Innovations as Key to the Green Revolution in Africa

Andre Bationo · Boaz Waswa ·
Jeremiah M. Okeyo · Fredah Maina ·
Job Kihara
Editors

Innovations as Key to the Green Revolution in Africa

Exploring the Scientific Facts

Volume 2

Editors
Andre Bationo
Alliance for a Green Revolution in Africa (AGRA)
Soil Health Program
6 Agostino Neto Road
Airport Residential Area
PMB KIA 114, Airport-Accra
Ghana
abationo@agra-alliance.org

Boaz Waswa
Tropical Soil Biology and Fertility Institute of the International Centre for Tropical Agriculture (TSBF-CIAT)
Nairobi, Kenya
bswaswa@yahoo.com

Jeremiah M. Okeyo
Tropical Soil Biology & Fertility (TSBF)
African Network for Soil Biology and Fertility (AfNet)
c/o ICRAF, Off UN Avenue
P.O. Box 30677-00100
Nairobi, Kenya
jmosioma@gmail.com

Fredah Maina
Kenya Agricultural Research Institute
Socio-economics and Biometrics
P.O. Box 14733-00800
Nairobi, Kenya
fredah.maina@yahoo.com

Job Kihara
Tropical Soil Biology & Fertility (TSBF)
African Network for Soil Biology and Fertility (AfNet)
c/o ICRAF, Off UN Avenue
P.O. Box 30677-00100
Nairobi, Kenya
j.kihara@cgiar.org

Please note that some manuscripts have been previously published in the journal 'Nutrient Cycling in Agroecosystems' Special Issue "Innovations as Key to the Green Revolution in Africa: Exploring the Scientific Facts". (Chapters 13, 14, 19, 20, 23, 36, 42, 57, 59, 78, 80 and 113)

Printed in 2 volumes
ISBN 978-90-481-2541-8 e-ISBN 978-90-481-2543-2
DOI 10.1007/978-90-481-2543-2
Springer Dordrecht Heidelberg London New York

Library of Congress Control Number: 2011930869

Printed on acid-free paper

Springer is part of Springer Science+Business Media (www.springer.com)

Contents

Volume 1

Part IV Innovation Approaches and Their Scaling Up/Out in Africa

Part III
Limitations to Access and Adoption of Innovations by Poor Farmers

Some Facts About Fertilizer Use in Africa: The Case of Smallholder and Large-Scale Farmers in Kenya

P.F. Okoth, E. Murua, N. Sanginga, J. Chianu, J.M. Mungatu, P.K. Kimani, and J.K. Ng'ang'a

Abstract It is argued that for a green revolution to take place in Africa, fertilizer use must be increased from the current average of 8 kg ha^{-1} to around 50 kg ha^{-1} by 2015. This was a major issue tackled by the African Fertilizer Summit and endorsed by the African Heads of States at Abuja, Nigeria, in June 2006. This chapter assesses fertilizer situation in Kenya versus the agreed milestones which include that (i) the African governments should take appropriate policy measures to reduce price of fertilizers by mid-2007; (ii) African Union Member States should take concrete measures to increase access to fertilizers by scaling up the network of agro-input dealers and community networks by mid-2007; (iii) African Union Member States should specially address the needs of the farmers and (iv) African Union Member States should work with development partners to tackle issues relating to subsidies that favour the fertilizer sector including infrastructure. In order to provide insight into some of these, we carried out a survey in *Kiambu*, *Thika* and Nairobi districts to establish the extent of fertilizer use in the area with a view to determining if the African Dream of the "*green revolution*" is realizable as planned. The study area was selected due to its strategic contribution to the agricultural economy of Kenya through coffee and tea. Results indicate that only large-scale coffee and tea farmers apply sufficient quantities of fertilizer. High price of fertilizers and overall access to it still remain unresolved issues among smallholder farmers.

Keywords Fertilizers · Africa · Kenya · Use · Policy

Introduction

Role of Fertilizers in Soil Fertility

Low-inherent soil fertility in the highly weathered and leached soils largely accounts for the low and un-sustained crop yields in most African countries (Bationo et al., 2006; Okalebo et al., 2006). The major nutrients (nitrogen N and phosphorous P) are commonly deficient in African soils. There is ample evidence that increased use of inorganic fertilizers has been responsible for an important share of worldwide agricultural productivity growth. It has been observed that fertilizers are as important as seeds in green revolution (GR), contributing as much as 50% to yield growth in Asia (Hopper, 1993). Several studies have found that one-third of the cereal production worldwide is due to fertilizer and related production factors (Bumb, 1995, citing FAO). Van Keulen and Breman (1990) and Breman (1990) stated that the only real cure against land hunger in the West African Sahel lay in increased productivity of land through the use of external inputs, especially inorganic fertilizers. There are numerous cases of strong fertilizer response for maize in East and Southern Africa (Byerlee and Eicher, 1997). Soil fertility depletion in smallholder farms is another fundamental biophysical root cause of declining per capita food production in Africa. Over 132

P.F. Okoth (✉)
Tropical Soil Biology and Fertility Institute of the International Centre for Tropical Agriculture (TSBF-CIAT), UN Avenue, Gigiri, Nairobi, Kenya
e-mail: p.okoth@cgiar.org

A. Bationo et al. (eds.), *Innovations as Key to the Green Revolution in Africa*,
DOI 10.1007/978-90-481-2543-2_86,

million t of N, 15 million t of P and 90 million t of K have been lost from cultivated land in 37 African countries in 30 years (Smaling, 1993). Nutrient loss is estimated to be 4.4 million t N, 0.5 million t P and 3 million t K every year from the cultivated land (Sanchez et al., 1997). These rates are several times higher than Africa's (excluding South Africa) annual fertilizer consumption of 0.8 million t N, 0.26 million t P and 0.2 million t K. The loss is equivalent to 1400 kg ha^{-1} urea, 375 kg ha^{-1} triple super phosphate (TSP) and 896 kg ha^{-1} KCl during the last three decades. Figure 1 shows the annual average nutrient depletion in Africa measured in kilograms per hectare per year between 1993 and 1995.

The scarcity of data on the nature of fertilizer demand at country levels in Africa makes it difficult to assess how much fertilizer small farmers use compared to what is used by the commercial farmers. This kind of information is important for the design of interventions that improve fertilizer use among the smallholders as the basis for achieving an Africa's GR. This concern is the main motivator for the authorship of this chapter.

In Africa generally, fertilizer consumption is highly concentrated, with Egypt and South Africa alone accounting for about 40% of the total consumption (Camara and Heineman, 2006). Farmers in 33 African countries applied an average of <10 kg ha^{-1} of fertilizer in the 1970s and about 22 kg ha^{-1} between 1970 and 2002. In 2002, Africa accounted for only 3% (4 million t) of the world consumption, compared to 9% (13 million t) for Latin America and 54% (77 million t) for Asia and the Pacific.

Kenya is currently using an average of about 31 kg ha^{-1} of fertilizers which is seen as better than many African countries (Camara and Heineman, 2006). However, it is still unclear how fertilizer use in Kenya is partitioned between smallholders and commercial farmers. There is scanty information on the amounts of fertilizers that service the food sector compared to what goes into cut flower production (mostly for export). The quantity of fertilizers being used by smallholder farmers for food crops needs to be assessed. Such an analysis will help to match fertilizer use in Kenya with the goals of African GR set by the African Heads of States. The information generated will provide a yardstick for gauging the situation in Kenya specifically and Africa in general since the majority of the countries are still below what is needed to turn

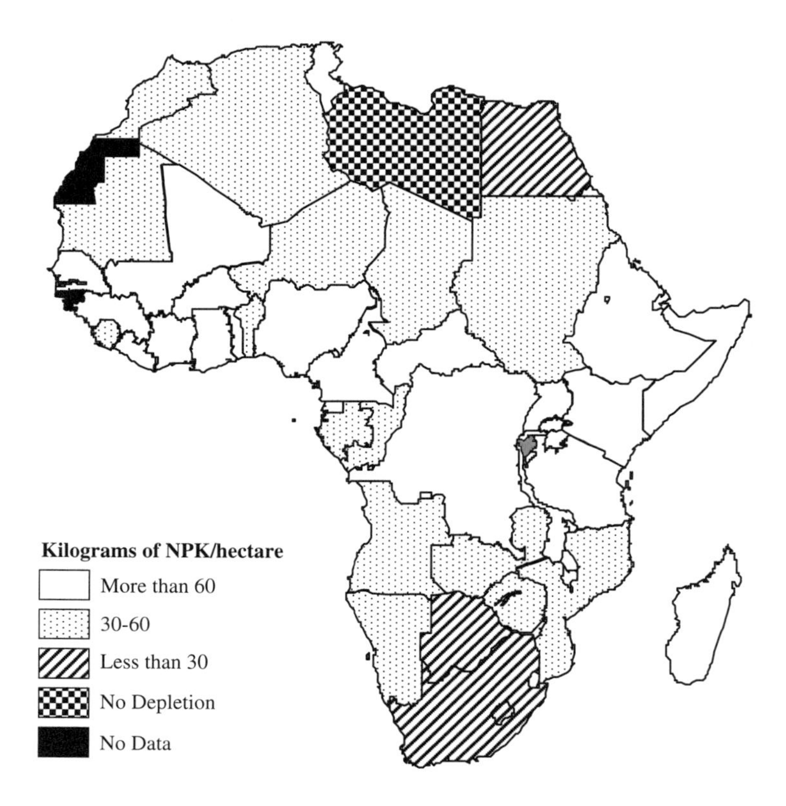

Fig. 1 Average annual nutrient depletion (NPK) in Africa between 1993 and 1995. *Source*: Henao and Baanante (1999)

around Africa's agriculture and achieve the desired GR. In order to meet the African GR by 2015, the African Heads of States highlighted some milestones that each country was supposed to meet to be on-course. They included the following: (i) the African governments should implement appropriate policy measures to reduce the price of fertilizers by mid-2007; (ii) African Union (AU) Member States should take concrete measures to increase access to fertilizers by scaling up the network of agro-input dealers and communities by mid-2007; (iii) AU Member States should specially address the needs of smallholder farmers and (iv) AU Member States should work with development partners to tackle issues related to subsidies that favour the fertilizer sector including infrastructure. These would be used to gauge the response of the Kenyan and other African governments in meeting the 2006 Abuja declaration.

Objectives of the Study

The objectives of the chapter were to provide insight into the fertilizer situation in Kenya and thereby assess the country's preparedness for the proposed Africa's green revolution and to use a selected high potential area in Kenya in a case study to provide insight into the above objective.

Materials and Methods

The Study Area

Data were collected mainly from *Kiambu* and *Thika* districts and parts of Nairobi Province covered by latitudes 0°46′ and 1°31′ south of the Equator and longitudes 36°30′ and 37°20′ east of Greenwich. Figure 2 shows the location of the study area. The rainfall regime is bimodal with the long rains falling in April and May. The rainy season is followed by a cool dry season during the months of July and August. The short rains fall from October to December. The mean annual rainfall of the area is 1100 mm. The study area is mainly composed of volcanic rocks of varying ages (Saggerson, 1967). To the northwest of Kiambu town, the geology varies from Miocene to

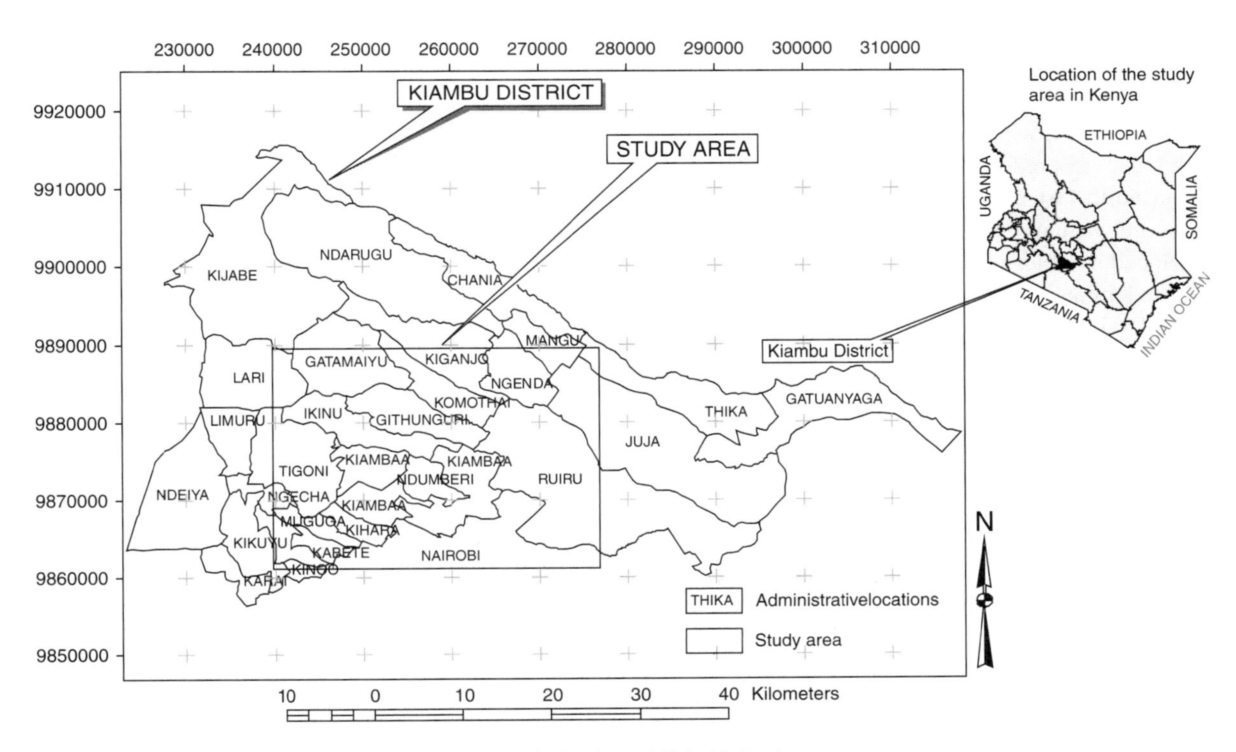

Fig. 2 Location of the study area in Kenya and parts of Kiambu and Nairobi districts

Table 1 Average values of soil properties in the study area (150 samples covering the entire area)

	Land use					
Soil properties	Coffee	Coffee and Macadamia	Fallow	Forest	Intercrop (annual and perennials)	Tea
Sand (%)	22	25	18	31	23	20
Silt (%)	20	26	22	32	25	20
Clay (%)	58	49	60	37	52	60
Silt/clay ratio	0.3	0.5	0.4	0.9	0.5	0.3
SOC (g kg^{-1})	2.2	2.5	2.9	3.8	2.9	3.5
pH–H_2O (1:2.5)	5.3	4.9	4.9	5.2	5.2	5.5
S-P (mg kg^{-1})	51	122	138	128	87	47
S-K (mg kg^{-1})	1.1	1.0	0.4	0.9	1.0	1.1
S-Ca (mg kg^{-1})	5.7	5.9	5.3	9.3	6.4	6.3
SEC (mmhos cm^{-1})	0.2	0.3	0.2	0.3	0.2	0.3

SOC, soil organic carbon – Walkley and Black method (Walkley and Black, 1934); S-P, soil phosphorous (Mehlich, 1984); S-K, soil potassium (Flame photometer method); S-Ca, soil calcium (ammonium acetate method, pH 7); SEC, soil electrical conductivity (1:1.5 soil:H_2O ratio)
Source: Okoth (2003)

Pleistocene volcanic. The northwestern part of the area is characterized by a high-level mountain range (the Aberdare Mountains). Fifty percent of the study area is dominated by Nitisols (well-drained, dark red to dark reddish brown clay soils with good structures and high water infiltration rates) (Shitakha, 1983). The depths vary between extremely to moderately deep soils in the steeply sloping parts. Most of the soils have a high concentration of organic matter in the topsoil. The soils on the toe slopes currently supporting large-scale coffee (*Coffea arabica* and *Coffea canephora*) are well-drained, strongly weathered, extremely deep and dark red to dark reddish brown friable clay soils. Areas to the northwest of Kiambu town where large-scale tea is grown have Andosols (low bulk density soils developed on volcanic ash and tuffaceous deposits) as the predominant soils and are extremely deep, well drained with poorly drained pockets in bottomland areas. Table 1 shows some of the chemical properties of the soils.

Data Collection Methods

Simple random sampling methods were used to select respondents (smallholder farmers, large-scale tea and coffee farmers and stockists). Sampling frames of all in the eligible populations were constructed. On policy matters, the sampling was purposive and targeted officials involved in agricultural policy making. Data were collected using structured questionnaires. They were administered by face-to-face interviews.

Issues addressed in the farm-level questionnaire included the type and quantity of fertilizers used by farmers, availability and timely access to fertilizers, sources of technical agronomic knowledge and its availability. The data collected from stockists included sources, quantities and types of fertilizers stocked and prices at which they sold fertilizers. Fertilizer policy issues concentrated on how the Kenyan government was responding to the Abuja declaration.

Secondary data were collected from the Kenyan Central Bureau of Statistics, Ministry of Planning, and based on publications pertaining to fertilizer imports and monetary value from 1997 to 2006. This was to assess the trend in supply and availability in the country. Data were entered in Microsoft Excel. The data was analyzed after cleaning using Microsoft Excel and SPSS softwares.

Results and Discussions

Fertilizer Use

Over the years, fertilizer application rates among the smallholder farmers are on the decline. The farmers have been supplementing fertilizer with manure and some have stopped using fertilizers altogether. The fertilizer types that dominate smallholder farms are

Table 2 Percentage of farmers using different types of fertilizers in the survey area

Farmer type	DAP	TSP	SSP	CAN	NPK	Urea	SA	Dolomite
Smallholder (%)	77.4	5.7	0	64.2	28.3	11.3	1.9	0
Large scale (%)	40	20	20	90	100	30	100	20

DAP and CAN with 77% of the farmers using DAP and 64.2% using CAN and mostly on food crops (Table 2). Most farmers used combined applications, not exclusively one fertilizer type. Other types of fertilizers include NPK (28.3% of smallholder farmers), urea (11.3%) and SA (1.9%). SSP and dolomite lime were not encountered among smallholders.

Among large-scale farmers, fertilizer application rates have been on the increase despite the high fertilizer prices. Majority obtain fertilizers from importers directly or from distributors, the major problems being late deliveries and unreliable supply. The large-scale farmers have good knowledge of fertilizer rates and types to use with recommendations made by experts based on regular soil tests. The most commonly used fertilizers in the large-scale farms were the NPK and the CAN (Table 2). All the farmers used NPK while 90% used CAN. The large-scale farmers also use a combination of fertilizers, although more intensely than smallholder farmers. Other fertilizers are DAP (40% of large-scale farmers); urea (30%); and TSP, SSP and dolomite lime (20% each).

The differences in the types of fertilizer used by the large-scale farmers could mostly be attributed to the fact that fertilizers used by the large-scale farmers are applied on coffee and tea that require good nutrition (mainly of N, P, K). The sulphate of ammonia used by large-scale farmers provides sulphur that is a micro-nutrient that is not provided by most of the other fertilizers but an important nutrient for coffee and tea. Large-scale farmers use dolomite to control soil acidity whereas smallholders do not use it at all (Table 3). Soil acidity is a major problem in the area (see Table 1).

Table 3 shows the major constraints cited by smallholder farmers (53). The most important among these are high prices and transport problems. The second most important problem is transportation. Transport was either expensive or not readily available in some localities both for purchase of fertilizer and for the sale of farm produce. Despite this setback, fertilizers are generally accessible with the majority of farmers using fertilizers and obtaining them from local shops within their localities. Some farmers felt that the distances to markets were sometimes inhibitive implying high transportation costs with isolated cases of inaccessible transport.

Findings by other researchers in Africa bear similar conclusions or provide different insights on the low levels of fertilizer use in Africa. For example, according to Mwangi (1997), in Africa, policy issues are important and they particularly need to address the issue of subsidies since African farmers are faced with a myriad of farming problems. Policies should address issues like credit and the need to support appropriate agricultural research and the need to develop and maintain infrastructure as well as foster development of the private sector alliances for increased farmer access to fertilizer. Demeke (1997) points out that there is much debate in the agricultural development

Table 3 Constraints that smallholder farmers cite for inadequate use of fertilizers

Major issues raised	Number of respondents	Percentage of respondents out of 53
High prices (ever increasing and variable)	41	77.36
Transport problems (high cost and reliability)	9	16.98
Unreliable supply	6	11.32
Long distances to markets	6	11.32
Lack of money and credit	3	5.66
Unsure of quality	2	3.77
High cooperative interest rates	1	1.89
Corrupt suppliers	1	1.89
Wrong recommendations by stockists	1	1.89
No problem	1	1.89

literature about whether fertilizer use in Africa is profitable. According to him, Africa's use of fertilizers is constrained primarily by poor input distribution systems, farmers' lack of knowledge and incorrect use of fertilizer and lack of effective demand because fertilizer use is not sufficiently profitable. Smaling (2006) points out that until 1990, many African governments were involved in the fertilizer sector. Fertilizer subsidies were widespread – placing a heavy burden on the limited financial resources of the countries. During the 1980s the World Bank imposed financial adjustment programmes in most countries, but the privatization of the distribution systems in countries with low levels of fertilizer consumption on food crops was not very successful as demand remained low.

Another factor that constrained the smallholder farmers was the aspect of cost advantage. Table 4 shows the mean price of fertilizers by farmer type. It is evident that smallholder farmers purchased fertilizers at higher prices than large-scale farmers. It was, for example, 41% more expensive for the smallholder farmer to buy TSP from retailers compared to the large-scale farmer who purchased directly from wholesalers.

The major constraints cited by large-scale farmers are high prices, late delivery and availability at source. Many problems (e.g. high cooperative interest rates, lack of credit, corrupt suppliers) never featured among large-scale farmers (Table 5).

Nutrient Application Rates

Table 6 shows the comparison between nutrients used by the large and small-scale farmers. Though the large-scale farmers were using the nutrients on cash crops, the difference between them and the smallholder farmers (growing maize and other food crops) was enormous. For instance, the 22 kg of N applied by the smallholder farmers on maize is well below the commonly recommended rates (about 120 N ha^{-1}) for maize and for grain yields of over 4 t ha^{-1} in similar soils. Good yields require not less than 150 kg P ha^{-1} if supplemented with adequate applications of K (Jensen et al., 2003; Kwabiah et al., 2003; Schmidt and Adriaanse, 2003). The amount commonly applied by the smallholder farmers bears a negative difference

Table 4 Mean price of fertilizers by farmer type in Kenya

	Cost per kilogram of fertilizer in Kenya Shillings					
Farmer type	DAP	TSP	SSP	CAN	NPK	Urea
Smallholder farmers	41	45	–	35	36	33
Large-scale farmers	35	32	26	28	32	28
Percentage difference	17	41		25	13	18

Table 5 Constraints that large-scale farmers cite for inadequate use of fertilizers

Major issues raised	Number of respondents	% of respondents out of 10
High prices (ever increasing and variable)	5	50
Late delivery/availability	4	40
Transport problems (high cost and reliability)	1	10
Unreliable supply	1	10
Long distances to markets	0	0
High cooperative interest rates	0	0
Lack of money and credit	0	0
Corrupt suppliers	0	0
Unsure of quality	0	0
Wrong recommendations by stockists	0	0
No problem	2	20

Table 6 Mean nutrients applied by smallholder and large-scale farmers in Kenya

Farmer type	Nitrogen (kg ha^{-1})	Phosphorous (kg ha^{-1})	Potassium (kg ha^{-1})	Calcium (kg ha^{-1})	Sulphur (kg ha^{-1})
Large scale	254 ± 43	127 ± 44	86 ± 18	59 ± 29	6 ± 6
Smallholder	22 ± 5	26 ± 6	3 ± 2	3 ± 1	0

of about 100 kg N and 120 kg P. According to the Coffee Research Foundation of Kenya (CRF, Coffee nutrition, Technical Circular No. 23, Coffee Research Foundation of Kenya, unpublished), the recommended N for a good coffee crop that yields 1500–2000 kg ha^{-1} of clean coffee is 100–150 kg ha^{-1} $year^{-1}$ which falls within what the large-scale farmers are currently applying. In reality, the large-scale coffee producers are on average applying 100 kg N more than what is recommended. According to Dang (2005), tea requires about 150–200 kg N ha^{-1} $year^{-1}$ that conforms with the application rate of the large-scale tea farmers.

In Kenya, fertilizer distribution is based on two types of private sector groups (Wanzala et al., 2002). Importers, wholesalers and retailers constitute the first group and smallholders who purchase fertilizer directly from private importers constitute the second group. Based on their assessment of demand, weather conditions, credit availability and the policy environment, importers decide on the quantity of fertilizers to import. The structural adjustment programmes (SAP) by the World Bank and the IMF of the 1980s introduced turbulence in the Kenyan fertilizer sector with a high rate of instability in fertilizer imports (see Figs. 3 and 4). Figure 3 shows fluctuations in all the fertilizers imported. Figure 4 shows the percent fluctuations in the individual imported fertilizers between 1997 and 2006. Table 7 shows the amount of fertilizers imported into Kenya between 1997 and 2006 and the associated expenditure. Though fertilizer import fluctuations could be a factor caused by changing policies, it could also be as a result of relying on importers whose main concerns are driven by commercial gains rather than what the country needs. Table 1 shows that fertilizer imports increased between 1997 and 2006. Figure 3 shows that non-nitrogenous and non-phosphate fertilizers dominated the Kenyan fertilizer market making it possible that the fertilizer imports were servicing other sectors rather than the food sector that is normally dominated by the nitrogenous and phosphatic fertilizers.

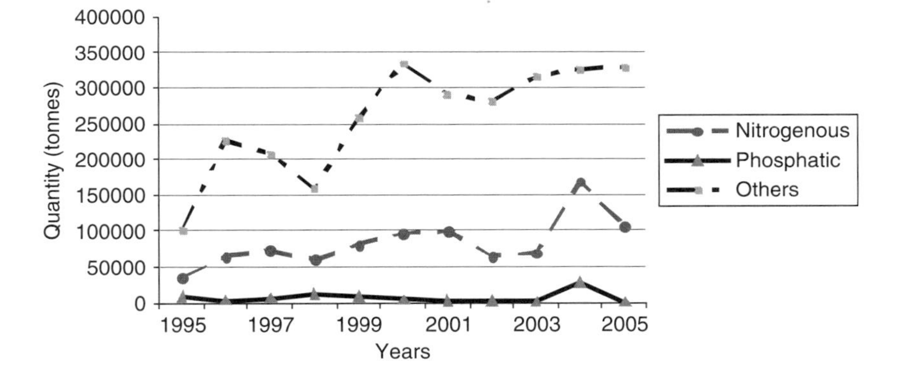

Fig. 3 Fertilizer quantity and type imported in Kenya over the years 1995–2005 (by category). *Source*: Central Bureau of Statistics, Ministry of National Planning, Kenya: Economic Survey (2006)

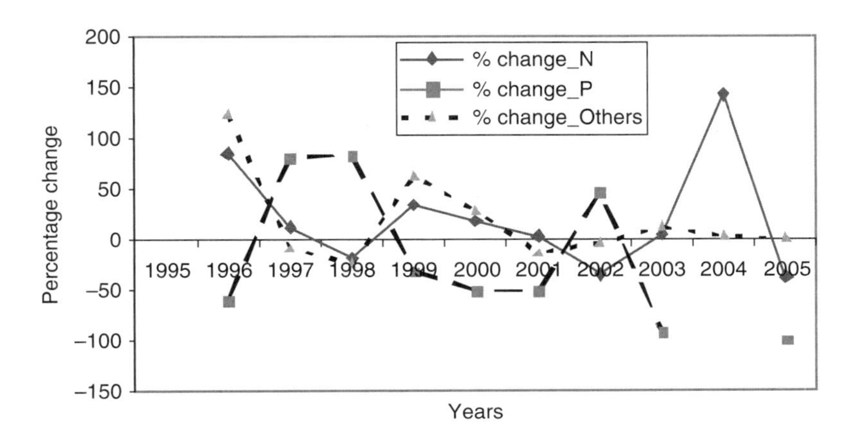

Fig. 4 Percentage change in the quantity of fertilizer imports (by category). *Source*: Central Bureau of Statistics, Ministry of National Planning, Kenya: Economic Survey (2006)

Table 7 Types of fertilizers imported into Kenya (1997–2006)

Year	Nitrogenous (metric tons)	Phosphatic (metric tons)	Others (metric tons)	Total (metric tons)
1995	35, 067	8195	100, 634	143, 896
1996	65, 155	3277	226, 824	295, 256
1997	73, 392	5963	207, 547	286, 902
1998	59, 965	10, 968	158, 181	229, 114
1999	81, 064	7570	258, 150	346, 784
2000	96, 305	3768	333, 756	433, 829
2001	99, 738	1860	290, 933	392, 531
2002	65, 265	2742	280, 510	348, 517
2003	69, 122	221	315, 311	384, 654
2004	167, 986	27, 950	326, 486	522, 422
2005	35, 067	117	330, 013	365, 197

Source: Central Bureau of Statistics, Ministry of National Planning, Kenya: Economic Survey (2006)

Policy Issues

Result shows that there was a comprehensive policy on fertilizer use in Kenya under the Ministry of Agriculture. The main objectives of the policy framework were (i) to promote efficient and sustainable use of natural resources; (ii) to provide framework for developing and applying appropriate management techniques; (iii) to facilitate research and dissemination of appropriate technologies; (iv) review soil fertility-related legislation; (v) to provide enabling environment for public/private sector investment and (vi) proper coordination of soil fertility activities.

However, the major issues of concern included farmers' access to fertilizers; extension information relating to fertilizer use and the enforcement of inputs quality. We also dealt with the question as to which strategies the Ministry of Agriculture employed to get millions of smallholder farmers to use fertilizers in terms of both appropriate types and consistent use. The response was that "*the ministry provided extension services to the farmers and that the ministry had developed a National Accelerated Inputs Access Program that aims at providing needy farmers (smallholders) with input grants to enable them start using fertilizers*". The system aims at using a voucher system. The very needy farmers are given vouchers which they use to access fertilizers from stockists who are then refunded by the government. Besides, the government through the Kenya Cereals Board is stabilizing the buying process of produce so that the farmer is guaranteed a steady price to enable him/her buy own fertilizer. According to Gregory (2006), input vouchers can play a vital role in "jumpstarting" market inclusion or millions of smallholder farmers in sub-Saharan Africa.

Fertilizer Use Problem

Whether the strategy of the government on fertilizer use and access works is doubtful since the impact is not directly felt on the ground by the smallholder farmers. The average smallholder farmer is only using on average 22 kg N, 26 kg P and 3 kg K which are all below what is recommended and only accounts for about 18% of what is recommended for maize. The implication is that the food systems are only performing at 20% of their true potentials and need to be propped up either through policy interventions (implementation) or through strong awareness campaigns or effective knowledge dissemination. The Baraza method in Kenya does not seem to work for the smallholder farmers. The introduction of the voucher system might address the issue of access by poor smallholder farmers.

The Kenyan government imports other types of fertilizers, most likely fertilizers with micro-nutrients and other types of nutrients for use in the flower, tea, coffee and horticulture sectors which generate a lot of revenue for the four economic sectors in Kenya. It is, however, not clear whether the imbalance in imports is deliberate and based on market forces or whether it is based on strategic thinking that takes into consideration the fertilizer requirements to adequately meet the food sector. Going by what the smallholder farmers are currently applying this seems to be remote.

High Price and Poor Infrastructure

The issue of high prices is consistently cited by all classes of farmers. The high prices are due to a variety of reasons including poor roads, VAT on road transport, high port charges, inefficiencies in port clearance and lack of financial infrastructures in rural areas (Poulton et al., 2006). The strategy by the Kenyan government that the producer prices should determine the fertilizer demand that ensures purchase of fertilizers by the farmers is still questionable since it is not linked with successful use of adequate amounts of fertilizer. The strategy has not translated to farmers accessing adequate quantities of fertilizer or nutrients and should therefore be re-examined for appropriateness.

Stockists and Distribution Networks

Most of the fertilizer stockists obtain their stocks from major fertilizer importers who are easily accessible to them and ensure that fertilizer is readily available. They, however, feel that the fertilizer prices are too high and unaffordable to their clientele and that the government's intervention is not evident. The increased use of fertilizers by large-scale farmers has almost been negated by the decline in consumption by small-scale farmers and hence reflecting a marginal change in fertilizer demand from the stockists. Since some farmers cite distance as a possible bottleneck, the issue of distribution and stockists networks should as well be addressed.

Conclusions

Overall, Kenya does not appear to have met most of the milestones agreed upon by the African Heads of States Summit to enable the country propel itself towards the green revolution. The input voucher system when applied will be a sure way forwards but it is clear that there is still work to be done. If price incentives are introduced, there could be an upturn on fertilizer use and a scenario change in smallholder farms for Africa's green revolution. Policy should address issues relating to access, affordability, infrastructure, distribution and levies charged on fertilizer-related enterprises.

Acknowledgements The authors would like to thank the Director of TSBF Dr. Nteranya Sanginga for facilitating the data collection that made the authorship of this chapter possible. We also recognize the support of the farmers in the study area who availed time for the interviews and filling in the questionnaires. The Kenyan Ministry of Agriculture is also acknowledged for providing information on the policy situation in Kenya. We are greatly indebted to all of them.

References

Bationo A, Hartemink A, Lungu O, Naimu N, Okoth PF, Smaling E, Thiombiano L (2006) African soils: their productivity and profitability for fertilizer use. Background paper prepared for the African Fertilizer Summit, Abuja, Nigeria, 9–13th June 2006

Breman H (1990) No sustainability without external inputs. Sub-Saharan Africa; Beyond Adjustment. Africa Seminar. Ministry of Foreign Affairs, DGIS, The Hague, The Netherlands, pp 124–134

Bumb B (1995) Global fertilizer perspective, 1980–2000: the challenges in structural transformations: technical bulletin T-42. International Fertilizer Development Center, Muscle Shoals, Alabama

Byerlee D, Eicher C (eds) (1997) Africa's emerging maize revolution. Lynn Rienner, Boulder, CO

Camara O, Heineman E (2006) Overview of the fertilizer situation in Africa. Background paper prepared for the African Fertilizer Summit, Abuja, Nigeria

Dang MV (2005) Soil-plant nutrient balance of tea crops in the northern mountainous region, Vietnam. Agric Ecosyst Environ 105:413–418

Demeke M, Kelly V, Jayne TS, Said A, Le Vallée JC, Chen H (1997) Agricultural market performance and determinants of fertilizer use in Ethiopia. Department of Agricultural Economics, Michigan State University in its series International Development Collaborative Working Papers with number ET-FSRP-WP–10

Gregory I (2006) The role of input vouchers in pro-poor growth. Background paper prepared for the African Fertilizer Summit, Abuja, Nigeria

Henao J, Baanante C (1999) Estimating rates of nutrient depletion in soils of agricultural lands in Africa. International Fertilizer Development Centre, Muscle Shoals, Alabama, USA

Hopper WD (1993) Indian agriculture and fertilizer: an outsiders observations: keynote address to the FAI Seminar on Emerging Scenario in Fertilizer and Agriculture: Global Dimensions. FAI, New Delhi

Jensen JR, Bernhard RH, Hansen S, McDonagh J, Moberg JP, Nielsen NE, Nordbo E (2003) Productivity in maize based cropping systems under various soil-water-nutrient management strategies in a semi-arid, Alfisol environment in East Africa. Agric Water Manage 59:217–237

Kwabia AB, Stoskopf NC, Palm CA, Voroney RP, Rao MR, Gacheru E (2003) Phosphorus availability and maize response to organic and inorganic fertilizer inputs in a short term study in western Kenya. Agric Ecosyst Environ 95:49–59

Mehlich A (1984) Mehlich 3 soil test extractant: a modification of Mehlich 2 extractant. Commun Soil Sci Plant Anal 15:1409–1416

Mwangi WW (1997) Low use of fertilizers and low productivity in sub-saharan Africa. Nutr Cycl Agroecosyst 47:135–147

Okalebo JR, Othieno C, Woomer P, Karanja N, Semoka J, Bekunda M, Mugendi DN, Muasya R, Bationo A, Mukhwana E (2006) Available technologies to replenish soil fertility in East Africa. Nutr Cycling Agroecosyst 76(2–3):153–170

Okoth PF (2003) A hierarchical method for soil erosion assessment and spatial risk modelling: a case study of Kiambu district in Kenya. PhD thesis, Wageningen University and Research Centre, Wageningen, The Netherlands

Poulton C, Kydd J, Dorward A (2006) Increasing fertilizer use in Africa: what have we learned? Background paper presented to the African Fertilizer Summit. Agriculture and Rural Development Discussion Paper, Abuja, Nigeria. World Bank Paper, 25

Republic of Kenya (2006) Economic survey 2006. Central Bureau of Statistics, Ministry of Planning and National Development. Government printer, Nairobi

Saggerson EP (1967) Geological map of the Nairobi Area. (To accompany Geological Report No. 98). Geological Survey of Kenya

Sanchez PA, Shepherd KD, Soule MJ, Place FM, Buresh RJ, Izac AMN, Mokwunye AU, Kwesiga FR, Ndiritu CG, Woomer PL (1997) Soil fertility replenishment in Africa: an investment in natural resource capital. In: Buresh RJ, Sanchez PA, Calhoun F (eds) Replenishing soil fertility in Africa. SSSA Special Publication No. 51, Soil Science Society of America, Madison, WI, pp 1–46

Schmidt CJJ, Adriaanse FG (2003) Nitrogen fertilizer guidelines for small-scale farming on an irrigated duplex soil based on soil analyses. S Afr J Plant Soil 21(1):31–37

Shitakha FM (1983) Detailed Soil Survey of the Coffee Research Station, Ruiru, Kiambu District. Kenya Soil Survey Soils Report

Smaling EMA (1993) An agro-ecological framework of integrated nutrient management with Special Reference to Kenya. Wageningen Agricultural University, Wageningen, The Netherlands

Smaling E, Toure M, De Ridder N, Sanginga N, Breman H (2006) Fertilizer use and the environment in Africa: friends or foes? Background paper prepared for the African Fertilizer Summit, 9–13th June 2006

Van Keulen H, Breman H (1990) Agricultural development in the West African Sahelian region: a cure against land hunger? Agric Ecosyst Environ 32:177–197

Walkley A, Black AI (1934) An examination of the Degtjareff method for determining organic carbon in soils: effect of variations in digestion conditions and of inorganic soil constituents. Soil Sci 63:251–263

Wanzala M, Jayne T, Staatz J, Mugera A, Kirimi J, Owuor J (2002) Fertilizer markets and agricultural production incentives: insights from Kenya. Working Paper No. 4. Egerton University, Tegemeo Institute, Nairobi

Farm Input Market System in Western Kenya: Constraints, Opportunities, and Policy Implications

J. Chianu, F. Mairura, and I. Ekise

Abstract Widespread and increasing rural poverty in sub-Saharan Africa (SSA) has been of great concern to development community. Compared to other developing regions of the world, low use of inputs by small farmers is one of the factors responsible for the gap between potential and actual yields. Market constraints reduce profitability in use of inputs, increasing production risks. This study interviewed 130 agro-input dealers in Kenya to analyze trends, inputs stocked, distance to markets, services to farmers, and constraints and suggests how to improve input delivery to farmers. Results indicate that although the number of agro-dealers is still small relative to farmer population, there has been a steady annual increase (2–22%, with mean of 16% across inputs) in their number from 2003 to 2005. DAP fertilizer (stocked by 92% of respondents) was most commonly stocked. Others are CAN fertilizer (84%), urea (78%), and NPK (40%). Other services provided by agro-dealers are input information (75% of respondents), credit (13%), bulk breaking (8%), and spraying (4%). Selling price of inputs increased with distance to markets. High transport cost (53%), low demand (30%), lack of market information (21%), lack of storage facilities (13%), and limited business knowledge (12%) were the most important constraints faced by agro-dealers. Policies and institutional frameworks suggested by dealers to streamline agro-input trade were associated, and government was the main institution proposed. The study concludes with suggestions on how to enhance efficiency of agro-dealers in input delivery – timely since SSA governments are presently creating structures to enhance input use.

Keywords Farm input delivery · Input dealers · Kenya · Market constraints · Poverty and yield gap

Introduction

Widespread and increasing rural poverty in sub-Saharan Africa (SSA) has been of great concern to the development community. Compared to other developing regions of the world, the low use of farm inputs by smallholder farmers in SSA is responsible for the gap between potential farmers' yields and actual crop yields at farm level. A comparison of fertilizer consumption trends in SSA and developing countries of Asia shows that while average annual fertilizer consumption increased by 182% in the latter between 1980–1989 and 1996–2000, it increased by only 16% in the former (FAOSTAT, 2003). The slow growth in the use of modern agricultural inputs in the farming systems of SSA has resulted in missed opportunities to increase Africa's agricultural production, productivity, and household incomes and welfare. Fertilizer use in SSA is the lowest in the world and is actually less than 10% of the global mean (about 93 kg ha^{-1}) (IFDC, 2006) (see Fig. 1).

This chapter examines constraints and challenges limiting the expansion of farm input use by smallholder farmers in Western Kenya by assessing input

J. Chianu (✉)
Tropical Soil Biology and Fertility Institute of the International Centre for Tropical Agriculture (TSBF-CIAT), Nairobi, Kenya
e-mail: jchianu@yahoo.com

A. Bationo et al. (eds.), *Innovations as Key to the Green Revolution in Africa*,
DOI 10.1007/978-90-481-2543-2_87,

Fig. 1 (**a**) Global mean fertilizer use (93 kg ha^{-1}); (**b**) fertilizer recommendation (50 kg ha^{-1} by the African Fertilizer Summit recommendation, Abuja, Nigeria, 2006). *Source*: Adapted from IFDC (2006)

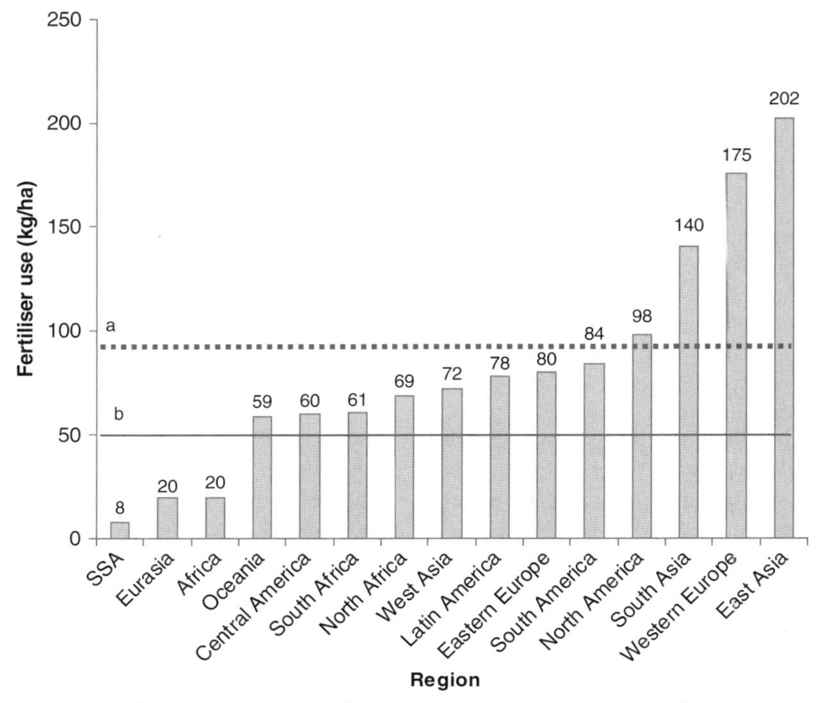

a = Global mean fertilizer use (93 kg/ha); b = Fertilizer recommendation ((50 kg/ha) by the African Fertilizer Summit recommendation, Abuja (Nigeria), 2006

supply side issues. This is critical in creating conducive atmosphere for agricultural intensification and to enable farmers (especially smallholders) produce for markets and lift them out of poverty. The study surveyed agro-input dealers in order to ascertain why their farm and other services are not reaching many farmers, especially those in remote rural areas. An important element of the study was to assess the main farm services provided by agro-input dealers, the constraints and challenges they face, the policy and institutional frameworks they would want to see implemented in order to enhance the environment for a sustainable expansion of their areas of coverage, and the access of smallholder farmers to farm inputs and the other services that they provide. This type of study has a great potential to contribute in the attainment of the goal of the Comprehensive Africa Agricultural Development Program (CAADP) that calls for a 6% annual growth in agricultural production as a framework for restoring agricultural growth, food security, and rural development in Africa and a key step toward attainment of first Millennium Development Goal of halving poverty by 2015.

Materials and Methods

Study Area

The study was carried out in 13 districts in Western Kenya, a densely populated region of Kenya with high levels of hunger, extreme poverty, and disease. Over 21% of the region's children under 5 years of age are underweight (malnourished). Adult HIV/AIDS prevalence is estimated at 30% in much of the region, leaving large numbers of orphans. Western Kenya is also characterized by low crop yields and low household cash incomes (Kelly et al., 2003). Potable water, paved roads, electricity, and telephone landlines are all scarce in this region. Most residents are subsistence farmers. However, farm sizes are so small and often in the neighborhood of 0.1 ha. Farm inputs (such as fertilizer, improved seeds, or water pumps) are also very scarce. Rainfall is so unpredictable often resulting in low crop and livestock productivity. Many families have difficulty producing enough food to meet their needs. Those who manage to produce a surplus have difficulty finding buyers or getting good prices.

The altitude in the study area ranges from low-medium (1000–1300 m.a.s.l) in places such as *Kasewe* in *Rachuonyo* district, through medium (1270–1320 m.a.s.l) in places such as *Mabole* (in *Butere* district) and *Akiites* (in *Teso* district), to high (1500–2100 m.a.s.l) in places such as *Oyani* (in *Migori* district) and *Riana* (in *Kisii* district) (Jaetzold and Schmidt, 1982). Average annual rainfall ranges between 1200 and 2100 mm for *Riana* area, 1400 and 1600 mm for *Oyani* area, 1300 and 2000 mm for *Mabole* area, 900 and 2000 mm for *Akiites* area, and 1000 and 2200 mm for *Kasewe* area (Jaetzold and Schmidt, 1982). Annual mean temperature ranges between 16.2 and 20.5°C for *Riana* area, 20.5 and 21.7°C for *Oyani* area, 22 and 27°C for *Mabole* area, 21 and 22°C for *Akiites* area, and 20 and 21.5°C for *Kasewe* area (Jaetzold and Schmidt, 1982).

Survey Method

A cross-section survey using structured questionnaire was used to collect data from 130 agro-input dealers randomly selected from 40 markets from 13 districts. Data collection was carried between the months of February and November 2005.

Parameters Investigated

Among others, the main variables on which data were collected were the agro-inputs sold, quantity of different inputs (different types of fertilizers, seeds, farm tools, etc.) stocked, prices at which different inputs were sold, distances from where different inputs were sourced, additional farm services offered to farmers, constraints and challenges faced by the agro-input dealers, and the policies and institutional frameworks which the agro-input dealers would like to see implemented in order to enhance their efficiency and areas of coverage in timely provision of farm services to the smallholder farmers, including those in far rural areas.

Data Entry and Analysis

Microsoft Excel was used for data entry. Online distance calculator (ODC), based on the World Geodetic System WGS84 (DMA, 1991) ellipsoid, was used to estimate distances between input selling and purchasing points. Data were analyzed using SPSS version 11.5 (SPSS, 2002). The World Geodetic System reference coordinate frames were established more than a decade ago to facilitate mapping, charting, positioning, and navigation applications.

Results and Discussion

Trends in the Number of Agro-input Dealers in Western Kenya

Between 2003 and 2005, the increase in the number of agro-input dealers ranged from 2% (for seed treatment chemicals) to 22% (mineral fertilizers) with a mean of 16% across agro-inputs (Table 1). The difference in the magnitude of percent increase (between 2003 and 2005) in the number of agro-input dealers selling different inputs reflects the demand for different agro-inputs in the farming systems of Western Kenya. Although many farmers are yet to be reached with the various farm services of the agro-input dealers, the trend in the growth of the number of agro-input dealers is encouraging, especially considering the current low level of infrastructure development in support of good network of private agro-input dealers and the usual period of slow, gradual growth before experiencing a period of relatively dramatic and rapid growth that characterizes a typical technology adoption process (Surrey, 1997). According to Encyclopedia Britannica (2003), the kilometers of paved roads per million people in selected African countries are 59 for DR-Congo, 66 for Ethiopia, 94 for Uganda, 114 for Tanzania, 141 for Mozambique, 230 for Nigeria, 494 for Ghana, 637 for Guinea, 1402 for South Africa, and 1586 for Zimbabwe compared with 20,987 for USA, 12,673 for France, and 9,102 for Japan and with 1064 for Brazil, 1004 for India, and 803 for China.

Farm Inputs Stocked

Altogether, the agro-input dealers surveyed sold about 2357 bags (or about 118 t) of mineral fertilizers annually (Table 2). The corresponding figure for improved

Table 1 Number of agro-input dealers selling different farm inputs in Western Kenya: 2003–2005

Agro-input	Number of input dealers[a]			% increase (2003–2005)
	2003	2004	2005	
Mineral fertilizers	245	276	299	22
Insecticides	314	351	372	19
Farm machinery	69	78	82	19
Herbicides	61	67	72	18
Fungicide	234	255	275	18
Improved seeds	176	189	198	13
Storage chemical	244	263	271	11
Seed treatment chemicals	43	43	44	2
Total across inputs (*not mutually exclusive*)	1498	1643	1742	16

[a]Responses were collated for different input brands within each agro-input category. Hence, for some inputs, the number of dealers is more than the sample size because of the grouping together of similar inputs

Table 2 Mean quantity of different farm inputs sold by agro-input dealers in Western Kenya: 2003–2005

Agro-input	Quantity	n^a
Mineral fertilizers (bags)	2357.0	383
Improved seeds (kg)	10,144.0	186
Fungicide (liters)	416.5	299
Insecticides (liters)	671.0	273
Storage chemicals (liters)	868.0	223
Herbicides (liters)	491.2	69
Seed treatment chemicals (liters)	1616.0	46
Machinery (units of different tools)[b]	236.3	17
Other farm implements (units of different tools)[c]	187.7	17

[a]Each input category included several specific input brands; hence n was larger than the number of agro-input dealers
[b]Machinery here includes tractors, sprayers, and spare parts
[c]Other farm implements here include long hoes, short hoes, and cutlasses

seeds was about 10 t. The average for the liquid agro-chemicals ranges from about 490 l (for herbicides) to about 1600 l (for seed treatment chemicals) (see Table 2).

Table 2 clearly shows the low demand which the agro-input dealers ranked second to high cost of transport among the constraints and challenges that they face in trying to provide inputs to smallholder farmers. Coupled with the usually high prices of most agro-inputs, these constrain the development of efficient farm input distribution systems and are fed into by farmers' inability to sell their farm's surplus produce at high prices, especially immediately after harvest. This situation contributes to poor land stewardship, accelerated land degradation, decline in household welfare, and negatively affects farmer investments in farm inputs and returns to agricultural production (Bashaasha, 2001). To break the cycle of high input price and low input demand, the need to stimulate a huge increase in input demand is critical. This requires reduction in input prices (through economies of size, new institutional arrangements, etc.) at the farm level, credit availability to farmers for the purchase of agro-inputs, and attractive prices for farm produce.

Other Farm Services That Input Dealers Provide to Farmers in Western Kenya

Only 2–11% of the agro-input dealers surveyed provided other services to small-scale farmers. Farm services most commonly provided by the farm input dealers were information [especially related to agrochemicals (24% of respondents) and improved varieties of seeds (22%)]. The least important farm services that dealers provided farmers included credit and spraying services. Twenty-seven agro-input dealers (or 18% of the respondents) provided farmers with credit services. However, only three agro-input dealers (about 2% of all respondents) provided farmers with spraying services. It is little surprising that only few agro-input dealers provided credit to smallholder farmers since many of the traders themselves lacked access to financial resources beyond their own savings and income (Yanggen et al., 1998). This is critical especially considering that farm inputs such as fertilizer are expensive to market (high storage and transport costs) and to purchase and underscores the fact that both the input dealers and the farmers lack access to critical service markets (especially credit markets). This is a typical example of how the private sector growth is constrained by a weak enabling environment.

Table 3 Number of farmers who benefited (according to dealers) from different other farm services given by agro-input dealers in Western Kenya

Other agro-input service	Number of farmers that benefited from service[a]	
	Female	Male
Input packaging	3100 (2)	15,000 (2)
Soil suitability information	796 (7)	3650 (7)
Soil fertility information	787 (7)	3645 (7)
Seed variety information	656 (13)	2656 (12)
Agrochemicals information	735 (14)	2328 (13)
Credit facilities	255 (12)	298 (12)
Spraying[b]	30 (3)	75 (3)

[a]Values in parentheses are effective sample sizes of agro-input dealers who offered services
[b]Free and at cost

It is critical that credit guarantees are used to link farm input manufacturing companies with agro-input dealers. Successful instances of smallholder farmer adoption of technologies (e.g., in Malawi) have demonstrated the importance of credit as an important enabling factor.

The above information was gender disaggregated to provide a better understanding of the distribution of farm services by input dealers. Agro-input buyers (mostly farmers) mostly benefited through input packaging. Data from the agro-input dealers show that of about 18,100 customers (farmers) that benefited from this service, about 83% were male farmers. Although only a few of the surveyed agro-input dealers gave an estimate of the farmers that benefited from their other farm services, it is striking that in 100% of the cases, more male farmers than female farmers benefited from the farm services of agro-input dealers (Table 3). This was the case for both free farm services (e.g., some aspects of the spraying services) and the services that were paid for. This is generally expected because, compared to their male counterparts female farmers are less likely to have contact with agro-input dealers due to their more limited financial resources than male farmers.

Business Constraints

The agro-input dealers surveyed (in western Kenya) faced five major constraints. In a decreasing order of importance, these were high transport cost (mostly due to poor infrastructure), low demand, lack of market information, lack of storage facilities, and limited business knowledge (Table 4). The high cost of transport services must have been due to the long average distance covered (and many hours' drive away) by agro-input dealers to source their goods. This shows how poor infrastructure undermines farmers' access to necessary farm inputs and hence food security. Poor domestic infrastructure (in much of Africa) and limited access to agricultural credit (including seasonal credit) also undermine the effect and equitable participation of many African countries in world trade. Africa's road density was less than India's in 1960 (Sanchez, 2005). All these explain why many smallholder farmers use few purchased inputs. Kelly et al. (2003) earlier identified the serious deficiencies in roads, education, market information systems, and supportive institutions as the major limitations that need to be addressed by governments to expand farm input use, increase agricultural productivity, and improve the livelihoods of smallholder farm families in SSA. Recently (in June 2006), an Africa Fertilizer Summit declared that the Africa Union Member States should, among others, take specific action to improve farmer access to market information. This is in line with the Comprehensive Africa Agriculture Development Program (CAADP)'s Pillar II agenda, which seeks to improve Africa's infrastructure and trade-related capacities for market access.

Table 4 Business constraints faced by agro-input dealers in Western Kenya

Constraint	% of respondents[a]	Effective n
High transport cost	52.8 (67)	127
Low demand	30.3 (37)	122
Lack of market information	21.2 (24)	113
Lack of storage facilities	13.0 (15)	115
Limited business knowledge	12.3 (14)	114

[a]Figures in parentheses are proportion of effective sample size affected by constraint; responses are not mutually exclusive

In western Kenya, the distance from agro-input marketing points to paved roads ranged from 0 (where market is next to a paved road) to 3 km. However, the distance (km) that agro-input dealers had to travel to source the farm inputs sold (i.e., from the market where interviewed to where agro-input dealers sourced or purchased specific inputs) ranged from 20 to over 300 km, indicating a more or less inaccessible distance. For specific farm inputs, the mean agro-input sourcing distance ranged from about 80 km (inorganic

Table 5 Mean distances (km) traveled by agro-input dealers to acquire selected farm inputs

Input	Distance (km)
Improved seeds	112.6 (97)
Storage chemicals	99.7 (22)
Farm machineries	92.8 (350)
Other agro-chemicals (herbicides, pesticides, etc.)	86.3 (71)
Mineral fertilizers	79.5 (182)

Figures in parentheses refer to the number of sourcing distance included in determining the statistics in the table

Table 6 Correlation between distances from where traders sourced farm inputs and the selling price

Input	Correlation
Muriate of potash	1.000**
Improved millet seeds	1.000**
Tractor	1.000**
Thiram	0.574*
NPK fertilizers	0.543
Improved sorghum seeds	0.209
Urea fertilizer	0.19
Improved seeds of common bean	0.174
Improved maize varieties	0.014

Source: Computed from survey data (2005)
** for 5%; * for 10%

fertilizers) to 113 km (improved seed varieties) (see Table 5). All these, coupled with the poor state of most of the road and market infrastructure in Western Kenya, explain the high cost of transport that ranked highest among the business constraints facing agro-input dealers in the area. It also explains the lack of access to timely and affordable farm inputs by most of the smallholder farmers. Under these situations, the inputs can be there but still far from the reach of the farmers because of high prices caused by many factors including "excessive" profit margins being asked for by agro-input dealers. All these again underscore the need to improve the existing rural infrastructure and build and develop new ones in most parts of SSA, a prerequisite for increasing the access of rural farmers to farm inputs at affordable prices and for increasing agricultural productivity, farm incomes, and general improvements in the livelihoods of the people.

Our analysis shows a positive correlation between the price at which agro-input dealers were willing to sell farm inputs and input sourcing distances. However, the magnitude of the correlation coefficient varied with agro-inputs. For instance, the sourcing distances and selling prices of muriate of potash (MoP), improved millet seeds, and tractors had a perfect correlation. Other correlations range from 0.014 (for improved varieties of maize) to 0.574 (for Thiram) (see Table 6). This result shows that increasing the networks and percent coverage by the main distributors of different agro-inputs are important ways of getting farm inputs closer to the poor and has a great potential to reduce the unit price at which farm inputs are sold to the resource-limited farmers in rural communities of SSA. This is an important step toward developing rural small-scale agro-input dealers. The question then is "how to provide the necessary incentives for the main agro-input distributors to increase the networks through which they supply agro-inputs to the retailers that ultimately sell to the smallholder farmers in distant and far-flung rural communities." Both policy support and infrastructural development are critical issues that must be addressed.

Frameworks Agro-input Dealers Would Want to See Implemented

Sustainable expansion of the areas of coverage by agro-input dealers for increased access of smallholder farmers to farm inputs is critical. How to arrive at this desirable situation was assessed from the point of view of the agro-input dealers. They were asked about the policy and institutional frameworks they would want to see implemented to enhance the environment for a sustainable expansion of their areas of coverage and the access of smallholder farmers to farm inputs. The most important areas where agro-input dealers would want to see improvements were training on agro-input business (29%); enhanced access to credit and loans to enable them purchase more goods (21%); agricultural extension, research, and infrastructure development (15%); tax reduction (14%); illegal trade and adulteration (13%); and input supply management (8%). These would provide some incentives for them to take the risk implied in supplying farm inputs to many farmers in remote rural communities. The major institutions that agro-input dealers felt should intervene to remove their business constraints were the government (48%) (facilitate access to credit and loans, tax reduction, and input quality control) and the universities and research institutions (26%) (provide various training support).

Overall, it is important to note that the government is extremely important when it comes to price (including farm input price) reduction mechanisms.

We also examined the relationship between all the policy-related interventions (training of farmers, agro-input dealers, and agricultural extension personnel; giving credit to farmers and agro-input dealers; price and tax reduction; curbing illegal trade and licensing; supply management; provision of agricultural extension services; supporting market research; etc.) and all the institution-related interventions (market cleanliness; customer care, agro-vet services in market, government framework and institutions such as the Ministry of Agriculture and KEPHIS; formation of cooperatives; financial institutions, etc.) proposed by the agro-input dealers using multivariate correspondence analysis. However, it is important to note that the division is our own creation aimed at increasing our understanding of the data. Correspondence analysis helps to find a dimensional representation of the dependence between categories of variables in a two-way contingency table (Hair et al., 1995). Multivariate correspondence analysis was conducted to relate the areas of policy interventions suggested by the agro-input dealers and institutions that they felt were in the best position to handle the situation. As expected, the result shows that both the policy-related interventions and the institution-related interventions were significantly correlated (Pearson chi-square = 208.504, $df = 100$, $p = 0.000$). The agro-input dealers who proposed training and credit and loans as a means to improve agro-input marketing also proposed government departments as the expected main actor. Government was also the main organization that was suggested to deal with illegal trading and adulteration of agro-inputs and improve marketing efficiency. The agro-input dealers who proposed the need to improve transportation suggested road improvement policies, and they were also associated with those who proposed extension policies and chemical distribution institutions and improved supply policies.

Agro-input dealers who proposed tax reduction and the curbing of illegal agro-input trade and adulteration proposed Kenya Revenue Authority as the best institution that could deal with the situation. Government role is important in promoting the expansion of input use (Kelly et al., 2003). The first dimension (horizontal axis) of the solution is related to road improvement and the curbing of illegal input trading and adulteration with contributions from extension and the university. The second axis (vertical axis) is related to training policies and favorable loan policies with government and Kenya Revenue Authority intervention as indicated by contribution of points to inertia of dimensions, which are equivalent to the interpretation of loadings in a principal component analysis (PCA) for numeric data reduction (data not shown). The suggestions of input dealers were found to be in tandem with those proposed by Kelly et al. (2003), including the need of government to invest in rural infrastructure (roads, markets, storage facilities, etc.), education, agriculture research and extension, and market information systems. Government, training, and loan provision were closely related as policies and institutions.

Conclusions and Way Forward

The study shows that although the number of agro-input dealers in Western Kenya has been growing, the growth is still a far cry from what is needed to ensure that smallholder farmers, especially those in far away rural communities, have adequate access to agro-inputs. Besides, agro-input dealers in Western Kenya still face enormous problems (e.g., infrastructure, low demand) in their business. Most of the agro-input dealers still cover long distances to source different agro-inputs – a situation that has continued to result in high farm-level prices for farm inputs. The problem of high unit price of agro-inputs is compounded by the fact that credit services were rarely provided to the smallholder farmers by the agro-input dealers who themselves lack access to credit and loans required for increased stocking of goods. Besides, only very few agro-input dealers (2–11%) were in a position to provide small-scale farmers with other services (especially input-related information).

The fact that agro-input dealers faced numerous constraints (high transport costs, low demand, lack of market, limited market information, lack of storage facilities, and limited skills and knowledge) is particularly worrisome because most of these are serious problems of infrastructure that require strong political will to address. Policy and institutional development are, therefore, of paramount importance for efficient agro-input market development in Western Kenya and similar environments in SSA. A similar measure is also needed on limited business skill and knowledge, especially considering that knowledge reduces risk and increases rewards.

As a way forward, government policy and institutional intervention (in areas such as reducing the risks that the private sector faces in rural markets, improving road and other infrastructure, developing and extending credit and loans to agro-input dealers and their networks) are critical in stimulating the input supply sector and in effectively ushering in sustainable green revolution in Africa.

Acknowledgment We acknowledge the agro-input dealers for giving us their attention during data collection.

References

Bashaasha B (2001) The evolution and characteristics of farming systems in Uganda. Department of Agricultural Economics, Makerere University. A resource paper for the Policies for Improved Land Management in Uganda Project. A collaborative project of International Food Policy Research Institute, German Center for International Development, Makerere University, Agricultural Policy Secretariat, and National Agricultural Research Organization

DMA TR 8350.2 (1991) Department of defense world geodetic system 1984, its definition and relationships with local geodetic system, 2nd edn. Defense Mapping Agency, Washington D.C.

FAOSTAT (Food and Agricultural Organization Statistical Database) (2003) http://apps.fao.org/page/collections?subset=agriculture

Hair JF, Anderson RE, Tatham RL, Black WC (1995) Multivariate data analysis, 4th edn. Prentice Hall, Upper Saddle River, NJ

IFDC (International Center for Soil Fertility and Agriculture) (2006) Report. 31(1). June

Jaetzold R, Schmidt H (1982) Farm management handbook of Kenya, vol II – natural conditions and farm management information – part A West Kenya (Nyanza and Western, Provinces). Ministry of Agriculture, Kenya, in Cooperation with the German Agricultural Team (GAT) of the German Agency for Technical Cooperation (GTZ). Mald, Kenya, p 397

Kelly V, Adesina AA, Gordon A (2003) Expanding access to agricultural inputs in Africa: a review of recent market development experience. Food Policy 28:379–404

Sanchez P (2005) Implementing the hunger task force recommendations in Africa. PowerPoint presentation. Hayes Memorial Lecture, University of Minnesota, 25 May 2005

SPSS (Statistical Package for Social Sciences) (2002) Statistical package for social sciences release 11.0. Prentice Hall, Chicago

Surrey DW (1997) Diffusion theory and instructional technology. Paper presented at the Annual Conference of the Association for Educational Communications and Technology (AECT), Albuquerque, NM, 12–15 Feb. Available at www.gsu.edu/wwwitr/docs/diffusion/

Yanggen D, Kelly V, Reardon T, Naseem A (1998) Incentives for fertilizer use in sub-Saharan Africa: a review of empirical evidence on fertilizer response and profitability. MSU International Development Working Paper No. 70. Departments of Agricultural Economics and Economics, Michigan State University, East Lansing, MI, 48824

Gender Differentials in Adoption of Soil Nutrient Replenishment Technologies in Meru South District, Kenya

E.G. Kirumba, D.N. Mugendi, R. Karega, and J. Mugwe

Abstract Understanding gender differentials in adoption of soil nutrient replenishment technologies is critical to their successful implementation by farmers. This study was conducted first to examine gender differentials in choices of technologies adopted at intrahousehold level. Second, to investigate socio-economic, institutional, and demographic factors influencing adoption, and finally to examine gender differences in the frequency of participation in project activities. The results indicated gender differences in the choice of cattle manure and inorganic fertilizer. Gender differentials were also observed in the frequency of participation in project activities. A logistic regression model developed revealed that different factors influenced adoption at intrahousehold level. In male-headed households, adoption was positively influenced by the number of cattle owned, the access to credit, the number of adults working on farm, and farmer group membership. For female-headed households, adoption was positively influenced by the area under cash crops, the number of goats owned, the number of adults working on farm, participation in project activities, and farmer group membership. There is a clear need for strategies and policy to address gender disparities in adoption of soil improvement technologies and to encourage women's participation in agricultural training activities.

E.G. Kirumba (✉)
Department of Environmental Science, Kenyatta University, Nairobi, Kenya
e-mail: gathoni_edith@yahoo.com

Keywords Adoption · Choice · Gender · Participation · Technologies

Introduction

Soil fertility depletion on smallholder farms has been cited as the fundamental biophysical root cause responsible for the declining per capita food production in Africa (Sanchez et al., 1996). Studies of soil nutrient balance across countries in Africa show evidence of widespread nutrient mining leading to severe nutrient deficiencies across ecological zones. Soil fertility decline is made all the more alarming, given that recurring devaluations and removal of subsidies have made inorganic fertilizers unaffordable to most small-scale African farmers (Mukhopadhyay and Pieri, 1999).

In the Central Kenya Highlands, farmers themselves have persistently expressed that low soil fertility is a major constraint to food crop production (Sanchez and Jama, 2000). One of the major factors contributing to this decline is soil impoverishment caused by continuous cropping without adequate fertilizers and/or manure, mostly due to lack of readily available resources to replenish the soil. As a result, deliberate efforts by scientists have led to the introduction of several soil nutrient replenishment technologies that integrate nutrient management which encompasses organic materials and mineral fertilizers. These technologies have been found to be both technically feasible and socially acceptable (Sanchez and Jama, 2000).

Despite these impressive advances over the last three decades, soil improvement projects commonly

A. Bationo et al. (eds.), *Innovations as Key to the Green Revolution in Africa*,
DOI 10.1007/978-90-481-2543-2_88, © Springer Science+Business Media B.V. 2011

suffer from inadequate rates of adoption and/or abandonment soon after adoption (Franzel et al., 2002). To address this situation, Sanchez (1995) highlighted "the need to develop a predictive understanding of how farm households make decisions regarding land use," as others argued for more socio-economic research on adoption of soil improvement technologies. Understanding the factors affecting farmers' adoption of improved technologies is vital to the success of implementing soil innovation programs. An important factor affecting adoption is gender.

The criticality of gender issues in adoption of soil innovations is increasingly gaining global recognition and there is a strong call for its integration in development projects and programs (Gladwin, 2002). Several studies have reported the existence of gender disparities in adoption of soil improvement technologies (Fort, 2007; Mwangi et al., 1996). For example, Franzel and Scherr (2002) found that gender differences in adoption of innovations reflect men and women's differential ability to independently decide on how resources will be used and allocated. The power differentials between men and women lay the foundation for gender bias from the household-level decisions to policy-level decisions.

Further, research indicates that participation in project activities related to dissemination of knowledge on soil improvement technologies encourages adoption. However, women have been found to participate in lesser numbers compared to men. The reasons why women do not participate include lack of time, lack of self-confidence, lack of information, or misunderstanding of project goals (Sarin et al., 2006).

Past research work in the study area primarily focused on biophysical aspects of adoption with limited work on socio-economic factors influencing adoption. However, there has been very little mention of gender differentials in adoption of soil improvement technologies at household level. Also lacking is the information on the choices of technologies for adoption between male- and female-headed households and the socio-economic, institutional, and demographic factors influencing adoption. Further, information was limited on how the frequency of participation in project activities varied among male and female farmers. This study focused on addressing these concerns. As such, the analysis will not only be helpful in predicating choices of technologies adopted by gender, factors influencing adoption but also provide gender-disaggregated data frequency of participation in project activities. These results could inform gender-sensitive policy making and enhance gender targeting in technology dissemination.

Methodology

The Study Area and Background

The study was carried out between November 2006 and January 2007, in 14 villages in Mukuuni location, Chuka division, Meru South District. The area is located in upper midland 2 and 3 (UM2 and UM 3), a predominantly maize growing area, which is also referred to as a coffee agro-ecological zone (Jaetzold and Schmidt, 1983). The altitude is 1200–1600 m above sea level with an annual mean temperature of 20°C. Annual rainfall varies between 1200 and 1400 mm. The rainfall pattern is bimodal, falling in two seasons, with the long rains between March and June and short rains between October and December. The soils are mainly Humic Nitisols (Jaetzold and Schmidt, 1983). The farming system in the area is characterized by integration of both crops and animals. A wide variety of species and breeds of livestock, which include cattle, goats, sheep, and poultry, are found in the area. The major cash crops grown in the area are coffee (*Coffee arabica*) and tea (*Camellia sinensis*) and food crops like maize (*Zea mays*) and beans (*Phaseolus vulgaris*).

Maize is the main staple food, which is cultivated from season to season (Muriu, 2005, Farmers decision making process in their preference for soil nutrient replenishment technologies in the Central Kenya Highlands, unpublished masters thesis presented to Kenyatta University). As a result of soil fertility decline leading to low yields and food insecurity, a project was initiated in the area in 2003. One of the objectives of this project was to promote adoption of newly introduced soil fertility improvement technologies through participatory approaches. To start with, a demonstration trial was set up and the technologies were demonstrated to farmers during field days, which were organized every growing season. The demonstration trials comprised of soil fertility improvement technologies, which were those technologies that farmers had been introduced to with a view to improving soil fertility and food security. They included adoption of *Tithonia diversifolia*, *Calliandra calothyrsus*, *Leucaena trichandra* and improved management of inorganic fertilizers and cattle manure. By the time the study was conducted in November 2006, the farmers had visited the demonstration plots eight times during the short and long rain seasons. During field days, village training workshops and nursery groups, farmers were taught on how to implement the technologies on their farms by scientists and extension agents. The farmers also exchanged information and technologies among themselves.

Sample Selection

A sampling frame of all 350 households in the area was obtained from the provincial administration office and availed by the area chief. All the households in the area practice farming as the major economic activity. Ninety of the households were female-headed households while 260 were male-headed households. A sampling frame of adopters already existed in the project records. From these records, 160 households had adopted at least one of the technologies by the time the study was conducted. When crosschecked against the chief's list, it was found that 108 male-headed households had adopted at least one technology while 152 had not adopted any. Fifty-two female-headed households had also adopted at least one technology, while 38 female-headed households had not adopted any of the introduced technologies. Male-headed households were sampled by the use of simple random sampling while female-headed households were purposively sampled because they were fewer in number. The sample size comprised of 140 households, 70 who were adopters and another 70 who were non-adopters. Of the 70 adopters, 35 were male-headed households while another 35 were female-headed households. Of the 70 non-adopters sampled, 35 were male-headed households while another 35 were female-headed households.

Data Collection

The study relied on both primary and secondary sources of data. Primary data were obtained by the use of semi-structured interview schedules, which were administered to the 140 respondents. The interview schedules generated information on the frequency of participation in project activities and choices of technologies for adoption by gender, demographic, institutional, farm, and socio-economic factors influencing adoption. In all households, household heads were interviewed. Secondary data were obtained from journals, text books, and periodicals.

Analytical Approaches and Empirical Model

The data collected were analyzed using the statistical software for social scientists (SPSS). Data summary was done by the use of descriptive statistics and presented as frequencies and means. Chi-square tests ($p < 0.05$) were run to determine significant statistical relationships between categorical variables. Independent sample *t*-tests ($p < 0.05$) were run to test for equality between two means.

The logistic regression model was used to analyze the gender differences in factors influencing adoption of soil nutrient replenishment technologies. The logistic regression model is a non-linear regression model that has a binary response variable. As such, it has a binomial distribution with parameter (probability of success). According to Pampel (2000), the model equation is as follows:

$$\text{Logit}(E[Y]) = \text{Logit}(P) = X^{\mathrm{T}}\beta$$

The model is based on the binomial logistic probability function. The Logit($E\,[Y]$) is the binary response variable. It represents the probability of the number of events being successful. It is an index reflecting the combined effects of factors that predict adoption. The Logit(P) is the natural log of the odds of success. It is defined in the open unit interval (0,1). $X^{\mathrm{T}}\beta$ is the product of the transpose of the column matrix X of explanatory variables and the unknown column matrix β of regression co-efficient. The logistic regression model accounts for both categorical- and dichotomous-dependent variables and has been widely used in adoption studies (Wekesa et al., 2003). The dependent variable was a dichotomous variable depicting adoption of a technology and took the value of 1 if the farmer had adopted at least one technology and 0 if none. The independent variables included demographic, socio-economic, farm characteristics, and institutional factors and are as described in the following section.

Demographic Factors

HHSIZE was a continuous variable that depicted the size of the household. Labor constraints often limit farmers' use of soil improvement innovations. Due to the high labor demand of biomass transfer technologies such as *T. diversifolia* (Gladwin et al., 2002), it was expected that the larger the household size, the greater the labor availability. As such, it was hypothesized that the household size would be positively related to adoption.

AGE, a continuous variable, indexed the age of the household head. Age may or may not affect potential adoption. Rogers (1983) observes that age has no definite direction on adoption whereas Obonyo (2000, The adoption of biomass transfer technology in Western Kenya, unpublished MSc thesis, Kwame Nkrumah University of Science and Technology, Ghana) reported a positive relationship between age and potential adoption of *Tithonia* biomass transfer technology. Consequently, the net effect on age could not be determined a priori.

Farm Characteristics

FARMSIZE was a continuous variable referring to the variable on the size of the farm in acres. Farm size dictates the amount of crops grown and input levels. Small farms have a greater likelihood of adopting improved varieties as they are more intensively managed (CIMMYT, 1993). Farm size has been found to influence adoption decisions positively (Sullivan, 2000, Decoding diversity: strategies to mitigate household stress, unpublished master's thesis, University of Florida). Accordingly, it was anticipated that farm size would have a positive influence on adoption.

FOODCROP indexed the area of land under food crops in acres. Wekesa et al. (2003) reported that a larger area under maize is considered to increase a farmer's interest in new technologies and therefore area under food crops was hypothesized to be positively related to adoption. CASHCROP was a continuous variable that indexed the area of land under cash crops. Irungu et al. (1998) in a study conducted among dairy farmers in Kenya found that the higher the household income, the greater the financial ability to adopt. As a result, it was conjectured that the area of land under cash crops would have a positive influence on adoption.

CATTLE indexed the variable on cattle ownership. It takes on the value of 1 if the farmer owns cattle and 0 if otherwise. GOATS indexed the variable on ownership of goats. It took the value of 1 if the farmer owned goats and sheep and 0 if otherwise. SHEEP refers to the variable on the number of sheep owned. It took the value of 1 if the farmer owned sheep or 0 if otherwise. Fodder tree technologies were more likely to be adopted by households with cattle, goats, and sheep (Neupane et al., 2002). It was thus postulated that ownership of cattle, goats, and sheep was likely to influence adoption positively.

Socio-economic Factors

EDUCATION measured the farmer's level of education. The variable took the value of 1 if the farmer had an education and 0 if otherwise. A higher level of education increases a farmer's ability to obtain, process, and use adoption information (Ouma et al., 2002). Education was thus posited to be positively associated with adoption.

ACCLAND indexed access to land resources. The variable took the value of 1 if there was access and 0 if there was no access. CONTLAND indexed control of land resources. The variable took the value of 1 if there was control and 0 if otherwise. According to Sanchez (1995), ability to access and control resources is likely to influence adoption positively. It was therefore predicted that access to and control of land were likely to influence adoption positively.

PARTCPTN was a dichotomous variable that indexed the variable on participation in project activities. The variable took on a value of 1 if the farmer participated and 0 if otherwise. It was hypothesized that participation would be positively related to adoption as it exposes farmers to information about the soil fertility improvement technologies (Kariuki and Place, 2005).

TENURE indexed whether the farmer had security of tenure. Security of tenure took a value of 1 if the farmer purchased or inherited land and 0 if otherwise. It is expected that borrowed, rented/leased will be negatively related to adoption, as renting or leasing land is a sign of land scarcity. Inherited or purchased land is likely to be used to take up technologies that include a tree component (Adesina and Chianu, 2002). This may have a positive influence on adoption.

HIRELABOR indexed the variable on the ability of the household to hire labor. The dichotomous variable took on the value of 1 if labor was hired and 0 if otherwise. Studies conducted in West Africa cited labor availability as a major limiting factor to adoption of soil fertility replenishment technologies. Others said that due to labor shortage they have not been able to adopt technologies that require extensive labor investments (Enyong, 1999). Consequently, availability of labor was expected to have a positive influence on adoption.

PERCEPTION indexed whether farmers perceived a soil fertility problem in their farms. It took the value of 1 if yes and 0 if otherwise. If farmers' perceptions are that soil fertility is not a problem, labor and capital resources will not be channeled toward this cause (Nabifo, 2003). As a result, farmers' perception of soil fertility as a problem was hypothesized to have a positive relationship with adoption.

OFFINCOME was a dichotomous variable that indexed the variable on the household's access to off-farm income. It took the value of 1 if yes and 0 if otherwise. Wekesa et al. (2003) reported that an employed household head who was permanently employed had an assured income and was more likely to hire labor and adopt recommended maize technologies. It was thus posited that off farm would influence adoption positively.

Institutional Factors

CREDIT depicted access to credit. The dichotomous variable took the value of 1 if there was access and 0 if otherwise. The statistically significant positive coefficient for access to credit was established in West Africa in a study identifying credit as a major constraint to adoption of planted forages (Elbasha et al., 1999). Access to credit was postulated to affect adoption positively.

EXTENSION was a dichotomous variable that measured if the farmer had contact with extension agents, taking on the value of 1 if there was such contact and 0 if none. Extension services are a major source of technical information for farmers. Njuki (2001, Gender roles in agroforestry: a socio-economic analysis of Embu and Kirinyaga districts, Kenya, unpublished PhD thesis presented to Sokoine University, Tanzania) reports that contact with extension agents is one of the most important factors that determine adoption and hence was expected to influence adoption decisions positively.

GROUP was a dichotomous variable that indexed whether the farmer was a member of any farmers' group, taking on the value of 1 if the farmer belonged to a farmer's group and 0 if he did not belong to any group. It has been found that soil improvement innovations have higher success rates in adoption when soil fertility management projects work through farmers' groups (Adesina and Chianu, 2002). It was thus theorized that membership in a farmers' group would be positively related to adoption.

Results and Discussion

Sample Characteristics

Demographic and Farm Characteristics

Results revealed gender differences in demographic and farm characteristics between adopter and non-adopters (Table 1). In male-headed households, adopters had larger farm sizes, larger household sizes, higher number of adults working on the farms, and higher number of cattle owned than non-adopters. For female-headed households, adopters had larger areas of land under cash crops, more adults working on the farm, and a higher number of goats. The age of the household head, area under food crops, area under cash crops, and the number of goats and sheep owned were not significantly different between adopters and non-adopters for male-headed households. Similarly, for female-headed households, there were no significant differences between adopters and non-adopters in relation to age of household head, farm size, area under

Table 1 Farmers' demographic and farm characteristics disaggregated by gender

	Male-headed households ($N = 70$)			Female-headed households ($N = 70$)		
	Mean			Mean		
Variable	Non-adopters	Adopters	*T*-statistic	Non-adopters	Adopters	*T*-statistic
Farm size (acres)	2.60	3.25	–2.845**	1.89	2.04	0.888 NS
Area under food crops (acres)	1.33	1.56	–1.532 NS	1.04	1.31	1.421 NS
Area under cash crops (acres)	0.46	0.63	–1.599 NS	0.11	0.30	3.199***
Age of HH head (years)	45.23	46.91	0.654 NS	47.51	47.14	–1.49 NS
Household size	3.26	4.86	4.149***	3.51	3.17	–1.249 NS
Number of adults working on farm	1.86	2.66	4.866***	1.23	1.74	4.085***
Number of cattle	2.80	4.11	4.464***	1.89	2.20	1.460 (NS)
Number of goats	9.31	10.91	–1.367 NS	4.09	5.94	–2.706***
Number of sheep	1.60	1.71	–0.344 NS	1.09	1.43	–1.142 NS

Legend: NS = not significant; ** = significant at $p < 0.05$; *** = significant at $p < 0.01$

cash crops, household size, the number of cattle and sheep owned (Table 1).

Socio-economic and Institutional Factors

For male-headed households, factors that were significant were participation in project activities, access to credit, and membership in a group. Perception of soil fertility as a problem, ability to hire labor, secure land tenure, access to land, control of land, education, off-farm income, and access to extension services were not significant for both male- and female-headed households. For female-headed households, significant factors were participation in project activities and membership in a group (Table 2).

Frequency of Participation in Project Activities by Farmers in Meru South District

Significant gender differences were found in the frequency of participation in project activities, where men were found to participate in larger numbers than women. Participation in problem diagnosis meetings, field days, and village training workshops was significantly higher among male-headed households in comparison to female-headed households (Table 3). Most female household heads (80%) reported that they had no time to attend project meetings as they were either busy in their farms or performing domestic chores. Njuki (2001, Gender roles in agroforestry:

Table 2 Farmers' socio-economic and institutional factors disaggregated by gender for farmers in Meru South District, Kenya

	Male-headed households ($N = 70$)			Female-headed households ($N = 70$)		
	Mean			Mean		
Variables	Non-adopters	Adopters	Chi-square value	Non-adopters	Adopters	Chi-square Value
Farmer's perception of soil fertility as a problem	77.1	68.6	0.650 NS	71.4	68.6	0.068 NS
Ability to hire labor	82.9	88.6	0.072 NS	71.4	74.3	0.402 NS
Has secure land tenure	94.3	97.1	0.952 NS	77.1	85.7	0.324 NS
Has access to land	85.7	94.3	1.263 NS	82.9	94.3	0.729 NS
Has control of land	94.3	91.4	0.402 NS	85.7	88.6	0.128 NS
Participates in project activities	51.4	77.1	5.040**	42.9	77.1	8.571***
Has education	97.1	100	0.514 NS	77.1	80	0.701 NS
Receives off-farm income	22.9	31.4	0.348 NS	20	25.7	0.324 NS
Has access to extension	82.9	82.9	0.094 NS	71.4	74.3	0.280 NS
Has access to credit	17.1	57.1	9.130***	20	37.1	2.520 NS
Belongs to a group	74.3	22.9	18.529***	31.4	71.4	20.696***

Legend: NS = not significant; ** = significant at $p < 0.05$; *** = significant at $p < 0.01$

Table 3 Frequency of participation in project activities by farmers disaggregated by gender in Meru South District, Kenya

Project activity	Mean number who participated		
	MHHs ($N = 70$)	FHHs ($N = 70$)	*T*-statistic
Farmers who attended problem diagnosis (yearly)	0.60	0.41	–2.158**
Farmers who attended field days (twice in a year)	0.96	0.63	–2.673***
Farmers who attended village training workshops (twice in a year)	1.04	0.74	–2.195**
Farmers who attended nursery group meetings (once a week)	1.51	1.54	0.101 (NS)

Legend: NS = not significant; ** = significant at $p < 0.05$; *** = significant at $p < 0.01$.

a socio-economic analysis of Embu and Kirinyaga districts, Kenya, unpublished PhD thesis presented to Sokoine University, Tanzania) reported that the men mainly participate in project activities; thus, women who provide much of the labor do not receive sufficient training required to implement the technologies on their farms.

Choices of Technologies for Adoption Disaggregated by Gender for Farmers in Meru South District, Kenya

The findings indicated significant gender differences in the adoption of cattle manure and inorganic fertilizer (Table 4). Adoption of cattle manure was higher among male-headed households (65.7%) in comparison to female-headed households (40%). Similar results were noted for adoption of inorganic fertilizers: 74.3% for male-headed households and 37.1% for female-headed households. These results coincide with the results reported by Lekasi et al. (2001), who reported women's lack of animal and pasture land limits their access to manure, and although organic fertilizer is important for maize in addition to inorganic fertilizer, women can rarely use it as they lacked animals as well as cash to procure it. However, no significant gender differences were found in adoption of *T. diversifolia*, *C. calothyrsus*, and *L. trichandra*.

Factors Influencing Adoption in Female-Headed Households

The logistic regression model was significant at $p < 0.05$ and correctly predicted 84.3% for both adopter and non-adopter female-headed households. This implies that the model predicted 84.3% of the total variations in adoption of soil nutrient replenishment technologies and was therefore very reliable. The exponential beta (β) or odds ratio indicated the proportion with which adoption could occur, while the beta (β) sign predicted whether the variable influenced adoption decisions positively (+) or negatively (–). The results indicate that adoption by female-headed households was significantly influenced by the number of goats owned, membership in a group, area under cash crops, participation in project activities and the number of adults working on the farm.

The number of goats owned influenced adoption positively. The model predicted that a unit increase in the number of goats owned by a female household head was 1.37 times likely to increase adoption of soil nutrient replenishment technologies. This may have been so because most female household heads could not afford cattle and mainly fed their legume technologies to goats. Goats were cheaper to purchase and more risk averse. A study of high-potential areas in Kenya showed that on small farm sizes where farmers had no access to credit and inputs, dairy goats were

Table 4 Choices of soil nutrient replenishment technologies for adoption by farmers disaggregated by gender in Meru South District, Kenya

Technology adopted	Frequency of adopters (%)		
	MHHs ($N = 35$)	FHHs ($N = 35$)	Chi-square value
Cattle manure	65.7	40	4.644**
Tithonia	82.9	80	0.094 NS
Leucaena	22.9	17.1	0.357 NS
Calliandra	71.4	54.3	2.203 NS
Inorganic fertilizer	74.3	37.1	9.785***

Note: NS = not significant; ** = significant at $p < 0.05$; *** = significant at $p < 0.01$

more profitable and less risky than dairy cattle (IFAD, 2007). Generally, men and women tend to own different animal species. In many societies, cattle and larger animals are usually owned by men, while smaller animals, such as goats and backyard poultry which are kept near the house, are more women's domain (Fort, 2007).

A farmer's members into a group increased the chances of adoption. The more farmer's that joined groups, the higher the possibility of taking up technologies. This may have been because farmers were able to access legume seedlings and also receive adequate training and skills in groups. These results were supported by Adesina and Chianu (2002), who found that soil improvement innovations had higher success rates in adoption when soil fertility management projects work through farmers' groups.

Farmers with larger areas of land under cash crops were more likely to adopt than those with smaller areas. Indeed, a unit increase in the area of land under cash crops was 23.16 times likely to increase adoption. This may have been because income generated from cash crops could be used to hire labor and purchase required inputs. Wekesa et al. (2003) also found that the area of land under cash crops dictated a farmer's income and thus his ability to invest in soil fertility improvement technologies.

Farmers' participation in project activities significantly and positively influenced adoption. This could be attributed to the fact that through participation, farmers became aware of the technologies, were trained, and also could visit demonstration plots to learn more. These results are corroborated by Njuki (2001, Gender roles in agroforestry: a socio-economic analysis of Embu and Kirinyaga districts, Kenya, unpublished PhD thesis presented to Sokoine University, Tanzania), who reports that farmers who participate in farmer training courses are more likely adopters than those who do not.

An increase in the number of adults working on the farm was likely to have a resultant increase in adoption. This was probably given that most households in the area relied on family labor and with the free primary education initiative by the government; labor from children was unreliable as more children enrolled in schools. These results coincide with Mwangi et al. (1996), who found that households with a higher number of adults working on farm were more likely to adopt compared to those with fewer adults (Table 5).

Logistic Regression Results for Male-Headed Households

The logistic model estimate was significant at $p < 0.05$ and correctly predicted at 90% for both adopter and non-adopter male-headed households, suggesting that the model's precision in prediction was very high at 90% and was thus dependable. The results revealed that the number of adults working on farm, access to credit, member in a farmer's group, and the number of cattle owned significantly and positively influenced adoption.

A unit increase in the number of adults working on farm was likely to increase adoption by 8.19 times. This may have been due to the fact that most households in the area mainly relied on family labor, which was readily available and cheap. Studies conducted in West Africa cited labor availability as a major limiting factor to adoption of soil fertility replenishment technologies (Enyong, 1999).

Farmers who accessed credit were more likely to adopt than those who did not. This may have been possibly because financial access translated to access to purchasable inputs, including hired labor. Mwangi et al. (1996) reported that access to credit by male-headed households encouraged adoption of improved maize varieties in Tanzania.

Table 5 Logistic regression parameter estimates for female-headed households

Variable	β	S.E.	Wald	Sig.	exp(β)
Number of adults working on farm	2.656	0.832	10.192	0.001	14.246
Farmer belongs to a group	2.116	0.785	7.257	0.007	8.297
Number of goats owned	0.315	0.146	4.665	0.031	1.370
Area of land under cash crops (acres)	3.142	1.683	3.486	0.062	23.160
Farmer participates in project activities	1.447	0.721	4.030	0.045	4.251
Constant	–7.938	2.009	15.615	0.000	0.000

Table 6 Logistic regression parameter estimates for male-headed households

Variable	β	S.E.	Wald	Sig.	exp(β)
Number of adults working on the farm	2.103	0.754	7.777	0.005	8.190
Access to credit	2.475	1.100	5.061	0.024	11.880
Farmer belongs to a group	3.107	1.075	8.347	0.004	22.355
Number of cattle owned	1.323	0.437	9.180	0.002	3.754
Farm size in acres	0.307	0.440	0.488	0.485	1.359
Farmer participates in project activities	1.387	1.005	1.907	0.167	4.005
Constant	−13.336	3.419	15.212	0.000	0.000

A farmer's membership into a group was a motivation for adoption. A unit increase in group membership was found to possibly increase the adoption by 22.4 times. The reason for this may have been that farmers who were group members could access legume seedlings and receive training and short loans with which they could purchase inputs. These findings are supported by Kariuki and Place (2005) who found that a farmer's membership in a group increased adoption as farmers exchanged information and obtained resources through groups.

Farmers who owned more cattle were likely adopters in comparison to those who did not own as many. The explanation for this could probably be that cattle ownership is a sign of wealth and thus well-to-do farmers could also afford to buy other inputs and also hire labor. Sullivan (2000, Decoding diversity: strategies to mitigate household stress, unpublished master's thesis, University of Florida) found that the number of cattle owned influenced adoption decisions in male-headed household. Farm size in acres and participation in project activities influenced adoption positively though not significantly (Table 6).

Conclusions, Implications, and Recommendations

This study affirms the need for gender consideration in adoption of soil nutrient replenishment technologies. The choices of technologies for adoption were differentiated by gender, with female-headed household adopting fewer technologies than male-headed households. Significant gender differences were observed in the choices of cattle manure and inorganic fertilizers for adoption, with female-headed households using these technologies at lower levels in comparison to male-headed households.

Factors significantly influencing adoption in male-headed households were found to be access to credit, membership in a farmer's group, the number of adult household members working on farm, and the number of cattle owned. Those that significantly influenced adoption in female-headed households were found to be the following: participation in project activities, membership in a farmer's group, area of land under cash crops, the number of goats owned, and the number of adult household members working on farm. Further, significant gender differentials were also noted in the frequency of participation in project activities such as village training workshops, field days, and problem diagnosis meetings. There is a clear need for researchers to target technologies by gender to encourage adoption by both male- and female-headed households.

Further, successful implementation of projects will need to take into account the equal participation of women and men in all project activities. Interventions targeted at women need to consider women's available time, not only for new activities but also for participating in meetings and committees. Based on the fact that adoption by male- and female-headed households is influenced by different factors, it is vital that these factors are considered as a reference point when conducting studies on adoption and gender. In addition, future researchers need to investigate gender differentials at interhousehold level and also determine whether farmers who participated in project activities disseminated the knowledge acquired to their spouses and community members. Finally, gender literacy for policy makers and project implementers requires immediate attention.

Acknowledgments The authors wish to thank RUFORUM for the financial support. They also wish to appreciate the collaboration of Kenyatta University (Department of Environmental Science). The cooperation and contributions by researchers and

Chuka community members (farmers, administrators, and agricultural extension staff) are acknowledged.

References

Adesina AA, Chianu J (2002) Determinants of farmers' adoption and adaptation of alley farming technology in Nigeria. Agroforest Syst J 55:27–35

CIMMYT Economics Program (1993) The adoption of agricultural technology: a guide for survey design. CIMMYT, Mexico, DF

Elbasha E, Thornton PK, Tarawali G (1999) An expost economic assessment of planted forages in West Africa In: ILRI Impact Series, vol 2. ILRI, Nairobi, Kenya

Enyong LA (1999) Farmers' perceptions and attitudes towards introduction of soil fertility enhancing technologies in Western Africa. ICRISAT, West Africa

Fort L (2007) Collecting gender data on access and ownership of economic assets. The World Bank Gender and development programme, Washington DC, USA

Franzel S, Arimi H, Murithi F (2002) Calliandra calothyrsus: assessing the early stages of adoption of a fodder tree in the highlands of Central Kenya. In: Franzel S, Scherr SJ (eds) Trees on the farm: assessing the adoption potential of agroforestry practices in Africa. CAB International, Wallingford, UK, pp 125–33

Franzel S, Scherr S (2002) Trees on the farm: assessing the adoption potential of nutrient replenishing technologies in Africa. CAB, UK

Gladwin CH, Peterson JS, Uttaro R (2002) Agroforestry innovations in Africa: can they improve soil fertility in women fields. J Afr Stud 6:12–18

International Fund for Agricultural Development (2007) Livestock development in East and Southern Africa: some features if IFAD's policies and programmes. IFAD, Rome, Italy

Irungu P, Mbogo S, Munei K, Staal S, Thorpe W, Njubi D (1998) Factors influencing adoption of napier grass in smallholder dairy farms in the Highlands of Kenya. In: Food, lands and livelihoods: setting the research agenda for animal science, KARI, Nairobi, Kenya

Jaetzold R, Schmidt H (1983) Farm management handbook of Kenya, Natural conditions and farm information vol 11/C East Kenya. Ministry of Agriculture, Kenya

Kariuki G, Place F (2005) Initiatives for rural development through collective action: the case of household participation in group activities in the highlands of central Kenya. CAPRi Working Paper # 43

Lekasi JK et al (2001) Managing manure to sustain smallholder livelihoods in the east African highlands. HYDRA Publications, Kenilworth, UK

Mukhophadhyay M, Pieri C (1999) Integration of women in sustainable land and crop management in Sub-Saharan Africa. Collaborative effort between Gender in Rural Development (GENRD) team and Sustainable Land and Crop Management (SLCM) Thematic Team

Mwangi W et al (1996) Gender differentials in adoption of improved maize production technologies in Mbeya region of the southern highlands of Tanzania. CIMMYT/ United Republic of Tanzania

Nabifo P (2003) The acceptance and profitability of alternative soil improvement practices in Tororo District. Published Masters Thesis Submitted to Makere University, Uganda

Neupane RP, Sharma KR, Thapa GB (2002) Adoption of agroforestry in the hills of Nepal: a logistic regression analysis. Agric Syst 12:177–196

Ouma N et al (2002) Adoption of maize seed and fertilizer in Embu District, Kenya. CIMMYT, Nairobi

Pampel FC (2000) Logistic regression: a primer. Sage quantitative applications in the social sciences series #132. Sage, Thousand Oaks, CA, pp 54–68 provide a discussion of probit

Rogers EM (1983) Diffusion of innovations. Free press, New York, NY

Sanchez PA (1995) Science in agroforestry. Agroforest Syst 30:5–55

Sanchez PA, Izac A-MN, Valencia I, Pieri C (1996) Soil fertility replenishment in Africa: a concept note. In: Breth SA (ed) Achieving greater impact from research investments in Africa. Sasakawa Africa Association, Mexico City, pp 200–207

Sanchez PA, Jama BA (2000) Soil fertility replenishment takes off in East and Southern Africa. ICRAF, Nairobi, Kenya

Sarin M et al (2006) An example from India: a women's groups' suggestions for women's participation in project activities. People's Education and Development Organization (PEDO, India

Wekesa E et al (2003) Adoption of maize production technologies in the coastal lowlands of Kenya. Kenya Agricultural Research Institute (KARI) and International Maize and Wheat Improvement Center (CIMMYT, Mexico, DF

Enhancing Agricultural Production Potential Through Nutrition and Good Health Practice: The Case of Suba District in Kenya

O. Ohiokpehai, T. Hongo, J. Kamau, G. Were, J. Kimiywe, B. King'olla, D. Mbithe, L. Oteba, G. Mbagaya, and O. Owuor

Abstract Several studies have shown that HIV and nutrition operate in tandem. Moreover, it has been shown that the two greatly affect agricultural production due to reduced energy to work, inability to purchase agricultural inputs, low labor, and eventual death. The link between agricultural productivity, malnutrition, and HIV can therefore not be overlooked. People who are inadequately nourished are more susceptible to diseases and poor health. In an attempt to achieve optimal nutrition and good health among vulnerable groups, various intervention programs have used food supplementation and especially the plant-based food products to achieve this. Such programs have proved to be effective in restoring the nutrition and health status of the people. However, much more value would be achieved if such programs were complemented with basic health services such as deworming, water, sanitation, malaria control, hygiene. This chapter explores the benefits of research on nutrition as the basis for improving threatened rural communities' nutria-health and potential economic performance. The premise is that good nutrition and preventive measures will reverse some of the human health problems associated with HIV, hunger, and/or malnutrition. Emphasis is placed on food preservation, processing, nutrition intervention, and education. Micronutrients through agronomic fortification/fertilization is recommended as an intervention with the benefit of improving the nutritive quality of food and thereby providing essential elements needed by the human body to combat malnutrition and poor health. The conclusion is that better nutrition will contribute to better health and increase productivity and production on the farm.

Keywords Corn–soy blend · Nutria-health · Processing · Soybeans · Threatened communities

O. Ohiokpehai (✉)
Tropical Soil Biology and Fertility Institute of the International Centre for Tropical Agriculture (TSBF-CIAT), Nairobi, Kenya
e-mail: oohiokpehai@yahoo.com

Introduction

The HIV pandemic has grossly affected development and nutrition security in sub-Saharan Africa (SSA). The poor health conditions associated with HIV have resulted in people living in poor socio-economic conditions and poor health. Due to this problem, people are unable to grow adequate food for their nutritional needs leading to food insecurity (UNAIDS, 2001). Inadequate household food security or prolonged lack of availability of food in the home results in poor nutrition and malnutrition. Undernutrition and growth retardation in young children are associated with reduced physical activity, impaired resistance to infection, impaired mental development, reduced educational capacity, and increased morbidity and mortality (FAO/WHO, 1992).

HIV has made it difficult for households to be food secure; yet food insecurity and malnutrition may hasten spread of HIV by increasing people's vulnerability and exposure to the virus and the risk of infection following exposure (Gillespie and Kadiyala, 2005). Food insecurity along with poor health and care may lead to increased malnutrition rates which can lead to increased transmission by lowering immunity. Gillespie and Kadiyala (2005) showed that a drought in 2002 in Southern Africa interacted with HIV in

A. Bationo et al. (eds.), *Innovations as Key to the Green Revolution in Africa*,
DOI 10.1007/978-90-481-2543-2_89,

high-prevalence areas to cause rapid deterioration of nutrition in children. The study showed large changes in child underweight prevalence which increased from 5 to 20% in Maputo (Mozambique 1997–2002), 17 to 32% in Copperbelt (Zambia 1999–2001/2002), and 11 to 26% in Midlands province (Zimbabwe 1999–2002).

The relationship between HIV and food insecurity can be demonstrated further by the severe drought that happened in South Africa in 2002 where many households without active adults earned 31% less income than those with active adults and those with chronically ill adults had 66% less income than households without chronically ill adults. In Zambia, households with a chronically ill head planted 53% less and were 21% more food insecure than those without a chronically ill head of household (De Wagt and Connoly, 2006).

Further, HIV lowers household food production as shown by a study in Kenya whereby the death of a household head reduced the value of the household crop production by 68%. A similar study in Rwanda showed that 53% of households had a less nutritious diet when the father had died and 23% when the mother had died. About 42% of households had a less nutritious diet when the father was ill while 34% had a less nutritious diet when the mother was ill. A study by World Bank (1993) showed that among the poorest 50% of households in Kagere, Tanzania, food expenditure was reduced by 32% and food consumption by 25%. About 47% of Kenyans lack adequate food to meet their nutritional needs leading to stunting and underweight rates of 31 and 20%, respectively, among children below 5 years of age (GOK/MOH, 2006). Suba district experiences food shortages due to factors such as the high disease burden in the area including HIV which has a prevalence of 31%, the highest in Kenya (MOH/NASCOP, 2006). In addition, the community is traditionally fishermen and therefore gives little importance to agricultural production leading to food insecurity.

According to FAO/WHO (1992), nutritionally balanced food aid is important for those who are food insecure or at risk. Mortality and morbidity data show that AIDS has a significant impact on ability of people to produce, transport, sell, and buy food leading to loss of income and malnutrition. Since malnutrition leads to increased vulnerability to HIV infection and hastens progression to AIDS, life-sustaining foods and safe fluids are the only realistic cost-effective approaches to manage the disease until medication becomes available to all and accessible to people living with AIDS (Kraak, 2001). In addition, the global school feeding campaign by World Food Program promotes policies that make food aid conditional on girls' participation in education, an essential package that includes school sanitation and water and environmental improvement; nutrition education that improves the quality of students' diets; and HIV prevention education and nutrition services that include food, deworming, and alleviation of short-term hunger (Donald et al., 2005).

School meals programs have been successful in supporting national nutrition and education goals (McGuffin, 2005). However, to provide high-quality, safe, nutritious foods to children requires many types of trained personnel in content such as food safety, sanitation and storage, foods and nutrition, leadership and administration, and record keeping. According to Probate et al. (2003), successful concepts in nutrition and health education carried out in Burkina Faso were those that addressed specific needs of target audience, those that showcased successful programs and practices, and those that used precise, accurate, relevant, and compelling training materials (Probate et al., 2003). The aim of this chapter is to bring out the link between nutrition and health interventions and enhancement of agricultural productivity.

The entry point and main objective of this chapter is to expound on the benefits of nutrition research as a means of understanding the place of nutrition in human health and especially those infected/affected by HIV and also as the basis for improving threatened rural communities (fishing and farming) nutria-health. The premise is that good nutrition will reverse some of the human health problems associated with HIV, hunger, and/or malnutrition and thereby improve their potential economic performance and food production. Suba district in the Nyanza province of Kenya with a HIV prevalence of 31% (MOH/NASCOP, 2006) has been selected to make and prove our case.

Materials and Methods

The study employed a descriptive experimental research design with experimental groups and controls. The study targeted HIV infected and affected children from Mbita, Lambwe, and Central divisions,

Suba district. A baseline survey was first carried out to establish socio-demographic characteristics, food consumption patterns and nutritional status of children, and morbidity patterns. HIV affected and vulnerable school children aged 6–9 years were enrolled into a feeding trial using corn–soy blend. The school children took the porridge at school during the mid-morning break. Water for the preparation of porridge was treated with water guard. All the children in the study were supplemented with vitamin A and given anti-helminthic drugs for expulsion of worms and an insecticide-treated mosquito net for malaria control.

Data were collected using a structured interview schedule where the primary caregivers were interviewed to obtain socio-demographic data, dietary intake, and morbidity. Food frequency table and 24 h recall were used to gather the children's food consumption patterns at baseline and every month of the feeding trial. Anthropometric assessments, mainly weight, height, and mid-upper arm circumference controlled for age, were performed on the children at baseline and after every month of feeding using standard anthropometric tools to establish the children's nutritional status. Observation checklist was used to collect information on any confounding factors that may affect the nutrition and health status of the children. The instruments were pre-tested in a nearby division which was exempted in the main project. Data were analyzed using SPSS and Nutri Survey computer packages and expressed in descriptive and inferential statistics. A p value of < 0.05 was considered significant.

Nutrition and health education was given to the parents/guardians of all the children in the study area and the school children themselves. This package focused on issues such as nutrients, their functions and sources, importance of good nutrition, and barriers to adequate nutrition. The participants were also enlightened on management of nutritional deficiencies and food security including factors that promote food security in the community. Parents/guardians were educated on the importance of soybeans in the diet and their nutritional value. Under health education, personal, food, and environmental hygiene was taught. Water treatment, latrine usage, and the importance of sanitation were also emphasized.

Soybean processing at the household and community level in Suba district focused on training mostly women groups, HIV support groups, and community-based organizations on the benefits of soybean in the household food consumption. Farmers who had soybean trials on their farms were also trained on processing and utilization following the harvest. Food fortification and incorporation of soybeans into locally consumed foods such as maize, millet, and sorghum were carried out to improve the nutritional value of the local dishes. Soybean handling and storage and thus preservation were taught to encourage sound post-harvest practices which have been lacking in this district leading to food insecurity. Due to the sensitive nature of microorganisms especially on the background of HIV infection, food hygiene and sanitation were given paramount importance and included in the training.

Results and Discussion

Food and Animal Production and Consumption

Baseline results indicate low level of land ownership in Suba with mean total land area of 1.8 acres owned by a single household and a mean 1.2 acres used for food crop cultivation. There is low food crop production with nearly all the households (91.7%) growing little maize with an annual mean production of 450 kg (five 90 kg sacks). Whereas this situation persists, there is a high consumption rate of maize, millet, and fermented porridges with 76.4, 46.1, and 36.2% inhabitants, respectively, consuming the grain-based foods. It is important to note that most of the food consumed is bought as opposed to own production. About 31.1% of the respondents reported not to be cultivating land at the time of data collection. Out of those who were cultivating, 48.4% of them reported to grow maize, millet, and beans while 10.2, 3.2, and 7.1% reported to grow maize and sorghum, tubers (cassava and sweet potatoes), and vegetables (kale and tomatoes) and fruit, respectively. Table 1 shows that 37% of the average household food consumed was purchased. About three-quarters of the population (72.4%) reported not to use soybean in any form. Among those that used soybean, 2.8 and 21.7% purchased the product from shops and open air markets, respectively. The quantities bought were low with an average of half a kilogram once a month. Nearly all households (94.1%) were not growing it. After offering nutrition

Table 1 Sources of food by the households in Suba district

How food was obtained	Sub-location Township (%)	Kayanja (%)	Kaswanga (%)	Total (%)
Own produce and assistance	0.4	0	0.8	1.2
Purchase	31.3	4.5	1.5	37.4
Own farm and purchase	23.4	13.6	17	54
Assistance and purchase	2.6	0	0	2.6
Own produce, assistance, and purchase	0.8	0.4	0.8	1.9
Others (food for work)	2.6	0.4	0	3

education and training on soybean processing and utilization, there is a growing interest in the crop with more people including it in their diets. Farmers are also interested in planting soybeans and three farmer groups have already harvested their crop and intend to continue growing it annually. Results indicated that cereals (maize, millet, and sorghum) and cassava were the most consumed foods on a daily basis by children. These were consumed in the form of mixed flour porridges. The most consumed vegetable was kale (50.9%) followed by African leafy vegetables mostly from the family pot. Fresh milk (47.2%), large fish (41.9%), and omena (29.4%) were the most consumed proteins. Except for avocado, oranges, lemon, and pawpaw, other fruits were not consumed 7 days prior to the survey because they were not in season. Further, most children consumed fruits once a week because they were possibly bought once a week on a market day. During periods of food insufficiency, majority of the population (81.6%) purchase food (Table 2). The focus group discussions confirmed these results.

Farming equipment use reflected low levels of farming implements with 46.9% using hand *pangas*/hoes while 26.8, and 0.8% using oxen and tractors, respectively. This can possibly be explained by the low ownership of land and the fact that the community is traditionally fishermen. About three-quarters of the respondents (75.2%) reported to own livestock with 36.2, 10.2, 42.9, 65.4% reporting to have cows, sheep, goat, and poultry, respectively. However, households owned just a few animals with an average of two cows, a sheep, three goats, and seven chickens per household. It is expected that the nutrition intervention will motivate people to improve on agricultural productivity including rearing livestock that will provide more food to the community while supplementing the traditional foods that are grown in the area. This is already taking effect with the renewed interest of people on modern agricultural activities and adoption of new high-quality crops such as soybeans.

Income Levels

The mean household monthly income of the study population was approximately Ksh. 5,000 while the maternal income was Ksh. 1,550. This converts to about US $100 per month and US $3.3 per day. This income is too low to meet all the needs of the household including dietary requirements thus placing this population in the low socio-economic status category. These results further agree with the Geographic Dimensions of Wellbeing in Kenya (2005) which shows that 64–74% of the population lives below the poverty line (Central Bureau of Statistics, 2005).

Table 2 Coping strategies in times of food insufficiency and by sub-location in Suba district

Coping strategies	Sub-location Township (%)	Kayanja (%)	Kaswanga (%)	Total (%)
Purchase	49.4	16.1	16.1	81.6
Assistance	1.1	0.4	0.8	2.3
Food for work	2.6	0.8	0.8	4.2
Selling assets	1.2	0	0.4	1.6
Purchase reduced amounts	6.8	1.6	1.9	10.3
Total	61.1	18.9	20	100

Nutritional Status of Study Children and Effect of the Intervention

The baseline survey showed very high malnutrition levels especially among the 6- to 9-year-old children. Stunting, underweight, and wasting on average stood at 8, 16, and 7%, respectively. These figures agreed with a study by UNESCO (2004) which identified stunting and underweight as a common occurrence in school-age children with stunting levels prevailing at 18% among Kenya's school-going children and growth retardation at 34%. By the end of 3 months feeding trial on corn–soy blend, there was notable improvement in the children's nutritional health with regard to weight-for-age (underweight), height-for-age (stunting), and weight-for-height (wasting) (Table 3). The children were reported to be more active in class and eager to learn. School absenteeism among experimental schools in Suba dropped from 18 to 4%. This agrees with various studies which have shown that girls who are better nourished are more attentive and more involved during class and boys have improved classroom behavior and increased activity levels. Stunted children enroll in school later than other children. School foodservice programs have been successful in improving school attendance before (Donald et al., 2005). Further, it has been established that increase in school attendance and enrollment is directly related to school feeding (McGuffin, 2005).

The need for a package of health and nutrition services including micronutrient supplementation (vitamin A, iron, and iodine), issuance of bed nets, deworming, health and nutrition education, and improved hygiene and sanitation cannot be overemphasized whether in schools or in a community setting. Micronutrient deficiencies including iron deficiency, the most common form of micronutrient deficiency in school-age children, are caused by inadequate diet and infection, particularly by hookworm and malaria. More than half the school-age children in low-income countries are estimated to suffer from iron deficiency anemia. Children with iron deficiency score 1–3 SD worse on educational tests and are less likely to attend school. Iron supplementation reduces these deficits (Donald et al., 2005). Iron supplementation of children in Burkina Faso supported by the Helen Keller Foundation led to a drop in anemia from 50% before the intervention to 21% (Sifri et al., 2003).

Although we did not collect on micronutrient and vitamin A deficiency among the children and adults, vitamin A deficiency affects an estimated 85 million school-age children. The deficiency, which causes impaired immune function and increases risk of mortality from infectious disease, is an important cause of blindness. Recent studies suggest that this deficiency is also a major public health problem in school-age children. Vitamin A supplementation of school children in Suba improved their general health. Studies show that multiple micronutrient supplements have improved cognitive function and short-term memory in schoolchildren and have reduced absenteeism caused by diarrhea and respiratory infections.

Worm infestation especially among children is a common phenomenon in Kenya and most parts of Africa. The most intense worm infections and related illnesses occur at school age and account for some 12% of the total disease burden and 20% of the loss of disability-adjusted life years (DALYs) from communicable disease among schoolchildren (World Bank, 1993). Infected schoolchildren perform poorly in tests of cognitive function; when they are treated, immediate educational and cognitive benefits are apparent only for children with heavy worm burdens or with concurrent nutritional deficits. Deworming of school children in Suba led to a decline in absenteeism and better

Table 3 Effect of feeding intervention on the nutritional value of school children in Suba

	Before intervention (%)			3 months after intervention (%)		
Malnutrition level	Sindo school (fishing area)	Lambwe school (farming area)	Control school (Ong'ayo)	Sindo school (fishing area)	Lambwe school (farming area)	Control school (Ong'ayo)
Stunting[a]	28.9	20.4	28.5	16.7	7.1	21.5
Underweight	10.2	9.8	11.4	6.2	2.2	14.3
Wasting	5.6	9.8	8.7	3.4	0.0	9.5

[a]Stunting is the measure of malnutrition of a population. WHO recommends that if 20% and above of a population is stunted then the population is considered malnourished

health. This agrees with a study done elsewhere in Kenya where treatment of worm infestations reduced absenteeism by one-fourth, with the largest gains for the youngest children who suffered the most ill health (Miguel and Kremer, 2004).

School health and nutrition programs that help children complete their education and develop knowledge, practices, and behaviors that protect them from HIV infection as they mature have been described as a "social vaccine" against the disease. In Kenya, primary school students miss 11% of school days because of malaria, equivalent to 4–10 million days per year (Brooker et al., 2000). Issuance of treated mosquito nets to children in Suba reduced the prevalence of malaria from 22 to 15%. This agrees with a study carried out in Tanzania where the use of insecticide-treated bed nets reduced malaria and increased attendance. Further, oral anti-malarial treatment reduced school absenteeism by 50% in Ghana. These basic health services can be provided to communities while emphasizing the need for growing quality crops that are nutritious and high in micronutrients such as vitamin A.

In light of the above interventions, there is a positive response from parents/guardians of the study children in Suba and the general population. Everyone has learnt to appreciate the good health that their children have achieved leading to a higher motivation to enhance agricultural production and thus promote good nutritional health. Various mechanisms can be used to achieve higher agricultural production such as fertilization and sustainable agriculture that will enhance nutria-health and improve the economy.

Sustainable agriculture should add value to family livelihoods in terms of markets, nutrition, and health. At the same time it is expected to improve soil fertility and income generations and facilitate availability of markets to farmers. Adoption of multipurpose crops such as promiscuous soybeans in many parts of Western Kenya, including Suba, will lead to better soil quality (striga control) and increase of yields of inter and subsequent crops due to N-fixation by soybean. Family incomes will improve due to the multiple benefits of soybean as a food and a cash crop. The farmers are also able to process some of the soybeans for household consumption thus improving the nutrition of household members, especially children who are more exposed to malnutrition that is high in the area (Table 3).

Soybean processing and utilization and training of trainers, farmers, support groups, and community-based organizations in Suba district are expected to cause a ripple effect whereby it is expected that there will be more diversity in peoples' diets in the household thus benefiting from the high protein value and other nutrients in soybean diets. It is also expected that more soybean will be grown in the area to supplement the traditional staple foods such as maize and sorghum. This is especially so because the feeding trial using soybean has served as a demonstration to parents/guardians and community members of the high nutritional value of soybean and its effect on nutria-health. Table 3 has shown a drop. With time surplus in production will be realized which can be sold or processed into products that can be sold to generate income. This will boost the economic well-being of the households in the region. Consequently, this will lead to enhanced agricultural production of good-quality crops that can address HIV, hunger, and malnutrition in Kenya and SSA in general.

Conclusions

The feeding program improved the children's health in both the farming and the fishing communities. Good nutrition can thus improve the health of the general population and result in improved labor productivity, enhanced income, and overall well-being of the community. The data show that even in the fishing areas crop production must be advocated. Similarly, in farming areas, improved food/nutrition security can be achieved through production and consumption of high-quality foods. An African Green Revolution, which will lead to improved agricultural production, should result in improved nutrition and health of the populace. The private sector has a role to play in nutritional enhancement especially for kids. Not only is an increase in production required but also proper transformation to enable populace harness the goodness of the foods. Improved health will diminish the susceptibility of the population to infectious disease especially HIV infection. Reduction in infectious diseases will decrease expenditure in health care. The surplus money gained can then be utilized in other productive activities. As scientists we should give robust information to farmers so that they can act purposefully.

Acknowledgments The authors sincerely acknowledge Rockefeller Foundation for funding this research and all those who took part in the study, including the parents/guardians of study children and enumerators. The authors are also thankful to the Ministry of Health staff, Suba District, for their cooperation and collaboration during the study.

References

Brooker S et al (2000) Situation analysis of malaria in school-aged children in Kenya: what can be done? Parasitol Today 16:183–186

Central Bureau of Statistics (2005) Geographic dimensions of well- being in Kenya. Who and Where are the poor? A constituency level profile, vol ll. CBS, Nairobi

De Wagt A, Connoly M (2006) Orphans and the impact of HIV/AIDS in Sub-Saharan Africa. In: Food Nutrition and Agriculture (FAO, 2006/34)

Donald A et al (2005) Disease control priorities. School-based health and nutrition. In: Nutrition Matters. No 79. (92007)

FAO/WHO (1992) Major issues for Nutrition strategies: preventing and managing infectious diseases. International conference on Nutrition. Theme paper No. 3. Rome

Gillespie S, Kadiyala S (2005) HIV/AIDS and food and nutrition security. From evidence to action, food policy review. International Food Policy Research Institute (IFPRI), Washington, DC

GOK/MOH (2006) Kenya national guidelines on nutrition and HIV/AIDS. Ministry of Health. Kenya

Kraak V (2001) Soybean farmers offer information in conjunction with world AIDS day. http://www.PRNewswire.com

McGuffin R (2005) IP supports school feeding in Kenya. F:/IP supports school feeding in Kenya. WFP-newsroom.photo-galleries

Miguel E, Kremer M (2004) Worms: identifying impacts on education and health in the presence of treatment externalities. Econometrica 72:159–217

MOH/NASCOP (2006) Kenya National Guidelines on Nutrition and HIV and AIDS. Ministry of Health, Government of Kenya, Government Printer, Nairobi, Kenya

Probate C et al (2003) Communication and education for training of school food service directors in Pennsylvania, United States. In: Food, nutrition and agriculture, FAO 2003

Sifri Z et al (2003) School health programmes in Burkina Faso: the Helen Keller International experience. In: Food, nutrition and agriculture, FAO 2003

UNAIDS (2001) Nutrition and HIV/AIDS. Nutrition policy paper No. 20. Report of 28th session symposium. Nairobi, Kenya

World Bank (1993) World development report: investing in health. Oxford University Press, New York, NY

Linking Policy, Research, Agribusiness and Processing Enterprise to Develop Mungbean *(Vigna radiata)* Production as Export Crop from Senegal River Valley

M. Cisse, M. Diouf, T. Gueye, and A. Fall

Abstract Mungbean (*Vigna radiata*) is a short-duration and high nutritive value leguminous pulse crop. To allow Senegalese farmer to access the increasing mungbean market, different stakeholders, policy makers and research, private and industry sectors were brought together. First phase was screening varieties and evaluating yield performance. In an experiment carried out in 2004, 34 varieties were evaluated for their potential, whereas in 2005, 10 best accessions were submitted to a second screening, including grain technological analysis for marketability. Considering mean yield level and stability criteria, Line 4, VC 6123 B-11, VC-6173 B-10, CDH-A, VC 6123 A, Line 7, Line 6, VC 6379 (23-21), KPS 7 and Line 5 were retained following the first screening process. Global aboveground biomass average was 3677 kg ha^{-1}, ranging from 2707 kg ha^{-1} by Line 5 to 4736 kg ha^{-1} by VC 6379 (23-21). Harvest index varied between 0.47 by Line 6 to 0.59 in VC 6123 A, with global biomass average standing at 0.54. All varieties yielded higher than 1500 kg ha^{-1} with a peak at 2222 kg ha^{-1} by VC 6123 B-11. The performance of varieties was established as follows: VC 6123 B-11 > Line 4 >KPS 7 > VC 6379 (23-21) > Line 5 > Line 7 > CDH-A > VC-6173 B-10 > VC 6123 A > Line 6. Using technological scoring criteria, Line 4, Line 6 and KPS 7 were chosen as best varieties for export purpose. Other varieties should be devoted to local use to enhance farmer's food security.

Keywords Mungbean · Senegal River Valley · *Vigna radiata*

Introduction

Senegalese agricultural production comes primarily from rainfed cropping, covering nearly 96% of cultivated area. Managed irrigated area represents only 69,679 and 3930 ha in the left bank of the Senegal River Valley (SRV) and the Anambé Basin, respectively (Dia et al., 1998). The Senegalese agricultural production comes primarily from the rainfed crops which represent nearly 96% of cultivated area. The managed field on left bank of the Senegal River Valley (SRV) represents 69,679 ha in 1996 against 3930 ha in the Anambé Basin in Kolda, with a total of 73,609 ha. From the left bank of the Senegal River to the Falémé, big dams were realized for the development of irrigated crop which will contribute to the economic and social development of the bordering areas of the SRV. This is a part of the land use planning, the area management and agriculture development and diversification. It is within the framework of this diversification of agricultural activities that our study on mungbean crop in the Valley is recorded. However, on the Delta of the Senegal River Valley, the essential characteristic of agriculture is the prevalence of rice cropping system. However, this rice growing does not manage to make profitable the whole of managed area because of some difficulties related to low productivity level, trade difficulties and liberalization of rice

M. Cisse (✉)
Institut Sénégalais de Recherches Agricoles (ISRA), BP 240
Saint-Louis, Senegal
e-mail: sbamand@yahoo.com

A. Bationo et al. (eds.), *Innovations as Key to the Green Revolution in Africa*,
DOI 10.1007/978-90-481-2543-2_90, © Springer Science+Business Media B.V. 2011

sector. In the presence of this alarming situation, farmers became aware to diversify agricultural activities and make incomes safe. Thus in 2000, in terms of contribution to sales turnover, rice preceded diversification crop (onion, tomato, sweet potato, gumbo, water melon, corn, sorghum, groundnut and cotton) which gradually seeks to integrate irrigated farming system. Research and development institutes have been requested to work on identifying mungbean varieties adapted to SRV conditions, on optimizing farming techniques, productivity and economic profitability of the crop. As consequence, growing demand of mungbean in international market, will allow farmers to get new export opportunities and the country to gain some market parts.

Mungbean is a short-duration leguminous pulse crop cultivated extensively in Asia and that can be grown in different soil conditions, in rotation with rice and other main crops such as corn, sesame and cotton (Huijie et al., 2003; Ahn et al., 1985). Mungbean has a very high nutritive value. It is prized for its considerably high protein content ranging from 28.5 to 79.5%, about three to four times that of cereals, making it an excellent source of cheap but high-quality and easily digestible protein. It can be processed into vermicelli, flour, candy and sweets, boiled dry bean, bean sprouts, green beans, noodles and bean cakes, both at the household and at the industrial levels. For this reason it is increasingly been asked in European market due to its suitability to replace soybean, with use of genes modified ones. For this use, suitable varieties must be chosen (He et al., 1988; Ahn et al., 1985).

To fulfil the requirements for safety food for consumers and to develop competitive bean sprout in the European market, the company Evers Special Taugé (processing industry) has developed strategy for mungbean grain supply through diversification of suppliers through other sources of bean that meet quality and phytosanitary requirements, because of difficulties to satisfy increasing inner demand in China, the main grain provider.

Materials and Methods

In the first experimental set, plant material includes 30 entries of mungbean from Asian Vegetables Research Development Centre (AVRDC) and four varieties also from AVRDC which already was a subject of preliminary tests in Senegal, namely a total of 34 accessions or entries (Table 1).

Table 1 List of varieties submitted to preliminary screening

Genotype	Code	Origin
VC E17 (B-10)	V1	Netherlands/AVRDC-Taiwan
Barimung 7	V2	Netherlands/AVRDC-Taiwan
VC 3960-88	V3	Netherlands/AVRDC-Taiwan
ML 613	V4	Netherlands/AVRDC-Taiwan
SML 134	V5	Netherlands/AVRDC-Taiwan
KPS 1	V6	Netherlands/AVRDC-Taiwan
KPS 7	V7	Netherlands/AVRDC-Taiwan
ML 762	V8	Netherlands/AVRDC-Taiwan
VC 6310 (70-65)	V9	Netherlands/AVRDC-Taiwan
VC-6327 (45-8-1)	V10	Netherlands/AVRDC-Taiwan
VC-6173 B-10	V11	Netherlands/AVRDC-Taiwan
VC-6153 B-70 P	V12	Netherlands/AVRDC-Taiwan
NIMB 101	V13	Netherlands/AVRDC-Taiwan
VC 6123 A	V14	Netherlands/AVRDC-Taiwan
NM 97	V15	Netherlands/AVRDC-Taiwan
NM 94	V16	Netherlands/AVRDC-Taiwan
NM 54	V17	Netherlands/AVRDC-Taiwan
VC 6123 B-11	V18	Netherlands/AVRDC-Taiwan
Line 1	V19	Netherlands/AVRDC-Taiwan
Line 2	V20	Netherlands/AVRDC-Taiwan
Line 3	V21	Netherlands/AVRDC-Taiwan
Line 4	V22	Netherlands/AVRDC-Taiwan
Line 5	V23	Netherlands/AVRDC-Taiwan
Line 6	V24	Netherlands/AVRDC-Taiwan
Line 7	V25	Netherlands/AVRDC-Taiwan
Line 8	V26	Netherlands/AVRDC-Taiwan
Line 9	V27	Netherlands/AVRDC-Taiwan
VC 6379 (23-21)	V28	AVRDC-ARP Tanzanie
VC 6379 (45-8)	V29	Tanzania /AVRDC-ARP
VC 6148 (50-12)	V30	Tanzania/AVRDC-ARP
MungPB1	V31	Netherlands
MungPB2	V32	Netherlands
CDH-A	V33	Tanzania/AVRDC-ARP
CDH-B	V34	Tanzania/AVRDC-ARP

This first test was carried out on sandy soil at Ndiol research station, 25 km from Saint-Louis (Table 2). A randomized complete block design with two replications was used, because of the limited seeds quantity. Sowing took place on March 19, 2005, therefore in hot dry season. Emergence began 5 days after sowing (DAS). Two seeds were placed at 3–5 cm depth per hole with 50 cm spacing between lines and 50 cm between holes, giving 80,000 plants ha^{-1} density. Irrigation was performed through sprinkler.

In the second trial, ten entries resulting from the first screening were subject to a second performance

Table 2 Electric conductivity, pH, C_{total}, N_{total} and soil of the different locations

	Locations	
	Ndiol	Fanaye
Soil type	Sandy clay	Sandy
CE (mS cm^{-1})	1.1	0.5
pH_{H_2O}	7.2	7.3
C_{total} (‰)	3.6	3.7
N_{total} (%)	0.31	0.22

evaluation in rainy season 2005. Field trial was conducted in representative location of the Senegal River Valley, the Fanaye experimental research station, 150 km from Saint-Louis (16°30′N, 15°18′W). This experiment differs from the first one in season, soil type (sandy clay), plant density and geometry, structure (on ridge) and irrigation mode.

A RCBD with three replications and ten entries was set up. Varieties were Line 4, VC 6123 B-11, VC-6173 B-10, CDH-A, VC 6123 A, Line 7, Line 6, VC 6379 (23-21) and Line 5. Sowing took place on August 25, 2005, in rainy season, with 80 cm distant ridges (ridged cultivation), with two sowing lines positioned at 2/3 height on ridge sides. Plants are 25 cm distant in sowing line, giving 200,000 plants ha^{-1}. Furrow irrigation was done soon after sowing to boost seed vigour after emergence and was carried out ten times during growth.

In both the experiments, fertilizer was applied as top dressing at 150 kg ha^{-1} NPK (9-23-30) ha^{-1} at sowing. First side dressing was applied at a rate of 33 kg ha^{-1} urea as starter, 7 days after emergence (DAE), and the second one at 150 kg ha^{-1} rate of NPK (9-23-30) at 26 DAE. There applications correspond to the use of 42 kg N, 30 kg P and 75 kg K per hectare. Seeds protection was ensured with a mixture of insecticides and fungicides (Spinox).

Mixed insecticides (Deltamethrine and Dimethoate) against insects, *Chlorothalonil* against fungi and bacteria and Dicofol against *Aculops lycopersici* were sprayed six times during cropping period. The seeds protection was achieved through Spinox T. application. Manual weeding was ensured at request.

Data were collected on sowing date, days to first flowering, days to first pod emergence and maturity, plant height, number of plants per plot, pod number per plant, grain number per pod, pod length, 100-grain weight and yield (kg ha^{-1}). Other important parameters are precocity and plant survival rate.

Results and Discussions

Bringing Together Policy Maker, Research, Private and Processing Industry

To help Senegalese farmer to have access to mentioned available market, the ministry in charge of agriculture, the Senegalese Institute for Agricultural Research (ISRA) and a private agribusiness enterprise (Touba Agribusiness International, TABI) undertook a working mission in the Netherlands to evaluate these market opportunities, important for crop and farmer income diversification. The objective was to comfort the company Evers Specials to support a short-term action plan to lead the mungbean production emergence chain and build partnership (Fig. 1). As a result of the visit, Evers Specials confirms its availability to import mungbean grains from Senegal.

That's why, it was suggested to initiate a research program primarily based on identification of varieties adapted to local conditions and secondly on development of appropriate cultivation techniques. In addition, this research will make it possible to build crop production budget and define conditions of economic profitability of this newly introduced crop for Senegalese farmers. At the same time, a seed production programme has to be set up to allow grain quantity, quality and sanitary requirement, to start an industrial production.

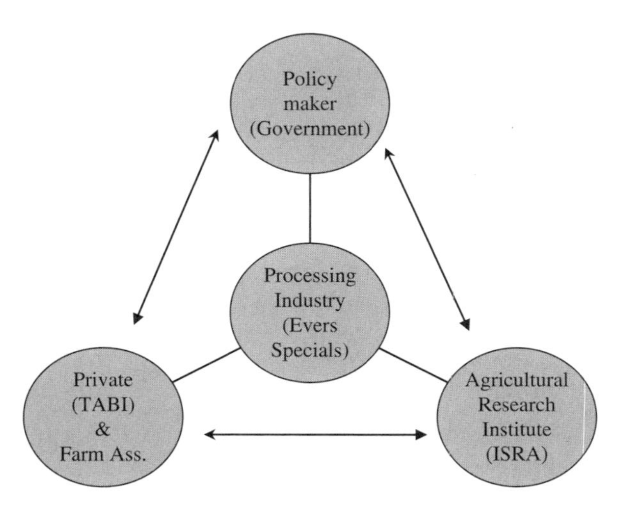

Fig. 1 Linking institution relation to support mungbean production in the Senegal River Valley

Screening and Evaluating Varieties and Yield Performance

Following a mission of the Ministry for Agriculture in Netherlands (Evers Special), the contribution from Agricultural Research Institute (ISRA) was to undertake a series of trials, to allow decision making for possible public investments to this support-related crop. The work presented within this study framework reveals agronomic and technological performance following varieties screening process through a series of tests carried out in 2004 and 2005.

Screening for Varieties

Because of limited number of repetition (2) and high treatments (34), the conventional data statistical analysis was not carried out to sort varieties. As a result, average yield performances of indicated varieties with their specific performance at block level could be taken into account as selection criteria to evaluate variety performance and stability. To discriminate varieties, following aspects were studied: general average yield, specific yield level in each individual block and higher yield compared to total block-specific yield. The first selection criterion (CRT1) is based on yield general average, whereas CRT2 corresponds to stability and performance level in individual blocks. In this case, yield of varieties must be equal or higher than general yield average of considered block.

Following these criteria, the following results were obtained (Table 3). Thirteen varieties were directly eliminated because of general mean yield limitation while ten varieties were regarded as approved through first criterion because of performance level of their general average yield (Group 1). However, these varieties must be confirmed by stability criterion (CRT2). The ten first varieties approved following CRT1 must pass through the second stability criterion, namely yield of a variety in individual block must be higher than intervarietal yield on the block (i.e. the average yield of all varieties in considered block). Eleven varieties were regarded as acceptable because of their general level of performance following immediately the ten preceding varieties. These varieties will be allowed only if one of the ten varieties of Group 1 fail following the introduction of stability criterion in block level and that these varieties succeed with this stability criterion, while satisfying general yield performance criterion (Table 3).

Varieties NM 97 and VC 3960-88 of Group 1 were eliminated because of their result accordingly to criterion 2. KPS 7 and Line 5 of the second group were then selected to replace them. Table 3 indicates selected varieties with indication of their plot yield level and rank (Table 4).

It should, however, be specified that the 34 varieties were cultivated under relatively stressing conditions, namely hot dry season, sandy soil with low organic matter and nutrients content and low plant density. It would not be surprising that varieties having been eliminated at this first screening process could get

Table 3 Performance and stability criteria used for selecting varieties

Varieties eliminated through CRT1	Varieties selected according to CRT1 and rank: first group	Varieties selectable according to CRT1 and rank: second group	Varieties admissible according to CRT2: second group	Definitively selected
E1 Line 8	S1 Line 4	S11 KPS 7	KPS 7**	Line 4
E2 ML 762	S2 VC 6123 B-11	S12 VC-6153 B-70 P	VC-6153 B-70 P	VC 6123 B-11
E3 VC E17 (B-10)	S3 VC-6173 B-10	S13 VC-6327 (45-8-1)	VC-6327 (45-8-1)	VC-6173 B-10
E4 VC 6310(70-65)	S4 CDH-A	S14 Line 5	Line 5**	CDH-A
E5 KPS 1	S5 NM 970*	S15 Line 9	NM 94	VC 6123 A
E6 VC 6148(50-12)	S6 VC 6123 A	S16 NM 94	**: Selected due to CRT2 to replace NM 970 and VC 3960-88	Line 7
E7 VC 6379 (45-8)	S7 Line 7	S17 Barimung 7		Line 6
E8 MungPB1	S8 Line 6	S18 Line 3		VC 6379 (23-21)
E9 SML 134	S9 VC 6379 (23-21)	S19 VC-6153 B-70 P		KPS 7
E10 ML 613	S10 VC 3960-88*	S20 CDH-B		Line 5
E11 Line 2	*: eliminated following CRT2	S21 NM 54		
E12 MungPB2				
E13 Line 1				

CRT1: Criterion 1, CRT2: Criterion 2

Table 4 List of selected varieties

Genotype	Plot yield (g plot^{-1})	Rank
Line 4	102	1
VC 6123 B-11	97	2
VC-6173 B-10	81	3
CDH-A	80	4
VC 6123 A	73	5
Line 7	72	6
Line 6	69	7
VC 6379 (23-21)	67	8
KPS 7	65	9
Line 5	55	10

relatively greater yield under favourable conditions, particularly with good choice in cropping calendar and good soil fertility conditions. Organic and mineral fertilization must be investigated, especially in relationship with the various soil types of the SRV. Related yield in this first screening process was relatively feeble and by far noncomparable to what was obtained in Fanaye in rainy season, with intense increase in yield. That illustrates the necessity to adapt crop fitting to the three cropping seasons of the region.

Agrophysiological Characteristics of Varieties

In the second experiment, the ten best accessions from the first experiment, including Line 5, VC 6379 (23-21), VC-6173 B-10, Line 6, VC 6123 A, Line 4, Line 7, CDH-A, KPS 7 and VC 6123 B-11, were submitted to a second evaluation process including agronomic characterization. During this trial, yield components (density at harvest, number of pods/plant, number of grain/pod, hundred grain weight) were investigated to allow predictability of crop production (Table 5). Above-grown biomass was also evaluated to prospect livestock feeding potential of different varieties. In average, varieties come to flowering 32 days after sowing (DAS). VC-6173 B-10 and KPS7 was the first with 29 and 30 DAS, whereas Line 4, Line 5 and Line 6 come to flowering at 35, 35 and 36 DAS, respectively. By 41 DAS, 50% of plant population has already flowered with the first place occupied by KPS 7, VC-6173 B-10 and CDH-A all with 38 DAS. By evaluating 14 accessions, Qiao (1998) found that days to flowering varied between 32 and 35 DAS. In the USA, planted in early June, mungbean crop begins to flower in 50–60 days and then continues flowering for a few weeks (McAlavy, 2006).

If mungbean has to be cropped in different season, good varietal choice is important because research results show that (a) mungbean strains differ in their flowering response to photoperiod and to mean temperature; (b) increasing photoperiod or reducing mean temperature delayed flowering, the amount of delay varying with strain; (c) variations in mean temperature may alter photoperiod effect on flowering in particular strains (Qiao, 1998). This author mentioned variety failing to flower within 105 days in specific photoperiod-temperature treatment.

Days to maturity varied between 47 and 50, whereas total maturity appears between 52 and 54. Other authors report days to maturity varying between 58 and 64. In the USA, authors reported that some varieties need 105 days to reach maturity (McAlavy, 2006).

Table 5 Some physiological characteristics of varieties including days to emergence of flowering, to pods and to maturity

	Days to					
Varieties	Flowering	50% flower	First pod	50% pod	Ripening pod	Maturity
KPS 7	30	38	35	42	48	54
VC-6173 B-10	29	38	36	42	47	52
VC 6123 A	30	39	39	43	50	53
VC 6123 B-11	32	41	38	44	49	53
Line 4	35	42	39	44	49	53
Line 5	35	45	40	46	51	53
Line 6	36	44	41	46	51	53
Line 7	32	41	38	43	48	52
VC 6379 (23-21)	31	39	37	42	48	54
CDH-A	32	38	38	43	48	53
EI	32	41	38	44	49	53

EI, environmental index

Yield Components and Aboveground Biomass

The germination was almost complete for all varieties. Only on block 3, insufficient ridge height generated overloading which was prejudicial to germination on incriminated ridge. The objective density was of 200,000 plants ha^{-1}. The density at harvest varied between 18,2638 plants ha^{-1} by variety KPS 7 and 13,4027 plants ha^{-1} by CDH-A. The survival rate (SVR) of plants varied between 91% by KPS 7 and 67% with CDH-A (Table 4). Eight varieties out of the ten tested had a number of grains per pod varying between 8 and 9. Only the varieties KPS 7 and CDH-A have produced 7 grains pod^{-1} (Line 5).

Lowest aboveground biomass was registered by Line 5 with 2706 kg ha^{-1} while VC 6379 (23-21) showed the highest aboveground biomass with 4736 kg ha^{-1} (Table 5). For pods yield, the varieties VC 6123 B-11, VC 6379 (23-21) and Line 4 with respective yield of 5934, 5403 and 4548 kg ha^{-1} gave higher averages than environmental index of 4424 kg ha^{-1}. Only variety Line 6 has a harvest index (HI) lower than 0.50. The maximum index was recorded by Line 5 and VC 6123 A with 0.59. The general average for this index was 0.54. VC 6379 (23-21), VC 6123 A, VC 6123 B-11 and CDH-A have lowest shelling yield. This weakness explains why certain varieties like VC 6123 A and the CDH-A, despite a good pods yield level, do not belong to the six best varieties relating to grain yield (Tables 6 and 7).

Evaluation of Yield Performance

All varieties yielded higher than 1500 kg ha^{-1} grains with a peak at 2222 kg ha^{-1} and a deviation of regression of +25% by variety VC 6123 B-11. Lowest deviation was registered by Line 6. The performance succession of varieties was established as follows: VC 6123 B-11 > Line 4 >KPS 7 > VC 6379 (23-21) > Line 5 > Line 7 > CDH-A > VC-6173 B-10 > VC 6123 A > Line 6.

Table 6 Plant density at harvest (PDH), survival rate (SVR), grain number per pod (GNP) and hundred grain weight (HGW) of tested varieties

Varieties	PDH (plants ha^{-1})	SVR (%)	GNP	HGW (g)
Line 5	166,666	83	7	9.4
VC 6379 (23-21)	152,083	76	9	8.7
VC-6173 B-10	147,916	74	8	8.2
Line 6	161,805	81	8	8.4
VC 6123 A	168,402	84	8	8.2
Line 4	170,138	85	9	7.8
Line 7	178,472	89	8	8.3
CDH-A	134,027	67	7	8.4
KPS 7	182,638	91	9	6.1
VC 6123 B-11	136,111	68	8	7.2
EI	159,826	80	8	8

Table 7 Aboveground biomass (AGB), pod yield (PY), harvest index (HI), grain yield (GY), shelling yield (SHY) and deviation of the regression of yield mean (DREI)

Varieties tested	AGB (kg ha^{-1})	PY (kg ha^{-1})	HI	GY (kg ha^{-1})	DREI (%)	SHY (%)
Line 5	2706	3823	0.59	1820	+03	47.6
VC 6379 (23-21)	4736	5403	0.53	1836	+04	34.0
VC-6173 B-10	3471	3929	0.53	1595	–10	40.6
Line 6	4254	3803	0.47	1515	–14	39.8
VC 6123 A	3061	4434	0.59	1541	–13	34.8
Line 4	3722	4548	0.55	1851	+05	40.8
Line 7	3156	3965	0.56	1782	+01	44.9
CDH-A	3955	4270	0.52	1636	–07	38.3
KPS 7	3086	4130	0.57	1848	+05	44.7
VC 6123 B-11	4626	5934	0.56	2222	+26	37.4
EI	3677	4424	0.54	1765		0.42

Among the 14 entries tested by Qiao (1998), VC 3960A-89 exhibited good agronomic characteristics (early flowering, early maturing) and had the highest yield of 1500 kg ha^{-1}. This indicated good behaviour of mungbean crop in Senegal River Valley.

Line 5 and Line 7 with relatively low pods yield belong to the six leading varieties, in terms of grain yield.

Screening for Marketability

As productivity and marketability of bean are key factors, private processing industry and agribusiness enterprise were involved as agribusiness enterprise-funded research trials and processing industry contributed to grain technological analysis. In fact, mungbean market demands a clean, fresh sprout with particular length and diameter size that is 6–7 cm long and 4–5 mm in diameter. Other characteristics are sprout white brilliance with crusty texture.

From ten different varieties of mungbean received from Senegal, Evers Specials could make a selection using grains selection criteria, namely hard grains rate, presence of the fungi and bacteria harmful diseases, hundred grain weight, sprout size and diameter, sprout conservation duration and grain size and yield (Table 8). Hard grain rate varied between 0% by Line 6 and 23% by VC 6379 (23-21). The hundred grain weight increased from a general average of 6.3–6.9 by grain with diameter higher than 4 mm. The rate of grain with grain diameter higher than 4 mm is an important character, as only these grains are allowed to be sprouted. Using this information, the real yield for variety appropriateness to sprouting is depicted in Table 9.

Other important characters include sprout diameter and length and conservation ability as indicated in Table 10. The reason for rejecting varieties VC 6123 B-11, VC-6173 B-10, VC 6123 A and Line 7 was existence of dark-coloured cotyledons.

Using these scoring criteria, results from Evers Specials show that Line 4, Line 6 and KPS 7 were the

Table 8 Hard grain rate (%), hundred grain weight (HGW), HGW and rate (%) for grain with diameter superior to 4 mm

Varieties	Hard grains (%)	HGW (g)	HGW for grains > 4 mm (g)	Rate of grain > 4 mm (%)
Line 4	3	6.7	6.8	92
VC 6123 B-11	13	6.2	7.3	81
VC-6173 B-10	8	6.5	7.1	68
CDH-A	14	6.1	7.1	54
VC 6123 A	11	6.5	7.3	75
Line 7	3	6.6	6.6	83
Line 6	0	6.6	7.4	87
VC 6379 (23-21)	23	6.1	6.8	49
KPS 7	2	5.8	6.2	58
Line 5	5	5.9	6.7	52
EI	8.2	6.3	6.9	70

Table 9 Grain yield as influenced by rate of grain with diameter superior to 4 mm

Varieties	Whole yield (kg ha^{-1})	Rate of grains > 4 mm (%)	Yield of grains > 4 mm (kg ha^{-1})
Line 4	1851	92	1705
VC 6123 B-11	2222	81	1797
VC-6173 B-10	1595	68	1092
CDH-A	1636	54	876
VC 6123 A	1541	75	1157
Line 7	1782	83	1487
Line 6	1515	87	1315
VC 6379 (23-21)	1836	49	897
KPS 7	1848	58	1076
Line 5	1820	52	940
EI	1765	70	1234

Table 10 Sprout diameter (mm), length (mm), conservation ability

Varieties	Sprout diameter (mm)	Sprout length (mm)	Conservation after 7 days
Line 4	4.28	51.7	Good (2)
VC 6123 B-11*	4.27	48.5	Mean (7)
VC-6173 B-10*	4.35	57.3	Mean (7)
CDH-A	4.08	56.1	Good (1)
VC 6123 A*	4.34	57.9	Mean (7)
Line 7*	4.22	53	Mean (7)
Line 6	4.00	55	Good (3)
VC 6379 (23-21)	3.91	56	Good (5)
KPS 7	4.09	52.1	Good (4)
Line 5	3.96	46.3	Good (6)
EI	4.15	53.4	–

*: indicate rejected varieties

best varieties, because Line 4 had a good score on all technological criteria; Line 6 had a good score on all criteria, but suffers from a lower yield, but this variety does not contain hard grains, whereas KPS 7 had smaller grains and could be interesting for some Evers Specials customers.

Conclusion

Looking for the Future of Mungbean Production

Results show good yield potential for mungbean production and technological analysis using processing industry criteria confirms interesting varieties suitable for export purpose. These varieties are good in several agronomic and technological characteristics. Next steps are (i) continuing research action to increase productivity and seed availability; (ii) to scale up technology to farmer and private enterprise through training; (iii) to begin market study and to better facilitate decision making from policy maker, private agribusiness enterprise and farmers.

With a global average of 1765 kg ha^{-1}, mungbean production in the Senegal River Valley can be considered to a large extent with optimism, as additional technical progress is still achievable. As farming practices are important for crop yield level (ARC-AVRDC, 1997), choosing appropriate soil type and cropping season, good crop density and geometry, suitable mineral and organic fertilization, an integrated plant health protection and a good protection against weeds, will help in improving the above mentioned yield level.

Acknowledgements Authors would like to express their gratitude to Touba Agribusiness International (TABI) for the financial support. Authors are very grateful to Mr Amadou Ba, manager of TABI, for the valuable contribution and encouragement during this study.

References

Ahn CS, Chen JH, Chen HK (1985) International Mungbean nursery: performance of the 9th and 10th IMN. AVRDC, Tainan

ARC-AVRDC (1997) Mungbean KPS I & II: cultural practices. Kasetsart University, ARC-AVRDC, Bangkok

Dia M, Gaye I, Sylla MM, Gueye AA, Diallo Y (1998) Rentabilité et compétitivité de la filière rizicole du Sénégal. Unité de Politique Agricole: Ministère de l'Agriculture

He XH, He TY, Xiong YIV, Jiao CH (1988) Research and use of Mungbean germplasm resources in Hubei, China. In: Shanmugasundaram S, Maclean BT (eds) Mungbean: proceedings of the 2nd international symposium. AVRDC, Tainan, pp 35–41

Huijie Z, Ninghui L, Xuzhen C, Weinberger K (2003) The impact of Mungbean research in China. AVRDC – The World Vegetable Center, AVRDC publication no. 03-550, Working paper no. 14, Shanhua, Taiwan, 26 pp

McAlavy TW (2006) Texas Mungbeans have sprout potential. AgNews. Texas A&M University System Agricultural Research and Extension Center 2112:30

Qiao X (1998) Mungbean Varietal Trial AVRDC-ARC Research Report 1998

Prioritizing Research Efforts to Increase On-Farm Income Generation: The Case of Cassava-Based Farmers in Peri-urban Southern Cameroon

J.W. Duindam and S. Hauser

Abstract The extent to which agricultural technologies have had an impact in the humid forest zone remains questionable as adoption levels have been low. The technologies developed emphasized on maintaining soil fertility and crop yields in short fallow systems. This chapter defines the problems and opportunities of commercialization of cassava (*Manihot esculenta*) production in the forest margins of peri-urban Cameroon. Cassava is the chief subsistence staple and mainly produced extensively in traditional mixed food crop fields in a short fallow rotation. The fallow period is mostly around 2–4 years with natural regrowth typically dominated by *Chromolaena odorata* (Ngobo et al., 2004). The urban demand for cassava products is currently higher than the supply which improves cassava income generation potential and justifies the development of more commercially orientated fields. Yet intensification levels are low and yields are generally far below the potential attainable. Observed production increments have been mainly based on increased cassava growing area. Data from farmer interviews and group discussions in three villages in peri-urban Yaoundé indicate that technologies for sustainable intensification of cassava production should target both pre- and post-harvest activities. They should focus on reduced labour requirements and pest and disease management. The technology proposed therefore re-emphasizes on returns to labour as a parameter of success and includes the involvement of farmers in technology testing. A commercial cassava field with a rotational *Pueraria* fallow system is discussed as a basic design. To ensure appropriateness and subsequently a higher adoption potential and hence higher impact, the system has built-in flexibility for further on-farm adaptations.

Keywords Cassava · Impact · Poverty reduction · Labour saving · Technology targeting

Introduction

A smallholder farm in the humid forest zone of southern Cameroon is still largely based on the traditional slash and burn system, where the land holding consists of a mosaic of manually slashed patches of cropped land and areas of natural regrowth vegetation of different successional stages. Labour and land are the principal inputs in the system. Soil fertility is generally considered to drop rapidly during the process of burning and subsequent cropping (Hauser and Norgrove, 2001), especially on low clay activity soils (acid Ultisols and Oxisols) prevailing in southern Cameroon. However, this system can provide sustained crop yields if sufficiently long fallow periods are respected. Sustainability of this system is often related directly to population density and associated land use intensity. Maximum population numbers estimated range from 7.8 (Nye and Greenland, 1960) to 20 people km^{-2} to allow for a sufficient fallow period of 12 years for the Congo Basin (Landelout, 1990). However, recent figures show that the mean fallow period in areas around Yaoundé with a population density of 70–80 people km^{-2} is 3.9 years (Gockowski and Tonye, 2004). In the absence of inputs, this is expected to cause productivity declines in the long term.

J.W. Duindam (✉)
International Institute of Tropical Agriculture, Yaoundé, Cameroon
e-mail: jelleduindam@hotmail.com

A. Bationo et al. (eds.), *Innovations as Key to the Green Revolution in Africa*,
DOI 10.1007/978-90-481-2543-2_91, © Springer Science+Business Media B.V. 2011

Based on the above theoretical argumentation, agricultural research in short fallow systems in the humid forest zone has focused on soil fertility and many attempts over recent decades have been undertaken to develop a system that can halt the perceived decline that is associated with land use intensification (e.g. IITA, 2004; Sanches and Benites, 1999). Research primarily focused on the integration of shrubs and trees within the system, either planted in hedgerows between crops (e.g. Hulugalle and Ndi, 1993; Nolte et al., 2003) or as improved (planted) fallows (e.g. Tian et al., 2005; Koutika et al., 2004). None of the systems developed and tested is currently used at a large scale by farmers (Hauser et al., 2006b). One reason for the poor adoption is that, overall, tree-based fallows were generally not capable of attaining higher crop yields or did so only at the expense of unacceptably high labour requirements (Hauser et al., 2006b). Besides, the systems were targeted and promoted to improve the soil fertility, whereas the farmers may not perceive this as the main constraint within their cropping systems. In many cases evaluation by farmers of the suitability of an improved fallow technology involves other aspects than solely crop yields but also labour requirements and by-products (Kanmegne and Degrande, 2002).

Declining soil fertility may, in the case of cassava cultivation in the humid forest margins of southern Cameroon, not be a major driver of land management decisions. Farmers in the urban periphery of Yaoundé most often practise much shorter fallow periods (3–5 years) than strictly necessary based on the age of fallows available in their land holdings. Various other decision criteria like protection of land use rights and labour requirements were of higher importance than the soil fertility related to fallow age (Brown, 2006). The contradiction between theory and reality may justify a change in approach of agricultural research without emphasis on the importance of reducing soil fertility and development of pre-defined technological improvements targeted to stop this decline.

Cassava (*Manihot esculenta*) is the main subsistence staple for farmers and a preferred food item for urban consumers in Cameroon (Lesaffre, 2004; Bricas and Seck, 2004). Based on rural migration, urban populations have been growing rapidly in Cameroon, which increased the local demand for cassava products. Although between 1983 and 1996 the number of inhabitants of Yaoundé doubled and the consumption of cassava more than tripled, this did not result in major shocks in consumer price (Dury et al., 2004 and references cited herein). The growing urban demand was simply compensated by an increased rural supply represented by the flexibility of the number of producers. More farmers were selling a part of their production. However, this flexibility has reached its limits (Dury et al., 2004). Prices of both fresh and transformed cassava doubled in recent years (Dury et al., 2004) which undermines food security for a growing segment of urban poor. Nevertheless, only low levels of intensification in cassava production have been observed in the peri-urban zone, where it is mostly cropped in subsistence fields. FAO statistics on cassava yields are comparable and indicate that past increases in production have been mainly achieved by increasing the area under cultivation rather than increases in yield (Table 1).

If Cameroon wants to reduce its dependence on costly food imports, every farmer needs to produce for an increasing number of city dwellers. Based on proximity to main market and high transportation costs, it is most logical to target commercial cassava production in peri-urban areas. Currently, fresh and processed cassava is already significantly a more important revenue source in peri-urban areas than in more isolated rural areas (Gockowski and Tonye, 2004).

Table 1 Statistics of cassava root (fresh and dry) production, consumption and producer prizes over the past 17 years in Cameroon

	1990	1995	2000	2005
Production quantity (1000 Mg)	1,587.87	1,780.00	1,918.30	2,138.80
Yield per hectare (kg/ha)	16,438.60	11,866.70	13,299.20	6,196.60
Producer price (CFA/Mg)	36,958.00	36,000.00	42,582.00	70,000.00
Food consumption quantity (1000 Mg)	1,172.64[a]	1,344.90	1,449.14	1,441.43[b]

[a]Estimate in 1991
[b]Estimate in 2003
FAOSTAT | © FAO Statistics Division 2007 | 24 April 2007

The overall objectives of this chapter are to (i) diagnose the main constraints to intensification in the current system in the peri-urban areas around Yaoundé, (ii) identify appropriate agronomic practices for intensified cassava production and (iii) develop a strategy for on-farm testing and adoption of these practices.

Materials and Methods

Individual Farmer Surveys

In April–June 2007, a survey was carried out in three villages (Nkolmeyang2 (03°52′N, 11°39′E), Nkometu (04°04′N, 11°34′E) and Ekoumdouma (04°06′N, 11°32′E)) in the periphery of Yaoundé. All sites were located within 30–40 km from the main road axis. A total of 63 randomly selected farmers, equally distributed in the three sites, were interviewed in a formal survey discussing cassava production and marketing issues. Interviews were conducted on farmers' fields when possible. Data were analysed using the SPSS 12.0 software package.

Group Discussions

In addition, group discussions in the same villages were organized. Average attendance was 21 persons, mostly women. Main issues discussed were problems faced during cassava production and marketing. This was done by listing all problems and subsequent prioritization by individual participants by providing ten matchsticks to each participant with the instruction to divide them proportionally according to their importance between the listed items. This served to quantify and rank the relative importance of the problem factors. Mentioned problems were grouped into five classes and percentages averaged for the three villages.

Results and Discussion

The average age of respondents during the survey was 49 years and 63% were female. Most (71%) respondents depended on farm produce for the major proportion of their income. Nearly all (98%) respondents indicated that all food requirements were met by their own farming activities throughout the year. Cassava is primarily (99% of fields) cropped in association with many other food crops. These mixed food crop fields, locally called *afub owondo*, have groundnut, maize and cassava as major components with the primary objective of providing food for the family. This system was observed to be very common throughout the forest margins in Cameroon (Gockowski and Tonye, 2004). The field is grown in a short fallow rotation with often always some residual cassava in the field, providing a certain level of food insurance for the household. As women are traditionally responsible for the provision of food to the farming household, these fields are principally managed by women (Guyer, 1984). Men are involved with family cash income. They take care of traditional cash crops including cocoa and crops that require forest fallow clearing, like plantain and melon fields (Gockowski and Tonye, 2004). From the group sessions, primary constraints were identified as (i) lack of labour, (ii) pests and diseases, (iii) lack of good quality planting material and (iv) low soil fertility (Fig. 1).

Data from the interviews show that farmers primarily rely on family labour with mean family size of 10.3 persons with an average of 5.9 members permanently working on the farm. Nevertheless labour scarcity resulted in several non-family labour arrangements. Temporary extra labour was hired by 52% of farmers at seasonal peaks in farm operations. This purchased labour was mostly used for clearing fallow fields (45%), seed bed preparation and planting (32%) and, to a lesser extent, weeding and harvesting (both 11%). Seasonality of labour needs coincides very well with the bimodal rainfall pattern which allows two rain-fed annual cropping seasons starting in March and September, respectively. Cash needs follow this trend, but peak sharply in September, when school fees have to be paid.

Cassava was sold when there were surpluses by 94% of the farmers, either as fresh roots or as on-farm processed products. None of the cassava fields had a clear commercial orientation, although some farmers referred to cassava as "cacao des femmes" indicating its cash generation potential for women. Produce was either sold locally to middlemen or first transported to an urban market where higher prices could be obtained. Farmers additionally indicated increasing

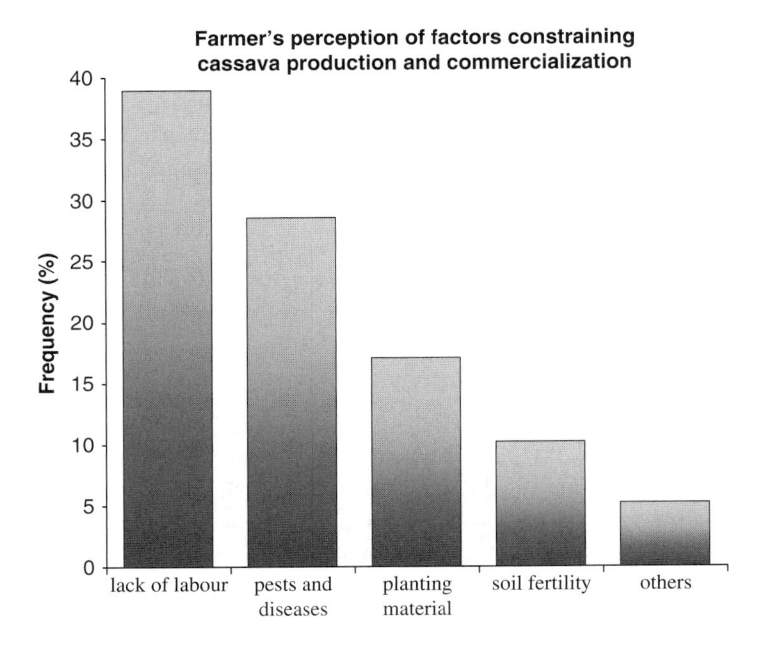

Fig. 1 Constraints to cassava production ranked by farmer (*bars* represent averages of ranked constraints by farmers in peri-urban Yaoundé)

sales to international traders from neighbouring countries such as Equatorial Guinea and Gabon which have substantial oil revenues and a net food deficit. Quantitative data on yields, marketing and income revenues of cassava produce remained largely incomplete during the survey. This was because farmers had difficulties to estimate yields from a particular field. Harvesting takes place over a long period and it proved difficult to estimate the ratio between consumed and marketed cassava. However, as cassava was exclusively produced in fields with many other crops, it can be assumed that productivity is generally low. The average plant density in cassava fields was estimated at 3,200 plants ha^{-1} (Gockowski and Tonye, 2004) and fresh root yields range typically from 5 to 10 Mg ha^{-1} fresh matter (IITA, 2004 and references cited herein). Numben (1998) estimated even a 3 Mg ha^{-1} fresh root mass in typical intercrop fields.

During the group discussions, farmers attributed the low yields to most visible pests and diseases including root rots caused by fungal infection and African root and tuber scale (*Stictococcus vayssierei*) (ARTS). Both affect yields considerably and reduce the time cassava roots can be stored in the ground beyond the normal harvesting time (Ngeve, 2003; Onyeka et al., 2005). Among the foliar diseases, farmers only perceived cassava mosaic disease (CMD) as a problem. Although some farmers were aware of the existence of CMD resistant varieties, they did not know where to obtain this material. In terms of marketing, processed products were favoured by farmers due to increased storability and better prices which are not subject to seasonal price fluctuations. Nevertheless, high labour requirements for processing discouraged farmers to increase production beyond basic family needs. The need for on-farm transformation is confirmed by findings in a recent marketing survey discussing the sustainability of local food supply system of Yaoundé (Dury et al., 2004). The paper identified the bulky and perishable nature of fresh cassava roots compared to cash crops like cocoa as a main constraint in the development of more commercial cassava production.

Cocoa is the most dominant cash crop in the region and almost half of the farmers in the survey indicated having at least one of their fields planted with cocoa. The authors of the chapter foresee the recent increases in cocoa prices becoming a threat to urban food supplies, as farmers may prefer to allocate resources to their cocoa fields to meet their cash needs, at the expense of starchy staples. Processing before transport adds value to the product and reduces the volume that needs to be transported. However, processing comprises several labour-intensive operations including peeling and especially the grating of raw roots and pounding of fermented roots. Considering the high labour requirements of processing (grating and pounding) and the unavailability of grinding machines at village level, processing for external marketing will only be a viable option when grinding machines are available at village level.

Soil fertility was not identified as a serious problem during the group discussion. The farmers' response during interviews also did not indicate that they deal with severe soil nutrient deficiencies. Although 17% of farmers claimed to use chemical fertilizers on their food crop fields, nutrient recycling through household waste, crop residues or compost was practised only by 10% of farmers.

Identification of Technological Improvement

Currently production of cassava is in short fallow rotation systems with low levels of intensification. However, demographic growth in the city has reached the point where rural supplies no longer meet urban demands. The observed increase in retail prices may justify the development of a cassava field where chief production target is cash generation instead of subsistence food production. By changing the prime objective of the production system, major yield improvements may already be achieved by altered agronomic practices, like increasing plant density, using improved varieties, timed weeding and increased land use intensity. However, labour requirements, representing the main current constraint, are likely to be unacceptable, as farmers are likely to continue with their subsistence fields for family food requirements in addition to cassava cash crop field. In this respect, an herbaceous legume *Pueraria phaseoloides* covering the soil both during and after cropping phase (also termed live mulch system) may prove to be an appropriate technology. Compared to natural fallow or other improved fallow species it has many potential positive features in terms of reduced labour requirements like easy establishment, rapid weed suppression and labour efficient biomass management (Hauser et al., 2006b). In the current system, clearing new fields after a fallow period was identified as the most labour-intensive farm operation. The typical *Chromolaena odorata* fallow regrowth produces biomass with high lignin content. This makes clearing laborious and limits biomass management options only to burning. When *P. phaseoloides* is allowed to replace *C. odorata* in the fallow, fallow biomass can be retained on the field without obstructing activities (Hauser et al., 2006a). As *Pueraria* is a perennial with a deep root system (FAO, 2007), re-sowing is not necessary and it is able to survive burning of slashed vegetation on the soil surface. Thus it offers both options: slash and burn and slash and mulch. In addition, a planted *P. phaseoloides* fallow system has shown potential in replacing weeds like *C. odorata* (Banful et al., 2007) and other "nuisance" weeds which are associated with the presence of *C. odorata* in short fallow systems (Ngobo et al., 2004). Even though the cassava–*Pueraria* system has not been further investigated, the few examples have nevertheless shown great potential, especially in short fallow systems for weed suppression (Chikoye et al., 2001; Ekeleme et al., 2000, 2004), maintaining soil biology and fertility (Koutika et al., 2001) and sustaining yields (Akobundu et al., 1999; Tian et al., 2005). In addition, permanent soil cover by *P. phaseoloides* has several advantages including N and C gains through N fixation and biomass production, protection from soil erosion and tight nutrient recycling (Hauser et al., 2006a; Leihner, 2002).

Strategy for On-Farm Testing and Conditions for Adoption

The proposed technology is so far a conceptual design combining a few management factors with built-in flexibility. As such it leaves options to farmers to adapt it to their local conditions and preferences, which need to be determined in a participatory testing phase. The testing phase will be executed on individual farmers' fields. In this particular case, farmers will have several management factors that can be tested. These include the following:

Fallow Biomass Management

Farmers have the choice to retain the *P. phaseoloides* mulch cover after slashing or burn biomass as usual, allowing farmers to either focus on saving labour or manage soil fertility.

Maize Intercropping

Maize is a principal component of their subsistence fields, but is appreciated as a cash crop for its short cropping cycle and high demand for grains from breweries (Hauser et al., 2006a). Based on the fact that in a pure cassava stand the nitrogen accumulated by the *P. phaseoloides* may be under-utilized or may even cause excessive leaf growth at the expense of storage root development (Howeler, 2002), it is likely to be attractive to farmers to include an N-responsive crop. Maize–cassava could be intercropped either simultaneous or relay. The choice of farmers may be based on cash needs during the cassava crop cycle (simultaneous) or by labour availability (relay). Maize, when planted in a *P. phaseoloides* fallow, can produce about 80% higher dry grain yield than in natural fallows (Hauser et al., 2006a). Additionally, a maize–cassava cropping system has been shown to yield the highest land equivalent ratio and reduced stem borer infestation rates in southern Cameroon (Chabi-Olaye et al., 2005).

Pueraria Climbing Control

P. phaseoloides is a vigorously growing legume (FAO, 2007) that can be a threat to cassava if it is not controlled appropriately (Leihner, 2002; Chikoye et al., 2001). The tools by which the cover crop needs to be kept from climbing on the cassava stems can be tested together with farmers. Several options are available, like the use of herbicides which nowadays might be cheaper than manual labour. If low doses are applied, *P. phaseoloides* may be damaged partially and growth may be arrested or slowed down during cropping phases. Hence, without killing *P. phaseoloides*, soil cover will be retained, but competition for nutrients, light and water will be reduced and the decomposition of the killed portion will release nutrients to the crops. Because *P. phaseoloides* is only a threat when it climbs on the cassava other simple mechanical means could be developed to prevent it from climbing.

Conclusion

Insufficient supply of cassava (products) to urban markets represents a threat to urban food security. At the same time this poses an opportunity for farmers to commercialize their cassava produce. This chapter has discussed the design of a technology to assist farmers to scale up their cassava production, based on existing concepts, with a more flexible way of testing several elements on farmers' fields. Although flexibility for farmers in adapting an existing technology has been shown to improve adoption rates (Kanmegne and Degrande, 2002), the effectiveness depends also on pre-conditions that farmers may not control (Dormon et al., 2007). A technology to increase productivity of cassava beyond household food needs is unlikely to be adopted if there is no synchrony with improvements at processing level. Appropriate processing equipment is available but purchasing these machines is constrained by cash accessibility, partially caused by the failure of formal rural credit institutions in Cameroon (Gockowski and Tonye, 2004). Besides, the operation of a machine requires a certain level of organizational capacity that currently does not exist within the fragmented smallholder producers. Investments in establishment and organizational capacity of producer groups and better linkages with the post-harvest market chain may additionally open up the industrial markets for cassava products like starch and unfermented flower. Moreover, this reduces the risk of price reductions when consumer markets get saturated.

To make this innovation complete, it may require external assistance from developmental aid organizations or donor institutions to promote group formation and micro-credits in order to allow farmers to benefit from the potential that cassava has to offer in this region.

Acknowledgements The authors gratefully acknowledge the assistance provided by all members of the involved farmers' organizations in the villages Nkomeyang2, Ekoumdouma and Nkometu and the field technicians of IITA station Cameroon. This research is partly sponsored by the APO project of the Dutch Ministry of Foreign Affairs.

References

Akobundu IO, Ekeleme F, Chikoye D (1999) Influence of fallow management systems and frequency of cropping on weed growth and crop yield. Weed Res 39:241–256

Banful B, Hauser S, Ofrori K et al (2007) Weed biomass dynamics in planted fallow systems in the humid forest zone of southern Cameroon. Agroforest Syst 71:49–55

Bricas N, Seck PA (2004) L'alimentation des villes du Sud: les raisons de craindre et d'espérer. Cah Agric 13:10–14

Brown DR (2006) Personal preferences and intensification of land use: their impact on southern Cameroonian slash-and-burn agroforestry systems. Agroforest Syst 68:53–67

Chabi-Olaye A, Nolte C, Schultess F et al (2005) Relationships of intercropped maize, stem borer damage to maize yield and land-use efficiency in the humid forest of Cameroon. Bull Entomol Res 95:417–427

Chikoye D, Ekeleme F, Udensi UE (2001) Cogongrass suppression by intercropping cover crops in corn/cassava systems. Weed Sci 49:658–667

Dormon ENA, Leeuwis C et al (2007) Creating space for innovation: the case of cocoa production in the Suhum-Kraboa-Coalter district of Ghana. Int J Agric Sustain 5:232–246

Dury S, Medou J, Tita DF et al (2004) Sustainability of the local food supply system in sub-Saharan Africa: the case of starchy products southern Cameroon. Cahier d'études et de Recherches Francophones/Agricultures 13(1):116–124

Ekeleme F, Akobundu IO, Isichei AO et al (2000) Influence of fallow type and land-use intensity on weed seed rain in a forest/savanna transition zone. Weed Sci 48:604–612

Ekeleme F, Chikoye D et al (2004) Impact of natural, planted (*Pueraria phaseoloides, Leuceana leucocephala*) fallow and land use intensity on weed seedling emergence pattern and density in cassava intercropped with maize. Agric Ecosyst Environ 103:581–593

FAO (2007) Factsheet *Pueraria phaseoloides* (Roxb.) Benth. Available via URL: http://www.fao.org/ag/AGP/AGPC/doc/Gbase/DATA/Pf000058.HTM, Cited 17 Dec 2007

Gockowski J, Tonye J et al (2004) Characterization and diagnosis of farming systems in the forest margins benchmark of Southern Cameroon. Social science working paper series no. 1, International Institute of Tropical Agriculture, Ibadan, Nigeria

Guyer J (1984) Family and farm in Southern Cameroon. Afr Res Studies 15, Afr Studies Center. Boston University, Boston, MA

Hauser S, Bengono B, Bitomo OE (2006a) Maize yield response to *Mucuna pruriens* and *Pueraria phaseoloides* cover crop fallow and biomass burning versus mulching in farmer managed on-farm experiments. Paper presented at the Conference on International Agricultural Research for Development (Tropentag), University of Bonn, 11–13 Oct 2006

Hauser S, Nolte C, Carsky RJ (2006b) What role can planted fallows play in the humid and sub-humid zone of West and Central Africa? Nutr Cycling Agroecosyst 76:297–318

Hauser S, Norgrove L (2001) Slash-and-burn agriculture, Effects of. Encycl Biodiv 5:269–284

Howeler RH (2002) Cassava mineral nutrition and fertilization. In: Hillocks RJ, Thresh JM et al (eds) Cassava: biology, production and utilization. CABI, Wallingford, CT, pp 115–147

Hulugalle NR, Ndi JN (1993) Effects of no-tillage and alley cropping on soil properties and crop yields in a Typic Kandiudult of southern Cameroon. Agroforest Syst 22:207–220

IITA (2004) Annual report of program E: enhancing livelihoods in the humid and sub humid zones of West and Central Africa through profitable and sustainable intensification of diverse agricultural systems. International Institute of Tropical Agriculture, Ibadan, Nigeria

Kanmegne J, Degrande A (2002) From alley cropping to rotational fallow: farmers' involvement in the development of fallow management techniques in the humid forest zone of Cameroon. Agroforest Syst 54:115–220

Koutika L, Hauser S et al (2004) Comparative study of soil properties under *Chromolaena odorata, Pueraria phaseoloides* and *Calliandra calothyrsus*. Plant Soil 266:315–323

Koutika L, Hauser S, Henrot J (2001) Soil organic matter assessment in natural regrowth, *Pueraria phseoloides* and *Mucuna pruriens* fallow. Soil Biol Biochem 33:1095–1101

Landelout H (1990) La jachère forestière sous les tropiques humides, Unité des Eaux et Fôrets, Centre de Recherches Forestières de Chimay, Université Catholique de Louvain, Louvain-la-Neuve

Leihner D (2002) Agronomy and Cropping systems. In: Hillocks RJ, Thresh JM et al (eds) Cassava: biology, production and utilization. CABI, Wallingford, CT, pp 91–113

Lesaffre B (2004) L'alimentation des ville: de nouveaux défis pour la recherche. Cah Agric 13:9–15

Ngeve JM (2003) The cassava root mealybug (*Stictocuccus vayssierei* Richard): a threat to cassava production and utilization in Cameroon. Int J Pest Manage 49(4):327–333

Ngobo M, MacDonald M et al (2004) Impacts of typo of fallow and invasion by *Chromolaena odorata* on weed communities in crop fields in Cameroon. Ecol Soc 9(2):1. Available via URL: http://www.ecologyandsociety.org/vol9/iss2/art1, Cited 17 Dec 2007

Nolte C, Tiki-Manga T et al (2003) Effect of Calliandra planting pattern on biomass production and nutrient accumulation in planted fallows of southern Cameroon. Forest Ecol Manage 179:535–545

Numben ST (1998) Soil productivity potential under indigenous fallow systems (Afub owondo) in the humid forest zone of southern Cameroon. PhD thesis, Faculty of the Graduate School, Cornell University, Ithaca, 226 pp

Nye PH, Greenland DJ (1960) The soil under shifting cultivation. Commonwealth Agric Bureaux, Harpenden

Onyeka TJ, Dixon AGO et al (2005) Field evaluation of root rot disease and relationship between disease severity and yield in cassava. Exp Agric 41:357–363

Sanchez PA, Benites JR (1999) Improved fallows come of age in the tropics. Agroforest Syst 47:3–12

Tian G, Kang BT et al (2005) Long-term effects of fallow systems and lengths on crop production and soil fertility maintenance in West Africa. Nutr Cycling Agroecosyst 71:139–150

Policy Framework for Utilization and Conservation of Below-Ground Biodiversity in Kenya

C. Achieng, P.F. Okoth, A. Macharia, and S. Otor

Abstract The reasons for the lack of inclusion of below-ground biodiversity in the Kenyan policy and legal framework were sought through a purposeful survey. Gaps were identified in the relevant sectoral policies and laws in regard to the domestication of the Convention on Biological Diversity (CBD). Below-ground biodiversity had no specific schedule in any of the sectoral laws. Most sectoral laws were particular about the larger biodiversity and soils but had no specific mention of below-ground biodiversity. Material Transfer Agreements and Material Acquisition Agreements that are regarded as tools for the domestication of the CBD to guide transfers, exchanges and acquisition of soil organisms lacked a regulating policy. The lack of regulating policy could be attributed to the delay in approval of draft regulations by the Ministry of Environment while the lack of specific inclusion of below-ground biodiversity in Kenya's legal and policy framework could be as a result of lack of awareness and appreciation among stakeholders.

Keywords Below-ground biodiversity · Convention on biological diversity · Material acquisition agreement · Material transfer agreement · Policy framework

C. Achieng (✉)
Department of Environmental Science, Kenyatta University, Nairobi, Kenya; School of Environmental Studies and Human Sciences, Kenyatta University, Nairobi, Kenya
e-mail: cellineoduor@yahoo.com

Introduction

Concerns about the loss of soil biodiversity related with land use change have been raised in many studies. Even as agricultural intensification becomes a real occurrence, it has received relatively little attention in Kenya (Okoth, 2004). The reduction of below-ground biodiversity (BGBD) decreases agricultural productivity (resulting in a higher number of hectares having to be used to get the same yield) (Altieri, 1999) and also decreases the resilience of agricultural ecosystems so that they are more vulnerable to erosion, pests, diseases, and the general degradation of the land (Ibid). Despite these concerns, BGBD has been given very little attention by stakeholders (UNEP, 1995).

There is a need to keep agricultural ecosystems as healthy and sustainable as possible so that the biodiversity loss, which is usually so high in agricultural systems, is reduced (Swift, 1997). Nevertheless, if higher priority is to be given to the conservation and management of the soil and its associated biota, then policy makers need a better understanding of the soil-based ecosystems' services and of their commercial values (Giller et al., 1997). In many countries, enacting of laws formulated to protect the soil has not kept pace with measures intended to protect other natural resources, such as air and water. The same case applies to Kenya where there is no specific policy or legislation to guard soil organisms and yet they are affected by land use change and agricultural intensification. What exist are sectoral laws which are not encompassing.

A policy framework that stipulates appropriate land use systems and provides guidelines on transfer, exchange, acquisition, commercialization, utilization, and conservation of BGBD is needed. Such a policy framework would ensure abundance and diversity,

A. Bationo et al. (eds.), *Innovations as Key to the Green Revolution in Africa*,
DOI 10.1007/978-90-481-2543-2_92,

which would in turn increase BGBD functions in the soil. Increased activities result in adequate nutrient supply for crop growth, hence increased food production (Swift, 1997). Increased food production ensures enough food for subsistence and commercial purposes, thus helping to alleviate poverty. The objective of this study is to propose an improved policy framework for the enhancement of biodiversity and sustained utilization of its components. Sustainable utilization of biodiversity is stipulated in the Environmental Management and Co-ordination Act (1999).

Materials and Methods

Several methods were used in the data collection, ranging from reports and publications, survey data from field visits to Gatunduri and Kibugu sub-locations of Embu District. Use of meta-data to relate land to below-ground biodiversity was done. A survey was carried out with 120 farmers in the two sub-locations; interviews were conducted with 30 scientists handling BGBD and 12 institutions dealing with the same in Nairobi and Embu. Key informants were also interviewed to get specific information about different aspects of BGBD.

Use of Meta-data to Relate Land to Below-Ground Biodiversity

The use of meta-data in this research was to obtain general information on the effects of land use systems and intensities on the diversity and abundance of soil organisms in Embu. Such data are found in literature, in published papers, and in gray literature existing in institutions. In the case of this study, current information was obtained from work done by an ongoing project (the conservation and sustainable management of below-ground biodiversity) in Kenya and in six other tropical countries (Indonesia, Brazil, Mexico, Uganda, India, and Côte d'Ivoire).

Farmer Interviews and Focus Group Discussions

Another method employed was a survey of farmers' current knowledge on BGBD through interviews and focus group discussions. One hundred and twenty farmers from both Gatunduri and Kibugu villages were interviewed and the purpose of the questionnaires was to generate perceptions of the farmers on the effect of agricultural intensification on the abundance and diversity of BGBD. The questionnaires were intended to find out if farmers had any knowledge on management and their view of such undertakings in future. A total of three focus group discussions were held. The groups were organized based on gender with the first group constituted mainly by women, the second by men, and the third by both men and women. Each group had a membership of 10 persons, where the last group consisted of 5 women and 5 men. The purpose of the focus group discussions was to further analyze and understand the previously obtained results from individual farmer interviews. The groups comprised farmers of all ages including the youth, middle age, and the elderly farmers.

Institutional Studies

Institutional studies were carried out in Nairobi and Embu Districts. Among the institutions interviewed were government ministries, lead agencies, non-governmental organizations, universities, and private firms dealing with soil organisms. The use of open-ended questionnaires for this part of the study was to probe the institution for information that would otherwise be restricted, as is the case with close-ended questionnaires. The use of questionnaires enabled the researcher to present questions in a uniform manner, thus reducing bias.

Individual Scientists' Study

Thirty individual scientists were issued with questionnaires to understand their experiences with BGBD. Such experiences included management, transfers, exchanges, acquisition, and commercialization of BGBD, as guided by existing policies and regulations in Kenya.

Key Informant Interviews

Interviews with individuals with knowledge on specific issues on BGBD that the researcher sought to find were

Table 1 Farmers' knowledge of various aspects of BGBD

Knowledge	Numbers	Percentage (%)
Commercialization	Yes: 38/120	31.7
	No: 46/120	38.3
	Needed education first: 36/120	30
Transfers	Yes: 70/120	58.3
	No: 50/120	41.7
	Had witnessed soil samples being picked from their farms: 30/50 did not know	60
Exchanges	Understood as exchange between farmers: 44/120	36.7
	Sought knowledge first: 54/120	45
	Needed an explanation on exchanges first: 22/120	18.3

carried out. Such issues included policies relevant to BGBD, revision and improvement of policies to incorporate BGBD issues, intellectual property rights, and individual handling of BGBD.

Results and Discussions

The Role of Policy in BGBD Conservation, Management, Preservation, Maintenance, Acquisition, Transfer, Commercialization, Benefit Sharing and Intellectual Property Rights (IPR)

Farmers' Perceptions

Farmers were asked if they knew if soil organisms can be used in trade, if properly identified for type and function in the soil. The purpose of this question was to get insights on farmers' knowledge in commercialization with BGBD. The following responses were obtained: 31.7% (38 out of 120) said yes, 38.3% (46 of 120) said no, while 30% (36 out of 120) said that they would be in a better position to answer this question, when educated about such issues before hand. Out of the 120 farmers interviewed, a significant 70 had an idea of what transfer of soil organisms was. Out of the 70, a significant 36 said it was the removal from one farm to another, while 40 said, it was the movement of soil organisms from one farm to another. The remaining 50 farmers had no idea about transfers at all. This implies that farmers lack knowledge in BGBD issues, hence the need to educate them. These results are shown in Table 1.

From the last column of the row on transfers in Table 1, more insights were obtained from farmers during focus group discussions (Kreuger, 1988) and the results are as shown in Box 1. The transfers that farmers were more familiar with were those in which soil organisms are carried in manure, for example, the case of the dung worm – locally known as "marindi." Other organisms transferred in manure were enumerated as small black ants, big beetles, millipedes, centipedes, termites, earthworms, and nematodes. The other significant transfer was from their farms to the University of Nairobi and the Kenya Agricultural Research Institute (KARI), for research. One farmer shared his experience, as in Box 1.

Box 1

"We do not know where they were taken since people in a team came and took samples from our plots and never returned. They even put pegs which are still here up to now, but did not explain, what the pegs were for! Soil samples have been picked on a number of occasions, but we have no idea what was extracted from such samples."

Source: Interview with one of the farmers in Gatunduri sub-location

This is an outcry that transfers of BGBD occur mainly in form of soil samples being taken away from farmers' farms. When this is done, the farmers are not made aware of what is obtained from such samples and whether they are beneficial or not. Therefore, they do not get any benefits in case such organisms are used for commercial purposes. This act is unethical because it overlooks the advice given by the CBD on prior informed consent, when handling biological resources from one individual to another or one institution to another or one country to another (http://www.biodiv.org). National regulations/guidelines on prior informed consent would guide such a process and serve all stakeholders equitably. Regulations on prior informed consent have been formulated, but are still in draft form and have not been approved by the Minister for Environment. They are found in schedule II of the (conservation of biological diversity and biological resources, access to genetic resources and benefit sharing, and the protection of environmentally significant areas) regulations (Republic of Kenya, 2005, Third National Report).

Also, farmers' opinions on future exchanges of BGBD were requested. The purpose of this question was to gather information that would enable the inclusion of farmers, as relevant stakeholders in exchanges of BGBD. Such information would also be useful, when formulating guidelines on BGBD issues. The results obtained are shown in Fig. 1.

Institutions' and Scientists' Perceptions

Material Transfer Agreements (MTA) and Material Acquisition Agreements (MAA)

Information about familiarity with MTA and MAA was sought from institutions and scientists dealing with BGBD. The purpose of this question was to find out how many institutions know about MTA and MAA. Familiarity with MTA was at 83.3% (10 out of 12), while unfamiliarity was 16.7% (2/12). For the MAA, it was 50% (6 out of 12) for familiarity against 50% (6 out 12) for lack of familiarity. Most of the agreements termed as MTA and MAA were actually Memorandum of Understanding (MoU). This shows that transfers were more commonly taking place than acquisition. This explains the fact that soil organisms were mostly used for research between and among institutions rather than such institutions claiming to own the organisms through acquisition. The signatories to these agreements were given as lead agencies and governments 66.7% (8) and lead agencies only 33.3% (4); the rest were as shown in Fig. 2. The difference in the percentages explains the fact that both agreements were considered at the same time without separating them. It can also be explained that a lead agency in a government would sign an agreement with a lead agency in another government or from the same government without the governments of such countries being part of such agreements. It was found out that there were a number of flaws in the existing MTA/MAA/MoU. First, they were not strictly adhered to, because there were no legislative appendages in such agreements. The draft regulations on access to genetic resources and benefit sharing outline various forms of punishment to offenders who violate the draft MTA regulations (Draft regulations, Part V, Section 35.1, 35.2, 36, and 37). These regulations once approved by the Minister for Environment and gazetted will ensure proper adherence to MTA and

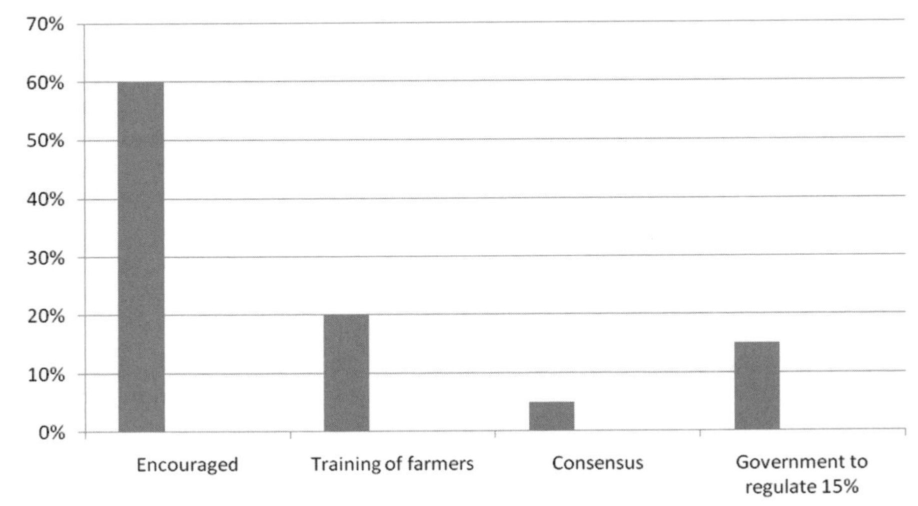

Fig. 1 The *horizontal axis* shows farmers opinions on future exchanges of soil organisms among farmers and with institutions dealing with BGBD. The *vertical axis* shows the percentage of farmers. Consensus: consensus needed with farmers; encouraged: exchanges among farmers and with institutions should be encouraged in future; Farmers T: farmers to be trained; G regulate: government to regulate

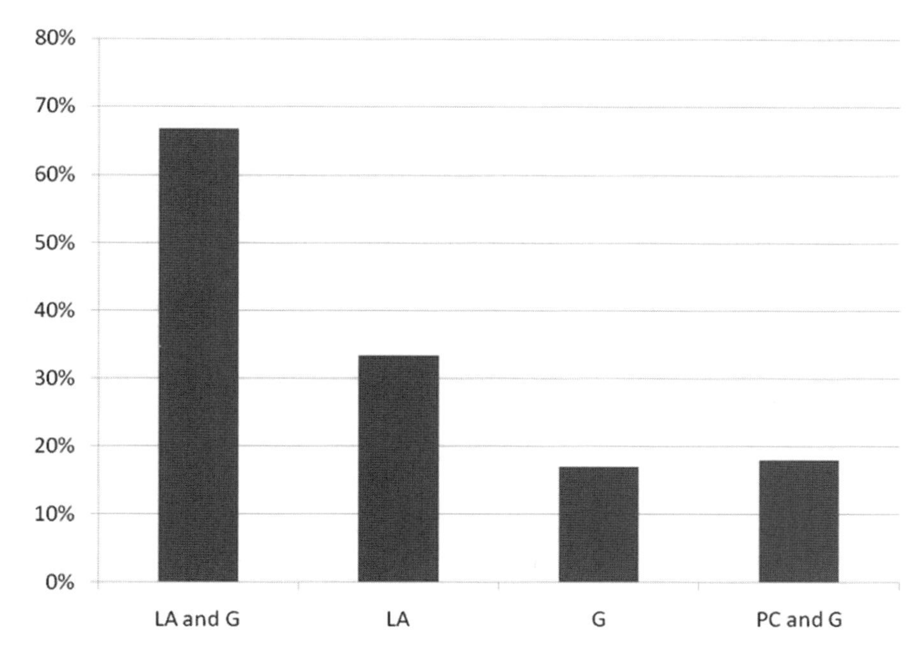

Fig. 2 MoU signed between institutions dealing with biological resources in Kenya. The *horizontal axis* shows the type of institutions while the *vertical axis* shows the percentage of agreement types signed by different institutions. LA and G: lead agencies and government; LA: lead agencies; G only: governments only; PC and G: private companies and governments

MAA. Secondly, MTAs and MAAs accessed lacked an outline on benefit sharing, hence it would be difficult for the stakeholders involved to claim any benefit accruing from the genetic material in question.

Conclusions

From the information obtained from the three groups of stakeholders (farmers, scientists, and institutions dealing with genetic resources), it is clear that there is need to include all the key stakeholders in the process of BGBD policy formulation. This is because all have different and important stakes in BGBD issues. In this regard, the proposed ways and means include, first, having prior informed consent of all the stakeholders involved when handling BGBD and in particular the farmers. Farmers also need to be educated on various aspects of BGBD. Secondly, having incentive measures to ensure proper utilization and conservation of BGBD have to be considered. Such incentive measures need to be outlined in a national policy document or in regulations, formulated by relevant authorities. Thirdly, there is a need to have specific BGBD regulations on equitable benefit sharing. Lastly, the guidelines stipulating the formulation of MTAs and MAAs should be approved sooner than later to guide institutional arrangements in relation to transfers, exchanges, acquisition, and commercialization of BGBD in Kenya.

Acknowledgments I wish to express my sincere appreciation to Kenyatta University and to all my supervisors: Dr. Ayub Macharia, Dr. Samuel Otor and Dr. Peter Okoth for their guidance, advice, patience, moral, and professional support during the entire study period. My appreciation also goes to Ms. Ritu Verma of TSBF-CIAT who advised me on various aspects of social science that were relevant to my research; Prof. Albert Mumma of the University of Nairobi, for his advice regarding environmental policy issues. Special thanks go to the Below-Ground Biodiversity (BGBD) project's global coordinator Dr. Jeroen Huising for providing me with financial support to undertake part of my research. Much gratitude goes to the African Network in Agroforestry Education (ANAFE) for providing me with a grant to undertake my research work.

Many thanks go to all the farmers, research scientists, and employees of institutions that were interviewed for their collaboration and willingness to give information. I would also like to acknowledge the hard work undertaken by my research assistants in Embu: Rachel Ireri, Munyi, Njoka, and Muchangi. They worked tirelessly in translating the questionnaires to the farmers as well as interviewing them. To the World Agroforestry Centre (ICRAF) library staff for their assistance in providing me with relevant research material and reading space. Special thanks go to Ms. Jacinta Kimwaki, the ICRAF chief librarian for her good coordination. Lastly, much gratitude to my classmates Naomi Njeri, Salome Muriuki, Mukera, Kahunyo, Weru, and Felista Muriu for their moral support.

References

Altieri MA (1999) The ecological role of biodiversity in agro ecosystems. Agric Ecosyst Environ 74:19–31

Environment Management and Coordination Act (EMCA) (1999) Kenya Gazette supplement No. 3 (Acts No. 1)

Giller KE, Beare MH, Lavelle P, Izac AMN, Swift MJ (1997) Agricultural intensification soil biodiversity and agro ecosystem function. Appl Soil Ecol 6:3–16. http://www.biodiv.org/convention/article.asp. Accessed 26 Feb 2007

Kreuger RA (1988) Focus groups: a practical guide for applied research. Sage, London

Okoth SA (2004) An overview of the diversity of microorganisms involved in decomposition in soils. Trop Microbiol 3(1):3–13

Republic of Kenya (2005) Third national report to the conference of parties to the convention on biological diversity on the implementation of the convention, 2005

Swift MJ (1997) Agricultural intensification, soil biodiversity, and agro ecosystem function on the tropics. Appl Ecol 6: 1–2

UNEP (1995) Global biodiversity assessment. Cambridge University Press, Cambridge

Policy Issues Affecting Integrated Natural Resource Management and Utilization in Arid and Semi-arid Lands of Kenya

J.W. Munyasi, A.O. Esilaba, L. Wekesa, and W. Ego

Abstract This chapter analyses policy issues affecting utilization and management of natural resources in Kenya's arid and semi-arid lands (ASALs) that constitute 84% of the total landmass and support 25% of human population, 54% of livestock and almost all wildlife. The ASALs have fragile natural resources, which require appropriate management strategies. Focus group discussions and single subject interviews were conducted involving various policy institutions and lobby groups to identify policy flaws in management and utilization of natural resources and recommend to policy makers for review. Privatization policy on land ownership was found to be unsustainable for resource development and utilization. It denies communities access to communally utilizable resources and nomadic grazing systems. Conventional farming systems have failed to address soil and water conservation practices contributing to land degradation. The policy is unclear on appropriate areas for farming and livestock keeping. Wildlife conservation policies have impacted negatively on communities living in wildlife dispersal areas due to few direct benefits from wildlife. Yet, the communities bear all the costs of living with wildlife. Existing marketing policies are unfavourable in guaranteeing markets for ASAL products. Water management and utilization policies favour large-scale farmers at the expense of pastoralists. Policies on energy conservation are unclear leading to excessive charcoal burning, which degrades the environment. The government needs to review its national policies and invest more in development of human capital in rural areas to release pressure from the environment. There is also a need to involve many stakeholders in policy formulation.

Keywords Arid and semi-arid lands (ASAL) · Kenya · Natural resource management · Policies

Introduction

Natural resource degradation in Kenyan rangelands has reached alarming levels resulting in decreased livestock productivity and loss of biodiversity. This situation had led to ecological imbalance, food insecurity, stagnation in economic growth and conflicts over resource use by various interested parties (Wambugu and Kibua, 2004). This has been due to continued mismanagement of fragile natural resources, application of inappropriate technologies, inadequate involvement of stakeholders in policy formulation and failure to enforce sound policies for enhancing integrated natural resource management and utilization. It has, however, been shown that pastoralism is the most efficient means of production in ASALs of Africa and if enhanced can be used to meet the growing demand for livestock products (Omiti and Irungu, 2002). It is against these concerns that a desk research and consultations with various players in policy analysis and advocacy was conducted with the support of Desert Margins Programme (DMP) and Kenya Agricultural Research Institute (KARI) to identify policy options that will enhance integrated natural resource management practices in arid and semi-arid lands of Kenya. This will lead to formulating and circulating policy guidelines

J.W. Munyasi (✉)
Kenya Agricultural Research Institute, Kiboko Research Centre, Makindu, Kenya
e-mail: munyasijoseph@yahoo.com

A. Bationo et al. (eds.), *Innovations as Key to the Green Revolution in Africa*,
DOI 10.1007/978-90-481-2543-2_93,

to appropriate stakeholders and initiating policy dialogue with policy makers using empirical evidence to influence the adoption and enforcement of sound-integrated natural resource management and utilization in Kenyan ASALs. Specific aims of the study were to identify and document conflicting policy issues in the management and utilization of the natural resources in ASAL and establish the contribution of community involvement in policy formulation and implementation in integrated natural resource management and utilization in ASALs.

Materials and Methods

Study Sites

The study focused on policy issues influencing integrated natural resource management and utilization in southern Kenya rangelands. Two administrative districts of Kajiado and Makueni were purposively selected for the study as they represented two distinct production systems in the region. Makueni District occupies an area of about 7965 km^2 and Kajiado 21,902 km^2. The two districts are in ecological zones IV and VI and lie between latitudes 1°44′S and 2°26′S and longitudes 36°50′E and 37°45′E (Kajiado District Development Plan, 2002–2008; Makueni District Development Plan, 2002–2008). They receive a bimodal rainfall, with Kajiado receiving a mean annual rainfall of 300–450 mm while Makueni 600–700 mm (Jaetzold and Schmidt, 1983). The general altitude of the actual study areas is between 900 and 1100 m (Kajiado District Development Plan (2002–2008; Makueni District Development Plan (2002–2008). Vegetation types of the study areas consist of woodland, grassland and bushland (Pratt and Gwynne, 1977). The soils are deep and fine textured with hill masses being volcanic while the plains range from stony Cambisols on the upper slopes to dark, cracking Vertisols in the bottomlands and valleys. Most of these soils are fragile and prone to wind and water erosion, where conservation measures are neglected. Kajiado District is largely pastoral production systems characterized by livestock keeping, communal grazing (though in some parts land tenure is partially individual) and inhabited by the Maasai. Makueni District is agro-pastoral production systems characterized by mixed livestock and crop production, sedentary life style on individually owned land and inhabited by the Kamba (Munyasi, 2004).

Data Collection Methods

Since natural resource management and utilization has been a complex issue affected by many factors, and by policies operating at different scales and through different mechanism, a multiple data collection approach was used for the purposes of triangulation and reliability of the findings (Scherr et al., 1996). Thus two levels of data collection were applied. The first level was literature review to establish existing policies on integrated natural resource management and utilization. The second level involved purposively identifying policy institutions and lobby groups and conducted focus group discussions on policy issues affecting natural resource use and management. These institutions and lobby groups were Institute of Policy Analysis (IPAR), Kenya Institute for Public Policy Research Analysis (KIPPRA), Olkejuado Pastoral Development Organization (DUPOTO-e-MAA), Masongaleni Community Organization for Sustainable Development (MACOSUD), Resource Conflict Institute (RECONCILE) and Natural Resource Department, Egerton University.

Results and Discussion

Extension Policy in ASAL

Extension policy in the country is moribund and ineffective for development of the ASALs. However, its development is historical. In the Session Paper No. 4 of 1981 on National Food Policy, the Kenya government emphasized the flow of information to farmers through extension services. However, over time, this policy has changed leading to limited extension services to stimulate development and sound management practices of the existing natural resources in rural areas (Kiara et al., 1997). This has been exacerbated by the on-going structural adjustment programme that forced the government to retrench the civil service, affecting a substantial number of extension agents, particularly in ASAL (Muok et al., 2000). Given this scenario, the advisory role and implementation of range improvement programme are seriously threatened and the government ought to review this policy with the purpose of increasing advisory services to rural communities in ASAL. Besides limited extension staff, poor infrastructure and inadequate facilitation by the government need to be addressed further so as to experience remarkable adoption rate of range improvement technologies for enhancement of livestock production in ASALs.

Interaction between farmers, extension agents and researchers is vital in addressing farmers' needs. This relation is achieved through existence of clear institutional guidelines and policies that encourage participatory research and technology dissemination (Chamala and Pratley, 1987). For instance, a pastoralist might require a Zebu breed that has fewer challenges for survival in ASALs, while researchers through extension agents advocate for a Sehiwal breed that has high risks and is less adapted to arid climatic conditions. Such technologies would also need large quantities of finance for adoption (Lee et al., 2001), which are mostly unavailable among resource-poor farmers. In such circumstances, a policy that enhances collaboration between technology developers and other stakeholders hastens uptake rate of appropriate natural resource management practices. Maranga (1998) recommended involvement of pastoralists in national development, while Vanclay (1992) suggested that policies geared to developing research technologies that are compatible with potential adopters' needs are more relevant than those that are not demand driven.

Incompatible Land Tenure and Land Use Policy

Land ownership and land use in Kenya are administered and regulated by the constitution (Trust lands) and over 50 statutes (Wamukoya and Situma, 2000). Despite the existence of these legal instruments, there has not been proper and comprehensive land use planning and coordination in Kenya (Wamukoya and Situma, 2000). In the ASALs, the Land Act (Cap 280) of the Laws of Kenya, which classifies all pastoral land under communal ownership whose use is governed by Customary Land Law (Omiti and Irungu, 2002), is pertinent. This law provides every member of the community equal rights to access and utilize land and is usually governed by a range of social factors including kinship, ethnicity, status and residence that have been established historically as a result of alliance, collaboration and competition between groups (Omiti and Irungu, 2002). However, in the early 1980s, the Kenyan government started changing its land tenure policy from communal land ownership to individual after the pastoral communities felt threatened by immigrants taking their land and government allocating part of it to game parks (Kimani and Pickard, 1998). Currently, however, the call is to revert to communal land ownership with the Olkejuado Pastoral Development Organization (DUPOTO-e-MAA) arguing that land privatization policy denies some community members access to some common resources such as watering points, saltlicks, plants of cultural values and historical sacred sites. The policy also interferes with the traditionally coordinated grazing systems that are characterized by migration in times of scarcity as opposed to sedentarized lifestyle that is more applicable in the high potential areas. There is thus a call for a land tenure policy that is not generalized across production systems countrywide, but is flexible and recognizes cultural differences and potential uses.

Transition of landholdings from indigenous communities to migrants in ASAL through land sales has contributed to loss of natural resource biodiversity and land degradation (DUPOTO-e-MAA members', 2006, personal communication). Musimba and Nyariki (2003) observed that introduction of nontraditional land use systems such as crop farming and mining in traditionally dry grazing areas has contributed to environmental degradation, in cases where appropriate soil and water management practices were not observed. DUPOTO-e-MAA maintained that Land Control Board membership in each district should comprise prominent indigenous people who have historical background of their land, local land politics and cultural attachment. This will avoid disposal of land to elite migrants whose interest is to maximize on available natural resources with little emphasis placed on ecological sustainability. In this regard, DUPOTO-e-MAA recommended that the government should review and abolish the special Land Control Board that had contributed to the selling of land in the pastoral community to outsiders or elite kinsmen without involving the entire board members.

In addition, the group argued that the current farming system policy should be reviewed to match with the ecological zones. For instance, clear policy should be provided for the growing of crops such as maize, tea, coffee, and wheat, specifically in areas where rainfall is reliable, while beef cattle, goats and sheep should be the domain of the ASAL areas. This policy will encourage successful investment and exchange of goods locally without importation of the same from other countries.

Land ownership and control are usually linked to production and creation of employment opportunities. However, during the dry season, land in ASALs is low in primary production leading to food shortage and unemployment due to inadequate rainfall to raise crops. Consequently, people have turned to alternative livelihood options, which impact negatively on the stability of the environment. These activities include felling of trees for charcoal burning and construction, sand harvesting and wildlife poaching (Munyasi and Nichols, 2007). Therefore, DUPOTO-e-MAA group members suggested that a clear policy and legislation should be developed and enforced to address the main causes of communities turning onto the environment for survival while also addressing the need to conserve biodiversity in ASAL for posterity.

Policy on Income Sharing from Wildlife and Its Related Activities

Given that most of Kenya's game parks/reserves are found in the ASALs, a large proportion of Kenya's wildlife tourism income comes from the ASALs. Adventure tourism is also focused on the ASALs because of their spectacular scenery and a sense of wilderness. However, the interaction between wildlife and livestock in these areas has been a major source of antagonism between livestock keepers and the Kenya Wildlife Service (KWS) which manages these parks and reserves. Under the current Wildlife (Conservation and Management) Act (Cap 376) of 1985, wildlife is largely seen by communities as a liability imposed upon them by the government since it is classified as a national heritage held in trust for the benefit of the public. For instance, communities bordering wildlife protected areas in Kenya receive less than 1% of the total foreign exchange earnings from wildlife-based tourism annually. Yet, most of these communities have to bear the brunt of many of the direct negative effects of wildlife invasion (Obunde et al., 2005; Ashley and Elliot, 2003; Kinyua et al., 2000). Furthermore, with the ever-increasing human population, Mwaloma (2007) asserted that it was shameful to seek famine relief food from donors while potential agricultural areas remain under wildlife conservation. Consequently, the local communities remain poor while having resources whose potential they have not exploited. Therefore, efforts to promote integrated wildlife and livestock management need to ensure that the local communities receive adequate benefits from wildlife resources. The building of the community capacity to actualize the potential of the wildlife resources will be necessary. In addition, mechanisms to ensure access for pastoralists to seasonal grazing and water and to limit the negative aspects of wildlife integration such as disease transmission, predation and crop damage (Boyd et al., 1999) are needed.

Policy on Water Resource Use

As a result of inadequate rainfall for cropping and to some extent also livestock production, water resource availability in ASALs is one of the main factors limiting the pace of development (GOK, 2004; Omiti and Irungu, 2002). The legal framework for managing water resources in Kenya is the Water Act (Cap 372). Despite the existence of this legislation, large-scale farmers and companies have been allowed to tap the scarce water resources for large-scale uses to the detriment of pastoralists (Pastoral Civil Society, 2006; Wamukoya and Situma, 2000). According to DUPOTO-e-MAA group members, water management and utilization policy should be enforced for equitable benefits among the stakeholders. Although the policy of developing earth dams, water pans, wells and boreholes is supportive, it should be handled with care by avoiding environmental degradation through consideration of the mobility of pastoral groups, seasonal use of pastures and stock routes (Omiti and Irungu, 2002). Thus, management of water resources should be placed under pastoral associations whose role should be backed up by clear legislation and enforcement. In addition, measures should be taken to rehabilitate the rangeland through forage reseeding, bush management and soil and water conservation with the aim of optimizing grazing pressure both spatially and temporally (Omiti and Irungu, 2002).

Animal Disease Quarantine Policy

Disease transmission across regions due to livestock movements has been identified as a major constraint to range productivity. Although the quarantine policy exists, lack of enforcement has led to prevalence of communicable diseases in ASAL environment. Therefore, enforcement of quarantines, disease surveillance, vaccination and livestock movement rules and regulations will reduce livestock mortalities and cost of management (Pastoral Civil Society, 2006). This will involve regional cooperation of the national veterinary services (Mugunieri and Omiti, 2004). Other important structures that have collapsed and require revitalization are community cattle dips, holding grounds, stock routes and vaccination centres.

The role played by Community-Based Animal Health Workers (CBAHW) in the marginal areas of Kenya cannot be underrated (Omiti and Irungu, 2002). However, the existing legal and policy framework on veterinary service delivery prohibits veterinary practice by 'unqualified' individuals (IPAR, 2002). In the policy reform, the CBAHW should be encouraged to offer services to livestock and also be integrated in formal veterinary services. IPAR (2002) suggests that this will be achieved through formation of associations, which will monitor, regulate and encourage training of CBAHWs. It is, however, proposed that backstopping of CBAHWs by veterinarians should be strengthened.

Credit Facilities, Resource Support and Capacity Building Policy

The risk and uncertainty of investment in new agricultural technologies in arid and semi-arid lands cannot be over-emphasized, and policies aimed at reducing these risks through financial, technical and manpower assistance are imperative. In a study to assess the factors enhancing new technology adoption by smallholders, Cramb (1999) noted that farmers' decisions towards technology adoption were prompted by interventional projects that provided resource support. The scarce resources that pastoralists possess are jealously protected against investment in technologies that take too long period to realize tangible outcome.

The price of inputs determines the level of adoption in a heterogeneous society in terms of income, land size and property rights. New technologies at times are too expensive for rural resource-poor farmers to afford. Allub (2000) and Barr and Cary (2000) suggested that policies that provide subsidies to ease burdens on smallholders help them gain access to material resources and enhance the adoption rate. In addition to acquiring new technologies, training on natural resource management and conservation is essential (Musimba and Nyariki, 2003) and policy makers should incorporate workshop trainings and seminars about new technologies, which should improve the income and living standard of the rural people (Allub, 2000). According to Barr and Cary (2000), cost sharing arrangement is one of the most important incentives that could improve the adoption rate of new practices since such endeavours are risk associated. These incentives should be channelled through existing community structures and should be combined with training of potential adopters.

Marketing Policy on Range Products and Value Addition

The guarantee by the government of access to market outlet for livestock and livestock products motivates livestock keepers to improve management practices and livestock utilization. While developing a model for pasture improvement in southeastern New South Wales, Australia, Vere and Muir (1986) found that favourable livestock prices and input costs were among important incentives to farmers adoption of pasture improvement strategies while adverse livestock markets acted as disincentive. In ecological conditions, such as rangelands, dependency on livestock is more relevant since crop production has always been unpredictable. However, the high grading systems set by the Kenya Meat Commission (KMC) Act (Cap 363) of 1990 on the quality of animals required for slaughter has limited livestock keepers from full benefits of the commission's facilities. The government should intervene and review KMC policy to favour livestock keepers so that they could benefit more from this market opportunity. This policy review should emphasize contracting livestock keepers to raise beef animals according to KMC requirements. The contracted farmers should be facilitated by KMC in the livestock management technologies to attain market weight faster and according to desired specifications. Similar arrangements have been demonstrated in the crop sector such as tea, coffee and sugarcane enterprises in Kenya. In addition, the current location of KMC factory is quite far from most livestock keepers. The intervention strategy in this problem should be a policy put in place to require KMC establish mobile abattoirs and satellite slaughterhouses to cater for livestock in far places.

The main source of livelihood for the ASAL communities is livestock and livestock products, which are significantly affected by risks inherent in dryland production systems. However, markets for livestock products especially in their raw form are not always readily available and when available, the prices are not economically appealing to the producers (Musimba and Nyariki, 2003). In addition, during wet season, pastoralists produce more livestock products such as milk than the demand hence the prices are low leading to losses and wastage. Similarly, during drought periods there are lots of hides and skins as well as bones going to waste as animals succumb and die in large numbers. These seasonal surplus products need to be processed with a view to adding value that will fetch higher market prices and increase shelf life. However, the major constraints to value addition to livestock products include availability of small-scale processing plants and supporting infrastructural policies. Therefore, a policy on value addition to livestock and livestock products should be formulated and enforced so that farmers could maximize profits from these products. For instance, milk storage facilities (cooling stores) and processing plants should be introduced at farm level in order to increase milk shelf life, improve nutritional level, consumption appeal and market value. Livestock manure available in abundance should be processed into energy (cooking fuel) for local consumption as has been done in Ethiopia (Mnene, 2007, personal communication) while bones should be processed into buttons, necklaces, belts, shoes, animal feeds at farm level.

Taxation and Market Liberalization Policy

In developing nations, taxation on output has had disastrous effects on technology adoption. Government policies that minimize taxation on input and output provide incentive to adoption of high yielding technologies, which in turn enable farmers to increase production (Just et al., 1988, cited in Sunding and Zilberman (2000). In the Argentinean agricultural industry policy, the slow growth rate between 1940 and 1973 was attributed to heavy output taxation imposed by the government (Just et al., 1988, cited in Sunding and Zilberman (2000). Therefore, the government should emulate the Argentinean policy to realize rapid economic growth in ASAL.

Trade liberalization and exchange rates in the agricultural sector have significantly affected the adoption of techniques for natural resources management and utilization due to high costs of inputs and low prices of products. Just et al. (1985) cited in Sunding and Zilberman (2000) observed that macroeconomic policies with high interest rates reduced the adoption of new technologies due to the competitive local and international market forces that come into play. In a liberal market economy like Kenya, the livestock market has been flooded with cheap livestock and livestock products from neighbouring countries such as Tanzania, Ethiopia and Somalia. As a result, pastoralists have been discouraged from investing in the industry and have not fully adopted livestock improvement and environmental conservation technologies. This implies that market policies that are unfavourable to local industry discourage farmers from adopting technologies such as pasture improvement, ago-forestry, and improved livestock breeds.

Environmental Policy in ASAL

According to the Masongaleni Community Organization for Sustainable Development (MACOSUD) manager (personal communication), conducting environmental impact assessment for new and existing projects affects mostly private developers but not government projects. Therefore, it implies that government projects have substantially contributed to environmental degradation. For instance, settling of squatters on government land without first conducting comprehensive environmental impact assessment and later environmental audit has led to land degradation in Makueni District through overgrazing, soil erosion, inappropriate farming and felling of trees for charcoal burning (MACOSUD, personal communication, 2007). Implicitly, policy enforcers should ensure that no institution is above the law.

Charcoal production in rural ASAL set-up has been taken as a source of income and urban areas are major markets. Currently there is no clear policy on rural energy source to reduce pressure on fuelwood. Therefore, the government should commit substantial resources on research for developing affordable alternative sources of energy, which should be accompanied by community training on how to apply the new technologies in view of the high levels of illiteracy. In addition, more efficient charcoal stoves should be developed and provided to local communities at an affordable price or free of charge.

The cost of transport has increased lately due to increased fuel prices. This cost is passed on to rural people who pass the buck to the environment. In other words, environmental degradation is influenced by increased poverty among the people. Therefore, the government should invest in the rural people to release pressure from the environment rather than concentrating in urban centres and neglecting the rural ASAL set-up.

Conclusions

Nonparticipatory and poorly formulated policies have direct impact on the fragile ASAL development and management, which in turn exacerbate poverty levels among communities living in this area. Implementation of policies that were not derived in consultative manner fails to address the real problem of the inhabitants of that particular area. Study has showed that pastoral communities regard some recommended policy guidelines for integrated natural resource management and utilization as inapplicable to their current situation and should be reviewed in a participatory manner. Policies which stem from the grass root are valued and defended by the communities than the imposed policies which sometimes require force from the government to be adopted.

Acknowledgements The authors wish to thank the staff of institutions and lobby groups that were engaged in discussions that led to the success of this project. We would also like to appreciate the role of Kenya Agricultural Research Institute, Kiboko, especially the Centre Director for providing enabling environment. We deeply appreciate the United Nations Environmental Programme (Global Environmental Facility) under the Desert Margins Programme and the Kenya Agricultural Research Institute for funding this project successfully. We also acknowledge Dr. Mnene W.N and Ms. Muthiani E.N for reviewing the draft of this chapter.

References

Allub L (2000) Attitudes towards risk and the adoption of new technologies among small producers in arid, rural regions: the case of San Juan, Argentina. http://www.general.uwa.edu.au/u/dpannell/seaos8.htm Cited 5 Apr 2002

Ashley C, Elliot J (2003) Just wildlife? Or a source of local development? In: Natural resource perspectives No. 85. ODI, London

Barr N, Cary N (2000) Influencing improved natural resource management on-farms. A guide to understanding factors influencing the adoption of sustainable resource practices. http://www.affa.gov.au/corporate_docs/publications/pdf/rural_science/nat_resource_mgt.pdf. Cited 5 Apr 2002

Boyd C, Blench R, Bourn D, Drake L, Stevenson P (1999) Reconciling interests among wildlife, livestock and people in Eastern Africa: a sustainable livelihoods approach. Natural Resource Perspectives No 45

Chamala PS, Pratley JE (1987) Adoption process and extension strategies for conservation farming. Tillage: new direction in Australian agriculture. pp 400–419

Cramb RA (1999) Process influencing the successful adoption of new technologies by smallholders. In: Stur WW, Horne PM, Hacker JB, Kerridge PC (eds) Working with farmers: the key to adoption of forage technologies. Proceedings of an international workshop held in Cagayan de Oro City. Mindanao, 2–15 Oct 1999

GOK (2004) National Policy for the sustainable development of the arid and semi-arid lands of Kenya: Draft report, Nairobi

IPAR (2002) How public policy can improve the situation of Kenyan pastoralists. Policy Brief 8(3):1–4

Jaetzold R, Schmidt H (1983) Farm management handbook of Kenya Vol/II/B Central Kenya. Ministry of Agriculture, Nairobi

Kajiado District Development Plan (2002–2008) Ministry of planning and national development. Government Printing Press, Nairobi

Kiara JK, Skoglund E, Eriksson A (1997) Development of soil conservation extension in Kenya. In: Sombatpanit S, Zobisch MA, Sanders DW, Cook MG (eds) Soil conservation extension: from concepts to adoption. Science, Enfield, pp 35–45

Kimani K, Pickard J (1998) Recent trends and implications of group ranch sub-division and fragmentation in Kajiado district, Kenya. Geogr J (London) 164(Part 2):202–213

Kinyua PID, van Kooten GC, Bulte EH (2000) African wildlife policy. Protecting wildlife herbivores on private game ranches. Eur Rev Agric Econ J 27(2):227–244
Lee AC, Conner JR, Mjelde JW, Richardson JW, Stuth JW (2001) Regional cost share for rancher participation in bush control. J Agric Resour Econ 26(2):478–490
Makueni District Development Plan (2002–2008) Ministry of planning and national development. Government Printing Press, Nairobi
Maranga EK (1998) A review of range production and management extension activities in Kenya. J Agric Environ Ethics 11(2):131–144
Mugunieri LG, Omiti JM (2004) Strategies for improving the contribution of livestock sector to food security and increased incomes: the case of red meat. Discussion Paper No. 042/2004. IPAR Discussion Paper Series
Munyasi JW (2004) Adoption of weed management technologies for natural pasture improvement in the South-eastern Kenyan Rangelands. MSc thesis, Southern Cross University, Lismore, Australia
Munyasi J, Nichols JD (2007) Communities and contrasting values attached to pasture weeds: the case of the Maasai and Kamba peoples in South-east Kenyan Rangelands. Agroforest Syst J 70:185–195
Muok BO, Owuor B, Dawson I, Were J (2000) The potential of indigenous fruit trees: results of a survey in Kitui District, Kenya. Agroforest Today 12(1):13–16
Musimba NKR, Nyariki DM (2003) Development of and policy on the range and pastoral industry with special reference to Kenya. Anthropologists J 5(4):262–267
Mwaloma AG (2007) Only homegrown policies can prevent human-wildlife conflict. Daily Nation, Wednesday June 13 2007, Nation Media group, Nairobi
Obunde PO, Omiti JM, Sirengo AN (2005) Policy Dimensions in human-wildlife conflicts in Kenya: evidence from Laikipia and Nyandarua districts. Discussion Paper No. 065/2005. IPAR Discussion Paper Series
Omiti J, Irungu P (2002) Institution and policy issues relevant to pastoral development in Kenya. Discussion Paper No. 031/2002. IPAR Discussion Paper Series
Pastoral Civil Society (2006) Quarterly newsletter of RECONCILE/IIED programme on reinforcement of pastoral society in East Africa. Issue No. 14
Pratt DJ, Gwynee MD (1977) Rangeland management and ecology in East Africa. Hodder and Soughton, London
Scherr JS, Bergeron G, Pender J, Barbier B (1996) Policies for sustainable development in fragile lands: methodology overview. Fragile lands programme environment and production technology division. International Food Policy Research Institute, Washington, DC
Session Paper No. 4 of 1981 on National Food Policy. Government printing press, Nairobi
Sunding D, Zilberman D (2000) The agricultural innovation process: research and technology adoption in a changing agricultural sector. (For handbook of agricultural economics). Department of Agricultural and Resource Economics, University of California at Berkeley. http://are.berkeley.edu/~zilber/innovationchptr.pdf 8 am 01 Mar 2011
The Kenya Meat Commission Act (Cap363) (1990) Laws of Kenya. Government Printer, Nairobi
The Wildlife (Conservation and Management) Act (Cap 376) (1985) Laws of Kenya Government printer, Nairobi
Vanclay F (1992) The social context of farmers' adoption of environmentally sound farming practices. In: Lawrence G, Vanclay F, Furze B (eds) Agriculture, environment and society. Macmillan, Melbourne, VIC, pp 94–121
Vere DT, Muir AM (1986) Pasture improvement adoption in South-Eastern New South Wales. Rev Mark Agric Econ 54(1)
Wambugu NN, Kibua TN (2004) Economic growth & property in Kenya: a comparative analysis of effects of selected policies. Discussion Paper No. 058. Institute of Policy Analysis and Research, Norfolk Towers, Nairobi
Wamukoya GM, Situma FDP (2000) Environmental management in Kenya. A guide to the Environmental Management and Coordination Act 1999

Stakeholder Characterisation of the Biophysical and Socio-economic Potential of the Desert Margins in Kenya

J.W. Onyango, A.O. Esilaba, and P.K. Kimani

Abstract The arid and semi-arid environment in eastern Kenya is composed of a range of socio-economic and biophysical conditions that have not been characterised. The human population in such areas is on the increase making the area unsustainable for human, wildlife and natural resource utilisation. A participatory characterisation was carried out to provide guidance towards sustainable utilisation of this environment. It was observed that the area has low fertility with stocks for major nutrients being either zero or negative. Soils are also highly susceptible to severe wind and water erosion. The vegetation consisting of *Acacia* spp. and other wooded bushland is under constant decline due to deforestation for firewood and building materials and increase in demand for agricultural land due to low level of agricultural intensification. The households have less than 50% of the total area as farmland with over 70% of them living below the poverty line of US $1/day. The rest of the area is rangeland and is conserved for wildlife. Pastoralism and arable farming practices are often in direct competition for land with wildlife while the number of livestock per unit pastureland usually disproportional. Crop failures of rain-fed agricultural production are common and occur in 3 out of every 5 years. Since the area suffers from serious moisture deficits, irrigation is practiced at various places but sometimes these offer inadequate methods leading to salinisation and sodification and ultimate abandonment. Kenya Agricultural Research Institute has undertaken, through three designated transects, some GIS mapping, soil fertility studies and socio-economic characterisation. Appropriate soil and water management technologies under prevailing socio-economic conditions have been carried out in the area to enhance agricultural productivity.

Keywords Arid and semi-arid lands · Biophysical and socio-economic characterisation · Kenya

Introduction

At least 485 million Africans are affected by land degradation and agricultural productivity with an estimated $42 billion in income and 6 million hectares of production lost annually (AfNet, 2007). Continentally, between 1990 and 2000 alone, 14.6 million hectares were lost annually against the planting rate of 5.2 million hectares due to associated human activities. Thirty-four percent of Africa is under threat of desertification affecting 73% of agricultural dryland, 74% of rangelands, 61% of rain-fed croplands and 18% of irrigated lands (Christensen et al., 2002). In East Africa deserts are dominated by food insecurity; degradation of soils, water and woodlands; deforestation arising from drought and mismanagement (Toulmin, 2005) caused by population growth; and expansion of agricultural land. Darkoh (1993) explains that drought in Africa can be expected to continue recurring at unpredictable intervals and is caused by mistakes and mismanagement occurring from within and without the continent. Unganai (1997) explained that in southern Africa two severe droughts are expected every 10 years while in 4 out of 10 years the area receives

J.W. Onyango (✉)
Irrigation and Drainage Research Programme, Kenya Agricultural Research Institute (KARI), Nairobi, Kenya
e-mail: joabwamari@yahoo.com

A. Bationo et al. (eds.), *Innovations as Key to the Green Revolution in Africa*,
DOI 10.1007/978-90-481-2543-2_94,

below normal rainfall. Terrestrial rains in these arid lands are detrimental causing soil erosion and vegetative destruction. No durable economic development is expected unless based on sustainable management of Africa's land, soils, forests and water.

Desertification is due to climate change and proposals in both economic development and environmental security should be employed as measures to combat the process (Toulmin, 2005). Possibilities of reversing the negative trends should be determined in resource projects along with communities through activities such as reclamation, sustainable harvesting, wise use of the scarce woody reserves and biotechnology (Kasusya, 1988). Failure to these would lead to abandonment of land and an increase of the sand by denuded areas where regeneration is slow due to surface sealing properties of the soil (Gachimbi, 1990). Focus should be on aspects of sustainability including productivity, protection, resilience, economic viability and social acceptability. Changes in rainfall patterns and increasing periods of droughts have negative effects not only on the agricultural production but also on the rich biodiversity (Climate Network Africa, 1994). Climate change and climate variability can worsen desertification by altering temperature and rainfall patterns (Christensen et al., 2002). Since desertification is both natural and anthropogenic it may also lead to environmental migration and is caused by occurrence of drought and desiccation through human activities. Despite the awareness of the dangers of these magnitudes, government branches still operate independently with little or no coordination to permit integrated approaches. To end hunger in Africa there is a need to increase the level of fertilizer use from the current average of 8 kg ha^{-1} to an average of 50 kg ha^{-1} by 2015 as agreed by the African Heads of State and Government following the Africa Fertilizer Summit held at Abuja, Nigeria, in June 2006. This has to go alongside with soil and water management, improved agroforestry species development and bio-socio-economic conditions.

The Desert Margins Programme (DMP) is a project in the sub-programme of the Environmental Science and Research of the United Nations Environmental Programme (UNEP) whose objectives include arresting land degradation in the arid and semi-arid areas focussing especially on poverty alleviation increase in agricultural productivity and environmental protection through sustainable use of biodiversity (Kenya Agricultural Research Institute – KARI, 2005). DMP is inter-institutionally implemented by KARI, Kenya Meteorological Department (KMD), Kenya Forestry Research Institute (KEFRI), Kenya Wildlife Services (KWS), National Museums of Kenya (NMS), Department Remote Sensing and Surveys (DRSS), University of Nairobi (UoN) and National Environmental Management Authority (NEMA). DMP is expected to develop methodologies and approaches for monitoring and evaluation, promotion of biodiversity conservation, capacity building of stakeholders, improving natural resource management (NRM) policies, promotion of alternative livelihood systems and encouraging participation of stakeholders (Njoka, 2005).

The Long-Term Ecological Monitoring Project (ROSELT/OSS) has identified the Kiboko-Kibwezi Observatory covering parts of Makueni and Kajiado Districts (Fig. 1) as one of the 14 pilot observatories in Africa (ROSELT/OSS, 2001). In subscription to the Rio declaration and to Agenda 21 of the United Nations Charter, the proposed activities in the observatory are intended to utilise existing scientific and technical collections of information to assist in observing the causes, consequences, mechanism and extent of desertification and establish an elaborate monitoring system adapted to conditions of arid zones for better decision making (CRIESEL, 1983). In both Kajiado and Makueni Districts, the observatory has low rainfall which is unreliable and erratic. Soils are fragile, low in fertility and are susceptible to erosion and leaching (Gachimbi et al., 2003). Also prevalent especially in irrigated agricultural land is the problem of soil salinity and sodicity often leading to abandonment of areas previously conceived to be highly productive. Some water sources in the area are saline due to urban effluence and environmentally unfriendly methods of irrigation to natural causes. To achieve such an inventorisation, the UN convention to combat desertification emphasises the development of benchmark procedures and sites and indicators in order to monitor and assess desertification and hence establishment of operational, cost-effective, early warning systems for drought and desertification. A participatory

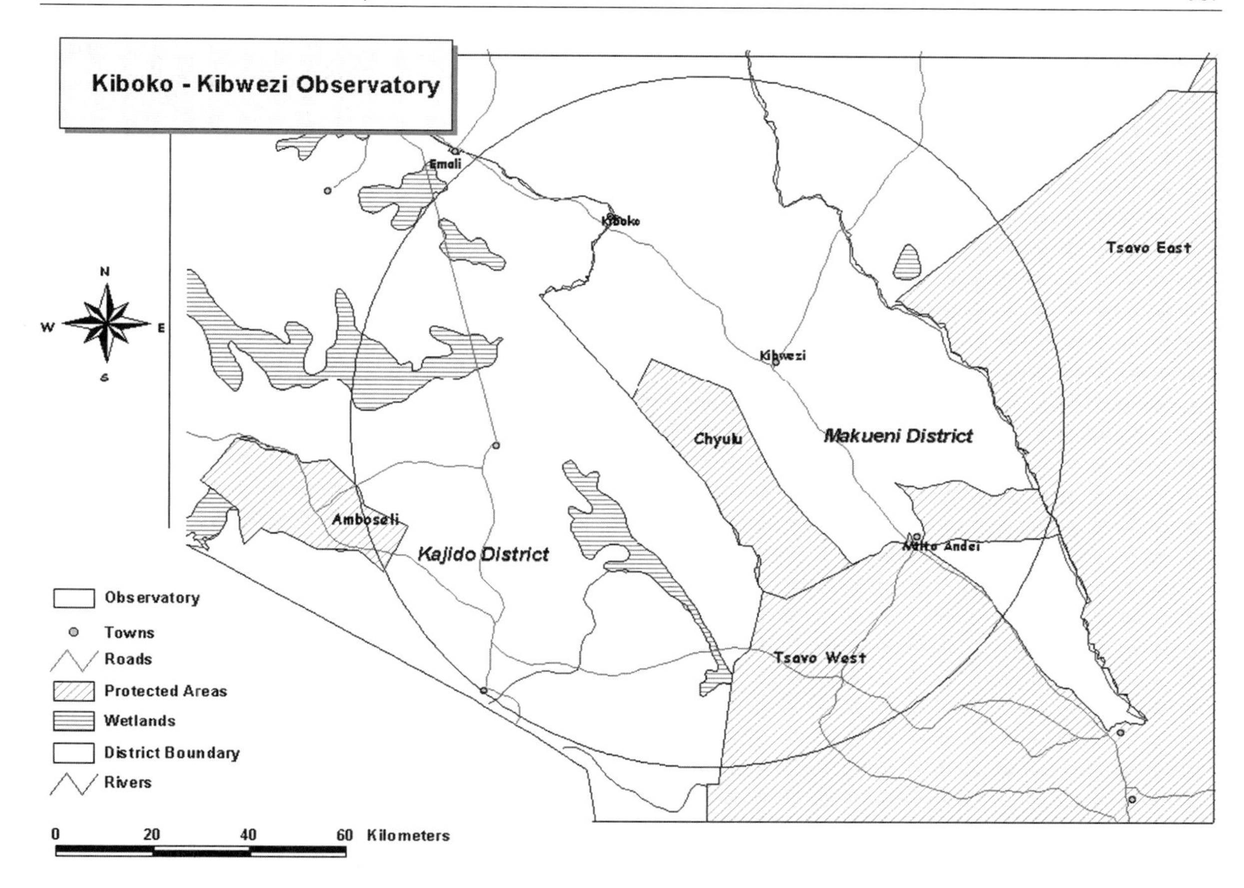

Fig. 1 Boundaries of the Kiboko-Kibwezi observatory

characterisation by relevant stakeholders was carried out to provide guidance towards sustainable utilisation of the Kiboko-Kibwezi observatory in south eastern Kenya.

Materials and Methods

The observatory lies at the intersection of Kajiado, Makueni and Taita-Taveta Districts with land degradation impacts emanating from farming, pastoralism and wildlife conservation. Two districts, Makueni and Kajiado, were considered for characterisation based on surveys (reconnaissance, diagnostic and soil) and stakeholder workshops. Samples were taken from 16 points on these transects (7 points in the Kiboko-Kitise, 4 on the Kibwezi and 5 on the Mtito Andei) considered to cover the agro-ecological characteristics of the observatory (see Fig. 1). The soil samples from the three transects were analysed at National Agricultural Research Laboratories (NARL) for all chemical and physical characteristics. A digital camera was also used to obtain information of the situation as it was at the time of the visit.

A semi-structured questionnaire was administered at a total of four points where inhabitants were interviewed. Three permanent transects each running from Athi River to Chyulu hills were identified and marked. These are the Athi River–Kitise–Nganani Hill Transect, the Athi River–Kibwezi–Umanyi Spring Transect and the Athi River–Mtito Andei–Kitingisyo Transect. An additional transect within the Tsavo West National Park running from the offices at Komboyo to Mzima springs was marked. Each transect was proposed to have three permanent sampling points (PSPs) for monitoring changes in vegetation. The variables on which data were collected were agricultural crops, frequency of drought, livestock composition, uses of vegetation, deforestation processes, population characteristics and urbanisation. Translation into local dialect was done through the assistance of staff members of KARI Range Research Centre – Kiboko. Scientists from KARI headquarters, NARL Kabete, KARI Kiboko, KEFRI and KWS were involved in identification and setting up of permanent transects for the Kiboko-Kibwezi DMP/ROSELT/OSS observatory. Literature review was done on information of work previously carried out in the area. The relationship between drought and desertification was drawn from a wide literature search on documentation by UNEP and the Climate Network Africa publications on the relevant subjects (FAO, 1993). A number of reports were perused from pre-colonial era to date in this area, but in Makueni District in particular. Literature review of previously recorded information was also carried out for relevant data to avoid repetition and actually utilised to ensure that the content of the report was synergistic.

Results and Discussions

Biophysical Characteristics (Makueni)

After the Chyulu there is an almost featureless plain lying between 150 and 1000 m above sea level. On higher areas is the marginal cotton zone with some maize and sunflower (Jaetzold and Schmidt, 1983). Natural vegetation is dominated by wooded bushland consisting of *Enteropogon macrostachyus* and *Chloris roxburghiana* (30% of grasses) with *Acacia brevispica*, *Combretum exallatum*, *Cammiphora spp.* and grasses such as *Premma holstii*, *Ocinum basilicum* and *Grewia* spp. Baobab (*Adansonia digitaria*), some estimated to be 3000 years old, grows in all the ASAL areas and has many uses including fodder, food, drink, medicine

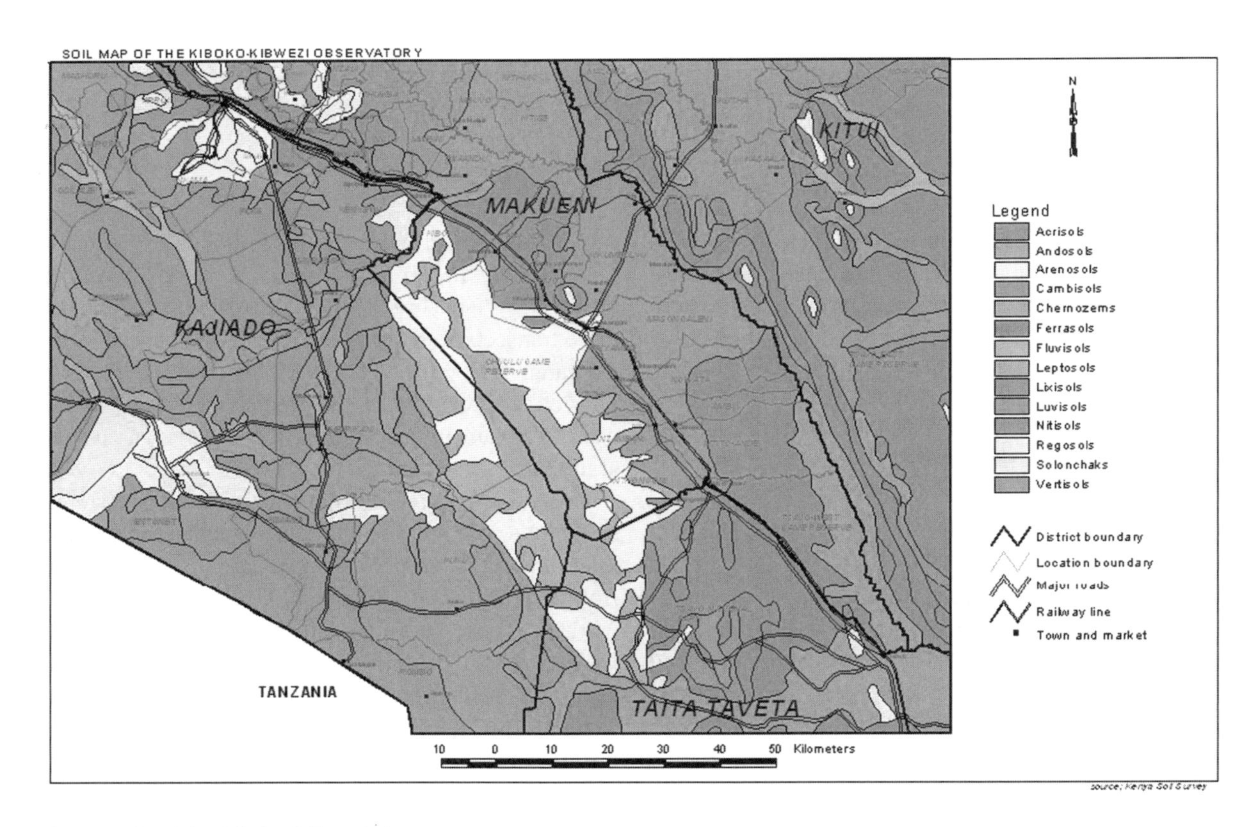

Fig. 2 Soils of the Kiboko-Kibwezi observatory

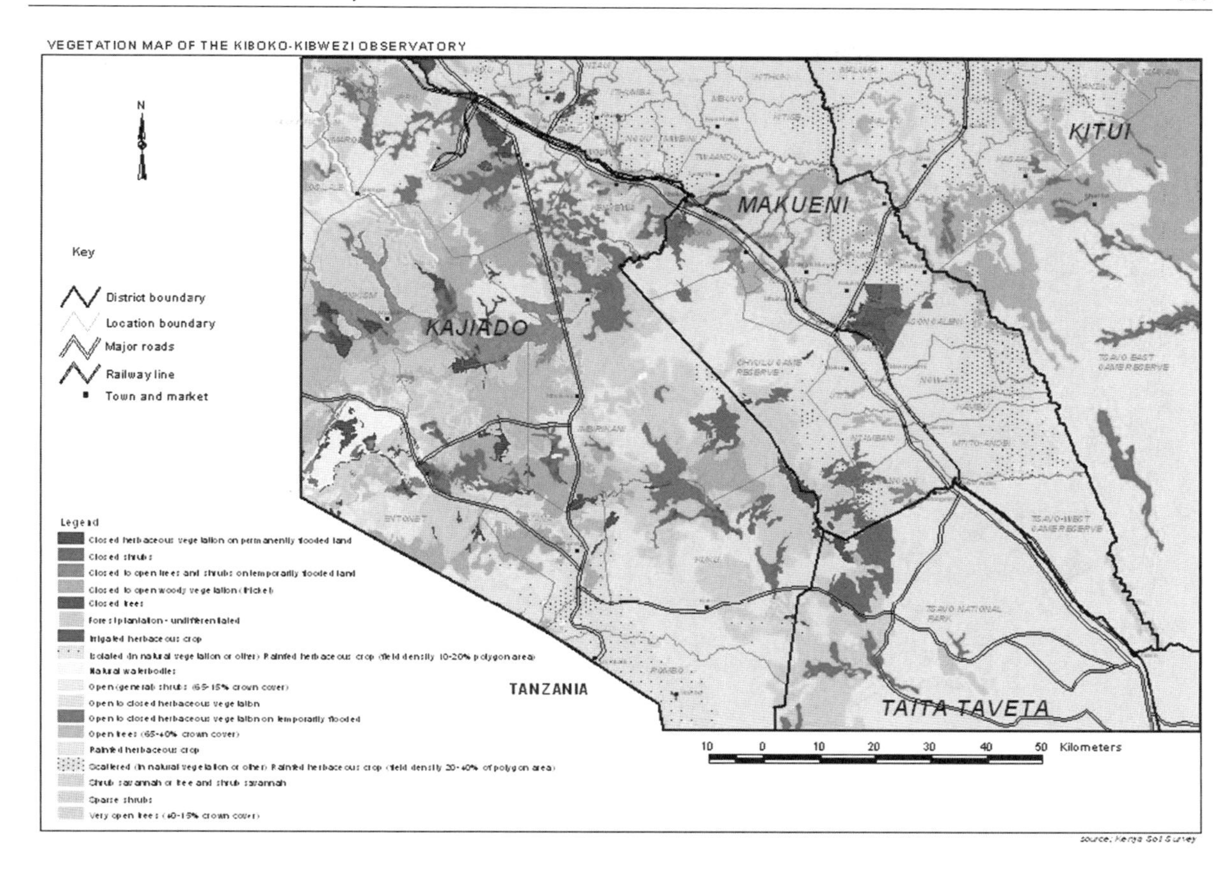

Fig. 3 Vegetation of the Kiboko-Kibwezi observatory

and dyes. Baobabs also store water in the dry seasons in their hollow trunks. Soil erosion close to the trees is restricted and its exposed roots indicate land degradation (Figs. 2, 3 and 4).

The tree also encourages infiltration; serves as houses, prisons, storage barns; and refuges from marauding animals. It is also a rainfall-onset indicator since it sheds its leaves at onset of rains (Gachimbi, 1990). Soils here are developed on sandstones rich in ferro-magnesium minerals and are well drained, deep to very deep, red to dusky red and sandy clay to clay (Ekirapa and Muya, 1991). These are represented around Kiboko and west of Kibwezi by *chromic* Luvisols and interfluves of *orthic* and *xanthic* Ferralsols (Ministry of Agriculture, 1987). The Chyulu hills were formed between the Pleistocene to recent era but lateral eruptions in the surroundings occurring in the last century. In the Chyulu are lava flows of olivine basalt and pyroclastic material of rather course texture (Muchena and Njoroge, 1984). The Tsavo National Park lies on basement system rocks where fertility is fairly low and is dominated by surface sealing leading to very low infiltration susceptible to erosion. Around Kiboko, infiltration is considered high (49.30 cm/h) while in the semi-abandoned areas and completely abandoned areas infiltration rates of between 5.13 and 16.3 cm/h, respectively, occur (Omulabi et al., 2000). The University Dryland Farm is dominated by *chromic* Luvisols and *haplic* Lixisols with a *sodic* phase. Comparable pH values for Kibwezi town, TARDA and KARI farm are 9.53, 7.70 and 9.93, respectively, with 78.9% of the area susceptible to severe erosion, 5.3% susceptible to between moderate to slight rates of erosion and only a mere 2.6% not susceptible. Soils vary in depth depending on parent material; slopes are generally low in organic matter and deficient in nitrogen, phosphorus and potassium. The district covers five agro-ecological zones with a small portion within zone III (Table 1). The climate of the target observatory area is semi-arid to arid with mean annual rainfall ranging

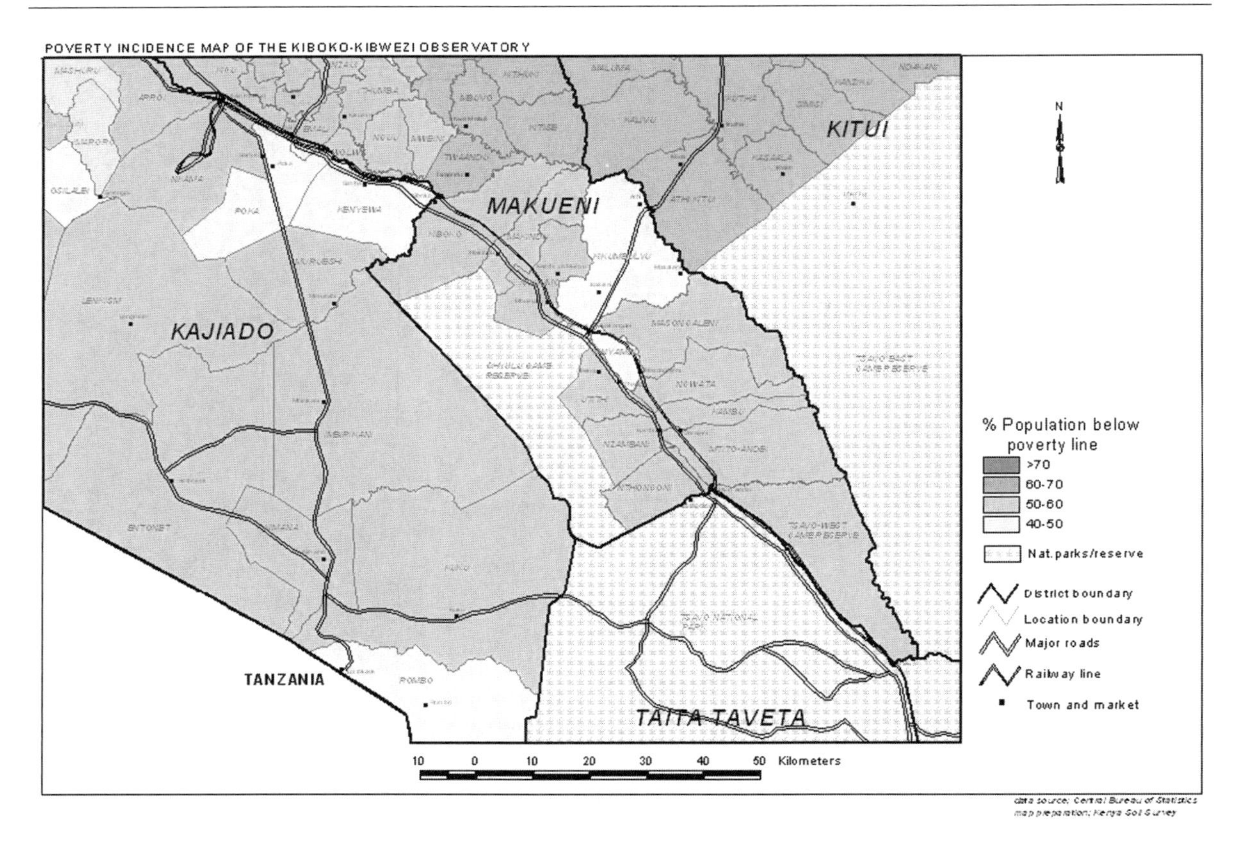

Fig. 4 Poverty incidence at Kiboko-Kibwezi observatory

between about 600 and 750 mm as opposed to potential evaporation of between 2100 and 2150 mm (Achieng and Muchena, 1979). Over the record period these rainfalls have ranged between 38–830 and 18–511 mm occurring between October to December and March to May, respectively. In the National Park annual rainfall varies between 300 and about 900 mm and increases towards the south (Table 1). Distribution of the rainfall is very poor with unreliable onset especially during the first season (March to June). Deficit soil moisture is prevalent between January and March and between May and October (Table 2).

Socio-economic Activities (Makueni District)

The population consists of some 43,377 households, a density of about 10–150 people/km^2 and farm sizes averaging 3.5 ha/household (De Jager et al., 2005). About 48.5% of the total farmland is under cultivation while 71% of the inhabitants live below the poverty line of US $1/day. Increased pressure on land has led to land use intensification often without the necessary external inputs to sustain its productivity. A wide variety of crops including maize, green grams, cowpeas, sorghum and pigeon peas are grown for food, although green grams and cowpeas are sold in local markets. Surrounding Tsavo National Park the inhabitants practice bush–fallow cultivation (Van Wijngaarden and van Englen, 1984). In Kibwezi average farm sizes per household are about 3.3 ha/household. About 1.8 ha of this is used to grow crops with the rest under pasture. Nutrient (K and P) flows are zero while nitrogen (N) is negative at about –12 kg/ha/year, with leaching accounting for the highest losses (–8 kg/ha/year or about 67%) (Gachimbi et al., 2002). Livestock are also kept but their quality is low due to poor breeds, feed supply and feed rations. Use of farmyard manure as a source of nutrients is very low due to poor manure management, inefficient collection methods,

Table 1 Agro-ecological characteristics of the observatory within Makueni District

AEZ	Alt (m)	r/Eo (%)	Mean annual rainfall	Potential evaporation	Average seasonal rainfall March to May	Average seasonal rainfall Oct to Dec	Example
III	1830–1350	50–65	1000–1250	1900–2000	400–500	400–500	–
IV	1830–1160	40–50	750–1000	2000–2100	300–400	300–400	Makueni
V	1710–990	25–40	600–750	2100–2150	200–300	300–400	Kibwezi
VI	990–790	15–25	350–600	2150–2500	150–200	200–300	Nthongoni market
VII	790–650	<15	<350	>2500	50–150	100–200	N/Park

Source: Achieng and Muchena (1979) and Jaetzold and Kutsch (1980)

Table 2 Monthly climatic information for selected stations within observatory

Month	Temp. (°C)	Rainfall (mm) Makindu 997 m	Rainfall (mm) Kiboko 975 m	Rainfall (mm) Kibwezi 914 m	Eo (mm)	Effective rainfall 2/3Eo=Et (mm)	Effective rainfall r−Et Makindu	Effective rainfall r−Et UoN farm
Jan	23	39	40	37	171	114	−75	−89
Feb	24	30	40	30	181	121	−91	−96
Mar	25	91	65	91	198	132	−41	−48
Apr	24	123	122	123	169	113	+10	+11
May	23	25	29	25	154	103	−78	−73
Jun	21	1	3	5	150	100	−95	−93
Jul	20	1	1	1	145	97	−96	−97
Aug	21	2	1	2	164	109	−107	−110
Sep	22	4	3	4	193	129	−125	−122
Oct	23	22	24	22	210	140	−118	−177
Nov	23	161	157	161	273	115	+46	+43
Dec	23	140	113	140	252	101	+35	+28
Mean	22.7	641	598	641	2060	1374	−735	−780

Source: Njeru (1980) and Jaetzold and Schmidt (1983)

Table 3 Some spatial and temporal chemical changes along Loituleilei River

Analytical item	Intake in July 1976	Canal in November 1977	Nature of change
pH	7.0	8.6	Getting alkaline
EC	275	264	Decreasing
Na (me/l)	0.96	1.90	Reducing
K (me/l)	0.21	0.21	Constant
Ca (me/l)	0.79	0.48	Reducing
Mg (me/l)	1.27	0.70	Reducing
CO_3 (me/l)	–	0.54	–
Bicarbonates (me/l)	3.06	2.46	Reducing
Chlorides (me/l)	0.60	0.30	Reducing
Sulphates (me/l)	Trace	Nil	Minimal
Sodium adsorption ratio (SAR)	0.95	2.5	Increasing
Adjusted SAR	1.62	3.75	Increasing

Source: Mugai and Kanaka (1978)

poor storage and haphazard application. Soil moisture storage and hence growing periods for rain-fed crops are greatly reduced within the annual calendar (Table 3). Irrigation is operated in the University farm, KARI Katumani and Kiboko sub-centre, the CARE-Kenya farms and within TA but inadequate methods of irrigation (usually sprinkler) often lead to soil salinisation and sodification and eventual abandonment of such lands. Saline soils have a pH ranging from 5.2 to 8.5 while alkaline soils have a pH ranging from 7.5 to 12.4 (Radiro et al., 2003). The Athi River traverses the entire western longitudinal parts of the observatory with other rivers including Kibwezi and the Thange. Athi River has a large basin with seasonal flow discharge, peaking between February and March, of 0.5 m^3/S. Kibwezi River is free of sodicity but has medium salinity problems (i.e. 650 mm/cm EC, 3.48 me/l Na and 3.3 SAR) (Ekirapa and Mare, 1991). The Mzima springs form a major source of domestic water in many urbanised areas along the Mombasa highway. Except the Athi River which is acidic the rest of the rivers had alkaline waters. Livestock is an integral component of livelihood with the hardy donkeys forming a special component in Kambaland, especially east of Emali since they are used to convey water (for domestic and other uses) over long distances. Cattle are used for ploughing agricultural land and are a major source of household cash income. Medium cash income is obtained from the sale of goats while poultry keeping forms an important source of food.

Biophysical Characteristics (Kajiado)

The district mainly lies in the livestock-millet zone (LM5) and is dominated by both highly variable rainfall and negative nutrient balance and covers some 2,210,600 ha which is largely rangeland (only 7% (167,000 ha)) available for agricultural production. Three agro-ecological zones are recognised in the park with zone VI occupying over 50% and are composed of *Commiphora–Lannea* group with wooded grassland structure. The grass species are dominated by the *Combretum zeyheri* in the wetter areas. Agro-ecological zones IV and V occupy about 30 and 10% of the park, respectively. Several tree species are a source of livelihood as can be seen by piles of charcoal sales by the road side. The *Balanites* spp. and *Acacia mellifera* are used for making charcoal while *Commiphora africana* and *Cammiphora baliensis* are major sources of timber and other building material. Grasses are mainly cattle pastures while goats browse on some specific tree species. Large portions of the soil are *rhodic* Ferralsols within the parks along with pockets of laval soils up the Chyulu hills. Water erosion in these areas can sometimes be very severe with about 65% of the wind erosion associated with overstocking. Water losses occur through runoff, deep percolation and evaporation. Infiltration rates in the region are between 2.0 and 38.0 cm/h with very high external flow of nutrients from the farms, especially of N which is usually lost through leaching. Nutrient status measured from Enkorika, for example, Ca, N and P were 36.5, 55.5 and 99.1% below critical values, respectively. However, K values were in high supply (175% above critical). The mean annual rainfall at Oloitoktok is around 800 mm but in some years and places (e.g. Enkorika) the actual amounts may be only 10% of this (Gachimbi et al., 2003). Loitokitok receives the long rains between October and December and the short ones between March and May with May to October being usually very dry. Mean annual temperatures are 22°C with cold months in July to August and warm months being February to March. Rainfall is only 40% of the potential evapo-transpiration causing serious moisture deficits between July and September (Ndaraiya, 1996).

Socio-economic Activities (Kajiado)

Average household size is about 13 persons often without any form of off-farm income source with about 87% of the households falling below poverty line of US $1/day/person (see also Fig. 2). Livestock consists of large herds on relatively large areas of communally owned grazing land and is largely inhabited by the Maasai who are largely pastoralists keeping a very large number of cattle (i.e. >28.4 livestock per livestock unit). Farmers are not market oriented with an average distance from local market of about 35.6 km. Kimana and Tikondo springs are the major water sources used at Kimana irrigation scheme. Table 3 shows the chemical spatial and temporal changes along the Loitulelei River at two points. Infiltration rates within these schemes varied from moderate (2.0–6.5 cm/h) to rapid (12.5–25.5 cm/h) (Mugai and Kanaka, 1978).

Loituleilei River provides water for minimal irrigation within the district (e.g. Kimama irrigation scheme). The Kimama irrigation scheme occupies an area of some 177 ha and has loam to sandy clay-loam soils developed on alluvial and colluvial deposits of volcanic origin. High-input rain-fed agricultural production is confined to production of maize and beans which yields about 1.8 and 0.5 t/ha, respectively. The agricultural areas are confined to Loitoktok and Ngong divisions. In some areas (e.g. Enkorika) farm sizes are

Table 4 Crop types, hectarages and yields in Kajiado District

Food and cash crops	Hectare	Yields (t/ha)	Horticultural crops	Hectare	Yields (t/ha)
Maize	30,453	1.8	Tomatoes	165	45
Wheat	140	0.5	Onions	80	9.0
Sorghum	37	0.5	Cabbages	54	10
Millet	39	0.5	Kales	62	11
Beans	56,367	0.5	Asian vegetables	230	6.0
Cowpeas	27	0.6	Spinach	24	15.0
Pigeon peas	7	0.6			
Green grams	8	0.4			
Irish potatoes	531	4.0			
Sweet potatoes	24	5.0			
Cassava	9	6.0			
Groundnuts	29	0.6			
Soya beans	5	0.4			
Cotton	10	0.2			
Coffee	74.5	14.0			

Source: Mwangi (1996)

large (about 52 ha/household) but with only 0.2% of this land area being under cultivation (Gachimbi et al., 2005). Table 4 shows the crops grown in the district along with their hectarages and yields. Livestock is by far the most important activity while arable farming is found only in small pockets (Gachimbi et al., 2003). Goats are kept in areas considered to be too dry for cattle. Maasai keep large herds of cattle and is the measure of their wealth hence a display of their characteristic nomadic lifestyle, usually in search of pasture.

Conclusions and Recommendations

Soil characteristics are related to diversity in plant species with the Acrisols dominating *Acacia senegal* and *Commiphora*, while *Acacia tortilis* occur mainly on Lixisols, Arenosols and Solonetz. *Acacia seyal*, *Acacia mellifera* and *Acacia reficiens* were found on Vertisols (Wekesa et al., 2007). The soils were low in fertility with major nutrients being at zero or negative (i.e. $N = 0$, $P = 0$ and $K = -18$ kg ha^{-1}), although in parts of Kajiado K being in high supply (i.e. 175% above critical levels). Soil moisture deficits for agriculture and livestock development could be countered by interventions from KARI of appropriate technologies (i.e. use of drip irrigation; raising of dryland food, feed and high-value crops; fertility; and water conservation). In both districts over 70% of the households live below the poverty line arising from natural conditions such as the prevalently low and erratic rainfall and distribution coupled with wind and water erosion on predominantly sandy soils. Under these conditions vegetation dwindles and soils status deteriorates outstripping the increased resource demand by human, livestock and wildlife populations. Kenya Agricultural Research Institute recommends the need to invest in the rural people in order to ease pressure on environment rather than concentrating in urban centres and neglecting the rural set-up (KARI, 2005). Survey analysis showed that overgrazing was widespread in the observatory except within the Chyulu National Park. The major livestock raised were cattle, goats, poultry and donkeys but cattle were reported to be declining due to shortage of feed and decreasing sizes of farmland. Forests were dwindling while soil erosion was on the increase mainly due to human activities. Land use changes over time included increased crop farming that was negatively impacting on the vegetation as farmers cleared more land for cultivation. Charcoal burning was a source of income and this requires to be addressed with a view to reduce pressure on fuel wood. Suggested coping strategies in response to long periods of drought and declining land quality included use of plant species tolerant to very severe and adverse environmental conditions which are also nutritious for the animals. Considerable quantity of biological resources and potential for water harvesting exist and could be exploited for improved livelihoods.

Acknowledgements Personnel from the KARI Kiboko Range Research Centre, the Kenya Soil Survey (KSS) and Irrigation and Drainage (IDRP) (i.e. from the National Agricultural Research Laboratories (NARL)) are acknowledged for assistance with analysis of the soil samples and cartography. Financial assistance for the field visit and additional soil analysis was provided by the DMP-KARI project and KASAL-KARI and the Director KARI accepted to have the chapter published.

References

Achieng NJM, Muchena FN (1979) Land utilisation types of the Makueni area. Miscellaneous soil paper no. M20, Kenya Soil Survey, Kabete

African Network for Soil Biology and Fertility (AfNet) (2007) International symposium organised by AfNet of TSBF institute of CIAT in collaboration with soil fertility consortium for Southern Africa. (SOFECSA), TSBF, Nairobi

Christensen J, Kesten T, Sheperd G, Haddad AM, Lesinge FE (2002) How UNEP can help dry-land countries address land degradation, climate change and biodiversity loss in hotter dry-lands. United Nations Environmental Programme (UNEP). www.unep.org

Climate Network Africa (1994) Poverty and climate change. Tiempo 14:20–22

Comprehensive Resource Inventory and Evaluation System (CRIESL) (1983) Kenya natural resource assessment. The CRIESL project. Michigan State University

Darkoh MBK (1993) Desertification: the scourge of Africa. Tiempo 8:1–6

De Jager A, van Keulen H, Mainah F, Gachimbi LN, Itabari JK, Thuranira EG, Karuku AM (2005) Attaining sustainable farm management systems in semi-arid areas in Kenya. Few technical options, many policy challenges. Int J Agric Sust 3:189–205

Ekirapa AE, Mare EM (1991) Detailed soil survey of part of the University of Nairobi Dryland Field Station, Kibwezi Machakos district. Detailed soil survey report no. D75. Kenya Soil Survey

Ekirapa AE, Muya EM (1991) Detailed soil survey of part of the University of Nairobi dryland field station, Kibwezi. Machakos District. Detailed soil survey report no. D75. Kenya soil survey

Food and Agricultural Organisation (FAO) (1993) Agro-ecological assessment for national policy. The example of Kenya. Soil resources, management and conservation service, FAO. Land and water development division and international institute of applied systems analysis, Rome

Gachimbi LN (1990) Land degradation and its control in the Kibwezi area, Kenya. MSc thesis, Department of range management, University of Nairobi

Gachimbi LN, De Jager A, van Keulen H, Thuranira EG, Nandwa SM (2002) Participatory diagnosis of soil nutrient depletion in semi arid areas of Kenya. In: International Institute for Economic Development (IIED) (ed) Managing Africa's soils No. 26. IIED, London, 15pp

Gachimbi LN, van Keulen AH, De Jager A, Thuranira EG, Karuku AM, Itabari JK, Nguluu SN, Gichangi EM (2003) Participatory land use analysis for semi arid regions in Kenya. A nutrient monitoring approach. In: Proceedings of the international systems for sustainable dryland agricultural systems, ICRISAT, pp 37–48, Dec 2003. www.icrisat.org

Gachimbi LN, van Keulen H, Thuranira EG, Karuku AM, de Jager A, Nguluu S, Ikombo BM, Kinama JM, Itabarai JK, Nandwa SM (2005) Nutrient balances at farm level in Machakos (Kenya), using a participatarory nutrient monitoring (NUTMON) approach. Land Use Policy 22:13–22

Jaetzold R, Kutsch H (1980) Climatic data bank of Kenya. Department of Cultural and Regional Geography, University of Trier, Germany

Jaetzold R, Schmidt H (1983) Farm management handbook of Kenya, vol. II. Natural conditions and plant growth. Part C, East Kenya (Eastern and Coast). Ministry of Agriculture, National Agricultural Research Laboratories, Kenya

Kasusya P (1988) Combating desertification in northern Kenya (Samburu) through community action. A community care experience. J Arid Environ 39(2):325–329

Kenya Agricultural Research Institute (KARI) (2005). Annual report of the Kenya DMP/GEF project for January–December 2005. Desert margins programme. KARI Headquarters, Kenya Agricultural Research Institute

Ministry of Agriculture (MoA) (1987) Description of the first priority sites in the various districts. Fertilizer use recommendation project (FURP) 26, Machakos District, NARL-Kabete

Muchena F, Njoroge CRK (1984) Soils of the Makueni area. (Quarter degree sheet 163). Set of Printed maps. Kenya soil survey

Mugai ENK, Kanake PJK (1978) Detailed soil survey of Kimana irrigation scheme, Kajiado district. Detailed soil survey report no. D11. Kenya soil survey

Mwangi SM (1996) Crop production and its constraints in Kajiado district. Miscellaneous paper no. M46. Kenya soil survey

Ndaraiya FM (1996) The role of climatology in land use planning for agriculture. In: Kilambya DW, Gicheru PT (eds) Proceedings of workshop between Kenya soil survey and other key players concerned with land resources of Kajiado district held at Moran Hotel Kitengela. Miscellaneous paper no. M46. Kenya Soil Survey, Nairobi

Njeru EB (1980) Semi detailed soil survey of Tikondo irrigation scheme. Oloitoktok division, Kajiado District. Semi detailed soil survey report no. S10, Kenya soil survey

Njoka JT (2005) Proposed summary of biophysical indicators in ROSELT observatory based sustainable land management criteria. Report on monitoring and evaluation framework for sustainable land management in Africa; Appendix on biophysical indicators. Food and Agricultural Organisation, Rome

Omulabi JE, Kinyali SM, Tirop SK (2000) The influence of soil physical properties on the infiltration rates in the salt affected soils in Makindu-Makueni District. In: Proceedings of the 15th soil science society of East Africa held on 19th–23rd Aug 1996, Nanyuki

Radiro MO, Onyango JW, Maingi PM, Sijali VI (2003) Salt affected soils of Kibwezi. In: Proceedings of workshop held in Kibwezi from 29th August to 3rd September 2003 at

DANIDA conference hall, Kibwezi. Irrigation and Drainage Research Programme, NARL

ROSELT/OSS (2001) Observatory network for long term ecological monitoring. ROSELT/OSS collection- scientific document no.2

Toulmin C (2005) Climate changes and Africa and climate change. Tiempo 57:12–15

Unganai L (1997) Regional drought monitoring. Tiempo 23: 17–19

Van Wijagarden W, van Englen VWP (1984) Soils and vegetation of the TSAVO area (Mtito Andei-Voi). Reconnaissance Soil Survey report no. R7. Kenya Soil Survey, Nairobi

Wekesa L, Mulatya J, Esilaba AO (2007) Bio-socioeconomic factors influencing tree production in south eastern drylands of Kenya. Paper presented at the Symposium on Innovations as Key to the green revolution in Africa: exploring the scientific facts held. Arusha, Tanzania, 16–22 Sep 2007

Soil Fertility Management in the Region of Gourma, Burkina Faso, West Africa

M. Traoré, T.G. Ouattara, E. Zongo, and S. Tamani

Abstract Meeting the millennium development goals in Africa, such as poverty reduction, alimentary self-sufficiency, sustainable natural resources management etc., can only be achieved by intensification and increasing crop yield per unit area. Despite the research efforts made in the field of soil fertility management and the proposition of technology packages, crop yields are still below population needs in Africa. How can African soil adequately nourish its population? To answers this question, field investigations using a rapid and participatory method of research were carried out in 15 villages of the region of Gourma. The investigations aimed to determine farmers' knowledge on soil resources: soil classification, soil degradation using local indicators, local soil fertility management, and smallholder farmers' capacity to adopt new technologies in soil fertility management. The results of the investigations show that the main parameter for soil classification was texture. Fallow, organic manure and crop rotation were the main soil fertility management methods; soil degradation was judged according to the agricultural output, and the apparition of certain species of weeds. Concerning the villages where modern soil fertility management technologies were introduced, the lack of tools and information were the main limiting factors for the adoption of these technologies at the smallholder farmers' level. Our investigations have shown that stone belt construction in the field was essential for limitation of soil erosion processes and additionally they represent a starting point of intensification encouraging farmers from the practice of shifting agriculture.

Keywords Local soil classification · Region of Gourma · Smallholder · Soil fertility management · Stone belt

Introduction

Land degradation can be considered in terms of the loss of actual or potential productivity or utility as a result of natural or entropic factors; it is the decline in land quality or reduction in its productivity. In the context of productivity, land degradation results from a mismatch between land quality and land use (Beinroth et al., 1994).

Land degradation will remain an important global issue for the 21st century because of its adverse impacts on agronomic productivity, environment, and its effect on food security and quality of life (Eswaran et al., 2001). African farmers are more vulnerable to land degradation because of their poor soil management system, the low agricultural inputs, the high dependence of yield on climatic hazard, and the large number of persons depending directly on agricultural output. The history of farming in Sub-Saharan Africa (SAA) is characterised by low input–low output as farm income is often too low for farmers to purchase enough mineral fertilisers and organic inputs to compensate for nutrient outflows. This consequently leads to soil degradation (Badiane and Delgado, 1995).

M. Traoré (✉)
Bureau National des Sols (BUNASOLS), Burkina Faso
e-mail: madouchef@yahoo.fr

A. Bationo et al. (eds.), *Innovations as Key to the Green Revolution in Africa*,
DOI 10.1007/978-90-481-2543-2_95, © Springer Science+Business Media B.V. 2011

In SAA in general and Burkina Faso particularly, it is still a daily struggle for agricultural production to provide enough food to meet the basic needs of an increasing population. The investigation areas are characterised by particularly low agricultural productivity due to the low adoption rate of modern technologies, nonavailability of capital for off-farm input, and land tenure. There is a definite amount of land suitable for food production while the population is growing fast (3.2% growth rate per year); there is also an input of migrants from the central part of the country. As a result, the ratio of cultivated land per capita is declining rapidly with sometimes the cultivation of marginal lands leading to degradation of natural resources.

Aware that land degradation seriously impacts population life with unexpected social and economic consequences, the Government of Burkina Faso, with the aid of partner institutions has initiated a community soil fertility management program. Farmers are fully involved in the steps of identification of the limitation factors for soil fertility management and actions are proposed by them to mitigate soil degradation.

Materials and Methods

The study was carried out in 15 villages of the region of Gourma located in the eastern part of Burkina Faso (Fig. 1).

The investigation was made by a pluri-disciplinary team of: one soil scientist, a rural economist, and a specialist in rural sociology; the investigation period for each village was 4 days. The investigation procedure was a rapid and participatory method of research (FAO, 2002). Table 1 shows the tools and procedures for data collection.

Results and Discussion

Results

Soil Taxonomy

For the 15 investigated villages soil taxonomy was mainly based on soil textural properties. That is why for the three main classes of soil identified the names are given in the main Goulmatcheman and mooré languages of the investigation area:

Tanciaku or Zinca (stony soils): brown colour (10YR5/4). Textural composition was sandy-loamy; with low development of soil structure; the aggregates were blocky subangular with moderate consistence. This class of soil was in general located at the summit, near hills leading to an important quantity of gravels and rock fragments spread out over the soil surface. The soil had good porosity due to the high content of gravel which does not retain water infiltration; poor biological activities; low vegetation cover; and was acidic (pH = 6.0). Human activities noted on these soils were greasing, wood collection, and agriculture. Agriculturally, the main crops grown on this class of soil were: sorghum, millet, beans, cotton and peanuts. Soil degradation indices were erosion threats at the surface, fire, and compaction due to animal trampling and the apparition of "*penpelgu*", a denuded surface. These soils were recognised to have medium fertility but they were very sensitive to water shortages.

Tambima or Biisga (sandy soil): light yellowish brown (10YR6/4). Textural composition was sandy-loamy; with low development of soil structure; the aggregates were blocky subangular and of weak consistence. Because of the textural porosity, this class of soil is characterised by good porosity with high water infiltration rates. In general *Tambima or Biisga* (sandy soil) was located on the middle slope; had high biological activities; low water retention capacity; low vegetation cover; and was lightly acidic (pH = 6.5). Human activities noted on these soils were greasing, wood collection, and agriculture. Agriculturally, the main crops grown on this class of soil were: millet, peanuts, Vouandzou, sesame, cotton, and cassava. Soil degradation indices were erosion threats at the surface, fire, and compaction due to animal trampling and the apparition of *Striga sp.*, and a light colour.

Tambouali or Baongo (clay soils): located on the lower slopes, light brownish gray (10YR6/2). The textural constitution was loamy clay; with low development of soil structure; the aggregates were blocky subangular with strong consistence. This class of soil is characterised by poor porosity due to the high clay content; it therefore also has high water-holding capacity. The infiltration rate is high at the beginning of the rainy season because of surface cracks but gradually

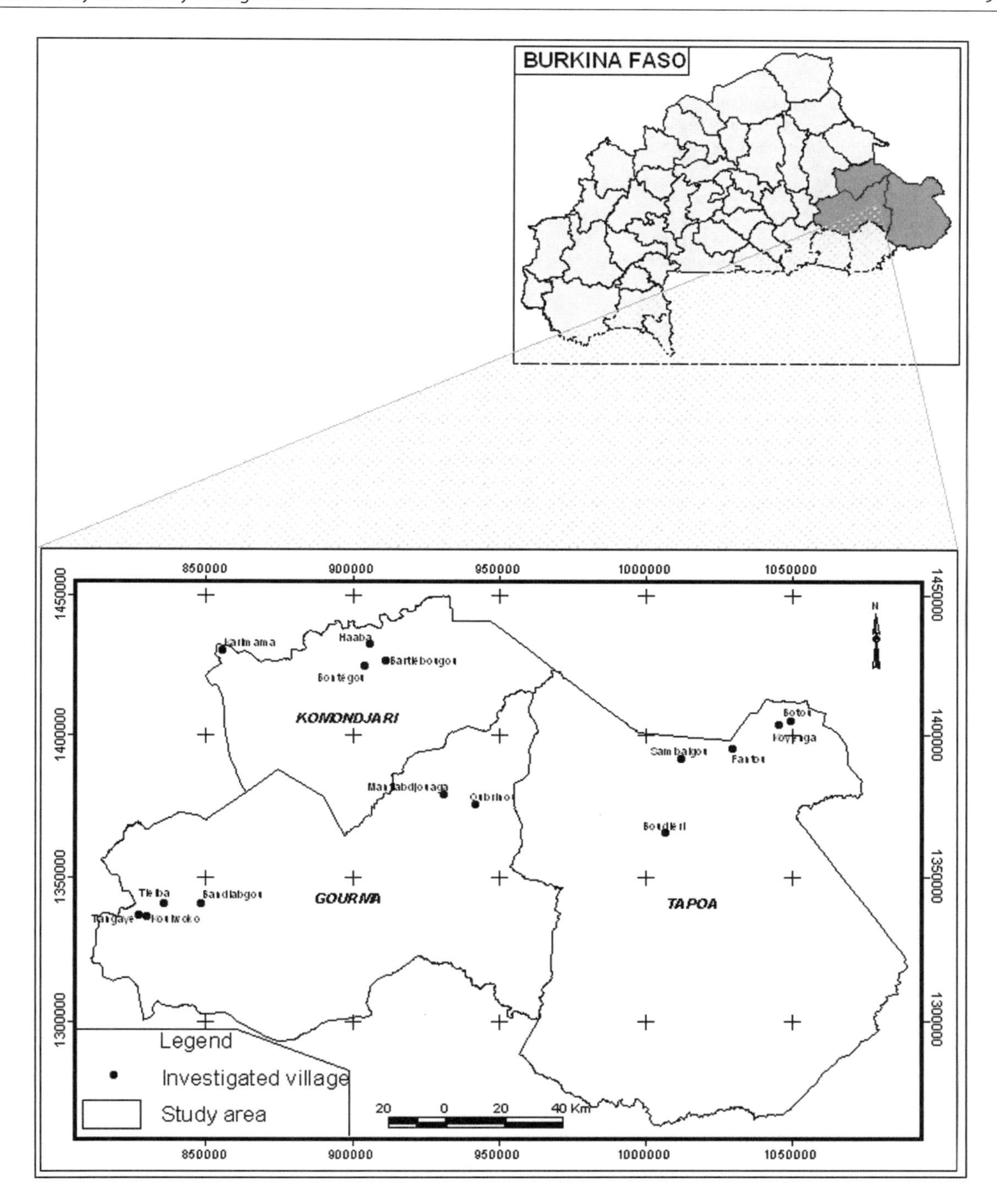

Fig. 1 Investigation area

decreases and becomes low as the soil becomes wet and the surface cracks disappear. This class of soil had high vegetation cover due to its long conservation of humidity; they have medium acidity (pH = 6.0). Human activities noted on these soils were greasing, wood collection, and agriculture. Soil degradation threats were water erosion threats at the surface, fire, and compaction due to animal trampling.

Table 1 Tools and procedures for data collection

Tools	Participants	Analysed parameters	Interactions
1. Thematic maps	All the members of the community (men, women, old and young persons)	– Soil taxonomy according to the local classification – Soil fertility status – Land use (infrastructure, natural resources, etc.) – Indicators for fertile or poor soils	– Transect – Social map – Interview
2. Transect	Focus groups (local organisation leaders, old persons, and member of the family in charge of the land question), mixed groups, old persons.	– Basic natural resources – Soil type and land use – Location of fields and size of fields – Infrastructure and services – Other activities	–Thematic map – Interviews (in general assembly and with focus groups)
3. Interview	– All the members of the community (men, women, old and young persons) – Focus groups: association leaders, vulnerable groups (women, young and migrants), old persons in charge of traditions, etc.	In addition to the information given by the thematic maps and the transect, the interview allowed one to obtain information on: – Land tenure – Social organisation for land management – Actual knowledge on soil fertility management at the village level, etc.	–Transect – Social map
4. Venn diagram	– Focus groups – The entire village	– Institutions and local associations – Interaction with the different partners, etc.	– Social map – Institutional background of the village

The investigations have shown the importance of local soil taxonomy for the local communities because the varying crop production on the land was mainly based on the local soil taxonomy. Also soil fertility status and its vulnerability to degradation were defined according to the local soil taxonomy. Also, the different technology packages proposed by the population for soil fertility management were done with regard to the soil class.

Local Indices for the Determination of Poor Soils

Ten parameters were used to characterise poor soils within the 15 investigated villages; the frequencies of expression of the parameters are given in Fig. 2 .

Soil Degradation Factors

Investigations were also carried out in order to test farmers knowledge on the degradation factors; the results of these investigations are expressed in Fig. 3 and, the investigation shows that degradation is mainly due to human activities through mismanagement of soils.

Technology or Technology Packages Proposed by Farmers for Mitigating Soil Fertility Decline

According to the social and economic standing of the investigated villages, the combination of two or three technologies outlined in Fig. 4, were proposed to farmers for soil fertility management.

Discussion

There is some knowledge at the community level on soil fertility management. First of all, each investigated village had local soil taxonomy; this soil taxonomy is mainly based on soil textural properties and land use for the production of different crops is guided by this local soil classification.

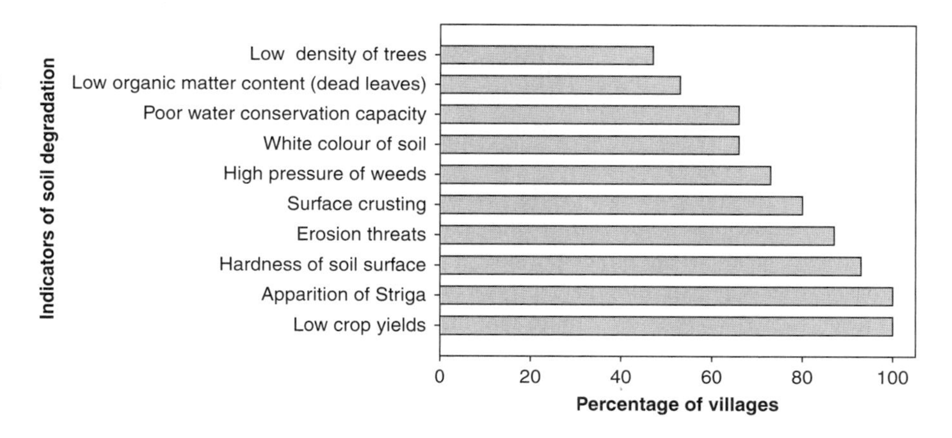

Fig. 2 Frequencies of characterisation of poor soil using local indices within the investigated villages

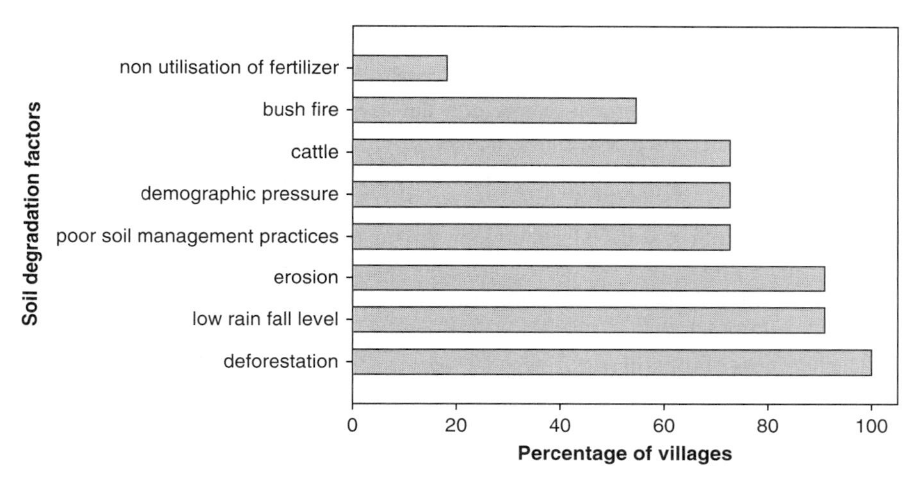

Fig. 3 Distribution of soil degradation factors as expressed by farmers of the investigated villages

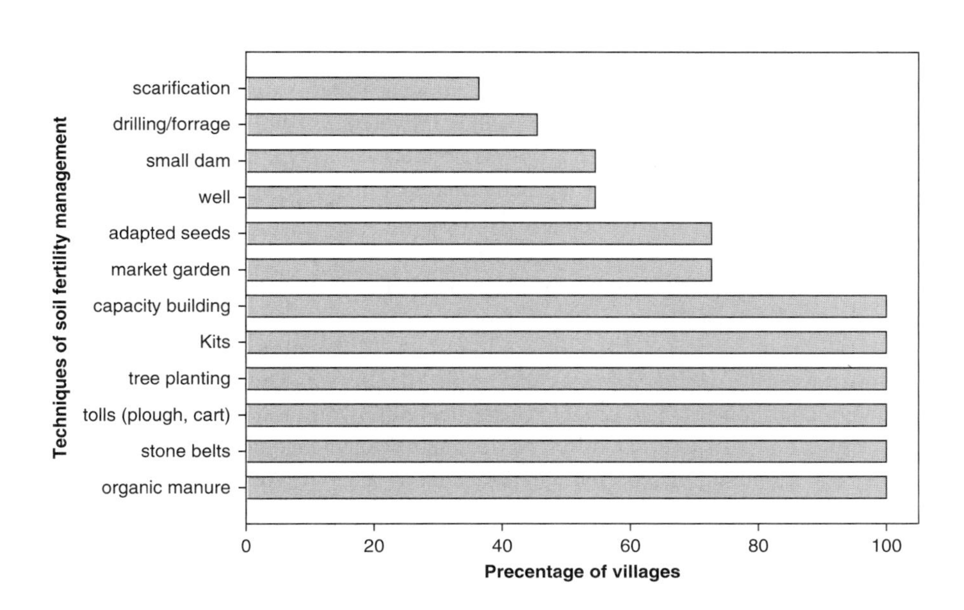

Fig. 4 Ratio of demand of soil management techniques in the investigated villages

Soil degradation has been determined using ten indices (Fig. 2). In the 15 investigated villages, the presence of *Striga sp.* and the yield decline have been pointed out as land degradation indicators. *Striga sp.* is a parasitic weed infecting poor soils, and during its development cycle, the roots of *Striga* are fixed to those of the host plants and form connections diverting nutrients to the detriment of the host plants. This seriously affects the development of the host plants and leads to yield reduction. This traditional method of assessment of land degradation has a scientific basis because many authors have also based land assessment on its capacity to perform its functions in a sustainable manner. And, for Eswaran et al. (2003), the importance of land degradation among other global issues is enhanced because of its impact on food security and the quality of the environment. Therefore, the use of yield decline as an indicator for and the assessment of soil degradation by local communities is an acceptable methodology.

The hardness of the soil surface, erosion threats and the impermeability of the soil surface were pointed out as indicators of soil degradation in the respective rates of 93, 87 and 80%.

The hardness of the soil surface and surface crusting are mainly the results of soil compaction which is a worldwide problem. In Africa, between 40 and 90% of yield reduction is due to hardness of the soil surface or soil compaction (Kayombo and Lal, 1994).

High weed pressure, white colour of soils, poor water conservation capacity and low organic matter content were pointed out at the respective rates of 73, 66, 66 and 53% as soil degradation indicators.

High weed pressure: the growing of crops is to favour the developmental conditions of the cultivated plants to the detriment of weeds. Poor soils characterised by low nutrient content lead to difficulties in nutrient uptake by cultivated plants. The weeds are more resistant and are better able to overcome such conditions. That is why they always grow more successfully and faster than cultivated plants on poor soils.

Soils take on a white colour as the result of poor organic matter content. Soil organic matter has different origins and can exist at different decomposition levels. The mixture gives the soil a dark colour. When organic matter content is low, the soil becomes fragile for the following reasons: (1) organic matter represents a nutrient pool that is indispensable for plant development; (2) beyond its role as a nutrient pool, organic matter contributes greatly to improving and stabilising soil structure. Brady (1990) and Stevenson (1982) have found that organic matter and clay, either directly or indirectly, influence various chemical, physical and biological reactions leading to greater soil aggregation and increasing aggregate stability and water conservation in soil. This avoids the soil compaction discussed above as an indicator of degraded soil.

Among the eight degradation factors given by the local communities (Fig. 3), deforestation was the most important; 100% of the investigated villages recognised deforestation as the main factor of soil degradation. It was followed by soil erosion and low rainfall level (91%). In the investigated area, shifting cultivation and transhumance pastoral systems are practiced. Because of the low-input low-output farming system, land vulnerability and the demographic pressure, farmers are clearing forest using traditional methods with slash-and-burn as an integral way to remove vegetation and reduce pest. That reduced considerably the possibilities for trees to regenerate. Bridges and Oldeman (2001) have documented the negative impact of food production on the environment through removal and modification of the vegetation. Another reason for forest clearing in the investigated zone is the increasingly high demand for fire wood to meet the needs of cities like Ouagadougou, Fada NGourma and Poytenga.

The erosion process is at the origin of lateral transport of arable soil and nutrients; it is strengthened by poor soil management practices. In Sub-Saharan African farming systems, the share of erosion in nutrient depletion is predominant. Stoorvogel and Smaling (1990), FAO (1986) have shown that 38% nitrogen, 40% phosphorus and 39% of potassium were lost through erosion.

In 73% of the investigated villages, poor soil management practices, demographic pressure and the increase in the animal population were pointed out as soil degradation factors. Low rainfall level and population pressure were pointed out by local communities as factors of land degradation. Concerning population pressure, a clear relationship is not present between land degradation and population pressure. Case studies in West African savannah showed that in agricultural areas with similar soil, climate, and agricultural methods, land may be more subject to degradation where the rural population is of the order of 30 persons km^{-2} than it is with 80 persons km^{-2} (Drechsel et al., 2001). According to Henao and Baanante (1999); FAO (1986), population density ranged of 10–500 inhabitants km^{-2}; in the investigated area, the population density is 22–50 persons km^{-2}. Therefore high population density can not be considered as causes of land degradation.

The low rainfall level leads to soil water stress and this has been pointed out as a soil degradation factor. The rainfall level of the Gourma region lies between 600 and 900 mm. If this quantity of water was conveniently used with adapted crop varieties, soil water stress should not be a limiting factor for farming activities. Eswaran et al. (1997a, b) have shown that about 14% of African soil under water stress could be free of this constraint if convenient water conservation practices were applied.

Due to the magnitude of soil degradation in the Gourma region, the actual land management system performed by farmers are unsuitable mainly because of the lack adequate land management tools. From these results we can conclude that land degradation is mainly due to failure to implement new technologies rather than the draught and population pressure.

Cattle pressure is expressed by overgrazing and destruction of soil aggregates leading to surface compaction which facilitates the erosion process because of low permeability of soil.

Fifty-five and 18% of the investigated villages have respectively recognised bush fire and nonutilisation of fertiliser as causes of land degradation. Bush fire is one of the most important causes of soil fertility decline. Continuous burning of the soil has a negative impact on soil fauna. Bush fire is at the origin of soil organic matter destruction and leads in the long term to surface crusting on clay soil because of the exposition to high temperatures.

Nonutilisation of fertiliser leads for sure to soil degradation even if all the technology packages of land conservation were practiced. The degradation arises from the loss of nutrients through harvest and exportation of crop residues; and it has to be compensated by external inputs. Herweg and Ludi (1999) have shown that yield impact is barely observed on soil fertility if no other inputs are provided simultaneously. In the investigated area, the use of organic matter from local material is recommended because farmers do not have enough money to invest in mineral manure. Also, the low economic values of yields (cereals) do not allow them to meet their basic needs and reinvest in soil fertility management. In this case using mineral manure places them under permanent dependence on donators (projects); this is not a sustainable system because the collaboration stops at the end of the project and the problem of soil fertility management will remain.

There is a global awareness that urgent soil conservation methods are needed to mitigate land degradation. The principal actions recommended by the local population are: stone belts, organic manure, tree planting, kits for soil fertility management and soil conservation, agricultural tools and capacity building.

Even if the farmers didn't mention it, one of the crucial problems of soil fertility management in the investigated villages is land property right. In the Gourma region, land belongs to the community; that is a limitation for soil fertility management by foreigners and even for some natives of the village because they are not sure they will receive the future benefit of their investment. Tree planting is a sign of appropriation of land; it cannot be recommended at an individual scale of the soil fertility management issue because it can lead to social conflicts. Only the community plantations on extremely degraded lands and the lands under high risk of degradation can be recommended. Also for the extremely degraded land, tree planting can be preceded by scarification or the Zaï system; these techniques were proposed by farmers in 73% of the investigated villages.

Application of organic manure is necessary to stabilise the soil nutrient pool as stated above. The benefits of this practice have to be maximised by using stone belts. The stone belts reduce the lateral transport of organic manure due to erosion. Also, the stone belts are important because they are the starting point of intensification in the smallholder farming system. It represents investment encouraging farmers to stay on the same plots for many years. In addition to these advantages, the stone belts help to maximise water infiltration reducing moisture stress. All these actions need to be supported by capacity building and acquisition of adapted tools for farming and soil conservation activities. Without this support it will be impossible for farmers to mitigate soil degradation in view of their economic and social standing.

Adapted seed: one of the identification parameters of poor soil is yield decline; in some cases yield decline is due to water stress. Sub-Saharan

Africa is characterised by general decline and irregular repartition of rainfall. This problem can be mitigated using adapted seed because the local varieties with long vegetation cycles do not become fully developed before the end of the rainy season.

A permanent water source is indispensable for convenient soil fertility management. The production of 10.8 m^3 of organic manure requires 1,000 L of water per week. Providing this quantity of water for a large-scale production of manure is not assured without a permanent water source, which is why a small dam was demanded by 55% and drilling by 73% of the investigated villages.

Adoption of new technologies by villagers depends on their economic standing. Market gardening allows farmers to keep busy during the dry season and the vegetables produced during this period are an additional source of earnings which can improve farmers' living standards and help them to adopt new technologies for soil fertility management.

Conclusion and Recommendations

Farmers have empirical knowledge of soil; in every investigated village, they had local soil taxonomy, identification factors for poor or fertile soils, and knowledge of the parameters and causes leading to soil degradation. In this article we have given a scientific explanation for most of this empirical knowledge. There is a global awareness concerning land degradation and its danger to the smallholder farming system because soil degradation has a particularly negative impact on farmers' life due to their greater dependence on agriculture. Because of their low economic income, smallholder farmers cannot afford cereal to compensate for yield decline due to soil degradation. This will lead to food insecurity. Agcaoili et al. (1995) have shown that increased land degradation could increase the number of malnourished children in rural and urban areas in 2020 by 3–4%.

The investigation has shown that sustainable soil conservation has to be practiced using local materials and it has to include farmers' access to markets which can increase the economic value of the agricultural products. These parameters are important because they allow farmers to meets their basic needs and invest in sustainable soil management activities. Sanders and Cahill (1999) have shown that in addition to technology packages, financial support, local capacity building of soil conservation projects, guaranteed output prices are the real achievement of soil conservation.

Acknowledgement First I thank BUNASOLS, my home institution, for the facilities offered during the field investigation, and PICOFA for funding the investigation. My gratitude goes to the population of the 15 investigated villages and the decentralised authorities for their collaboration and for hosting the investigation team during the field work.

References

Agcaoili M, Perez N, Rosegrant M (1995) Impact of resource degradation on global food balances. Paper prepared for the workshop on "Land degradation in the developing world: implication for food, agriculture, and environment to the year 2020". Annapolis, MD, IFPRI, Washington, DC, 4–6 Apr 1995

Badiane O, Delgado CL (eds) (1995) A 2020 vision for food, agriculture, and the environment in sub-Saharan Africa. Food, agriculture, and the environment discussion paper 4, World Bank, Washington, DC

Beinroth FH, Estwaran H, Reich PF, Van der berg E (1994) Land related stresses in agroecosystems. In: Virmani SM, Katyal JC, Eswaran H, Abrol IP (eds) Stressed ecosystems and sustainable agriculture. Oxford and IBH, New Delhi, 441 pp

Brady NC (1990) The nature and properties of soils, 10th edn. Macmillan, New York, NY

Bridges EM, Oldeman LR (2001) Food production and environmental degradation. In: Bridges EM et al (eds) Responses to land degradation. Science Publishers Inc, Enfield, NH

Drechsel P, Penning DE, Vries FWT (2001) Land pressure and soil nutrient depletion in sub-Saharan Africa. In: Bridges, M et al (eds) Response to land degradation. Science Publishers, Inc. Enfield, NH, pp 55–64

Eswaran H, Almaraz R, Vendenberg E, Reich PF (1997a) An assessment of the soil resources of Africa in relation to productivity. Geoderma 77:1–18

Eswaran H, Lal R, Reich PF (2001) Land degradation: an overview. In: Bridges EM, Hannam ID, Oldeman LR, Pening de Vries FWT, Scherr SJ, Sompatpanit S (eds) Responses to land degradation. Proceedings of 2nd international conference on land degradation and desertification, Khon Kaen, Thailand. Oxford Press, New Delhi, India

Eswaran H, Reich PF, Beinroth FH (1997b) Global distribution of soils with acidity. In: Moniz AZ, Furlani AMC, Schaffert RE, Fageria NK, Rosolem CA, Cantarella H (eds) Plant-soil interactions at low pH: sustainable agriculture and forestry production. Proceedings of the 4th international symposium on plant-soil interactions at low pH, Belo Horizonte, Minas Gerais, Caminas: Brazilian Soil Science Society, pp 159–164

FAO (1986) Atlas of African agriculture. African agriculture: the next 25 years. ARC/86/3, Rome, 72 pp

FAO (2002) Guide dápplication au niveau terrain. ASEG. Programme dánalyse socioéconomique selon le genre. Viale delle Terme di Caracalla 00100 Rome, Italie, 118p

Henao J, Baanante C (1999) Estimating rates of nutrient depletion in soils of agricultural lands of Africa. IFDC Technical Bulletin T48. Muscle Shoals, IFDC, Alabama, 76pp

Herweg K, Ludi E (1999) The performance of selected soil and water conservation measures – case studies from Ethiopia and Eritrea. Catena 36(1/2):99–114

Kayombo B, Lal R (1994) Response of tropical crops to soil compaction. In: Sloane BD, Van Ouwerkkerk C (eds) Soil compaction in C production. Elsevier, Amsterdam, pp 87–315

Sanders DW, Cahill D (1999) Where incentives fit in soil conservation programs. In: Incentives in soil conservation: from theory to practice. Science Publishers, Enfield, NH, Science Publishers Inc., 11p

Stevenson FJ (1982) Humus chemistry: genesis, composition, reactions. Wiley, New York, NY

Stoorvogel JJ, Smaling EMA (1990) Assessment of nutrient depletion in sub-Saharan Africa 1983–2000, vol 1, Report 28. Winand Starting Centre, Wageningen, 137p

Understanding Cassava Yield Differences at Farm Level: Lessons for Research

A. Babirye and A.M. Fermont

Abstract Cassava is an important food crop in Africa. A range of improved varieties and Integrated Pest Management (IPM) practices have been developed to increase and maintain its yields in response to emerging pest and disease problems. Much less attention has been given to Integrated Soil Fertility Management (ISFM) and agronomic practices. Wide cassava yield disparities between farmers exist. Using a farm typology approach and six case study sites in East Africa, yield differences between sites and farm types were quantified and factors contributing to the differences evaluated. Overall, yield differences were twice as large between sites (6 t/ha) as between farm types (3 t/ha), whereas within sites, cassava yield differences between farm types varied from 1.5 to 7.5 t/ha. While differences in agro-ecological conditions explain part of the variation found at site level, differences in management were important in explaining cassava yields between farm types. Richer households obtained significantly ($p < 0.001$) higher cassava yield (+3.2 t/ha) than poorer households. Hired labour input ($p < 0.01$–0.05), monocropping ($p < 0.05$–0.15) and timing of first weeding ($p < 0.01$–0.1) significantly explained yield differences between sites and farm typologies. Use of improved varieties was rarely linked to higher yield levels. Manure and/or inorganic fertilizer use was rarely targeted to cassava. To improve cassava production in Africa, more emphasis should be given to the development and dissemination of appropriate management practices and higher yielding varieties. Research for development efforts in cassava should take into account and benefit from differences in cassava production that exist between farm types, while not ignoring agro-ecological differences.

Keywords Agro-ecologies · Farm typologies · Management · *Manihot esculenta Crantz* · Sub-Saharan Africa

Introduction

Cassava (*Manihot esculenta* Crantz) is the largest non-cereal carbohydrate source for human food in the world, with Africa being the largest centre of production. In Africa, cassava is the most important food crop and is grown primarily for local food consumption by small-scale farmers. It is grown either in monoculture or in association with the main cereals and/or legumes. Most soil types are suitable for cassava growing, apart from hydromorphic soils (Fauquet and Fargette, 1990). Production in Africa has increased mostly due to increases in area under cultivation rather than increase in yield per hectare (Hillocks, 2001). Average farmer yield in Africa is 8.9 t/ha, but its potential yield at research stations exceeds 40 t/ha in Eastern Africa (NAARI, 2000). Throughout Africa, cassava is being commercialized into diverse human food products (gari, flour), animal feeds and industrial products such as starch and ethanol (FAO, 2004; Anonymous, 2006).

In spite of its vast importance, cassava is the least researched crop among the major crops of the world. Main research themes in Africa focus on breeding in the context of Integrated Pest Management (IPM).

A. Babirye (✉)
International Institute of Tropical Agriculture (IITA-Uganda), Kampala, Uganda
e-mail: annetbabirye2@yahoo.com

A. Bationo et al. (eds.), *Innovations as Key to the Green Revolution in Africa*,
DOI 10.1007/978-90-481-2543-2_96,

Limited research has been conducted on Integrated Soil Fertility Management (ISFM) and other agronomic practices in cassava. In order to develop appropriate management practices, this study aims to contribute to the understanding of what management factors influence cassava yields at farm level by studying differences in cassava yield and selected management parameters between sites and farm types. Implications will be drawn out for the development of cassava management technologies for smallholder farmers in the cassava-based farming systems of Eastern Africa.

Materials and Methods

Study Sites

A household survey was conducted in the major cassava-growing regions of western Kenya, central and eastern Uganda. From these regions, a total of six sites were randomly selected for the study. The Kenyan sites were Kwang'amor (0°29′N; 34°14′E), Mungatsi (0°27′N; 34°18′E) and Ngunja (0°10′N; 34°18′E) parishes in Teso, Busia and Siaya districts, respectively. The Ugandan sites were Kisiro (0°67′N; 33°80′E), Bujjabe (1°40′N; 32°38′E) and Chelekura (1°14′N; 33°62′E) parishes in Iganga, Nakasongola and Pallisa districts, respectively. Sub-humid climatic conditions prevail across all the six sites. The whole region has a bimodal rainfall distribution and hence two cropping seasons running from March to June and September to November. Annual rainfall varies between 750 and 1770 mm (Anonymous, 2007; Fischler and Wortmann, 1999; Shepherd et al., 1997).

Farm Selection and Characterization

At each site one lowest level administrative unit was randomly selected for the survey. Three local civic leaders with jurisdiction over the selected administrative unit were used as key informants. The three key informants were separately interviewed to identify locally applicable criteria for wealth ranking and to generate a complete list of household heads in the administrative unit, grouped by wealth class (poor, medium and rich) according to the identified criteria. Among the household wealth ranking criteria variously used across the six sites were farm size, number of animals, off-farm income and production orientation. At each site, a total of 20 households (except for Kisiro parish with 21 households) were proportionately selected by wealth class strata but ensuring a minimum of three households per wealth category in the final subsample for the site.

Data were collected using a structured questionnaire during farm visits to the selected households at each site. The questionnaire was designed to collect general household information, labour aspects, average income in the previous 2 years and cassava cropping system characteristics. The household heads were interviewed and their responses cross-checked with other adult household members. Sensitive information on household income, farm size and use of hired labour was cross-checked on farm visits subsequent to the maiden interview using probing techniques geared towards triangulation with the initial information provided. Hired labour data for cassava were obtained by multiplying the total acreage grown using hired labour per year by the number of persons hired per activity.

Cassava Yield Determination

Average fresh cassava yield level in tons per hectare for each farm was calculated from farmer estimates of the number of bags of dry cassava obtained from a given area (either 1 acre or local area unit). Although yield estimates from farmers often give unreliable information (Wortmann and Kaizzi, 1998) and are considered specifically difficult for cassava (Fresco, 1986), as the majority of farmers harvest their cassava fields piecemeal wise, for most households cassava yield estimations corresponded reasonably well to the other cassava production data obtained in the survey.

Farm Types

Households were categorized into two wealth categories based on their average annual household income: households with a medium to good resource endowment and households with a medium to poor resource endowment. The first group of 'rich' households consisted of households with an annual income of more than $750/year, while the 'poor' households earned less than this. In addition we categorized households along their production objectives for cassava: (a) households producing cassava both for the market and for their own food security and (b) households producing cassava mainly for home consumption. The first group of 'market'-oriented cassava farmers earned more than $50/year from cassava commercialization, while the second group of 'home consumption'-oriented farmers earned less than that amount.

Statistical Analysis

Data were analysed using SPSS version 10.0. Analysis of variance was used to test for significant differences in yield and management practices between countries, sites, country by farm type and site by farm type.

Results and Discussion

With average yield levels of 11.3 and 7.0 t/ha, the overall mean cassava yield in this study was significantly ($p < 0.01$) higher in Uganda than in Kenya (Table 1). Yield levels were somewhat lower than the average yield levels for Uganda (14.4 t/h) and Kenya (9.1 t/ha) according to the FAO (2007). Highest average yields in Uganda and Kenya were obtained in Kikooba (12.7 t/ha) and Kwang'amor (7.7 t/ha), respectively. Higher yield levels in Uganda can be partially explained by the significantly ($p < 0.01$) larger part of the cassava acreage under monocropping (+21%), the higher use of improved varieties (+16%) and the higher amount of hired labour used on cassava (+52 man days/year) in Uganda compared to Kenya (Table 1). Variations in agro-ecological conditions between sites may also have partially contributed to yield differences.

Richer households obtained significantly ($p < 0.01$) higher cassava yields than poorer households in both Uganda and Kenya (Table 2). In Uganda, higher yield levels of rich farmers can partially be explained by the larger part of the cassava acreage they planted with monocropped cassava (+16%), the higher amount of hired labour used on cassava (+119 man days/year) and possibly a better division of weed operations within the growing season as rich farmers postponed the last weeding in cassava by 1 month compared to poor farmers. In Kenya, higher yield levels of rich farmers can be explained by a multitude of factors such as a higher use of improved varieties, three times more hired labour on cassava, more manure and fertilizer use and a later first weeding (Table 2). Analysing the data set by site × wealth class shows that in five out of the six study sites rich farmers tended to obtain ($p < 0.05$–0.1) higher (2.3–4.5 t/ha) cassava yields than poor farmers (Table 3). Higher yield levels can be related to more hired labour used on cassava by rich farmers in four sites ($p < 0.1$) and a later first weeding in two sites. Nonetheless, the management parameters included in this study cannot explain higher yield levels in all sites.

Semi-commercial cassava farmers obtained significant ($p < 0.05$) higher cassava yields than farmers producing cassava for home consumption in both Uganda and Kenya (Table 4). However, both types of farmers manage their cassava fields rather similarly.

In this study, higher yields were frequently correlated with more monocropping, with Ugandan farmers growing relatively more monocropped cassava than Kenyan farmers and rich Ugandan farmers growing relatively more monocropped cassava than poor Ugandan farmers. Cassava when planted solely gives higher yields than when intercropped with other crops. Its yield can be drastically reduced if the intercrop is planted earlier than cassava, creating strong competition for light, water and nutrients (Olasantan et al., 1996). In the Kenyan sites, farmers planted cassava on average 4 weeks after planting maize, while Ugandan farmers planted cassava on average 2 weeks before maize, creating more favourable growing conditions for cassava. Farmers estimated that intercropping reduced cassava yields on average by 30–50%.

The findings that higher yields were obtained in farms where more hired labour was used are confirmed by Nweke (1996). He found that hiring labour increases cassava yields considerably, as it facilitates the timeliness in the performance of critical farm operations. Farmers faced with a lack of labour are more likely to delay farm operations of cassava, as cassava is more tolerant to late planting and weeding than most other crops. Nonetheless, delaying farm operations in cassava has adverse effects on cassava yields.

Table 1 Cassava yield and selected management parameters by country

Country	n	Cassava yield (t/ha)	Monocropping (% cassava acreage)	Improved varieties (% cassava acreage)	Time first weeding (weeks after planting)	Total no. of weedings (–)	Time last weeding (months after planting)	Hired labour on cassava (man days/year)	Manure use on farm (kg/ha)	Fertilizer use on farm (kg/ha)
Uganda	52	11.3	70	46	4.0	3.9	8.4	72	167	2
Kenya	59	7.0***[a]	49***	30**	3.9	5.1***	8.0	20**	582***	28***

[a]*** $p < 0.01$; ** $p < 0.05$

Table 2 Cassava yield and selected management parameters by country and wealth category

Country	Wealth[a]	n	Cassava yield (t/ha)	Monocropping (% cassava acreage)	Improved varieties (% cassava acreage)	Time first weeding (weeks after planting)	Total no. of weedings (–)	Time last weeding (months after planting)	Hired labour on cassava (man days/year)	Manure use farm (kg/ha)	Fertilizer use farm (kg/ha)
Uganda	Rich	22	13.0	79	40	4.2	3.8	9.2	139	123	4
	Poor	29	9.8***[b]	63*	51	3.9	4.0	8.0*	20***	207	0
Kenya	Rich	18	8.5	51	40	4.5	5.3	7.8	42	945	63
	Poor	41	6.3***	48	25*	3.6***	5.0	8.0	11***	423**	14***

[a]The cut-off point between 'rich' and 'poor' farmers is an annual income of $750/year
[b]*** $p < 0.01$; ** $p < 0.05$; * < 0.1

Table 3 Cassava yield and selected management parameters by country and main objective of cassava production

Country	Objective cassava production	n	Cassava yield (t/ha)	Monocropping (% cassava acreage)	Improved varieties (% cassava acreage)	Time first weeding (weeks after planting)	Total # weedings (–)	Time last weeding (months after planting)	Hired labour on cassava (man day/year)	Manure use farm (kg/ha)	Fertilizer use farm (kg/ha)
Uganda	M-HC[a]	16	13.3	81	26	4.6	3.4	9.6	111	9	0
	HC[a]	35	10.2**[b]	65	56***	3.7	4.2**	8.0**	54	244*	3
Kenya	M-HC	18	8.5	51	33	3.9	5.1	7.9	23	308	29
	HC	41	6.4***	48	28	3.9	5.1	8.0	19	703	28

[a]M-HC: Cassava produced both for home consumption and for the market; HC: Cassava produced primarily for home consumption
[b]*** $p < 0.01$; ** $p < 0.05$; * $p < 0.1$

Table 4 Cassava yield and management parameters by site and wealth category

	Uganda								Kenya			
Country Site	Kisiro		Kikooba		Chelekura		Kwang'amor		Mungatsi		Ngunja	
Wealth	Rich[a]	Poor[a]	Rich	Poor	Rich	Poor	Rich	Poor	Rich	Poor	Rich	Poor
N	7	6	9	9	6	14	5	15	8	10	5	15
Cassava yield (t/ha)	9.7	6.6**[b]	14.9	10.5*	14.1	10.7*	8.9	7.4	8.0	5.7**	8.9	5.8**
Parameters that vary significantly ($p < 0.1$) between wealth category	Hired labour		–		Hired labour Time first weed		–		Hired labour Time first weed Manure use farm Fertil. use farm		Hired labour	

[a]The cut-off point between 'rich' and 'poor' farmers is an annual income of \$750/year
[b]*** $p < 0.01$; ** $p < 0.05$; * $p < 0.1$

The relative acreage under improved varieties is one of the management factors that could explain higher yield levels in two cases. However, in Kikooba where cassava yields were highest, farmers rarely grew improved varieties. Poor Ugandan farmers grew twice as much improved varieties as their rich counterparts, but harvested much less cassava (Table 2). Numerous studies in Uganda and Kenya (IITA, 2006; NARO, 2003; Bua et al., 1997) have demonstrated the higher yield levels of improved versus local varieties. In this study, most of the improved varieties found in the field were Nase 3 and SS4, two of the early released varieties in Uganda and western Kenya to combat the cassava mosaic disease (CMD) epidemic that hit the area from the mid-1990s onwards (Obiero and Ndolo, 2005). Many of the local varieties currently found are the result of a natural/farmer-led selection process and have a certain level of tolerance to CMD (Legg et al., 2006). Their yield levels are often similar to those of Nase 3 and SS4 (Obiero and Ndolo, 2005).

Ugandan and Kenyan farmers weed cassava on average four to five times, which is much more than the recommended three times (Melifonwu et al., 2000) or the two to three weed operations carried out in Asia (Howeler et al., 2001). This may explain why a higher number of weed operations was never associated with higher yield levels in this study. Cassava can withstand weed pressure and its weeding requirements are flexible (Fresco, 1986) but when fields are weeded too late, cassava growth is reduced and additional weed operations are still required at a later growth stage. Farmers in this study were found to weed their cassava until 8–9 months after planting while on-station work showed that in well-maintained trials weeding after 3–4 months after planting did not increase root production (Doll et al., 1982, cited in Leihner, 2002). By timing their first weeding (not too early) and subsequent weed operations better, farmers may be able to reduce the total number of weed operations and therefore labour requirements for weeding.

Richer households obtained higher cassava yields than poorer households in both Uganda and Kenya. This is contrary to the findings of Baijukya et al. (2005) who did not find any difference in cassava yield across wealth categories. However, this may be explained by the extremely low yield levels in his study (1.2 t/ha dry matter). Considerable differences in management practices between rich and poor farmers were found which could explain the cassava yield differences observed. Much less differences in management practices were found between farmers growing cassava with the objective of selling part of their production and those growing cassava mainly for home consumption.

Conclusion

This study shows that management practices that contribute to higher cassava yields at farm level in Eastern Africa include among others monocropping, the use of hired labour on cassava and timely weeding. Resource endowment influences management practices much more than the production objective of cassava farmers. By optimizing management farmers can both increase cassava yields and reduce labour requirements. At farm level the use of better agronomic practices currently pays off in 2–4.5 t/ha more yield on average. With more attention given to developing and recommending improved agronomic practices, in combination with the dissemination of higher yielding improved varieties to farmers, there is huge scope to boost cassava production in Eastern Africa.

Acknowledgements We acknowledge IITA and the Dutch Ministry of Development Cooperation through its programme for Associated Professional Officers for providing the necessary financial support for this study. We also thank Joseph Kizimi and Josephine Lubondi (IITA) for their excellent contribution to the survey work in Kenya and Uganda and Hannington Obiero (KARI) and several staff of NARO for their assistance in selecting field sites and sharing their experience. Great appreciation goes to all farmers and civil leaders who participated in this study.

References

Anonymous (2006) First draft roadmap for a commercial cassava sector in Africa by 2020 that satisfies the domestic and export demands for food, feed, ethanol and starch. The outcome of a cassava meeting held in Bellagio, 2–5 May 2006

Anonymous (2007) 2002 Population and housing census analytical report, Iganga district, Nakasongola district and Pallisa district. Local Government, Kampala, Uganda

Baijukya FP, de Ridder N, Masuki KF, Giller KE (2005) Dynamics of banana-based farming systems in Bukoba district: changes in land use, cropping and cattle keeping. Agric Ecosyst Environ 106:395–406

Bua A, Otim-Nape GW, Acola G, Baguma YK (1997) The approaches adopted for and the impact of cassava multiplication in Uganda. In: Otim-Nape GW, Bua A, Thresh JM (eds) Progress in cassava technology transfer in Uganda. Proceedings of the national workshop on cassava multiplication, Masindi, 9–12 Jan 1996

Doll JD, Piedrahita CW, Leihner DE (1982) Métodos de control de malezas en yuca (Manihot esculenta Crantz). In: Yuca: Investigación, Producción y Utilizació. Referencia de los cursos de capacitación sobre yuca dictados por el Centro Internacional de Agricultura Tropical, CIAT, Cali, Colombia. CIAT and Programa de las Naciones Unidas para el Desarrollo (PNUD), Cali, Colombia, pp 241–249

FAO (2004) The global cassava development strategy and implementation plan, Vol 1. Proceedings of the validation forum on the global cassava development strategy. FAO, Rome, 26–28 Apr 2000. Reprint from 2001

FAO (2007) FAO Statistics. The Food and Agricultural Organisation of the United Nations Statistics. http://faostat.fao.org/site/367/Default.aspx. Cited 4 March 2007

Fauquet C, Fargette D (1990) Cassava mosaic disease virus: etiology, epidemiology and control. Plant Dis 74(6):404–411

Fischler M, Wortmann CS (1999) Green manures for maize-bean systems in eastern Uganda: agronomic performance and farmer's perceptions. Agroforest Syst 47:123–138

Fresco LO (1986) Cassava in shifting cultivation: a systems approach to agricultural technology development in Africa. Royal Tropical Institute, Amsterdam

Hillocks RJ (2001) Cassava in Africa. In: Hillocks RJ, Thresh JM, Bellotti A (eds) Cassava: biology, production and utilization. CABI, New York, NY, pp 41–54

Howeler RH, Oates CG, Allem AC (2001) Strategic environmental assessment, an assessment of the impact of cassava production and processing on the environment and biodiversity, in FAO and IFAD. Proceedings of the validation forum on the global cassava development strategy, vol 5. FAO, Rome, 26–28 Apr 2000

IITA (2006) Cassava mosaic disease pandemic mitigation in East and central Africa, a system-wide whitefly IPM affiliated project, quarterly technical report October–December 2005, International Institute of Tropical Agriculture, Kampala, Uganda, Jan 2006

Legg JP, Owor B, Sseruwagi P, Ndunguru J (2006) Cassava mosaic virus disease in East and central Africa: epidemiology and management of a regional pandemic. Adv Vir Res 67:355–418

Leihner D (2002) Agronomy and cropping systems. In: Hillocks RJ, Thresh JM, Bellotti A (eds) Cassava: biology, production and utilization. CABI, New York, NY, pp 91–113

Melifonwu A, James B, Aihou K, Weise S, Awah E, Gbaguidi B (2000) Weed control in cassava farms, IPM field guide for extension agents. International Institute of Tropical Agriculture, Cotonou

NAARI (2000) Annual report. Namulonge Agricultural and Animal Research Institute 1999–2000. Namulonge

NARO (2003) On-farm evaluation of cassava mosaic disease resistant varieties in Teso and Lango farming systems National Agricultural Research Organization/DFID, CORF45 Terminal Report, Namulonge, Uganda

Nweke FI (1996) Cassava: a cash crop in Africa. COSCA working paper no. 14. Collaborative study of cassava in Africa. International Institute of Tropical Agriculture, Ibadan, Nigeria

Obiero HM, Ndolo PJ (2005) Accelerated multiplication and distribution of improved healthy planting materials of cassava varieties in Western Kenya, Technical Report for the Year 2004 Kenya Agricultural Research Institute, Nairobi, Kenya, 33p

Olasantan FO, Ezumah HC, Lucas EO (1996) Effects of intercropping with maize on the micro-environment, growth and yield of cassava Agriculture. Ecosyst Environ 57: 149–158

Shepherd KD, Ndufa JK, Ohlsson E, Sjögren H, Swinkels R (1997) Adoption potential of hedgerow intercropping in maize-based cropping systems in the highlands of western Kenya. 1. Background and agronomic evaluation. Exp Agric 33:197–207

Wortmann CS, Kaizzi CK (1998) Nutrient balances and expected effects of alternative practices in farming systems of Uganda. Agric Ecosyst Environ 71:115–129

Organic Matter Utilisation and the Determinants of Organic Manure Use by Farmers in the Guinea Savanna Zone of Nigeria

A. Bala, A.O. Osunde, and A.J. Odofin

Abstract A survey of 450 farmers was conducted in nine localities within the Guinea savanna zone of Nigeria to examine organic matter usage and to determine the factors that affect organic manure use by farmers. A multi-stage sampling technique was used to administer 50 questionnaires in each locality, and descriptive statistical tools were used to analyse the results. The respondents had an average age range of 37–50 years and a family size of 3–9. The highest literacy rate among the localities was 58%. All the respondents used some form of organic amendment for soil fertility management, although the bulk of the organic resource is acquired from external sources. Farmers that produce both crops and livestock constituted an average of 67%. However, less than half of this group apply the manure generated by animals on their farms. The average per capita farmyard manure generation and consumption were 1.05 and 0.38 t yr^{-1}, respectively. Cattle dung and crop residues were the most commonly used organic resources. There were distinct locality and farmer variations in the use of organic manure. The major factors that determine farmers' use of organic manure include access to mineral fertiliser, crop–livestock integration, number of farms owned by the farmer, family size and educational background of the farmer. In one locality with a large number of migrant farmers, use of organic amendment was significantly lower on rented lands than on lands owned by the farmers.

Keywords Guinea savanna · Organic matter resource · Organic matter use · Soil fertility management

Introduction

Soils of the Nigerian Guinea savanna are inherently low in soil fertility and generally characterised by low organic matter and base cation contents. However, farmers in this region have limited access to fertilisers with which they could backstop the nutrient supply required for optimal production. The low fertiliser use is occasioned by inadequate supply, inefficient distribution and high cost of the fertilisers and the poor resource base of most farmers (Manyong et al., 2001). Against this background, organic matter could be used to complement fertiliser use in the region. In addition to the improvement of soil fertility (Sanchez et al., 1989), organic matter application promotes soil aggregation, improves moisture infiltration and increases the water-holding capacity of the soil (Oades, 1984; Lal, 1986; Lavelle, 1988).

The major sources of organic matter in Nigeria include animal manure, crop residues and municipal wastes (Adetunji, 2004). Animal manure, or farmyard manure (FYM), is one of the most common traditional organic inputs used in Africa, and there are several traditional links between arable farming and livestock husbandry in the continent. The integration of crop and livestock for maintaining soil fertility in the savanna ecological zone of Nigeria was introduced in 1922, and subsequent studies have shown that modest applications of FYM often result in substantial improvement in soil chemical and physical properties (Bache and Heathcote, 1969; Mokwunye, 1980). In spite of its

A. Bala (✉)
School of Agriculture and Agricultural Technology, Federal University of Technology Minna, Minna, Niger State Nigeria
e-mail: Abdullahi_bala@yahoo.com

A. Bationo et al. (eds.), *Innovations as Key to the Green Revolution in Africa*,
DOI 10.1007/978-90-481-2543-2_97, © Springer Science+Business Media B.V. 2011

obvious benefits, farmers in some parts of the region are yet to fully embrace the use of organic manure to support crop production. Reasons adduced for this apathy include bulkiness and difficulty in handling of the product as well as its unavailability (Agboola and Unamma, 1991; Adebayo and Ajayi, 2001). However, there is a dearth of information on some of the socio-economic considerations that farmers in the Nigerian Guinea savanna face in deciding whether or not to use organic manure. This study, therefore, examines the use of organic amendments by farmers in localities within the Nigerian Guinea savanna with the specific objective of identifying some of the determinants of organic manure use.

Materials and Methods

Study Sites

The study was carried out at nine sites in the north-central region of Nigeria. A multi-stage sampling technique was used for the selection of the study sites and the respondents. Three states (Federal Capital Territory, Kwara and Niger) out of the seven in the region were selected. Administratively, each state in Nigeria is divided into three senatorial zones. Dabi, Kilankwa West and Kilankwa East were selected to represent the three zones in the Federal Capital Territory (FCT), while Alapa, Isapa and Lade were chosen for Kwara State. The Gidan Mangoro, Nassarawa and Lioji sites represent the three zones in Niger State. All the sites are located in the southern Guinea savanna agro-ecological zone, except Lioji which is in the northern Guinea savanna zone. These are characterised by annual rainfall of 1200–1500 mm. Rainfall is unimodally distributed in both ecologies with a growing season of 150–200 days. The Guinea savanna zone is characterised by daily mean temperatures above 22°C.

Sample Collection and Data Analysis

Three villages/hamlets within each study site were randomly chosen for the survey and 50 farmers per site selected for the administration of questionnaires. Descriptive statistical tools were used to explain the results. To assess the determinants of organic manure use, chi-square (χ^2) was used to test the following null hypotheses:

(i) There is no significant relationship between the use of organic manure and the crop–livestock integration practised.
(ii) There is no significant relationship between the use of organic manure and the size of the farmer's household.
(iii) There is no significant relationship between the use of organic manure and the educational level of the farmer.
(iv) There is no significant relationship between the use of organic manure and the land tenure status of the farmer.
(v) There is no significant relationship between the use of organic manure and the number of farms owned by the farmer.
(vi) There is no significant relationship between the use of organic manure and the major source from which the farmer secures mineral fertiliser.

The frequency of organic matter use was rated as 'regularly', 'occasionally', 'rarely' or 'not at all'. Crop–livestock integration was grouped as 'mixed' or 'crops only'. Farmer's household size was classified as small (1–3), medium (4–6), large (7–10) and very large (>10). Farmer's educational level was stratified into 'not educated', 'primary' and 'secondary/tertiary' levels. Land tenure status was 'rented' or 'owned' and the number of farms owned by the farmer had three classes (1, 2–3 and >3). The major sources of fertiliser were government agency, cooperative, market or none.

Results

Socio-economic Characteristics of Respondents

The respondents from the Dabi, Kilankwa West and East in the Federal Capital Territory were aged 37–45 years with family size of 6–9 and a literacy rate of 33–58% (Table 1). The respondents in Dabi had the

Table 1 Socio-economic characteristics of farmers in the Guinea savanna zone of Nigeria

	Survey sites								
		Kilankwa							
Characteristics	Dabi	West	East	Alapa	Isapa	Lade	Gidan Mangoro	Lioji	Nassarawa
Age range	32–80	25–60	30–65	28–65	27–79	24–60	24–73	28–70	22–83
Mean age (yr)	37	37	45	42	50	40	37	46	41
Mean family size	6	8	9	7	5	3	5	3	5
Literacy rate	58	33	40	12	32	12	22	50	39
Mean number of farms per farmer	3	3	4	3	3	5	5	5	4
Percentage of farmers that fallow their lands	57	90	100	75	90	92	100	69	96
Average length of fallow (yr)	3	3	3	5	6	6	3	3	5
Percentage of farmers that use mineral fertilisers	100	100	100	90	33	100	100	100	100

highest literacy rate of the three sites and indeed all of the sites surveyed. The Dabi site also had the lowest proportion (57%) of farmers that fallow farmlands. In Kwara State, farmers in Alapa, Isapa and Lade have a mean age range of 40–50 years, an average family size of 3–7 and literacy rate of 12–32% (Table 1). The average age (50 years) of farmers in Isapa was the highest of all the nine sites. Literacy rate (12%) in Alapa and Lade was the lowest in any of the nine sites. Farmers in Gidan Mangoro, Lioji and Nassarawa were aged 37–41 years, with a mean family size of 3–5 and literacy rates of 22–50% (Table 1).

Farmers' Perception

Farmers' perception of organic matter effect on soils is presented in Fig. 1. Except for the Alapa site, most of the farmers at all the sites were aware of the positive effect of organic matter on soil fertility and crop yields. Less than half of the respondents in Dabi, Alapa, Isapa and Lade were aware of organic matter effect on soil organic carbon accumulation (as expressed through soil colour) and soil moisture retention. Fewer than half of the respondents in Dabi were aware of organic matter effect on soil stability. More than half of the farmers at all the sites also mentioned other effects of organic matter, chief of which was a cooling effect due to application of mulches.

Types of Organic Amendments Used by Farmers

The major organic matter resource used by farmers in the Federal Capital Territory was FYM (Fig. 2). This was followed by crop residues and green manure. The organic resource commonly mentioned under the 'others' category was dry grasses used for mulching of yam heaps. None of the farmers in the FCT or at any of the other sites use compost. In Kwara State, other organic materials, mainly cassava peels and dry grass used for mulching of yam heaps and ash from the burning of crop residues, were mentioned as the most commonly used organic resource in Alapa (Fig. 2), although ash is strictly more of a mineral resource than organic matter. These were followed by crop residue and FYM in that order. Only 7% of the farmers in Asapa apply FYM. About 20% of the farmers in this locality use each of dry grass/cassava peels, crop residue and green manure for soil management. The commonest organic resource in Lade is FYM. About 16 and 12% of the farmers in this locality also use green manure and crop residues, respectively, while 4% also use waste from municipal dumps. FYM and crop residues are the predominant organic resources used in Gidan Mangoro of Niger State (Fig. 2). Dry vegetation and ash are also used by up to 30% of the respondents at this site. At the Lioji site, FYM is the most commonly used organic matter at about 40%, followed by crop residues (16%) and municipal waste (13%). Equal proportions (about

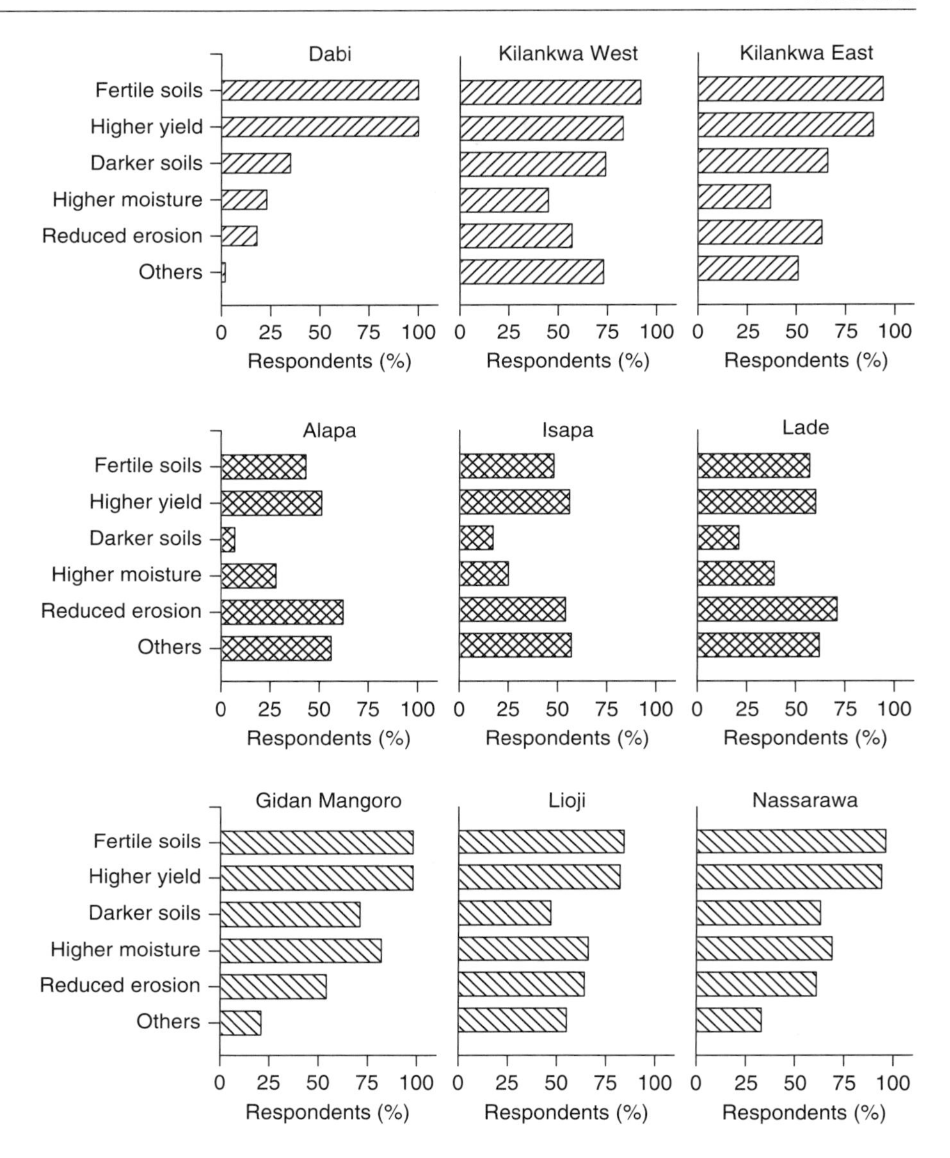

Fig. 1 Distribution of farmers' perception on the effect of organic matter on soils in the Guinea savanna zone of Nigeria

50%) of farmers use both crop residues and ash soil amendments in Nassarawa, while 42% use FYM.

occasionally. Twenty two percent of the respondents in Lioji do not use manure at all.

Frequency of Use of Animal Manure

There were more farmers in Dabi that do not use FYM at all than those that use it either regularly or occasionally (Fig. 3). Most farmers in Kilankwa West and East use FYM regularly. In Kwara State, 37% of the farmers in Alapa and 42% in Isapa do not use FYM, while the proportion in Lade was 23%. Most farmers in Gidan Mangoro, Lioji and Nassarawa use FYM regularly or

Production and Collection of FYM

Twenty five percent of the farmers at the Dabi site raise livestock in addition to being engaged in crop production, while all the farmers in Alapa and Lade combine crop and livestock production (Table 2). Only 4% of the farmers that raise livestock in Isapa apply the droppings collected from their animals on their farms as against about 70% in Gidan Mangoro and Lioji.

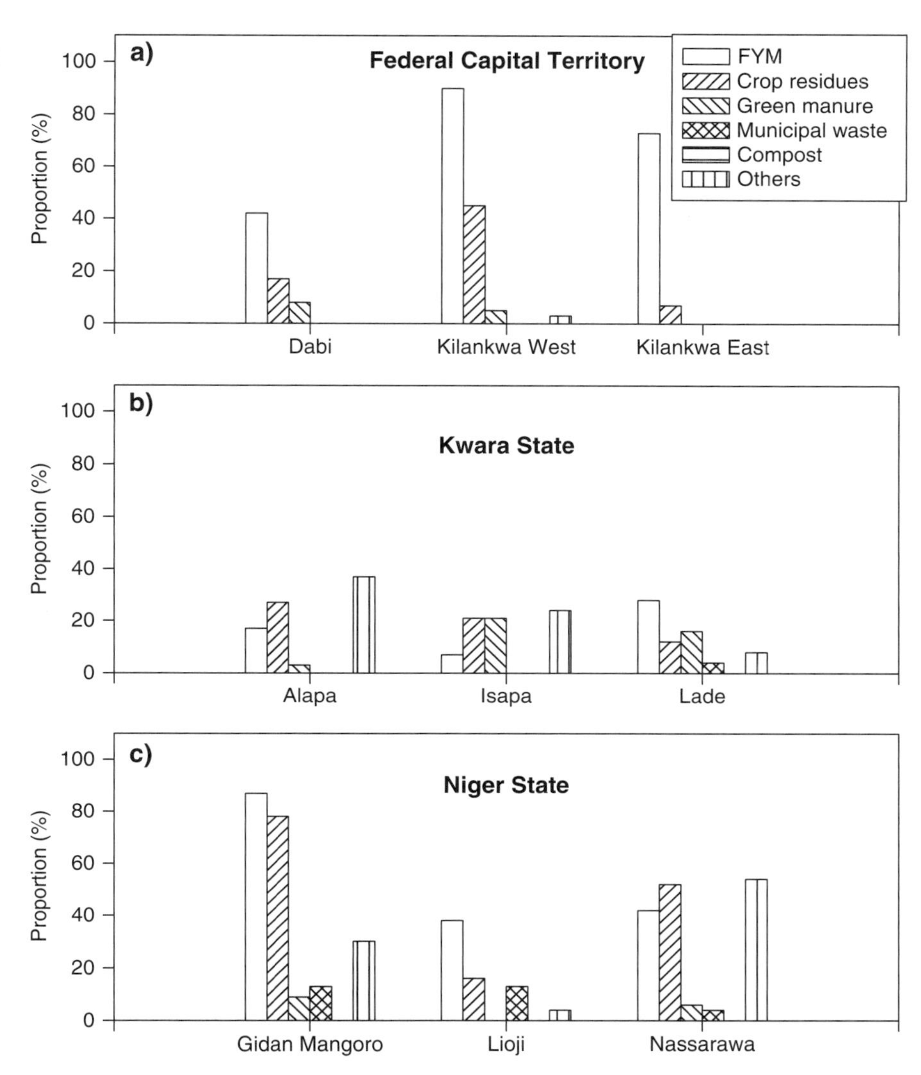

Fig. 2 Distribution of farmers using various types of organic amendments in (**a**) Federal Capital Territory, (**b**) Kwara State and (**c**) Niger State in the Guinea savanna zone of Nigeria

The approximate per capita quantity of FYM collected annually in the nine localities ranged from 50 kg in Dabi to about 2.7 t in Alapa. About 5% of the manure generated in Lade and Asapa is applied on farms, while about 90% is used on farms in Kilankwa West and Gidan Mangoro.

Determinants of Organic Manure Use

The relationships between FYM use and selected socio-economic factors at the various sites as defined by chi-square tests are presented in Table 3. In Dabi, there was a significant relationship ($p \leq 0.01$) between FYM use and crop–livestock integration, educational level of the farmer and land ownership. At this site, it was observed that farmers that keep animals tended more to use FYM than those that did not. The result also showed that those that use FYM were more likely to have had no formal education than those that do not use FYM. Additionally, farmers were found to be less likely to apply FYM where the land was rented than if the land was owned. In both Kilankwa West and Kilankwa East, the size of the farmer's household and the number of farms owned by the farmer were significantly ($p \leq 0.05$) linked to the farmer's use of FYM. Farmers with at least seven family members were more likely to use FYM than those with fewer members. In addition, farmers with more than three farms tended less to apply FYM than those with fewer farms.

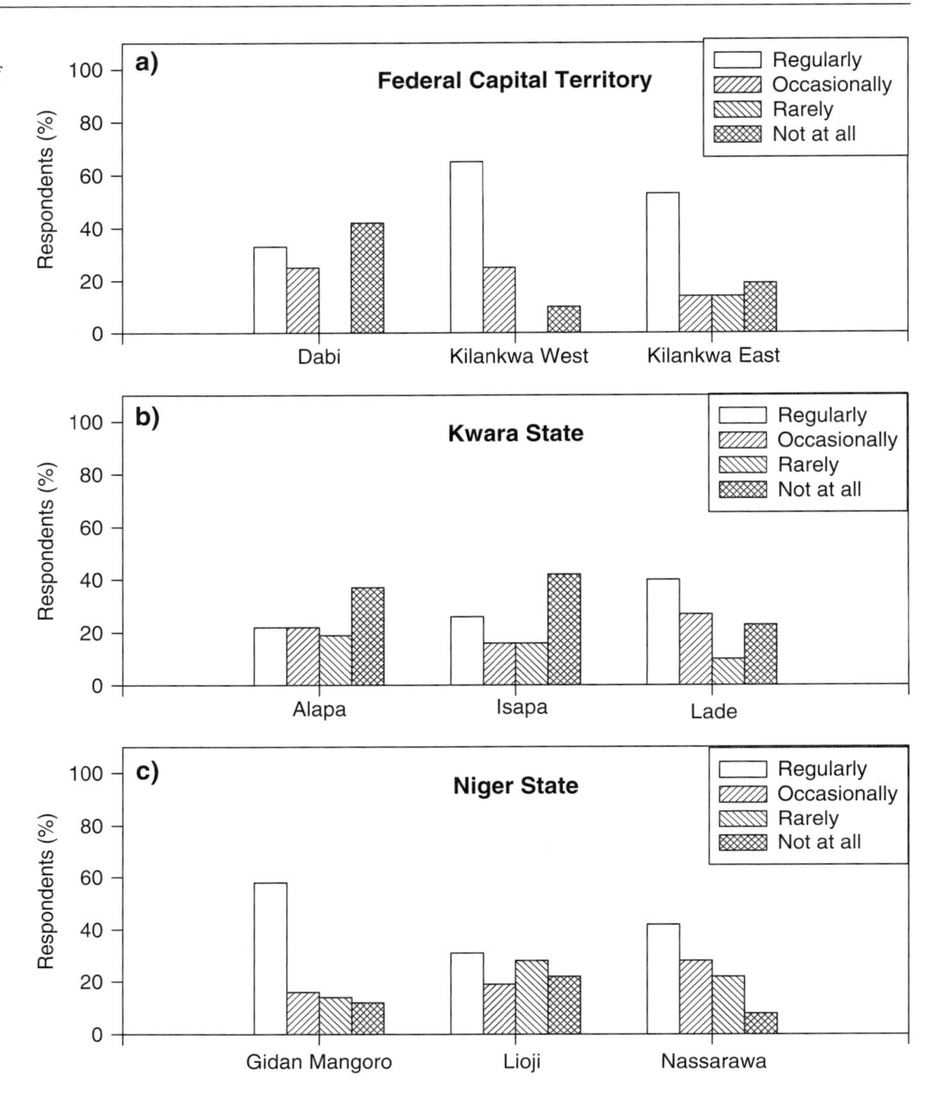

Fig. 3 Distribution of farmers based on frequency of animal manure use in (**a**) Federal Capital Territory, (**b**) Kwara State and (**c**) Niger State in the Guinea savanna zone of Nigeria

Table 2 Distribution of farmers based on the generation and use of farmyard manure in the Guinea savanna zone of Nigeria

Site	Farmers raising livestock (%)	Quantity of FYM generated per capita (t yr^{-1})	Quantity applied on farms per capita (t yr^{-1})	Farmers using FYM collected from their livestock (%)
Dabi	25	0.05	0.02	67
Kilankwa West	30	0.46	0.41	33
Kilankwa East	60	0.37	0.28	56
Alapa	100	2.69	0.82	10
Isapa	90	2.02	0.09	4
Lade	100	0.85	0.03	12
Gidan Mangoro	76	1.77	1.52	74
Lioji	53	0.59	0.31	71
Nassarawa	72	1.03	0.75	36

Table 3 Chi-square (χ^2) values establishing the relationship between farmyard manure use and some socio-economic factors

Site	Crop–livestock integration (df = 2)	Household size (df = 6)	Educational level (df = 4)	Land tenure (df = 2)	Number of farms (df = 4)	Major source of fertiliser (df = 6)
Dabi	17.11**	3.48	15.20**	21.34**	1.21	5.41
Kilankwa West	1.49	11.28*	0.06	0.03	17.68**	3.23
Kilankwa East	3.65	9.82*	0.17	0.07	12.53**	2.51
Alapa	0.44	3.86	2.69	0.11	18.93**	7.18
Isapa	0.02	1.13	7.56	0.24	14.08*	26.61**
Lade	1.04	7.54	1.06	1.21	10.37*	4.05
Gidan Mangoro	2.13	10.21*	24.65**	1.77	3.45	14.63*
Lioji	13.42**	4.44	5.71	0.09	13.16*	12.71*
Nassarawa	4.39	2.98	9.13	2.12	15.17*	15.45*

*Significant at 5%; **Significant at 1%

The number of farmlands owned by a farmer was also significantly related to organic manure use in Alapa, Isapa and Lade (Table 3). Farmers at these locations are more likely to use FYM if they have three or fewer farms than if they have more than three farms. Additionally, organic manure use had significant relationship ($p \leq 0.01$) with the source from which farmers in Isapa obtain mineral fertilisers. Farmers that get their supplies from cooperatives were less likely to apply FYM than those that buy from open markets or from government agencies.

In Gidan Mangoro, farmers' household size, their educational level and the major source from which they obtain mineral fertiliser all have significant relationship ($p \leq 0.05$) with organic manure use (Table 3). Farmers with fewer than four family members were less likely to use FYM than those with greater number. Farmers with no or primary school education were more likely to use FYM regularly on their farms than those that had secondary or tertiary education. Additionally, farmers that rely on government agencies for their fertiliser supplies tended to apply FYM on their farms more than those that purchase theirs from the open markets.

In Lioji, combining cropping and livestock management, the number of farms owned by the farmer and the major source of fertiliser significantly ($p \leq 0.05$) influence organic manure use. Farmers that keep livestock were more likely to use organic manure than those that do not, while farmers with more than three farms were less likely to use FYM than those with fewer farms. Farmers that get their fertilisers from cooperatives were less likely to use manure than those that buy from the market or from government agencies. Both the number of farms owned by the farmer and the major source of fertiliser are also significantly related to organic manure use with similar trends as in Lioji.

Discussion

The average age of the respondents in eight of the nine sites was less than 50 years. This result is in contrast to that of Adebayo and Ajayi (2001) who reported an average age of 59 years for farmers within the derived savanna and forest zones of Nigeria. However, consistent with the results of these authors, our study shows that the average household size of the farmers at most sites range from five to nine. The survey showed that most of the family members participate in some tasks on the farms during the season. On average, women contribute at least 46% of the labour inputs on family farms (Dixon, 1982; Gladwin and McMillan, 1989). The literacy rates at most sites were low, and the case of Dabi in the FCT is different because of the large number of migrant community from the southern part of the country where literacy rates are much higher than those in the north. Bush fallowing is still being practised by most of the farmers at most of the sites surveyed. This is possible because land is not a major limitation for cropping in this region; hence, the average number of farms owned is three to five. However, the length of the fallow averages 3–6 years, which is shorter than the minimum of 10 years required for the land to recover (Agboola and Unamma, 1991). Other than at the Isapa site, virtually all the farmers use inorganic fertilisers. However, consistent with other studies

(Badiane and Delgado, 1995; Larson and Frisvold, 1996), we observed that they are unable to obtain adequate supplies in quantity and in time due to delivery delays, high cost of the fertilisers and low purchasing power.

The proportion of farmers that are aware of the positive effects of organic matter on soils varied depending on the location. In general, farmers in Niger State (Gidan Kwano, Lioji and Nassarawa sites) and those of Kilankwa West and East in the FCT showed a high degree of awareness in this respect. Fewer farmers at the other sites (Dabi, Alapa, Isapa and Lade) had a good grasp of the organic matter effect. This may be responsible for the low usage of organic materials at these sites or it may be the result of low organic matter use.

Consistent with practices in other parts of Africa (Palm et al., 1997), FYM and/or crop residues were observed to be the most commonly used organic resources at most of the sites surveyed. Use of FYM in Alapa and Isapa was relatively low in spite of the large numbers of farmers that raise livestock. The animals are largely raised on free range and the little manure collected in pens overnight is disposed at communal waste dumps. A large majority of the farmers questioned do not consider manure production as their main reason for keeping livestock. On closer enquiry, some of the farmers said most of them grew vegetables on their farms in addition to staple crops, and nobody would buy vegetables that were treated with manure. Lack of information and training for farmers on collection and use of manure appears to be a major problem in these parts of north-central Nigeria. In some parts of the Eastern Highlands of Africa, it is reported to be a taboo to use manure for growing crops (Kihanda and Gichuru, 1999). Competing uses for conventional residues, such as cereal stovers and legume haulms (Agbim and Adeoye, 1991), have led to the emergence of agricultural processing wastes, such as cassava peels and rice bran, as a major source of organic inputs in crop production (Adediran et al., 2003), hence the use by some farmers of cassava peels in Alapa and Isapa. Ash obtained from burning of crop residue was also often mentioned as soil amendment especially in Nassarawa. Studies have shown that application of ash causes an increase in soil pH, exchangeable cations and yield of various crops (Adetunji, 1997; Ojeniyi and Adejobi, 2002). Green manure was considered a major organic resource in some of the communities, contrary to the suggestion that it is not popular with farmers because it does not fit into farmers' traditional mixed cropping system (Agboola and Unamma, 1991). In spite of its potential as a good organic resource (Adetunji, 2004), municipal waste is not used in considerable amounts except in Niger State and parts of Kwara State. This may be because of the perception that it may contain heavy metals, parasites and pathogens that may not be beneficial to soils and plants or the consuming public (Adetunji, 2004). None of the farmers use compost and the survey showed that they were not conversant with this resource, although they were willing to learn its production. This, therefore, is a major gap in extension service that needs to be addressed.

A large number of farmers in parts of the FCT and in Niger State were found to apply FYM regularly or occasionally as opposed to a greater proportion of non-users observed in Dabi and parts of Kwara State. The frequency of manure application relates to whether to apply large amounts of manure to last a long time or to apply small amounts frequently (Kihanda and Gichuru, 1999). In our study, we observed four major types of manure sourcing. The first type is where farmers collect manure from their animals at home in small amounts and make frequent trips to the farm for dumping of the manure into heaps to be worked into the soil at the beginning of the rains. Most farmers in this category manage their animals free range during the day and the manure collected is what is generated by the animals in their pens overnight. The second group usually obtain manure from dumps at neighbouring settlements of nomadic herdsmen either through purchase or free of charge. This is usually done just before the rains, and large quantities of manure are often secured and transported to the farms by trucks or other means of transportation. The third type of FYM sourcing is where farmers come to an agreement with herdsmen for the animals to forage on crop residue on farmers' fields during off-season. In this arrangement, the animals are not stationed on the farms as against the fourth type of manure sourcing where herdsmen and their animals are permanently stationed on farmers' field throughout the off-season period. In this case, the herds may often go out for grazing, sometimes for days, but do come back to station on the farm. These variants of manurial contracts were observed at the sites where manure use is common but not at sites with low use. These are approaches that could be used to facilitate farmers' access to manure as agreements between farmers and pastoralists are of greater importance in many areas than the integration of crops and livestock within a single production unit (Bayer and Watersbayer, 1990).

The proportion of farmers that use FYM obtained from home varies across sites (Table 2). For sites with high percentages, such as Dabi, Kilankwa East, Gidan Mangoro and Lioji, raising small ruminants and poultry plays a key role for farmers' use of manure in the farm. In spite of this, vast amounts of FYM and nutrients are lost because the bulk of the livestock are managed free range or through lack of farmers' diligence in collecting the manure for use in the farm. As a result, there is a huge gap between the quantities of manure generated and the amounts applied on farms annually. The gaps are narrowed at sites such as Kilankwa West and East and Gidan Mangoro because of the preponderance of farmers that secure FYM from pastoralists. Even in areas with high per capita applications, such as Gidan Mangoro, the amounts applied are very modest in comparison with the recommended manure levels of 5–10 t ha^{-1} yr^{-1} for West African savanna (Jones, 1971).

Other than crop–livestock integration, the size of the farmer's household was also significantly related to FYM use in three locations. This may be related to the number of extra hands available to provide labour in the farm given that ready availability is a major requirement for organic manure processing and application (Schleich, 1983). The influence of educational level of the farmer on manure use appeared to be less of appreciating the benefit of FYM, but rather due to the fact that the two sites where this factor was significant are close to urban centres and farmers with higher education have paid employment, which provides them with some money to buy fertilisers from open markets. Their position may also confer on them some influence as to more easily secure fertiliser from government agencies. Land tenure was significant only at the Dabi site perhaps because most of the migrant farmers at this site had rented lands and may feel less inclined to invest in organic matter use especially since many could afford fertilisers from open markets. The number of farms available to the farmer was significantly related to FYM use at seven of the nine locations surveyed. Given that most of the farmers practise bush fallowing, availability of extra farms could be a disincentive for investing in the collection, transportation and application of FYM. Although fertiliser supply is a major limitation in agricultural production in smallholder farms in Africa (Larson and Frisvold, 1996), many farmers consider fertiliser ahead of organic manure because of the ease of handling and application. Therefore, where a farmer has access to fertiliser, the chances of using organic manure are minimised. Thus, even in areas where livestock production is traditional (such as in Lioji), it is difficult to improve the collection and use of manure. In the case of Isapa, Lioji and Nassarawa, the availability of subsidised fertilisers through cooperatives makes it more difficult to promote the use of farmyard manure.

Conclusion

This study shows that the average farmer age in the nine localities was less than 50 years but literacy rates were generally lower than 50%. Most farmers appreciate the positive effect of organic matter on soil and crop yield; hence, organic matter use in the areas surveyed is widespread. FMY is among the most commonly used organic resource, although its use is relatively low in areas within Kwara State. Even in areas where FYM use is high, the amounts applied are low in comparison to the quantities required for appreciable crop yield. Depending on the locality, the decision by farmers to use FYM on their farms is influenced by a combination of factors especially the number of farms a farmer owns, accessibility of farmers to mineral fertilisers and the size of farmers' households. Land tenure was only significant as a determinant of FYM use at the Dabi site because most of the farms were rented. Farmers at all sites but most especially those in Kwara State need better information and training on the collection, management and application of organic manure.

References

Adebayo K, Ajayi OO (2001) Factors determining the practice of crop-livestock integration in the derived savanna and forest zones of Nigeria. ASSET Series A 1:101–107

Adediran JA, Taiwo LB, Sobulo RA (2003) Effect of organic wastes and method of composting on compost and yields of two vegetable crops. J Sust Agric 4:95–109

Adetunji MT (1997) Organic residue management, soil nutrient changes and maize yield in a humid Ultisol. Nutr Cycling Agroecosyst 47:189–195

Adetunji MT (2004) Integrated soil nutrient management options for Nigerian agriculture. In: Salako FK et al (eds) Managing soil resources for food security and sustainable environment. 20th Annual Conference of the Soil Science Society of Nigeria, Abuja, Nigeria, pp 189–195, 326 pp, Dec 2004

Agbim NN, Adeoye KB (1991) The role of crop residues in soil fertility maintenance and conservation. In: Lombin G et al (eds) Proceedings of national organic fertilizer seminar. Federal Ministry of Agriculture, Abuja, Nigeria, pp 21–42, 125 pp

Agboola AA, Unamma RPA (1991) Maintenance of soil fertility under traditional farming systems. In: Lombin G et al (eds) Proceedings of national organic fertilizer seminar. Federal Ministry of Agriculture, Abuja, Nigeria, pp 51–66, 125 pp

Bache BW, Heathcote RG (1969) Long-term effects of fertilizers and manure on soil and leaves of cotton in Nigeria. Exp Agric 5:241–247

Badiane O, Delgado CI (1995) A 2020 vision for food, agriculture and the environment in sub-Saharan Africa. Food, agriculture and the environment. discussion paper 4. International food policy research institute, Washington, DC, USA

Bayer W, Watersbayer A (1990) Beziehungen zwischen Ackerbau und Tierhaltung in traditionellen Landnutzungssystemen im tropischen Afrika. Der Tropenlandwirt 91:133–145. Oktober 1990

Dixon R (1982) Women in agriculture: counting the labour force in developing countries. Pop Dev Rev 8:558–559

Gladwin CH, McMillan D (1989) Is a turnaround in Africa possible without helping African women to farm? Econ Dev Cult Change 37:279–316

Jones MJ (1971) The maintenance of soil organic matter under continuous cultivation at Samaru. Nig J Agric Sci 77: 473–482

Kihanda FM, Gichuru M (1999) Manure management for soil fertility improvement. Manure management mini workshop, Nairobi, Kenya, Jul 1999

Lal R (1986) Soil surface management in the tropics for intensive land use and high and sustained productivity. In: Steward BA (ed) Advances in soil science, 5:1–105. Springer, New York, NY

Larson BA, Frisvold GB (1996) Fertilisers to support agricultural development in sub-Saharan Africa: what is needed and why. Food Policy 21:509–525

Lavelle P (1988) Earthworm activities and the soil system. Biol Fertil Soils 6:237–251

Manyong VM, Makinde KO, Sanginga N et al (2001) Fertilizer use and definition of farmer domains for impact-oriented research in the Northern Guinea savanna of Nigeria. Nutr Cycling Agroecosyst 59:129–141

Mokwunye U (1980) Interactions between farmyard manure and fertilizers in savanna soils. FAO Soils Bull 43:192–200

Oades JM (1984) Soil organic matter and structural stability: mechanisms and implications for management. Plant Soil 76:319–337

Ojeniyi SO, Adejobi KB (2002) Effect of ash and goat dung manure on leaf nutrient composition, growth and yield of amaranths. Nig Agric J 33:46–57

Palm CA, Myers RJK, Nandwa SM (1997) Combined use of organic and inorganic nutrient sources for soil fertility maintenance and replenishment. In: Buresh RH et al (eds) Replenishing soil fertility in Africa. Soil Sci Soc Am Special Publication Number 51, Madison, WI, p 251

Sanchez PA, Palm CA, Szott LT et al (1989) Organic input management in tropical agroecosystems. In: Coleman DC et al (eds) Dynamics of soil organic matter in tropical ecosystems. University of Hawaii Press, Honolulu, Hawaii, 23–41, 247 pp

Schleich K (1983) Die Anwendung von Rinderdung im Norden der Elfen-beinküste. Zentrum für Regionale Entwicklungsforschung (Hrsg). JLU, Giessen, pp 47

Innovativeness of Common Interest Groups in North Rift Kenya: A Case of Trans-Nzoia District

L.W. Mauyo, J.M. Wanyama, C.M. Lusweti, and J.N. Nzomoi

Abstract One measure of group evaluation is the classical structure, conduct and performance (SCP) analysis approach. Ignoring the inter-relationships among actors in the agricultural value chain leads to poor intervention strategy designs. Common interest groups (CIGs) are farmer associations established with the purpose of promoting special interest of communities. The groups were formed to alleviate poverty, enhance food security and improve health status through income generation. Kenya Agricultural Productivity Project (KAPP) recognized the need to use CIGs as key actors in agricultural sector value chain for intervention. However, there was limited information on the structure and performance of the CIGs. This study aimed at examining the structure, conduct and performance of the CIGs. A survey was conducted in Trans-Nzoia district in 2006. Fifty seven randomly selected CIGs were interviewed using a semi-structured questionnaire. SCP and logit models were used in data analyses. The results show that groups interviewed were women (23%), research (9%), health (9%), youth (5%) and mixed (54%). There was evidence that most of the groups had devolution of power to various sub-committees. The lifespan of the groups was 5 ± 4.8 years with membership of 26 ± 15.1. However, most (89%) of the groups were not registered. From logit regression results, the major factors significantly influencing group external technical and financial support were gender, lifespan, group type and group special projects undertaken. This implies, for group sustainability, these factors may be considered. The CIGs form favourable target for KAPP interventions for enhanced impact and improved community welfare.

Keywords Common Interest Groups · Conduct · Performance · Structure

Introduction

Smallholders in rural farming communities produce the bulk of the regional food requirements, yet their production system has many limitations including poor access to information on improved production technology such as new high-yielding varieties, better crop management strategies, access to inputs and markets. As such agricultural production is often for subsistence, productivity levels tend to stagnate and sometimes decline as their natural resource base degrades. Agriculture is directly linked to poverty eradication, sustainable consumption and production of agricultural commodities in most of the developing economies, like the Kenyan one. Agricultural growth is crucial to Kenya's overall economic and social development. The sector directly contributes about 26% of gross domestic product (GDP) and a further 27% through linkages with manufacturing, distribution and service-related sectors (GOK, 2004). About 80% of the population live in the rural areas and depend mainly on agriculture and fisheries for livelihood. About 50% of Kenyans are food insecure while significant potential for increased production remains largely unexploited (GOK, 2004). Efforts to promote growth and development in this sector have had limited success.

L.W. Mauyo (✉)
Masinde Muliro University of Science and Technology,
P.O. BOX 190-50100, Kakamega, Kenya
e-mail: lmauyo@yahoo.com

A. Bationo et al. (eds.), *Innovations as Key to the Green Revolution in Africa*,
DOI 10.1007/978-90-481-2543-2_98, © Springer Science+Business Media B.V. 2011

To significantly reduce the high levels of poverty and unemployment in Kenya, a green revolution-type action is required. In Kenya, smallholders produce most of their own food and also contribute about 68% of the nation's total marketed output (GOK, 1998; Kinyua, 2004). Worldwide, the attraction to small farms lies in their economic efficiency relative to larger farms and the fact that they can create large amounts of productive employment (Hazell, 2003). They can also reduce rural poverty and food insecurity, support a more vibrant rural non-farm economy and help to contain rural-to-urban migration. Smallholders in Kenya are, however, known to be resource poor and therefore operate below their potentials (Nyikal, 2000).

Combined action among stakeholders is extensively used as a positive instrument for rural development in many economies (Braun et al., 2000; Nyameino, 2007). Group innovativeness empowers farmers to make rational decisions to solve their own problems and source funds from interested donors for enhanced livelihoods (Ajili, 2000). Such collective activities have succeeded in cereal banking and farmer field schools in Western Kenya (Mukhwana et al., 2005; Nyameino, 2007). This is why many change agents use groups as a tool to decision making. However, before utilizing the group, there is need to know how it is performing. One measure of group evaluation is the classical structure, conduct and performance (SCP) analysis approach. Ignoring the inter-relationships among actors in the agricultural value chain leads to poor intervention strategy and designs.

Common interest groups (CIGs) are farmer associations established with the purpose of promoting special interests of communities Sarkar, 2003). Kenya Agricultural Productivity Project (KAPP) recognized the need to use CIGs as key actors in agricultural sector value chain for intervention. CIGs act as participatory platforms for improving decision-making capacity and stimulating local innovation for sustainable agriculture of the farming communities (Braun et al., 2000). However, there was limited information on the structure and performance of the CIGs. The objectives of the study were to examine the structure, conduct and performance of the common interest groups in Trans-Nzoia district, North Rift, Kenya. This was perceived to contribute to effective and efficient implementation of the KAPP activities. Subsequently, the group approach would lead to a more pronounced impact.

The key activities within the CIGs include the following:

- Formation, partner identification as well as their roles and norms for running the activities
- Constraint identification, analyses and prioritization of possible solutions and the development of intervention agendas
- Regular CIG meetings to discuss problems and exchange information and possible solutions to problems
- Farmer-led research and field demonstrations of improved technologies, field visits and training including farmer exchange visits
- Synergistic partnering with other stakeholders along the agricultural value chain on improvement of access to markets
- Farmer training in business awareness for better farm profit engagements

Methodology

Conceptual Framework

Community members come together for collective action aimed at improving welfare. Groups undertake a number of activities to achieve set objectives and goals. In order to know how the group is functioning, there is need to evaluate them. One way of evaluating them is to understand the structure, conduct and performance of those groups. This will bring forth issues leading to group success or failure. Group performance is influenced by heterogeneity of the groups, technical and financial support received, devolution of power, benefits group members receive, rules/constitution, committees and frequency of electing officials.

Survey Design

A survey team conducted a survey in Trans-Nzoia district in 2006. After establishing a district sample frame by listing the groups in divisions, a total of 57 randomly selected common interest groups were interviewed using a semi-structured questionnaire. Interview dates were fixed with groups after discussions. Focus group discussions were held and some of the questions were targeted at group officials.

Data Analysis

Descriptive and regression statistics were used in data analyses. These include means, standard deviation and standard error. The regression statistics included qualitative and ordinary regression analyses which were conducted using the SPSS software Version 12 (Field, 2002).

Regression Models

The data were then analysed using descriptive statistics, SCP, logit and ordinary linear regression models. Discrete choice econometric models have been widely used in estimating models that involve discrete economic decision problems. Guerre and Moon (2004) observe that when the economic decision is affected by fundamental macroeconomic or financial variables, many of which are known to show non-stationary characteristics, one may need to consider a discrete choice model with non-stationary explanatory variables. In situations where the phenomenon being investigated is discrete rather than continuous, we develop models with discrete dependent variables. In this study, to participate in group activities involves a discrete choice and is dependent on the individual characteristics of the participant which influence him or her in making the choice. Thus we construct models that link the decision or outcome to a set of factors peculiar to the participant. Greene (1990) suggests an approach akin to the general framework of probability models in which

$$\begin{aligned} &\text{Prob(event' } j\text{' occurs} = \text{Prob}(y = j) \\ &= F(\text{relevant effects:parameters})) \end{aligned} \tag{1}$$

In choosing the model it is appreciated that both the logit and the probit models are frequently utilized in discrete choice studies, although the logit model is often preferred over the probit model on both the statistical and theoretical grounds (Amemiya, 1981). According to Amemiya (1981) the two models have statistical similarities that often pose difficulties to researchers when exercising choice of one model over the other. Although choosing between the probit and the logit models may be evaluated a posteriori on both a statistical and theoretical basis, in practice there are no strong grounds for choosing one model over the other. This study involves the use of a qualitative choice model since the dependent variable involves two choices: to participate or not to participate in group activities. In this study, the logit model was most appropriate, mainly on grounds of theoretical tractability. The logistic regression, like the log-linear model, is part of a category of statistical models called generalized linear models. Logistic regression enables a researcher to predict a discrete outcome such as group membership from a set of variables that may be continuous, discrete, dichotomous or a mixture of any of these. In a functional form, we therefore developed the following simple reduced-form participation model:

$$\begin{aligned} Y_i^* = f(&\text{AGE,GENDER,DOSUP,ASPENT,} \\ &\text{SPACT,ELECOFF,REGMET,} \\ &\text{GRPTYPE,GHAVCOMM)} \end{aligned}$$

where AGE = age of group in years, GENDER = gender considerations (1 = male; 0 = female), DOSUP = donor support (1 = yes; 0 = no), ASPENT = aspire for post-production enterprises, SPACT = special activities undertaken by group (1 = yes; 0 = no), ELECOFF = elected official periodically (1 = yes; 0 = no, REGMET = regular meetings attendance held, GRPTYPE = group type (1 = women, 0 = otherwise), GHAVCOMM = groups have effective communication channels (1 = yes; 0 = no).

Following Gujarati (1995), the specific model estimated was further assumed to take the log-linear form, which tends to give the best results. It is also very convenient for elasticity calculations. The log-linear model is one of the specialized cases of generalized linear models for Poisson-distributed data. It is an extension of the two-way contingency table where the conditional relationship between two or more discrete, categorical variables is analysed by taking the natural logarithm of the cell frequencies within a contingency table. Log-linear models are commonly used to evaluate multivariate contingency tables involving three or more variables. Since the study utilized cross-sectional data, we specified the participation equation without time lags as shown:

$$Y_i = B_0 + B_1 \ln X_1 + B_2 \ln X_2 + B_3 \ln X_3 + B_4 \ln X_4 + B_5 \ln X_5 + B_6 \ln X_6 + B_7 \ln X_7 + B_8 \ln X_8 + B_9 \ln X_9 + e. \quad (2)$$

where X_1 = age, X_2 = gender considerations, X_3 = donor support, X_4 = aspire for post-production enterprises, X_5 = special activities undertaken by group, X_6 = elected official periodically, X_7 = regular meetings attendance held, X_8 = group type, X_9 = group having effective communication channels, e = error term.

The dependent variable is the natural log of the probability of participating in group activities (P) divided by the probability of not participating ($1 - P$). From Eq. (2) we note that the value of the dependent variable is a linear combination of the values of the independent variables plus an error term which is assumed to be normally distributed with zero mean and a constant variance. The model was estimated using the Omnibus logistic regression model of the SPSS software Version 12.

The Logit Model

This model is based on the cumulative logistic probability function. The logit model is often used instead of the probit model because it is computationally easier to use. The specification of the model takes the following form:

$$P_i = F(Z_i) = F(\alpha + \beta X_i) = 1/(1 + e^{-zi}) = 1/(1 + e^{-(\alpha+\beta X_i)}) \quad (3)$$

In this notation e represents the base of natural logarithms which is approximated at 2.718. P_i is the probability that an individual will make a certain choice, whether to participate in group activities or not. In estimating Eq. (3), we multiply both sides by $(1 + e^{-z})$ $P_i = 1$ so that dividing by P_i and then subtracting 1 yields

$$e^{-zi} = 1/P_i - 1 = 1 - P_i / P_i$$

However, since $e^{-z} = 1/e^{zi}$, then $e^{zi} = P_i/1{-}P_i$ so that by taking the natural logarithm on both sides of the equation, we obtain $Z_i = \log P_i / 1{-}P_i$ (Pindyck and Rubinfeld, 1998) or from Eq. (3), we have

$$\log P_i / 1 - P_i = Z_i = \alpha + \beta X_i. \quad (4)$$

where the dependent variable is the log of the odds that a certain decision will be made.

α = The constant of the equation
β = The coefficient of the predictor variables

Unlike the probit model, one special feature of the logit model is that it transforms the problem of predicting probabilities within a 0.1 range of the real line. The slope of the cumulative logistic distribution is greatest at $P = ½$ implying that changes in independent variables will have their greatest effect on the probability of choosing a given option at the midpoint of the distribution. Because of the advantages it has over the probit model, the logit model was used in this adoption study.

The logit model was used to evaluate factors influencing group members to participate in group activities. It is a logistic distribution bound between 0 and 1. The model was specified as shown in Eq. (5) (Maddala, 1983; Rogers, 1983):

$$\log\left[\frac{\text{Prob(event)}}{\text{Prob(no} - \text{event)}}\right] = \beta_0 + \beta_1 X_1 + \cdots + \beta_k X_k \quad (5)$$

where β_is are estimated coefficients and X_i are independent variables such as farmer group characteristics. The independent variable was the active farmers participating in group activities. The ordinary regression model was specified as given in Eq. (6).

$$Y_i^* = \beta X_i + u_i \quad (6)$$

For ordinary linear regression model Y^* is the dependent variable (number of group members), X is a set of independent members and group factors, while logit model is a vector of parameters, including a constant, which are estimated using the maximum likelihood method, and u is the error term (Table 1). Table 1 describes the variables in the models and specific hypotheses.

Table 1 Variable definition and description

Variable name	Nature of variable
Logit independent variable (1 = yes; 0 = otherwise)	Active participants in the group (1 = yes; 0 = otherwise)
Ordinary regression independent variable	
Group size	
AGE	Age of group in years
GENDER	Gender considerations (1 = male; 0 = female)
DOSUP	Donor support (1 = yes; 0 = no)
ASPENT	Aspire for post-production enterprises
SPACT	Special activities undertaken by group (1 = yes; 0 = no)
ELECOFF	Elected official periodically (1 = yes; 0 = no)
REGMET	Regular meetings attendance held
GRPTYPE	Group type (1 = women, 0 = otherwise)
GHAVCOMM	Group having effective communication channels (1 = yes; 0 = no)

Structure, Conduct and Performance Model

In structure–conduct–performance (SCP) model there are five force aspects to be considered and they include barriers to entry, intensity of competition, substitution, bargaining power and strategic groups. The SCP model was utilized to understand the relationship between the group's environment, the group's behaviour and the group's performance. The structure of groups influences their conduct, which subsequently influences the performance. In this context the group structure refers to the organizational characteristics of the groups (power sharing, offices established). The characteristics may also include number and size of the groups and their distribution and conditions of entry/registration by members or the group. Group conduct is the competitive tactics the groups use to promote itself in the society (e.g. type of groups and institutions the group works with). It also includes the operational policies/rules that influence its products or services. Finally, group performance means the economic results of the structure and conduct of the group (Mose, 2007). The parameters that measure this may include capital accumulation of the group and its members, profits accruing from group activities and number of activities within the group. Overall performance in the SCP model looks at the productive efficiency of the group and resource use within the groups. The measurement of social performance involves investigating the structure of an organization (i.e. mission, ownership, management principles, relation to and care for its staff) and its conduct in the market and local and wider community (services, products, market behaviour, other relations with clients and other stakeholders, community and social/political organizations) (Zeller et al., 2003).

Results and Discussions

Types of Groups Interviewed

The results indicate that different types of groups were interviewed as shown in Fig. 1. Out of the 57 groups interviewed, 23% of them were women groups, 9% were research groups, 9% were health groups, 5% were youth groups and 54% were mixed (combinations of one to four) groups. This implies that the groups were relatively heterogeneous in terms of age, sex and interests. The diverse groups also imply that change agents have a higher chance of using different groups depending on the population the project targets.

Demographic Characteristics of Groups

The group and individual member demographic factors influence the performance of groups. Performance may be influenced by the period the group has been in existence, membership number per group and gender balance.

As shown in Table 2, there were significant differences in the period the CIG had been in existence,

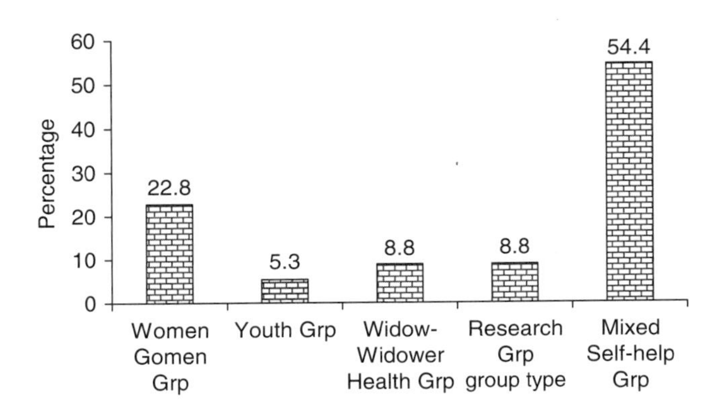

Fig. 1 Types of groups interviewed – 2006

Table 2 Demographic characteristics of CIGs in Trans-Nzoia district, Kenya

Variable	Women group	Youth group	Widow_widower health group	Research group	Mixed self-help group	*F* test
Age of CIG	8.7 ± 1.523	5.00 ± 0.01	5.60 ± 3.94	1.60 ± 0.93	3.55 ± 0.62	3.713***
Number of active male	2.00 ± 0.764	8.00 ± 4.16	4.20 ± 1.85	10.80 ± 2.48	28.22 ± 3.00	4.454***
Number of active female	14.92 ± 1.930	4.33 ± 2.19	16.20 ± 2.82	19.40 ± 7.51	12.26 ± 1.40	1.876
Number of active youth	1.69 ± 0.91	11.33 ± 4.37	2.40 ± 1.03	6.00 ± 2.43	4.68 ± 1.53	1.395
Total membership	18.23 ± 2.38	23.67 ± 2.03	22.80 ± 1.32	36.20 ± 9.54	25.93 ± 2.00	1.773
Percentage males	7.26 ± 3.19	31.17 ± 15.86	18.62 ± 8.20	34.21 ± 9.24	38.63 ± 4.17	5.660***
Percentage females	87.67 ± 8.45	16.98 ± 8.49	70.81 ± 11.53	48.66 ± 9.79	46.45 ± 4.50	5.697***

Note: *** = 1%

number of active male members and percentage of male and female members in the group.

Membership in Groups

Starting special groups with common interest and common challenges requires screening of membership. The groups interviewed gave a number of factors considered for membership in CIGs. They included registration fee, age, mandatory share contributions, place of residence, type of work and commitment. The cut-off levels of the factors are unanimously decided in special and/or annual general meetings.

Activities Undertaken by Groups

The overriding aims of group formation were poverty alleviation, food security and improvement of health status of members. Out of 57 groups interviewed, 12% were engaged in dairy, 29% in horticulture, 10% in shoats, 27% in poultry production, 3% in HIV/AIDS, 6% in credit and 13% in other activities like small businesses (Fig. 2).

Factors Influencing Group Size

From the ordinary linear regression analysis shown in Table 3, the significant factors that influence the group size are special activities the groups are engaged in after project/donors withdrawal and gender considerations. The groups with special consideration on men reduce the group size, while those special considerations on women increase the group size. This implies that women groups are more likely to maintain larger group sizes than men. In addition groups with a vision to have special activities initiated by the groups are more likely to sustain the group for a longer period than those without.

From Table 3, we can generate the following estimated equation:

$$
\begin{aligned}
Y_i = {} & 60.138[\text{Constant}] - 0.34[\text{AGE}] \\
& - 10.322[\text{GENDER}] - 2.212[\text{DOSUP}] \\
& + 10.235[\text{ASPENT}] + 5.84\,[\text{SPACT}] \\
& + 9.27[\text{ELECOFF}] - 19.209[\text{REGMET}] \\
& + 1.854[\text{GRPTYPE}] + e
\end{aligned}
$$

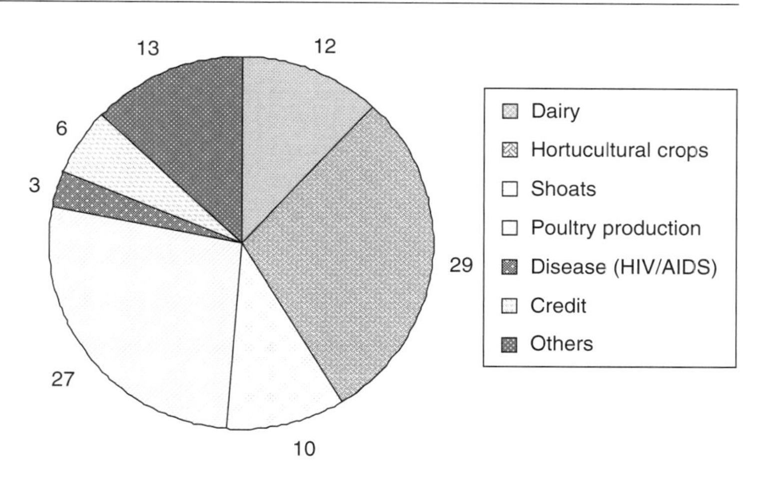

Fig. 2 Potential benefits from group activities

Table 3 Factors influencing group size (dependent variable = total number of members per group)

Variable	Coefficient	SE	*t* Value	Significance
Age of group	–0.34	0.486	–0.7	0.488
Gender considerations	–10.322	5.91	–1.746*	0.088
Donor support	–2.2127	4.53	–0.47	0.641
Aspire for post-production enterprises	10.235	4.633	2.209**	0.032
Special activities	5.84	4.507	1.296	0.202
Elected official	9.27	15.602	0.594	0.555
Regular meetings	–19.209	15.396	–0.174	0.219
Group type	1.854	1.432	1.294	0.202
Constant	60.138	26.802	2.244	0.03
R square	0.49			
F value	1.74			
Number of cases	57			

Note: * = 10%; ** = 5%

Factors Influencing Farmers to Participate in Group Activities

The groups that had effective communication channels positively and significantly influenced the likelihood of members participating in group activities. The odds in favour of participating in group activities increase by 0.01 for groups with effective communication channels. This demands effective communication channels and skills among group members and also implementers of projects like KAPP (Table 4).

Discussion

Novel ideas and farming practices spread, and evolve, through interpersonal interaction and communication in rural communities and subsequently facilitating cooperation and exchange among farmers becomes a core objective of most research-extension interventions (Godtland et al., 2004; Darr and Pretzsch, 2006). The results of this study imply that farmers' groups can play proactive roles in promoting diffusion of agricultural knowledge and technologies and thus improve the efficiency and efficacy of the technology extension dissemination. This is evident from the group's mission, vision and objectives of alleviating poverty, improving food security and promoting good health. These aspects are not only in project objectives but also spelt out in government strategic plan (GOK, 2001, 2004, 2005). As indicated in the study research statement the effect of structural and functional variables of farmers' groups on the spread of agricultural innovations can be enhanced through identifying relevant groups that are cohesive, have on-going activities and motivate their members to actively get involved in group activities. The results showed that most of the groups

Table 4 Factors influencing farmers to participate in group activities

Variables	Coefficient	SE	Wald	Significance	Exp (*B*)
AGE	−0.017	0.065	0.068	0.794	0.983
TOTMEMB	0.011	0.021	0.282	0.596	1.011
CIGREG	7.077	36.661	0.037	0.847	1184.369
GENDISSU	−0.71	0.945	0.564	0.453	0.492
DONSUPP	−0.698	0.595	1.377	0.241	0.497
GHAVCOMM	1.357*	0.781	3.018	0.082	3.884
Constant	−7.218	36.723	0.039	0.844	0.001
Predicted correct	72.4				
Overall predict	61.0				
2 log likelihood	0.171				
Cox and Snell	0.128				

Note: * = 10%

selected and interviewed are potential vehicles for disseminating agricultural innovations among the farming communities because they are actively involved in agricultural activities. However, there could be demand to have some support structures put in place to support them technically and also in financial and administrative management (Sarkar, 2003). This is perceived to enhance group stability. The longer the group stays, the more the spread of the technology. Additionally inter-group interaction needs to be enhanced in terms of visits and tours for group members to learn and implement what others are doing.

From the responses there were a significant proportion of the groups that were aware of some of the agricultural technologies implying that these groups seek new agricultural information to have the common intervention agenda. This signals the synergistic options if KAPP uses the groups to spread other agricultural innovations to the group members and the wider community at large.

According to Kitetu (2005), the group as an approach of community intervention is not new. She indicated that the informal groups popularly known as 'merry go round groups' are key examples of common interest groups working for a common goal. The results of this study showed that a number of women groups were sampled and interviewed. This could probably indicate that women groups could be actively used in technology transfer.

Intra-communication among group members reflects the state of relations between officials and other group members, development workers and ordinary members (Kitetu, 2005). The results from this study showed that communication significantly influences members' participation in group activities. This implies that effective communication between the officials of the groups and members need to be taken into account in order to sustain the groups. One of the key weak links in the groups is lack of access to markets and value addition. It is indicated that linking farmers to markets can enhance research-extension impacts on farming communities (Sanginga et al., 2004). This means that in the course of using the groups as vehicles of technology dissemination in KAPP, there could be demand to strengthen these areas.

Conclusions and Recommendations

The results of the study indicate that there are a variety of groups in the district. The groups engage in activities aimed at alleviating poverty, enhance food security and improve health status which is in line with the aims and objectives of the Kenyan government (GOK, 2002, 2004). The activities that groups are engaged in are agricultural, health and social in nature. A number of factors were observed to affect group performance as documented by HIMSS (2006). They include heterogeneity of the groups, technical and financial support received, devolution of power within the group sub-committees, benefits group members receive from engagements, rules and norms the groups constitute and follow, committees established within the group and frequency of electing officials. This implies that for group sustainability and if one wants to use groups as entries into communities, these factors may be considered. The CIGs may form favourable target groups for KAPP interventions for enhanced impact and improved community welfare.

Acknowledgements The authors acknowledge the financial support from World Bank through KARI. Special thanks go to group members and extension staff who were key players in the generation of information for this study.

References

Ajili A (2000) Aspects of traditional versus group extension approaches on farmer behavioral change in an extensive grazing environment in the Bathurst district of New South Wales, Australia. Department of Wool and Animal Science. The University of New South Wales, Sydney, NSW, pp 1–310

Amemiya T (1981) Qualitative response models: a survey. J Econ Lit 19:1483–1536

Braun AR, Thiele G, Fernández M (2000) Farmer field schools and local agricultural research committees: complementary platforms for integrated decision-making in sustainable agriculture. Agricultural Research & Extension Network, ODI, UK, pp 1–20

Darr D, Pretzsch J (2006) The spread of innovations within formal and informal farmer groups: evidence from rural communities of semi-arid Eastern Africa. Conference on international agricultural research for development, University of Bonn

Field A (2002) Discovering statistics using SPSS for windows. Advanced techniques for the beginner. SAGE, London

Godtland EM, Sadoulet E, de Janvry A, Murgai R, Ortiz O (2004) The impact of farmer-field-schools on knowledge and productivity: a study of potato farmers in the Peruvian Andes 1, Economic Development and Cultural Change, CA, USA

GOK (1998) Government of Kenya. Economic survey 1998–1999. Ministry of economic planning and development, Nairobi, Kenya

GOK (2001) Poverty reduction strategy paper for the period 2001–2004. Prepared by the people and government of Kenya, June 2001. Government printers. Nairobi, Kenya

GOK (2002) National development plan 2002–2008. Government Printers, Nairobi

GOK (2004) Strategy for revitalizing agriculture. Prepared by ministry of agriculture ministry of livestock and fisheries development ministry of cooperative development and marketing, Government printer, Nairobi

GOK (2005) MDG status report for Kenya 2005. Nairobi, Kenya. Ministry of planning and national development in partnership with UNDP, Kenya and government of Finland

Greene WH (1990) Econometric analysis. Macmillan, New York, NY, pp 10022

Guerre E, Moon HR (2004) A study of a semiparametric binary choice model with integrated covariates. LSTA, Universite Paris 6 and CREST, Los Angeles, CA, USA

Gujarati DN (1995) Basic econometrics. McGraw-Hill, New York, NY

Hazell PBR (2003) Is there a future for small farms? In: Proceedings of the 25th international conference of agricultural economists (IAAE). International Food Policy Research Institute (IFPRI), Washington, DC

HIMSS (2006) How to establish a HISS special group. Healthcare Inf Manage Syst, 2006

Kinyua J (2004) Priorities for action: perspectives for East and Central Africa. Assuring food and nutrition security in Africa by 2020: prioritization, strengthening actors, and facilitating partnerships, IFPRI, Kampala, Uganda, 1–3 Apr 2004

Kitetu CW (2005) Farmer groups as a way of mobilising citizen participation in development: an example from Kenya. Paper presented at the 11th general assembly, Maputo

Maddala GS (1983) Limited-dependent and qualitative variables in econometrics. Cambridge University press, New York, NY

Mose LO (2007) Who gains, who loses? The impact of market liberalization on rural households in North Western Kenya. PhD thesis, Wageningen University, The Netherlands

Mukhwana EJ, Nyongesa M, Ogemah V (2005) Facilitating small scale farmer collective marketing activities in Africa. The case of cereal banks in Kenya. A Filed manual for government and NGO extension workers facilitating collective marketing activities. The sustainable agriculture center in Africa. SACRED Africa, Nairobi, pp 1–32

Nyameino DM (2007) Cost of doing farming as a business farmer group experience in Kenya. Presentation for the 2nd African grain trade summit, Intercontinental Hotel-Nairobi, Kenya

Nyikal RA (2000) Financing smallholder agricultural production in Kenya: an economic analysis of the credit market. PhD thesis, University of Nairobi, Nairobi, Kenya

Pindyck RS, Rubinfeld DL (1998) Econometric models and economic forecasts, 3rd edn. McGraw-Hill, New York, NY

Rogers EM (1983) Diffusion of innovations, 3rd edn. The Free Press, New York, NY, pp 1–29

Sanginga PC, Best R, Chitsike C, Delve R, Kaaria S, Kirkby R (2004) Linking smallholder farmers to markets in East Africa empowering mountain communities to identify market opportunities and develop rural agroenterprises. Mt Res Dev 24(4):288–291, Nov 2004

Sarkar R, West A (2003) A handy guide to getting involved for voluntary and community groups, London. First published in Great Britain in 2003 by the community development foundation, London

Zeller M, Sharma M, Lapenu C, Henry C (2003) Measuring social performance of micro-finance institutions: a proposal for social performance indicators initiative (SPI) Final Report. Göttingen, Argidius foundation and consultative group to assist the poorest (CGAP), pp 1–18

Economic Analysis of Improved Potato Technologies in Rwanda

R.J. Mugabo, D. Mushabizi, M. Gafishi, J. Chianu, and E. Tollens

Abstract The Rwanda Agricultural Research Institute (ISAR), in collaboration with the international potato research center (CIP) and PRAPACE, generated and disseminated over the last three decades improved potato-based technologies. Conducted in two major potato-producing agroecological zones in Rwanda (*hautes terres de laves* and *hautes terres du Buberuka*), this study aimed at identifying the best-bet technological packages out of five alternatives. Based on the minimum acceptable marginal rate of return criterion, results from a two-season on-farm trial revealed that three technological packages, T_2 (improved seeds + fertilizer "NPK" + fungicide "dithane"), T_3 (improved seeds + fertilizer "DAP" + fungicide "dithane"), and T_4 (improved seeds + fertilizer "NPK" fungicide "ridomil+ dithane"), were profitable in both zones. T_5 (improved seeds + fertilizer "DAP" + fungicide "ridomil + dithane") was profitable only in the *terres de laves* zone. T_1 (improved seed + farmer practices) would be attractive to farmers only in *Buberuka* zone. The sensitivity analysis showed that all the treatments have almost the same trend and are very sensitive to a fall in potato prices. However, T_5 was the least sensitive to changes in fertilizer prices. T_2 was least sensitive to changes in pesticide prices in the *terres de laves* zone. The sensitivity analysis in *Buberuka* zone revealed that although T_4 was superior to the other options, it became most sensitive to change in potato price beyond a 25% increase.

R.J. Mugabo (✉)
Institut des Sciences Agronomiques du Rwanda (ISAR), Musanze, Rwanda
e-mail: mugabojosa@yahoo.fr

Keywords Technological packages · Partial budget · Stochastic dominance · Marginal analysis · Sensitivity analysis

Introduction

The Irish potato was introduced in Rwanda late in the 19th century by German missionaries. It was considered for a long time as white colonialists' food, a reason why its integration in the traditional agricultural systems went very slow (Scott, 1988). Its production was very low until the creation of the Rwanda Agricultural Research Institute (ISAR) in 1962. The production then increased from 33,000 t per year to 170,000 t per year in 1979 at the start of the national potato program (PNAP) in ISAR with the help of experts from Holland. It later increased by an annual rate of about 8% due to big efforts made by the new potato program in collaboration with the international research center of potato "CIP" (Nezehose, 1990; Munyemana and von Oppen, 1999).

Over the last three decades, ISAR in collaboration with CIP and the Regional Potato Research Network (PRAPACE) generated improved potato-based technologies that were widely disseminated in Rwanda by various partners and collaborators. Components of the disseminated technologies included improved varieties, fertilization techniques, pest and disease control techniques, and other potato management practices. Most of the earlier potato research in Rwanda had focused on production issues.

The purpose of this study is to identify the best-bet options from different technological package

A. Bationo et al. (eds.), *Innovations as Key to the Green Revolution in Africa*,
DOI 10.1007/978-90-481-2543-2_99,

alternatives and recommend those that would meet the socio-economic conditions of farmers and command widespread adoption as a result. It was assumed that some of the potato technological packages were more profitable than the current farmer potato production practices.

Materials and Methods

Materials

Farmer-managed trials under the monitoring of ISAR scientists were conducted in two of the three major potato-producing agroecologies in Rwanda: the "Birunga highlands" and the "Buberuka highlands." The Birunga highlands zone is located in the north and northwest of Rwanda. Soils in that zone are very rich but shallow and result from the degradation of lava. They have a pH between 5.1 and 5.9. The rainfall is greater or equal to 1500 mm/year. The slope of lands is not in general steep on the average altitude of 2300 m. Located in the northern region of Rwanda the Buberuka highlands zone has steep slope hills on the altitude of 1900–2300 m. Soils are in general acidic with pH between 3.5 and 4.5 but have good physical properties. Most of the agricultural lands are on steep slopes making soil erosion a complex problem in the region. The rainfall is between 1100 and 1300 mm/year.

The two zones are characterized by two rainy seasons and two dry seasons that correspond to two agricultural seasons. Season A covers the period from September to February, the last two months being the small dry season of the year. Season B goes from March to August and the bigger dry season of the year is made up by the last 3 months of the season.

Eighteen farmers conducted trials during two seasons: 2006A and 2006B. They were selected in six sites (three farmers per site), three sites being located in each of the two agroecologies. Two types of seeds (improved seeds of the variety called *Mabondo* and farmer seeds), two pesticides (dithane and ridomil), and three chemical fertilizers (NPK 17-17-17, DAP, and urea) were used in different combinations that correspond to different research recommendations. Each trial was made up of six treatments (including the control).

The description of the trial is as follows:

T_0: Control – farmer's seeds and farmer's agricultural practices[1]

T_1: *Mabondo* + farmer's agricultural practices

T_2: Improved seeds + fertilizer "NPK" (300 kg/ha) + fungicide "dithane (25 g/are/week for about 8 weeks)"

T_3: Improved seeds + fertilizer "DAP (100 kg/ha) + urea (50 kg/ha)" + fungicide "dithane (25 g/are[2]/week for about 8 weeks)"

T_4: Improved seeds + fertilizer "NPK" (300 kg/ha) + fungicide "ridomil (25 g/are/week the first 2 weeks) + dithane (25 g/are/week for about 5 weeks)"

T_5: Improved seeds + fertilizer "DAP" (research recommendations) + fungicide "ridomil (25 g/are/week the first 2 weeks) + dithane (25 g/are/week for about 5 weeks)

Statistics and Data Analysis

Three analytical tools were used to identify the technological packages that were not only profitable but also exhibited good margin and remained profitable in different situations of input and output prices. First was the partial budget analysis which generates the net benefits from the alternative technologies and the control. Second was the marginal analysis that compares the net benefits from the partial budget by considering the magnitude of corresponding variable costs. Technologies that produced attractive margins were submitted to sensitivity analysis, the third analytical technique. The sensitivity analysis helps detect technologies that better withstand different input and output price shocks (CYMMIT, 1988; IRG 65, Partial budget analysis for on-farm research. http://www.iita.org/cms/details/trn_mat/irg65/i65.pdf).

[1] His current way of planting potato, rates of fertilizers, and rates of pesticides

[2] 1 are = 0.0 ha

Results and Discussion

Partial Budgets

The partial budgets computed for Birunga highlands and Buberuka zones (Tables 1 and 2) reveal that seeds have the highest share of the production cost, indicating the relatively high costs of seeds compared to other inputs. The net benefits of all improved technological packages were higher than that of the control in both zones.

Stochastic Dominance and Marginal Analyses

Results from stochastic dominance and marginal analyses are presented in Table 3 for the *Birunga highlands*

Table 1 Partial budget analysis for the *Birunga* agroecological zone

	Treatment[a]					
	T_0	T_1	T_2	T_3	T_4	T_5
Average yield (kg/ha)	15,794	19,883	22,233	21,128	23,928	21,961
Adjusted yield (kg/ha)	15,321	19,287	21,566	20,494	23,210	21,302
Gross field benefits (Frw/ha)	1,057,122	1,330,792	1,488,077	1,414,082	1,601,486	1,469,857
Cost of seed	217,528	405,104	405,104	405,104	405,104	405,104
Cost of fertilizer	68,422	68,422	102,978	49,969	102,978	49,969
Cost of pesticides	66,960	66,960	37,333	37,333	78,750	78,750
Cost of hired labor	8259	8259	9911	8722	8967	7778
Opportunity cost of family labor	5851	5851	6724	5836	5817	4929
Total cost that vary (Frw/ha)	367,020	554,596	562,050	506,964	601,616	546,530
Net benefits (Frw/ha)	690,102	776,196	926,027	907,118	999,870	923,327

[a]T_0 (control): Farmer seeds and his agricultural practices
T_1: *Mabondo* + farmer's agricultural practices
T_2: Improved seeds + fertilizer "NPK"+ fungicide "dithane"
T_3: Improved seeds + fertilizer "DAP" + fungicide "dithane"
T_4: Improved seeds + fertilizer "NPK" + fungicide "ridomil + dithane"
T_5: Improved seeds + fertilizer "DAP" + fungicides "ridomil + dithane"

Table 2 Partial budget analysis for the *Hautes terres du Buberuka* agroecological zone

	Treatment[a]					
	T_0	T_1	T_2	T_3	T_4	T_5
Average yield (kg/ha)	7388	12,845	15,535	13,914	16,665	14,244
Adjusted yield (kg/ha)	7166	12,460	15,069	13,496	16,165	13,816
Gross field benefits (Frw/ha)	551,782	959,420	1,160,313	1,039,192	1,244,705	1,063,832
Cost of seed	110,250	325,875	325,875	325,875	325,875	325,875
Cost of fertilizer	0	0	103,650	41,767	103,650	41,767
Cost of pesticides	17,059	17,059	21,050	21,050	72,881	72,881
Cost of hired labor	3542	3542	3951	3735	3639	3423
Opportunity cost of family labor	4424	4424	6405	6352	5949	5896
Total cost that vary (Frw/ha)	135,275	350,900	460,931	398,779	511,994	449,842
Net benefits (Frw/ha)	416,507	608,520	699,382	640,413	732,711	613,990

[a]T_0 (control): Farmer seeds and his agricultural practices
T_1: *Mabondo* + farmer's agricultural practices
T_2: Improved seeds + fertilizer "NPK"+ fungicide "dithane"
T_3: Improved seeds + fertilizer "DAP" + fungicide "dithane"
T_4: Improved seeds + fertilizer "NPK" + fungicide "ridomil + dithane"
T_5: Improved seeds + fertilizer "DAP" + fungicides "ridomil + dithane"

Table 3 Dominance analysis and marginal rate of return in the *Birunga* zone

Treatment	Total costs that vary (Frw/ha)[a]	Net benefits (Frw/ha)	MRR[b] (%)	MRR from T_0 (%)
T_0 (control): Farmer seeds and his agricultural practices	367,020	690,102		
T_3: *Mabondo*+ fertilizer "DAP" + fungicide "dithane"	506,964	907,118	155	155
T_5: Improved seeds + fertilizer "DAP" + fungicides "ridomil + dithane"	546,530	923,327	41	130
T_1: Improved seeds + farmer's agricultural practices	554,596	776,196 D[c]	–	–
T_2: Improved seeds + fertilizer "NPK"+ fungicide "dithane"	562,050	926,027	17	121
T_4: Improved seeds + fertilizer "NPK" + fungicide "ridomil+ dithane"	601,616	999,870	187	132

[a]Frw: Rwanda Francs (US $1 = 546 Frw)
[b]MRR = marginal rate of return
[c]D = dominated

and Table 4 for the *Buberuka highlands*. One treatment was dominated and then excluded from marginal analysis in each agroecological zone. T_5 that dominated T_1 in *terres de laves* zone happened to be the only treatment that was dominated in the *hautes terres du Buberuka* zone.

As long as the marginal rate of return between two treatments exceeds the minimum acceptable rate of return, a switch from one treatment to the other should be attractive to farmers. It was not possible to get an exact figure of the minimum rate of return acceptable to the farmers in the two recommendation domains. It is accepted, however, that for the majority of situations the minimum rate of return acceptable to farmers will be between 50 and 100% (CYMMIT, 1988). Considering that all the improved technological packages in our experiment were really slight modifications of farmer's practice, especially in the *terres de laves* agroecological zone, a minimum rate of return of 70% would be acceptable.

In the two zones all the improved packages, except those dominated, met the minimum acceptable rate of return (see Tables 3 and 4). These technological packages appear to be highly attractive to farmers in the *terres de laves* zone for having marginal rates of return above 120%, whereas they would be slightly attractive to farmers in the *Buberuka* zone because their marginal rates of return are just higher than the minimum acceptable rate of return and below 90%. Improving farmer practices by using improved seeds combined with fertilization with DAP and fungicide control by dithane constitutes the more attractive alternative in *terres de laves zone*. A simple replacement of farmer seeds by improved seeds in the farmer practices is revealed to have a higher impact in the *Buberuka* zone.

Table 4 Dominance analysis and marginal rate of return in the *Buberuka* zone

Treatment	Total costs that vary (Frw/ha)[a]	Net benefits (Frw/ha)	MRR[b] (%)	MRR from T_0 (%)
T_0: Control – farmer seeds and his agricultural practices	135,275	416,507		
T_1: *Mabondo* + farmer's agricultural practices	350,900	608,520	89	89
T_3: Improved seeds + fertilizer "DAP" + fungicide "dithane"	398,779	640,413	66	85
T_5: Improved seeds + fertilizer "DAP" + fungicides "ridomil + dithane"	449,842	613,990 D[c]	–	–
T_2: Improved seeds + fertilizer "NPK"+ fungicide "dithane"	460,931	699,382	95	87
T_4: Improved seeds + fertilizer "NPK" + fungicide "ridomil+ dithane"	511,994	732,711	65	83

[a]Frw: Rwanda Francs (US $1 = 546 Frw)
[b]MRR = marginal rate of return
[c]D = dominated

Sensitivity Analysis

With a decrease of about 25% in the current price all the improved potato technological packages are no longer attractive to farmers in the *Buberuka* zone (Figs. 1 and 2). Three of the improved potato technologies, T_3, T_4, and T_5, were still surviving the marginal rate of returns criterion in *Birunga highlands* zone with a decrease of 25% in the current price. A further decrease of the price by about 37.5% makes, however, the three mentioned treatments not attractive to farmers in that zone due to higher total variable costs associated with these treatments.

Change in pesticide price does affect net benefits in both zones but in different ways. In the *terres de laves* zone, improved packages become more and more attractive due to higher pesticide cost for the control treatment, with the exception of T_5 which was still better than the control treatment but not as in the current price (Fig. 3). In the *Buberuka* zone, three from the four undominated treatments (T_1, T_2, and T_3) were as attractive to farmers as they were at the current price; the fourth (T_4) survived the sensitivity analysis test only up to an increase of 50% of the current price (Fig. 4). This situation could be explained by the fact that treatment T_4 has the highest net benefit in Buberuka zone, although it has the highest cost associated with the use of pesticides.

In the *terres de laves* zone, T_4 was not only the most profitable alternative at current fertilizer prices but also the most sensitive to changes in fertilizer price along with T_2 (see Fig. 5). In this zone, as price of fertilizer increased, T_3 and T_5 that had lower fertilizer cost at the current prices became more attractive to farmers.

Increase in fertilizer price would negatively affect more potato production in *Buberuka* zone (see Fig. 6). In fact, with an increase of about 50% of the current price of fertilizer none of the improved packages would be attractive to farmers, all of them generating marginal rates of return below the minimum rate of return acceptable to farmers and so would not be acceptable to them.

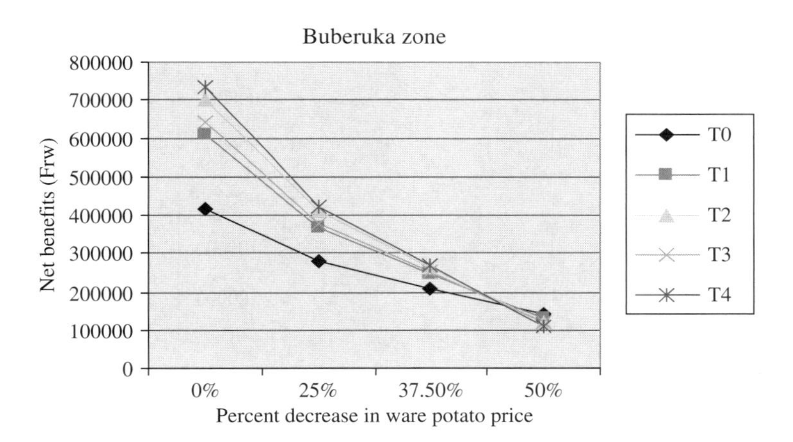

Fig. 1 Change in potato price in *Buberuka* zone

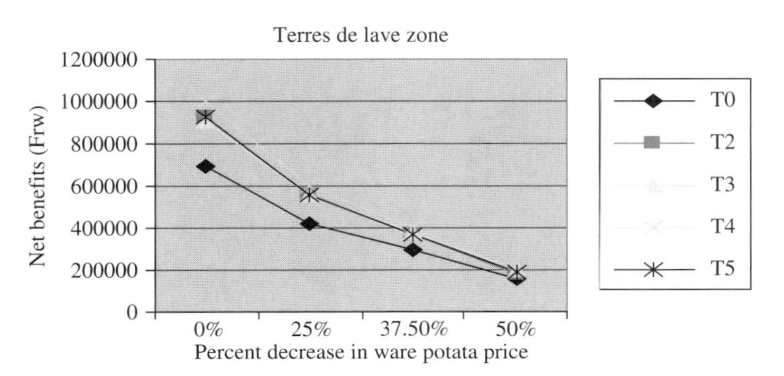

Fig. 2 Change in potato price in *Birunga* zone

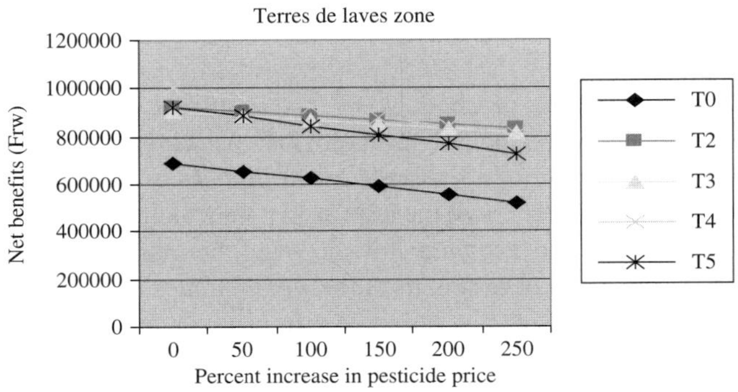

Fig. 3 Change in pesticide price in *Birunga* zone

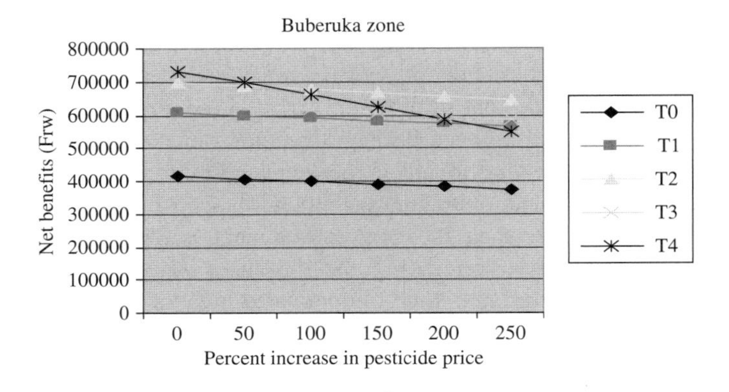

Fig. 4 Change in pesticide price in *Buberuka* zone

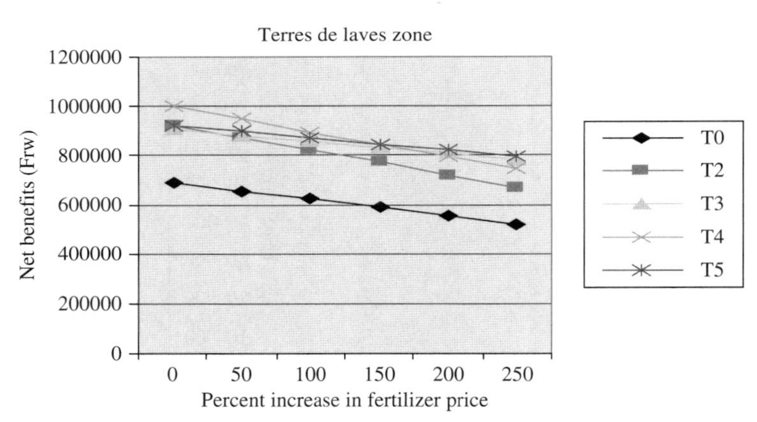

Fig. 5 Change in fertilizer price in *Birunga* zone

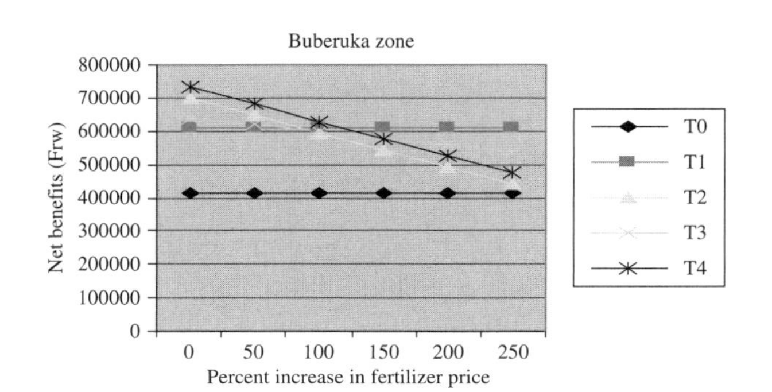

Fig. 6 Change in fertilizer price in *Buberuka* zone

Conclusion and Recommendations

The major objective of this study was to identify the best-bet options from five most known improved potato technological packages and recommend those that would meet the socio-economic conditions of farmers in the two major potato-producing agroecologies: *terres de laves* and *hautes terres du Buberuka*. In the former zone, all the non-dominated packages have marginal rate of return of at least 120%, whereas their counterparts in the latter have marginal rate of return of at most 89%. These results can be explained partially by the fact that soils in the *Buberuka* zone are acidic and need to be improved by applying lime.

Four improved potato technological packages survived both the dominance analysis and the minimum acceptable marginal return test in the *terres de laves* zone when compared with the control treatment. In the *Buberuka* zone four improved technological packages survived also the minimum acceptable marginal returns test not only when compared with the control treatment but also when marginal analysis considers pairs of adjacent treatments. Based on the minimum acceptable marginal rate of return criterion, T_1 (*Mabondo* and farmer's agricultural practices) was more appropriate in *Buberuka* zone while T_3 (improved seeds + fertilizer "DAP" + fungicide "dithane") would be the first to be recommended in the *terres de laves* zone.

All the profitable treatments have almost the same trend and were very sensitive to a fall in potato prices in both zones. Measures to stabilize potato prices around the current price are then very important in order to sustain the adoption and use by farmers of improved technologies in *terres de laves* zone.

Changes in pesticide and fertilizer prices have less effect on net benefits in both zones compared to the change in ware potato price. The treatment that includes "ridomil" and "NPK," that is, T_4, although highly profitable at the current price would be the last to be recommended when prices of pesticides and fertilizers increase.

References

CYMMIT (1988) From agronomic data to farmer recommendations: an economics training manual. Completely revised edition. International Maize and Wheat Improvement Centre, Mexico

Munyemana A, von Oppen M (1999) La pomme de terre au Rwanda: Une analyse d'une filière a hautes potentialités. Centre International de la Pomme de Terre, Lima

Nezehose JB (1990) Agriculture Rwandaise: Problématique et Perspectives. Inades Formation Rwanda, Kigali

Scott GJ (1988) Potatoes in central Africa: a study of Burundi, Rwanda, and Zaire. International Potato Centre, Lima

Assessment of Occupational Safety Concerns in Pesticide Use Among Small-Scale Farmers in Sagana, Central Highlands, Kenya

P. Mureithi, F. Waswa, and E. Kituyi

Abstract Small-scale farmers in Sagana area of central Kenya constitute a population at risk due to intensive use of pesticides in the production of mainly horticultural crops for commercial purposes. This chapter examines the main causes of pesticide hazards and risks, barriers to taking risk reduction measures and cues to adopting safety behaviour when dealing with pesticides. Data were collected by the use of interviews conducted in 2006/2007 from a sample of 140 farmers. Perception scales were developed from interview items and were ranked along a modified three-point Likert scale. Analysis of the items and scales showed that farmers had fairly high levels of perceived risk, perceived severity and perceived benefits of taking action to mitigate pesticide hazards. Results from this study showed that farmers are still susceptible to pesticide-related dangers notably due to resignation to fate, perceived high cost of purchasing protective gear and lack of adequate training in the use and handling of pesticides. Further, contrary to conventional thinking, farmers' education had limited positive effect to safety behaviour when handling pesticides. The challenge to policy and practice towards safe use of pesticides lies in issues of farmers' economic survivability, perceptions and attitudes, along the whole chain from pesticide procurement, storage, farm application and disposal.

Keywords Intensive farming · Pesticides · Occupational safety · Environmental health

Introduction

Due to decreasing land sizes commensurate with population pressure in small-scale agro-ecosystems, agricultural intensification and thus use of agro-chemicals seems to be the only viable option farmers have to maximise yields in pursuit of survival. As a result, preventing, minimising and controlling occupational health/safety and environmental hazards associated with pesticide use remains an important concern in such agro-ecosystems. According to UNEP (1985), about 2–3 million tonnes of pesticide products are scattered all over the environment each year in an attempt to control pests and diseases. On the other hand, ILO (1999) estimates that there are some 3 million acute cases of pesticide-related poisoning in the world each year, 70% of which take place in developing countries and 340,000 results into death. According to WHO (1989) about 3% of agricultural workers in developing countries on average suffer an episode of pesticide poisoning every year, which means that of the 830 million agricultural workers in the developing world (ILO, 1999), there are about 25 million cases of occupational pesticide poisoning. With about 350,000 cases in the 1980s, Kenya is among African countries with the highest cases of pesticide poisoning per year.

With pesticide imports of about 7,600 tonnes, worth Kenya shillings 3.114 billion in 1998, Kenya is among the highest pesticide users in sub-Saharan (NEMA,

P. Mureithi (✉)
Department of Environmental Studies and Community Development, Kenyatta University, Nairobi, Kenya
e-mail: petmukariuki@yahoo.co.uk

A. Bationo et al. (eds.), *Innovations as Key to the Green Revolution in Africa*,
DOI 10.1007/978-90-481-2543-2_100,

2005). NEMA further estimates that pesticides poison 7% of agricultural sector's population in the country every year. However, these figures are mainly derived from large-scale agriculture where workers seek poisoning compensation from their employers. In small-scale rural areas like Sagana with poor monitoring and reporting systems, more cases could be going unreported.

Widespread use of pesticides in Kenya has made them ubiquitous in the environment. Wandiga et al. (2002) reported pesticide residues in water and sediment samples from several places at the Kenyan coast where River Tana and River Athi drain after traversing intensively cultivated areas in Kenya's highlands including Sagana. These included dichloro-diphenyl-ethane (DDE), dichloro-diphenyl-dichloro-ethane (DDD), dichloro-diphenyl-trichloro-ethane (DDT), Lidane, Dieldrin, with concentrations often exceeding 200 mg/g.

Although many human exposure cases go unreported, especially non-acute cases in rural farming, high pesticide levels have been found in human milk among Kenyans compared to corresponding levels in other countries (Kanja et al., 1988). This places high economic and health burden to households who are already resource poor. Kenya's horticultural sector has been under special pressure from international governments, especially Europe where it has been the largest exporter of horticultural produce for a decade to address pesticide hazards (Jaffee et al., 2005). According to the UK Pesticide Residue Committee (2005), random samples of Kenyan French beans and passion fruits tested in Europe in the years 1999–2004 indicated the presence of high levels of pesticides well above recommended maximum residue levels (MRLs). Farmers seem to have accepted pesticide hazards because of the perceived economic gains derived from farming with little consideration of the documented dangers (Moses et al., 1993; Garcia, 1998), especially when pesticides are mishandled or used for the wrong purpose.

This chapter is based on a study carried out at Sagana, in the central highlands of Kenya, with the aim of informing policy decisions towards safe use and handling of pesticides. This area was chosen because it forms an island of an intensively cultivated region (0.6 ha per household; average population density of the area is 202 persons per kilometre squared) in the country (GOK, 2001, 2003). The specific objectives of the study were as follows:

i. To determine the farmers' attitudes, perceptions and behaviour towards pesticides
ii. To assess the potential occupational safety hazards associated with such attitudes, perceptions and behaviour
iii. To suggest measures that would inform policy interventions towards safe use of pesticide at the farm level

Materials and Methods

The study design entailed a life cycle approach (LCA), which examines selected variables at all stages of the life of a product or an event. In this case farmers' attitudes, perceptions and behaviour when dealing with pesticides were assessed starting from choice of pesticides, purchase, transport and storage, mixture preparation, use and disposal. The unit of analysis was the individual farming household. Simple random sampling was used to select a representative sample of 140 households consistent with standard statistical procedures as discussed by among others (Saleemi 1997). Purposive sampling was used to select key informants who included local leaders, both public and private agricultural extension officials, sales agents of firms dealing in pesticides, provincial administration and local health officers.

Data were collected using researcher-administered questionnaires and observation checklists. Data on unreported cases of pesticide poisoning as well as the factors that make the farming community vulnerable to pesticide hazards were obtained through multi-stakeholder focussed group discussions. Secondary data were obtained from such sources as environmental impact assessment reports from the area and health records from two health centres serving the area, i.e. at Mutaga and Kiamariga shopping centres. Both descriptive and inferential statistics were used in data analysis. These included frequency distributions, modified Likert scale for analysing farmers' attitudes and perceptions towards pesticides and Pearson correlation to test relationships between variables such as education level and attitudes towards pesticide-associated hazards.

Results and Discussion

Out of the sampled population, 135 (96%) used pesticides as the first choice in pest management (Table 1). Only about 64% of the total respondents had appropriate pesticides application equipment while others borrowed or used twigs. Only 17% of the total respondents knew of and had the recommended list of pesticides for the crops they grew. This means that up to 83% of respondents used any pesticides including banned and highly persistent brands as long as they appeared to solve the pest problem at hand. This served to increase the hazards and risks associated with the use of these agro-chemicals.

As far as education and training levels were concerned, results indicated that training and not level of formal education seemed to influence safe use of pesticides (Table 2). Lack of training was identified as the single most important factor determining adoption of safety behaviours when handling pesticides. Over 69% of the respondents felt that they had no adequate information regarding pesticide hazards to be able to make safe use decisions. Pesticide promoters were biased in their delivery of information regarding pesticides. A lot of emphasis was put on the economic gains associated with pesticides use, while their hazards to human health and the environment were avoided.

Regarding timing of application, at least 64% of the respondents sprayed pesticides in any weather conditions. This was common among farmers with relatively large farm sizes. Such farms required higher volumes of pesticides, which had to be done within a tight time schedule. This tended to ignore prevailing weather conditions. Further, up to 53% of the farmers believed that using higher dosages of animal-based pesticides to control crop pests was more effective. Similarly, due to misconceptions, farmers also applied crop pesticides to get rid of ticks. In this way it would be difficult to use recommended rates, with the consequence of increased health and environmental hazards.

Table 1 Pest control methods most frequently used in Sagana

Pest control method	% frequency
Synthetic chemicals	96
Cultural	2
Integrated pest management (IPM)	2

Only about 11% of the farmers interviewed were involved in calibration of their application equipments (Fig. 1). Most knapsack sprayers used spilled their contents, had loose fitting caps and others leaked from below. Up to 84% of interviewed farmers lacked proper pesticides measuring containers raising questions on the accuracy of pesticides measurement in the area. About 52% of the farmers interviewed disposed empty containers to the environment. Bearing in mind that 24% of all the containers were rinsed less than three times before disposal, it can be deduced that these empty containers had pesticide residues. Surplus application mixes were disposed to the environment, while knapsack sprayers were washed near streams thus contaminating the water.

Pearson correlation analysis showed that low-income farmers were more vulnerable to pesticide hazards than higher income farmers who engaged less in risky behaviour when dealing with pesticides (Table 3).

There was a positive correlation between farmers spraying during windy weather and practicing other risky behaviours (Table 4). These included wetting of self with pesticides ($r = 0.83$, $n = 138$, $p = 0.01$), smoking or eating while handling pesticides ($r = 0.37$, $n = 138$, $p = 0.01$), not changing work clothes ($r = 0.46$, $n = 138$, $p = 0.01$) and not bathing after work ($r = 0.46$, $n = 138$, $p = 0.01$). This means that farmers engaging in one hazardous pesticide handling

Table 2 Pearson's correlation of farmers' training and education against various behaviours associated with human health and the environmental safety

Behaviour	Farmers' training *r* value	Farmers education *r* value
Spray during windy weather	–0.18*	–0.1
Touch pesticides with bare hands	–0.21*	–0.09
Wet themselves with pesticide mixtures	–0.26**	–0.14
Touch crops after spraying	–0.18*	–0.05
Use unlabelled pesticides	–0.21*	–0.09

* (r) values significant at $p = 0.05$; ** (r) values significant at $p = 0.01$

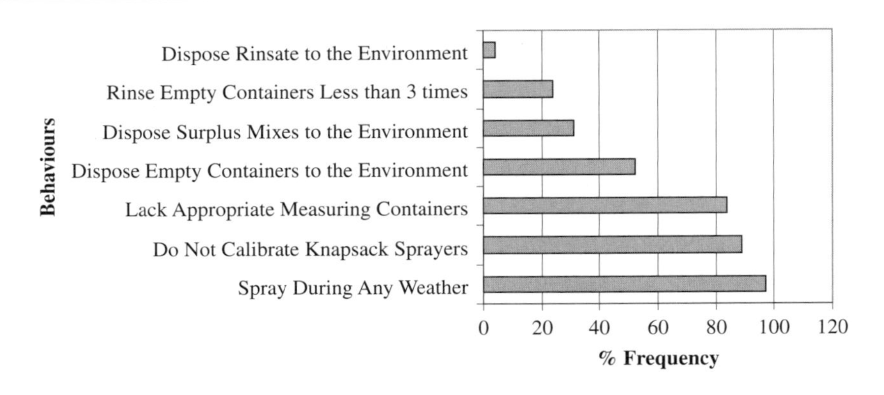

Fig. 1 Farmers' pesticides handling behaviour hazardous to the environment

Table 3 Pearson's correlation coefficient (r) for farmers income and practicing risky pesticide handling behaviours

Behaviour	Pearson's correlation coefficient (r) $n = 140$
Use of pesticides without labels	–0.17*
Reuse of work clothes	–0.18*
Mixing work clothes with other clothes	–0.22**
Smoked, ate or drank while handling pesticides	–0.22**
Not taking bath after work	–0.20*
Spray during windy weather	–0.17*

*(r) values significant at $p = 0.05$; **(r) values significant at $p = 0.01$

Table 4 Pearson's correlation coefficient (r) for farmers spraying during windy weather and practicing other risky behaviours

Behaviour	Pearson's correlation coefficient (r) $n = 140$
Wet self with pesticides	0.83
Smoked or ate while handling pesticides	0.37
Used pesticides without labels	0.91
Touched pesticides with bare hands	0.91
Brought home unwashed crops from field	0.76
Reused work clothes	0.42
Stepped on pesticides with bare feet	0.86
Not bathed after work	0.50
Swallowed sweat during spraying	0.67
Mixed work clothes with other clothes	0.59

All (r) values significant at $p = 0.01$ level of confidence

behaviour were more likely to engage in another leading to multiple pesticide exposure. The majority of respondents (96%) sprayed during windy weather as shown in Fig. 1 which led to increased, risks of inhalation and wetting self with application mixtures. This was attributed to ignorance and limited access to protective gear.

Further, 96% of the respondents reported using pesticides without labels, which effectively denied farmers' essential information on safety requirements. This was because pesticide vendors repackaged products into plastic bags or bottles for sale in small quantities. Majority of the farmers stored pesticides in their living rooms, which posed poisoning danger to unsuspecting household members, particularly children.

More than 90% of the respondents admitted mixing pesticides with bare hands. In addition to this, spraying of pesticides without protective gear was common. More than 80% of respondents did not change their clothes after work. Where changing occurred, the clothes were casually mixed with other family wear thus exposing other household members like children to pesticide hazards. Over 83% of the interviewed respondents ate food, smoked or carried out both activities during pesticides handling. Reuse of empty containers as receptacles for household food products (sugar, milk and salt) was practiced by 75% of all respondents. This increased the risk of food contamination, poisoning and endangered health to the households.

More than 90% of respondents indicated that pesticides caused harm to the environment and could also harm people. This was, however, based on indicators that communicated lost opportunity for timely interventions such as cases of death or long-term illnesses and livestock deaths after pesticide exposure. Due to limited use of respiratory masks, it was no surprise that the highest number of respondents (25%) experienced breathing problems after exposure to pesticides. Other symptoms included skin problems (burns and abrasions) and eye problems, which they attributed to

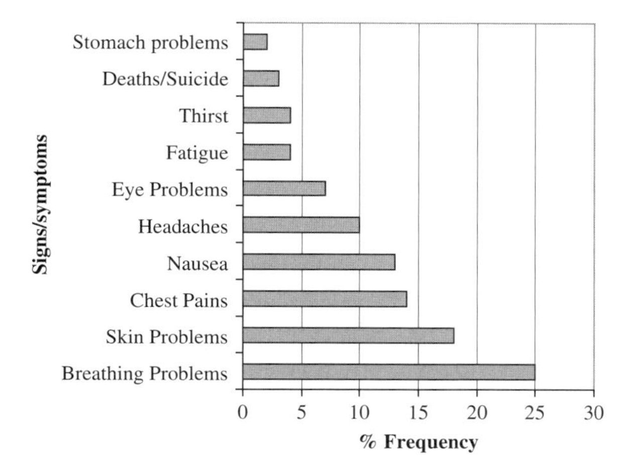

Fig. 2 Symptoms experienced by farmers after handling pesticides

chronic exposure to pesticides. Chest pains, nausea, headaches, fatigue thirst and stomach problems were also reported (Fig. 2).

The use of protective gear in the area was very low (Fig. 3). Over 69% of the total respondents took no action to protect themselves or the environment from pesticide hazards. Individuals who took no action to protect themselves were either ignorant of pesticide hazards or believed that they had no ability to protect themselves from the hazards leaving everything to God.

Among the farmers who used protection gear, none used a complete set. At 26%, gumboots were the most frequently used protective gear notably to protect the farmers from mud during rainy seasons and not necessarily from pesticides exposure through the feet. Hand gloves were used by 19% of the respondents. Overalls were used by 16% of the respondents, in many cases these overalls were torn and in bad condition to guarantee protection. Respiratory masks were used by only 12% of the respondents.

Up to 85% of the respondents indicated that the recommended pesticides were too costly. The cost of purchasing personnel protective gears and construction of prescribed storage and incineration facilities were equally prohibitive. Combined, these factors reduced the adoption of safety enhancing behaviour in pesticide use.

There was very little awareness of the long-term effects (chronic) of pesticides exposure. Victims sought medication for either acute poisoning cases or where chronic exposure led to serious illnesses that prevented the victims from going on with their day-to-day activities. Not going to seek medical attention was the misconception that the symptoms are inconsequential, last for a short time and are not serious. Milk was often used in treating exposure, while traditional herbs were used to treat later symptoms.

Conclusions

Although aware of the environmental and human safety hazards associated with pesticides use, the community and its environment remained vulnerable to pesticide hazards largely due to lack of adequate information about safe use, handling, appropriate doses, and handling procedures. Perception of the protective gear as expensive made farmers to generally avoid using them. In addition, messages that reached farmers from pesticide manufacturers were only intended to promote sales at the expense of potential environmental and human health risks associated with pesticides use.

A significant proportion of the farming population deliberately exceeded recommended pesticide

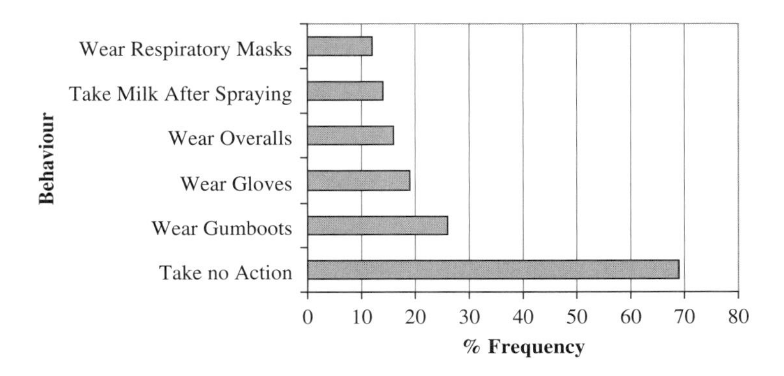

Fig. 3 Sagana farmers' personal protective practices

dosage in an attempt to increase their effectiveness. They failed to calibrate application equipments, lacked appropriate pesticide measuring containers to guarantee recommended dosages, sprayed during any weather and disposed surplus chemicals and empty containers anywhere. Hazards to human health included failure to use personal protective equipments, eating and smoking while handling pesticides, reuse of empty containers to store food products and mixing pesticides with bare hands. There was prolonged exposure as few used overalls, boots, respiratory masks and gloves while majority wore unwashed and contaminated clothing.

Recommendations

- Safe use and handling of pesticides can be achieved with increased level of safety awareness through on-farm training among the users. This awareness should address users' perceptions, attitudes and behaviour towards pesticides at all stages of pesticide life cycle.
- Principles of IPM, scouting and use of non-chemical measures to keep pest populations below damaging levels should be promoted so as to enhance the safety of resource poor farmers.
- The government should provide leadership in revising existing pesticide legislation with the special conditions of small-scale farmers in mind such as the need for smaller quantities affordable by farmers. This would reduce repacking at the farm level.
- Strategic extension services involving agricultural officers and environmental officers are needed to continuously influence farmers positively on pesticide use and handling.
- Adequate participatory monitoring systems should be put in place so that hazardous pesticides and handling practices are eliminated before a lot of damage to the environment or a large number of people are affected.

Acknowledgements Many thanks are extended to the Sagana community for their overwhelming support and cooperation during the study and to all family members, friends and colleagues who assisted in all ways towards the completion of this study.

References

Garcia AM (1998) Occupational exposure to pesticides and congenital malformations: a review of mechanisms, methods, and results. Am J Ind Med 33:232–240

Government of Kenya (2001) The 1999 population and housing census. Central Bureau of Statistics, Vol. 1 Jan 2001

Government of Kenya (2003) Nyeri District Development Plan 2002–2008, Effective Management for Sustainable Economic Growth and Poverty Reduction. Central bureau of statistics, Nairobi

ILO (1999) The ILO programme on occupational safety and health in agriculture. International Labour Office, Geneva

Jaffee S, van der Meer K, Henson S, de Haan C, Sewadeh M, Ignacio L, Lamb J, Lisazo MB (2005) Food-safety and agricultural health standards: challenges and opportunities for developing countries export. The World Bank, Washington, DC

Kanja L, Skaare JU, Nafstad I, Maitai CK, Lokken P (1988) Organochlorine pesticides in human milk from different areas of Kenya 1983–1985. J Toxicol Environ Health 19:449–464

Moses M, Johnson ES, Anger WK, Burse VW, Horstman SW, Jackson RJ, Lewis RG, Maddy KT, McConnell R, Meggs WJ et al. (1993) Environmental equity and pesticide exposure. J Environ Ind Health 9:913–959

National Environment Management Authority (2005) State of Environment Report 2004. NEMA Secretariat, Nairobi

Saleemi NA (1997) Statistics simplified. NA Saleemi Publishers, Nairobi

UK Pesticide Residue Committed (2005). Minutes of the meeting of the pesticide residues committee (PRC) on 2 February 2005. Pesticide safety directorate, London

UNEP (1985) Africa environment outlook, past, present and future perspectives. UNEP Secretariat, Nairobi

Wandiga SO, Yugi PO, Barasa MW, Jumba IO, Lalah JO (2002) The distribution of organochlorine pesticides in marine samples along the Indian Ocean Coast of Kenya. J Environ Technol 23(11):1235–1246

WHO (1989) Public health impact of pesticides used in agriculture, world health organization document 86.926. WHO, Geneva

Variation in Socio-economic Characteristics and Natural Resource Management in Communities with Different Potato Market Linkages in the Highlands of Southwestern Uganda

R. Muzira, B. Vanlauwe, S.M. Rwakaikara, T. Basamba, J. Chianu, and A. Farrow

Abstract Many countries in Africa have started implementing economic reforms that could lead to rapid growth and improved socio-economic conditions of growing populations. Uganda has been undergoing major economic reforms through stabilization, economic recovery, and structural adjustment programs. Adoption of market liberalization policies has favored growth of private sector in the country demanding increased production at farm level, which in turn favors marketability of farm produce in competitive manner. A study was conducted in *Kamuganguzi* sub-county to assess variation in socio-economic characteristics among households and level of natural resource management in potato production using two communities with and without farmer field schools. Farmers used livelihood indicators to place households in different wealth categories. It was observed that most female-headed households were in wealth categories III and IV. Most main houses were semi-permanent and farmers strived to use iron sheets to roof their houses due to scarcity of thatching material. It was in community without farmer field school that had grass thatched household in both wealth categories III and IV. Most potato fields had no soil conservation structures to combat land degradation in form of soil erosion. In both communities trenches and newly planted grass bunds were the most common soil conservation technologies. Beans and potato are the two main crops grown for household incomes. Potato and beans production take the biggest acreage in communities with and without farmer field school, respectively, and level of production and utilization depends on the market drives. It was observed that communities with farmer field school had higher potato yield disintegrated into different usages and applied more fertilizers on potato enterprise as compared to community without farmer field school. The demand of the market to be supplied with potato tuber of not more than 80 cm in girth and the higher price offered to farmers compared to open market is an incentive for farmers to invest in soil fertility management.

Keywords Economic reforms · Socio-economic · Natural resource management · Farmer field school

Introduction

Uganda has been undergoing major economic transformations since 1986 through economic stabilization, recovery, and structural adjustment programs toward economic growth (UNDP, 2000) and prosperity for all. This led to reduction of poverty levels from 56% in 1992 to 34% in 2000 (Benin, 2004). Cash crop farmers benefited most dramatically in the improved economy that reflected increase in cash crop prices due to extensive liberalization of export trade and removal of trade barriers (Abdalla and Egesa, 2005). Poverty levels in this group dropped from 60 to 44% between 1992 and 1996. However, poverty levels among the food crop farmers in the country dropped slightly in the same period from 64 to 62% (UBOS, 2005).

Although the level of poverty at household kept on reducing in the country, there were some sections

R. Muzira (✉)
National Agricultural Research Organization, Mbarara, Uganda
e-mail: nrmuzira@yahoo.com

A. Bationo et al. (eds.), *Innovations as Key to the Green Revolution in Africa*,
DOI 10.1007/978-90-481-2543-2_101,

of population still experiencing abject poverty (Benin, 2004), more so in rural communities (Muzira et al., 2004). To respond to the urgency of poverty reduction in households, the Government of Uganda developed a strategic plan known as Plan for Modernization of Agriculture (PMA) program, which aimed at addressing food security and sustainable agricultural development, trade, and investments (Abdalla and Egesa, 2005). This plan was designed to boost economic growth, restructure and revitalize Uganda's exports, curb environmental degradation, and enhance food security at the household level. Since most Ugandans are self-employed, mainly in agriculture, PMA became a central role in poverty eradication.

It was against this background that International Center for Tropical Agriculture (CIAT) in collaboration with Africare-Uganda trained farmer groups in agronomic skills of potato production in three parishes (*Buranga*, *Katenga*, and *Kicumbi*) of *Kamuganguzi* sub-county, Kabale district, using Farmer Field School (FFS) approach. Potato production was chosen as community enterprise from the list of alternative options put forward by farmers. The choice of potato production was obtained after conducting participatory market research (PMR), scrutinizing the market options, and subjecting alternative enterprise options to critical cost–benefit analysis. Farmer participatory research (FPR) was conducted in soil fertility management (SFM) and plant spacing to achieve the potato quality in terms of tuber girth required by the market, let alone optimizing yields. After understanding agronomic requirements in potato production to achieve the needed marketable tuber quality, participating communities started supplying 4.7 t of potato to urban fast food restaurant every fortnight under agreed contract which was renewable on annual basis. By the time this research was conducted farmers had supplied potato to the restaurant for 10 years.

Therefore, the study was conducted to investigate the differences in socio-economic characteristics and levels of natural resource management (NRM) in terms of SFM in potato production between FFS and non-FFS communities. Non-FFS community comprising of *Kasheregyenyi, Kyasano*, and *Mayengo* is in the same sub-county with the FFS community and both have similar biophysical characteristics. The non-FFS community was not involved in the training in potato production and management and had no link to definite market.

Materials and Methods

Study Area and Characteristics of Farming System

Kamuganguzi sub-county is located in Kabale district, southwestern Uganda. It is located at 1.35°S and 30.02°E with an altitude ranging between 1,791 and 2,000 m above sea level. It has a population of 50,312 (KDLG, 2002). The area experiences bimodal rainfall of 1,800 mm/year, with the first rains occurring in March to May, which are also considered as short rains, while the second rains which are long occur in September to January. A major dry season occurs during June to August while rainfall peaks are in April and November (Wortmand and Eledu, 1999).

The soils on hill slopes are ferralitic in nature having low pH and productivity while most valleys have Histosols with thick topsoil that contains high organic matter (Wortmand and Eledu, 1999). Soils in the valleys have medium to high productivity due to abundant organic matter. In general, however, most soils on hilltops and slopes are deficient in N and P due to over-cultivation and erosion (Muzira et al., 2004). The potential of erosion is high due to long and steep slopes, which are intensively and extensively cultivated exposing the soil to erosion at the beginning of the rain season. The topography is very undulating with most slopes ranging between 34 and 75%. Most of the land is farmed and is interspersed with woodlots on individual plots (Wortmand and Eledu, 1999).

The farming systems are biologically and agronomically diverse with numerous small plots of land intercropped at varying planting seasons among other characteristics. The average number of plots per household varies between 6 and 10, which are intensively cultivated. The rich middle-class farmers possess many plots and can afford to fallow some of the plots to allow natural soil fertility regeneration. The plots are highly fragmented and distributed on different hill slopes and parishes (Lindblade et al., 1996). The proximity of plots to farmers' homestead varies and thus some farmers stay far leading to longer time taken to reach the plots before starting to work. The main food crops grown are beans, sweet potato, sorghum, and field peas while potato and *Artemisia* are the main cash crops. Livestock kept by the community is comprised of mainly cattle, goats, and sheep, grazed on free range that includes natural fallow or crop residues (Mbabazi et al., 2003).

Site Selection

Three parishes (*Buranga, Katenga*, and *Kicumbi*) had farmers organized in groups, trained in potato production and natural resource management using FFS approach. The three parishes had definite market chosen from the list of possible options. For comparison purpose three parishes (*Kasheregyenyi, Kyasano*, and *Mayengo*) in the same sub-county with similar biophysical characteristics were selected as control (Fig. 1). Biophysical characteristics considered included similarity in landscape, soil physical properties, and nature of vegetation. The soils are sandy loam, which are well drained with some patches of sandy clay on hill slopes and valleys. These communities grow potato as cash crop as well as food crop.

Farm Selection and Characterization

Approximately 10 key informants that included community leaders having wide knowledge about well-being of fellow farmers in each parish were invited in the focus group discussions (FGDs) to categorize households by wealth. Using identified indicators of wealth possession, participants categorized households accordingly. The indicators included types of main houses, number of plots, management practices, and labor use. Rapid rural appraisal survey was conducted among 97 households in the study area using semi-structured questionnaire. A plan was drawn showing the location of each plot in the community to enhance identity of homestead. The interview targeted both male- and female-headed households though in male-headed households, women were also interviewed since they spend most of their time in the field and could understand the history of each plot very well.

Statistical Analysis

The information derived from the semi-structured interviews was expressed as average values for socio-economic indicators (e.g., mean number of plots under cultivation in different wealth categories) and percentage distribution (distribution of plots under potato in different parishes or communities).

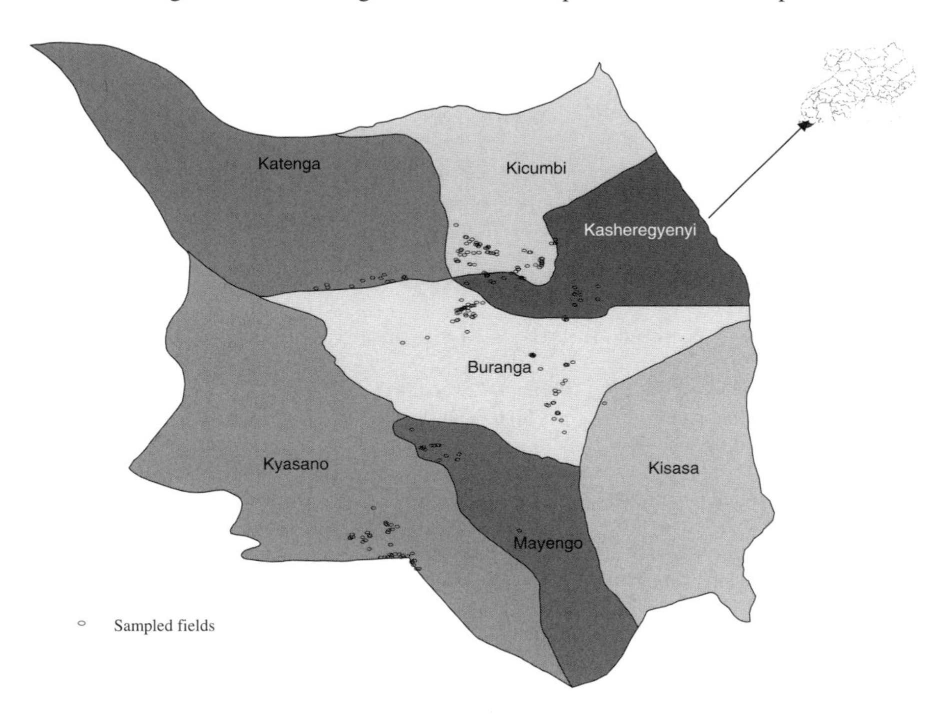

Fig. 1 Distribution of sampled potato fields in Kamuganguzi sub-county

The means and percentage distributions were calculated using survey analysis of group summaries in Genstat Release 6.

Results and Discussion

Description of Households in Different Wealth Classes

Socio-economic factors differentiating households in *Kamuganguzi* sub-county were explored during FGDs based on the well-being of the household members and resources in their possession (Table 1). Most households in wealth categories (WCs) III and IV had neither permanent nor semi-permanent main houses. Households in both categories with permanent or semi-permanent main houses are those that were doing well some years back but family members who were supportive died leading to deterioration of conditions of well-being among household members. Children of school-going age belonging to households in WC IV were not attending any form of primary education including universal primary education (UPE) due to lack of basic scholastic requirements such as books, pens, and pencils. In FGDs, children from such households were said to show protein deficiency symptoms such as brown hairs and protruded abdomen. This category of households has few plots (1–4) situated near homesteads. Plots that are far from homesteads are often located on very steep and stony slopes and are often low in productivity. Such households rarely keep any livestock and therefore hardly use any farmyard manure (FYM) as soil inputs. This category of households does not have means of transport such as bicycles and walk long distances on foot. These households are known to have no woodlots as source of firewood. Their main source of fuel is crop residues such as sorghum stover which is normally available after harvest. They are known to have one meal a day in most cases after exchange for labor and rarely have money in excess of US $2 at any one time.

Households in WCs III and IV are not actively involved in commercial potato production due to high costs involved that include using new and clean seed,

Table 1 Farmers' own criteria of categorizing different households

	Class I	Class II	Class III	Class IV
Nature of main house	Good and clean permanent	Good but semi-permanent	Grass thatched or roofed with old iron sheets	Small and poor grass thatched
Means of transport	Has a vehicle or bicycle	Has a bicycle	Most of the time uses foot	Most of the time uses foot
Possession of money	Has US $1,571–2,619 on bank account	Has US $104–523 on bank account	Has US $10–26 in the house	Has US $0–2 in the pockets
No. of plots	Has 15–100 plots	Has 10–14 plots	Has 2–4 plots	Has one or few plots around homestead
No. of heads of cattle	Has 4–10 improved cows	Has 1–3 local cows	None	None
No. of small ruminants	Has 10–20 goats and sheep	Has 2–10 goats and sheep	Has one goat or sheep	None
No. of woodlots	Has 3–5 woodlots	Has 1–2 woodlots	None	None
Nature of schools attended by children	All children are in boarding schools	Children attend good day schools	Universal primary or secondary schools	None due to need of scholastic materials
No. of workers	Uses hired labor 30–50 times a year	Uses hired labor 5–20 times a year	Does not use hired labor	Does not use hired labor
No. of meals taken per day	Four meals (breakfast, lunch, evening tea, and supper)	Two meals (lunch and supper)	One meal in most cases (one late lunch)	One meal and gets it for exchange for labor
Use of soil amendments	Use FYM and fertilizers	Use FYM and rarely fertilizers	Use little FYM and not fertilizers	Do not use FYM and fertilizers
No. of potato bags produced per season	100–500 from all plots cultivated	10–50 from all plots cultivated	2–5 from all plots cultivated	One if at all from all plots cultivated

purchase of fertilizers, and periodical application of pesticides to control pests and diseases.

Households in WC I are known to possess larger amounts of money (above US $1,000) on bank account and have more animals compared to other households. They can afford several meals a day and often incorporate sources of proteins in their diet. Just like households in WC II, most households in WC I possess at least a bicycle, which is used as a means of transport within and outside the community. Households in WC I can afford to open several plots of their own in addition to using hired labor. They normally use fertilizers in potato fields that are located far from the homesteads because organic inputs are bulky to carry which may increase production costs. Nearby plots are normally applied with FYM to improve potato yields. At the end of the growing season these households normally have 100–500 bags of 100 kg units to take to market.

Distribution of Households

Distribution of households showed that female-headed households ranged from 16.7% in *Kyasano* to 39.8% in *Buranga* while male-headed households ranged from 60.2% in *Buranga* to 83.3% in *Kyasano* (Table 2). Female-headed households were those of widows and wives whose husbands moved to distant districts in search of jobs and take long time without checking on their families. Husbands who are always away provide labor to tea plantations or casual work in neighboring districts or Rwanda making wives play the leading role in the homesteads. Some husbands are in other non-agricultural activities such as teaching, brick making, or car driving outside the communities. In *Mayengo* parish, high proportion of households is headed by males. This parish is situated at *Katuna* border with Rwanda where men are engaged in cross-border business and therefore commute from their homes.

Grass thatched houses were mostly common with households in WC III (37.5%) and IV (62.5%) in non-FFS community comprising *Kyasano*, *Mayengo*, and *Kasheregyenyi* parishes (Table 3). In FFS community, grass thatched houses were only found among households in WC IV. The presence of few grass thatched houses in FFS community was attributed to improved household incomes from potato sales and due to scarcity of thatching material like sorghum stover. Sorghum is no longer a priority crop for most households due to its low market value compared to potato (Mbabazi et al., 2003) in FFS community. Lack of thatching material has led farmers to purchase iron sheets for their houses. The presence of permanent houses in WCs II and III in non-FFS communitywas attributed to the support households used to get from relatives before they died of natural causes or during the war of Rwanda. Similarly, households in WCs III and IV with semi-permanent houses are mainly for widows or wives who were abandoned by their husbands resulting in deterioration of welfare of the households.

Acreage Distribution of Potato with Other Crops in FFS and Non-FFS Communities

Crops most common with farmers include potato, beans, field peas, and maize. These are grown in crop rotation with potato in non-systematic sequence whereby potato and beans become most common in FFS and non-FFS communities. Potato was developed as cash crop for the whole sub-county and its prevalence in the fields depends on market forces and NGOs promoting it as a source of income for farmers. On

Table 2 Distribution of households in different parishes under study

		Households		
Parish	Village	Number	Female headed (%)	Male headed (%)
Buranga	Rushebeya	83	39.8	60.2
Katenga	Nyabyumba	88	21.6	78.4
Kicumbi	Rushongati	71	31.0	69.0
Kasheregyenyi	Nyangoye	89	29.2	70.8
Kyasano	Isingiro	90	16.7	83.3
Mayengo	Rwakizamba	109	21.1	78.9

Table 3 Distribution of households in different wealth categories in FFS and non-FFS communities

	Number of household types in FFS community (n=46)			Number of household types in non-FFS community (n=51)		
Wealth class	Permanent	Semi-permanent	Grass thatched	Permanent	Semi-permanent	Grass thatched
I	71.4	5.7	0.0	100	17.0	0.0
II	14.3	22.9	0.0	0.0	36.6	0.0
III	14.3	37.1	0.0	0.0	26.9	37.5
IV	0	34.3	100	0.0	19.5	62.5
	(n=7)	(n=35)	(n=4)	(n=2)	(n=41)	(n=8)

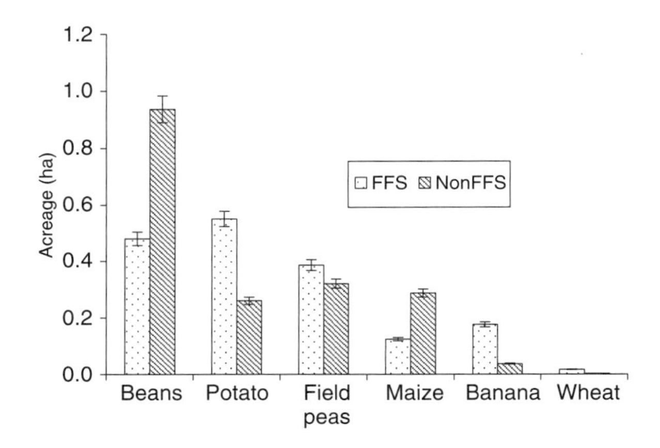

Fig. 2 Average land sizes under different crops in FFS and non-FFS communities

the other hand, beans are produced for both food security and household income. The two crops including field peas and other array of crops produced other than cereals such as maize and wheat are alternative host for bacterial wilt and is one of the major reasons for its high prevalence in both communities. In FFS community, acreage per household under potato production takes the lead of 0.55 ha followed by beans with average size of 0.48 ha (Fig. 2). On the other hand in non-FFS community, bean production takes the lead with average land size of 0.94 ha per household followed by field pea and maize average acreage of 0.32 and 0.28 ha per household. Acreage of potato in non-FFS community is 0.26 ha per household.

In non-FFS community, bean production is popular due to intervention by the Kabale District Farmers' Association (KADFA) to market beans on behalf of farmers. Bean producers under KADFA are normally given essential inputs such as seed, fertilizers, and pesticides on subsidized loans. They are also entitled for loans from KADFA when bean prices are low at the time of harvest so that the beans are sold when good prices prevail. Such friendly environment has made farmers in non-FFS community put most of the land under bean production. Potato is more delicate compared to beans in terms of disease attack. Potato crop is frequently attacked by diseases such as bacterial wilt and late blights which most farmers cannot endure if the crop is planted on a larger area. However, the availability of market for potato at steady price of US $25 per bag of 100 kg units throughout the period of contract has enabled farmers in non-FFS community put most cultivable land under potato production.

Variation in SFM in Potato Fields in FFS and Non-FFS Communities

Fields in both FFS and non-FFS communities are on continuous cultivation without fallow to break down disease and pest cycles and rejuvenate natural soil fertility (Fig. 3). High pressure on the land due to high population (361 people km^{-2}) has led to the change of crop rotation sequence with bush fallow (Lindblade et al., 1996; Grisley and Mwesigwa, 1994; Breyer et al., 1997). In the study, it was found that the period of continuous cultivation without fallow ranged from 12 to 34 years with households in WC IV in FFS and non-FFS communities, respectively. Fields

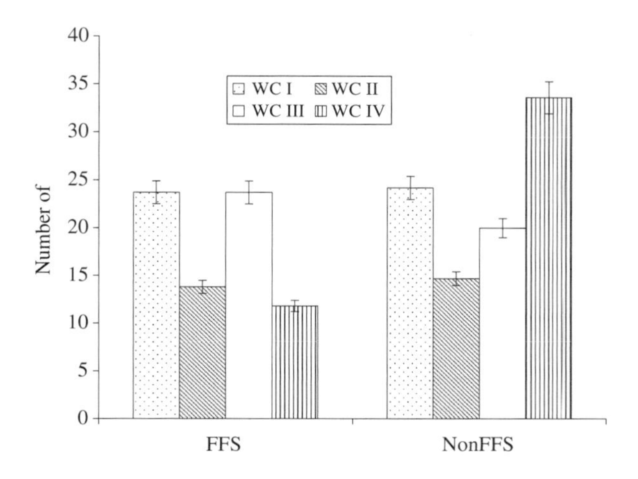

Fig. 3 Variation of the number of years plots under potato have been subjected to continuous cultivation without fallow

which normally rest for long (>5 years) under fallow and look abandoned are located on very steep slopes. Nonetheless, such fields have shallow soils and low productivity (Raussen et al., 2002) and farmers prefer using them for woodlots or grazing. Problems of land degradation in Kabale are normally blamed mostly on cultivation on the very steep slopes covering most of the land surface, accompanied with scant measures to prevent soil erosion (Bamwerinde et al., 2006).

This was evident in both FFS and non-FFS communities where grass bunds that were constructed during colonial periods were destroyed. Proportion of land without barrier is high with 28.2 and 27.4% in FFS and non-FFS communities, respectively, and is a cross-cutting issue in both communities (Fig. 4). Farmers destroy grass bunds when searching for fertile and deep soil that accumulate over time due to erosion from upper terrace positions. Destroyed grass bunds are normally replaced with trenches and new grass bunds. However, at times farmers will leave plots unprotected and this has resulted in soil erosion and fertility decline on hill slopes. The proportion (20.9%) of newly constructed grass bunds is lower in the FFS community and this is attributed to the urge farmers have to maximize the whole field for potato production without constructing soil conservation structures. Trash line as soil conservation technology is normally done after sorghum harvest. Nonetheless, the technology has competition with need for fuel, animal feed, and mulch in vegetable production. Some farmers have planted trees such as grevillea and eucalyptus scattered in the field to hold soil and also improve its fertility through drop-litter. Hedgerow and dugout ditches are not common in both communities due to lack

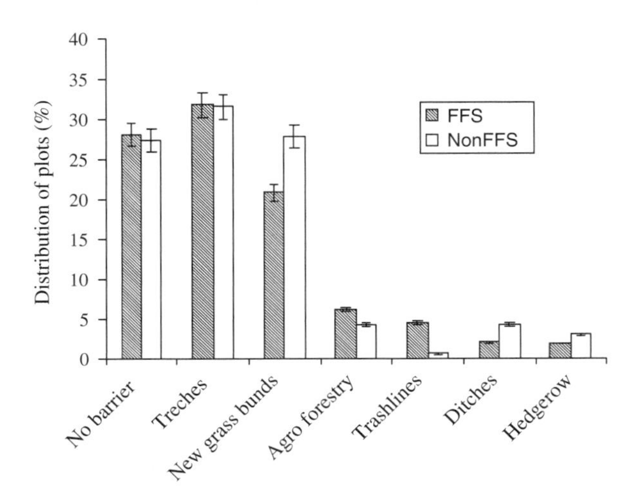

Fig. 4 Distribution of plots with and without barrier systems to reduce soil erosion and surface runoff

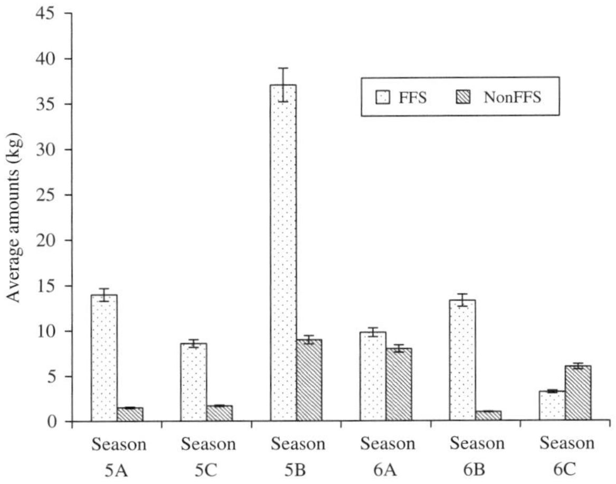

Fig. 5 Variation in quantity of fertilizer NPK used by FFS and non-FFS communities in different seasons

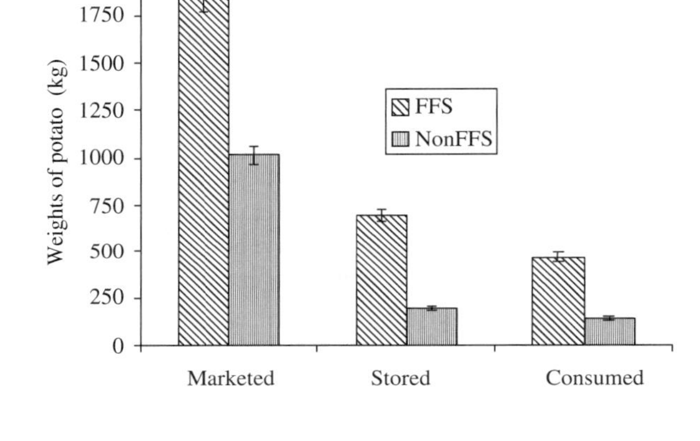

Fig. 6 Average amounts of harvested potato per household put under different uses in season 2006 B

of planting materials such as calliandra and localized effect of dugout ditches on surface runoff and soil erosion. Trenches that spread across the fields along the contours are preferred to dugout ditches as they are more effective in controlling surface runoff and soil erosion.

Although the level of using NPK fertilizer is not consistent, it is generally high in FFS community (Fig. 5). This is mainly attributed to the demand of large tuber size needed in the market. Lack of consistency in the level of the amounts of fertilizer used is mainly attributed to changing prices that are always high (US $77–88 for a 50 kg bag of NPK) and its availability in retail shops. When prices shoot up higher compared to the returns expected from potato sales, farmers will reduce the level of fertilizer application.

There is a close link between increased land productivity and fertilizer application in most parts of Africa (Vanlauwe and Giller, 2006). Potato yields per household disaggregated into the different uses put to by farmers are lower in non-FFS community and it is mainly attributed to the low level of fertilizer applied (Fig. 6). Though market drive may facilitate farmers to use improved technology to combat land degradation and improve potato yields, NRM practices may be hampered by scarcity and increased prices of inputs (Nkonya, 2003). Farmers are willing to invest in natural resource management (NRM) when there is positive return.

Conclusions

Farmers are very knowledgeable of their communities and local factors that could be used to categorize households in different WCs. Female-headed households are mainly for widows or females who were abandoned by their husbands. Such households are mostly in WCs III and IV. Most of the main houses are semi-permanent and belong to households in WCs I–III. The level of crop production for household income depends on several factors. Community that went through FFS and have linkages to profitable market put most of the land under potato production. On the other hand, in non-FFS community, though growing potato for household income, most of the land is under bean production due to intervention by KADFA with favorable policies of supplying inputs at subsidized loans and marketing beans on behalf of farmers on gainful profits. In both communities land is under pressure with continuous cultivation without fallow and most of the pieces of land have had soil conservation structures destroyed in search of fertile soil and also the need for expanding the area of cultivation. However, construction of trenches and planting new grass bunds is becoming common in both communities as a way of controlling soil erosion and surface runoff. The setback is that trenches are not always protected with grasses or shrubs and therefore this requires farmers to keep on de-silting. However, the distribution of plots with newly planted grass bunds is lower in FFS community because most farmers cannot afford to leave a stretch of 0.5 m to be planted with grasses or shrubs. Most of the soil in the plot is put to use for potato production. The level of fertilizer used for potato production is higher in FFS community compared to non-FFS community. This is due to the market demands of high-quality tubers in terms of girth. In non-FFS community, farmers use fertilizers on beans and this is because of the intervention by KADFA to improve household income through marketing beans. Fertilizers and other inputs are given to farmers on subsidized loans. Therefore, policies such as favorable policies and availability of profitable markets for farmers produce are incentives to fertilizer use and have an impact on the level of production.

Acknowledgments The authors would like to acknowledge the Belgium Technical Cooperation (BTC), International Center for Tropical Agriculture, and Tropical Soil Biology and Fertility Institute of the International Centre for Tropical Agriculture (TSBF-CIAT) for funding the study. Sincere thanks go to farmers in both FFS and non-FFS communities for their invaluable time accorded to the authors during after this study. Lastly, the authors would like to extend their gratitude to the enumerators Charity, Edmond, Damalie, Grace, Patience, and Dustan for having persevered the terrain while looking for farmers in their plots that were scattered all over the different hills.

References

Abdalla YA, Egesa KA (2005) Trade and growth in agriculture: a case study of Uganda's export potential within the evolving multilateral trading regime. Bank of Uganda working paper, p 44

Bamwerinde W, Bashaasha B, Ssembajjwe W, Place F (2006) Determinants of land use in the densely populated Kigezi highlands of southwestern Uganda. Contributed paper prepared for presentation at the International Association of Agricultural Economists conference, Gold Coast, Australia, 12–18 Aug 2006

Benin S (2004) Enabling policies and linking producers to markets. Uganda J Agric Sci 9:863–878

Breyer J, Larsen D, Acen J (1997) Land use cover change in South West Uganda. African Highland Initiative, ICRAF, Nairobi

Grisley W, Mwesigwa D (1994) Socio-economic determinants of seasonal cropland fallowing decisions: Smallholder in South-western Uganda. J Environ Manage (42):81–89

Kabale District Local Government (KDLG) (2002) Population and housing census, District Planning Unit, Kabale

Lindblade K, Tumahairwe JK, Carswell G, Nkwiine C and Bwamiki D (1996) More People, More Fallow – Environmentally favorable land-use changes in southwestern Uganda. (Report prepared for the Rockefeller Foundation and CARE International, 1996). In: Lindblade K, Carswell G and Tumahairwe JK (eds) (1998) Mitigating the relationship between population growth and land degradation: Land-use change and farm management in southwestern Uganda. *Ambio* 27(7):565–571

Mbabazi P, Bagyenda R, Muzira R (2003) Participatory land degradation assessment in the highlands of Kabale district, southwestern Uganda. Report submitted to African highlands initiative (AHI)

Muzira R, Sanginga P, Delve R, Kabale farmers' groups (2004) Farmers' participation in soil fertility management research process: opportunities for rehabilitating degraded hilltops in Kabale. In: Advances in integrated soil fertility management in sub-Saharan Africa; challenges and opportunities

Nkonya E (2003) Soil conservation practices and non-agricultural land use in the southwestern highlands of Uganda: a contribution to the strategic criteria for rural investments in productivity (SCRIP) program of the USAID Uganda mission. The International Food Policy Research Institute (IFPRI), Washington, DC, 31p

Raussen T, Place F, Bamwerinde W, Alacho F (2002) Report on a survey to identify suitable agricultural and natural resource–based technologies for intensification in South-Western Uganda. A contribution to the strategic criteria for rural investment in productivity (SCRIP) policy framework of the USAID Uganda mission. International Food Policy Institute, Kampala

Uganda Bureau of Statistics (UBOS) (2005) Precise maps of Ugandan poverty. A report on poorest people at the county level, pp 1–4
United Nations Development Programme (UNDP) Poverty Report (2000) Uganda country assessment from decentralization to participation
Vanlauwe B, Giller KE (2006) Popular myths around soil fertility management in sub-Saharan Africa. Agric Ecosyst Environ 116:34–46
Wortmand CS, Eledu CA (1999) Uganda's agro ecological zones: a guide for planners and policy makers. Centro Internacional de Agricultura Tropical, Kampala

Crop Rotation of Leguminous Crops as a Soil Fertility Strategy in Pearl Millet Production Systems

L.N. Horn and T.E. Alweendo

Abstract Crop rotation represents one strategy of the crop and soil fertility management practices being implemented in Namibia to improve soil fertility and crop yield per area of land cultivated. Experiments are being conducted at four research stations in the northern Namibia communal area where Pearl millet is predominantly grown. During the 2005/2006 cropping season, different crops used in rotation with Pearl millet became well established and 3.82 t/ha of Lablab (*Lablab purpureus*) was recorded at Bagani Research station with 0 t/ha recorded at Okashana and Omahenene. Different rainfall figures were recorded per location. Bagani recorded the highest average rainfall of 890 mm in 2005/2006 and 540 mm in 2006/2007, while Okashana recorded the lowest with 63 mm during the same year. When all the plots were treated with Pearl millet, the yield of Pearl millet dropped slightly from 2.70 to 2.18 t/ha at Omahenene in the plots where Pearl millet was planted previously and 2.63 t/ha was recorded in the plot where Lablab was previously planted. At Bagani a higher yield of Pearl millet was recorded in plots where Lablab was planted and a lower yield where Sorghum was planted with 2.49 t/ha and 1.56/ha, respectively. A higher yield of Pearl millet at Mannheim was recorded in plots where nothing was planted (fallow) with 1.63 t/ha and a lower yield was recorded where Pearl millet was grown previously with 1.40 t/ha. Soil samples taken for fertility analysis revealed that the soils at all four research stations lack major elements such as P and K. This project is expected to end during the 2009/2010 growing season.

Keywords Soil fertility · Pearl millet · Leguminous · Crop rotation · Crop yield

Introduction

Pearl millet (*Pennisetum glaucum*) is the most important staple food for most people living in northern Namibia with *Sorghum bicolor* the second most important (Labetoulle, 2000). This is true as Irving et al. (1993) showed that Pearl millet is the principal crop in this area and represents up to 90% of the cultivated land in northern Namibia. Even though most farmers in the north central regions also cultivate Sorghum, the cultivated areas for Sorghum are less important than those reserved for Pearl millet yet larger than for maize. Despite the cultural practices and effort made by farmers to produce Pearl millet and Sorghum in northern Namibia, average yields are still low with a national average of 100–360 kg/ha of the improved cultivars (MAWRD, 1994). In addition to Pearl millet, Sorghum and Maize, farmers grow Cowpea (*Vigna unguiculata*) and Bambara groundnut (*Vigna subterranean*) (Mendelsohn et al., 2000). These crops are usually grown in isolation on separate portions of land; however Cowpea is mostly intercropped with both Pearl millet and Sorghum. Sorghum and Bambara groundnut are usually not cultivated in rotation with other crops. They are grown in isolation on different piece of land. Such practices coupled with the increasing human population have placed

L.N. Horn (✉)
Division of Plant Production Research, Ministry of Agriculture, Water and Forestry, Government Office Park, Luther Str. Windhoek, Namibia
e-mail: lnhorn@yahoo.com

A. Bationo et al. (eds.), *Innovations as Key to the Green Revolution in Africa*,
DOI 10.1007/978-90-481-2543-2_102, © Springer Science+Business Media B.V. 2011

immense pressure on the cultivated land, and resulted in depletion of soil fertility levels (Nyagumbo, 1993; Keyler, 1995). The practice of annual tilling of the land is also one of the factors contributing to soil depletion hence fertility loss (Mendelsohn et al., 2000). The Ministry of Agriculture, Water and Forestry through its efforts in crop improvement and diversification have released new cultivars of Pearl millet, Sorghum and Cowpea. The released cultivars are early maturing, high yield due to their drought-escaping mechanisms. The benefits of improved cultivars could only be realized if combined with other soil fertility enhancement practices that not only improve the soil structure but also retain moisture and other needed soil nutrients. In addition, other soil management practices such as the use of organic manure and crop residue incorporation into the soil must be encouraged. It has been recognized that experience elsewhere indicates that when issues of crop rotation and crop protection are addressed, crop yield increases (Matanyaire, 1995).

On the contrary, increased shortage of agricultural land and the demand for higher agricultural production systems to improve household food security and income, pose a challenge to traditional soil fertility management strategies such as shifting cultivation and fallow. As the population increases, pressure on the land and its resources has also increased, causing rapid decline in soil fertility and crop yield. This problem is not unique to Namibia but has also been observed in other SADC countries (Bationo et al., 1992). Despite the fact that they do not bring any economic benefits and any loss is largely attributed to poor soil fertility management strategies farmers continue to carry out their traditional farming activities (Matanyaire, 1995). Aware of the problems facing farmers, this project was developed to (1) determine the effects of crop rotation on Pearl millet yield and other yield components, (2) identify appropriate crops that are beneficial for soil fertility improvement and maintenance, and (3) determine the effects of legume crops and crop residues on Pearl millet yield.

Materials and Method

The experiments involve a completely randomized design with four replications and are being carried out concurrently at each site. The experiments are being conducted at four different locations in different political regions in Namibia namely: Okashana Research and Training Centre, Mannheim Research Station (Oshikoto Region), Omahenene Research Station (Omusati Region) and Bagani Research Station (Kavango Region). This is a dry-land experiment and depends on rainfall. The project started during the 2004/2005 cropping season and is expected to run until the 2009/2010 growing season. The trials are being located on poor sandy soil selected for this purpose and soil samples were taken before planting. During the first growing season 2004/2005, all the plots were planted with Pearl millet and on the alternating year 2005/2006 different crops including legumes and fallow (Cowpea, Lablab, Unplanted and Sorghum) were planted in different plots. For the purpose of this experiment, no chemical fertilizers are used; however, crops residues are being incorporated into the soil. Soil samples are collected before planting each season for analysis of elements like P, K, Na, pH, organic matter and other variables at depths of 0–30 and 30–60 cm; however, the results presented in this report are those from the 0–30-cm depth interval only as listed in Tables 1, 2, 3, and 4.

Table 1 Soil physical and chemical properties after the second harvest for Bagani Research Station

	P	K	Ca	Mg	Na	pH-H_2O	ECw (μS/cm)	OM%	$CaCO_3$		Sand	Clay	Silt
Treatment	ppm									Texture	%		
Fallow	6.97	75.5	216	24	31	6.51	33.5	0.32	None	Sand loamy	66.15	5.85	28
Pearl millet	5.63	38	1175	81	41.5	7.13	75	0.375	None	Loamy sand	80.85	11.25	7.9
Lablab	4.93	43	136	48	48	7.81	29	0.32	None	Loamy sand	82.1	5	12.9
Cowpea	6.37	65.5	406	28	38	7.535	37.5	0.64	None	Sand	87.15	7.45	5.45
Sorghum	14.11	85	380	52	43	7.39	38	0.29	None	Sand	87.2	6.2	6.6

Table 2 Soil physical and chemical properties after the second harvest for Omahenene Research Station

Treatment	P	K	Ca	Mg	Na	pHw	ECw (μS/cm)	OM %	$CaCO_3$	Texture	Sand	Clay	Silt
	ppm										%		
Fallow	37.18	79	578	92	15	6.81	39	0.54	None	Sand Loamy	74.9	11.8	13.35
Pearl millet	22.65	81	575	85	86	6.76	33.5	0.32	None	Loamy sand	79.4	12.45	8.1
Lablab	15.4	48	557	63	18.5	6.66	41.5	0.47	None	Sandy loam	85.9	10.9	3.3
Cowpea	21.6	70	494	316	28	6.94	26.5	0.55	None	Loamy sand	87.15	8.05	4.8
Sorghum	19.5	82	798	88.5	13.5	6.72	37.5	1.23	None	Loamy sand	84.45	10.7	4.55

Table 3 Soil physical and chemical properties after the second harvest for Okashana Research Station

Treatment	P	K	Ca	Mg	Na	pHw	ECw (μS/cm)	OM%	$CaCO_3$	Texture	Sand	Clay	Silt
	ppm										%		
Fallow	2.64	60	270	110	9	7.55	25	0.25	None	Sand	92.6	4.3	3
Pearl millet	3.87	58	226	92	8	6.49	30	0.17	None	Sand	94.7	4.5	0.8
Lablab	3.32	52	290	106	8	6.43	30	0	None	Sand	94.4	4.2	1.4
Cowpea	3.53	64	244	100	9	6.24	25	0.23	None	Sand	94.6	5	0.4
Sorghum	4.34	78	272	108	9	6.66	25	0.21	None	Sand	92.9	5	2

Table 4 Soil physical and chemical properties after the second harvest when for Mannheim Research Station

Soil ID	P	K	Ca	Mg	Na	pHw	ECw (μS/cm)	OM %	$CaCO_3$	Texture	Sand	Clay	Silt
	ppm										%		
Fallow	3.32	54	5124	500	32	7.24	110	1.02	None	Clay loam	45.6	31.3	23.1
Pearl millet	5.15	132	4688	504	13	7.13	125	1.13	None	Sandy, clay loam	48.7	29.6	21.6
Lablab	7.9	108	5016	564	15	7.31	150	0.16	None	Sandy, clay loam	51.9	28.4	19.7
Sorghum	6.17	84	5480	636	25	7.17	125	1.47	None	Sandy, clay loam	54	28	18
Cowpea	8.41	112	5268	588	18	7.09	145	1.27	None	Sandy, clay loam	50.6	27.5	21.8

Data Collection

Data such as days to bloom, head weight and grain yield were recorded and entered into Microsoft Excel. A Sigmastat 2.0 statistical package was used to analyze the data. The bar graphs presented were drawn using Microsoft Excel however, in this report only grain yield, rain fall and soil sample data are reported.

Results and Discussion

During the 2005/2006 cropping season, crop establishment and development was good due to sufficient rainfall with the highest recorded at Bagani (100–890 mm) and the lowest at Okashana (between 18 and 169 mm). Lablab did not produce any yield at Mannheim and Omahenene because planting was carried out late, however the plant produced much litter and remained green during winter, 3.8 2 t/ha was recorded at Bagani Research station. Pearl millet and Sorghum produced 2.67 and 3 t/ha respectively at Omahenene. The Cowpea yield was 1.10 t/ha at Omahenene and low at Bagani with 0.08 t/ha recorded. These data are presented in Fig. 1. Variations in yield could be attributed to differences in rainfall, soil types as well as planting times at different locations.

During the 2006/2007 cropping season, all plots were treated with Pearl millet following the previous alternating crops as indicated in Fig. 2. The yield of Pearl millet dropped slightly from 2.70 to 2.18 t/ha at Omahenene in the plots where Pearl millet was planted previously and rose to 2.63 t/ha recorded in the plot where Lablab was previously planted. No difference

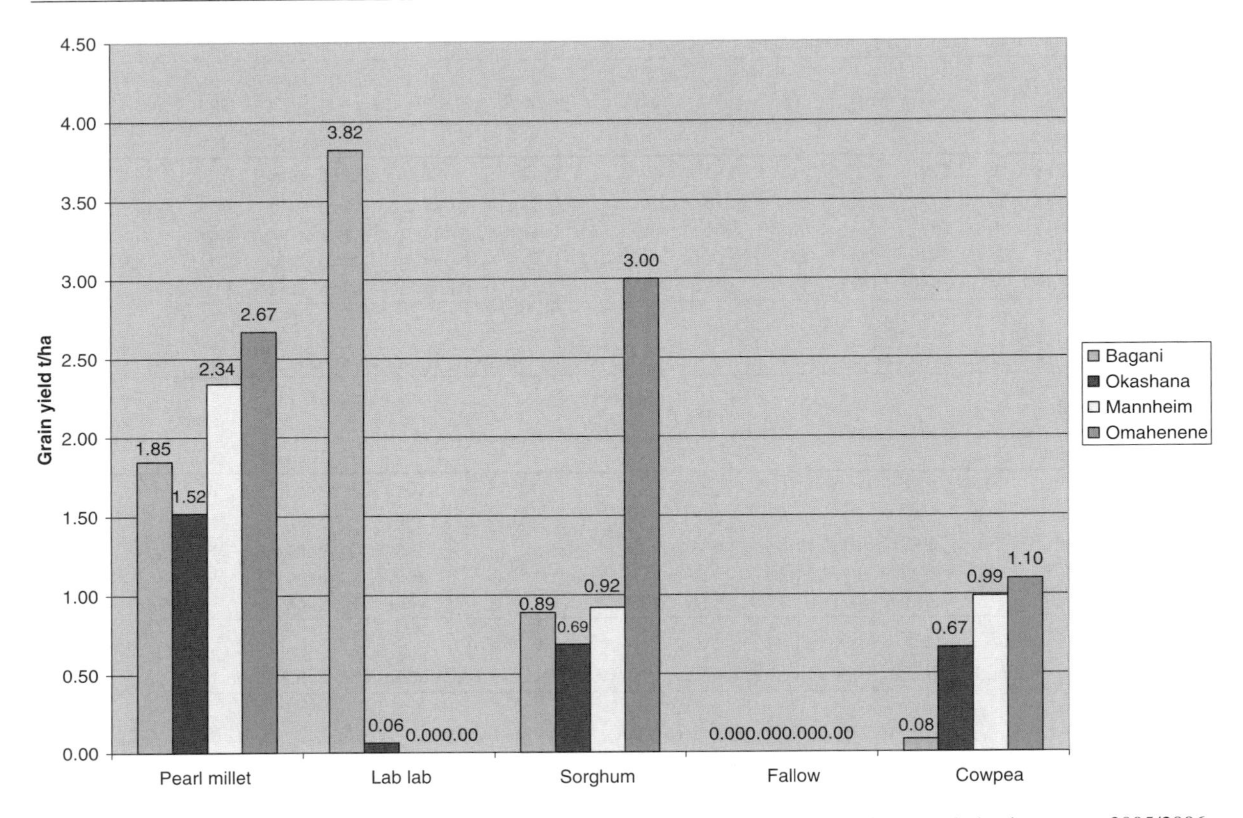

Fig. 1 Grain yield of different crops used in this experiment to improve soil fertility during the second planting season 2005/2006

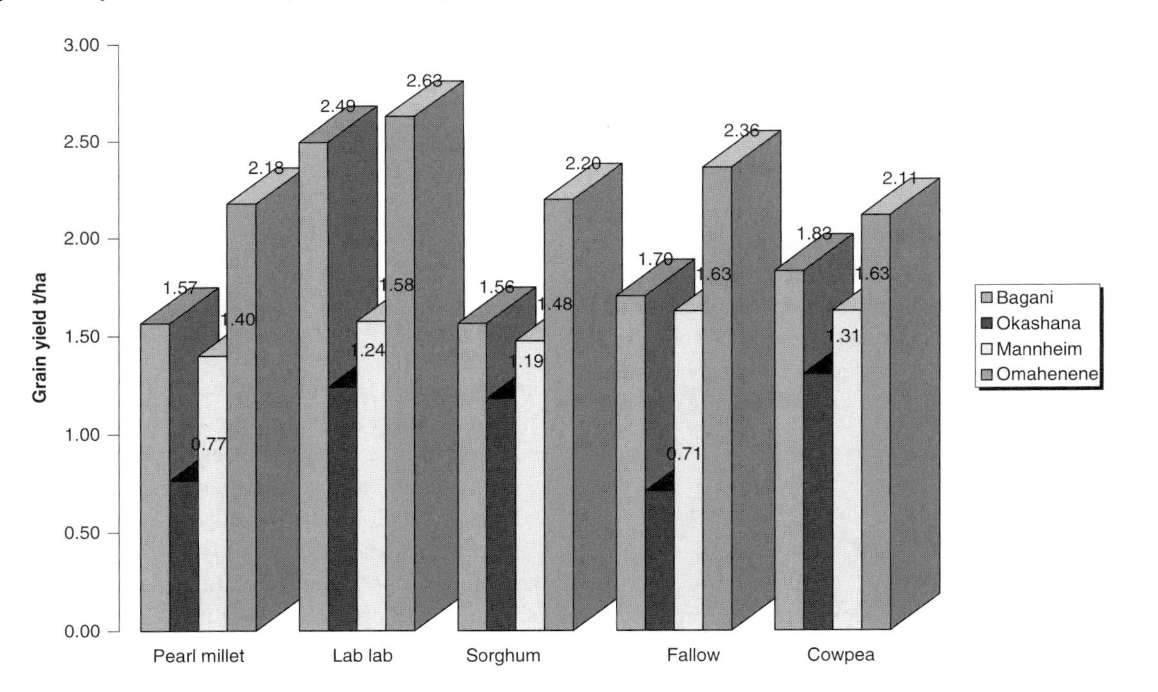

Fig. 2 Yield of Pearl millet in t/ha during 2006/2007 harvested from the plots where different leguminous crops where planted during the third 2005/2006 cropping season

could be detected with the other crops in rotation. The lowest yield was recorded at Okashana in plots under fallow with 0.71 t/ha; this could be as a result of the low rainfall recorded during the 2006/2007 growing season, ranging between 17 and 63 mm, and the very poor soil quality at that location. At Bagani a higher yield of Pearl millet was recorded in plots where Lablab was planted previously and a lower yield where Sorghum was planted with 2.49 and 1.56 t/ha, respectively. A higher yield of Pearl millet at Mannheim was recorded in plots where nothing was planted previously (fallow) with 1.63 t/ha and a lower yield where Pearl millet was grown previously with 1.40 t/ha (see Fig. 2).

Effects of Crop Rotation on Soil Physical and Chemical Properties

During the second harvest, when all plots were treated with Pearl millet, the soil was sampled and analyzed for organic matter content as well as other physical and chemical properties indicated in Tables 1, 2, 3, and 4. Samples were collected from the 0–30-cm depth range in all plots. The results obtained from each research station are shown in Tables 1, 2, 3, and 4. It was noted that the soils at some locations lack some major elements such as P and K. Nitrogen % was not measured due to a lack of analysis tools. For Bagani recorded P was between 3 and 14.11 ppm. This is regarded as low because according to Mylavarapu et al. (2007), the recommended P level in agronomic soil is low when it is <15 and high when it is between 31–60 ppm. At Omahenene the recorded P levels began just above the lower level limit, ranging from 15.37 ppm in the Lablab plot to 37.18 ppm in the fallow plot, as shown in Tables 1 and 2, respectively. Low P levels were also observed at Okashana and Mannheim, where the P level noted at Okashana was 2.64 ppm in fallow and 4.33 ppm in Sorghum plots while at Mannheim the P level was 3.32 ppm in fallow and 8.41 ppm in Cowpea plots, as shown in Tables 3 and 4, respectively. It was, however, noted that the K level was high at Bagani, with recorded values ranging from 38 to 85 ppm. The K level is defined as high as according to Mylavarapu et al. (2007) K levels are defined as low from 20–35 and high from 61–125 ppm. Magnesium is low at <15 and high at >30 ppm. Most soils were found to be slightly acidic to neutral at all locations with values between 6 and 7.55 pHw. It was noted that the soils were very sandy with values of 94% for Okashana and 87% for Omahenene and Bagani, however the soils at Mannheim were less sandy with only 54% sand. Organic matter content noted was very low at all locations but variable at Okashana with 0% in Lablab plots to 0.25% in fallow plots, while at Mannheim and Omahenene organic matter content was above 1%. These results could also be the result of rainfall variation as recorded from the various locations as the average rainfall recorded at Okashana was 164 mm during 2005/2006 and 64 mm during the 2006/2007 cropping season. While at Omahenene 230 mm was recorded during 2005/2006 and 140 mm during 2006/2007. At Bagani the average rainfall recorded was 890 mm during 2005/2006 and 540 mm during the 2006/2007 season. The average rainfall recorded at Mannheim was 230 mm during 2005/2006 and 130 mm during the 2006/2007 season. It is clear from the rainfall data that Bagani received the highest amount of rainfall and this correlates with the high yield of Lablab recorded at Bagani during 2005/2006.

Soil properties for Bagani Research Station showed some acidic to neutral pH levels ranging between 6.51 and 7.81. The level of P recorded was also low, below the critical level of <15 ppm (Table 1).

Conclusions

The study showed the great potential of using Lablab (*Lablab purpureus*) for soil fertility improvement and soil conservation when compared to the other crops used in this study. Lablab showed greater growth and contribution to soil coverage and organic matter when compared to any other crop used during the past two years. Bearing the small-scale farmers in Namibia in mind, Lablab could be incorporated into the Namibian crop farming system as a soil fertility improving crop. Moreover, Lablab could be a good source of soil fertility and moisture retention in the soil. In addition to the information presented herein, ongoing experimentation may produce new data and lead to more conclusions on this topic.

Acknowledgements I would like to thank the organizers of the International Symposium on Innovations for the Green

Revolution in Sub-Saharan Africa and AFNET for their active role in information dissemination to its members. I would also like to thank the Ministry of Agriculture, Water and Forestry for continuous financial support to enable this project to continue.

References

Bationo A, Christianson WE, Mokwunye AU (1992) A farm level evaluation of nitrogen and phosphorus fertiliser use and planting density for pearl millet production in Niger. Fertil Res 31:175–184

Irving TF, Marsh A, Ryhn IV (1993) An environmental assessment of Uukwaluudhi. University of Namibia, Windhoek

Keyler S (1995) Economics of the pearl millet sub-sector in northern Namibia. A summary of baseline data ICRISAT. Southern and Eastern Africa Region working paper 95/03

Labetoulle L (2000) Nutritional survey of rural communities in the Northern central regions of Namibia. A research report prepared for the French Embassy in Namibia and for MAWRD. Windhoek, Namibia

Matanyaire C (1995) Progress report, research projects of the division of plant production on research 1994/1995, MAWRD, Namibia

MAWRD (1994) Workshop report of the yearly planning meeting of the plant production division at Tsumeb, Windhoek, Namibia

Mendelsohn J, Obeid S, Roberts C (2000) A profile of Northern Namibia. Gamsberg Macmillan, Windhoek

Mylavarapu R, Wright D, Kidder G, Chambliss CG (2007) UF/IFAS Standardized fertilization recommendations for agronomic crops.

Nyagumbo I (1993) Farmer participatory research in conservation Tillage: part 2: practical experiences with no-till tied ridging in the communal areas lying in the sub-humid North of Zimbabwe. Windhoek, Namibia, pp 236–245

Participatory Variety Selection of Pulses Under Different Soil and Pest Management Practices in Kadoma District, Zimbabwe

L. Rusinamhodzi and R.J. Delve

Abstract A study was initiated in Kadoma district, Zimbabwe to select the most preferred varieties of cowpea (*Vigna unguiculata* L. Walp), common beans (*Phaseolus vulgaris* L.) and soybean (*Glycine max*) based on farmers' criteria. These three crops had earlier been selected to address food insecurity and improved income through a participatory diagnosis process. Farmers developed their own criteria for evaluating the different crop varieties under soil fertility (fertilizer vs. manure), time of planting (early vs. late) and pesticide use (sprayed vs. unsprayed). The first experiment was established in late November 2005 and the second in late November 2006 based on results of the previous season. In 2006, the experiments tested botanical crude extracts on selected varieties. Farmers' selection criteria included disease tolerance, yield, grain characteristics, cooking time and taste. Pesticide application was the most important factor for production of cowpea and bean yield underlying the fact that pest control is still crucial for optimum yields. Across all criteria, CBC1 (cowpea), UBR92/25 (bean) and Solitaire (soya bean) were considered to be the best pulse varieties that could be grown in Kadoma district with conventional pesticide applications.

Keywords Farmer evaluation · Food security · Grain legumes · Participatory research

L. Rusinamhodzi (✉)
Tropical Soil Biology and Fertility Institute of the International Centre for Tropical Agriculture (TSBF-CIAT), Harare, Zimbabwe
e-mail: l.rusinamhodzi@cgiar.org

Introduction

Farmer participation in on-farm experimental work is an important component in the evaluation of new technologies for sustainable agricultural development. These evaluations when combined with formal assessments in experimental plots can provide a basis for making suitable recommendations about new interventions (Abeyasekera et al., 2002). Participatory methodologies have revealed intrinsic selection criteria for new technologies, preferences for established technologies as well as constraints (Chianu et al., 2006). Farmers' participation and performance in the research process is directly related to their human capital endowment, which includes both inborn and learned skills (Anderson and Feder, 2004).

The emphasis on farmer participation in on-farm variety selection research is to broaden the recommendation domain for crop breeders and other researchers so that the final product is resilient enough under local conditions and meets the preferences of the ultimate beneficiaries to enhance widespread adoption (Johnson et al., 2003). By inviting farmers to make decisions in the research process, it is also assumed that they will not only adopt but also adapt the available technology to their own needs and environment (Ashby, 1991).

After a participatory diagnosis exercise was conducted in Kadoma district in 2005 to investigate opportunities for addressing food security and increasing market-oriented production, farmers chose soya bean, cowpea and bean (Sanginga and Chitsike, 2005). Farmers recognized that these crops had different varieties and as such they needed to find varieties that were suitable to their local climatic conditions, that could yield substantially with minimum inputs and that meet

A. Bationo et al. (eds.), *Innovations as Key to the Green Revolution in Africa*,
DOI 10.1007/978-90-481-2543-2_103,

their food and market preferences. The objective of this chapter is to report on a participatory variety screening process to select the most suitable varieties of bean, soya bean and cowpea.

Materials and Methods

Study Area

This field study was conducted in Kadoma district, western Zimbabwe (29° 53′E, 18°19′S, altitude 1156 masl). Soils in Kadoma are mostly well-drained, reddish brown fersiallitic (5E.2) (Zimbabwe), Ferralic Cambisol (FAO, 1995) and Oxic Ustropept (USDA, 1999), (Thompson and Purves, 1978; Nyamapfene, 1991). The area receives 680–800 mm of rainfall annually, distributed in a unimodal pattern between November and April (Vincent and Thomas, 1960). Rainfall received in 2005/2006 and 2006/2007 seasons of 650 and 440 mm, respectively, was below the long-term average of 740 mm. The main crops are predominantly cotton and maize. Others are groundnuts, cowpea and soya bean. The miombo woodlands are a primary source of energy, in the form of firewood and charcoal, and crucial source of essential subsistence goods such as herbs for curing minor ailments, building materials and grass for thatching.

Treatments Screened

In the year 2005/2006, three cowpea varieties (CBC1, CBC2 and CBC3, all developed from IT18), seven soya bean varieties (Bimha, Solitaire, Soma, Storm, Roan, Nyati and Mhofu) and seven bean varieties (Pan 148, Iris, Local, UBR92/25, Natal Sugar, CIM9314 and Sugar 131) were screened under three nutrient management strategies: (i) no external nutrients, (ii) basal fertilizer (8:6:6) at 250 kg ha^{-1} and (iii) manure at 10 t ha^{-1}. In total, six treatments were used: (i) control (no pesticide, no fertilizer), (ii) F+P (fertilizer and pesticide), (iii) F–P (fertilizer, no pesticide), (iv) P (pesticide only), (v) M+P (manure plus pesticide) and (vi) M–P (manure, no pesticide).

The evaluation in 2005/2006 identified pesticide application to be important for legume production, although it was considered to be expensive. The opportunity was therefore to try alternatives for pest control. Varieties that performed well and selected by farmers in 2005 were further screened using bio-pesticides in 2006. These were cowpea (CBC1 and CBC2), bean (Pan 148, UBR 92/25, Natal Sugar and Sugar 131) and soya bean (Solitaire, Storm, Soma and Roan). The four treatments were no pesticide application (control), conventional pesticide application (P), extract of cactus (Bio-1), and chilli and tobacco mixture (Bio-2).

Evaluation Criteria

The characteristics considered by farmers to be important for evaluation of legumes were similar across crops (Table 1). The vegetative stage considered the whole crop, while the harvest stage evaluation considered the grain characteristics. Amount of leaves or biomass in cowpea was important because cowpea leaves and immature pods are important for the diet. Gender had no effect on preference scoring and was not included in the tabulated summaries. The overall preference was computed by averaging the scores across the criteria (Maxwell and Bart, 1997).

Farmer Evaluation of Legume Varieties

The evaluation process for legumes was in two stages: at vegetative and at harvest stages. Farmers would go around the plots during the season to look for characteristics they felt were important for the legume varieties being screened. The characteristics or indicators were then recorded and used to develop the criteria for evaluations based on these observations. Men and women developed their criteria separately and these were merged during the scoring exercise. Actual yield in this last stage was measured. Appropriate time of planting of the legumes was evaluated by comparing the same crops after harvest planted at different times and then the criteria were developed based on the differences noted.

Table 1 Farmer criteria for evaluation of cowpea, soya bean and bean crops at mid-season and end of season

Technology component	Evaluation criteria: Mid-season (vegetative)	End of season
Cowpea	Number of pods Biomass/leaves	Yield Size of grain Taste
Soya bean	Maturity period Resistance to shattering Resistance to lodging	Seed colour Seed size Yield
Bean	Disease resistance Maturity period Number of pods	Taste Seed colour Cooking time Grain size Yield
Management practices	Crop performance (yield, vigour, colour of leaves)	
Time of planting	Affordability Yield Seed appearance	

Scoring Procedure

Farmers were given 20 seeds to score each criterion across varieties; allocation of seeds was based on their preferences, with the most preferred criteria getting more seeds out of the possible 20. Farmers allocate the seeds based on their preferences, e.g. 'out of 20 seeds, how much would be assigned to the technology *x* for criteria *y*?' This means that each technology and criterion combination has an equal chance of getting any score between 1 and 20 (inclusive) from each farmer. Scoring was done with 20 farmers across four categories (5 young men, 5 young women, 5 mature men and 5 mature women). These farmers belonged to a farmer group involved in participatory research with legumes. The scoring was done individually and was independent as others were not allowed to watch the process (Abeyasekera et al., 2002; Njuki, 2007, unpublished).

Statistical Analysis

Variety evaluations were made in the form of scores rather than ranks; scores have numerical meaning and can be subjected to a variety of statistical analysis procedures (Agresti, 1996). The comparisons for effect of gender and variety characteristics on preference were done through analysis of variance (ANOVA) (Genestat, 2002).

Results and Discussion

Evaluation of Treatments

The important criteria for evaluating treatments in both seasons considered crop performance and affordability (Table 1). Jaiswal (1994) observed that farmers tend to use a different set of criteria for the evaluation of new varieties to the scientists who develop them. In particular, farmers tend to use a wider range of criteria (including, for example, consumption and cooking criteria) than those prioritized by plant developers (such as yield and resistance to pests). Although performance was high in treatments that involved pesticide applications, pesticide prices were considered to be prohibitive and affordability had lower scores compared to crop performance and conversely. Farmers did not prefer treatments with manure as they seldom use manure on legumes and they also use less fertilizer due to biological nitrogen fixation (BNF) which they believe is sufficient for high yields. The fertilizer and pesticide application treatment was regarded as the best followed by manure plus pesticide (Table 2). Other factors constraining production of grain legumes are directly linked to farmers' preferences, as they favour the main staple crop maize, over grain legumes, to ensure food security instead of income generation (Giller et al., 1997; Svubure, 2000). As farmers lack adequate nutrient resources to fertilize all crops, they prefer to apply fertilizers to maize, the staple food security crop, and rarely apply fertilizers directly to

Table 2 Evaluation of treatments used to screen legume varieties in 2005/2006 season

Treatment	Evaluation criteria		
	Crop performance	Affordability	Mean score
Control	9.8	81.5	45.7
P	57.8	48.8	53.3
F + P	91.5	47.3	69.4
F–P	20.5	37.8	29.2
M + P	72.0	49.8	60.8
M–P	36.8	47.3	42.1
$SED_{0.05}$	2.0	2.8	2.5
F-test probability (variety)	< 0.001	< 0.03	< 0.002
F-test probability (farmer category)	0.6	0.4	0.2

grain legumes that are mostly grown on residual fertility. Farmers also reserve the largest areas on the most fertile plots, often closest to homesteads, to maize and allocate small portions of outfields poor in fertility to production of grain legumes (Waddington and Karigwindi, 2001; Zingore et al., 2008).

Increasing the proportion of land under cash crops appears to reduce the amount of manure used, supporting the observation that manure is applied mainly for food crop production using household labour. The results imply that manure use is more important to the production of food crops (than of cash crops), and this is critical to low-income households (Heisey and Mwangi, 1996).

Farmers argued that in intensive production systems, high outputs are a result of high investments, and they would favour that option for maximum production. The criteria for evaluation were crop performance and affordability. The bio-pesticides tested were poorly rated compared to conventional pesticides, with the alternatives less preferred than the control (Table 3). Farmers argued that preparing the bio-pesticides was difficult and time consuming, yet they were not effective on the identified legume pests, and crop performance was similar to that of control. The results showed distinct preferences, farmers' favoured treatments that included pesticide application as this would affect the crop appearance. The clear message coming out of this was that pest control remains critical to derive maximum benefits from pulses.

Cowpea Variety Evaluation

Evaluation of cowpea was for green leaves as relish and grain as food, therefore crop appearance in the field and yield at harvest were very crucial (Table 4). In terms of biomass and leaf shape (considered it important for relish), there was no difference between CBC1 and CBC2, while CBC3 was rejected. The number of pods which directly reflects the yield was highest in CBC1 followed by CBC2 and lowest in CBC3. After harvest, grain size and taste were important evaluation criteria as yield and cooking time were not significantly different. In terms of grain size and taste, CBC1 was way above the other two varieties making it the best cowpea variety among those tested and CBC3 was rejected. Biomass and leaf shape was important because cowpea leaves are used extensively as a vegetable throughout the African continent, being incorporated into a variety of dishes, soups and sauces. In Kenya, the leaves are consumed as a side dish for *ugali* (a paste prepared from maize meal) and to a lesser extent as part of the main meal where leaves are cooked with maize kernels, legumes and either green bananas or potatoes (Oomen and Grubben, 1977).

Soya Bean Variety Evaluation

Important attributes were lodging resistance and shattering resistance as maturity period was considered to be similar (Table 5). Solitaire and Soma were considered to be the ideal varieties as they were resistant to both shattering and lodging. Ranking was distinct with four varieties showing positive attributes and being accepted by farmers and three being rejected (Table 5). Solitaire was regarded by farmers as the overall best variety because of its high resistance to shattering. Mhofu, Nyati and Bimha were clearly rejected by

Table 3 Evaluation of treatments used to screen legume varieties in 2006/2007 season

Treatment	Evaluation criteria		
	Crop performance	Affordability	Mean score
Control	5.3	92.5	48.9
P	94.8	19.8	57.3
Bio-P1	12.5	33.5	23.0
Bio-P2	10.5	36.8	23.8
$SED_{0.05}$	1.5	1.4	2.1
F-test probability (variety)	< 0.001	< 0.001	< 0.001
F-test probability (farmer category)	0.3	0.4	0.5

Table 4 Farmer preference scores for the different cowpea varieties during mid-season evaluation

Variety	Evaluation criteria						
	Mid-season			End of season			
	Biomass/ leaf shape	Number of pods	Yield	Grain size	Taste	Cooking time	Mean score
CBC1	83.0	90.8	92.8	89.5	100	99.2	92.6
CBC2	79.5	64.8	89.8	64.3	78.5	99.8	79.5
CBC3	36.3	54.5	33.2	55.3	53.0	85.5	53.0
$SED_{0.05}$	1.9	0.8	2.5	1.8	3.2	2.8	4.0
F-test probability (variety)	< 0.001	< 0.001	< 0.001	< 0.001	< 0.001	0.003	< 0.001
F-test probability (farmer category)	0.4	0.4	0.5	0.3	0.4	0.4	0.4

Table 5 Farmer preference scores for different soya bean varieties during mid-season evaluation

Variety	Evaluation criteria						
	Mid-season			End of season			
	Maturity period	Lodging resistance	Shattering resistance	Yield	Grain size	Colour	Mean scores
Storm	85.3	55.2	48.3	89.5	55.2	54.8	64.7
Solitaire	86.3	83.8	96.8	98.5	98.8	94.5	93.1
Soma	85.3	72.8	85.5	95.0	96.5	93.8	88.2
Roan	82.3	52.5	64.5	81.0	52.5	64.5	66.2
Mhofu	90.3	22.8	46.5	85.0	22.7	46.5	52.3
Nyati	92.0	20.8	41.3	83.8	20.7	41.3	50.0
Bimha	90.1	19.2	31.5	84.5	21.7	28	45.8
$SED_{0.05}$	1.7	4.45	2.8	2.2	5.4	2.1	3.8
F-test probability (variety)	< 0.001	< 0.001	< 0.001	< 0.001	< 0.001	< 0.001	< 0.001
F-test probability (farmer category)	0.5	0.3	0.3	0.7	0.6	0.7	0.7

farmers on both evaluation occasions in the first season evaluation. They all had negative attributes and were more vegetative and were regarded as fit for fodder not for marketable grain. In the second evaluation, colour and grain size were important; some varieties such as Roan and Mhofu had the green colour and were not preferred even with the best seed size. The results agree with the evidence provided by studies of consumer demand showing that consumers critically evaluate characteristics of a product and that demand is affected by consumers' subjective assessments of product attributes (Lin and Milon, 1993). Farmers are the consumers of the products of agricultural research and their subjective preferences for characteristics of new agricultural technologies affect their adoption decisions.

Bean

Disease resistance and number of pods were more important in mid-season evaluation, whereas yield, seed size, cooking time, colour of seed and taste were important at end of season (Table 6). Maturity period was an important criterion but during scoring, no differences were observed among the varieties evaluated. UBR 92/25 variety was the best followed by Sugar 131, while the local variety was among those rejected. This clearly shows that there is a desire by farmers to adopt improved varieties if they are available; they keep the local variety for many season due to lack of options exacerbated by high prices for hybrid seed every season. The clear picture coming out is that research should focus on releasing varieties that have multiple attributes as shown by farmers' preference.

Time of Planting

Late planted cowpea and bean were preferred over the early crops, whereas early planted soya bean was preferred over the late crop (Table 7). These observations agreed with those by Hildebrand and Nosenga (2000), who noted that beans are not suited to production during the normal summer rainfall period as they tend to be susceptible to a range of bacterial, fungal and viral diseases. Flower drop can also be excessive if flowering occurs during the height of the rainy season. The appropriate time to plant according to farmers is end of December to end of January for beans and cowpea, while appropriate time to plant soya bean is the start of the season in early November. Seed appearance was important in that it was used to reflect the particular quality of each crop. Early planted bean and cowpea

Table 6 Farmer preference scores for different bean varieties during both mid-season and end of season evaluations

	Evaluation criteria								
	Mid-season				End of season				
Variety	Maturity period	Disease resistance	Number of pod	Colour	Taste	Yield	Size	Cooking time	Mean score
Natal Sugar	89.5	55.2	54.8	64.8	58.3	61.3	63.0	61.5	63.6
Sugar 131	85.3	83.8	73.0	74.0	64.3	83.8	80.5	70.0	76.8
UBR 92/25	85.0	96.0	92.5	91.8	96.3	94.3	97.0	88.3	92.7
Pan148	81.0	52.5	64.5	30.8	42.5	73.0	57.8	48.8	56.4
Iris	85.0	22.8	46.5	12.8	25.8	4.5	91.5	47.3	42.0
CIM 9314	83.3	20.8	41.3	34.3	20.8	18.3	57.8	48.8	40.7
Local	86.0	25.5	29.5	26.5	32.0	10.3	91.5	47.3	43.6
$SED_{0.05}$	2.3	5.3	1.8	2.8	3.3	1.7	2.0	2.1	
F-test probability (variety)	0.06	< 0.001	< 0.001	< 0.001	< 0.001	< 0.001	< 0.001	< 0.001	< 0.001
F-test probability (farmer category)	0.6	0.6	0.6	0.4	0.4	0.4	0.5	0.4	0.6

Table 7 Farmer preference scores for different time of planting of each of the three crops at the end of season evaluations

		Evaluation criteria		
Crop	Time of planting	Yield	Seed appearance	Mean score
Cowpea	Early	89.4	56.3	72.9
	Late	91.2	98.2	94.7
	$SED_{0.05}$	0.8	13.3	15.4
Soya bean	Early	93.4	93.6	93.5
	Late	63.2	36.4	49.8
	$SED_{0.05}$	9.5	18.1	30.9
Bean	Early	48.3	54.8	51.6
	Late	88.1	94.7	91.4
	$SED_{0.05}$	12.7	13.0	28.1

flowered during the peak of the rainy season, flowers were lost and there was more grain rot at harvest.

Conclusions

This research documents the increasing importance of farmer participation in setting the research agenda in agricultural research and development. While it is important to consider farmer inputs, caution should be exercised on some of their recommendations as there is a tendency to explore 'dead ends' as shown by the biopesticide results. This study places emphasis on factors that are qualitative in nature and represents a challenge to breeders who have to release varieties that meet both quantitative and qualitative aspects. Farmers' preferences are mainly driven by the desire to minimize investment costs, but it is also clear that they have no intentions to compromise on crop yield. Farmers noted that any of the top three varieties for bean and soya bean and two for cowpea was good enough for production, providing themselves with options in periods of seed shortages. CBC1 (cowpea), UBR92/25 (bean) and Solitaire (soya bean) were considered to be the best pulse varieties that could be grown in Kadoma district with conventional pesticide applications. Given the local nature of the project, a macro-level study on this to cover most of the agro-ecological zones is still desirable for wider dissemination of results. An advantage of this approach is that it is possible to identify the key characteristics of technologies that need to be targeted for improvement if adoption is to be achieved. The drawback of this approach is that it is repeatable only with the same sampled farmers.

Acknowledgements The authors are grateful to E. Ndoro and S. Makaza (AREX-Kadoma), F. Mukoyi (AREX-Harare) for setting up meetings with farmers and facilitation skills, and Tayambuka Farmers Group for taking part in the process. This chapter is an output of the project 'Using Market-Led Approaches to Drive Investments in Soil Fertility Management and Improve Production and Incomes of Rural Communities in Selected Areas of the Central Watershed of Zimbabwe' hosted by Agricultural Research and Extension (AREX), Kadoma, and managed by CIAT-TSBF, Harare. This project was financially supported by CIDA-Harare Office. The opinions expressed in this chapter are those of the authors and they do not necessarily reflect the views of AREX or CIDA.

References

Abeyasekera S, Ritchie JM, Lawson-McDowall J (2002) Combining ranks and scores to determine farmers' preferences for bean varieties in southern Malawi. Exp Agric 38:97–109

Agresti A (1996) An introduction to categorical data analysis. Wiley, New York, NY

Anderson J, Feder G (2004) Agricultural extension services: good intentions and hard realities. World Bank Res Observer 19(1):41–60

Ashby J (1991) Adopters and adapters: the participation of farmers in on-farm research. In: Tripp R (ed) Planned change in farming systems: progress in on-farm research. Wiley, New York, NY, pp 273–286

Chianu J, Vanlauwe B, Mukalama J, Adesina A, Sanginga N (2006) Farmer evaluation of improved soybean varieties being screened in five locations in Kenya: implications for research and development. Afr J Agric Res 1(5): 143–150

FAO (1995) The digitized soil map of the world including derived soil properties. (version 3.5) FAO land and water digital media series 1. FAO, Rome

Genestat (2002) Release 6.1. VSN International, Oxford, UK

Giller KE, Cadish G, Ehaliotis C, Adams E, Sakala WD, Mafongoya PL (1997) Building soil nitrogen capital in Africa. In: Buresh RJ, Sanchez PA, Calhoun F (eds) Replenishing soil fertility in Africa. Soil Science Society of America Special Publication No. 51, Madison, Wisconsin, WI, pp 81–95

Heisey PW, Mwangi W (1996) Fertilizer use and maize production in sub-Saharan Africa. CIMMYT economics working paper 96-01. CIMMYT, Mexico, DP

Hildebrand GL, Nosenga AZ (2000) Growing beans in Zimbabwe. Seed Co Limited. Harare, Zimbabwe

Jaiswal JP (1994) Increasing the relevance of farmers' participation in generating appropriate technology for the hill areas of Nepal. Unpublished MSc dissertation, Department of Agriculture, The University of Reading, Reading, UK

Johnson N, Lilja N, Ashby J (2003) Measuring the impact of user participation in agricultural and natural resource management research. Agric Syst 78:287–306

Lin CTJ, Milon JW (1993) Attribute and safety perceptions in a double-hurdle model of shellfish consumption. Am J Agric Econ 75:724–729

Maxwell S, Bart C (1997) Beyond ranking: exploring relative preferences in P/RRA. PLA Notes 22:35–39

Nyamapfene KW (1991) Soils of Zimbabwe. Nehanda, Harare, pp 75–79

Oomen HAPC, Grubben GJH (1977) Tropical leaf vegetables in human nutrition, Communication 69. Department of Agriculture Research. Koninklijk Instituut voor de Tropen, Amsterdam

Sanginga P, Chitsike C (2005) The power of visioning. A handbook for facilitating the development of community action plan. Enabling rural innovation guide 1. CIAT, Kampala, Uganda

Svubure O (2000) Contributions of biological nitrogen fixation (BNF) by selected grain legumes to sustainabilty of the maize-based cropping systems in communal areas of Zimbabwe. MPhil thesis, University of Zimbabwe, Harare, Zimbabwe

Thompson JG, Purves WD (1978) A guide to the soils of Rhodesia. Department of Research and Specialist Services. Rhodesia Agric J Tech Handbook No 3, pp 64

USDA (1999) Soil taxonomy, a basic system of soil classification for making and interpreting soil surveys. US Government Printing Office, Washington, DC

Vincent V, Thomas RG (1960) An agricultural survey of southern Rhodesia: part I: agro-ecological survey. Government Printer, Salisbury

Waddington SR, Karigwindi J (2001) Productivity and profitability of maize plus groundnut rotations compared with continuous maize on smallholder farms in Zimbabwe. Exp Agric 37:83–98

Zingore S, Murwira HK, Delve RJ, Giller KE (2008) Variable grain legume yields, responses to phosphorus and rotational effects on maize across soil fertility gradients on African smallholder farms. Nutr Cycling Agroecosyst 80(1):1–18

Economic Returns of the "MBILI" Intercropping Compared to Conventional Systems in Western Kenya

M.N. Thuita, J.R. Okalebo, C.O. Othieno, M.J. Kipsat, and A.O. Nekesa

Abstract In Kenya, smallholder farmers practice maize–bean intercropping. Low nutrient levels in soils result in low yields for both crops. Farmers plant both maize and beans in the same hill or between maize rows, which results in low yields. In the MBILI (Managing Beneficial Interactions for Legume Intercrops) system, the spatial rearrangement gives high legume and cereal yield. An on-farm experiment was carried out in four districts (Bungoma, Siaya, Trans Nzoia, and Uasin Gishu) of western Kenya in a randomized complete block design giving three intercropping systems (MBILI, hill, and conventional), maize and two legumes (bean, soybean, or groundnut) and two fertilizer levels (0 and 150 kg of DAP/ha) with three blocks. The aim was to compare grain yields and economic returns of the MBILI and conventional intercropping systems. Treatment effects were determined by ANOVA analyses using the general linear model of the SAS system. The Bungoma site had the highest groundnut yields for long rains of 2005 and 2006; bean yields under MBILI intercropping (1.4 t/ha). Kitale gave the highest soybean yields of 573 kg/ha from MBILI with fertilizer. The MBILI intercropping gave the highest maize yields (4 t/ha) in all the sites except Sega, while the controls gave low yields (1 t/ha) in all the sites compared to the fertilized intercrops. Economic analyses showed that MBILI gave the highest returns on capital. The distinct finding is that MBILI gave increased maize and legume yields compared to conventional intercropping systems.

M.N. Thuita (✉)
Department of Soil Science, Moi University, Eldoret, Kenya
e-mail: thuitam@yahoo.com

Keywords MBILI · Intercropping · Legume · Cereal · Yield

Introduction

Agricultural output in the 1970s in sub-Saharan Africa (SSA) decreased by 1.3%, while population rose by 2.7%. It was also observed that crop yields are lower in SSA than elsewhere globally (Meerman and Cochran, 1982). The main cause of declining agricultural productivity in SSA is the reduction in traditional systems of shifting cultivation due to change of land tenure systems (Bouldin et al., 1980). Each farmer continuously farms the same area and as each generation subdivides their land, farm sizes keep decreasing (Bouldin et al., 1980). Very limited use of manures and fertilizers and increased nutrient removal lead to soil fertility decline and lower yields per unit land area (Oram, 1981). Organic manures are not available in the qualities and quantities that are required to impact positively on agricultural production (Bouldin et al., 1980; Oram, 1981). Fertilizers are expensive to most smallholder farmers and are thus rarely used in the right quantities.

Among the solutions being recommended to reverse the declining agricultural productivity are the adoption of cropping systems that conserve soil fertility and the close integration of livestock and arable farming (Pratt and de Haan, 1979). Okigbo and Greenland (1976) observed that increasing agricultural output in SSA requires improved cropping systems, which necessitate the integration of traditional and modern technologies as farmers are more likely to adopt modifications to existing farming systems than completely new ones. One of the entry points for improving agricultural productivity could be the modification of cropping

A. Bationo et al. (eds.), *Innovations as Key to the Green Revolution in Africa*,
DOI 10.1007/978-90-481-2543-2_104, © Springer Science+Business Media B.V. 2011

systems such as intercropping that leads to higher or diversified production per unit land area.

Intercropping is defined as the growing of two or more crops simultaneously on the same field (Sullivan, 2003). The most common form of cropping pattern in SSA is mixed intercropping, the growing of two or more crops simultaneously with no distinct pattern. Other types of intercropping are row, strip, and relay cropping. Traditionally, agricultural systems in Kenya are based on growing of crops in mixtures. For most of the smallholder farmers, it is justified that a variety of crops should be grown to allow for a varied diet and food security. The farmer sees intercropping as a means of ensuring this diversification.

In western Kenya, it is common practice to plant both maize (staple) and beans in the same hill or between maize rows (conventional). Competition for light by crops in such practices significantly contributes to overall low yields (Mukhwana et al., 2002). The shortcoming of both intercropping systems is the low legume and maize grain yields, thus not allowing the farmer to grow different legumes for wider market penetration and reduce biotic pressures such as pests and diseases due to continuous cropping of the same legume. In the newly introduced "MBILI" (Managing Beneficial Interactions in Legume Intercrops) intercropping system (Tungani et al., 2002), two rows of maize and two rows of beans, or suitable food grain legume, are planted in alternating rows. Maize rows are spaced as 50 cm pairs that are 100 cm apart (the gap). Two rows of legumes are planted within the gap of 33 cm row spacing. By staggering the rows of maize into a two-by-two arrangement, the same plant populations as conventional intercrops are maintained, but the legumes are better able to compete favorably with maize for moisture, nutrients, light, or solar radiation (Woomer and Mukhwana, 2000). Thus plant populations and nutrient inputs are the same for both MBILI and conventional arrangements. The MBILI intercropping system also reduces leaching of nitrates and better nutrient content in grains compared to conventional intercropping systems (Thuita et al., unpublished data). The MBILI system also provides the flexibility for choice of the legume crop, focusing the cultivation of high value or economic legumes (groundnuts, green grams, and soybeans) for the smallholder farmer.

Nutrient inputs from any other soil fertility restoration technology (e.g., FURP, PREP-PAC) may also be applied to the two forms of intercropping, or spatial arrangements (Mukhwana et al., 2002), but the economics of using these technologies also needs to be studied. It is also stressed here that farmers in western Kenya plant maize and bean seed in the same hole, while others plant beans between maize rows.

Materials and Methods

Description of the Study Sites

The study was conducted as an on-farm trial in four districts (Siaya, Bungoma, Uasin Gishu, and Trans Nzoia) of western Kenya. Siaya and Bungoma districts experience a bimodal rainfall season. Average annual rainfall ranges from 800 to 2000 and 1200 to 1800 mm in Siaya and Bungoma, respectively (Republic of Kenya, 1994, 1997). The main soil types in the districts are Nitisols, Ferralsols, and Acrisols. Uasin Gishu and Trans Nzoia districts have one cropping season with rainfall fairly well distributed throughout the year, averaging 800–1300 mm (Jaetzold and Schmidt, 1983). The main types of soils are Nitisols, Ferralsols, and Acrisols for Uasin Gishu district. In Trans Nzoia, one finds deep red, friable clays and sandy clays derived from the basement complex and black cotton soils (Vertisols) (Jaetzold and Schmidt, 1983).

Trials

In each district, one trial was planted in the long rains of 2005 and 2006. Each trial was a randomized complete block design with three replicates and twelve treatment combinations as shown in Table 1. The plots were 4.5 m × 5 m giving six rows of legume and cereal each. Four inner rows were harvested as samples, while the outer two rows and the end plants in the inner rows were not sampled. Land preparation was done by hand hoes in February 2005.

Soil samples were taken in all the plots for initial chemical and physical characterization before treatment application. In all sites, planting was done in March during the long rains. Fertilizer treatments were applied by broadcasting the fertilizer evenly within the plots and then incorporated using hand hoes. P was applied as DAP at a rate of 150 kg/ha (30 kg P/ha) at planting, while N was applied at a rate of 75 kg N/ha (FURP, 1994) as a split with 27 kg/ha DAP at planting and 48 kg/ha as top dressing at legume flowering stage.

Table 1 Treatments applied in the field study (12 treatment combinations)

		Nutrient level (kg DAP/ha)			
		Kuinet and Kitale sites		Bungoma and Sega sites	
Row arrangement	Intercrops	0 (O)	150 (P)	0 (O)	150 (P)
MBILI (M)	Maize/beans (B)	MBO	MBP	MBO	MBP
	Maize/soybean (S)/groundnut (G)	MSO	MSP	MGO	MGP
Conventional (C)	Maize/beans (B)	CBO	CBP	CBO	CBP
	Maize/soybean (S)/groundnut (G)	CSO	CSP	CGO	CGP
Hill (H)	Maize/beans (B)	HBO	HBP	HBO	HBP
	Maize/soybean (S)/groundnut (G)	HSO	HSP	HGO	HGP

Key Three row arrangements: MBILI (M), conventional (C), hill (H)
Two intercrops for each site; maize/beans (B), maize/groundnuts (G) for Bungoma and Sega, and maize/soybean (S) for Kitale and Kuinet with one long growing season
Two nutrient levels: 0 kg DAP/ha (O) and 150 kg DAP/ha (P)
Beans were grown in all the four sites

Maize (IR *Striga*-resistant maize variety, the farmers' favorite "Wairimu" bean seed, and groundnut (red Valencia) were planted in Siaya and Bungoma sites, while H614D seed maize and soybean TGx variety were planted at the Uasin Gishu and Trans Nzoia sites as follows.

Within the "MBILI" system, maize row spacing was 100 cm, with an intrarow spacing of 50 cm. Two seeds per planting hole were used and later thinned to one at first weeding stage. Two rows of legumes were planted between rows with a row spacing of 33 cm and an intrarow spacing of 15 cm, giving maize plant population of 44,000 plants/ha and legume population of about 88,000 plants/ha. Within the conventional system, maize was planted with a row spacing of 75 cm and an intrarow spacing of 30 cm, resulting in a plant population of 44,000 plants/ha. A legume row was planted between the maize rows (37.5 cm) and an intrarow spacing of 15 cm, giving a legume population of about 88,000 plants/ha. In the hill system, maize and legume seeds were planted in the same row, using a row spacing of 75 cm. Maize was planted 30 cm apart within rows and two legume seeds were planted in-between two maize seeds at a spacing of 10 cm apart such that the plant population was maintained in all the intercropping systems.

Soil Sampling and Data Analysis

All fields were sampled before trial layout and a second soil sampling for initial soil characterization was carried out at maize harvesting in each plot (or field). A composite sample from 12 sampling points taken at random (0–15 cm) per plot was taken for laboratory analysis according to procedures in Okalebo et al. (2002). All data were subjected to analysis of variance (ANOVA) with the general linear model (GLM) procedure of the SAS system V 8.1 (SAS, 1999), while an economist was consulted for economic evaluation.

The gross margin (GM) for each treatment was computed as the difference between total revenue and total variable costs. Labor and capital productivities indicate the returns to labor and capital, respectively. Labor productivity represents the gross margin per shilling invested in labor, while capital productivity refers to gross margin per shilling invested in capital. Value-to-cost ratio (VCR) was computed as the total revenue divided by the total variable costs. Net change in income refers to change in income as a result of the introduction of the treatment. It is the difference between net income for specified treatment and net income for the corresponding control treatment.

Results and Discussion

Soils

Initial soil characterization was done for each field. Table 2 shows a summary of the soil parameters analyzed. The soil pH (H_2O) in all the sites, apart from Bungoma, was below the recommended 5.5–6.5, meaning the sites were acidic, especially Kuinet site which was far below the critical level (Othieno, C. O. pers comm.) for most food crops.

The amounts of available P (bicarbonate P) were very low in all the sites except Kuinet which was the

Table 2 Soil characterization for the top (0–15 cm) for all the sites before planting

Site	pH	% C	% N	Olsen P (mg/kg)	C mol/kg of soil K	Na	Ca	Mg	Fe	Mn	Zn	Cu
Kuinet	4.27	2.24	0.34	12.88	1.14	0.69	1	0.24	0.35	0.86	0.001	0.0007
Kitale	5.17	2.84	0.49	8.32	1.48	1.39	3.64	0.86	0.17	0.15	0.001	0.0014
Sega	4.86	2.01	0.26	1.35	0.4	0.69	2.44	0.11	0.21	0.08	0.0008	0.0014
Bungoma	5.58	1.37	0.18	4.16	0.25	1.47	0.04	0.37	0.18	0.09	0.0005	0.004

Soil texture	% Sand	% Clay	% Silt	Textural class
Kuinet	52	24	26	Sandy clay loam
Kitale	66	16	18	Sandy loam
Sega	64	32	4	Sandy clay loam
Bungoma	72	16	12	Loamy sand

only site with P above the critical amount of 10 mg kg^{-1} (Okalebo et al., 2002), while the amounts in Sega and Bungoma sites were far below this level. This justified the application of P. The organic carbon content in all sites was medium except for Bungoma, in which it was low (Okalebo et al., 2002). This related well with N percentage in all the sites whereby the site with the lowest N content also had the lowest carbon content and the site with the highest N content also had the highest N or organic carbon content. Potassium (K) was high in Kuinet and Kitale, but it was medium in Bungoma and low in Sega. Calcium (Ca) was very low in Kuinet and Bungoma and was below the critical level of 1.75 cmol/kg of soil (Wortman and Eledu, 1999). Mg was very high in Kitale, high in Sega, medium in Kuinet, and low in Bungoma (Okalebo et al., 2002). This initial characterization shows that soil fertility in our study sites was too low to support high crop production levels

Maize Yields

Tables 3 and 4 show mean maize grain yields during the long rains of 2005 and 2006 cropping season. Overall, maize yields were highest in Kitale (4.4 t ha^{-1}) and Bungoma (4.2 t ha^{-1}) and lowest in Kuinet (3.3 t ha^{-1}) and Sega (2.72 t ha^{-1}) ($P < 0.001$). In

Table 3 Unfertilized and fertilized maize grain yields (t ha^{-1}) in Kitale and Kuinet during the long rains in 2005 in three maize–legume intercropping systems, with beans or soybeans

		Kitale			Kuinet		
		DAP (kg/ha)			DAP (kg/ha)		
Intercropping	Legume	0	150	Mean	0	150	Mean
Conventional	Beans	1.71	3.71	2.71	2.00	2.86	2.43
	Soybean	2.48	4.71	3.60	1.37	3.88	2.62
	Mean	2.10	4.21	3.15	1.68	3.37	2.53
Hill	Beans	2.08	4.89	3.49	0.81	2.55	1.68
	Soybean	3.61	4.16	3.88	1.21	2.53	1.87
	Mean	2.84	4.53	3.69	1.01	2.54	1.78
MBILI	Beans	1.30	3.60	2.45	1.82	3.26	2.54
	Soybean	2.61	5.16	3.89	1.58	3.94	2.76
	Mean	1.96	4.38	3.17	1.71	3.60	2.65
SED legume (legu)				0.05			0.04
SED intercrop (inter)				0.07			0.05
SED fertilizer (fert)				0.05			0.04
SED inter × legu				0.09			0.07
SED legu × fert				0.08			0.06
SED inter × fert				0.09			0.07
SED inter × fert × legu				0.13			0.1

Table 4 Average fertilized and unfertilized maize grain yields (t/ha) in Bungoma and Sega during LR of 2005 and 2006 in three maize–legume intercropping systems, with beans or groundnuts

		Bungoma 2005			Bungoma 2006			Sega 2005			Sega 2006		
		DAP (kg/ha) 2005			DAP (kg/ha) 2006			DAP (kg/ha) 2005			DAP (kg/ha) 2006		
Intercropping	Legume	0	150	Mean	0	150	Mean	0	150	Mean	0	150	Mean
Conventional	Beans	1.89	4.99	3.44	2.20	4.18	3.19	1.36	2.88	2.12	1.09	2.33	1.71
	Groundnut	2.38	3.40	2.89	1.80	3.39	2.59	1.27	2.80	2.03	1.09	2.52	1.81
	Mean	2.13	4.19	3.16	2.00	3.78	2.89	1.31	2.84	2.08	1.09	2.42	1.76
Hill	Beans	1.53	3.8	2.67	1.76	3.63	2.70	1.50	3.79	2.65	0.92	2.82	1.87
	Groundnut	1.86	3.61	2.74	1.48	3.57	2.53	1.44	3.23	2.34	1.03	2.20	1.62
	Mean	1.7	3.71	2.70	1.62	3.60	2.61	1.47	3.51	2.49	0.97	2.51	1.74
MBILI	Beans	1.62	5.00	3.31	1.79	4.01	2.9	1.91	2.98	2.45	1.47	2.89	2.18
	Groundnut	1.95	3.69	2.82	2.32	4.35	3.33	2.07	2.65	2.36	1.53	2.94	2.23
	Mean	1.79	4.35	3.07	2.05	4.18	3.12	1.99	2.82	2.40	1.50	2.92	2.21
SED legume (legu)				0.05			0.02			0.03			0.02
SED intercrop (inter)				0.06			0.03			0.04			0.03
SED fertilizer (fert)				0.05			0.02			0.03			0.02
SED inter × legu				0.09			0.04			0.03			0.04
SED legu × fert				0.07			0.03			0.06			0.03
SED inter × fert				0.09			0.04			0.06			0.04
SED inter × fert × legu				0.12			0.06			0.08			0.05

all sites, fertilizer significantly increased maize yields ($P < 0.001$). Without fertilizer use, maize yields vary between intercropping systems and sites.

Overall, maize yields in the MBILI system outperformed maize production in the other two intercropping systems, either with or without fertilizer. Only in Sega was the performance of maize in the MBILI system significantly lower than was in the other two systems, while in Kuinet, Kitale, and Bungoma, maize yields in the MBILI system outperformed the other two systems when fertilized.

The best performing systems with respect to maize yields were the hill system in Kitale and Sega, the MBILI system in Kuinet and Bungoma, and the MBILI system in. The hill performed well in combination with beans. This was attributed to the very poor performance of beans in the hill system in both sites such that the maize was almost similar to a maize monocrop in terms of competition for nutrients. The MBILI performed best in combination with groundnuts for Bungoma and Sega, while soybean was better in the MBILI system for Kitale and Kuinet. Under this two-crop combination in MBILI row arrangement, maize seemed to benefit from the interaction as well as the reduced competition. Thuita et al. (2005) found reduced NO_3–N leaching under the MBILI intercropping system. This may have benefited the maize, especially because it is the system where the highest yields were obtained.

Legume Yields

Soybeans

Overall soybean yields were significantly higher in Kitale site than in Kuinet (Table 5; $P < 0.001$). Soybean yields were significantly higher in the MBILI system compared to the other two systems, independent of fertilizer application ($P < 0.001$). Soybean yields were lowest in the hill system ($P < 0.001$), with fertilized yields being similar to unfertilized soybean yields in the MBILI system.

Bean Yields

Bean yields were highest in the Bungoma site in the MBILI and conventional systems (1395 and 1018 kg/ha, respectively) ($P < 0.05$). Without fertilizer use, bean yields in the MBILI system significantly outperformed bean yields in the other systems in Sega and

Table 5 Soybean grain yields (kg/ha) for Kitale and Kuinet sites in 2005

	Kitale			Kuinet		
	DAP (kg/ha)			DAP (kg/ha)		
Intercropping	0	150	Mean	0	150	Mean
Conventional	264	414	339	182	252	217
Hill	180	349	265	133	216	175
MBILI	364	573	469	223	310	267
Mean	269	445	358	179	259	220
SED intercrop (inter)			9.19			6.94
SED fertilizer (fert)			7.51			5.67
SED inter × fert			13.00			9.81

Kuinet sites ($P < 0.05$), while with fertilizer use, bean yields in the MBILI system outperformed bean yields in the other systems in Sega, Kuinet, and Bungoma sites. Bean yields were lowest in the hill systems in all sites, independent of fertilizer ($P < 0.001$).

The conventional system with fertilizer also had bean yields above 1 t/ha. Kuinet had the second highest yields again from the MBILI with the highest yield of 800 kg/ha, which was higher than what was obtained in the long rains of 2005. The hill intercropping had the lowest yields at the Kuinet site as with the other sites. These were significantly lower than the other intercropping systems ($P < 0.001$) (Table 6).

Groundnut Yields

Groundnut yields in Bungoma were significantly higher than in Sega. Fertilizer use increased groundnut yields in all intercropping systems ($P < 0.001$). Groundnut yields were significantly higher in the MBILI system than in the hill system, independent of fertilizer application, but groundnut yields did not vary significantly between the MBILI and the conventional systems during the long rains of 2005 in the Bungoma site but were significantly higher during the long rains of 2006 in all the sites (Table 7). Highest groundnut yields were obtained in Bungoma with the MBILI system plus fertilizer (564 and 680 kg/ha in 2005 and 2006, respectively).

Economic Analyses of Grain Yields as Affected by Intercropping Systems

From an economic point of view, MBILI intercropping with soybean and beans (MSP, MBP) and hill system with bean (HBP) treatments performed best in Kitale, Kuinet, Bungoma and Sega sites, respectively. The results in Table 8 show that the best treatment in Kitale, MSP, had a GM/ha of Ksh. 783,409.20 and labor and capital productivities of Ksh. 3.39/Ksh. and Ksh. 1.82/Ksh., respectively. The treatment (MSP) had a value-to-cost ratio of 2.62 and a net change in income of Ksh. 74,659.85. Table 8 indicates that it was profitable to produce under all the specified techniques except under hill system controls for both soybean and bean (HSO and HBO, respectively) that had negative GM/ha, labor and capital productivities, and value-to-cost ratios that were less than 1.

All the production techniques in Bungoma were profitable with MBP having a GM/ha of Ksh. 101,856.15 and net change in income of Ksh. 75,875.10; HBO had the lowest GM/ ha of Ksh. 11,988.90 (Table 8). The economic evaluation of the techniques in Sega indicated that HBP had the highest GM/ha of Ksh. 41,927.2. The production under HBO was not economical at Sega site as it resulted in losses of Ksh. 1215.69/ha being incurred.

Discussion

Initial Soil Fertility Status

The soil of the four sites showed different fertility levels with nutrient deficiencies and low soil pH being more prevalent in all the sites. N and P were the main limiting nutrients in all the sites and this necessitated the application of both DAP and CAN as the main source of both nutrients. The high rainfall in these areas and continuous cropping have also led to high

Table 6 Bean grain yield (kg/ha) for Kuinet, Sega, and Bungoma during the LR of 2005 and 2006

	Kuinet 2005			Kuinet 2006			Sega 2005			Sega 2006			Bungoma 2006		
	DAP (kg/ha)			DAP (kg/ha)			DAP (kg/ha)			DAP (kg/ha)			DAP (kg/ha)		
Intercropping	0	150	Mean	0	150	Mean	0	150	Mean	0	150	Mean	0	150	Mean
Conventional	215	360	288	225	467	346	344	430	387	225	467	346	580	1018	799
Hill	127	208	168	58	407	233	183	335	259	58	407	233	240	270	255
MBILI	228	414	321	386	570	478	320	533	427	386	570	478	709	1395	1052
Mean	190	327	259	223	481	352	282	433	358	223	481	352	510	894	702
SED intercrop (inter)			3.89			26.18			4.88			6.19			11.31
SED fertilizer (fert)			3.18			21.38			3.99			5.5			9.24
SED inter × fert			5.51			37.03			6.9			8.75			1.6

Table 7 Groundnut grain yield (kg/ha) for Bungoma and Sega sites during long rains

	2005						2006					
	Bungoma			Sega			Bungoma			Sega		
	DAP (kg/ha)			DAP (kg/ha)			DAP (kg/ha)			DAP (kg/ha)		
Intercropping	0	150	Mean	0	150	Mean	0	150	Mean	0	150	Mean
Conventional	360	461	411	197	362	280	302	456	379	283	292	288
Hill	240	369	305	123	343	233	252	383	318	182	253	218
MBILI	356	564	460	221	416	319	378	680	529	290	448	369
Mean	319	465	392	180	374	277	311	506	409	252	331	292
SED intercrop (inter)			10.94			3.3			1.35			2.26
SED fertilizer (fert)			3.41			2.69			1.10			1.86
SED inter × fert			5.9			4.67			1.19			3.2

Table 8 Value-to-cost ratio of three maize–legume intercropping systems, with beans and/or groundnut or soybean and with or without fertilizer in four sites

Treatment	Kitale	Kuinet	Bungoma	Sega
Conventional + beans + no DAP (CBO)	0.29	1.00	0.42	0.30
Conventional + beans + DAP (CBP)	0.37	0.87	0.39	0.39
Conventional + soybean/groundnuts + no DAP (CSO/CSO)	0.32	0.95	0.39	0.26
Conventional + soybean/groundnuts + DAP (CSP/HGP)	0.32	1.01	0.43	0.28
Hill +beans + no DAP (HBO)	0.18	0.73	0.25	0.17
Hill +beans + DAP (HBP)	0.16	0.77	0.33	0.20
Hill + soybean/groundnut + no DAP (HSP/HGP)	0.21	0.87	0.30	0.19
Hill + soybean/groundnut + DAP (HSP/HGP)	0.25	0.80	0.31	0.21
MBILI + beans + no DAP (MBO)	0.43	1.27	0.51	0.43
MBILI + beans + DAP (MBP)	0.39	1.34	0.52	0.49
MBILI + soybean/groundnut + no DAP (MSO/MGO)	0.39	1.16	0.47	0.41
MBILI + soybean/groundnut + DAP (MSP/MGP)	0.47	1.05	0.54	0.47
Mean	0.32	0.99	0.41	0.32
Std. dev.	0.1	0.2	0.9	0.11
SE	0.03	0. 06	0.3	0.3

NB: Soybean for Kuinet and Kitale sites and groundnuts for Bungoma and Sega sites
Source: Computations derived from field trials in this study in 2005

losses of soluble cations and hence the low levels of Mg and Ca and the low soil pH.

The soils had high sand contents and low to medium organic matter content which predisposed the site to high leaching rates, hence losses of soluble nutrients, especially in the Bungoma site which had 72% sand content and this translates to the lowest N content of the four sites. This necessitates continuous nutrient replenishment as well as measures to reduce losses through runoff and leaching such as intercropping.

Maize Grain Yield

The MBILI intercropping with fertilizer resulted in increased maize yields compared to conventional intercropping systems ($P < 0.05$) in all the sites and seasons (Table 4) except for the Sega site during the long rains of 2005. This agrees with results of Tungani et al. (2002) that MBILI leads to higher yields of maize. It has also been reported that inclusion of legumes in intercropping systems reduces competition for N due to symbiotic N fixation by the legumes (Clement et al., 1992), and there is high potential for higher yields of intercrops when interspecific competition is less than intraspecific competition for a limiting resource (Francis, 1989). Similar levels of maize grain yields had been reported earlier by Ndung'u (2006) from modest P additions.

The better agronomic conditions offered by the rearranged cereal–legume rows under MBILI intercropping could also be a contributing factor. Tungani et al. (2002) observed that spatial arrangement of intercrops is an important management practice that improves radiation interception through more ground cover by both cereal and legume in the system. The MBILI intercropping had higher root length densities than did conventional intercropping systems ($P < 0.001$), which allowed it to explore a larger soil volume, especially if the nutrient was limiting, it became an important advantage for the intercrops in an MBILI system (Thuita, 2007).

Legume Yields

The hill intercropping system gave consistently the lowest yields in all the sites and seasons irrespective of whether fertilizer was applied or not for all the three legumes used. This is attributed to the negative interaction on the part of the legume such as competition for moisture, nutrients, and light among other factors that the legume in the hill intercropping system is subjected to. On the basis of these results, the hill system is an unsuitable form of intercropping, even though it is the most common form of intercropping practiced by farmers.

Soybean yields were generally low in both Kitale and Kuinet sites as higher yields under the MBILI intercropping have been reported in other studies (Thuita et al., 2005). This was attributed to poor variety choice as well as the low soil pH in the Kuinet and Kitale sites which could have affected the availability of other essential nutrients for legume growth mainly Mo as well as toxicity of Al and Mn as the soils in these areas are Fe rich. No nodulation of either soybean or bean was observed in the field in both seasons and this could have contributed to the low grain yields obtained. However, the MBILI intercropping system still gave the highest soybean yields in both sites (Table 5).

Application of fertilizers resulted in significantly higher bean yields than controls for all the intercropping systems and in all the sites and this agrees with findings of Kibunja et al. (2000), who reported that beans require external inputs to sustain high yields (over 1000 kg/ha) in low-fertility soils (in western Kenya, all crops need fertilizer to have high yields). This is contrary to what smallholder farmers practice, as they hardly use any fertilizers on the bean crop. The low bean yields in the hill intercropping system, even when fertilizer was applied, suggests that bean yields in the hill intercropping system are limited by other factors than soil fertility alone. Bean yields were low in the Kuinet soils for all the treatments. This was attributed to the soils in Kuinet site being more acidic (pH 4.27) compared to soils in Kitale (pH 5.17) and soil acidity may have affected the performance of the legume probably due to limitation in availability of nutrients such as molybdenum, which is essential in legumes but is affected by low soil pH (Tisdale et al., 1990). Most nutrients become less available at lower pH. This affects maize as well. This is in line with the conclusions of Woomer et al. (1997) and Tungani et al. (2002), who reported that although external nutrient input increases yields, the produce of the legume might only be modest.

The intercropping systems had a significant effect on the yields of groundnuts at both Bungoma and Sega sites ($P < 0.001$). In all the plots receiving fertilizers, in Bungoma site, there was increase in yields in 2006 compared to 2005, while in the Sega site, there was decrease in yields in 2006 compared to 2005 but it was not statistically significant. The increased yields were attributed to improved soil nutrient status that was observed at the end of the trials (Thuita, 2007). The marginal increased yields for plots could be due to the improved N status in the soil due to N_2 from the legumes and the residue that was always left on the plots at harvest for the three seasons that the trials were set.

The high yields of the MBILI system with addition of fertilizer were attributed to its row rearrangement which makes it suitable for growing groundnut as the legume is shade intolerant and does well only with adequate radiation (Tungani et al., 2002). This offers a viable option for increasing food production in the region as land sizes and the high food demands dictate that there should be increased food production.

The MBILI intercropping system has a future in this region as a cropping system but more work needs to be done to popularize it. Developing equipments either hand operated or animal drawn could go further by reducing the extra labor costs needed during planting.

Economic Returns from MBILI and Legume Grains

The *t*-test results showed that the net benefits from the treatments were significantly higher than those of controls at 95% confidence level. This means that the treatments led to significantly higher profitability than did controls in all the sites, emphasizing the need to apply external nutrients when doing any form of intercropping on low N and P soils in particular.

The higher returns achieved at Sega were attributed to the significantly high maize yields from hill intercropping with beans compared to the other systems, although its legume yields were significantly lower than MBILI and conventional intercropping systems. The conventional intercropping with beans at Bungoma was also highly ranked and this was because of the high bean yield (over 1 t/ha), thus significantly raising the economic returns from the treatment combination. At Kitale site the soybean yields were generally low and this resulted in the comparatively poor performance by MBILI system and conventional system with beans, while hill intercropping had high maize yields, resulting in the higher economic returns from the treatment.

Conclusions and Recommendations

Conclusions

The results from this study indicate that application of fertilizers in the low-fertility soils of western Kenya at the rate of 150 kg of DAP/ha significantly increased grain yields of maize, bean, groundnut, and soybean in all the nutrient-depleted sites.

Regardless of the intercropping system used, adding fertilizers resulted in increased maize grain yields. However, to increase the yields of legumes, smallholder farmers need to adopt the MBILI system as it gave high legume yields and even increased the maize yields as it probably benefited from N fixed by the legumes. However, MBILI alone will not ensure high yields without application of both N and P fertilizers in these low-fertility soils.

For each site, it was observed that the profitability of the intercrops depended on the legume yields. The marginal rate of returns were highest where the legume yield was highest and thus it was concluded that the choice of legume to be intercropped should be based on its suitability in the site but not just the market prices as the high yields easily outweighed any price difference between the legumes of choice.

Recommendations

Based on the observed high yields and rate of returns on invested capital obtained under MBILI intercropping systems, smallholder farmers should be encouraged to adopt MBILI but with suitable or high-value legumes for their region.

Using fertilizers (150 kg DAP/ha) and side dressing with CAN not only is profitable to the smallholder farmers but also leads to nutrient replenishment if applied for every cropping season and therefore farmers are encouraged to use it as it makes the profitability of intercropping much higher for their N- and P-depleted soils.

More work on MBILI intercropping system should now focus on diffusion, uptake, and adoption studies. The MBILI intercropping demonstration trials should include the required soil amendments needed in each area to enhance higher yields from the legumes.

Acknowledgment We wish to acknowledge the RUFORUM, who funded this study, and Ruth Njoroge and Scholarstica Mutua for their assistance in the field and lab works. I also wish to express my gratitude to all my co-authors for their part in the study and writing of this chapter. I could never have done it alone. To you all I say thank you.

References

Bouldin DR, Reid RS, Stangel PS (1980) Nitrogen as a constraint to non-legume food crop production in the Tropics. In: Drosdoff M, Zandstra H and Rockwood WG (eds) Priorities for alleviating soil-related constraints to food production in the tropics. IRRI, Los Banos, Laguna, Philippines, pp 319–337

Clement A, Chalifour FP, Bharati MP, Gendron G (1992) Effects of nitrogen supply and spatial arrangement on the grain yield of maize/soybean intercrop in a humid subtropical climate. Can J Plant Sci 72:57–67

Francis CA (1989) Biological efficiencies in multiple-cropping systems. Adv Agron J 42:1–42

FURP (1994) Fertilizer use recommendation project. Fertilizer use recommendations, vol 2. Kenya Agricultural Research Institute, Bungoma District, Nairobi, p 14

Jaetzold R, Schmidt H (1983) Farm management handbook Kenya; natural conditions and farm management information. Ministry of Agriculture, Nairobi

Kibunja CN, Gikonyo EW, Thuranira EG, Wamaitha J, Wamae DK, Nandwa SM (2000) Sustainability of long-term bean productivity using maize using maize stover incorporation, residual fertilizers and manure. Soil Science Society of East Africa, Nairobi

Meerman J, Cochran SH (1982) Population growth and food supply in sub-Saharan Africa. Finance Dev 193:1

Mukhwana EJ, Woomer PL, Okalebo JR, Nekesa P, Kirungu H, Njoroge M (2002) Best-bet comparison of soil fertility management recommendations in Western Kenya: independent testing by NGO's. Project proposal funded by the Rockefeller Foundation, 14

Ndung'u KW, Okalebo JR, Othieno CO, Kifuko MN, Kipkoech AN, Kimenye LN (2006) Residual effectiveness of Minjingu phosphate rock and fallow biomass on crop yields and financial returns in Western Kenya. Experimental Agriculture 42:323–336

Okalebo JR, Gathua KW, Woomer PL (2002) Laboratory methods of plant and soil analysis: a working manual, 2nd edn, Marvel EPZ (K) Limited

Okigbo BN, Greenland DJ (1976) Intercropping systems in tropical Africa. In: Papendick RI, Sanchez PA, Triplett GB (eds) Multiple cropping. Special Publication No. 27. American Society of Agronomy, Madison, WI, pp 63–101

Oram PA (1981) Production potentials in Africa: issues and strategies. In: Food policy issues and concerns in sub-Saharan Africa. IFPRI, Washington, DC, pp 45–80

Pratt DJ, de Hann C (1979) Crop and livestock integration in the semi-arid tropics. In: International symposium on development and transfer of technology for rain fed agriculture and the SAT farmer, ICRISAT, Hyderabad, India

Republic of Kenya (1994) Siaya District Development Plan (1994–1996). Office of the Vice President and Minister for Planning and National Development. Government Printer, Nairobi

Republic of Kenya (1997) Bungoma District Development Plan (1997–2001). Office of the Vice President and Minister for Planning and National Development. Government Printer, Nairobi

SAS Institute Inc (1999) Statistical analyses system/STAT user's guide version 8.1. SAS Institute Inc, Cary, NC

Sullivan P (2003) Intercropping principles and production practices. NCAT, ATTRA Publication #IP135. http://attra.ncat.org/attra-pub/PDF/intercrop.pdf

Thuita MN (2007) A study to better understand the "MBILI" and conventional intercropping systems in relation to root characteristics, nutrient uptake and yield of intercrops in western Kenya. MPhil thesis, Moi University, Eldoret

Thuita MN, Okalebo JR, Othieno CO, Kipsat MJ, Bationo A, Sanginga N, Vanlauwe B (2005) An attempt to enhance solubility and availability of phosphorus from phosphate rocks through incorporation of organics in western Kenya. 7th African crop science conference, Entebbe, 5–9 Dec 2005

Tisdale SL, Nelson WL, Beaton JD (1990) Soil fertility and fertilizers, 5th edn. MacMillan, New York, NY

Tungani JO, Mukhwana EJ, Woomer PL (2002) MBILI is number 1: a handbook for innovative maize–legume intercropping. SACRED, Bungoma, p 20

Wortman CS, Eledu CA (1999) Uganda's agro-ecological zones: a guide for planners and policy makers. Centro Internacional de International (CIAT), Kampala

Woomer PL, Bekunda MA, Karanja NK, Moorehouse T, Okalebo JR (1997) Agricultural resource management by smallhold farmers in East Africa. Nat Resources 34(4):22–33

Woomer PL, Mukhwana EJ (2000) A study on the effect of MBILI and conventional intercropping at on-farm level. A proposal forwarded and funded by the Rockefeller Foundation.

Bio-socio-economic Factors Influencing Tree Production in Southeastern Drylands of Kenya

L. Wekesa, J. Mulatya, and A.O. Esilaba

Abstract Empirical evidence on key biophysical, social and economic factors and their interplay for enhanced tree production in drylands of Kenya is scarce and scattered. The hypothesis for the study was that tree production in drylands was mostly practised by resource-poor farmers who based their decisions on a multiplicity of biophysical and socio-economic factors. Data were collected through a cross-sectional survey of 100 households using structured questionnaires. Sampling units were selected through multi-stage stratified random sampling procedures, and regression and descriptive statistics were applied in analysis. Findings suggest that tree production was used as a land use system mostly by those farmers who were young in age, poor and with modest levels of education. Apart from tree varieties, environmental and climatic conditions, tree survival was dependent on labour, germplasm, existing land uses, market access and availability of farmer, while tree technology adoption was dependent on farmer's age, capital assets, product price and availability of water and farmer. Household male heads were heavily involved in all tree operations. Marketing of tree products was dependent on price, labour, capital assets and technical skills. It was concluded that tree production in the southeastern drylands of Kenya was practised more with resource-poor farmers who based their decisions on a multiplicity of factors. Measures aimed at increasing farmer accessibility to water and markets for the tree products are crucial in enhancing tree production. Exploratory research on ways of enhancing growth rates and economic value of indigenous trees is urgent.

L. Wekesa (✉)
Kenya Forestry Research Station, Kibwezi, Kenya
e-mail: weknus@yahoo.com

Keywords Adoption · Land use system · Markets · Resource-poor farmers

Introduction

Trees are of great socio-economic importance to local economies in the drylands as well as the national economy of Kenya. Due to the energy crisis in the early 1970s, the government of Kenya realized the need to promote tree planting in areas outside gazetted forests and the Rural Afforestation Extension Scheme (RAES) was launched (Mukolwe, 1997). Since then, tree production especially in the drylands has been on the increase with resource-poor farmers taking the lead (GOK, 1986; World Bank, 1987; Kerkhof, 1990; ETC, 1995; Jayne et al., 1998; Muok and Cheboiwo, 2000). Resource-poor farmers are motivated into tree production because trees require low inputs and grant stable income generation (World Bank, 1986).

Farmers in drylands operate behind a background of biophysical and socio-economic constraints to tree production such as poor infrastructure, inappropriate germplasm, unfavourable land tenure and social cultural practices (Dewees, 1993; Muok et al., 2000; Wekesa et al., 2001). To sustain the increasing trend of tree production, pragmatic solutions over these hindrances are necessary. There was need for appropriately designed promotional arrangements focusing on biophysical as well as socio-economic dimensions of tree production in the drylands. Empirical evidence on key biophysical, social and economic factors affecting tree production in the drylands of Kenya is scarce. There have been minimal attempts to correlate tree

A. Bationo et al. (eds.), *Innovations as Key to the Green Revolution in Africa*,
DOI 10.1007/978-90-481-2543-2_105,

production with socio-cultural practices, accessibility to water and other resources, and capacities of households, making it hard to link tree production with community values, fears, preferences, capacities and farmer's decision mechanism. This study was aimed at filling this information gap by assembling empirical evidence on key factors that would form ingredients for a framework for enhancing tree promotion in the drylands.

The hypothesis for the study was that tree production in the drylands was mostly practised by resource-poor farmers who largely based their decisions on a multiplicity of bio-socio-economic factors. The main objective was to establish biophysical, social and economic factors affecting tree production in drylands. The specific objectives were the following:

(i) To characterize tree producers and their participation levels by gender and age characteristics;
(ii) To establish tree survival determinants in the drylands;
(iii) To identify factors affecting tree technology adoption in the drylands;
(iv) To establish marketing of tree products and their determinants.

Materials and Methods

Both primary and secondary data were used by the study. While secondary data were generated from existing literature through desk studies, primary data were obtained through a field survey covering 100 households. Data were collected using structured questionnaire. This was augmented with focus group discussions. Field study was conducted in three drylands districts of Kajiado, Kitui and Makueni located within the eastern and southern drylands of Kenya. Kajiodo district is found in the South Eastern part of Rift Valley Province, while Kitui and Makueni districts are found in the Southern part of Eastern Province. The districts were selected based on their high concentration of tree promotion projects.

Multi-stage stratified sampling procedures were applied in selection of sampling units for the survey. The data collected was coded and analyzed using Statistical Package for Social Scientists (SPSS) computer package for determination of frequencies, means, standard deviation and existence of relationships between variables. The *t*-test was used for significance testing at 90% levels.

Two-stage least squares (2SLS) regressions, which were consistent estimators, were applied to establish factors influencing tree survival, technology adoption and marketing as key components of tree production. It was hypothesized that tree survival was dependent on water, skills, labour, farm enterprise, market, education and germplasm. Tree technology adoption, on the other hand, was expressed in terms of total number of trees planted on a farm. Conventional wisdom is that rapid adoption of a technology is constrained by lack of income, limited access to extension, smaller farm size, inappropriate land tenure system, insufficient human labour and capital, absence of mechanized options to ease labour constraints, lack of access and untimely supply of farm inputs and water, and inappropriate transport and marketing facilities (Feder and Umali, 1993; Karanja et al., 1998). Finally, tree product marketing was hypothesized to be influenced by product price, technical know-how and labour. The functions applied were summed up as follows:

$$\text{Tree survival} = f(\text{WATER, SKILL, FAMILY, CROP, LIVESTOCK, MARKET, EDUC, SEED})$$

$$\text{Technology adoption} = f(\text{INCOME, FARM, EDUC, SKILL, ASSET, SEED, ROAD, TENU, FAMILY, WATER, MARKET, ENTERPRISE, PRICE, AGE})$$

$$\text{Market supply} = f(\text{PRICE, FAMILY, ASSETS, EDUC, SKILL})$$

The variable WATER was a measure of farmer's accessibility to water for tree production. Access to water variable was measured as the inverse of distance travelled to water source. The variable SKILL was used to reflect farmer's accessibility to technical skills through either technical training or extension services. The variable EDUC was a measure of education taken as a dummy for farmer's presence on the farm. It was hypothesized that those with higher levels of education were often working elsewhere from their farms as compared to those with limited levels of education whose off-farm employment opportunities were limiting. The variables CROP, LIVESTOCK, MARKET, FAMILY and SEED were used as a measure for crop production, livestock rearing, accessibility to markets, labour and availability of planting materials, respectively.

The variable INCOME was a measure of household's financial status in the form of credit, farm produce sales or accumulated savings. Income in the form of either accumulated savings or access to capital markets is needed to overcome fixed investment costs associated with adoption of new technologies (Lowdermilk, 1972; Lipton, 1976; Karanja et al., 1998). Tree production is associated with fixed investment costs because of long gestation period before returns are generated. The FARM was a measure of farm size. Several studies have demonstrated that adoption is related to farm size (Sharma, 1973; Barker and Herdt, 1978; Karanja, 1998). Where technology requires low investment, adoption is highest on smaller farms and vice versa (Karanja, 1998). In some cases, it has been observed that smaller farms initially lag behind large farms in adopting a technology but eventually catch up (Ruttan, 1997). Generally, smaller farms are more risk averse, thus failing to adopt new technologies. The variables AGE, ASSET, ROAD, TENU, ENTERPRISE and PRICE were a measure of age, capital assets, transport, tenure, other farm enterprises and price of products, respectively.

Results and Discussion

Key Tree Producers and Participation of Different Household Members in Tree Production

Tree production as a land use system was mostly adopted by farmers who were youthful in age, low-income earners and with modest levels of education. The number of trees planted per unit area decreased with age, income and education levels of farmers. Youthful farmers within 26–35 years of age category planted 64 trees/ha compared to those over 65 years who planted only 13 trees/ha (Table 1). Low-income earners with an annual income of less than Ksh. 24,000 planted 44 trees/ha unlike high-income earners with over Ksh. 96,000 earnings per annum planting only 9 trees/ha. Farmers without formal schooling planted 39 trees/ha compared to those with formal schooling at tertiary level planting a paltry 2 trees/ha.

Higher preferences for tree production by resource-poor farmers were attributed to low input requirement demanded and lower levels of failure by trees due to

Table 1 Number of trees planted per hectare by farmers of different age, education and income levels

Item	Category	Kajiado	Kitui	Makueni	Total mean
Age	Under 25 years of age	0	–	–	0
	26–35 years of age	3	170	20	64
	36–45 years of age	52	60	50	54
	46–55 years of age	10	57	15	27
	56–65 years of age	5	–	32	19
	Over 65 years of age	0	25	–	13
Education	Illiterate	18	95	4	39
	Primary	10	40	11	20
	Secondary	8	17	19	15
	Tertiary	4	0	3	2
Income	Lower (<24,000)		77	10	44
	Medium (24,000–96,000)	10	33	10	18
	Upper (>96,000)	1	15	11	9

Table 2 Levels of participation of household members in tree production operations

Tree operation	Percentage level of involvement			
	Husband	Wife	Sons	Daughters
Fruit seedling acquisition	62	19	20	10
Pitting for fruit planting	61	20	19	11
Fruit planting	61	20	20	12
Fruit weeding	61	25	22	17
Fruit watering	60	27	19	15
Fruit pruning	63	17	16	9
Fruit harvesting	61	14	20	9
Mean for fruits	*61*	*20*	*19*	*12*
Timber seedling acquisition	65	18	21	11
Pitting for timber tree planting	61	21	21	12
Timber tree planting	61	21	21	12
Timber tree watering	61	21	16	13
Timber tree weeding	60	19	21	11
Timber tree pruning	65	19	19	10
Timber tree harvesting	75	13	13	6
Mean for timber trees	*64*	*19*	*19*	*11*
Seedling acquisition of other trees	42	20	–	–
Pitting for other trees	39	27	–	–
Planting of other trees	40	27	–	–
Watering of other trees	40	27	–	–
Weeding of other trees	40	27	–	–
Pruning of other trees	30	20	–	–
Harvesting of other trees	50	0	40	0
Mean for others	*40*	*21*	*6*	*0*
Overall mean	*55*	*20*	*15*	*8*

vagaries of nature. Observations by Kerkhof (1990) showed that tree production was practised more by resource-poor farmers unable to meet their basic food needs and for whom it is a principal source of farm income. The World Bank (1986) pointed out that a decision to grow trees was influenced by their low labour and capital input requirements, and their stable income generation compared with food production. Dewees (1993) observed that tree cultivation (which required less labour and less capital investment than other types of crops) was an important source of income and land use option for the smallholder.

The level of participation of household members in execution of tree operations was based on gender and age. Level of involvement of male household heads in all tree operations was comparatively higher and was mean rated at 55% (Table 2). Daughters were least involved in execution of tree operations with their level of participation mean rated at a paltry 8%. In all operations the highest level of participation was with male household heads and least with children including sons and daughters.

Observations that male household heads were heavily involved in all tree operations exemplified their central role in decision making concerning tree production on the farm. Muok and Cheboiwo (2000) observed male heads of households to be main decision makers on tree planting in Kitui District. Chavangi (1993) noted that it was a taboo for women to plant trees when the husband is still alive in some communities, especially in western Kenya.

Tree Adaptability and Survival Determinants in the Drylands

Survival rates for tree species planted were variable. The rates were comparatively higher with indigenous fruit and woodlot tree mean rating at 81 and 74%, respectively. However, survival rates for exotic fruits and woodlot trees were comparatively lower with mean rating at 48 and 49%, respectively. Apart from climatic and environmental conditions, and tree types, tree

Table 3 Estimated parameters for determinants of tree survival

Dependent variable = Tree survival rate (%)	
Regressors	Regression coefficients
1. Constant	109.443 (3.393)
2. Household size (number of family members)	0.340 (0.784)
3. Education level	–6.657 (–0.704)*
4. Area under crops (ha)	7.738 (0.448)*
5. Area for livestock rearing	–37.520 (–2.043)
6. Distance to water source (km)	–6.394 (–1.008)*
7. Market distance (km)	–15.276 (–1.274)
8. Tree germplasm availability	176.296 (1.412)
R^2	0.991
$\check{R}^2$	0.932
F	16.644
N	80

The *t*-statistics are indicated in parentheses with * indicating significance at 10% significance level

survival was significantly influenced by farmer's level of education, crop cultivation and water accessibility (Table 3). Coefficient of determination (R^2) was 0.991, meaning that independent variables accounted for over 90% of variability in the dependent variable. Tree survival was positively related to area under crop cultivation and was statistically significant at 10% level. This meant, ceteris paribus, that increase in area allocated to crops would result in increased survival rates. Where intercropped with trees, crops acted as nurse plants for the young trees.

Tree survival rates were negatively related to education level and distance to water source. The sign of coefficients of variables was negative and was statistically significant at 10% level. This meant that increase in levels of education and distance to water source would result in reduced survival rates of trees planted. The negative impact of education on tree survival could be due to farmer's presence for supervision and execution of different tree operations. Farmers with higher levels of education were in better positions to take up off-farm employment, thus delegating most of the farm operations to workers or other members of their households. Such a move erodes farmer's supervisory role, thus compromising tree survival and establishment on the farm. The longer distances to water sources reduced household accessibility to water for supplementary watering of trees at initial stages of establishment.

The drylands are difficult environments full of vagaries of nature in the form of harsh climatic conditions and fragile environment, which negatively affect tree establishment on farms. Observations of high survival rates for indigenous trees indicate their high adaptability to these conditions. Studies conducted in the region on live fences showed that *Acacia mellifera* was the best in terms of both impenetrability and survival (Mohamed et al., 2000).

Tree survival depends on tree types, environmental and climatic factors, labour and site preparations. Labour is necessary for execution of tree operations including watering, weeding and general tree care. Crops act as nursing plants to the young trees planted as demonstrated under a *Shamba* (cultivation of crops in tree plantations) system. The survival rates could also be enhanced through increased supervision by farmer through his/her increased level of presence on the farm. Other measures like establishment of own nursery enhance tree establishment by enabling farmer make a choice to timely plant healthy and strong seedlings. Late planting in the drylands often compromises survival rates. Availability of ready markets also acts as an incentive to farmers to direct more resources in tree production.

Factors Affecting Tree Technology Adoption

The variables affecting tree technology adoption include price of products, age, capital assets, education, access to water and income status (Table 4). The coefficient of determination (R^2) was 0.849, demonstrating that independent variables accounted for over 80% variability in the dependent variable. The variables for capital assets, product price and access to water significantly and positively influenced tree technology adoption measured by number of farmers applying recommended technologies for tree production. Access to water variable was statistically significant at 1% level, whereas capital assets and price of product variables were significant at 10% level. The results demonstrate that, ceteris paribus, tree technology adoption is directly related to level of capital assets, access to water and price of tree products. Increase in levels of these variables would result in enhanced tree technology adoption by households in drylands. Capital assets including farm tools and transportation facilities like bicycles and vehicles

Table 4 Estimated parameters for tree technology adoption determinants

Dependent variable = number of trees planted per acre on a farm holding

Regressors	Regression coefficients
1. Constant	179.765 (0.974)
2. Price of tree product	2.385 (1.430)*
3. Age (combined age of husband and wife)	–1.013 (–2.393)**
4. Level of capital assets on the farm	3.229 (1.896)*
5. Education level (both husband and wife)	–13.766 (–2.793)**
6. Access to technical skills	4.289 (0.607)
7. Income status	–1.372 (–1.987)*
8. Access to water	44.922 (3.325)***
9. Farm size	0.463 (1.664)
10. Household family size	1.779 (0.605)
11. Market distance	–32.276 (–1.425)
12. State of road infrastructure	4.489 (0.278)
R^2	0.849
$-\breve{R}^2$	0.647
F	4.208
N	91

The *t*-statistics are indicated in parentheses with ***, **, * indicating significance at 1, 5 and 10% significant levels, respectively

are handy for tree production and those households with more capital assets were more inclined to adopt the technologies. In addition, households with easy access to water resource were more likely to adopt tree technologies because water is critical for tree establishment, especially at initial establishment stages and for enhanced productivity. Price as a distributor of resources positively spurs tree technology adoption by influencing households to allocate more resources towards tree production. Thus households accessing higher product prices were better adopters of technologies.

On the other hand, the variables for age, education and income significantly and indirectly influenced tree technology adoption. The sign of their coefficients was negative and was statistically significant at various levels. The variables for age and education were statistically significant at 5% levels, whereas income variable was significant at 10% level. The results demonstrate that, ceteris paribus, tree technology adoption is indirectly related to age, education and income. Any increment in levels of each of these variables would result in decreased tree technology adoption by households in the drylands. Households with couples advanced in age would not adopt tree technologies largely because of the length of time required to derive benefits from tree enterprises. Most people advanced in age would be reluctant to make such long-term investments, thus holding back on tree technology adoption. Education and income had negative impact on tree technology adoption probably because most households with higher education and incomes were mostly working off-farm and were more inclined to short-term, high-yielding farm enterprises.

Marketing of Tree Products from the Farms

The key products marketed were tree seedlings, fruits and charcoal. Average quantities marketed and prices were variable but fruits were the most highly marketed with mangoes taking the lead. The highly marketed seedlings were fruit tree seedlings including local mangoes and pawpaws. Factors significantly affecting market supplies of fruits to the market were price and access to technical skills (Table 5). However, the coefficient of determination (R^2) was only 0.20, meaning that the variables accounted for only 20% of variability in the dependent variable.

The coefficients of all the variables were positive, indicating their positive influence on market supply of products. The variable price and access to technical skill were statistically significant at 10% level.

Table 5 Factors affecting market supplies of fruits

Dependent variable = quantity of fruits supplied to the market in kilograms

Regressors	Regression coefficients
Constant	–37,971.793 (–0.843)
Price	38.097 (0.127)*
Family size	1740.637 (0.768)
Capital assets	767.500 (0.347)
Education	3836.515 (0.617)
Access to technical skills	176.217 (2.817)***
R^2	0.195
$\breve{R}^2$	0.087
N	43

The *t*-statistics are indicated in parentheses with ***, **, * indicating significance at 1, 5 and 10% significant levels, respectively

Commercialization of trees and their products like timber, poles and resins in the area was in the infancy stages of growth in terms of scale of production and marketed volumes. Low marketing of tree products in the region could be due to low levels of awareness and insufficient volumes of high-value tree products. Studies conducted by Wekesa et al. (2001) showed that charcoal and firewood were the commonly traded tree products in the region. The study observed that selling of such tree products was more often than not done when livelihood options became limiting in the area. Rather than viewing the selling of charcoal and firewood as a commercial venture, farmers engaged in their sale only as an off-season activity or when clearing bushes to create room for settlement or farming. A survey conducted by Muok and Cheboiwo (2000) identified charcoal and fruits as the major tree products sold by farmers in Kitui District.

Conclusions and Recommendations

Tree production in the southeastern drylands of Kenya was practised by resource-poor farmers who are youthful in age, low-income earners and with modest levels of education who based their decisions on a multiplicity of factors including biological, social and economic factors, and their interplay. The decision to engage into tree production was dependent on farmer's age, capital assets, price, income status and availability of water and the farmer. Indigenous tree species recorded better survival rates than did exotics in the region. Apart from environmental and climatic conditions, and tree types, survival of trees planted on farms in the region was dependent on type of farm enterprise, and availability of water and the farmer. Supply of tree products to the market was dependent on price and technical know-how. Although all household members were involved in different tree operations on the farm, male household heads played a major role in all tree operations on farms.

For the trend of tree production in the drylands to be invigorated and sustained, it is necessary that:

i. Resource-poor farmers who are highly receptive to tree planting be facilitated with information and financial support to access key inputs such as water in their tree planting efforts to enhance greening and conservation measures in the drylands;
ii. Sensitization and demystifying campaigns be held to enhance involvement levels of all household members in tree production for equity and fair sharing of resources;
iii. Exploratory research to identify technologies for enhancing growth rates and value of indigenous trees be initiated.

Acknowledgements Special thanks go to the director, KEFRI for funding the study. The National Programme Coordinator/Drylands is highly acknowledged for the timely release of required resources. The Desert Margins Programme (DMP) is acknowledged too for providing the much needed field logistical support. Special regards to farmers in the southeast drylands of Kenya who tirelessly laboured to plant trees and their contributions to this study. The contribution of the field staff including KEFRI Kibwezi Team is highly appreciated.

References

Barker R, Herdt RW (1978) Equity implication of technology changes. In: Interpretive analysis of selected papers from changes in rice farming in selected areas of Asia. International Rice Research Institute, Los Banos

Barrow EG (1996) The drylands of Africa: local participation in tree management. Initiatives Publishers, Nairobi, Kenya

Chavangi N (1993) Socio-cultural issues in agroforestry. Paper presented to district level agroforestry course. KEFRI, Muguga

Dewees PA (1991) The impact of capital and availability on smallholder tree growing in Kenya. DPhil dissertation, University of Oxford, UK

Dewees PA (1993) Social and economic incentives for smallholder treegrowing: case study for Murang'a District, Kenya. Community Forestry Case Study 5. Food and Agriculture Organization, Rome

ETC (1995) Agroforestry for integrated development in semi-arid areas of Kenya (ARIDSAK) project document

Feder G, Umali DL (1993) Adoption of agricultural innovations: a review. Technol Forecast Soc Change 43(3/4):215–239

Fridah MW (1999) Charcoal trade in Kenya. Working paper No. 5, RELMA/SIDA

Gerhart J (1975) The diffusion of hybrid maize in western Kenya. D.F. Centro International de Mejoramiento de Maiz y Trigo (CIMMYT), Mexico City

GOK (1986) Draft annual forest report. Forest Department, Ministry of Environment and Natural Resources, Nairobi, Kenya

Jamison DT, Lau LJ (1982) Farmer education and farm efficiency. John Hopkins University Press, Baltimore, MD

Jayne TS, Kodhek GA, Nyambane G, Yamano T (1998) Baseline characteristics of smallholder agriculture and non-farm activities for selected districts in Kenya. Paper presented at the conference on "Strategies for Raising Smallholder Agricultural Productivity and Welfare", Egerton University/Tegemeo Institute of Agricultural Policy and Development, Nairobi, 24 Nov 1998

Karanja DD (1998) An institutional and economic analysis of maize research in Kenya. MSU international development working paper No. 57. Department of agricultural Economics, Michigan State University, East Lansing, MI

Karanja DD, Jayne TS, Strasberg P (1998) Maize productivity and impact of market liberalization in Kenya. Paper presented at the conference on "Strategies for Raising Smallholder Agricultural Productivity and Welfare", Egerton University/Tegemeo Institute of Agricultural Policy and Development, Nairobi, 24 Nov 1998

Kerkhof (1984) Agroforestry in Africa: a survey of project experience. SIDA/CTA, Great Britain

Kerkhof P (1990) Agroforestry in Africa: a survey of project experience. Panos Publications Ltd, UK

Lipton M (1976) Agricultural finance and rural credit in poor countries. World Dev 4(July):543–554

Mbabu P, Wekesa L (2004) Status of indigenous fruits in Kenya. A report prepared for IPGRI-SAFORGEN in the framework of AFREA/FORNESSA

Mohamed AM, Mulatya J, Chikamai B, Gerkens M, Mengich E, Kimotho L, Muchiri D, Ochieng D, Wekesa L, Ego W (2000) Rehabilitation of degraded lands in sub-Saharan Africa: the case of the ARIDSAK project

Mukolwe M(ed) (1997) Proceeding of KEFRI/JICA conference on social forestry and tree planting technology in semi-arid lands at SFTP-KEFRI Headquarters, Muguga, Kenya, on 29th Sept to 2nd Oct 1997

Mulatya J, Misenya T (2004) *Melia volkensii* growth in the southern drylands of Kenya. A paper in the proceedings of the first national workshop on recent Mukau (*Melia volkensii* Gurke) research and development held at KEFRI Kitui from 16–19 Nov 2004

Muok B, Cheboiwo J (2000) Socio-economic and resource survey of Kitui district. A paper in the proceeding of the social forestry extension seminar for the promotion of tree planting in arid and semi-arid areas of Kenya held at KEFRI, Muguga, Mar 2000

Muok BO, Owuor B, Dawson IK, Were JM (2000) The potentials of indigenous fruit trees: results of a survey in Kitui district, Kenya. Agroforest Today 12:13–16

O' Kefee P, Raskin P, Bernow S (1984) Energy and development in Kenya: Opportunities and constraints. The Beijer Institute and the Scandinavian Institute of African Studies, Stockholm, Uppsala

Rosenzeweig MR (1978) Schooling, locative ability and green revolution. Paper presented in a meeting of the Eastern Economic Association. Washington, DC, 1978

Ruttan V (1977) The green revolution: seven generalizations. Int Dev Rev 19(December):16–23

Sharma AC (1973) Influence of certain economic and technological factors on the distribution of cropped area under various crops in the Ludhiana district. J Res 10(June)

Wekesa L, Mulatya J, Gerkens M (2001) Marketing of agroforestry products in the semi arid areas of South Eastern Kenya: the case of the drier parts of Makueni and Kajiado Districts. ARIDSAK project working paper no 3. Nairobi, Kenya

Wekesa L, Wang'ombe E, Mukolwe M, Yamada I, Sato Y (2005) Marketing study of ecological resource products. Intensified Social Forestry in Semi-arid Areas Project (ISFP). Can be found online at http://www.isfp-fd.org

World Bank (1986) Kenya: agricultural inputs survey. World Bank, Washington, DC

World Bank (1987) Kenya: forestry sub-sector. World Bank, Washington, DC

Economic Evaluation of the Contribution of Below-Ground Biodiversity: Case Study of Biological Nitrogen Fixation by Rhizobia

J. Chianu, J. Huising, S. Danso, P.F. Okoth, and N. Sanginga

Abstract Although it is common knowledge that soil microorganisms form an important constituent of below-ground biodiversity and provide ecosystem services, such knowledge does not often lead to formulation of policies to conserve and manage these soil microorganisms, or to strategies that lead to explicit use of these resources. Applying the knowledge gained from several experimental stations and from on-farm research [supplemented with necessary assumptions on FAO-sourced secondary data on soya bean (*Glycine max*) from 19 countries in Africa], this study attempts to increase the awareness on the importance of these microorganisms by quantifying the economic value of nitrogen fixation of legume-nodulating bacteria (LNB) associated with promiscuous soya bean. Computation of economic value (of nitrogen fixation) was based on the method of cost replacement or cost savings in terms of mineral nitrogen fertilizer that would have been required to attain the same level of nitrogen fixed biologically. Result shows that the economic value of the nitrogen-fixing attribute of soya bean in Africa, especially the promiscuous varieties, annually amounts to about US $200 million across the 19 countries. The study concludes with recommendations on various ways of increasing the chances of smallholder farmers benefiting from the nitrogen-fixing attribute of LNB, especially since many of them cannot afford adequate quantities of inorganic fertilizers required for increased crop productivity.

J. Chianu (✉)
Tropical Soil Biology and Fertility Institute of the International Centre for Tropical Agriculture (TSBF-CIAT), UN Avenue, Gigiri, Nairobi, Kenya
e-mail: jchianu@yahoo.com

Keywords Africa · Economic valuation · Favourable policies · Legumes · N_2 fixation

Introduction

Soil organisms contribute a wide range of essential services to sustainable functioning of agro-ecosystems (Swift, cited by Syaukat, 2005). However, few studies have been undertaken to measure the economic contribution these services represent and therefore underline the importance of these soil organisms as biological resources that need to be conserved and sustainably managed.

Biological nitrogen fixation (BNF) is one such process that is moderated by nitrogen-fixing bacteria in the soil, of which the benefits are evident as these bacteria fix nitrogen from the atmosphere that may then become available for plant growth. Of specific interest are leguminosae-nodulating bacteria (LNB) that form symbiotic associations with a leguminous host plant providing nitrogen to the host for which it gets carbohydrates in return as an energy resource. In agriculture, advantage is taken of this ability to fix nitrogen by growing legume crops like soya bean (*G. max*), cowpea (*Vigna unguiculata*), common bean (*Phaseolus spp.*), groundnut (*Arachis hypogea*) and others, all of which generally have the ability to nodulate with a wide variety of rhizobium strains in the soils (hence are generally referred to as being promiscuous). A substantial portion of the world's supply of organic nitrogen is fixed via the symbioses between root-nodulating bacteria and leguminous host plants (Simms and Taylor, 2002). Estimates available indicate that biologically fixed nitrogen in agricultural field is about double the

A. Bationo et al. (eds.), *Innovations as Key to the Green Revolution in Africa*,
DOI 10.1007/978-90-481-2543-2_106, © Springer Science+Business Media B.V. 2011

nitrogen produced industrially. For example, a mix of the legume cover crops (LCC), *Pueraria phaseoloides* and *Calopogonium caeruleum*, could supply approximately 133 kg N year^{-1} into the soil through its biomass (Siregar, 1984). This paper deals with the economic gains derived from the use of promiscuous soya bean cultivars in Africa as a particular case. The selection and breeding of high-yielding soya bean cultivars that maintain their promiscuous traits as done in Africa presents an interesting case in that it aims to make use of the variety of strains (biodiversity) that exist within the soil. The strategy is an alternative to the inoculation of more restrictive soya bean cultivars with selected rhizobium strains, as pursued in the USA and Brazil. The promiscuous soya beans cultivars are able to effectively nodulate with a wide variety of (brady)rhizobium strains. It was hypothesized that the diversity and efficiency of the existing 'native' strains is sufficiently large to allow achieving levels of N_2 fixation (and yield levels) similar to what is obtained with American type of soya bean under soil conditions prevailing in soya bean cultivation areas in the USA/Brazil.

Furthermore, nitrogen fixation is of particular relevance in providing an important source of N, especially in Africa where fertilizers are often not affordable to smallholder farmers and where fertilizer input remains at a very low level. Soya beans then represent a promising crop because of its dual-purpose character (produces saleable crop and contributes to fertility) and because of its high market potential (African demand is more than domestic supply).

The breeding programme at the International Institute of Tropical Agriculture (IITA) in the 1980s aimed to combine agronomic advantages (high yield) of US soya bean with the promiscuity of Asian cultivars so that the crop could be grown without bradyrhizobial inoculants (Kueneman et al., 1984). However, successful breeding lines could be developed only after considering the symbiotic effectiveness in combination with promiscuity (Sanginga et al., 1999). This resulted in one soya bean breeding line (1660-19F) that proved to nodulate effectively in various field locations. This soya bean line did not show response to bradyrhizobial inoculation and application of N (indicating effective N uptake), and appeared to be a suitable line of choice for the smallholder farmers who will not artificially inoculate their soya bean seeds (Sanginga et al., 1999). Additional benefits obtained in the form of residual N available for uptake by a second crop (companion or subsequent crop) range from 10 to 24 kg ha^{-1}. This represents 14–36% of the maize total N ha^{-1} (Sanginga et al., 2001) which is significant in the context of the low input use in subsistence farming in the tropics. Further efforts have been geared towards development of 'leafy' soya bean lines that produce more biomass while still maintaining acceptable yield levels to enhance the beneficial effect for soil fertility.

New soya bean varieties have had widespread adoption in Nigeria where production has increased from 60,000 (in 1984) to 405,000 t (1999), stimulated by industrial demand and home processing and utilization (Sanginga et al., 2003). Market access and expenditure on hired labour were important socio-economic determinants in the adoption of these varieties (Sanginga et al., 1999). Besides traits like early maturity, farmers appreciated the ability to give good yield without mineral fertilizer application.

In southern Africa (notably Zambia and Zimbabwe), the strategy was to select from available varieties bred from genotypes of east Asia origin (rather than breed), on the basis of yield potential and ability to nodulate with indigenous rhizobia strains (Mpepereki et al., 2000). *Magoye* and to lesser extent *Hernon 147* (with background of genotypes from different locations in east Asia) were found to nodulate effectively with isolates obtained from soils in southern Africa with no history of soya bean cultivation (which seems to suggest that these rhizobia bacteria are indigenous). *Magoye* also did not respond to inoculations (Mpepereki et al., 2000). In Zambia, *Magoye* and *Hernon 147* were released in 1981 as varieties recommended for production by small-scale farmers. In Zimbabwe, soya bean had been grown in small pockets since late 1950s, with varying successes. Promotion programmes have had temporary and local success in the 1980s. A large programme was initiated in 1996 to facilitate and accelerate the extension of promiscuous soya beans seed among smallholder farmers (Mpepereki et al., 1996). This led to an almost 10-fold increase in adoption of improved soya bean varieties. Farmers were offered seeds of both promiscuous and specifically nodulating soya beans. Most farmers opted for both the promiscuous and specifically nodulating varieties, citing the large biomass from promiscuous varieties for use as a soil amendment and fodder and the higher yields of specifically nodulating varieties for food and household cash income. The area under promiscuous soya bean decreased in 1998/1999 as a result of increase in the price of inputs (e.g. commercial soy seed). Training in processing for home use as well as good market prices was instrumental to the adoption and increased production of soya bean in Zimbabwe (Mpepereki et al., 2000).

This study deals with the economic gains derived from the use of promiscuous soya bean in sole cropping, intercropping and rotational cropping systems. Advantages like savings in fertilizer cost and enriching soil with nitrogen leading to better growth of a companion non-legume crop or of a succeeding non-legume crop could be evaluated economically. There are additional 'environmental' benefits like minimizing soil erosion, inhibiting weed growth, improving soil structure and maintaining microclimate, as well as the maintenance of the diversity and density of bacterial LNB populations that may be associated with the use of legume crops. These are, however, more difficult to quantify and certainly more difficult to attribute to these soil organisms. Analyses of economic gains in terms of benefits (increased yield, reduced fertilizer application) to the farmers therefore remain a partial analysis. So far, information on the economic value of N_2 fixed through BNF bacteria is scanty.

Methodology

Important elements in the approaches and methods used in this study are (i) review of articles, reports and methods for quantifying the N_2 fixed in various legume cropping systems (with particular reference to soya bean); (ii) review of papers and case studies on economic evaluation of N_2 fixation in legumes; (iii) use of existing data to compute the monetary value of some benefits from N_2 fixation in soya bean by rhizobia; (iv) outline of economic evaluation of N_2 fixation and (v) calculations of the economic value of the N fixed, N balance in soil and N made available to subsequent crops in legume–cereal rotation. The computed values indicate magnitudes of gross economic benefits generated by rhizobia through the fixing of N_2 from the atmosphere in association with promiscuous soya bean, since it was not possible to account for the related costs (fixed and variable) using existing data. A complete economic evaluation requires that all costs (fixed and variable) of all inputs into a system or into an enterprise and the value of all outputs from the system or from the enterprise be taken into account. However, most of the studies on N_2 fixation reviewed were not carried out with economic evaluation in mind. As a result, data were not collected on most of the inputs and many costs were not accounted for. Similarly, except for grain production and nitrogen, data on many outputs from N_2-fixing legumes (e.g. total biomass) were not supplied. The organic matter produced may be of more relevance than the N equivalent of this biomass. This notwithstanding, it is good to acknowledge that it is often very difficult to determine the monetary value of some inputs and related costs or outputs [e.g. cost of N_2 fixation as input into production of soil organic matter (SOM) and associated outputs, e.g. low soil erosion]. In this study, we settle, therefore, for partial economic analysis, ignoring the costs of many inputs and putting economic value on only those outputs that are clearly tangible.

There are various ways of tapping the benefits (e.g. economic) of N_2 fixation. *Direct tapping* describes a situation whereby the plant use part or the entire N_2 fixed. About 60% N within the biomass of LCC are produced through N_2 fixation (Mangoensoekarjo and Semangun, 2003). However, data from Sanginga et al. (2001) suggest that 60% might be difficult to obtain for soya bean. Sanginga et al. (2002) mention 42% for soya bean, and Sanginga et al. (2003) give a value of 30 kg N under promiscuous soya bean/maize rotation. *Indirect tapping* describes a situation whereby net N is contributed to the soil, subsequently used by a succeeding crop. In this case we need to know the differential N obtained from the soil, the N derived from the fertilizer and the N derived from the atmosphere. *Soil N* per se describes how through N_2 fixation farmers can save some money that would have been spent on N fertilizers to improve fertility, valued through cost replacement method. We quantified the economic benefits of N_2 fixation to fixing plant (*direct benefits*), to associated crops and non-legumes grown in rotation (*indirect benefits*) and to soil N per se. Aspects where data were easily accessible included N_2 fixation, total N and yield. We computed monetary equivalent of N fixed relative to urea N fertilizer that would have been purchased had no N been fixed (*cost replacement method*).

Direct Tapping

There is no single correct way of measuring N_2 fixation (Mafongoya et al., 2004). Several methods (e.g. N difference, N solute, isotope dilution, natural ^{15}N abundance) have been used. We adopted different approaches and formulae, giving examples.

To determine the cost of the equivalent amount of chemical N fertilizer (urea) that could achieve same effect as the N fixed in soya bean (*direct tapping*), we use *formula* 1 to compute the urea equivalent of N fixed and *formula* 2 to compute the corresponding monetary value of the urea equivalent.

$$Y = N_{\text{fixed}}(100/(\text{fNdfF})(100/\text{N avail CUrea}) \quad (1)$$

where

Y is the quantity of urea fertilizer equivalent of N fixed
N_{fixed} is the quantity of N fixed
fNdfF is the percentage efficiency with which urea is absorbed by soya bean
N avail CUrea is the percentage N available in a compound urea fertilizer

$$\begin{aligned} \text{M Value} &= Y \times \text{UreaPrice} \\ &= (N_{\text{fixed}} \times 100)/\text{fNdfF}\times \quad (2) \\ &\quad (100/\text{N avail CUrea}) \times \text{UreaPrice} \end{aligned}$$

where

MValue is the monetary value of N fixed
UreaPrice is the price per unit of urea fertilizer

When applying *formulae* 1 and 2, we assumed that N fixed was measured directly and actually known and 40% use efficiency for the urea fertilizer. Plants are incapable of absorbing the entire N applied through fertilizers. Data on nutrient use efficiency (the fraction of N applied in fertilizer that ends up in plant, fNdfF) are few in literature, with most reliable estimates obtained from experiments using ^{15}N fertilizers. The value of 40% use efficiency is based on the few available data and on the knowledge that legumes are poor scavengers of soil N. Urea was adopted as model fertilizer in all economic evaluations for all calculations where N_2 fixed was compared with N fertilizers. The available percentage of N in a urea fertilizer is 46.7%. For ease of calculation we took a fixed price for a unit of fertilizer of US $400 per ton of urea; prices may differ rather strongly from one place to the other in SSA (Kumar, 2007). This is representative of SSA farm gate price that accounts for all related transaction costs (i.e. FOB and transport cost). Based on this information, the following example can be given for the calculation of the equivalent value of urea needed to fix 50 kg of N ha^{-1}.

Example 1: The economic value of 50 kg N ha^{-1} of the N_2 fixed by natural soil biodiversity (un-inoculated) in a soya bean crop. Applying *formula* 2, we compute as follows:

$$\begin{aligned} \text{MValue} &= Y \times \text{UreaPrice} \\ &= [(N_{\text{fixed}} \times 100)/\text{fNdfF}] \times [(100/\text{N avail CUrea})] \times \text{UreaPrice} \\ &= [(50\,\text{kg} \times 100)/40] \times [(100/46.7)] \times \text{UreaPrice}] \\ &= [267.5\,\text{kg of Urea Fertilizer ha}^{-1}] \times \text{UreaPrice} \\ &= 267.5 \times \$0.40\,\text{ha}^{-1} \text{ or } \$\,107\,\text{ha}^{-1} \end{aligned}$$

From literature survey, we computed an average value (mostly on-station/on-farm trials) of 100 kg N ha^{-1} fixed in soya beans grown in Africa. We assumed that commercial farmers also grow promiscuous soya bean in most cases and that both smallholder and large-scale farmers attain similar yield levels. For farmers' fields, we accounted for the generally lower plant densities compared to trials and assumed a 15% reduction in N fixing levels as a consequence. Based on this, the N fixed in a country was obtained by multiplying the area harvested by 85 (15% downward adjustment from level of N fixation in trials). We assumed in all cases that the promiscuous soya bean has fixed N (even if this is not measured). See Example 2.

Example 2 Given that the total area grown to soya bean in a region is 1000 ha, we estimate the potential contribution of the N fixed and the economic value as follows:

$$\begin{aligned}\text{Total } N_{fixed} \text{ in equivalent N urea} &= 1000\,\text{ha}^{-1} \times 85\,\text{kg} \times 100/40\\ &= 212{,}500\,\text{kg N}\\ \text{Total economic or monetary value} &= 212500 \times 100/46.7 \times \$0.4\\ &= \$182013\end{aligned}$$

Indirect Tapping

When fixing N, legumes use less soil N than do cereals (*sparing N*). Also, legume residues are richer in N and decay faster than most cereals. Consequently, it is not uncommon that a cereal grown in rotation with legume obtains more N from soil than when cereal precedes cereal. The higher yield of the cereal in a legume–cereal rotation than in a cereal–cereal rotation has therefore been attributed to benefits derived from N_2 fixation (Klaij and Ntare, 1995; Bagayoko et al., 1996), even though there is much evidence that the benefits are not entirely attributable to the N_2 fixed only. Friesen et al. (2003) report that legume rotations in east Africa consistently produced higher maize grain yield (1.5–3.5 t ha^{-1} or 27–134%) than unfertilized maize in monocrop. Sanginga et al. (2003) state that the yield increases in subsequent maize crop are often larger than can be explained by the residual N from legumes. Sanginga et al. (2002) found that 10–22 kg N ha^{-1} was recovered from soya bean residues, representing 14–36% of the maize total N (Sanginga et al., 2001).

To derive the monetary value, formulae 1 and 2 as above were used, using urea as comparative N fertilizer (with assumptions). A more technical approach involves using N^{15} isotopes to assess extent to which unlabelled atmospheric N_2 derived by a preceding legume diluted the existing N^{15}/N^{14} ratio in soil. Cereal that follows legume, benefiting from N_2 fixed by earlier legume, would be expected to absorb N of lower N^{15}/N^{14} ratio than would the cereal that was preceded by a cereal. N contributed from fixation to next cereal is computed as follows:

Amount (kg N ha^{-1}) derived from residual N fixed is equivalent to 1–% ^{15}N atom excess in the cereal grown after a legume divided by % ^{15}N atom excess in the cereal grown after a cereal multiplied by kg ha^{-1} total N yield of cereal following legume. Again, formulae 1 and 2 are used to extrapolate to the equivalent amount of urea and its US$ value, again using urea as the comparative N fertilizer and with all the assumptions (or real cases where possible).

Contradictory reports have been issued on N fixed being made available to accompanying crop in legume–cereal intercrop. The usual high land equivalent ratio (LER) of an intercrop compared to respective sole crops is attributed, at least in part, to the sharing of the N fixed by both crop species. The ^{15}N methodology has provided evidence for N transfer. By deriving part of its N from the fixed N in a legume (of lower ^{15}N enrichment), the cereal in the intercrop is expected to contain N of lower $^{15}N/^{14}N$ ratio than is sole cereal. The amount of N transferred (kg ha^{-1}) is equivalent to 1–% ^{15}N atom excess in cereal in intercrop divided by % ^{15}N atom excess in sole cereal and multiplied by total N (kg ha^{-1}) of intercropped cereal.

Soil N Balance (Soil N Per se)

N-fixing legumes normally derive a portion of their N from soil N. Depending on relative amounts of N fixed and accumulated from soil, legume may deplete soil of its N. Relative distributions of N in seed versus the unharvested residues also influence the N balance in soil following the growth of legume. To compute N balance (kg N ha^{-1}), data are needed on (i) total N fixed (kg N ha^{-1}) in crop (including N in roots) and (ii) soil N (kg N ha^{-1}) in harvested seed. Soil N balance will be positive only if the % N fixed is greater than N harvest index (N in seed/N in residues), assuming that N in residues are returned to soil. Soil N could also be obtained if all inputs and outputs of N are measured and the difference computed. This approach was, however, not included in the calculations.

Results

Comparison of Soya Bean Cultivars in Terms of Economic Gains from N_2 Fixation

The extent of N_2 fixation by soya bean differs from cultivar to cultivar and as expected reflects on the magnitude of the economic value of the N_2 fixed. Using direct tapping, indirect tapping and soil N scenarios, we computed the economic value of N_2 fixed. Results based on data from Nigeria show that under 'direct tapping', the economic benefits of N_2 fixed by different soya bean cultivars range from US \$41 ha^{-1} (for IAC 100) to US \$288 ha^{-1} (TGx 1660-19F) with a mean of about US \$167 ha^{-1} (Fig. 1). When looking at the economic value of maize grain yield increment due to a preceding soya bean crop (so-called rotational effect), we found that the values range from about US \$65 ha^{-1} (IAC 100) to about US \$360 ha^{-1} (TGx 1456-2E) with a mean of about US \$205 ha^{-1} (Fig. 2). Another computation (Nigerian data also) based on mean of three maize cultivars (Sanginga et al., 2002) but only measured in terms of grain yields indicates that the rotational effect (on maize yields) ranges from about 239 kg ha^{-1} (TGx 1660-19F) to about 1802 kg ha^{-1} (TGx 1456-2E) with a mean of about 867 kg ha^{-1}, indicating huge N-fixing and rotational effects.

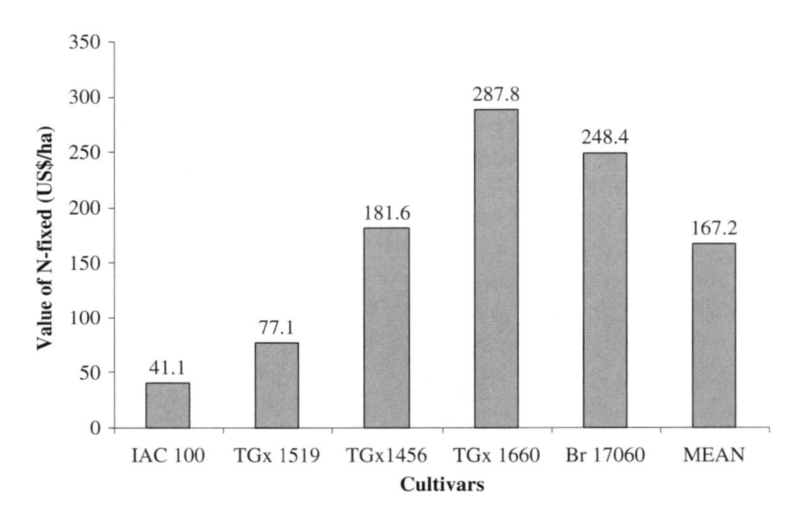

Fig. 1 Economic gains from natural biodiversity of *Bradyrhizobium* as influenced by plant genotype (cultivars)

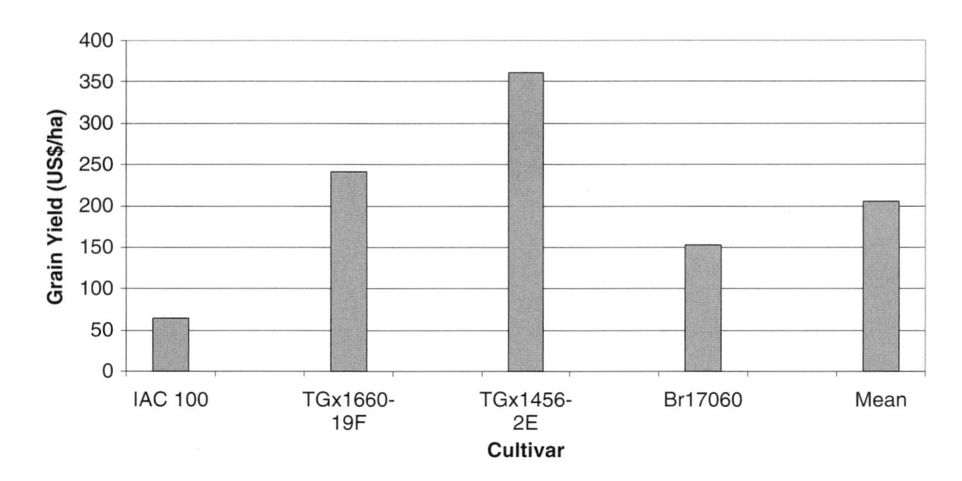

Fig. 2 Economic value of maize grain yield increment attributable to a preceding soya bean (Rotational Effect)

Country Estimates

Estimates of economic value of N_2 fixed per country were based on area harvested of soya bean (FAOSTAT) using 2002, 2003 and 2004 data available for 19 African countries (Benin, Burkina Faso, Burundi, Cameroon, Cote d'Ivoire, Egypt, Ethiopia, Gabon, Liberia, Madagascar, Morocco, Nigeria, Rwanda, South Africa, Tanzania, Uganda, Congo DR, Zambia and Zimbabwe). The estimated N_2 fixed, fertilizer N saved, urea fertilizer saved and values of urea fertilizer saved were computed for each country (table not shown). The estimated economic value (through replacement cost) of the urea fertilizer saved is depicted in Fig. 3 (for 2004) with the total value amounting to US $203 million across the countries.

There is little or no change in the area over the period (2002–2004) (table not shown). The total US$ value, however, increased by about 3%, from US $197 million (2002) to US $203 million (2004). Egypt and Rwanda show marked increases in soya bean area, although small in absolute terms. Nigeria and South Africa show moderate increases in soya bean area, contributing significantly in absolute terms. Uganda shows a marked decrease in soya bean area. An updated dataset of the area under soya bean cultivation (including data for 2005, 2006 and probably 2007) is needed to confirm the trend in area under soya bean cultivation. This will be undertaken in an expanded version of this document being planned for.

Inoculation and Nitrogen Fixation by Soya Bean

A comparison of crop yields and economic gain from N fixation under inoculated versus un-inoculated conditions was carried out using secondary data. Result from Zambia shows that for 75% (*Magoye*, *Hernon*, and *M30*) of the genotypes, *there was only a marginal increase in economic value of soya bean due to inoculation*, which, however, induced a significant difference in economic value of soya bean genotype *Kaleya* (Mpepereki et al., 2000) (Fig. 4). Results from Nigeria (Fig. 5) indicate a marked increase in N fixation and economic value especially for Br 17060. However, inoculation actually depressed N fixation and the economic value of the promiscuous genotypes TGx 1519-ID and TGx 1660-19F (contrary to some literature). The inoculated cultivar performed better than the un-inoculated highest producing promiscuous cultivar in none of the cases. These results show that response to inoculation is complex. Mpepereki et al. (2000) note a positive response of Hernon on acid oxisols in northern Zambia, whereas response in other sites was poor. There was no significant increase in yield. Increase in rainfall leads to increase in promiscuous nodulation (Mpepereki et al., 2000). Information on the extent of farmer application of inoculation in sub-Saharan Africa is limited. Using data from Tanzania (table not shown), the effect of inoculation

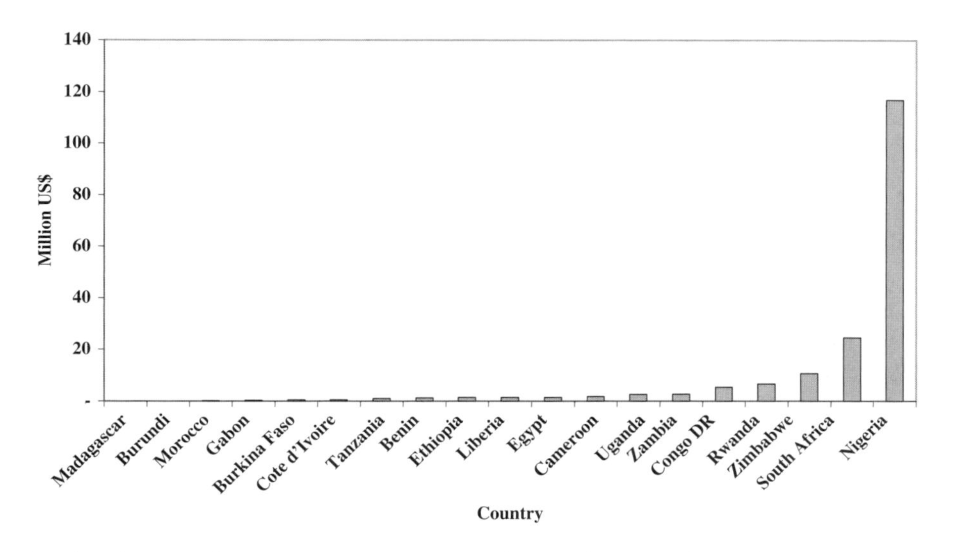

Fig. 3 Economic value of nitrogen fixed by soybean (*Glycine Max*) in 19 African countries in 2004

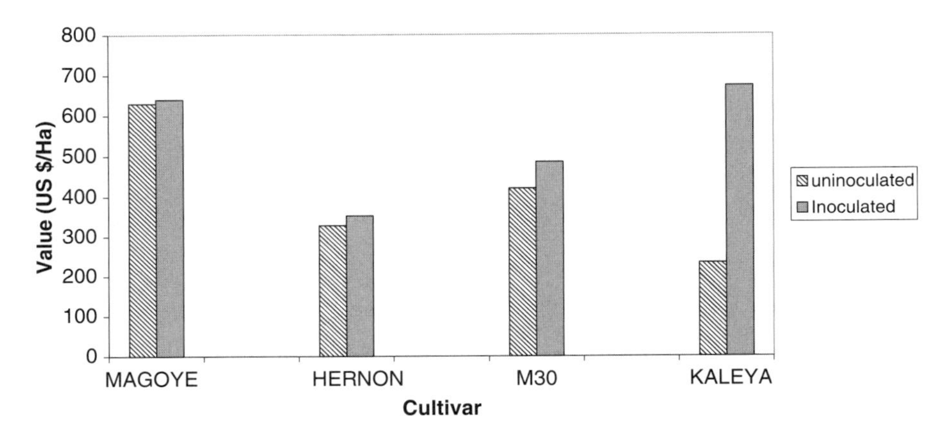

Fig. 4 Economic value of inoculated versus un-inoculated soya bean in Zimbabwe

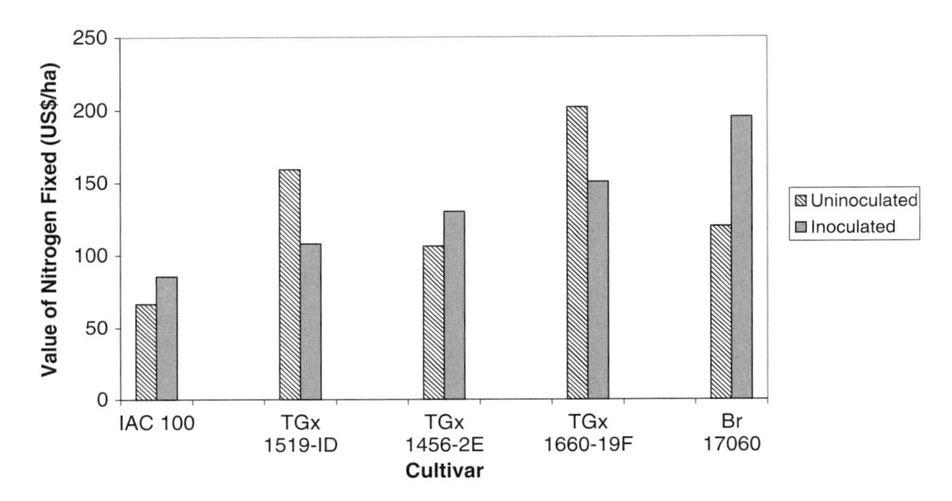

Fig. 5 Economic gains of nitrogen fixed in five inoculated (biodiversity enhanced) versus un-inoculated (Natural biodiversity)

with different *Bradyrhizobium* strains on N_2 fixation was determined. There was variable response of various cultivars to inoculation. While improved Pelican gave a negative yield response of −90 kg ha^{-1}, the cultivar 7H 101 gave a positive yield response of 740 kg ha^{-1}. Other cultivars gave intermediate yield figures. Overall, mean yield across cultivars was 375 kg ha^{-1}. The associated economic value ranges from US $36 ha^{-1} (Pelican) to US $296 ha^{-1} (7H 101) with a mean of US $150 ha^{-1}. Responses depend on the inoculant used. Nigerian data show soya bean yield increase from 170 kg ha^{-1} (using the *Bradyrhizobium japonicum* inoculant IRj 2123) to 468 kg ha^{-1} (using the strain IRj 2133) with a mean of 351 kg ha^{-1} (Olufajo and Adu, 1993). The corresponding economic value ranges from US $68 to US $187 ha^{-1} with a mean of US $132 ha^{-1}. Differences may also be found depending on location. Data from *Awka* (site in Nigeria) showed a marked difference in yield of 1180 kg ha^{-1} between inoculated soya bean and control using USDA 136 as inoculum, whereas, using the same inoculum, a negative response of 120 kg ha^{-1} was obtained from the other site (*Igbariam*) where the trials were carried out (Okereke and Eaglesham, 1993). Data from Kenya, Rwanda and Tanzania (Woomer et al., 1997) show that yield and economic value increase due to inoculation range from 592 kg ha^{-1} or US $ 237 ha^{-1} for Kabete (Kenya) to 1517 kg ha^{-1} or US$ 607 ha^{-1} for Mtwapa (also in Kenya) with a mean of 953 kg ha^{-1} or US$ 381 ha^{-1}. Rwandan data also show variation in response to inoculation from site to site.

Discussion

The BNF benefits (especially without inoculation) are examples of how below-ground biodiversity contributes to human welfare. Sanginga et al. (2001) mention that a 10% increase in legume (various) area in the cropping systems of the dry savanna Nigeria and 20% increases in yield translate to additional fixed N valued at 44 million per annum. Figures for the developing countries indicate that grain legumes (soya bean, common bean, groundnut, chickpea, cowpea, etc.) fix approximately 11.1 million metric tons of N per annum, representing about 6.7 billion US$ were this N to be supplied through inorganic fertilizers (Hardarson et al., 2003). However, this includes countries like Brazil, where soya bean area has reached over 13 million ha and soya bean is commercially being produced, exclusively using inoculum with economic benefits (fertilizer savings) of over US $2.5 billion per year (Alves et al., 2005). Effect of inoculation on N fixation in promiscuous soya bean varieties is inconclusive, needing more research to know when inoculation is best.

The results seem to indicate that the strategy that emphasizes the need to focus on promiscuous soya bean in SSA is a viable one. The total economic value of the N_2 fixed in soya bean for SSA is small compared to Brazil, but nevertheless substantial. The small figure suggests that there is enormous potential and opportunity to increase production of soya bean and tap into this important biological resource to expand the economic benefits, considering the rather uneven distribution of the area under soya bean cultivation among countries in SSA. Experience from Nigeria and Zimbabwe indicates that adoption rates are high if infrastructure and support functions are in place. Market access and processing for home utilization play key roles and are critical for success. In Malawi, *Magoye*, introduced in 1989 (with seeds distributed by Ministry of Agriculture), was readily adopted by smallholders. Stimulated by high demand from World Food Program, area cultivated rose to 90,000 ha in 1995, but production collapsed when demand dropped drastically. Malawi does not presently feature in FAO soya bean statistics. Zambia is another example where production rose steadily during the 1980s, with a parastatal organization *Litco* being responsible for promoting it among small farmers who depended on *Litco* for seeds and other inputs. With the bankruptcy of *Litco*, production declined. Zambia currently has 15,000 soya beans ha^{-1}, mainly commercial farms. Economic analyses of legume–cereal systems show 50–70% increase in income of adopters compared to farmers that embark on continuous maize cultivation (Sanginga et al., 2001).

Studies in various countries have indicated that African soils harbour a rich diversity of rhizobia bacteria, compatible and effective with a variety of soya bean cultivars. There does, therefore, not seem to be a major constraint to expansion of soya bean area in countries with marginal production presently. Of specific interest to SSA is the role that promiscuous soya bean might play in soil fertility restoration, because of possible net contribution to soil N (e.g. rotational effect of legumes in legume–cereal rotations). These promiscuous 'leafy' soya bean cultivars may be specifically suited for small farmers, since maintaining soil fertility through the activities of soil biota and symbiotic organisms is a natural process, does not involve huge cash outlays and is affordable to many farmers unlike mineral fertilizers. Indirect tapping has demonstrated LNB yield boosting in production of subsequent crop in rotation.

Exploitation of the economic benefits of below-ground biodiversity is influenced by plant genotypes. Selection is, therefore, an important issue to be addressed. Availability of the seed of selected varieties is next in importance. Also, regardless of the complexities in response, inoculation is a key factor, though investment in development and production of inoculum would at this stage not deserve priority giving the uncertainties in relation to the varied response in many places, and considering that we can get similar yields from promiscuous varieties without inoculation. Besides, all kinds of other difficulties are still associated with the production, maintenance and handling of inoculums and other live materials.

Conclusion

This chapter demonstrates the economic value associated with the fixing of N_2 by legume-nodulating bacteria (LNB), natively present in the soil, using promiscuous soya bean. It underlines the importance of the richness or diversity of LNB strains present in the soil and shows that there are ways of making use of that biological resource. Farmers should be able to choose between sourcing N through the LNB activity or the use of N fertilizers, and they would do so by choosing the proper crop (variety or cultivar) and by adopting the proper crop rotation systems. For effectiveness, such choice must not be hampered by factors such as lack of availability of the right soya bean seeds or market for produce. Through effective policies, the government must ensure that these constraints and distortions are removed.

Huge differences in the ability of different soya bean genotypes to fix N and generate economic returns have been seen. One key outcome of this chapter is the acknowledgement of such differences and the proper selection of the cultivars depending on the purpose for which it is being used (e.g. efficient sourcing of N), and putting farmers in the position that they are able to make this selection. In this sense, the contribution is an important step in the highly recognized recent need to start valuing the contributions of below-ground organisms to direct on whether or not efforts should be made to consciously conserve them.

References

Alves BJR, Zotarelli L, Araujo ES, Fernandes FM, Heckler JC, Medeiros AFA, Boddey RM, Urquiaga S (2005) Balanco de N em rotacao de culturas sob plantio direto em Dourados. Boletim de Prequita e Desenvolvimento 07. Embrapa Agrobiologia, Seropedica, Brazil

Bagayoko M, Mason SC, Traore S, Eskridge KM (1996) Pearl millet/cowpea cropping systems yield and soil nutrient levels. J Afr Crop Sci 4:453–462

Friesen DK, Assenga R, Bogale T, Mmbaga TE, Kikafunda J, Negassa W, Ojiem J, Onyango R (2003) Grain legumes and green manures in East African maize systems–an overview of ECAMAW network research. In: Waddington SR (ed) Grain legumes and green manures for soil fertility in southern Africa: taking stock of progress. Proceedings of a conference held on 8–11 October 2002 at the Leopard Rock Hotel, Vumba, Zimbabwe. SoilFertNet and CIMMYT—Zimbabwe, Harare, pp 113–118

Hardarson G, Bunning S, Montanez A, Roy R, MacMillan A (2003) The value of symbiotic nitrogen fixation by grain legumes in comparison to the cost of nitrogen fertilizer used in developing countries. In: Hardarson G, Broughton W (eds) Maximizing the use of biological nitrogen fixation in agriculture. Kluwer, Dordrecht, pp 213–220

Klaij MC, Ntare BR (1995) Rotation and tillage effects on yield of pearl millet (*Pennisetum glaucum*) and cowpea (*Vigna unguiculata*) and aspects of crop water balance and soil fertility in semi-arid tropical environment. J Agric Sci 124:39–44

Kueneman EA, Root WR, Dashiel KE, Hohenberg J (1984) Breeding soybeans for the tropics capable of nodulating effectively with indigenous Rhizobium spp. Plant Soil 8:387–396

Kumer S (2007) Dynamics of the global fertilizer market. PowerPoint presentation. Institute of Agriculture and Environment Research, Oslo, Norway

Mafongoya PL, Giller KE, Odee D, Gathumbi S, Ndufa SK, Sitompul SM (2004) Benefiting from N_2-fixation and managing rhizobia. In: Van Nordwijk M, Cadish G, Ong CK (eds) Below-ground interactions in tropical agroecosystems concepts and model with multiple plant components. CAB, Wallingford, CT, pp 227–242

Mangoensoe-Karjo S, Semangun H (2003) Manajemen agrobisnis kelapa sawit. Gadja Mada University Press, Yogyakarta, 605 hlm

Mpepereki S, Javaheri F, Davis P, Giller KE (2000) Soybeans and sustainable agriculture: promiscuous soybeans in Southern Africa, Field Crops Res 65:137–149

Mpepereki S, Makonese F, Giller KE (eds) (1996) Soybeans in smallholder cropping systems of Zimbabwe. SoilFertNet/CIMMYT, Harare

Okereke U, Eaglesham RJ (1993) Nodulation and nitrogen fixation by 70 ‘promiscuous’ soyabeans genotypes in soil in eastern Nigeria. Agronom Afr 2:123–136

Olufajo OO, Adu JK (1993) Nodulation of soyabeans grown under field conditions and inoculated with *Bradyrhizobium japonicum* strains. In: Mulongoy K, Gueye M, Spencer DSC (eds) Biological nitrogen fixation and sustainability of tropical agriculture. Wiley, New York, NY, pp 147–154

Sanginga N, Dashiell KE, Diels J, Vanlauwe B, Lyasse O, Carsky RJ, Tarawali S, Asafo-Adjei B, Menkir A, Schulz S, Singh BB, Chikoye D, Keatinge D, Ortiz R (2003) Sustainable resource management coupled to resilient germplasm to provide new intensive cereal–grain legume–livestock systems in the dry savanna. Agric Ecosyst Environ 100:305–314

Sanginga PC, Adesina AA, Manyong VM, Otite O, Dashiell KE (1999) Women get involved in soybean production. Siteresources.worldbank.org. Accessed 26 June 2009

Sanginga N, Okogun J, Vanlauwe B, Dashiell K (2002) The contribution of nitrogen by promiscuous soybean to maize based cropping in the moist savanna of Nigeria. Plant Soil 241:223–231

Sanginga N, Okogun JA, Vanlauwe B, Diels J, Dashiell K (2001) Contribution of nitrogen fixation to the maintenance of soil fertility with emphasis on promiscuous soybean maize-based cropping systems in the moist savanna of West Africa. In: Tian G, Ishida F, Keatinge JDH (eds) Sustaining soil fertility in west Africa. SSSA Special Publication Number 58, Madison, WI, pp 157–178

Sanginga N, Thottappilly G, Dashiel K (2000) Effectiveness of rhizobia nodulating recent promiscuous soybean selections in the moist savanna of Nigeria. Soil Biol Biochem 32: 127–133

Simms EL, Taylor DL (2002) Partner choice in nitrogen-fixation mutualisms of legumes and rhizobia. Integr Comp Biol 42:369–380

Siregar M (1984) Peranan tanaman penutup terhadap konservasi tanah dan pengaruhnya terhandap tanaman karet. Proseding Seminar Sehari Tanaman Penutup Tanah, Bogor, 34 hler

Syaukat Y (2005) Author of Chapter 8 of Indonesia book. Universitas Lampung, Bandar Lampung, Indonesia

Woomer PL, Karanja NK, Mekki EI, Mwakalombe B, Tembo H, Nyika M, Nkwiine C, Ndakidemi P, Msumali G (1997) Indigenous populations of rhizobia, legume response to inoculation and farmer awareness of inoculants in East and Southern Africa. Afr Crop Sci Conf Proc 3:297–308

Farmers' Perception of Soil Fertility Depletion and Its Influence on Uptake of Integrated Soil Nutrient Management Techniques: Evidence from Western Kenya

M. Odendo, G. Obare, and B. Salasya

Abstract Soil fertility depletion and the attendant declining agricultural productivity in western Kenya have led to many attempts to develop and popularize integrated nutrient management (INM) technologies that could restore soil fertility and improve productivity. INM bridges the gap between high external input agriculture and extreme forms of traditional low external input agriculture. The main components of INM are chemical fertilizers, animal manure, improved fallows and green manures. It is, however, not well understood why farmers who rely on agriculture for their livelihoods either do not adopt or adopt these technologies slowly. However, it is acknowledged that soil fertility depletion is an insidious and slow process, hence farmers' perception of severity of the problem and associated yield losses are critical in deciding adoption of soil fertility-enhancing technologies. The objective of this study was to evaluate farmers' perceptions of soil fertility depletion and assess its contribution to adoption of INM practices. Data were collected from a random sample of 331 households in Vihiga and Siaya districts and analysed by descriptive statistics and logit model. Results showed that most households (92.4%) perceived declining soil fertility to be responsible for the low crop yields and difference in perception between the two districts was insignificant ($P = 0.141$). From logistic analysis, farmers' perception of extent of depletion had significant effect on the adoption of animal manure and inorganic fertilizers. Other socio-economic factors militated against adoption of INM components.

Keywords Adoption · Households · Soil fertility · Perceptions · Integrated · Nutrients

Introduction

Declining soil fertility is behind the current low land productivity, food insecurity and poverty among most rural households in western Kenya (Jama et al., 1999). The concern for soil nutrient depletion and the attendant declining food productivity have led to many attempts in the past two decades to develop, test and popularize several soil fertility management technologies that could restore soil fertility and improve productivity in western Kenya. The attempts have been carried out by the World Agroforestry Centre (WAC) and Tropical Soil Fertility and Biology programme (TSBF) in collaboration with Kenya Agricultural Research Institute (KARI), Kenya Forestry Research Institute (KEFRI), the Ministry of Agriculture (MOA) and other agencies. This has been conducted mostly in Vihiga and Siaya districts (Rao et al., 1998).

The research and dissemination focused on integrated nutrient management (INM) approach, involving the use of organic and inorganic resources. This is because it is widely acknowledged that organic and mineral inputs cannot be substituted entirely by one another and both are required for sustainable crop production (Sanchez and Jama, 2002). INM bridges the gap between high external input agriculture and extreme forms of traditional low external input agriculture. It enables farmers to manipulate the organic and inorganic nutrient stocks judiciously and efficiently to save nutrients from being lost or to add nutrients

M. Odendo (✉)
Kenya Agricultural Research Institute (KARI), Regional Research Centre, Kakamega, Kenya
e-mail: Odendos@yahoo.com

A. Bationo et al. (eds.), *Innovations as Key to the Green Revolution in Africa*,
DOI 10.1007/978-90-481-2543-2_107,

to the farmland (Chianu and Tsujii, 2005). The main components of INM are chemical fertilizers, animal manure, improved fallows and green manures. It is, however, not well understood why farmers who rely on agriculture for their livelihoods either do not adopt or adopt these technologies slowly.

Household's decision making in soil fertility investment starts with perception of soil fertility depletion problem. It is acknowledged that soil fertility depletion is an insidious and slow process, hence farmers' perception of severity of the problem and associated yield losses are critical in deciding adoption of soil fertility-enhancing technologies (Ervin and Ervin, 1982). Past research (e.g. Ervin and Ervin, 1982; Norris and Batie, 1987) has highlighted the importance of perceptions in enhancing adoption of soil conservation technologies. Norris and Batie (1987) estimated perception as whether or not farmers believed that soil erosion was problem on their farms. The objective of this study was to evaluate farmers' perceptions of soil fertility depletion and assess its contribution to adoption of INM practices.

Methodology

The Study Area

The data used in this study were collected from households in the western Kenya highland areas of the larger Vihiga and Siaya districts. The two districts were selected for the study because both experience low soil fertility and high poverty levels, and INM technology was introduced in the districts. However, there are differences in agro ecological zones, agricultural potential, farming systems, culture, farm sizes and population density.

Vihiga is one of the eight districts in Western Province, whilst Siaya is one of the 12 districts in Nyanza Province. Much of western Kenya is considered to have good potential for agriculture, with medium elevation (1100–1600 m above sea level), deep, well-drained soils and relatively high rainfall (1200–1800 mm per year) that permits two growing seasons. The history of farming in the area, however, is characterized by low input–low output farming. Recent studies have found that crop productivity is very low (less than one ton of maize per hectare per year) and that nutrient balances are seriously in deficit (Jaetzold et al., 2005). As a result of the favourable climate, high population densities prevail, reaching over 1000/km^2 in some of the study divisions (GOK, 2001). The Luhya ethnic group inhabits Vihiga, while the Luo resides in Siaya.

The farming system incorporates crops, livestock and trees. Maize and beans are the most common agricultural enterprises. The food situation was reported as deficient by 89.5% of the households in Siaya and Vihiga, who had to buy food to supplement their own harvest (Wangila et al., 1999). Only 8.9% of the households were food secured from their own production.

Survey Design

The districts were purposively selected. In order to achieve fair representation of the population of reference in the districts, each of the survey districts was stratified on the basis of agro-ecological zones. One stratum comprised the high agricultural potential area (UM_1) and the second one consisted of low potential (LM_1–LM_3). The former comprised Vihiga district and the latter comprised Siaya district. In the first stage, all sub-locations in each stratum were listed as per the 1999 population census (GOK, 2001) and formed the sampling frame. In the second stage, lists specifying all households in each selected sub-location were constructed with the help of local administrators, from which a random sample of 331 households were sampled for interviews.

Data Analysis

Descriptive statistics and logit econometric model were used to analyse the data. Following Pindyck and Rubinfeld (1981) the functional form of logit model was specified as:

$$\log P_i/1 - P_i = \beta_o + \beta_i X_i + \cdots \beta_k X_k$$

where 1–P is the probability that household has adopted a named component of INM; P_i/1–P is odds of adoption; β_o is a constant; β_i is a vector of unknown

Table 1 Explanatory variables for logit model

Variable	Definition and units	Expected sign
D_RATE	Perception of extent of soil depletion (1–5 scale)	(±)
FSIZE	Size of the farm (acres) at time *t* of adoption	(±)
REMITTAN	1, Obtained remittances at *t*; 0, otherwise	(–)
CATTLENO	Number of cattle owned at time	(–)
EDUCHHD	Formal education of household head (years)	(–)
AGEHH	Age of household head (years)	(+)
HHSIZE	Total household members (number)	(±)
SEXHEAD	Sex of household head (0, male; 1, female)	(±)
HIRELAB	1, If hire labour; 0, otherwise	(+)
N_GROUP	1, household member belongs to group; 0, otherwise	(–)
CREDIT	1, access to credit; 0, otherwise	(–)
MARITAL	Marital status of household head: 1, married	(±)
ACRESCUL	Total area under cultivation (acres)	(±)

parameters to be estimated; X_i is a vector of socio-economic variables pertaining to a household i, which are explanatory variables for the probability (P_i) that the ith household has adopted a component of INM. The variables that were considered to affect the adoption and their expected signs are shown in Table 1.

Results and Discussion

Farmers' perceptions of the causes of declining crop productivity were assessed. Farmers often made maize as the reference crop. This was mainly because maize is the staple food crop in the study area and whenever fertilizers are applied, they are often applied on maize. Farmers perceived that declining soil fertility, shortage of labour, weeds and poor quality seed are the major reasons explaining declining trends of crop (Table 2).

Farmer Perceptions on Soil Fertility Status

Studies are increasingly showing that farmers clearly perceive and articulate differences in the levels of fertility of their farms and farm plots (Onduru et al., 2002). Majority of the farmers (94.6%) reported that soil fertility has been declining on their farms and listed a number of indicators explaining this phenomenon (Table 3). The key indicators of soil fertility decline were decline in crop yields, poor crop growth vigour and poor general crop health. Other indicators were presence of indicator weeds, especially *Striga* (thrive under low soil fertility), and fading of soil colour.

Onduru et al. (2002) report similar findings in eastern Kenya, where farmers identified indicators such as reduced yield, weed infestation, rocky outcrops and early wilting of crops to justify that soil

Table 2 Reasons advanced by farmers to explain the declining crop yields

Reason	Percentage of respondents reporting		
	Siaya	Vihiga	All
Declining soil fertility	90.3	94.6	92.4
Labour shortages at peak periods	88.5	94.0	91.2
Weeds	86.7	82.5	84.6
Poor quality of seeds	66.7	62.7	64.7
Low and erratic rainfall	68.5	45.2	56.8
Disease and pests	49.1	50.0	49.5
Inappropriate crop varieties	31.5	58.4	45.0
High cost of land preparation	24.8	21.1	23.0
Soil erosion	4.8	0.6	2.7
Do not know the reason	0.0	0.6	0.3

Responses were listed independently, hence percent does not sum to 100

Table 3 Farmers' indicators of soil fertility decline

Indicator	Percentage of respondents reporting		
	Siaya	Vihiga	All
Declining crop yields	83.6	89.2	86.4
Poor crop growth vigour	78.8	75.3	77.0
Presence of indicator weeds	30.3	47.0	38.7
Poor general crop health	59.4	61.4	60.4
Soil becoming lighter in colour	23.0	236.1	29.6

Responses were listed independently, hence percent does not sum to 100

fertility is declining. Farmers were also reported to be using soil colour, texture and certain physical characteristics to identify different soil types. Thus it appears that farmers' and scientists' perceptions on the decline in soil fertility may be congruous. However, Nandwa and Bekunda (1998) have indicated that there is little knowledge on the degree of correlation between parameters and indices used by farmers and researchers.

With regard to farmers' perceptions of the extent of soil fertility depletion in the past 5 years prior to this study, 41% of the respondents indicated that the rate of decline was moderate, about one-quarter reported that the rate soil nutrient depletion was high and only about 5% indicated that the rate was very high (Table 4).

Table 4 Farmers' perceptions of extent of soil fertility decline in past 5 years

Reason	Percentage of respondents reporting		
	Siaya	Vihiga	All
Very slow	10.5	21.3	15.8
Slow	2.5	7.5	5.0
Moderate	41.4	40.6	41.0
High	33.3	19.4	26.4
Very high	6.2	3.1	4.7

Determinants of Adoption of Selected Components of INM

The variables that influenced the adoption of the three INM components varied. The significant factors for the adoption of both animal manure and inorganic fertilizers were household size, whether household hired labour, area of land under cultivation and household perception of extent of soil depletion (Table 5). Adoption of animal manure was also significantly influenced education level of the household head, receipt of remittances, membership in local groups and number of cattle owned. None of the factors under investigation had significant effect on the adoption of compost manure. In contrast, Obare et al. (2004) show empirically that although perception of soil erosion

Table 5 Factors affecting adoption of manure, inorganic fertilizers and compost

Variable	Manure			Inorganic			Compost		
	β	S.E.	Sig.	β	S.E.	Sig.	β	S.E.	Sig.
SEXHEAD	36.754	59.499	0.537	−9.084	18.586	0.625	0.829	0.915	0.348
AGEHH	0.078	0.060	0.196	0.185	0.203	0.361	−0.010	0.010	0.342
EDUCHHD	−0.430	0.142	0.002	−0.227	0.219	0.300	0.022	0.041	0.593
MARITAL	−13.845	19.860	0.486	4.063	10.750	0.705	0.006	0.067	0.835
HHSIZE	1.835	0.342	0.000	0.822	0.212	0.000	0.118	0.255	0.662
HIRELAB	4.296	1.092	0.000	1.952	0.789	0.013	−0.320	0.265	0.126
ACRESCUL	4.215	0.860	0.000	1.765	0.536	0.001	0.039	0.265	0.560
REMITTAN	7.159	2.348	0.002	−5.681	11.791	0.630	−0.021	0.307	0.945
CREDIT	−19.649	4.836	0.000	8.450	24.929	0.735	0.287	0.499	0.897
D_RATE	5.127	1.398	0.000	4.726	1.523	0.002	−0.007	0.104	0.946
N_GROUP	2.591	1.700	0.128	2.169	2.250	0.335	0.263	0.357	0.321
CATTLENO	3.157	0.726	0.000	0.533	1.219	0.662	−0.008	0.075	0.912
Constant	−78.636	42.263	0.063	−60.662	46.403	0.191	−0.737	1.234	0.554

problem was relatively high in the marginal areas of Kenya, its effect on soil conservation was not significant.

Conclusions

Majority of the farmers (94.6%) reported that soil fertility is declining on their farms and listed a number of indicators explaining this phenomenon. The key indicators of soil fertility decline were decline in crop yields, poor crop growth vigour and poor general crop health. However, 41% of the respondents perceived the extent of soil fertility depletion to be moderate. The household perception of extent of soil depletion and other socio-economic factors influenced adoption of INM components, especially manure and chemical fertilizers. Effort should be devoted to enlighten farmers about the extent of soil depletion to enhance adoption to adoption of INM components.

Acknowledgements I am indebted to the World Bank and the Government of Kenya under the aegis of the Kenya Agricultural Productivity project (KAPP) for the financial support. Many thanks to director, KARI and the centre director, KARI-Kakamega for their support.

References

Chianu J, Tsujii H (2005) Integrated nutrient management in farming systems of the savannas of Northern Nigeria: what future? Outlook Agric 34(3):197–202

D'Emden FH, Llewellyn RS, Burton MP (2006) Adoption of conservation tillage in Australian cropping regions: an application of duration analysis. Technol Forecast Soc Change 73(6):630–647

Ervin CA, Ervin DE (1982) Factors affecting use of soil conservation practices: hypotheses, evidence and policy implications. Land Econ 58(3):277–292

GOK (Government of Kenya) (2001) 1999 Population and housing census. Counting our people for development. Volume I: Population distribution by administrative areas and urban centers. Ministry of Finance and Planning and National Development, Nairobi

Jaetzold R, Schimdt H, Hornetz B, Shisanya C (2005) Farm management handbook of Kenya, vol II. Natural conditions and farm management information, part a West Kenya, subpart A1 Western Province, 2nd edn. Ministry of Agriculture, Nairobi

Jama BA, Niang IA, Amadalo B, De wolf J, Rao MR, Buresh RJ (1999) The potential of improved fallows to improve and conserve the fertility of nutrient depleted soils of western Kenya. In: Proceedings of the 6th biennial KARI scientific conference (KARI), Nairobi, Kenya, pp 133–144

Nandwa SM, Bekunda MA (1998) Research on nutrient flows and balances in East and Southern Africa: state-of-the art. Agric Ecosyst Environ 71:5–18

Norris PE, Batie SS (1987) Virginia farmers' soil conservation decisions: application of tobit analysis. South J Agric Econ 19:79–90

Obare GA, Mwakubo SM, Ouma EA, Mohammed L, Omiti J (2004) Social capital and soil erosion control in marginal areas of Kenya: the case of Machakos and Taita districts. In: Proceedings of the inaugural symposium of African Association of Agricultural Economists, Grand Regency Hotel, Nairobi, 6–8 Dec 2004

Onduru DD, Gachimbi L, De Jager A, Maina F, Muchena FN (2002) Integrated nutrient management to attain sustainable productivity increases in East African farming systems. Report of a baseline survey carried out in Muthanu Village of Siakago Division, Mbeere district, Nairobi ETC—East Africa, KARI, LEI

Pindyck RS, Rubinfeld DL (1981) Econometric models and economic forecasts, 2nd edn. McGraw Hill, New York, NY

Rao MR, Niang A, Kwesiga A, Duguma B, Franzel S, Jama B, Buresh RJ (1998) Soil fertility replenishment in sub-Saharan Africa: new technologies and spread of their use on farms. Agroforest Today 10(2):3–8

Sanchez PA, Jama BA (2002) Soil fertility replenishment takes off in East and Southern Africa. In: Vanlauwe B, Diels J, Merckx R (eds) Integrated plant nutrient management in sub-Saharan Africa: from concept to practice. CABI, Wallingford, CT, pp 23–45

Wangila J, Rommelse R, deWolffe J (1999) Characterization of Households in the pilot project area of western Kenya. International Centre for Research in Agroforestry, Nairobi (Mimeo)

Taking Soil Fertility Management Technologies to the Farmers' Backyard: The Case of Farmer Field Schools in Western Kenya

M. Odendo and G. Khisa

Abstract Farmer field schools (FFSs) were introduced in western Kenya in 1995 to empower farmers with knowledge for informed decision making. Taking cognizance of diverse farming systems, the FFSs applied integrated production and pest management (IPPM) approach. IPPM involves training farmers on productivity-enhancing technologies, especially soil fertility and pest management, and how to access agricultural services – extension and credit to improve adoption. However, the influence of the FFSs on adoption of the technologies is not clearly understood. The objectives of this study were to assess adoption of soil fertility-enhancing and pest control practices by FFS and non-FFS households; analyze farmers' access to credit and extension services; and evaluate households' social capital. Data were collected from a random sample of 401 households and analyzed by descriptive statistics. The main soil fertility-enhancing technologies adopted were chemical fertilizers (64%), farmyard manure (56%), and compost (13%). The mean rate of chemical fertilizer application was lower than recommended, but slightly higher among FFS households than non-FFS households ($p = 0.149$). The main pest control method was use of local concoctions (50%). Sixty-nine percent of the households received extension advice and a paltry 3% accessed credit. Regarding social capital, 91% of the households had members in one or more social organizations/groups, of which the most important group for one-third of the households was "merry-go-round." We conclude that taking technologies to grass roots and improving farmers' knowledge requires concomitant improvement in access to complementary agricultural services for improved adoption and impacts.

Keywords Empowerment · Farmer field schools · Households · IPPM · Soil fertility

Introduction

Farmer field school (FFS) approach was introduced in western Kenya on a pilot basis in 1995 under the Food and Agriculture Organization's (FAO) special program for food security. A major scaling-up was undertaken between 1999 and 2002 under the East African sub-regional pilot project for FFS (financed by International Fund for Agricultural Development (IFAD)) and implemented by the Global Integrated Pest Management (IPM) Facility Project under the auspices of FAO. The model is not based on instructing farmers what to do but on empowering farmers with knowledge and skills regarding farm enterprises; sharpening the farmers' ability to make critical and informed decisions; sensitizing farmers in new ways of thinking and problem solving; and helping farmers learn how to organize themselves in farming in sustainable manner. Instead of focusing on using IPM as an entry point for the FFS as it is done in Southeast Asia, the program developed the notion of integrated production and pest management (IPPM), which is a broader and more holistic approach that accommodates other production-related issues (Khisa and Wekesa, 2003).

M. Odendo (✉)
Kenya Agricultural Research Institute (KARI), Regional Research Centre, Kakamega, Kenya
e-mail: Odendos@yahoo.com

A. Bationo et al. (eds.), *Innovations as Key to the Green Revolution in Africa*,
DOI 10.1007/978-90-481-2543-2_108,

In IPPM, farmers' priorities influence the program to add into the curriculum other aspects that have a direct bearing on production. The most important additions were HIV/AIDS prevention and management, basic principles of nutrition and environmental management, water and soil conservation, as well as basic financial management and marketing skills. All these activities aimed at improving the impact of agricultural research and technology transfer on rural livelihoods.

Methodology

The Study Area

The study was carried out in Kakamega, Bungoma, and Busia districts (Fig. 1), where FFSs were established. The districts mainly fall in the Lower Midland (LM) agroecological zone (AEZ), including LM_1, LM_2, and LM_3. Agriculture is the main economic activity in the study area Jaetzold et al., 2005.

In Bungoma district, maize is the staple food crop and in some parts it is also the cash crop. The other important crops are sugarcane, beans, sorghum, finger millet, cassava, tobacco, coffee, and groundnuts and a variety of horticultural crops. Zebu is still the most important cattle species. The other livestock are poultry, sheep, and goats. In Busia district, the main crops are maize, sorghum, cassava, sweet potato, beans, finger millet, cowpeas, soybean, groundnuts, and a variety of horticultural crops. Sugarcane is a major cash crop, particularly for the areas bordering Butere-Mumias district. Farmers in the district also keep livestock including cattle, poultry, sheep, and goats. Cattle are the most important livestock with the local zebus being the most common.

The district falls predominantly in LM_1 and LM_2. The survey covered the LM_1 and LM_2 zones (medium to high potential) where FFSs were established. The area receives a mean annual rainfall of between 1200 and 2200 mm per year. The rainfall is bimodal with the long rains falling between March and May and the short rains between August and October, hence allowing for two cropping seasons in a year.

Busia district falls within the Lake Victoria basin, with altitude varying from 1,130 on the shores of Lake Victoria to 1375 m above sea level in the central and northern parts. All the LM agroecological zones, ranging from LM_1 to LM_4, are found in the district. That means that the agricultural potential is also quite variable ranging from low potential in the LM_4 to LM_1. The survey was, however, in the LM_1 and LM_2 zones (medium to high potential) where FFSs were established. The area receives a mean annual rainfall of 1550 mm, which ranges between 760 and 1790 mm per year. The rainfall is bimodal with the long rains falling between March and May and the short rains between August and October, hence allowing for two cropping seasons in a year, particularly in the LM_1 and LM_2.

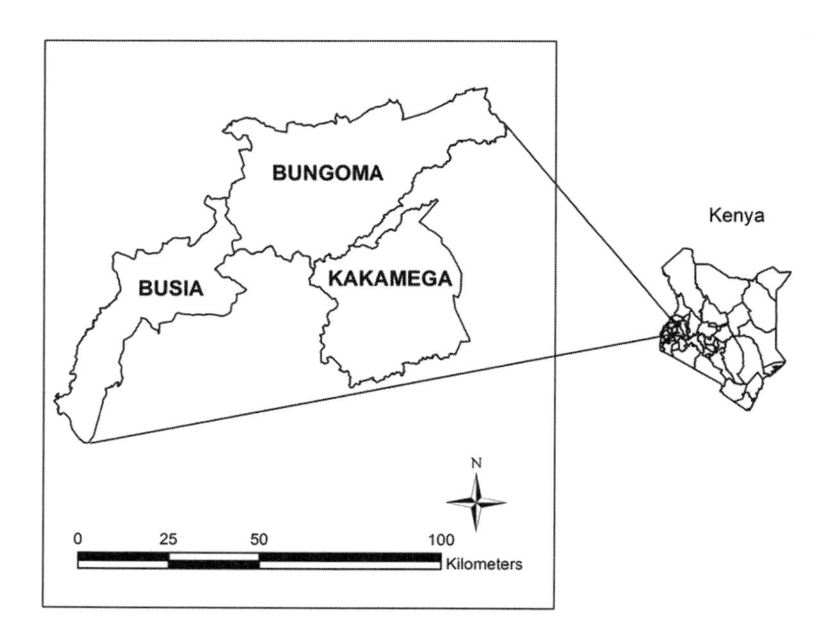

Fig. 1 Geographical location of the study sites

The mean maximum temperatures in the district range between 26 and 30°C and the soils are variable ranging from well drained to the heavy cotton soils (Jaetzold et al., 2005). Use of nutrient-replenishing inputs is low and farm households have moderate access to markets for both inputs and outputs with varying transaction costs. Kakamega district is located within the upper midland (UM) and LM agro-ecological zones (Jaetzold et al., 2005). The area receives a mean annual rainfall of between 1,000–2000 mm per year. The rainfall is bimodal with the long rains falling between March and May and the short rains between August and October, hence allowing for two cropping seasons in a year (Jaetzold et al., 2005). Maize is the staple food crop. The other important crops are beans, horticultural crops, sugarcane, tea, and coffee. Zebu is still the most important cattle species. As for the other districts poultry, sheep and goats constitute the major livestock. Use of nutrient-replenishing inputs is fair and farm households have moderate access to markets for both inputs and outputs.

Sampling Procedure

This study is based on face-to-face interviews of 401 households composed of 280 FFS households (households with FFS members) and 121 non-FFS households (households without FFS members). At the time of the survey, 34 FFSs had been registered, 14 in Kakamega district and 10 each in Bungoma and Busia districts. A two-stage random sampling procedure was used to select the study FFS households. In the first stage, a list of all registered FFSs in each of study districts was used as the sampling frame. In total, 20 registered FFSs were sampled for the survey and divided proportionally per district on the basis of the number of FFSs in the particular district and the diversity of agroecological zones. In the second stage, lists specifying all households in each selected FFS were used to randomly select the survey households. Again, the number of members selected for the interview was proportional to the total membership of a particular FFS.

In the case of non-FFS households, for each of the sampled FFSs, corresponding six non-FFS households were selected for interview. Lists of all villages in the sublocations where the selected FFSs were geographically located were obtained from the local administrators and extension agents. From the lists, the names of villages where FFS activities had never been carried out were identified, from which two non-FFS villages were randomly selected. Then lists specifying all households in each of the selected villages, with indication of membership or non-membership to FFS, were constructed with the help of village elders and extension agents. In relation to each selected FFS, two villages were randomly sampled and then six households were sampled from the two villages. The number of sampled households per village was proportionate to the total number of households in each village. In instances where FFSs were spread in most villages in the selected sublocation, different sublocations were chosen for the survey.

Data Collection and Analysis

A formal survey was conducted between August and September 2006. A structured questionnaire was designed and pre-tested to help restructuring of the questionnaire and paraphrasing questions that appeared ambiguous to the respondents. The questionnaire was administered by a team of 11 enumerators. The enumerators were trained on field survey techniques and the significance of each survey question and involved in pre-testing of the questionnaire. Among the non-FFS households, the household heads or in their absence, the most senior member available or the household member responsible for the farm management was interviewed, while for FFS households, the FFS member was interviewed. After the first day of interview, each filled questionnaire was reviewed by the authors for accuracy and completeness and discussed with the enumerators for improvement. The collected data were sorted and edited before analysis. Data were analyzed by descriptive statistics using combinations of Excel and Statistical Package for Social Scientists (SPSS) computer softwares.

Results and Discussion

Characteristics of Surveyed Respondents

A large majority of FFS members (59%) were married and living with their spouses. Over half (58%) were of primary school level, while 30% were of secondary level. With regard to occupation, 84% of the FFS members were farmers, while 8% were traders/business people. The rest were civil servants, private sector employees, casual laborers, retirees, or religious leaders. About half (51%) were male household heads, 44% were wives of male household heads, 4% were daughters of household heads, and 1% were husbands of household heads.

Households whose members hold leadership positions in any social organizations are likely to adopt technologies faster than those that do not (Dzuda, 2001). Asked about leadership positions held by household members in the community, a large majority of the households (69%) had no leadership positions. It was observed that non-FFS households had more leadership positions (76%) than did FFS members (66%). Of those who held leadership positions, about 18% held religious leadership, 6% belonged to school boards, 4% were village elders, and very few were assistant chiefs and councillors.

Soil Fertility Management Practices

About 86% of the households applied some soil-improving inputs in the past 1 year preceding this study. While there is ample evidence that judicious use of mineral fertilizers can bring about substantial crop yield increases (FURP, 1994), only 64% of the respondents applied chemical fertilizers (Table 1). Similar results were reported by Salasya et al. (2006) in a survey conducted in Bungoma, Busia, and Butere-Mumias districts, which shows that about 65% of the households applied some fertilizer in the long rains of 2004 and 2005 cropping seasons. The main basal fertilizer type used was diammonium phosphate (DAP) which was applied by all farmers who applied basal chemical fertilizers in 2004. Odongo et al. (2006) in a study conducted in western Kenya (Bungoma, Vihiga, and Siaya) reported that relatively few surveyed households used purchased inputs. Odendo et al. (2002) reported that unavailability of cash or credit was a major deterrent to farmers' access to purchased inputs.

Manuring is one of the traditional methods of maintaining soil fertility in Kenya (Omiti, 1998). In the sample studied, 56% of the farmers reported using manure. This high response indicates that manure is important for crop production in the study site. Other soil fertility maintenance practices included compost and green manure (Table 1). Green manure is not popular in the study areas as only 0.7% of the sample farmers practiced it. Onduru et al. (2002) reported a similar scenario in eastern Kenya, where only 2 out of 30 sampled farmers (7%) used green manure. Odendo et al. (2000) assert that the key factors that constrained the uptake of green manure were inadequate information on its use and technology for incorporating it into the soil, high labor demand at the time of planting, and unavailability of seed.

The mean amounts of fertilizer nutrients applied by the 250 farmers who applied DAP (18:46:0) were 19 and 48 kg of nitrogen (N) and phosphorous (P) ha^{-1} respectively (Table 2). However, the mean amounts applied by FFS households were slightly higher than those by non-FFS households, though there was no significant differences ($p = 0.149$). These amounts are less than the recommended rates of 60 kg P and 60 kg N ha^{-1} (FURP, 1994).

Disease and Pest Management

Recognition of the problem to be solved is one of the key factors that influence adoption of agricultural technology (Ervin and Ervin, 1982). About 84% of the respondents reported having experienced crop pests

Table 1 Soil fertility improvement options practiced by farmers

Soil management practices ($n = 401$)	All	Non-FFS	FFS
Applied some soil improvements	85.5	77.3	89.0
Chemical fertilizer	63.6	62.20	64.20
Farmyard manure	56.4	48.70	59.60
Compost	12.5	11.8	12.8
Green manure	0.7	0.8	0.70

Note: Multiple responses were taken independently and do not add up to 100%

Table 2 Quantities of fertilizer nutrients (N and P) applied

Nutrient type	*N*	Mean (kg ha^{-1})	Std. deviation
N	250	19.0	11.5
P	250	48.4	29.4
N	Non-FFS = 74	17.3	9.5
	FFS = 176	19.6	12.2
P	Non-FFS = 74	44.3	24.4
	FFS = 176	50.2	31.2

and diseases on their farms and 77 and 89% were non-FFS and FFS households, respectively. Of those, about half applied commercial pesticides and concoctions and 37% applied crop rotation.

Farmers' Access to Agricultural Services

Access to agricultural services was assessed in terms of access to credit and extension services and markets. The results of this study showed that only 30% of all the sampled households had ever applied for credit, with the proportion of FFS households being significantly higher than that of non-FFS households (Table 3). Although this study did not analyze factors likely to cause the differences, it is likely that individuals who join groups such as FFS are more exposed, and thus more likely to know the available sources of credit. The main reasons for applying for credit were household needs (43%), school fees (28%), and purchase of farm inputs (27%).

For the 281 households which had not applied for credit, 41% reported that they were not aware of credit application process, while 38% were not aware of credit sources (Table 4). Similarly, Nyangito et al.

Table 3 Reasons for credit application

Reasons	All	Non-FFS	FFS
Ever applied for credit ($n = 401$)	29.9	21.0	33.7
Reason for credit application ($n = 120$)			
Household needs	42.5	32.0	45.3
School fees	27.5	24.0	28.4
Purchase of farming inputs	26.7	16.0	29.5
Business	25.8	32.0	24.2
Medical bill	16.0	18.9	18.3

Table 4 Reasons for no credit application (percentage reporting)

Reasons for credit request	All ($n = 281$)	Non-FFS	FFS
Lack of awareness of application process	40.9	38.3	42.2
Lack of awareness of sources of credit	37.7	35.1	39.0
Fear of loss of collateral	32.1	12.4	19.5
Lack of security	23.1	26.6	21.4
Do not know	9.3	9.6	9.1
Bad past experience with credit	6.0	2.1	8.0
Had outstanding loan	0.4	0.0	0.5

(1997) showed that 45% of the households were ignorant of the available credit facilities. However, this study shows that 32% of the households did not apply for credit for fear of loss of collateral, while 23% lacked collateral. For those who applied for credit, 46% had been granted credit once, while 32 and 12% had been granted credit twice and thrice, respectively. Onduru et al. (2002) in a study conducted in eastern Kenya showed that most farmers had no access to credit because they could not marshal the required collateral to secure credit from financial institutions.

With regard to access to extension services, 69% of the whole sample reported having received some agricultural extension advice in the past 5 years, prior to this study. The major sources of new information on agricultural practices and technologies in the study area were research organizations (55%) and other farmers (44%) (Fig. 2). The other sources were NGOs and input suppliers, which represented 23 and 17% of the sample, respectively. Surprisingly, the number of households that received information from the government extension staff was discouragingly low (7%). These findings agree with the work of Rees et al. (2000), who reported that the major sources of agricultural knowledge for smallholders were local (neighbors, family, and CBOs), but contrast Rees et al.'s (2000) finding that between 40 and 70% of respondents obtain agricultural information from government agricultural extension agents.

Households' Membership to Organizations and Groups

It has been found in many studies of agriculture that social participation in formal or informal organizations influences adoption behavior. Social participation may expose individuals to a wide range of ideas, which may influence attitude toward an innovation (Alavalapati, 1990). The survey revealed that 91% of the whole sample had members in one or more social organizations/groups. The number of FFS households belonging to groups was significantly higher (59%) than that of non-FFS households (40%) ($p = 0.000$). Although the survey did not investigate explicitly the detailed activities of these organizations, it was observed that the organizations were carrying out a number of activities, including merry-go-round (micro-finance in kind

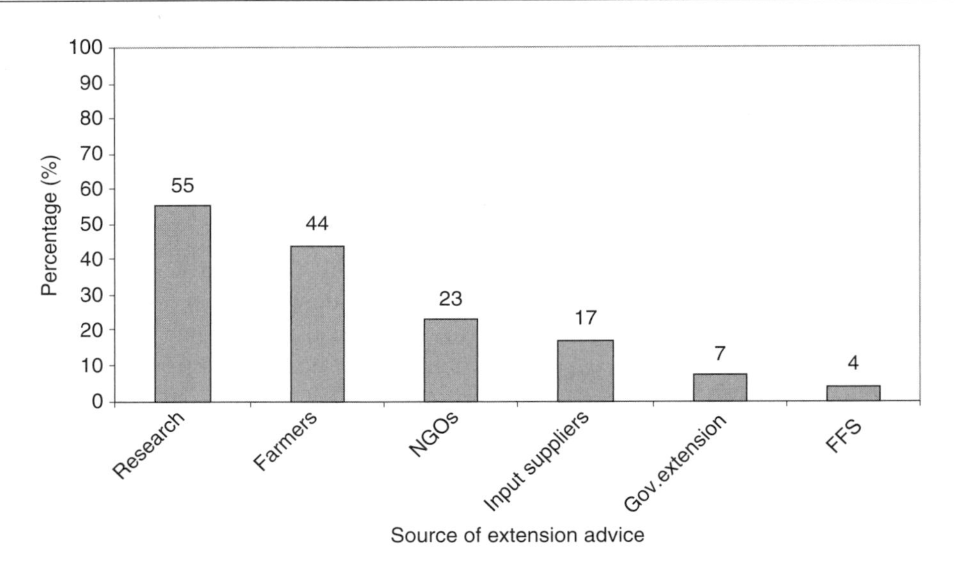

Fig. 2 Sources of technical agricultural advice for the whole sample

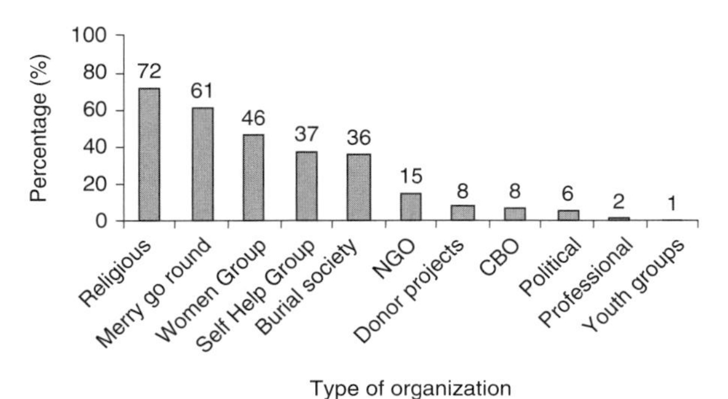

Fig. 3 Membership to organizations for the whole sample

or in cash), spiritual nourishment, provision of agricultural services, and offering platforms for learning new technologies. Most households had members who belonged to religious organizations (72%), followed by merry-go-round (61%) (Fig. 3). Other important organizations were women groups (46%), self-help groups (37%), and burial groups (36%). Considering positions held by household members in the organizations or groups, nearly 55% of the households had members who were leaders in at least one of the groups.

Conclusions

Farmers applied several soil fertility management technologies: 64% applied chemical fertilizers, 56.4% applied manure, and 0.7% used green manure. Most households (84%) experienced crop pests and diseases on their farms and 86% of them controlled diseases and pests. Half of the households applied concoctions and 46% applied commercial pesticides.

Regarding access to extension advice, 69% of the households received extension advice. The major sources of new information on agricultural practices and technologies in the study area were research institutes (55%) and other farmers (44%). On social capital, 91% of the households had members in one or more social organizations/groups and majority belonged to religious organizations (72%), followed by merry-go-round (61%). We conclude that taking technologies to grass roots and improving farmers' knowledge requires concomitant improvement in access to complementary agricultural services for improved adoption.

Acknowledgments We acknowledge IFAD-TAG through FAO for financial support. Special thanks to the enumerators for collecting high-quality data. We gratefully thank the staff from the Ministry of Agriculture for logistical support during data collection. Last, but by no means least, many thanks to the farmers for taking time to discuss with us.

References

Alavalapati JRR (1990) An analysis of factors influencing social forestry adoption: implications for forestry extension. Unpublished MSc. Thesis, Department of Rural Economy, University of Alberta

Dzuda LN (2001) Analysis of soil and water conservation techniques in Zimbabwe: a duration analysis. MSc thesis, University of Alberta

Ervin CA, Ervin DE (1982) Factors affecting use of soil conservation practices: hypotheses, evidence and policy implications. Land Econ 583:277–292

FURP (1994) Fertilizer use recommendations, vols 1–23. KARI, Nairobi

Jaetzold R, Schmidt H, Hornetz B, Shisanya C (2005) Natural conditions and farm management information (2nd Ed.), Part A. West Kenya, Subpart A1: Western Province in: Farm Management Handbook of Kenya Vol. 11. Ministry of Agriculture. Nairobi

Khisa SG, Wekesa RK (2003) Farmer field school feedback: a case of IPPM FFS programme. MOA, Kakamega

Nyangito HO, Mose LO, Mugunieri LG (1997) Farmer circumstances and their effects on fertilizer use in Maize. In: Proceedings of the international conference of the African Crop Science Society (ACSS), Pretoria, South Africa, Jan 1997. Makerere University, Kampala, pp 1413–1420

Odendo M, Barakutwo JK, Gitari NJ, Kamidi M, Kirungu B, Lunzalu EN, Mureithi JG, Nekesa CO, Ojiem J, Okumu M, Okwuosa E, Onyango R, Saha HM (2000) Potential for adoption of green manure legumes for soil fertility management in Kenya. In: Proceedings of the 2nd conference of the Soil Management Project (SMP) and Legume Research Network Project (LRNP), Mombasa, 26–30 Jun 2000. SMP and LRNP, pp 403–411

Odendo M, De Groote H, Odongo OM (2002) Assessment of farmers' preferences and constraints to maize production in moist mid-altitude zones of western Kenya. In: Proceedings of the 5th international conference of the African Crop Science Society (ACSS), Lagos, Nigeria, 21–26 Oct 2001. Makerere University, Kampala, pp 769–775

Odongo OM, Odendo M, Ambitsi N, Woyengo VW, Salasya BDS, Oucho P (2006) Farmers' production practices and variety selection criteria in maize-bean based farming systems of western Kenya. Kenya Agricultural Research Institute, Kakamega, Research report submitted to Rockefeller Foundation, Nairobi Office

Omiti J (1998) Micro-level strategies to improve soil fertility in semi and agriculture. Paper presented at 15th symposium African farming system-research extension. AFSRE, Pretoria

Onduru DD, Gachimbi L, De Jager A, Maina F, Muchena FN (2002) Integrated nutrient management to attain sustainable productivity increases in east African farming systems. Report of a baseline survey carried out in Muthanu Village of Siakago Division, Mbeere District, Nairobi ETC-East Africa, KARI, LEI

Rees D, Momanyi M, Wekundah J, Ndungu F, Odondo J, Oyure AO, Andima D, Kamu M, Ndubi J, Musembi F, Mwaura L, Joldersma R (2000) Agricultural knowledge and information systems in Kenya—implications for technology dissemination and development. Agricultural research and extension network. Network paper No. 107. Department for International Development, London

Salasya B, Odendo M, Mwangi W, De Groote H, Diallo A, Odongo OM, Saleem E (2006) An investigation of the factors influencing the adoption of WH 502 maize variety in Western Kenya. Research report. KARI and CIMMYT, Nairobi

Status and Trends of Technological Changes Among Small-Scale Farmers in Tanzania

E.J. Maeda

Abstract The purpose of this paper is to highlight technological changes in agriculture among small-scale farmers in Tanzania and how these changes have contributed towards poverty reduction and promotion of sustainable agriculture. Available evidence suggesting that most of the rural areas are faced with low productivity in agriculture and severe land degradation in the form of soil erosion and nutrient depletion, while food crop yields of less than a metric ton per hectare are common and deforestation and overgrazing are widespread. The paper discusses issues relating to status and trends affecting small-scale farmers in Tanzania including their use of technologies, land and gender, labor, income and resource allocation. The main objective of this paper is to demonstrate the current status and trends within agricultural production for small-scale farmers in Tanzania. While the specific objective is to illustrate how farmers use improved inputs efficiently in the current farming systems, a second objective is to identify sustainable requirements for more diversified systems of farming that include commodities which can be produced with more ecologically benign systems. Diversified farming systems traditionally have utilized crop rotations to control pest build up, conservation of soil (through agricultural conservation techniques) and maintenance of productivity. In addition, the paper attempts to determine the benefits of integrated cropping and livestock systems that have been used to reduce farm operational costs, recycle waste, and stabilize the incomes of the rural farmers. Finally, the paper also discusses the effects of technological changes on labor, production, income, and time allocation of members within a household to farm activities. To address these objectives, literature from past studies on the issues concerning technological changes were sought out and discussed. Results indicate that investment in agricultural technology is crucial in order to meet the growing demand for food at low cost. Current evidence provides support the view that such investment is, indeed, profitable and does contribute to improved productivity. However, there is still lack of empirical evidence derived from rigorously measuring the impact of technological change on household welfare, based on consumption and other factors. The few studies that are available show that technological change improves income and food consumption while increasing burden in terms of women's work hours in the field. The first part of the paper is an overview and introduction to the country, its location, typology, economy, and a general background. The second part describes agricultural status and trends. The final section provides a conclusion and defines the way forward towards poverty reduction.

Keywords Agricultural technologies · Farm labor · Gender-roles · Irrigation · Poverty · Resource allocation

Introduction

Agriculture is clearly the most important sector of the economy of Tanzania. It contributes 45.1% of GDP in 2000 (EarthTrends, 2003), while 80% of the population in the country rely on agriculture for their

E.J. Maeda (✉)
Ministry of Agriculture Food Security and Cooperatives, Tanzania
e-mail: elizabeth.maeda@kilimo.go.tz

A. Bationo et al. (eds.), *Innovations as Key to the Green Revolution in Africa*,
DOI 10.1007/978-90-481-2543-2_109,

livelihood. The statistics relating to the pattern of poverty are far from perfect but they show that poverty in Tanzania is pervasive and predominantly occurs in the rural areas where the majority of Tanzanians reside (Beier et al., 1990).

Rapid population growth relative to agricultural production continues to be a major challenge in Tanzania. This has resulted in Tanzania failing to increase food production to keep pace with the rapid annual growth of population (Braun and Kennedy, 1994). Poverty, low agricultural productivity, and natural resource degradation are severe and inter-related problems in the highland areas of Tanzania. These areas are characterized by less-favored areas including lands that have low agricultural potential because of limited and uncertain rainfall, poor soils, steep slopes, or other biophysical constraints, as well as areas that may have high agricultural potential but have limited access to infrastructure and markets, low population density, or other socioeconomic constraints.

Likewise the most-favored areas are found in small areas within the highlands of Northern Tanzania and are characterized by rapid population growth and large investments in terms of agriculture.

In order to provide food for the growing population there is a need for the country to upscale its dissemination of improved technologies to reach a much larger population in the rural areas.

It is estimated that the ratio of males to females in the agricultural sector is 1:1.5 with women in Tanzania producing about 70% of the food crops while also bearing substantial responsibilities for many aspects relating to export crops and livestock production. However, challenges that women face include access to productive resources (land, water etc.), supportive services (marketing services, credit and labor-saving facilities etc.), and income arising from agricultural production is severely limited by social and traditional factors (Jiggins, 1986).

Agricultural productivity in Tanzania has increased with the use of innovative technologies such as irrigation, agrochemicals, machinery, use of hybrid varieties and varietal blends (Braun et al., 1989). Moreover, other innovations used include multiple cropping, agroforestry, soil management methods which enhance organic matter and soil life, use of improved hand tools and animal traction, integration of cropping and animal husbandry, mixed cropping, crop protection by natural means, use of genetic diversity including those crops and animals which are regarded as unconventional by mainstream agricultural scientists and techniques of harvesting nutrients and water whether developed by research institutes or as part of farmers' own initiatives are among the technologies adopted and used currently by small-scale farmers.

But still Tanzania faces massive problems in terms of food security because of decreasing per capita food production. The majority of small-scale farmers still rely wholly or partly upon local plant genetic resources, local knowledge, experiences and skills to transform capital assets into meaningful and sustainable livelihoods.

Likewise, in areas using new or advanced technologies extreme poverty is widespread and massive environmental degradation is the norm. It is a direct consequence of a policy environment that resulted in large-scale nutrient mining that has converted natural habitats to areas where agricultural monocropping activities dominate through policies like fertilizer subsidies and encouragement of large-scale farms which indirectly encourage overuse of agrochemicals with detrimental effects on the environment.

In the case of agricultural research the overall aim is to promote sustainable food security, income generation, and employment growth and export enhancement through the development and dissemination of appropriate and environmentally friendly technology packages and increasing productivity and profitability in agriculture.

The development of new technologies and their delivery systems has been the key area of focus in order to remove identified constrains such as uncoordinated, fragmented, duplicated, and overlapping messages flowing from extension personnel, poor supervision of extension personnel and utilization. This in turn has discouraged farmers from the use of appropriate technologies because most of these technologies were not specific to a particular location but were broader in terms of their applications. Some technologies promoted at research stations were not appropriate to farmers.

Extension services are now demand driven and address livestock keepers, farmers and other beneficiary's needs. Although extension services are directed to all farmers/livestock keepers, special attention is given to women in recognition of the critical role they play in family household management and food production.

Status and Trends of Agriculture in Tanzania

The country has a dual agricultural economy with a smallholder subsector and a commercial/large-scale subsector with smallholder farmers dominating the agricultural sector estimated to be 4.8 million people according to the 2002/2003 agriculture sample census conducted by the Ministry of Agriculture. These agricultural holders carry out rain-fed agriculture, producing a variety of crops mainly for subsistence purposes. This is followed by temporary mixed crops (17.6%). Ninety-four percent of the total land available to smallholders is utilized (National sample survey of agriculture, 2002/2003 volume II). Food crops produced include maize, sorghum, rice, pulses, millets, cassava, wheat, sweet potatoes, bananas, oilseeds, fruits, and vegetables. Traditional export crops include coffee, cotton, cashew nuts, tobacco, tea, sisal, and pyrethrum. And other nontraditional crops include horticultural crops, flowers, spices etc (Teklu, 1996).

The agricultural sector over the last decade contributed an average of 45.8% to overall GDP and grew at an average rate of 3.7% per annum. The crop subsector contributed an average of 35% to GDP between the years 1993 and 2002 with an annual average growth rate of 3.8% (Kumar, 1994).

Crop production in Tanzania is severely affected by the overdependence on rain-fed agriculture. Rainfall is erratic and not distributed uniformly over the crop-growing season (Carloni, 1984; EarthTrends, 2003). Occurrence of severe drought in most parts of the country has been a common phenomenon in recent times. Tanzanian agriculture is dominated by subsistence farmers who operate 0.2–2.0 hectares and use approximately 85% of cultivated land because of a heavy reliance on the utilization of the hand hoe as the main cultivating tool, which sets obvious limitations on the area of crops that can be grown using family labor. The agricultural sector employs about 75% of the total labor force, of which 56% are women.

The development of new technologies and their delivery systems has been a key area of focus in order to remove identified constrains and increase productivity and profitability.

Land Expansion

Attention to poverty in rural areas has increased markedly during the last couple of decades as farmers are still struggling to improve their livelihood. In the case of the land under cultivation the size of the farms differ depending on many factors including the issue of altitude. Most of the lands under production are found within the high altitude area of the favored environment including lands that have high agricultural potential because of reliable rainfall, good soils, etc. Areas of low agricultural production are found within less favorable environments including marginal agricultural areas, forest, and arid and semi-arid areas.

These areas are characterized by low agricultural potential because of limited and uncertain rainfall, poor soils, steep slopes, or other biophysical constraints but also include areas that may have high agricultural potential but have limited access to infrastructure and markets, low population density, or other socioeconomic constraints. In high altitude areas where the population density is considerably high the average farm size in acres ranges from 0.2 to 4.8 while in the low altitude most marginal areas the average farm size is 2.5–12.6 acres per household.

The most important feature of the status of agriculture in Tanzania is the issue of land, not so much its location or the effect of climate or any of the other aforementioned criteria but the fact that land is an integral part of the rural economic, social and ecological system.

An example is given by a study carried out in Arumeru, Arusha, Tanzania. In this study a decreasing trend in farm size ownership per household was observed for both sites. In general a decreasing farm size trend was seen from 1988 to 1999. The average farm size of households in the low altitude zone was significantly higher than in the higher altitude zone in 1999 (Table 1).

Table 1 Farm size ownership per household ($N = 34$)

Zone	Village	Mean farm size (acres) in a year	
		1988	1999
High altitude	Ngiresi	4.8	2.8
	Olgilai	2.9	2.8
Low altitude	Kiseriani	12.6	6.9
	Sig. (0.005)	0.49	0.02

Source: Survey, 1999/2000 season

According to key informants a decreasing trend in average farm size per household was a reflection of population pressure increase over time due to an increase in family size, and in-migration at both sites.

Production Environment

Because of the heterogeneity of Tanzanian production environments, a variety of new technologies have been developed to correct the deficiency in soil nutrients and moisture, reduce rainfall-linked variability in production, and overcome the seasonal shortage of labor, through site-specific recommendations made by research scientists.

Additionally, with economic and environmental costs associated with land expansion increasing at the margin, rapid growth in technological change is critical in fostering agricultural growth. This is due to the improved adoption and efficient utilization of existing technologies, as well as continuous generation of new and appropriate frontier technologies that enhance productivity.

Most of the rural areas in Tanzania are faced with environmental problems associated with agricultural production. These problems include first-generation problems of soil erosion, deforestation, and habitat destruction which are associated with a dangerous second generation of problems including inadequate use of water and fertilizer and excessive reliance on pesticides (Paolisso et al., 1989). Thus, land degradation takes a number of forms including depletion of soil nutrients, salinization, agrochemical pollution, soil erosion, and vegetative degradation as a result of overgrazing and the cutting down of forests for farmland.

In the case of small-scale irrigation technology development is likely to yield the highest returns in areas with good market access and suitable soil conditions since they can enable high value production as well as intensified food crop production.

Technological Changes and Labor Distribution

Thus, the main effects of such changes are on the labor market to the extent that technological changes are accompanied by an outward shift in labor-market demand while at the same time there is an increase in both employment and wage rates. This may include positive or negative effects in the case where there are labor supply responses to technology-induced increased demand for labor as for example the need to perform more field work in the case of weeding, pesticide application, harvesting and processing before storage may vary.

Additional household members may enter into the work force at the same time as women's working time is increased in terms of an increase in field operations where they may work long hours and changes are also seen to the allocation of time across activities for example through a reduction in time for nurturing and child care. In a much broader perspective family labor supply may remain unchanged but additional labor is drawn from non-family labor such as hired labor.

Adoption of a technology that uses labor intensively could adversely affect the status of such households. That is, the potential gains from additional work and earnings due to technological change may not be adequately realized as some household members experience increased workload which is not fully compensated by the increase in income.

The increase in demand for farm labor was largely met through an increase in family labor. All categories of family labor (female adult, male adult, child labor) experienced an increase in level of employment, but the relative increase was much higher for male adult labor. A household survey in the fertile Ufipa plateau of the Rukwa region in Tanzania also showed that the increased profitability of crop production changed the gender-specific division of crop activities, especially the increased participation of men (Beier et al., 1990).

Another positive effect of technology changes is that the effect on employment is crop- and technology-specific. For example, the land-augmenting types of technology, which have been adopted in the irrigation technology of rice farms and improved seed, fertilizer and oxen technology, led to an increase in demand for farm labor (von Braun et al., 1989; Kumar, 1994). In the case of Tanzania in the irrigated rice lands an increase in the planting area, increased women's workload in weeding and transplanting, even though they could not own land in the irrigation schemes.

Such reallocation of labor may be due to a greater demand by these types of technologies for particular activities, which are usually gender-specific. For example, the introduction of mechanical technology for rice production increased the amount of time required for female labor to meet the increased demand for planting and harvesting, two activities usually carried out by female labor. However, the evidence shows that adult men are likely to replace females when they find farm work profitable. Not only were men more drawn to undertaking ox-drawn ploughing in Rukwa region, but they were increasingly engaged in weeding and harvesting, which were typically women's activities.

The positive effect of technology changes are indicated through the market. It is observed that an increase in production translates to a decline in food prices with the beneficiaries being net food buyers, which includes small-scale producers because they do not produce enough to provide for their households throughout the year.

In short, the agricultural technologies adopted contributed to an increase in demand for labor and time women spent in production activities compared to their reproductive activities. For family labor the increase was relatively larger, although the evidence is not conclusive in regards to allocation of the additional work within households, or on the effect on rural wages of a technology-induced shift in labor demand.

Effect on Production

Farmers tend to reallocate land to cash crops using new technology whereby in a few cases such reallocation of land may not necessarily be accompanied by a decline in the proportion of land allocated to food crops. This is because the areas grown under food crops are mostly areas that have been fallowed and brought back into crop production or are areas within the marginal land where food crop production has been practiced.

Secondly, where there is a decline in area allocated to food crop production, it does not necessarily translate into a decline in food production because where yield level is high, farmers can meet their subsistence needs without committing a large area of land while at the same time the adoption of yield-increasing technology has reduced the land area needed for subsistence production. This has been noted in areas that have best management practices as well.

The extent to which farmers choose to switch to non-staple commercial crops is not only governed by profitability considerations, but also by how much farmers weigh food security risk (due to uncertainties arising from production, off-farm employment, and remoteness from markets and services) and the subsistence orientation of farmers in these areas.

Evidence from studies conducted in various countries in Africa shows that farmers deliberately maintain food crops along with cash crops where the premium on food security is high. Where food security risks are low, farmers tend to specialize. Thus, the emphasis on a "food first" strategy becomes less of a priority in areas where there is a great deal of market integration (Teklu, 1996).

Effect on Income

In general, farmers adopting new technology experience an increase in income level due to an increase in factor productivity and a decline in unit costs of production. But the increased income comes with a shift in income control to men especially when it involves earning more money. For example, men tend to assume greater income control as farming becomes more profitable whether it is food or cash crop production. Such a shift in income control from women to men adversely affects distribution of food allocation within households.

Impact on Women's Time Relative to Child Care and Income

The common concern is that adoption of technology leads to an increase in demand for women's labor and, hence, to a reallocation of women's time away from child nurturing activities. Carloni (1984) reported that, time allocation studies from Tanzania indicate that women's time constraints during the peak agricultural season may interfere with breastfeeding. However, during the peak agricultural season, although, women devoted less time to child care they arranged for their older daughters to care for their younger siblings (Paolisso et al., 1989).

In the case of women's time, what happens in particular, especially the time allocated to child care depends on the relative valuation of time between income generation and child care. Thus, most rural women have to weigh the competition of time between earning income and child care especially in households that face an increased demand for their labor, but lack sufficient labor of their own or resources to hire labor. In addition, increased time demands from adoption of new technology might also lead to insufficient time devoted to child care.

On the other hand, there are studies that indicate an increasing share of women's time commitment after technological change such that most of these technologies double the labor time requirement for women's working hours in absolute terms (Jiggins, 1986).

Other factors that result in low agricultural productivity include cultural values and customs that can influence agricultural production because of the marked influence they have on the way people use the land.

Conclusions and Implications for Policy and Research

What is clear from the evidence presented so far is that technology adoption can affect household welfare through at least three different pathways (income-food consumption, income-nonfood expenditures-morbidity, and time allocation patterns-child care). Thus, the design of appropriate technology necessitates the evaluation of how these pathways can independently and jointly impact rural household livelihoods. The decision to intervene through a particular pathway needs to consider both independent and joint effects through the various links.

Thus, investment in agricultural technology is crucial in order to meet the growing demand for food at low cost and increase farmer's income. Current evidence provides support for the view that such investment is, indeed, profitable and sustainable and does contribute to improved productivity. What has currently been observed is the adoption by small-scale farmers of improved technology (for example, seeding of improved seeds, applications of fertilizer, animal traction technology and irrigation technology) has set in motion adjustments in household resources, especially labor, time and land, which are the primary factors of production and sources of income in most small-scale agricultural households.

Furthermore, research and technology alone will not drive agricultural growth. The interaction between technology and policy is crucial. Thus, distortions in input and output markets, asset ownership and other institutional distortions are unfavorable to the poor and must be minimized or removed.

Access by the poor to productive resources such as land and capital needs to be enhanced. Rural infrastructure and other institutions must be strengthened, for example expanded human capital investments in education, healthcare nutrition and sanitary environments.

In addition, the policy environment must be conducive to and supportive of poverty alleviation and sustainable management of natural resources in order to place agriculture at the forefront of reducing poverty. This will allow changes in agriculture which are geared towards helping farmers become more commercialized and increase their ability to operate in a liberalized market economy. With proper use and awareness of the appropriate technology befitting their particular environment and circumstances, farmers become more competitive and are able to comply with market standards and trade regulations.

Experiences within the field of technology development have clearly identified that new technologies have to be embedded in the local society, its ecological and physical environment, agricultural expansion and socioeconomic structures. Strategies for developing and disseminating technology must take into account the special characteristics and demand of the different areas because of the high degree of diversity in biophysical and socioeconomic conditions that prevail in rural Tanzania.

Other factors include changes in local agricultural agrobiodiversity causing the loss of diverse species and varieties by rapid replacement with a few genetically high-yielding uniform hybrid varieties developed by national and international institutes which are not available to most of the resource-poor farmers who comprise the majority of the farming community.

Thus, most resource-poor farmers use and maintain land race because modern varieties often provide yield advantages that results in their being sown over large areas.

Technological changes such as the use of fertilizers and irrigation also lower the demand for land races adapted to marginal growing areas.

References

Beier M, Falkenberg CM, Mjema N, Zewdie A (1990) The Southern highlands development and the food security needs of Tanzania: the case of Rukwa Region. A study paper prepared for GTZ, Division 426

Braun JV, Kennedy E (eds) (1994) Agricultural commercialization, economic development, and nutrition. Johns Hopkins University Press, Baltimore, MD

Braun JV, Puetz D, Webb P (1989) Irrigation technology and commercialization of rice in The Gambia: effects on income and nutrition. International Food Policy Research Institute, Washington, DC

Carloni A (1984) The impact of maternal employment and income on the nutritional status of children in rural areas in developing countries. United Nations Administrative Committee on Coordination – Subcommittee on Nutrition, Geneva

EarthTrends (2003) Economic indicators – Tanzania. Earth trends country profiles. http://earthtrends.wri.org/pdf_library/country_profiles/eco_cou_834.pdf. Accessed 24 Mar 2011

Jiggins J (1986) Gender-related impacts and the work of the international agricultural centers. Consultative Group on International Agricultural Research (CGIAR) study working paper no. 17. World Bank, Washington, DC

Kumar SK (1994) Adoption of hybrid maize in Zambia: effects on gender roles, food consumption and nutrition. Research report no. 100. International Food Policy Research Institute, Washington, DC

Paolisso M, Baksh M, Thomas JC (1989) Women's agricultural work, child care and infant diarrhea in rural Kenya. In: Leslie J, Paolisso M (eds) Women, work, and child welfare in the Third World. Westview Press, Boulder, CO, pp 217–230

Teklu T (1996) Food demand studies in Sub-Saharan Africa: a survey of empirical evidence. Food Policy 21(6):479–496

The Dilemma of Using Fertilizer to Power the Green Revolution in Sub-Saharan Africa

D.K. Musembi

Abstract Agriculture in over 84% of world farmlands is rainfed but yields just over half of the crops produced. In East Africa 60% of the landmass is arid and semi-arid land (ASAL), where rainfall is inadequate for arable agriculture. The ASAL environment is harsh. Rainfall is low, unreliable and bimodal, and the seasons have unequal production potential. In Kenya, the short rains contribute 55% of the annual rainfall, while the long rains contribute 35%. Frequent droughts cause crop failure raising the dilemma whether farmers can apply fertilizer to crops in either season. Fertilizer and hybrid seeds are expensive and farmers no longer keep sufficient livestock for manure due to small land units and lack of herdsmen occasioned by free primary education. Poor farmers plant inferior cultivars without fertilizer or manure and fail to apply pesticides or manage weeds. Illiteracy and low mechanization limit the ability to maintain required plant population, and planting is late due to inability to prepare land early. Crops are therefore unable to utilize all available moisture. Farmers must use recommended cultural practices including appropriate cultivars, fertility, seed, planting time, weeding and pesticide application. The produce must also attract competitive market prices. Investing in production, including use of fertilizer, is risky and can be done during the short rains. The government must be courageous in formulating enabling policies. Policy should regulate use of land based on size and potential. Planting grass for livestock, not maize for humans, gives better results in ASALs. Overcoming these dilemmas empowers the ASALs to power the green revolution.

Keywords ASALs · Bimodal · Mechanization · Production · Risk

Introduction

Agriculture in over 84% of world farmlands is rainfed but yields just over half of the crops produced, the rest are produced from irrigated farmlands. Africa has a total surface area of $3{,}010 \times 10^6$ ha of which 230×10^6 ha is water resources (FAO, 1978). Most of her soils are poor compared with other parts of the world due to weathering, erosion, leaching and lack of volcanic rejuvenation (Smaling, 1993). Ten percent of the soils, mainly vertisols and other clayey soils, have high to very high available water-holding capacity (AWHC) and 29% of the soils have medium AWHC; these are lexisols, ferralsols, loamy inceptisols and entisols. The rest are ferralsols with low AWHC. Fifty five percent of the lands are deserts, salt flats, dunes and rock lands and steep to very steep lands which are unsuitable for cultivated agriculture. These lands are considered fragile, erodible, easily degraded with bad management and only suitable for nomadic grazing (Eswaran et al., 1996).

The climate of Africa is tropical, and agriculture especially in the lowlands is controlled by rainfall. Most climatic elements, particularly temperature, are uniform in time and place but rainfall is much more variable (Nieuwolt, 1982). Lack of sufficient rainfall is the main limiting factor in agriculture and long-term predictions expected from agroclimatologists should

D.K. Musembi (✉)
Kenya Agricultural Research Institute (KARI), Kiboko Research Centre, Makindu, Kenya
e-mail: dkmusembi@yahoo.com

A. Bationo et al. (eds.), *Innovations as Key to the Green Revolution in Africa*,
DOI 10.1007/978-90-481-2543-2_110,

mainly be concerned with rainfall. However, its irregularity makes the task difficult.

There is need to quantify rainfall in relation to agriculture in simple methods that do not require the use of computers. Annual rainfall totals present a picture of endless variations with no clear trend or cycles but seasonal rainfall distribution is more regular as the large-scale weather patterns are controlled by monsoonal wind system. The day-to-day irregularity of rainfall affects the growth and yields of many crops, especially those with shallow root systems.

Sixteen percent of Africa's soils are of high nutrient quality and 13% have medium nutrient quality, both supporting 400 million people. The average use of fertilizer at 21 kg ha^{-1} is low with sub-Saharan Africa (SSA) averaging 10 kg ha^{-1} excluding South Africa (Henao and Baanate, 1999). Only Egypt, South Africa, Swaziland and Zimbabwe use 50 kg ha^{-1}. The low use of inorganic fertilizer leads to environmental degradation, while increased use benefits the environment by reducing the pressure to convert forests and other fragile lands to agricultural uses. Increasing biomass production helps to increase soil organic matter content (Wallace and Kneusenberger, 1997).

In the last 30 years, soil fertility depletion in Africa is estimated to have averaged 660 kg N ha^{-1}, 75 kg P ha^{-1} and 450 kg K from 200 million hectares of cultivated land in 37 African countries. An estimated $42 billion in income and 6 million ha of productive land are lost every year due to land degradation and declining agricultural productivity (UNDP/GEF, 2004). Soil fertility depletion in smallholder farms is a fundamental cause of declining per capita food production that has contributed to poverty and food insecurity. Nutrient loss from cultivated land is estimated at 4.4 million (M) tons (T) of N, 0.5 M T of P and 3 M T of K every year (Sanchez et al., 1997). These losses are higher than Africa's annual fertilizer consumption of 0.8 M T of N, 0.26 M T of P and 0.2 M T of K excluding South Africa.

Decline in inorganic matter threatens soil productivity. Most African soils are inherently low in organic carbon (<20–30 mg kg^{-1}) due to low crop and vegetation root growth and rapid turnover of organic matter due to high temperature and microfauna, especially termites (Bationo et al., 2006). Results from long-term soil fertility trials indicate that losses of up to 0.69 T of C ha^{-1} $year^{-1}$ in soil surface layers are common in Africa even with high levels of inputs (Nandwa, 2003).

Soils of the marginal ASALs in Africa are generally deficit in N whose enrichment is necessary in order to raise production (Mugabe, 1994). This is commonly achieved through the application of nitrogen fertilizers. However, most small-scale farmers, who constitute 60% of Africa's population, cannot afford them. Mugabe (1994) has observed that 75% of Africa's consumption of chemical fertilizers is imported. Kenya, for example, spends 40% of its foreign exchange on fertilizer imports, and this expenditure can be reduced through full exploitation of biological nitrogen fixation (BNF). Research has demonstrated that the common bean *Phaseolus vulgaris* has the potential to fix N up to 50 kg ha^{-1} $year^{-1}$(Mugabe, 1994).

During the 1960s and 1970s, an external input paradigm was driving the research and development agenda for agriculture. It was believed that inputs, i.e. fertilizers, lime, irrigation or water, were able to alleviate any crop production constraint. When this new approach was coupled with the use of improved germplasm, the "green revolution" was realized in Asia and Latin America where agricultural production increased tremendously. But only minor achievements were realized in Africa due to a variety of reasons (IITA, 1992). This was because removal of fertilizer subsidies due to imposition of structural adjustment programmes made fertilizers unaffordable for most subsistence farmers (Smaling, 1993), while in the 1980s there was a shift from mineral inputs only to low external input sustainable agriculture (LEISA), where organic resources were believed to enable sustainable production.

Fertilizer was as important as seed in the green revolution and contributed up to 50% of the increase in yield in Asia (Hopper, 1993), and several studies have found that one-third of cereal production in the world is due to use of fertilizers and related factors of production (Bumb, 1995). Van Kuelen and Breman (1990) and Breman (1990) stated that the cure for the land hunger in the West African Sahel lay in the increased productivity in the arable lands through the use of inorganic fertilizers, and maize trials in different agro-ecological zones (AEZ) have demonstrated the importance of fertilizers, especially NPK in the west, east and southern Africa.

Materials and Methods

Information is presented from published work obtained through literature search to highlight the problems facing rural African farmers using the experiences of the Kenyan subsistence farmers. The problems are environmental, socio-economical and related to government policy. Ways of enabling these farmers to contribute to the green revolution are discussed.

Results and Discussions

The Environment

Agriculture in over 84% of world farmlands is rainfed and yields just over half of the crops produced, the remainder coming from irrigated farmlands. In East Africa, 60% of the landmass is arid and semi-arid land (ASAL), where rainfall is inadequate for arable agriculture. In Kenya, the ASALs cover over 84% of the total land area (Dolan et al., 2004) and the environment is harsh, characterized by high temperature, low and unreliable rainfall (<800 mm) that is poorly distributed in time and space (Griffiths, 1972; Todorov and Pali-Shikulu, 1973).

The rainfall pattern is bimodal but only one season is reliable for agricultural production (Musembi, 1986; Kenworthy, 1964) as illustrated by rainfall data from Makindu Meteorological Station. The communities inhabiting the ASALs are either agropastoralists who suffer crop failure and famine during drought or pastoralists, whose livestock lack pasture and die when rains fail. The pastoralists are becoming more sedentary and turning to growing crops due to subdivision of group ranches (Pasha, 1986; Kimani and Pickard, 1998).

The rains are difficult to forecast (Nieuwolt, 1982; Stewart and Kashasha, 1984) and the lack of reliable seasonal forecasts makes it difficult for the farmer to make the right decisions on choice of seed, fertilizer use and other cultural practices. Musembi (1999) found that farmers in the pastoral districts of Kajiado and agropastoral district of Makueni were willing to make informed decisions on agricultural practices if accurate weather guidelines were available. The agropastoralists were willing to use appropriate crop cultivars and invest in farm inputs while the pastoralists were willing to reduce or increase their herds in drought or good years, respectively.

Similar observations were expressed in Burkina Faso, where farmers were interested in receiving seasonal forecasts (Ingram et al., 2002). In Zimbabwe, farmers responded to rainfall forecast information by altering crop distribution (Phillips et al., 2002) and in India, 14 marginal farmers wrote to researchers seeking their view on a forecast of poor 2000 monsoon rains. This was occasioned by large losses from investment in fertilizer and land leasing during the preceding poor rains of 1999 (Cadgil et al., 2002).

Socio-economic

Both the agropastoralists and the pastoralists are subsistent farmers, practising low-input and low-output agriculture with limited technology. The agropastoralists in Machakos and Makueni districts of Kenya are, for example, small landholders for whom cropping primarily is a means of feeding the family, with cash being of secondary importance (Stewart and Kashasha, 1984).

African farmers typically pay considerably higher for fertilizer than farmers in most other parts of the world (Jayre et al., 2003), and their cropping practice is non-use of fertilizers and low crop population densities (Collinson, 1978; Rodewald, 1978; Rukadema, 1981). The yields are low, ranging 442–2,434 kg ha^{-1} with no apparent relationship between plant population and rainfall or yield.

Farmers plant late for a variety of reasons. Some plant late because they cannot afford tractors and neither have the family labour nor the oxen to plough the land before the rains come and have to wait until those who own oxen prepare their farms before they can access them, often after the rains have started. Other farmers minimize cost of weeding by letting the rains to continue until the weeds emerge then plough them under before sowing. In such cases, the rains are wasted and the crops suffer moisture stress and produce low yields. Stewart and Kashasha (1984) have found a direct relationship between the date of planting and maize yield in the ASALs and established 180 mm of rainfall as the minimum for maize production with or without fertilizer.

Farmers in the dry areas lack precision technology for seed and fertilizer application. So they are unable to achieve the crop population densities and fertilizer levels recommended by extension. Where the attempt is made to achieve the recommended population density, this is not matched by appropriate nutrient application. Thus, although Stewart and Kashasha (1984) found increases of maize yield ranging from 680 to 2,400 kg ha^{-1} on farmers' fields during the good rains of 1982, there was no particular relationship between production and population.

Intercropping is commonly practised during the two rainy seasons even though the seasons have unequal production potential (Musembi, 1986). Legumes (beans, cow peas, pigeon peas, *Dolichos lablab*) are intercropped with cereals (maize, sorghum, pear millet). This is done due to limited land or in order to save on land preparation expenses. Stewart and Kashasha (1984) have found that the maize and beans intercrop is a system with higher water requirements and is only to be recommended for rainfall greater than 300 mm, above which maize yield is limited by fertilizer, not rainfall ICRISAT (1985–1988). Shapiro and Sanders (1988) found similar results and reported a water-use efficiency of 1.24 kg grain mm^{-1} without fertilizer and 4.14 kg grain mm^{-1} of water with fertilizer for the crop of millet in Sadore.

Policy

The government is able to affect agricultural activities through policy instruments or legislation. The draft constitution of Kenya (Kenya Gazette Supplement 2005 86(1)) had, for example, proposed prescription of minimum and maximum land holding acreage in agricultural areas. If the draft constitution had been passed, some of the problems associated with cultivation in the ASALs would have been sorted out through regulation of the type of agricultural activities acceptable in different ecozones and the question whether livestock production or arable agriculture or both should be practised in certain areas would have been answered.

Kenya, Ethiopia and Zambia have pursued official policy of encouraging private sector in fertilizer marketing but have taken different routes in reforming the fertilizer sub-sector (Jayre et al., 2003). In the 1980s, fertilizer prices in Kenya were controlled and the consumption of fertilizer in 1965–1989 varied from 90 to 250,000 T and was expected to rise above 250,000 T by 1993. The controlled pricing structure was designed to improve farmers' access to fertilizer but had the opposite effect in the more remote areas (Jayre et al., 2003).

Fertilizer prices were decontrolled in 1993 and the number of retailers rose from 5,000 to 7,000–8,000 by 2000. This market reform stimulated fertilizer use by improving farmer's access to inputs through expansion of private networks (Freeman and Omiti, 2003) Thus, the total number of small-scale farms using fertilizer increased from 61 to 65% but use varied considerably, ranging from <10% for households in the drier lowlands to 90% in the Central province and high potential maize zones in North Rift. This is because marketing costs remain high and as a result fertilizer application rates by small-scale farmers are below the levels recommended by the extension service.

Another significant problem to effective utilization of fertilizers has been the "pan-territorial/blanket" recommendations that fail to take into account differences in resource endowment (soil type, labour capacity, climate risk, etc). This is exacerbated by failure to revise recommendations following dramatic changes in input/output price rates due to subsidy removal and devaluation of currencies. Farmers have therefore adopted different approaches of applying fertilizers, e.g. the micro-dose technology in West Africa (Mali, Burkina Faso and Niger) where farmers apply small doses of fertilizer (4 kg P ha^{-1}) with the seed, which is one-third of the recommended rate (Tabo et al., 2007).

In Kenya, Ethiopia and Zambia, transport and handling costs accounted for 50% or more of the total domestic marketing margins, and for this reason, African farmers pay considerably higher prices for fertilizer than farmers in most other parts of the world. Jayre et al. (2003) have observed that for substantial increases in fertilizer use in Africa to take place, the high physical costs of exchange that impede marketing activities by all agents must be reduced, whether private, parastatal or cooperative.

One ton of urea is $90 in Europe, $500 in western Kenya and $700 in Malawi. The reason for this is the removal of subsidies, transport costs, poor infrastructure, poor market development, inadequate access to foreign exchange and credit facilities and lack of training to promote the use of fertilizers (Sanchez and Swaminathan, 2005; Sachs, 2005; Bationo et al., 2006). Jayre et al. (2003) have stated that the opportunities for reducing fertilizer prices lie in reducing port fees, coordinating timing of fertilizer clearance with upcountry transport, reducing transport costs through port, rail and road improvements, reducing high fuel taxes and reducing the uncertainty associated with government input distribution programmes that impose additional marketing costs on traders.

Conclusions

Over 84% of the farmers in Kenya are resource poor and are engaged in subsistence agriculture, and the price of fertilizers is unaffordable for them to meet the recommended rates of application. Fertilizer application is important if the full production potential of the farms is to be realized because without fertilizers, the full benefits of rainfall cannot be realized. In areas with marginal potential for crops, land size is too small to enable farmers to keep sufficient herds of livestock for adequate manure for their farms. The option open to them is under-dosing or micro-dosing of their farms with nutrients. Retailers cannot be expected to avail fertilizers to rural farmers outside profit considerations. Traders supply inputs if the following condition is satisfied by a sufficient number of farmers in an area:

$$E(\mathrm{WTP}i) > \mathrm{P}i$$

where

$E(\mathrm{WTP}i)$ = expectation of farmer's willingness to pay for the input
$\mathrm{P}i$ = competitive cost of input to farmer i, including transaction costs

The government must therefore put in place policies that reduce the price of inputs, including fertilizers, for there is no justification why fertilizers in poor African countries should be six to eight times the prices in the developed world. Affordable fertilizer prices will help to increase agricultural production and reduce food insecurity. Governments must also formulate bold policies to regulate the use of land for suitable agricultural activities. For example, agro-ecozone V in Kenya is better reserved for livestock production rather than cropping, which tends to degrade this important and fragile natural resource due to frequent rainfall failures.

More effort must be put into forecasting the notoriously difficult to forecast tropical rains and the forecasts should as much as possible be real time in nature so that advisories can be issued to farmers to evaluate their farm activity options. Farmers have indicated that they are willing to use agroclimatological information for decision making. The government must avail resources to develop the necessary capacity for this undertaking.

The farmers must also be empowered through capacity building to appreciate how their practices like late planting contributes to crop failure. Stewart and Kashasha (1984) have shown that there is a yield loss of 3.8 kg grain per every millimetre of rainfall wasted after the rains have started. More importantly, they must be capacity built to be able to undertake farming as a business. If these obstacles or dilemmas are removed, the African farmer will be able to participate in the green revolution.

Acknowledgements It is with great appreciation that the author acknowledges the support granted to him by the Director KARI and the Center Director at KARI Kiboko during the study leading to this chapter.

References

Bationo A, Hartermink A, Lungu O, Naimi M, Okoth PF, Smaling E, Thiombiano L (2006) African soils: their productivity and profitability of fertilizer use. In Africa fertilizer summit, Abuja, Nigeria, 9–13 June 2006
Breman H (1990) No sustainability without inputs. Sub-Saharan Africa beyond adjustment: African Seminar, Ministry of foreign affaris, DGIS, The Hague
Bumb B (1995) Global fertilizer perspective 1980–2000: the challenges in structural transformations, technical bulletin T-42, IFDC, Muscle Shoals, Alabama
Cadgil S, Rao PRS, Rao KN (2002) Use of Climate information for farm level decision making: rainfed agriculture in Southern India. Agric Syst 74:431–457
Collinson MP (1978) Survey of 132 farmers in parts of lower, drier areas of Northern Division, Machakos. CIMMYT/Medical Research Centre, Nairobi (mimeo)
Dolan R, Defoer T, Paultor G (2004) Kenya Arid and Semi-Arid Lands Research Programme, KARI. Final Feasibility Report
Eswaran H, Almaraz R, Van den berg E, Reich P (1996) Assessment of soil resources of Africa in relation to productivity. World Soil Resources, Soil Survey Division, USDA Natural Resources Conservation Service, Washington, DC

FAO (1978) Report on the agricultural zones project, methodology and results for Africa, vol 1. World soil resources project, 48, FAO, Rome, Italy, 158p

Freeman HA, Omiti JM (2003) A Freeman and Omiti, fertilizer use in semi-arid areas of Kenya: analysis of smallholder farmers' adoption behaviour under liberalized markets. Nutr Cycling Agroecosyst 66:23–31

Griffiths JF (1972) Climates of Africa. Elsevier, Amsterdam, pp 313–347

Henao J, Baanate CA (1999) Nutrient depletion in agricultural soils in Africa, 2020 Vision, No. 62. IFPRI

Hopper WD (1993) Indian agriculture and fertilizer: an outsider's observations, keynote address to the FAI seminar on emerging scenario in fertilizer and agriculture: global dimensions. FAI, New Delhi, India

ICRISAT (1985–1988) ICRISAT Sahelian Centre, Annual report, 1984, ICRISAT, Patancheru

IITA (International Institute of Tropical Agriculture) (1992) Sustainable food production in Sub-Saharan Africa No. 1, IITA's Contributions, Ibadan, Nigeria

Ingram KT, Roncoli MC, Kirshen PH (2002) Opportunities and constraints for farmers of West Africa to use seasonal precipitation forecasts with Burkina Faso as a case study. Agric Syst 74:331–349

Jayre TJ, Govereh J, Wanzala M, Demeke M (2003) Fertilizer policy development: a comprehensive analysis of Ethiopia, Kenya and Zambia. Food Policy Anal 28(4):293–3116

Kenworthy J (1964) Rainfall and water resources of E. Africa. Geographers and the tropics, liverpool essays. Longmans, Green and Co., London, 111–137

Kenya Gazette (2005) Kenya Gazette Supplement No. 63. Government Printer, Nairobi

Kimani K, Pickard J (1998) Recent trends and implications of group ranch sub-division and fragmentation in Kajiado District, Kenya

Mugabe J (1994) Research on biofertilizers: Kenya, Zimbabwe and Tanzania, Biotechnology and development monitor, No. 18, Mar 1994

Musembi DK (1986) The use of precipitation to identify soil moisture patterns and growing seasons in Eastern Kenya. J Agric For Meteorol 37:47–61

Musembi DK (1999) Survey report for Traditional indicators of rainfall in Kajiado

Nandwa SM (2003) Perspectives on soil fertility in Africa. In: Gichuru MP et al (eds) Soil fertility management in Africa, a regional perspective. Academy of Science Publishers (ASP) and Tropical Soil Biology and Fertility Institute, CIAT, Nairobi, pp 1–50

Nieuwolt S(1982) Tropical rainfall variability, the agroclimatic impact. Agric Environ 7:135–148

Pasha IK (1986) Evolution and individuation of group ranches in Masailand. In: Hansen RM, Woie BM, Childs RD (eds) Range management and research in Kenya. Winrock International Institute for Agricultural Development, Morrilton, pp 303–318

Phillips JG, Deane D, Unganai L, Chimeli A (2002) Implications of farm-level response to seasons climate forecasts for aggregate grain production in Zimbabwe. Agric Syst 74(2092):351–369

Rodewald GE (1978) Machakos Survey results. Analysis of 40 farmers during the Long rains of 1977, carried out by Development Planning Division, Ministry of Agriculture, Kenya (mimeo)

Rukadema M, Mavua JK, Audi PO (1981) Farming survey results from Mwala (100 farmers). UNDP/FAO/GoK Dryland Farming Research and Development Project

Sachs JD (2005) Investing in development: a practical method of achieving the millennium development goals. Report of the UN Millennium Development Project to the UN General Secretary, UN, New York, NY

Sanchez PA, Shepherd KO, Saule MJ, Place FM, Mokwunye AU, Buresh RJ, Kwesiga FR, Izac AMN, Ndiritu CG, Woomer PL (1997) Soil fertility replenishment in Africa: an investment in natural resource capital. In: Buresh RJ et al (eds) Replenishing soil fertility in Africa. SSA Special Pub. 51, SSSA, Madison, WI, pp 1–46

Sanchez PA, Swaminathan MS (2005) Cutting world hunger in half. Science 307:357–359

Shapiro BI, Sanders JH (1998) Fertilizer use in semi-arid West Africa: profitability and supporting policy. Agric Syst 56:467–482

Smaling EMA (1993) An agro-ecological framework of integrated nutrient management with special reference to Kenya. Waginengen Agricultural University, Wageningen

Stewart JI, Kashasha DAR (1984) Rainfall criteria to enable response farming through crop based climate analysis. E Afr Agric For J (Special Issue):58–78

Tabo R, Bationo A, Gerard B, Ndjeunga J, Machal D, Amadou B, Anou G, Sogodogo D, Taonda JBS, Hossane H, Dialo MK, Koala S (2007) Improving soil productivity and farmers' income using strategic application of fertilizers in West Africa. In: Bationo A, Waswa BS, Kihara J, Kimetu J (eds) In advances in integrated soil fertility management in Sub-Saharan Africa: Challenges and opportunities, vol XIV. Springer, Dordrecht. ISBN:978-1-4020-5759-5, 1097p

Todorov A, Pali-Shikulu J (1973) Agrometeorological assessment of temperature conditions. WMO No. 389, pp 170–180

UNDP/GEF (2004) Reclaiming the land sustaining livelihoods: lessons for the future. United Nations Development Fund/Global Environmental Facility, Nov, 2004

Van Kuelen H, Breman H (1990) Agricultural development in West African Sahelian: a cure for land hunger. Agric Ecosyst Environ 32:177–197

Wallace MB, Kneusenberger MI (1997) Inorganic fertilizer use in Africa: Environmental and economic dimensions, Environmental and natural resources policy and training (EPAT) project, Applied research, Technical assistance and training, September 1997, Winrock International Environmental Alliance, Arlington, Virginia

Overcoming Market Constraint for Pro-poor Agricultural Growth in the Eastern DR Congo, South Kivu

P.M. Njingulula and E. Kaganzi

Abstract The opportunities and constraints facing the marketing of the agricultural production have been poorly documented in DR Congo. To address this lack of information, a survey of the marketing opportunities and processing constraints facing agricultural markets was conducted in 11 rural markets in South Kivu. The objectives of the study were to assess the marketing opportunities, which may increase farmers' income, create technology demand and provide incentives for sustainable production, by investing in soil fertility improvement. The results have shown that (1) there are enormous marketing possibilities within and around the country: main national towns receive an average of 29,083 t (48.4%) of the beans produced by smallholder farmers in Kivu region and about 31,737 t (22.5%) is traded seasonally between DR Congo, Rwanda, Burundi and Uganda and (2) smallholders have few alternative sources of income and lack capacity to access and use these market opportunities. Most of the constraints are linked to limited market access and development, inhibiting economic and technological development, low economic activity, poor markets for agricultural input, output and finance, high transaction costs and risks and high unit costs, weak institutional and infrastructure environment, high cost and weak information access and property right. The study suggests that overcoming agricultural marketing constraints requires a strong information programme to sensitise farmers to take advantage of regional and national opportunities (potential market), a collaborative mechanism between public and private sectors, for improving access to agricultural services, including market information, instituting contractual enforcement measures and strategies for optimising the utilisation of processing capacity.

Keywords Market constraints · Agricultural market · Opportunities

Introduction

The last three decades have witnessed a dramatic increase in the amount of research carried out concerning food security in developing countries. A growing body of literature therefore exists on topics related to food production, consumption and nutrition. Alternatively, various authors have examined issues of food aid, international trade in agricultural commodities and food security. There is also a growing interest in the marketing of industrial and financial goods (Kindra, 1984; Kaynak, 1986; Kinsey, 1988; Roemer and Jones, 1991).

Studies on the internal distribution and sale of locally produced plant, livestock and fish products (domestic agricultural marketing) have tended to receive less attention for an assortment of reasons (Scott, 1995).

Following Shaffer (1986), a number of marketing studies were carried out in Latin America to emphasise the analysis of vertical linkages between participants in agricultural marketing activities. This issue has often been overlooked in previous research because of the limitations of the then prevailing analytical framework. These studies emphasised marketing activities from the producer through to the consumer for a particular commodity. Such an approach paid too much

P.M. Njingulula (✉)
Socio-economist INERA-Mulungu, Kivu, DR Congo
e-mail: pnjingulula@yahoo.fr

A. Bationo et al. (eds.), *Innovations as Key to the Green Revolution in Africa*,
DOI 10.1007/978-90-481-2543-2_111,

attention to the food requirements of poor, urban consumers, thus overlooking the production and marketing constraints of smallholder farmers (Bromley and Symanski, 1974).

The objectives of the study were to assess the marketing opportunities which may increase smallholder farmers' income, create technology demands and provide incentives for sustainable production by investing soil fertility improvement.

Materials and Methods

The study was conducted in two districts in the eastern part of the Democratic Republic of Congo (North and South Kivu). The study was based on nine rural markets (Mudaka, Katana, Mugogo, Kabamba, Kavumu, Musiru, Kankinda, Nyangezi, and Ninja) and two provincial markets (Bukavu and Goma). During the survey interviews were conducted with smallholder farmers, large-scale commercial growers, rural assemblers, transporters, wholesalers, retailers, consumers, researchers and extension agents.

A total of 175 individuals were interviewed: 15 individuals from rural market, including 5 smallholder farmers, 3 large-scale commercial growers, 2 rural assemblers, 2 transporters, 2 consumers and 1 public service agent. Given the economic importance of provincial markets, 20 people were interviewed for each market. These included five sellers, five transporters, four stockists, five buyers and one public administrative manager.

Bean market characterisation in North and South Kivu was the main concern of the first phase of the study. It was conducted from 2 February 2004 to 15 July 2004. Demands characterisation and agricultural market tendencies were carried out in the second phase from 1 March 2005 to 5 June 2005 in four rural markets (Katana, Mudaka, Mugogo and Musiru).

Field Data Collection

In view of the limited time available for the study, it was decided to rely on rapid appraisal techniques, i.e. informal interviews with key informants, direct observation of critical stages in the production–transformation–distribution sequence and reliance on secondary data sources whenever possible. Field data collection activities began with an informal reconnaissance survey. Itineraries were prepared for teams made up of two to three researchers, combining social scientists with technical scientists and extension agents (NGOs). The purpose of the surveys was to become familiar with important bean production zones, principal categories of farmers, input distribution networks, bean production technologies and bean utilisation patterns. As marketing of beans was of particular interest, an effort was made to directly observe the various components of the bean marketing system.

The study was designed to answer the following questions:

- What are the principal bean production zones in Eastern Congo?
- What are the predominant categories of bean producers?
- What are the most common bean production technologies?
- What are the principal usages of beans?
- Are beans planted as a food crop, cash crop or both?
- Where and to whom do farmers sell their beans?
- What are the principal marketing channels for beans?
- What type of marketing agents (intermediaries) buy and sell beans?

Results and Discussion

The Structure and Functioning of Kivu's Bean Marketing System

Based on information generated during the survey, as well as data collected through the survey of bean marketing agents, a fairly accurate picture was developed of the structure and functioning of Kivu's bean marketing system. Principal marketing channels are depicted in Fig. 1.

Figure 1 illustrates the commercial channels between North and South Kivu Provinces, which provide 60% of the national bean production in DR Congo according to National Extension Service (2004). The total seasonal average of beans used is

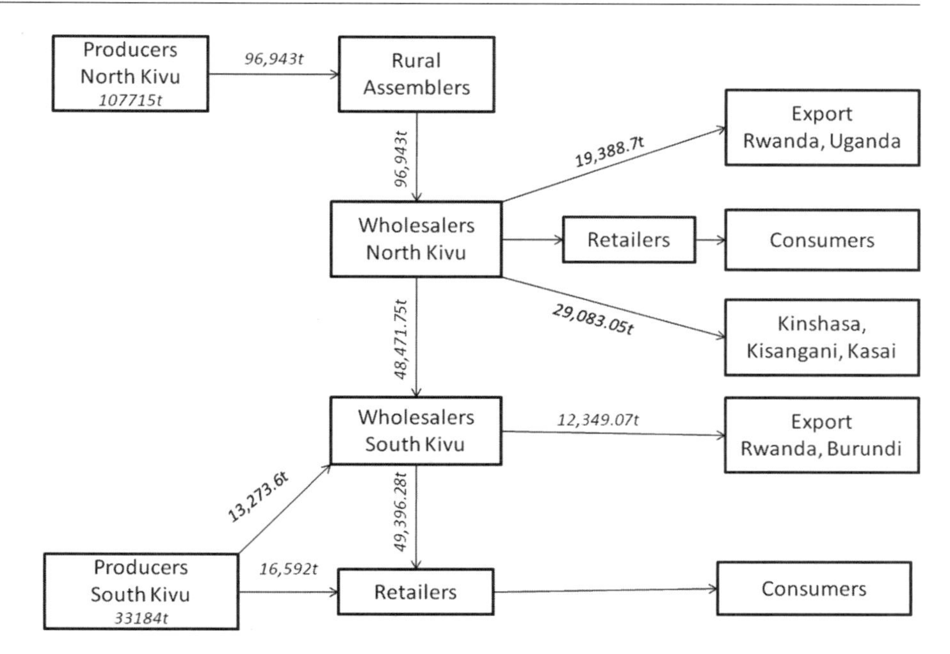

Fig. 1 Principal marketing channels for beans in Kivu

about 140,899 t. In addition Fig. 1 shows different marketing opportunities in Eastern Congo. There are enormous marketing possibilities within and around the country: The national towns like Kinshasa, Kisangani and Mbuji Mayi constitute principal consumption centres for these agricultural products. In total these towns receive an average of 29,083 t (48.4%) of the beans produced by smallholder farmers in Kivu region.

Figure 1 also shows the important commercial trans-border exchanges which exist between the east of the DR Congo and Rwanda, Burundi and Uganda. About 31,737 t (22.5%) of beans produced by Kivu's smallholder farmers is traded seasonally between these countries. Farmers who were interviewed, however, do not have the opportunity to exploit all these possibilities. National and regional markets for beans are dominated by middlemen, to whom most of the profit accrues, thereby serving as a disincentive to the smallholder farmers.

The majority of farmers interviewed (85%) said that they are unaware of procedures, market structures, penetration channels and rules that led these markets. Fifteen percent said that they knew how to proceed but lacked consequent financial resources to face the conditions of these markets. It is noted that the DR Congo Government has not invested in agricultural market development. The Statistics Marketing Services and Agricultural Inspection Services, which collect and disseminate agricultural market information, have not been adequately funded. This has limited the ability of information dissemination by these two services. In addition, the private sector is not developed enough to take over this process of setting up viable and organised structures, which could help smallholder farmers in market exploitation or creation.

Commercial Channel's Actors and Their Roles

A commercialisation circuit analysis identified six actor groups in total. These included producers, rural assemblers, transporters, wholesalers, retailers and consumers.

Producers are smallholder farmers using an average of 0.6 ha of available land for crop production. This group also includes agricultural farmers using more than 5 ha, representing less than 15% of the agricultural Kivu population. Collectors are mainly young boys in the villages who are recruited by sellers at the first set-off point in the village. Wholesalers are Goma market beans suppliers (all gender), who collect, transport and stock beans in a Goma market. Some individuals buy beans in a Goma market and supply big cities, such as Kinshasa, Kisangan and Mbuji Mayi, while at the same time organising export to neighbouring countries. Bean suppliers to Bukavu are mainly professional women or women who occasionally supply beans as a source of additional household livelihoods.

In general the relationship between different actors is informal and tends to change occasionally. However, Goma collectors and wholesalers tend to have reliable relationships, as each wholesaler works with specific collectors. The relationships of the rest of the agent groups are dictated by the market law.

Marketing Margin for Bean National Market

Marketing agents within each area gave remarkably consistent estimates of farm-gate prices, transport costs between various production and consumption points and wholesale prices for different market places.

Table 1 indicates the price increase for beans at each set-off point, as well as the cost and beneficiary margins at different stages of the marketing channel. The farm-gate price of a 100 kg bag of beans is $25. The same bag of beans is sold on the local market by the retailer for $40 with $15 for transport and handling. At the provincial market, the retailers sell a bag of beans at $70, with an additive cost of $30. Sent to Kisangani the bag is sold at $215 and costs $155. The transport and handling costs to Kinshasa (capital) are $195 and the bag of beans is sold by retailers at $265. From Table 1 it is evident that if they were connected to these markets, smallholder farmers in Kivu could take advantage of this linkage and benefit at least $80 per bag of beans that they sell. This would be sufficient to significantly increase their income.

Table 1 Estimated marketing margins and benefits for beans at the national market (US $/100 kg)

	Local market ($)	Provincial market ($)	Kisangani ($)	Kinshasa ($)
Farm-gate price of bean	25	40	60	60
Transport to market	+1	+10	+120	+140
Rural assemblers margin	+4	–	–	–
Wholesale price	30	50	180	220
Wholesale margin	+5	+10	+20	+25
Retailers buying price	35	60	200	245
Retailers margin	+5	+10	+15	+20
Retailers price	**40**	**70**	**215**	**265**

Note: The Value underlined and in bold indicate buying price in USD of a 100 kg bean bag at different stage of supply chain and in difference markets

Table 2 Use of produce generated from bean production

	North Kivu	South Kivu
Consumption	59.1%	65.5%
Sold	34.5	20.2
Seed	9.0	9.38

Source: Service National de Vilgarisation (2004)

Main Use of Agricultural Products by Households

The estimation of the quantity of produce sold is important for determining the behaviour of farmers. Data from the farm survey provided indicators of the percentage of produce sold. The study found that in a normal year the average household in South Kivu produced approximately 214.53 kg of beans, while in North Kivu the production is approximately 327.34 kg. The largest percentage of beans produced (60–65%) is consumed at the household level by the producers. The surplus (20–30%) is sold to other farmers and urban dwellers, and approximately 9% of beans produced are used as seed for the next season. Per capita bean consumption in Eastern Congo is about 35 kg/year. The major part of household production is consumed at this level, as bean prices at local markets are very low and less attracting to farmers. Producers will probably sell a big share of their harvest if they had access to an interested market.

Main Constraints of the Agricultural Product Market

The study revealed that poor farming areas in Eastern DR Congo are characterised by low total and monetary income for most people, with limited consumption and expenditure. In addition, a poor monetary economy with a narrow base and market (for agricultural input, output and finances, to consume goods and services), which are relatively 'thin' (with small volume traded, although for some items there may be very large numbers of people trading in very small volumes), and large seasonal variability in demand and supply also play a role. These conditions normally co-exist with poor roads and communication, poor information availability (particularly in agriculture; price, new technology, potential contracting partners), difficulties in enforcing impersonal contract and widespread rents.

Such conditions pose particular problems for the supply chain development needed for agricultural intensification, since this development needs significant simultaneous and complementary investments by new markets. These investments carry high risks of transaction failure and hence high transaction cost incurred in obtaining protection again such risks. The transaction risks (and costs) have three hand components: coordination risks (risk of an investment failing as a result of absence of a complementary investment by other role players in a supply chain); opportunity risks (which arise when contracting party, with monopolistic control over complementary investment, removes or threatens to remove itself from a supply chain after a player had made an investment that depended upon the party); and rent-seeking risk (where powerful government, political, criminal and other agents, who are not partial to the transaction, see associated investments and/or come back as opportunity to expropriate and threaten to expropriate income and assets from the investor).

Coordination, opportunity and rent risks (and the cost of protection against them) are closely related and where they are high compared to potential returns of investment, the investment required for the development of an agricultural intensification supply chain may be too risky. Therefore the supply chain might not develop even if it is potentially profitable.

Markets are generally unstable, partly because smallholder farmers and other role players have little storage and other capacities to absorb shocks, and the impact of production fluctuation is then largely transferred to poor producers through fluctuating prices. Intra-seasonal price variation has generally risen with market liberalisation. The fact that markets are generally poorly organised, prices being set by buyers who deal with the ones selling at low prices, also contributes to market instability. The existence of any important group of buyers on the market must be noted.

Sixty-five percent of small-scale traders complain of a lack of market information as well as obstructive and uniform rules and regulations (problems that large traders are more able to deal with). Adequate and accurate market information is a critical factor for farmers, traders and government officials and services to plan and make correct decisions. The local agricultural products, in particular those intended for the urban centre, are often in competition with imported products. The survey showed that services and organisations exist, who are responsible for awareness raising among farmers. Research is driven by INERA and the University (UCB), while public extension services and local and international NGOs are responsible for dissemination of information.

Respondents, however, indicated that only 30–67% of farmers received seed of improved varieties (IV), but none of them have been trained in improved bean management practices (IPM). In South Kivu approximately 3–9% of farmers had obtained bean seed and training, while 31–41% of farmers received seed, but only 2–10% had received training. It implies that farmers would have been able to improve the quality of their product if they had access to the range of technologies available and disseminated applicable services. This limited the possibility for farmers to produce a market competitive product.

Various other constraints have been identified:

- Tenure arrangements are not optimal and affect the ability of smallholder farmers to borrow, expand and exit with lump sum.
- Significant share of output for subsistence: output sold do not fully cover input and labour costs.
- Long production and marketing cycle (exacerbating risks) with seasonality in labour uses, cash flow and food availability.
- Discontinuous switches between technologies and crops, with threshold price and levels of performance above/below, while certain activities are viable or non-viable.
- Input purchases need seasonal finances (very difficult).
- Technical progress (and land pressure) needs financial input, which is often not available to smallholder farmers.

Overcoming Agricultural Market Constraints

The study suggested that overcoming agricultural marketing constraints requires strong information programmes to sensitise farmers to take advantage of regional and national opportunities (potential markets), building and managing efficient partnerships, liking technology development to market opportunities and facilitating participatory processes for policy changes. Some of the actions that need to be taken include the following:

a. Policy and institutional improvement
   To set up clear and consistent policy, there needs to be collaborative mechanisms between public and private sectors. There needs to be infrastructure investment for smallholder production, mechanisms for improving access to agricultural services, including market information, contractual enforcement measures and strategies for optimising the utilisation of processing capacity. Forward contracting also allows for direct bargaining between the buyers and the sellers, while simultaneously enhancing market stability and facilitating forward planning, especially for smallholder farmers. The setting up of contractual arrangements for commodity exchanges, involving fixed time frames, tonnages and specifications for commodities and legally binding contracts, could prove important-to-overcome market constraints. The contractual arrangements for out-grower designs led to smallholder farmers increasingly becoming viable and sustainable partners in out-growers' designs. This can be achieved through strengthened supervision of smallholder farmers by out-grower operators.
b. Invest in agricultural research and extension
   Market liberalisation has coincided with a sharp decline in state budgets and hence in public investment in key public goods, such as research, extension and infrastructure. It is necessary to invest in this area in order to develop and disseminate high-yielding technologies for increased smallholder farmer productivity. Demand-driven intervention and efficient participatory approaches in programme formulation, technology development and service delivery are critical factors of this investment.
c. Private sector development
   The private agricultural sector in Eastern Congo is relatively underdeveloped; there is an underutilised capacity as a result of high operation risk for large investments to transform parastatals and to splice regional markets. The development of the private sector will help

   - to diversify and promote community-based and private sector-led seed multiplication and distribution, to promote diversification and to develop a marketing system for agricultural input and products led by the private sector;
   - to contribute to the establishment of sustainable informal and formal farmer finance systems, to improve smallholders access to credit and other financial services;
   - to contribute to the establishment of a commodity exchange system, to provide centralised facilities for buyers and sellers of agricultural commodities and to ensure future prices based on market analysis; and

d. increase the role of NGOs.
   The outreach of national and international NGOs is fairly wide and they are able to service a high number of smallholder farmers around the country. The intervention of NGOs in the delivery of agricultural services to smallholders is really important until the private sector begins to take advantage of opportunities that exist in agricultural markets.
e. Crop diversification
   Food price instability and a high marketing margin encourage poor producers to prioritise staple food for their own consumption before diversifying into higher value crops. This reinforces the problem of low return of production activity and depressing low investment (including in soil fertility), with declining productivity. Efforts must be made to help farmers to move from producing crops for food security to producing cash crops. This is possible through the formation of farmers' associations as a strategy to attract production and marketing services, thereby stimulating and encouraging the development of farmer-owned and controlled organisations. The formation of farmer-owned and controlled organisations will improve crop marketing, available credit, training and advisory services to their members.

f. Establishment of agricultural market information services
   Agricultural market information services should be established to collect, analyse and disseminate market information. The information to be collected should include wholesale and retail prices on a variety of staple grain products, tube and vegetable prices and agricultural inputs. Information on prices, quantity and quality will help to reduce the risk of losses in market transactions, as a high risk of losses translates into high marketing costs. In an ideal competitive market, differences between markets only reflect costs. The process of the exchange of commodities in order to make a profit from prices that are more than market costs (spatial and temporal) is complete in a functioning market. Information flow to smallholders is constrained by logistical complexities, and the local farmers' organisations are not trained to analyse and use available market information. There is therefore room for improvement in collection, analysis and dissemination of market information as this will contribute to overcome market constraints.
g. The upgrading of infrastructure (roads, airport facilities)
   Upgrading the infrastructure and providing an efficient legal framework and system for contract enforcement will facilitate coordination to achieve quality. The lack of an adequate legal framework is an important constraint leading to the emergence of 'commodity raider insurance' (e.g. provision of common laboratory facilities). It is important at this meeting to create a forum for enhancing discussions between public services, the private sector and farmers' organisations, which will all be involved in implementation.
h. Removal of restrictions on to cross-country-edge
   Removal of restrictions on cross-country-edge trade by, for example, committing the chorus from export proclamations and taxes, working toward expansion of free trade area, while simultaneously minimising local licensing requirements.

Conclusion

The overcoming of agricultural market constraints in Eastern DR Congo should begin with strong information programmes designed to sensitise the potential farmers' organisation to appreciate and take advantage of existing and new market opportunities. Information packages on the implication and profits of agricultural markets should be developed for farmers, traders, the private sector, public services, technocrats and politicians.

The phenomena of 'bad markets' for local farmers should be overcome through the development of suitable institutions and frameworks that should improve farmers' organisation to entry existing and new markets. Special programmes are required to effectively involve the private sector in agricultural delivery services. These programmes should, however, be sustainable, promote private sector development and facilitate better access to markets by farmers with locational disadvantages.

The most cost-effective and technical sound tool for the improvement of delivery of agricultural services to farmers is through the stimulation and support of sustainable development of the private sector. This requires an intensive effort by both the private and the public sectors concerning infrastructure investment for smallholder production, mechanisms for improving access to agricultural services, including market information on a sustainable basis, instituting of contractual enforcement measures and strategies for optimising the utilisation of processing capacities.

References

Bromley RJ Symanski R (1974) Marketplace trade in Latin America. Lat Am Res Rev 9(3):3–38

Kaynak E (1986) Marketing and economic development. Praeger, New York, NY

Kindra GS (1984) Marketing in development countries. St Martin's Press, New York, NY

Kinsey J (1988) Marketing in development countries. Macmillan Education Limited, London

Roemer MR, Jones C (1991) Marketing in development countries, parallel, fragmented and black. Based on a workshop sponsored by Harvard Tea Institute for International Development, Cambridge, MA, USA, Nov 1988. San Francisco, CA, USA, International centres of Economic Growth

Scott GJ (1995) Agricultural marketing research in developing countries: old tasks and new challenges. In: Scott GJ (ed) Prices, products and peoples. Lynne Rienner Publishers, Boulder, CO, p. 5

Service National de Vulgarisation (2004) Rapport annuel

Shaffer JD (1986) A working paper concerning publicly supported economic research in agricultural marketing. Economic Research Service, Department of Agriculture, Washington, DC

Constraints in Chickpea Transportation in the Lake Zone of Tanzania

A. Babu, T. Hyuha, and I. Nalukenge

Abstract Chickpea (*Cicer arietinum*) is an important food legume providing food and income for farm families, is used as medicine, and provides the country with foreign earnings. Despite its importance, its marketing system still performs below par. The main objective of the study aimed at identifying the constraints facing traders in the chickpea marketing and determining the optimal quantities and costs of chickpea distribution from four supply sources to three markets. Eighty traders were interviewed using a structured questionnaire. The analysis was done using Statistical Package for Social Sciences and SOLVER–Microsoft Excel computer software. The findings indicate that the total cost of procuring chickpea was US $47,745 that was much higher than the optimal total cost calculated at US $37,710. Transport took up a large proportion of the marketing costs accounting for 45.3% of the total marketing costs. Transportation costs are high as a result of poorly maintained feeder roads, seasonal supplies of chickpeas, lack of information about prices, and haphazard choice of transportation routes. To reach the optimal solution, the distribution schedule was re-allocated, thus increasing the number of routes from 9 to 11 and changing the quantities transported in different routes. This change would result in a reduction of costs by 21%. Moreover improvement of road network, storage facilities, farmers' market education and information, and formation of traders' association would reduce marketing costs and hence increase marketing efficiency.

A. Babu (✉)
Agricultural Research Institute Ukiriguru, Mwanza, Tanzania
e-mail: adventinababu@yahoo.com

Keywords Chickpea · Constraints · Marketing costs · Optimal solution · Transportation

Introduction

Chickpea (*Cicer arietinum*) is an annual third most important legume crop grown globally (Ladizinsky, 1975). In Tanzania chickpea is mainly grown in Mwanza, Arusha, Mara, Shinyanga, Tabora, and Singida regions (Pundir, et al., 1996). Mwanza and Shinyanga regions contribute 80% of chickpea produced in Tanzania. Data from Bejiga and van der Maesen (2006) indicates that in Tanzania chickpea is cultivated on over 64,000 ha of land making an annual production of 25,000 MT (FAO, 2002). Chickpea production in Tanzania accounts for 6.3% of all the chickpea produced in Africa (Bejiga and van der Maesen, 2006). Chickpea export contributes 7% of foreign exchange earnings to Tanzania economy, from non-traditional agricultural export products (MAFS, 2001).

Ladizinsky (1975) reported that chickpea fixes nitrogen into the soil, thereby improving soil fertility. It does not compete for land with other crops, because it is cultivated in the dry season when other annual crops have been harvested (ICRISAT, 1996; Pundir et al., 1996). Chickpea is valued for its nutritive seeds with high protein content of 25.3–28.9%, carbohydrates 55%, and fats 5%; they are also rich in minerals and vitamins (Ladizinsky, 1975). Where chickpea is grown in the lake zone of Tanzania, malnutrition rates are low as compared to other areas (Roeleveld et al., 1994).

The husks, stems, and leaves are used as animal feed during the dry season in Tanzania. Despite their

A. Bationo et al. (eds.), *Innovations as Key to the Green Revolution in Africa*,
DOI 10.1007/978-90-481-2543-2_112,

importance, the marketing system of chickpea still performs below par. This is evidenced by high transportation costs as a result of poorly maintained feeder roads, seasonal supplies of chickpeas, haphazard choice of transportation routes where there is no guarantee that the transportation costs may be reduced, and lack of information about prices which further increases the marketing costs. Chickpea transport cost in the lake zone of Tanzania is 45.3% of the total marketing costs. Evidence from other areas in the sub-Saharan Africa shows that transport represents the largest component of marketing costs, accounting for 50–60% of the total costs (Omamo, 1998). High marketing costs from production to demand centers result in increase in domestic and border prices. If these are not checked, the impact will be to decrease the farm gate price and reduce export and farmers' income, hence reducing country's earnings. It is against this background that this study has been undertaken to identify the constraints facing traders in the chickpea marketing and to determine the optimal quantities and costs of chickpea distribution from four supply sources to three markets.

Materials and Methods

Sampling and Data Collection Methods

Misungwi and Magu districts were purposively selected as the chickpea supply sites on the basis that they are the major supply sources in the area, while Bukoba, Musoma, Bunda, and Mwanza towns were selected as the major consuming centers of marketed chickpea in the lake zone of Tanzania. Selection of the respondents is based on the tasks the trader performs (assembling, brokerage, retailing, wholesaling, and transporting), quantity handled, the supply source, and the time spent in chickpea trade. The list of traders was made and sequenced and a random sample was selected systematically until the required number was obtained. Selected traders were as follows: Misasi (13), Magu (10), Misungwi (13), Bunda (10), Mwanza (13), Bukoba (10), and Musoma (11); this made a total of 80 respondents. In addition 14 key informants were interviewed, 2 from each supply and demand center. To reach the number of respondents to include in the study the method of sample proportions was applied in calculating the sample size (Cooper and Emory, 1995).

The primary and secondary data were collected. The primary data were gathered by means of formal interviews using structured questionnaires with traders at rural and urban markets. The data collected included quantities supplied and demanded, marketing costs and constraints in the chickpea marketing, as well as traders' characteristics. Checklists were used to interview key informants on the general issues of chickpea marketing.

Data Analysis

To analyze marketing costs and identify the marketing channels, descriptive statistics like percentage, mean, and frequency tables were used. These tools were obtained using Statistical Package for Social Sciences (SPSS) computer software.

Linear programming (LP) transportation model helped to determine the optimal level of distribution of the chickpea transportation problem. The objective function involved minimizing the total transportation costs of delivering chickpea from Misungwi, Magu, Misasi, and Bunda supply centers to the chickpea demand centers of Mwanza, Bukoba, and Musoma. The model was developed using November and December 2005 total transportation cost for chickpea. The optimal level of chickpea was obtained using Microsoft Excel SOLVER computer package. The objective function was to minimize the total cost of transportation.

Algebraically the cost minimization LP model of the transportation problem according to Anderson et al. (1994) is presented as follows:

$$\text{Minimize: } \sum_{i=1}^{m}\sum_{j=1}^{n} c_{ij}x_{ij}$$

$$\text{Subject to: } \sum_{j=1}^{n} x_{ij} \leq S_i, \quad i = 1, 2, \ldots, m \text{ supply}$$

$$\sum_{i=1}^{m} x_{ij} \geq d_j, \quad j = 1, 2, \ldots, n \text{ demand}$$

$$x_{ij} \geq 0 \quad \text{for all } i \text{ and } j$$

where i is the index for origin, $i = 1, 2,..., m$; j the index for destinations, $j = 1, 2,..., n$; x_{ij} the number of units shipped from origin i to destination j; c_{ij} the cost per

Table 1 Transportation tableau for the chickpea transportation problem

Origins (m)	Destinations (n) 1	2	3	Origin supply (TS)
A	$XA1^{C_{A1}}$	$XA2^{C_{A2}}$	$XA3^{C_{A3}}$	SA
B	$XB1^{C_{B1}}$	$XB2^{C_{B2}}$	$XB3^{C_{B3}}$	SB
C	$XC1^{C_{C1}}$	$XC2^{C_{C2}}$	$XC3^{C_{C3}}$	SC
D	$XD1^{C_{D1}}$	$XD2^{C_{D2}}$	$XD3^{C_{D3}}$	SD
Destination demand (TD)	D1	D2	D3	TD = TS

TS = total supply; SA, SB, SC, SD = total supply in each of the supply centers A, B, C, and D; TD = total demand; D1, D2, D3 = total demand in each of the markets 1, 2, and 3; X_{ij} = variables representing the quantity (kg) of chickpea transported from source i to destination j; C_{ij} = transportation costs (Tzsh) of each kilogram of chickpea indicated in each cell as a superscript

unit of shipping from origin i to destination j; S_i the supply or capacity in units at origin i; d_j the demand in units at destination j.

Linear Programming Transportation Model

The transportation tableau was formulated to make easy the formulation of the objective function and the constraints for the model. In the tableau, columns 1–3 represent chickpea destinations/demand and rows A–D represent sources/supply (refer Table 1). The quantities X_{ij} of chickpea with the corresponding costs C_{ij} were entered in each cell. The total demand in each demand center is shown by D1–D3 in the last row while SA–SD indicates the total supply from each source in the last column. The objective function was to minimize the total transport cost for chickpea from sources (A, B, C, and D) to the markets (1, 2, and 3).

The *objective function* is the summation of the cost expressions:

$$\begin{aligned}\text{Minimize } & XA1^{C_{A1}}+XA2^{C_{A2}}+XA3^{C_{A3}}+XB1^{C_{B1}}+XB2^{C_{B2}}+XB3^{C_{B3}} \\ & +XC1^{C_{C1}}+XC2^{C_{C2}}+XC3^{C_{C3}}+XD1^{C_{D1}}+XD2^{C_{D2}}+XD3^{C_{D3}}\end{aligned} \quad (1)$$

Each supply source has limited supplies to satisfy demand at each destination. Therefore the objective function is subject to structural constraints.

Demand constraints:

$$XA1+XB1+XC1+XD1 \geq \text{D1} \quad (2)$$

$$XA2+XB2+XC2+XD2 \geq \text{D2} \quad (3)$$

$$XA3+XB3+XC3+XD3 \geq \text{D3} \quad (4)$$

Supply constraints:

$$XA1+XA2+XA3 \leq \text{SA} \quad (5)$$

$$XB1+XB2+XB3 \leq \text{SB} \quad (6)$$

$$XC1+XC2+XC3 \leq \text{SC} \quad (7)$$

$$XD1+XD2+XD3 \leq \text{SD} \quad (8)$$

$$\text{TD} = \text{TS} \quad (9)$$

Non-negativity of variables

$$S_{ij}, D_{ij}, C_{ij} \geq 0 \quad (10)$$

Sensitivity Analysis

This was done to find out how a change in a coefficient of the objective function affects the optimal solution and how will changes in the right-hand side value of a constraint affect the optimal solution. In order to find out how the optimal solution varies, the values in the limits report were added or subtracted to specific route solutions.

Results and Discussion

Characteristics of Chickpea Traders in the Lake Zone of Tanzania

The results indicated that men dominate the chickpea market at 70% in contrast to women who represented 30% (Table 2). The majority of traders in chickpea were between ages 31 and 40 (51.3%). Most traders had attained primary level education, 71.3%, whereas 27.5% have attended secondary education and a minority has not attended school. Majority of traders were sole proprietors at 78.8%, followed by partnerships and family business at 13.8 and 7.5%, respectively. These had been in the business for relatively longer periods of time, 40% had between 5 and 10 years of experience while 20.0% had over 10 years of experience in chickpea trading. Fifty percent of the traders earn between 10 and 25%, 44% earned between 26 and 50%, and 6% earned only 10% of income from chickpea trade.

Most traders (98.8%) deal in other commodities, mainly maize, green grams, rice, beans, and groundnuts. The main reason was the seasonal availability of chickpea; therefore, traders go for other businesses during the times of shortage. Retailers handled the largest share of chickpea followed by exporters as reported by 28.8 and 25.0% of the traders, respectively.

Table 2 Characteristics of chickpea traders in the lake zone ($n = 80$)

Characteristics	Category	%
Sex	Male	70
	Female	30
Age group	20–30 years	36.3
	31–40 years	51.3
	41 years and above	12.5
Marital status	Married	82.5
	Single	13.8
	Widow	2.5
	Divorced/separated	1.3
Education level	Primary	71.3
	Secondary	27.5
	Not attended school	1.3
Business ownership	Sole proprietorship	78.8
	Partnership	13.8
	Family	7.5
Trader category	Retailer	31.3
	Wholesaler	22.5
	Assembler	35
	Retailer/wholesaler	8.8
	Broker	2.5
Whether trades in other commodities	Yes	98.8
	No	1.3

Source: Babu et al. (2007)

The Chickpea Marketing Channel

Results show that 13 different types of marketing channels prevail in the chickpea marketing. The major routes include producer-wholesaler-retailer-local consumer; producer-assembler-retailer-local consumer; and producer-assembler-exporter representing 31.51, 19.22, and 14.01%, respectively. However, due to poor storage systems in rural areas producers buy seeds from urban retailers. From the farm gate chickpea passes through the farm level markets, local village markets, weekly markets, urban markets, and export markets.

Marketing Constraints

Table 3 shows constraints encountered by traders during marketing. The table shows that the most important constraint reported was poor road network as reported by 84% of respondents. Chickpea traders encounter a number of problems when transporting their chickpea, including poorly maintained roads with potholes, high vehicle hire charges (52%), long distances to procure chickpea (10%), and bribes to policemen during transportation (6%). Poor transport network restricts the access of traders to available chickpea in the market and the movement of chickpea by traders to market outlets owing to high transportation costs. Various researchers have emphasized the importance of road infrastructure, especially the rural feeder roads (Gabre-Madhin, 2001). Procurement of chickpea is efficient during harvesting season. Urban traders find the grain ready in collecting centers, farm gate, or weekly markets. Some wholesalers contract assemblers who collect the produce from all possible sources of supply ready for transportation. During off-season capacity utilization is very low not only for transportation equipment but also for other facilities like storage and processing plants.

Table 3 Percentage and severity of chickpea marketing constraints ($n = 80$)

Constraint	Percentage	Responses on the severity		
		Very severe	Severe	Slightly severe
Poor road network	84	31	29	7
Seasonal chickpea supplies	78	38	24	–
High tax rates	76	24	29	8
High transport costs	52	15	24	3
Poor storage facilities	50	12	26	2
High buying price of chickpea	24	12	5	2
Limited capital and credits	22	8	9	1
Lack of export market information	20	11	5	–
Fluctuation of chickpea price	14	7	2	2
Low quality of chickpea traded	14	6	2	3
High price of storage pesticides	12	7	1	2
Low selling price of chickpea	12	5	3	2
Long distances to the market	10	5	2	1
Few chickpea buyers	10	4	2	2
Unorganized chickpea markets	8	4	2	–
Bribes during transportation	6	3	2	–

Source: Babu et al. (2007)
1 = low severity, 2 = severe, and 3 = very severe

Seasonal chickpea supply was reported by 78% of traders as the problem in marketing of chickpea. The reason for this is that chickpea is grown annually during the dry season (June to September); therefore, during the off-season it becomes scarce. Large sales of chickpea are realized between the months of October and December while low sales are sustained between the months of June and August. On the other hand, high prices are realized between April and May compared to low prices, which are realized when the supply is high corresponding to the months of October to January. The planting season is May/June when the chickpea is scarce in the market, while the supply becomes high after the harvesting period in October. Expansion of the cropland, improvement of storage facilities, and irrigation will increase production and hence smooth supply. High tax rate (76%) was reported as a problem that hinders expansion of the marketing sector. Traders have to pay for the annual business license and the market fee that is paid every day the trader uses the market space and at each market where the produce passes (Table 4).

Poor storage facilities were mentioned by 50% of traders as one of the major problems in marketing of chickpea. Limited capital and credit, fluctuation of prices, problem of pest infestation (mainly weevils and rodents), high price of storage pesticides, and low-quality chickpea were cited as the major obstacles that limit the capacity to store chickpea (Table 5). Insect pests are the most severe on grain storage, and traders complained about the type of pesticides they use. The study showed that traders do not have the financial and managerial capability to store chickpea and be able to exploit seasonal price changes. Badiane and Shively (1998) observed the similar case and they suggested improvements in local storage to reduce food price variability.

Table 4 Correlation between low/high sales and prices (sales = dependent variable)

	Z	P	R^2
Large sales–low prices	–0.051	0.039	42.2%
Low sales–high prices	–1.647	0.067	32.0%

Source: Babu et al. (2007)
Z =................, P=.............. R^2=...............

Table 5 Mean length of time taken from buying to selling chickpea (months)

Retailer	Wholesaler	Assembler	Retailer/wholesaler
4.08 ± 0.60^a	2.56 ± 0.70^c	4.63 ± 0.61^a	7.14 ± 1.13^b

Source: Babu et al. (2007)
Superscripts with different letters depict significant difference at the 0.10 level

Lack of export market information was reported by 22% of chickpea traders that the information they receive is inadequate for decision making. All large-scale traders cited the problem; however, they are few to make the problem look big to other problems mentioned by small-scale traders. Traders pass information to each other only within the marketing chain. From wholesalers/exporters/processors who are few, information passes on the demand and price through brokers, and other middlemen towards farmers. Limited public information results in asymmetric information and is one of the institutional issues affecting efficiency of markets in Africa (Gabre-Madhin, 2001). Provision of market information is an area in which the government can play a key role in improving marketing.

Other marketing constraints reported by traders were high buying price (24%) of chickpea as mentioned by traders in urban markets and low selling price (12%) of chickpea mentioned by producers in rural markets. This implies that there is no compromise on the price of chickpea whereby each party is not satisfied. Local urban retailers also complained of getting few chickpea buyers (10%) during off-harvesting season. Local consumers are few and most of them are Indians. Therefore, considering the importance of chickpea to human nutrition, product development efforts are needed.

Chickpea Marketing Costs

Table 6 presents the mean costs of various marketing tasks. The table shows that transportation of chickpeas took up a large proportion of the marketing costs. In contrast, storage, packaging materials, and off-loading costs took the least share.

The Optimal Solution

Table 7 shows the volumes in kilograms of chickpea transported from sources to the markets and the superscripts are the unit transport cost. Table 8 shows the optimal quantities to be transported to different destinations. The total quantity to be transported remains the same (4,445,600 kg), though there are changes in quantities to be distributed to different routes. The transportation routes were increased from 9 to 11 as shown in Table 8.

Misasi–Musoma and Misungwi–Musoma routes have been added and therefore change the optimal quantities in all routes; however, the total quantities to be supplied and/or demanded remain the same. Mwanza demand quantities have been reduced by 50.95% (1,641,677 kg) and re-allocated to Bukoba and Musoma. Therefore, Musoma demand has been increased by 52.74% (1,484,536 kg) and Bukoba by 33.41% (1,319,388 kg). Traders in Musoma and Mwanza should collect chickpea from Bunda, Magu, Misasi, and Misungwi, while traders in Bukoba should collect produce from Magu, Misasi, and Misungwi as shown in Table 7.

Table 9 indicates the optimal costs of transporting a kilogram of chickpea for different routes. Bunda–Musoma route is the cheapest followed by Misasi–Musoma, while all routes to Bukoba are expensive; however, costs of all routes were reduced to reach the optimal costs. Re-allocation of chickpea flows would achieve the optimum pattern of chickpea transfer through reducing transfer costs. There would be a total transportation cost saving of 21%. In individual markets, transfer costs would reduce by 45.76% for Bunda–Musoma, 17.59% for Magu–Musoma, 46.9% for Bunda–Mwanza, 37.58% for Magu–Mwanza, 57.05% for Misasi–Mwanza,

Table 6 Mean marketing costs (Tzsh) ($n = 80$)

Marketing task	Mean cost	Percentage
Off-loading/loading costs at the market (Tzsh/kg)	1.03	2.89
Packaging materials costs (Tzsh/kg)	1.66	4.67
Storage costs per month (Tzsh/kg)	2.23	6.27
Total market fees (Tzsh/kg) sold	2.76	7.76
Chickpea transportation (Tzsh/kg)	16.13	45.33
Processing costs (Tzsh/kg)	11.77	33.08
Total marketing costs (Tzsh/kg)	35.58	100

Source: Babu et al. (2007)

Table 7 Quantity transported in different routes with their respective costs

	Musoma	Mwanza	Bukoba	Total supply
Bunda	534,553$^{4.720}$	300,047$^{16.929}$	0	834,600
Magu	248,467$^{8.179}$	533,480$^{7.849}$	159,213$^{30.517}$	941,160
Misasi	0	133,3347$^{9.662}$	126,693$^{34.151}$	1,460,040
Misungwi	0	1,054,826$^{6.769}$	154,974$^{30.422}$	1,209,800
Total demand	783,020	3,221,700	440,880	4,445,600

Data in superscript shows unit costs in Tzsh
Source: Babu et al. (2007)

Table 8 The optimal quantities (kg)

	Musoma	Mwanza	Bukoba	Total supply
Bunda	426,062.1	408,537.9	0	834,600.0
Magu	429,266.3	3991.512	507,902.1	941,159.9
Misasi	375,774.0	591,352.9	492,913.0	1,460,040.0
Misungwi	253,433.1	637,794.2	318,572.7	1,209,800.0
Total demand	1,484,536.0	1,641,677.0	1,319,388.0	4,445,600.0

Source: Babu et al. (2007)

Table 9 Optimal total costs of transporting chickpea (Tzsh)

	Musoma	Mwanza	Bukoba	Total
Bunda	920,294.1	3,672,756	0	4,593,050
Magu	2,893,255	19,558.41	8,344,832	11,257,645
Misasi	1,255,085	2,454,115	8,990,733	12,699,933
Misungwi	1,135,380	3,137,947	4,886,905	9,160,233
Total	6,204,014	9,284,376	22,222,470	37,710,860

Source: Babu et al. (2007)

17.32% for Misungwi–Mwanza, 46.17% for Magu–Bukoba, 47.6% for Misasi–Bukoba, and 49.58% for Misungwi–Bukoba routes.

These results may be difficult to implement because it requires high level of co-ordination between traders in various markets. However, traders should adopt group transport in order to offset the high cost of transferring the produces. Large trucks may also help if the distance is longer. With the chickpea market being competitive, the consumers and producers would enjoy the cost saving which would result from the optimal allocation of the chickpea grain. Only the small portion of the benefits may go to the intermediaries.

Sensitivity Analysis

Results from the analysis reveal that the optimal solution will not change if the transportation cost from Bunda to Musoma is allowed to increase by 2 Tzsh with no lower limit of allowable decrease (infinity). Bunda–Mwanza channel has no upper limit (infinity) while the lower limit of allowable decrease is 8 Tzsh. Findings from the study revealed that Magu–Musoma channel and Magu–Mwanza channel had no upper limit while minimum decreasable limits were 3.04 and 2.31 Tzsh, respectively. In addition, the Magu–Bukoba route had no upper limit while the lower decreasable limit was 12.73 Tzsh. It was further revealed that all the routes from Misasi had no lower bounds while the upper limits were observed to be 1.89, 3.92, and 13.65 Tzsh, respectively. Results from the sensitivity analysis revealed that the Misungwi–Musoma upper limit was 3.41 with no lower decreasable limit compared to Misungwi–Mwanza route that had an upper limit of 2.11 with no decreasable lower limit. Finally, the Misungwi–Bukoba channel had no upper limit while its lower limit was observed to be 13.29. These results imply that the optimal solution will not change significantly should the transportation costs vary by any of the above amount.

Conclusions and Recommendations

The study set out to identify the constraints facing traders in chickpea marketing and to determine the optimal quantities and costs of chickpea distribution from four supply sources to three markets. The following conclusions were drawn from the results of the study: The majority of traders had attained only the primary education level. This could be a cause of the inefficiency of the entire distribution system. A higher level of the farmer's education may help him/her to make better decisions, which would increase firm profitability. Also training through seminars or workshops and provision of information to producers and traders through radio or newspapers would help expand markets and thus improve the price.

The study reveals that majority of the traders are sole proprietors at 78.8%, followed by partnerships and family business at 13.8 and 7.5%, respectively. This solo-minded approach could have a negative impact on marketing since it has been well documented that when traders pool together they tend to perform more efficiently. Therefore, formation of groups will help them improve marketing efficiency. To empower producers to act as equal partners with the buyers, unionization and selection of leaders should come from the market participants themselves rather than being imposed from politicians. Association of traders and/or farmers has the potential of develop into representative organizations for their members.

Majority of traders receive information from fellow buyers and traders, through buyers and telephone. These are crude methods of information interchange and may not be reliable or fast or accurate. Given the importance of information, distribution of market information especially on demand by various export markets and prices would encourage farmers expand production of the crop for cash.

Results revealed that transport took up a large proportion of the marketing costs (up to 45.3% of the total marketing costs). The total cost of procuring chickpea in the lake zone is 47,745,013 Tzsh, which is much higher than the optimal total cost calculated at 37,710,860 Tzsh. There is a need to adjust the transportation plan in order to reduce total transportation cost by 21%. Adjusting the transportation plan to the optimal quantities and costs alone cannot make chickpea marketing efficient. Improvement of traders' capital, education, road infrastructure, and reduction of taxes will improve marketing of the crop.

Acknowledgments We would like to acknowledge all staff members and students in the Department of Agricultural Economics and Agribusiness, Makerere University, for their comments and support. Sincere thanks are extended to World Bank through Joint Japan/World Bank Graduate Program and Ministry of Agriculture and Food Security of Tanzania through its Department of Research and Development for sponsoring the course. The District Agricultural and Development Officers and farmers are thanked for the support during data collection.

References

Anderson DK, Sweeny DJ, William TA (1994) An introduction to management science: quantitative approach to decision making, 7th edn. West Publishing Company, New York, NY

Babu A, Hyuha T, Nalukenge I (2007) Evaluation of marketing efficiency of chickpea in the Lake zone of Tanzania. A thesis presented as a partial fulfillment of the master of science in agricultural economics of Makerere University, Uganda

Badiane O, Shively GE (1998) Spatial integration, transport costs, and the response of local prices to policy changes in Ghana. J Dev Econ 56:411–431

Bejiga G, van der Maesen LJG (2006) Cicer arietinum L. [Internet] Record from Protabase. In: Brink M, Belay G (eds) PROTA (Plant Resources of Tropical Africa/Ressources végétales de l'Afrique tropicale). Wageningen, The Netherlands

Cooper DR, Emory EW (1995) Business research methods, 5th edn. Richard D. Irwin Incorporation, Chicago, IL

Food and Agriculture Organization of the United Nations (2002) The global pulse markets: recent trends and outlook. Commodities and Trade Division, FAO Rome, Italy

Gabre-Madhin E (2001) Understanding how markets work: transaction costs and institutions in the Ethiopian grain market. IFPRI, Washington, DC

ICRISAT (1996) International Chickpea and Pigeon pea Newsletter. ICRISAT, Patancheru, Andhra Pradesh, India. ICPN 3, 1996

Ladizinsky G (1975) A new CICER from Turkey. Notes R Bot Garden Edinb 34:201–202

MAFS (2001) Annual report (1999/00). Ministry of Agriculture and Food Security, Dar es Salaam

Omamo SW (1998) Transport costs and smallholder cropping choices: an application to Siaya District, Kenya. Am J Agric Econ 80:116–123

Pundir RPS, Nyange LR, Sambai LM (1996) Bright Prospects for Chickpea Cultivation in Tanzania. In: International Chickpea and Pigeonpea Newsletter No. 3 (1996) ICRISAT. Pataccheru 503324, Andhra Prudish, India, pp 11–12

Roeleveld A, Wella E, Babu AK (1994) Livestock Survey in Kwimba District. LZARDI, Fieldnote No. 54, ARI Ukiriguru, Mwanza, Tanzania

Part IV
Innovation Approaches and Their Scaling Up/Out in Africa

Nutr Cycl Agroecosyst (2010) 88:3–15
DOI 10.1007/s10705-008-9200-4

RESEARCH ARTICLE

Micro-dosing as a pathway to Africa's Green Revolution: evidence from broad-scale on-farm trials

S. Twomlow · D. Rohrbach · J. Dimes · J. Rusike · W. Mupangwa · B. Ncube · L. Hove · M. Moyo · N. Mashingaidze · P. Mahposa

Received: 25 April 2008 / Accepted: 6 August 2008 / Published online: 28 August 2008

Abstract Next to drought, poor soil fertility is the single biggest cause of hunger in Africa. ICRISAT-Zimbabwe has been working for the past 10 years to encourage small-scale farmers to increase inorganic fertiliser use as the first step towards Africa's own Green Revolution. The program of work is founded on promoting small quantities of inorganic nitrogen (N) fertiliser (micro-dosing) in drought-prone cropping regions. Results from initial on-farm trials showed that smallholder farmers could increase their yields by 30–100% through application of micro doses, as little as 10 kg Nitrogen ha^{-1}. The question remained whether these results could be replicated across much larger numbers of farmers. Wide scale testing of the micro-dosing (17 kg Nitrogen ha^{-1}) concept was initiated in 2003/2004, across multiple locations in southern Zimbabwe through relief and recovery programs. Each year more than 160,000 low resourced households received at least 25 kg of nitrogen fertiliser and a simple flyer in the vernacular explaining how to apply the fertiliser to a cereal crop. This distribution was accompanied by a series of simple paired plot demonstration with or without fertiliser, hosted by farmers selected by the community, where trainings were carried out and detailed labour and crop records were kept. Over a 3 year period more than 2,000 paired-plot trials were established and quality data collected from more than 1,200. In addition, experimentation to derive N response curves of maize (*Zea mays* L.), sorghum (*Sorghum bicolor* (L.) Moench) and pearl millet (*Pennisetum glaucum* (L.) R.Br.) in these environments under farmer management was conducted. The results consistently showed that micro-dosing (17 kg Nitrogen ha^{-1}) with nitrogen fertiliser can increase grain yields by 30–50% across a broad spectrum of soil, farmer management and seasonal climate conditions. In order for a household to make a profit, farmers needed to obtain between 4 and 7 kg of grain for every kg of N applied depending on season. In fact farmers commonly obtained 15–45 kg of grain per kg of N input. The result provides strong evidence that lack of N, rather than lack of rainfall, is the primary constraint to cereal crop yields and that

S. Twomlow (✉) · D. Rohrbach · J. Dimes · J. Rusike · W. Mupangwa · B. Ncube · L. Hove · M. Moyo · N. Mashingaidze · P. Mahposa
International Crops Research Institute for the Semi Arid Tropics (ICRISAT), PO Box 776, Bulawayo, Zimbabwe
e-mail: s.twomlow@cgiar.org

Present Address:
D. Rohrbach
World Bank, Lilongwe, Malawi

Present Address:
J. Rusike
International Institute for Tropical Agriculture, Chitedze Research Station, Lilongwe, Malawi

Present Address:
B. Ncube
WATERnet, Department of Civil Engineering, University of Zimbabwe, Harare, Zimbabwe

This article has been previously published in the journal "Nutrient Cycling in Agroecosystems" Volume 88 Issue 1.
A. Bationo et al. (eds.), Innovations as Key to the Green Revolution in Africa: Exploring the Scientific Facts. © 2010 Springer.

A. Bationo et al. (eds.), *Innovations as Key to the Green Revolution in Africa*,
DOI 10.1007/978-90-481-2543-2_113, © Springer Science+Business Media B.V. 2011

micro-dosing has the potential for broad-scale impact on improving food security in these drought prone regions.

Keywords Fertiliser · Nitrogen · Semi-Arid · Smallholder · Sorghum · Maize · Pearl Millet · Southern Africa · Zimbabwe

Introduction

Throughout the 1980s and 1990s, the International Crops Research Institute for the Semi-Arid Tropics (ICRISAT) primarily targeted the development and dissemination of earlier maturing varieties of sorghum (*Sorghum bicolor* (L.) Moench) and pearl millet (*Pennisetum glaucum* (L.) R.Br.), as means to improve productivity and reduce the risks of drought in semi-arid agro-ecologies of southern Africa (Heinrich 2004). Farmers liked the new varieties for their early maturity and large grain size; adoption rates were favorable. However, limited gains were achieved in crop yields and productivity. This is because of the low inherent fertility of most soils in the region (Giller et al. 2006; Tittonell et al. 2005; Zingore et al. 2007). Even so, farmers are reluctant to risk investments in fertilizer, particularly at the recommended rates (Mafongoya et al. 2006). The main problem with most current fertility management recommendations is that they target maximization of yields or profits without consideration of the agricultural risks and resource constraints faced by many smallholder households. The levels of inorganic fertilizer, manure and rotations demanded are far beyond the capabilities of all but the wealthiest of households (Mapfumo and Giller 2001; Giller et al. 2006; Mafongoya et al. 2006; Zingore et al. 2007). Surveys in southern Zimbabwe, for example, indicated that less than 5% of farmers commonly used fertiliser (Ahmed et al. 1997; Rusike et al. 2003). Sixty percent of households owning cattle did not even use cattle manure as an amendment for crop production. Current and past use of inorganic fertiliser and manure and average rates of application for Malawi and Zimbabwe are summarized in Table 1. Similar data have been reported for elsewhere in Zimbabwe and other countries in sub-Saharan Africa (Hilhorst and Muchena 2000; Mafongoya et al. 2006; Morris et al. 2007; Zingore et al. 2007).

Table 1 Current and past use of inorganic fertiliser and manure and average rates of application, Malawi and Zimbabwe

Country	Practice	Proportion of farmers using technology (minimum and maximum for villages surveyed %)	Rate of application (kg N ha^{-1})	Official recommendation (kg N ha^{-1})
Malawi	Using inorganic fertiliser	4–31	17 kg/ha compound 23-21-0+4S (3.5 kg N ha^{-1})	100–150 kg ha^{-1} compound 23-21-0+4S soon after germination and 100–150 kg of CAN (28% N) or urea (42% N) two weeks after germination (50–105 kg N ha^{-1})
	Ever used inorganic fertiliser	99		
	Using manure	30–40	1.5 t ha^{-1}	10 t ha^{-1}
Zimbabwe	Using inorganic fertiliser	5–75	Less than 50 kg ha^{-1} mostly ammonium nitrate (34% N) (17 kg N ha^{-1})	150–200 kg ha^{-1} 1 compound D 8-14-7+6.5S and 100–150 kg ha^{-1} ammonium nitrate (55–79 kg N ha^{-1})
	Ever used inorganic fertiliser	21–50		
	Manure	6–60	Less than 4 t ha^{-1}	20–40 t ha^{-1}

Source: Twomlow and Ncube (2001)

In the late 1990s, ICRISAT began to use crop simulation models as a tool for more effective analysis of technology responses under conditions of high rainfall variability and low inherent soil fertility. In 1999, ICRISAT began a series of modeling workshops in conjunction with the International Maize and Wheat Improvement Center (CIMMYT) and the Agricultural Production Systems Research Unit (APSRU) in which research and extension officers used a simulation model (APSIM—Agricultural Production Systems Simulator model (Keating et al. 2003)) to evaluate the type of resource allocation questions faced by resource-poor farmers in semi-arid regions of southern Africa (i.e. under conditions of uncertain rainfall and with the objective of achieving household food security). A common theme started from the proposition that farmers may, at best, initiate investments in small quantities of fertiliser (Rohrbach 1999).

The robustness of the simulated responses to small quantities of nitrogen (N) fertiliser, was surprising, and contrary to much of the documented fertility research results in the region which start with at least 25 kg N ha^{-1} (Mafongoya et al. 2006; Mushayi et al. 1999). Simulation results for a 1951–1999 rainfall period in southern Zimbabwe, suggested that farmers could increase their average yields by 50–100% by applying as little as 9 kg N ha^{-1} (no spatial ability for N application in model). These results indicated farmers were better off applying lower rates of nitrogen on more fields, than concentrating a limited supply of fertiliser on one field at the recommended rates (Carberry et al. 2004). However, if the household could only afford a very small quantity of fertilizer, less than 25 kg of inorganic fertilizer, it should be targeted in the first instance on the homestead plots at a micro-dose rate. Unlike the fertility ring management systems of West Africa (Ruthenberg 1980), in smallholder farms of East and Southern Africa the homestead plots, irrespective of the resource status of the household, are the most fertile, with soil fertility declining as one moves away from the household (Giller et al. 2006; Tittonell et al. 2005; Zingore et al. 2007).

On-farm experimentation was then initiated with farmers on micro-dosing alone or in combination with available animal manures (Ncube et al. 2007). The on-farm trial results confirmed that farmers could increase their yields by 30–100% by applying approximately 10 kg N ha^{-1} (Rusike et al. 2006). Larger average gains could be obtained by combining the nitrogen fertiliser with a basal application of low grade manure (Ncube et al. 2007). The question remained whether these results could be replicated across much broader spectrum of farmers and soil types.

Scaling out of micro-dosing was initiated in 2003/2004 in the context of national drought relief programs. Donors were already distributing seed and fertiliser inputs to drought affected farmers. Support was obtained from the Department for International Development (DFID) and the European Commission Humanitarian Aid Office (ECHO) to encourage the application of the micro-dosing of ammonium nitrate (AN) fertiliser by more than 160,000 farmers (Rohrbach et al. 2005; Twomlow et al. 2007a).

This paper reports the results from three related studies on low-input soil fertility management practices for the cereal production systems in southern Zimbabwe. The first two studies were designed to provide direct field evidence to local extension staff on the benefits of small quantities of nitrogen compared to seed, as it is a commonly held belief amongst the relief and development communities that it is better to provide a vulnerable household with seed, rather than fertiliser (Rohrbach et al. 2005; Twomlow 2006). The third study was the wide scale testing of the micro-dosing concept across multiple locations in southern Zimbabwe through relief and recovery programs.

Materials and methods

Rainfall characteristics

On-farm trials were conducted across a total of 16 districts in southern Zimbabwe that covered Natural Farming Region III, IV and V (Vincent and Thomas 1961) from 2003 to 2006. These natural farming regions are characterized by semi-arid climatic conditions and annual uni-modal rainfall of between 450 and 750 mm. The duration of the rainy season is from October/November to March/April and is typically characterized by sporadic, heavy rainstorms, with periodic dry spells. It is followed by a cool to warm dry season from May to September. The length of a typical wet season is between 130 and 140 days for southern Zimbabwe, with Hwange District having the shortest at 107 days.

Soils

The soils of southern Zimbabwe range from deep (>150 cm) Kalahari sands (Eutric-Aridic Arenosol—93% sand, 4% clay, 3% silt, in the 0–11 cm layer) originating from aeolian sand parent material through granitic sands (Eutric Arenosol—93% sand, 3% clay, 4% silt, in the 0–11 cm layer) to clay loams (Eutric-Leptic Cambisol—61% sand 32% clay, 7% silt, in the 0–11 cm layer) (Moyo 2001). The typical pH-value (0.01 M $CaCl_2$) of the soils is slightly acidic (5.5 in the 0–11 cm layer and 5.8 in 11–30 cm layer), organic carbon content less than 1%, and cation exchange capacity (CEC) less than 5 cmolc kg^{-1}. Base saturation is typically less than 20% in the 0–11 cm layer, increasing to over 50% below 75 cm depending on the parent material (Moyo 2001).

Farming system

The farming systems in southern Zimbabwe are semi-extensive mixed farming, involving goat and cattle production, and cultivation of drought resistant crops. Both crop and livestock productivity in the smallholder-farming sector is poor, with farm sizes varying from less than 2 ha in the east of the country to more than 5 ha in the south west (Ahmed et al. 1997; Hikwa et al. 2001; Ncube et al. 2008). The farmers grow maize (*Zea mays* L.), sorghum (*Sorghum bicolor* (L.) Moench) and pearl millet (*Pennisetum glaucum* (L.) R.Br.) as the major cereal grain crops. Maize and sorghum are normally planted with the first rains from around mid-November and harvest from March onwards. Typical yields are frequently less than 500 kg ha^{-1} (FAOstat), with few if any households meeting basic households' food security needs (900 kg of cereal grain for an average household of six people) from one season to the next (Ahmed et al. 1997; Ncube et al. 2008; Zingore et al. 2007). Normal fertility management practice is to apply amendments (mainly manure) to the maize crop, and plant sorghum the following season (Carberry et al. 2004). Groundnut (*Arachis hypogaea* L.), bambara groundnut (*Vigna subterranea* (L.) Verdc.) and cowpea (*Vigna unguiculata* L. Walp. ssp. unguiculata) are the three legumes grown. But areas sown to legume each season are generally small (Ahmed et al. 1997; Twomlow 2004), and legumes receive less than 5% of the applied nutrients (Mapfumo and Giller 2001), and yields are less than 300 kg ha^{-1} (Hilderbrand 1996; Ahmed et al. 1997; Nhamo et al. 2003). To combat these low crop yields smallholder households pursue a combination of strategies/development pathways together or sequentially to meet their livelihood objectives and reduce their vulnerability. These include livestock enterprises, off farm employment and remittances—strategies common to smallholder communities throughout sub-Saharan Africa (for example: Twomlow 2004; Giller et al. 2006; Pender et al. 2006).

On-farm study 1—maize, sorghum and pearl millet nitrogen response curves

The N response curve for maize (var. SC403 or OPVZM421), sorghum (var. Macia) and pearl millet (var. PMV3 or Okashana) were determined using results from nine on-farm trial sites in southern Zimbabwe in 2003/2004, 2004/2005, and only for maize in 2005/2006. Nitrogen levels of 0, 8.5, 17, 25.5, 34 and 42.5 kg ha^{-1}, applied as ammonium nitrate (AN, 34%N), were evaluated to determine the response curve. The two highest nitrogen levels were applied as split dressing, the first at the 5-to-6-leaf-stage and the second three weeks later. Each on-farm site had a single set of treatments for each crop, with an individual plot size for each treatment of 100 m^2. The trials were located on the homestead field plot in each season, and the host farmer determined all management practices, including the date of planting. Composite soil samples were taken for the top 0.20 m of each on-farm trial site in 2006 and used as covariates in a pooled analysis.

On-farm study 2—maize variety by micro-dosing

Maize yields obtained from changing seed varieties, and adding a small dose of AN fertiliser (equivalent to 17 kg N ha^{-1}) were determined using results from nine on-farm trial sites in southern Zimbabwe in 2003/2004, 2004/2005, and 2005/2006. These were the same nine farmers that hosted trials for Study 1. Each season farmers were asked to prepare three 200 m^2 plots and plant their recycled maize seed, and open pollinated variety (var. Zm421) and a commercial hybrid (SC403). At the 5-to-6-leaf-stage the plots were split in half, one half receiving no nitrogen top dressing, and the other half receiving a small dose of AN fertiliser equivalent to 17 kg N ha[1]. The trials were located on the homestead field plot in each season, adjacent to the trials in study 1, and the host farmer determined all management practices, including the date of planting. Composite soil samples were taken for the top 0.20 m of each on-farm trial site in 2006 and used as covariates in a pooled analysis.

On-farm study 3—wide scale promotion and testing of micro-dosing

Between 2003 and 2006, under a series of recovery programs funded by DFID and ECHO, more than 160,000 farmers each cropping season, across 16 districts of southern Zimbabwe, Natural Farming Regions III, IV and V, were provided with 25 kg of AN along with a 1-page pamphlet in the local language advising on how to apply it to a growing crop. The fertiliser inputs were primarily distributed free, with the aim of improving food security of vulnerable households (see Table 2 for selection criteria for the vulnerability), typically 40–50% of households in most communities in southern Zimbabwe. The agricultural relief and recovery programs aimed at strengthening the capacity of these vulnerable households to produce their own food and produce some surplus for stabilization of national food supplies (Rohrbach et al. 2004).

In each of the three seasons between 300 and 1,200 farmers (more than 50% women in each season) were taught how to establish simple paired demonstration plots of approximately one acre (0.2 ha) in close collaboration with partner non-governmental organizations (NGOs) and local extension staff from the department of Agricultural Research and Extension (AREX). Half of the plot (0.1 ha) would receive approximately 10 kg of AN fertiliser and half of the plot received no fertiliser. The farmers applied the AN to any cereal grain they planted each season. They were advised to apply the AN using 1 beer bottle cap (4.5 g of AN fertiliser) for every three plants. This works out to a rate of about 17 kg N ha^{-1} (approximately 25% of recommended levels). It was recommended that this be applied when the cereal plant was at the 5-to-6-leaf-stage. All other crop management decisions (planting date and method, time of weeding, etc.) were the responsibility of the farmer. The total number of trials planned for each season along with the identity of the collaborating NGOs, and the trials that were successfully harvested in each season is summarized in Table 3. Throughout the on-farm evaluations women were encouraged to participate.

Table 2 Targeting criteria for beneficiaries under the Agricultural Relief and Recovery Programs in Zimbabwe

1. Households[a] without (or with limited) draught power and with limited small stock.
2. Female headed (dejure) households
3. Households with limited cash income, no pension, no formal employment and with little or no remittances
4. Households with high dependency ratio e.g. high numbers of children, orphans, handicapped, terminally ill and the elderly
5. Male headed households with limited assets

[a] Households were selected in public community meetings with representatives from donor NGOs, with the community leaders (village heads and chiefs endorsing the process). The recipients were deemed to be able to fully utilise the agricultural inputs they had received; source: Rohrbach et al. 2004)

Quantifying the long term sustainability of micro-dosing using simulation modeling

The simulation tool used was the Agricultural Production Systems Simulator (APSIM) model (Keating et al. 2003). The model is useful in capturing the interactions between climatic conditions, soil types and nutrient dynamics, and has been successfully used the in cereal based farming systems of southern Africa (Delve and Probert 2004; Robertson et al. 2005; Shamudzarira and Robertson 2002; Whitbread et al. 2004).

Analyses have been done for a sandy loam soil type typical of southern Zimbabwe using a 25 year weather record (1980–2005) record collected by the national Weather Bureau for Matopos Research Station that was extrapolated to 2015 by taking a random selection of weather records from the 45 year record (1960–2005). A short duration maize variety (SC403) was used to simulate maize growth and development to various crop production scenarios. The scenarios simulated are as follows:

1. Farmer practice-crop planted using overall spring ploughing in mid to late December, followed by

Table 3 Distribution of micro-dosing trials across southern Zimbabwe and collaborating NGO over the three seasons from 2003 to 2006

District	Natural region[a]	NGO	Number of paired micro-dosing plots targeted per seasons		
			2003/2004	2004/2005	2005/2006
Bikita	IV/V	CARE	80	ND	ND
Binga	IV/V	Save The Children	ND	ND	21
Buhera	III/IV				6
Chirumhanzu	III	OXFAM GB	ND	ND	15
Chivi	V	Zishavane Water Project	ND	ND	25
Gokwe	III	CARE	80	ND	ND
Hwange	IV/V	COSV	400	98	104
Inziza	IV	World Vision	ND	ND	17
Lupane	IV/v		ND	ND	22
Mangwe	IV/V	World Vision	ND	ND	9
Masvingo	III/IV	CARE	ND	13	ND
Matobo	IV/V	World Vision	400	ND	26
Mberengwa	IV/V	CARE	80	ND	ND
Nkayi	IV	COSV	ND	1	47
Zaka	III/IV	CARE	80	ND	ND
Zishavane	III/IV	OXFAM GB	80	ND	16
Total number of paired plots successfully harvested each season			915[b]	112[c]	308

ND = No demonstrations in that season

[a] Zimbabwe is divided into five agroecological regions, also known as Natural Regions I–V. Natural Region I and II receive the highest rainfall (at least 750 mm per annum) and are suitable for intensive farming. Natural Region III receives moderate rainfall (650–800 mm per annum), and Natural Regions IV and V have fairly low annual rainfall (450–650 mm per annum) and are suitable for extensive farming. Adapted from Vincent and Thomas (1960)

[b] 444 male, 471 female

[c] 49 male, 63 female

at least 2 weedings (Typical scenario for farmers with limited or no access to draught animals).

2. As for farmer practice with a micro-dose (17 kg N ha^{-1}) of fertilizer applied at the 5–6 leaf stage from 2005 onwards to show what contributions microdosing might have towards helping achieve the millennium development goal of increased food security (UN Millennium Project 2005).

For full details of the models parameterization for this soil type please refer to Carberry et al. (2004).

Data collection and analyses

Simple record books in the local vernacular (either Ndbele or Shona) were provided to each collaborating farmer that summarized the trial (Study 1–3) they were hosting and allowed them to record crop planted, date of planting, date and number of weedings, date of fertiliser application, yield information and any other observations they wished to make. We also collected data on basic household resource levels such as draught animal ownership. Field assistants were recruited in each locality to assist the farmers with record keeping, the collection of rainfall records from simple daily catch gauges located in each village for the host farms in Study 1 and 2, and at the individual farms in Study 3, harvesting of the plots and recording crop yields. Given the number of demonstrations undertaken in any one season, it was not possible to physically weigh the threshed grain yield from every plot. Where this was not possible, the yield from each sub plot was placed in 50 kg sacks, and the number to the nearest half sack was recorded. Spot checks were made throughout the districts where on-farm testing was undertaken in each season to quantify the weight of threshed grain that a 50 kg sack contained, in order to convert the number of sacks recorded into grain yield per ha on a dry weight basis. Typically, a 50 kg bag of maize cobs contained 21.6 kg of grain, a 50 kg bag of pearl millet heads contained 18.4 kg of grain, and with sorghum it was 20.7 kg of grain per 50 kg bag (Twomlow et al. 2007a).

Various national surveys have been undertaken since 2004, to assess impacts of the relief and recovery programs large scale distributions of seed and fertilisers. Full details of these surveys are reported in Rohrbach et al. (2005); Rohrbach and Mazvimavi (2005) and Woolcock and Mutiro (2007), and provide the necessary socio-economic inputs to allow a cost-benefit analyses of the micro-dosing intervention.

Statistical analyses

The cereal yield data was analyzed using the method of residual maximum likelihood (REML) included in the statistical software package GENSTAT version 9. The choice of REML was based on the fact that the model includes fixed and random factors, accounts for more than one source of variation in the data and provides estimates for treatments effects in unbalanced treatment designs. Season was included in the fixed model for Study 1 so that differences between seasons could be tested. Between seasons differences for Study 3 are not presented in this paper as locations of the trials varied from season to season, depending on the collaborating NGOs in that season (Table 3).

Gender, draught animal power ownership, household labour, number of weedings, field type (homestead plot/main field) and soils analyses (where available) were tested as fixed variables, but found to be not significant in accounting for any of the unexplained variability or significant interactions with fertiliser.

Therefore, the linear mixed model, used to analyze the seasonal effects on Studies 1–3, had the following components and terms:

Response variate: Yield
Fixed model: Constant + Fertiliser * (Season – included for studies 1 and 2)
Random model: District + Ward

Results

Rainfall patterns over the 3 years of observation varied considerably both within and between seasons, depending on location. Rainfall was found to have a statistically significant effect on cereal yield in the different districts ($P < 0.001$), but was however, found not to have any significant interaction with fertiliser application ($P = 0.697$) in each season. For the purposes of this paper it is sufficient to say that the 2003/2004 experienced below average seasonal rainfall (most districts receiving less than 550 mm), 2004/2005 experienced average seasonal rainfall (most districts receiving between 550 and 600 mm), whilst the 2005/2006 season experienced above average rainfall in all localities (Table 4).

On-farm study 1—maize nitrogen response curve

Figure 1 shows that 1.5–2 bags of ammonium nitrate (25–34 kg N ha^{-1}) are optimum for maize in dry regions, but also shows strong linear response at

Table 4 Maize yields obtained from changing seed varieties and adding a small dose of ammonium nitrate fertiliser (equivalent to 17 kg of N ha^{-1}) in semi-arid regions of Zimbabwe; measurements are the average from 9 farmers' fields, 2003/2004, 2004/2005 and 2005/2006 seasons

Season	Seasonal rainfall mm	Maize seed variety and nitrogen top dressing regime						e.s.e[a]
		Farmers retained seed		OPV ZM421		Hybrid SC403		
		Zero N	17 kg N ha^{-1}	Zero N	17 kg N ha^{-1}	Zero N	17 kg N ha^{-1}	
2003/2004	443	894	1,060	912	1,378	1,093	1,585	179.7
2004/2005	548	880	1,190	1,360	1,706	1,440	1,973	90.6
2005/2006	806	1,120	1,330	1,546	1,741	1,513	2,084	121.6

[a] e.s.e—experimental standard error

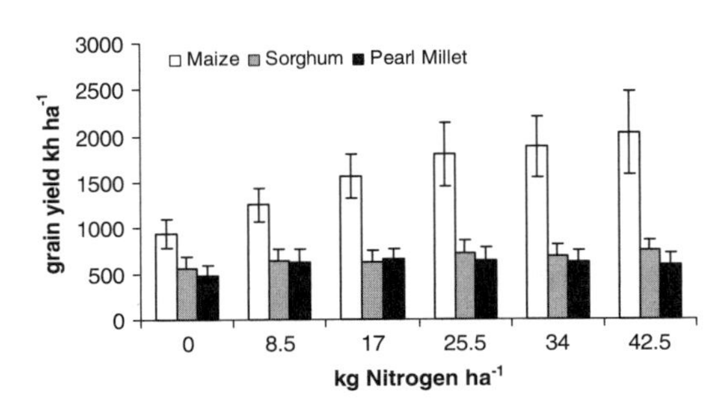

Fig. 1 Grain response of maize, sorghum and pearl millet to increasing levels of N fertiliser under farmer management. Mean of results from 9 sites since 2003 for three seasons for maize, and two seasons for sorghum and pearl millet. Error bars represent standard errors of differences between the predicted means of the nitrogen by crop yield by season

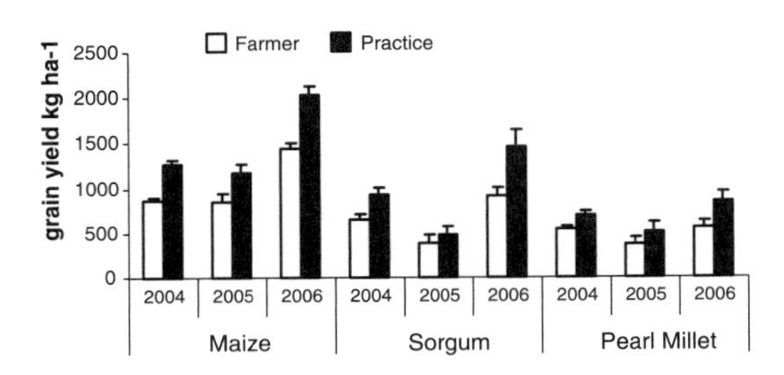

Fig. 2 Grain responses of cereals to a targeted application of 50 kg of ammonium nitrate fertiliser (17 kg N ha^{-1}) under farmer management. Mean of results across multiple sites for 2003/2004, 2004/2005 and 2005/2006 cropping seasons. (Grain increases due to each kg of N applied were between 18 and 35 kg for maize, 5 and 32 kg for sorghum, 8 and 16 kg for pearl millet). Error bars represent standard errors of differences between the predicted means of the micro-dosing by crop yields

lower application rates. It is worth noting that evidence for the linear maize response at low N rates is usually implied in published fertiliser response curves, which typically start at 30 kg N ha^{-1} or higher (e.g. Benson 1998; Mushayi et al. 1999). The slope of the maize response curve at the lower rates in Fig. 1 is about 11 kg of grain per kg of fertiliser input, and the economic returns to fertiliser investments at these sub-optimal levels have been shown to be quite profitable (Woolcock and Mutiro 2007). For example, the 25 kg of AN fertiliser commonly distributed through relief programs cost approximately US$ 2 kg^{-1} to deliver to the crop. This includes the estimated costs of labour used in applying this input. This compares with a post-harvest farm gate price for maize grain of US$ 0.4 kg^{-1}. In order to break even, farmers would have to obtain 5 kg of grain for every kg of fertiliser applied. This is easily surpassed by the grain response at low N rates in Fig. 1. In fact, at 11 kg of grain per kg of fertiliser input, the value cost ratio (VCR) exceeds 2:1, the commonly accepted threshold required to encourage risk-averse farmers to invest in fertiliser technology (Benson 1998; Morris et al. 2007). For N application rates above 25.5 kg ha^{-1}, the VCR falls below 2:1 and is approximately 1.6 for the highest rate applied in Fig. 1 (42.5 kg N ha^{-1}). However, this rate is well below the 46–76 kg N ha^{-1} promoted in current extension recommendations for these regions (Table 1).

The 3 years of on-farm experimentation in drier regions of Zimbabwe show consistent grain yield response and profitability of maize to low rates of nitrogen fertiliser, either alone (Fig. 1) or in combination with manure (Ncube et al. 2007). What is of concern, and requires more detailed study, are the poor responses of sorghum and pearl millet to nitrogen fertiliser shown in Fig. 1. It is speculated that some of these poor responses are due to poor root development and capability of sorghum and millet to extract P under low P conditions as observed in these soils (Vadez personal communication), despite the fact that the trials were located on homestead plots that are traditionally considered to be more fertile (Ncube et al. 2008). However, this lack of response by sorghum and pearl millet was not so evident in the broad-scale testing in farmers fields (Figs. 2 and 3).

On-farm study 2—maize variety by micro-dosing

Table 4 summarizes the three seasons' responses of different varieties of maize to micro-dosing. Improved OPV seed alone appears to give a significant increase in maize grain yield over the farmers retained seed in average to above average rainfall seasons, but not in below average seasons. In addition, the data suggests that the hybrid response to N was consistently about 500 kg, whereas that for farmer seed or OPV was less and more variable. The retained seed response to N was between 100 and 300 kg, whereas the OPV seed response was between 200 and 400 kg, depending on the rainfall received.

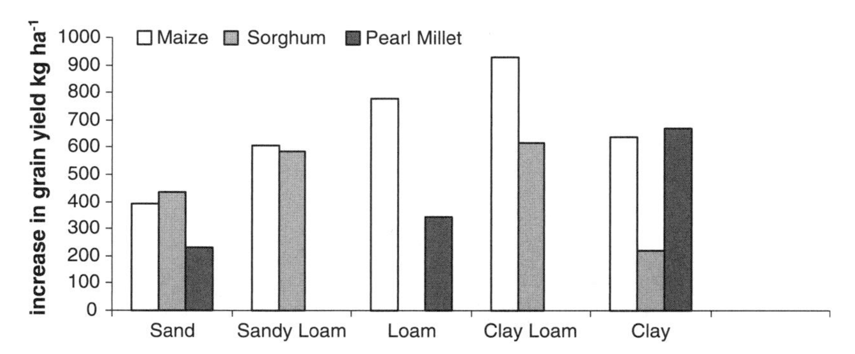

Fig. 3 Observed increases in cereal grain yield (kg ha^{-1}) for 323 households from 13 districts across southern Zimbabwe in response to a targeted application of 50 kg of ammonium nitrate fertiliser (17 kg N ha^{-1}) under farmer management for five different soil types in 2005/2006 season (Grain increases due to each kg of N applied were between 15 and 45 kg

On-farm study 3—wide scale promotion and testing of micro-dosing

Three years of wide scale testing in numerous districts across southern Zimbabwe with farmers has confirmed that small amounts of nitrogen fertiliser (17 kg ha^{-1} compared to recommended rates of 55 kg ha^{-1}) applied as targeted topdressing can give significant increases ($P = 0.001$) in cereal grain yield (Fig. 2), irrespective of farmers ability to manage the crop (Table 5) or soil type (Fig. 3). Despite the high variability shown in Table 5 for the timing of fertiliser application and weeding dates, which was observed to occur each season despite the flyers and training that were given, the response to small doses of N proved remarkably robust (Fig. 4). Only 7 of the 89 farmers (7.8%) in Fig. 4 either failed to obtain a yield gain with N micro-dosing or witnessed a decline in yield. A households' failure to achieve positive yield increases was, based on the farmers own record books and site visits attributed to either late planting, late or zero application of the fertiliser and poor weed management. At the same time, the few very high yield gains in Fig. 4 (those exceeding 850 kg or 50 kg grain/kg of N applied) are probably the result of unaccounted additional nutrient inputs (e.g., manure applications or extra fertiliser).

The observed efficacy of the grain response to low doses of N in this study is noteworthy and is an important result from the perspective of improving the food security of smallholder farmers in these dry regions. Even from the perspective of a breakeven yield (85 kg grain), the consistent gains exhibited in Fig. 4 are impressive-only 22 of the 89 farmers (25%) failed to achieve the necessary yield gain. In other words, 75% of farmers achieved a yield gain that would translate into a profit margin (over the input cost) when N was applied at a low rate.

Table 5 Timing of fertiliser application and weedings relative to planting dates for seven districts in southern Zimbabwe in 2003/2004

District	Days after planting		
	Fertilization (minimum–maximum)	First weeding (minimum–maximum)	Second weeding (minimum–maximum)
Bikita	58 (18–101)	27 (4–68)	57 (28–103)
Gokwe	42 (6–72)	22 (6–49)	38 (18–87)
Hwange	42 (0–74)	27 (3–105)	39 (21–97)
Matobo	52 (3–120)	33 (4–96)	50 (16–136)
Mberengwa	61 (25–111)	25 (1–80)	50 (16–96)
Zaka	54 (22–84)	21 (2–54)	25 (24–86)
Zishavane	39 (27–52)	25 (19–36)	75 (38–105)

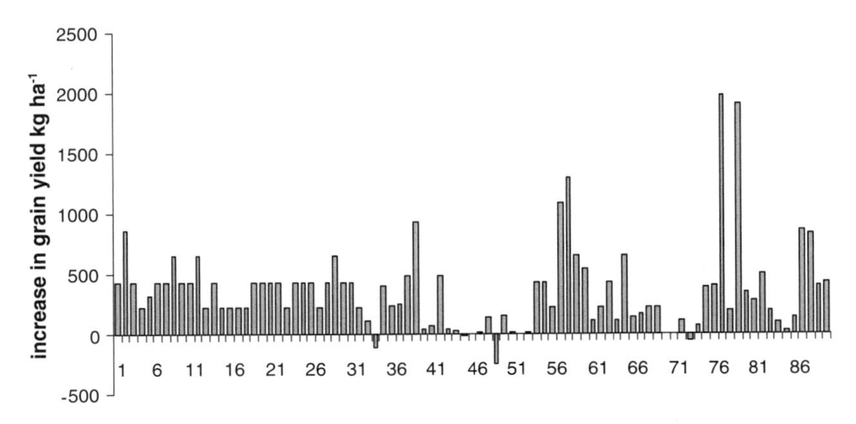

Fig. 4 Observed increases in maize grain yield (kg ha^{-1}) for 89 households from Hwange, Lupane, Masvingo and Nkayi Districts in response to a targeted application of 50 kg of ammonium nitrate fertiliser (17 kg N ha^{-1}) under farmer management in 2004/2005

Discussion

The drought relief program that was the platform for the research studies reported here facilitated widespread distribution of seed and fertiliser across southern regions of Zimbabwe. The innovation in the relief program was that it included fertiliser distribution into dry regions and that it promoted small doses of N fertiliser. The decision to do this was based on ICRISAT's results from a small number of on-farm trials in conjunction with output from crop simulation analysis. The question on whether the response to small doses of N could be replicated for much larger numbers of farmers with varied soil and management conditions and rainfall regimes remained. We pursued this question through trials that establish fertiliser response curves in dry regions (in the process helping to fill a research gap as such data are largely non-existent for these regions), comparing technology investments in N and improved seed, and broad-scale testing of small N doses under farmer management conditions.

Results from three seasons of extensive testing clearly show that response to small doses of nitrogen is measurable in on-farm trials for a wide range of soils, farmer management and seasonal rainfall conditions. This reflects the inherent low fertility (Mapfumo and Giller 2001; Ncube et al. 2007) of these cropping systems and the fact that nitrogen is more of a constraint to production than lack of soil moisture in most seasons. The grain yield increases achieved in the broad-scale studies are also consistent with the level of yield responses first suggested by the crop modeling analysis of the smallholder cereal production systems in southern Zimbabwe (Dimes et al. 2003; Carberry et al. 2004).

It is particularly remarkable that micro-dosing benefits accrued to almost all the farmers applying this technology, irrespective of season or resource status, as is shown in Fig. 4. Usually there are leaders and laggards in technology adoption. Often technologies are initially applied well by only a subset of better-than-average farmers. It is well known that fertiliser response depends on the application of complementary practices such as timely planting, timely weeding, timely fertiliser application, the starting quality of soils, and incidence of diseases and pests. Yet such a wide range of farmers have obtained significant yield gains from micro-dosing, even in drought years. The strong and consistent responses in Fig. 4 are further evidence of the inherently low N supply capacity of soils across the dry regions in Zimbabwe and that widespread yield responses to N can be generally expected (Mushayi et al. 1999; Mapfumo and Giller 2001; Zingore et al. 2007; Ncube et al. 2008).

The 2003/2004 and 2004/2005 seasons when micro-dosing was widely promoted were relatively poor rainfall years (Table 4), compared to the 2005/2006 season. Even so, the vast majority of fertiliser recipients achieved strong positive returns to this investment. With the aid of simulation modeling, Fig. 5 highlights the gains likely to be achieved if farmers continue to pursue micro-dosing in the

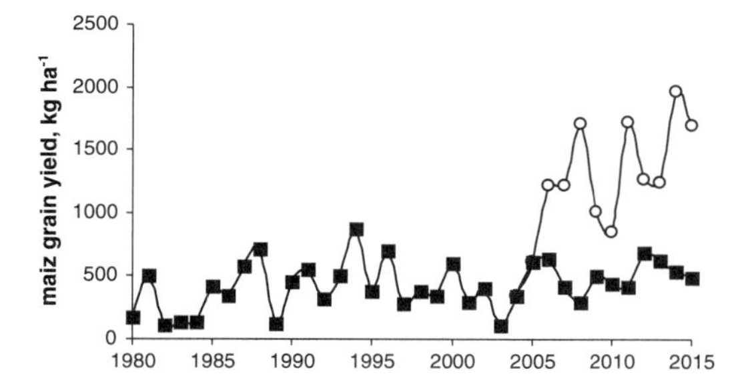

Fig. 5 Maize grain yields obtainable in drought prone semi-arid parts of Zimbabwe under current farmer practices without ammonium nitrate fertilizer (■) and with small doses of ammonium nitrate fertiliser (17 kg N ha^{-1}) post 2005 (○), based on crop simulation modeling using APSIM and confirmed by farmer managed demonstration trials

future. The initial data series (■) summarizes the levels of yields to be expected when farmers apply no fertiliser—the common current practice in semi-arid areas of the country. The second series (○) highlights the gains achievable with sustained use of as little as 17 kg N ha^{-1}, equivalent to one 50 kg bag of ammonium nitrate per hectare.

If the use of small quantities of AN can be continued after the relief programs stop handing out free fertiliser, these farmers can achieve a sustained set of higher grain yields and a sustained improvement in food security, thus meeting the first of the UN millennium goals (UN Millennium Project 2005). Even if severe drought occurs farmers will be better off than in previous drought years. On the other hand, if rains are more favorable, farmers will have appreciably higher yields as the N inputs contribute to higher water productivity in either situation.

The challenge remains, however, to move farmers from a dependence on free handouts toward a willingness to purchase fertiliser each year in a local retail shop. Currently, farmers are unaccustomed to purchasing fertiliser. Local retailers remain with the view that most of these farmers will not make this investment because it is too risky. Further, the willingness of fertiliser companies and retailers to pursue this market has been undermined by the continuing distribution of free seed and fertiliser directly to farm households. At a minimum, this sort of distribution should be through vouchers redeemable at local retail shops.

Some have questioned the logic of micro-dosing, claiming this is such a small quantity of fertiliser and that it is not sustainable. Some argue it is wrong to encourage farmers to adopt second best solutions. Some state that other nutrients such as phosphorous will quickly become limiting if only ammonium nitrate is promoted or low levels of organic matter will eventually restrict yield gains.

Yet the majority of farmers being assisted by the various donor programs in Zimbabwe did not use any fertiliser prior to the initiation of this effort. Extension recommendations calling for larger doses were consistently ignored as they were viewed to be impractical and too risky. The micro-dosing promoted by ICRISAT and many NGOs offers vulnerable households the first opportunity to lift their average yields to a new threshold. To apply only one 50 kg bag of AN offers a substantial improvement on food security that otherwise would not have been available. Extensive crop systems modeling data indicates this gain can be sustained in southern Zimbabwe for many years (Fig. 5). Importantly, however, the success of micro-dosing demonstrations has encouraged many farmers to begin to experiment with alternative improvements in crop management—combining organic and inorganic fertiliser, applying higher rates, and attempting conservation farming (Mazvimavi and Twomlow 2007; Twomlow et al. 2007b). In effect, this simple technology is renewing farmers' interest in exploring new options for technological change.

The 25 kg of AN fertiliser commonly distributed through the relief programs cost approximately US$ 2 kg^{-1} to deliver to the crop. This includes the estimated costs of labour used in applying this input. This compares with a post-harvest farm gate price for maize grain of US$ 0.4 kg^{-1}. In order to obtain a profit, farmers would have to obtain only 5 kg of grain for every kg of fertiliser applied. In fact, farmers more commonly obtained 15–45 kg of grain per kg of fertiliser input (Figs. 3 and 5).

Conclusions

This research set out to establish the efficacy of cereal crop responses to low doses of N fertiliser across dry regions of southern Zimbabwe. The results have provided strong evidence that N micro-dosing has the potential for broad-scale impact on food security for a large section of the rural poor. For example,

Rohrbach et al. (2005) estimate DFID's support for the distribution of 25 kg of ammonium nitrate fertiliser to each of 160,000 farm households contributed 40,000 additional tons of maize production, valued by the World Food Program at 5–7 million USD. A further question now arises for national research and extension agencies with a mandate for dry land cropping regions—is it rational and acceptable to recommend levels of fertiliser use lower than current recommendations? Our results from three years of observations in dry land areas say yes.

Acknowledgements We wish to thank the farmers and extension staff of Bikita, Binga, Buhera, Chirumhanzu, Chivi, Gokwe, Hwange, Inziza, Lupane Masvingo, Matobo, Mberengwa, Nkayi, Zaka and Zishavane Districts for their enthusiasm and collaboration, and the field staff of COSV, CARE, OXFAM UK and World Vision for assistance with farmer training, monitoring the trials program and record keeping. We also wish to thank ECHO, DFID and ICRISAT for the funding. The opinions expressed in this paper are those of the authors. We are grateful for the very useful comments made by two reviewers.

References

Ahmed MM, Rohrbach DD, Gono LT, Mazhangara EP, Mugwira L, Masendeke DD et al (1997) Soil fertility management in communal areas of Zimbabwe: current practices, constraints and opportunities for change. ICRISAT Southern and Eastern Africa Region Working Paper No. 6.ICRISAT. Bulawayo, Zimbabwe

Benson T (1998) Developing flexible fertiliser recommendations for smallholder maize production in Malawi. In: Waddington SR, Murwira HK, Kumwenda J, Hikwa D, Tagwira F (eds) Soil fertility research for maize-based farming systems in Malawi and Zimbabwe…. The Soil Fertility Network for Maize Based Cropping Systems in Malawi and Zimbabwe, Harare, Zimbabwe, pp 237–244

Carberry P, Gladwin C, Twomlow S (2004) Linking simulation modeling to participatory research in smallholder farming systems. In: Delve R, Probert M (eds) Modeling nutrient management in tropical cropping systems. ACIAR Proceedings no. 114. Australian Centre for International Agricultural Research, pp 32–46

Delve R, Probert M (eds) (2004) Modeling nutrient management in tropical cropping systems. ACIAR Proceedings No. 114. Australian Centre for International Agricultural Research

Dimes J, Twomlow S, Carberry P (2003) Application of APSIM in small holder farming systems in the semi-arid tropics. In: Bontkes T, Wopereis M (eds) Decision support tools for smallholder agriculture in sub-Saharan Africa: a practical guide. IFDC and CTA, pp 85–99

Giller KE, Rowe EC, de Ridder N, van Keulen H (2006) Resource use dynamics and interactions in the tropics: scaling up in space and time. Agric Syst 88:8–17. doi:10.1016/j.agsy.2005.06.016

Heinrich G (ed) (2004) A foundation for the future: the Sorghum and Millet Improvement Program (SMIP) in Southern Africa. Proceedings of the SMIP Final Review and Reporting Workshop, 25–26 Nov 2003, ICRISAT, Bulawayo, Zimbabwe

Hikwa D, Nyathi P, Mugwira LM, Mudhara M, Mushambi CF (eds) (2001) Integrated soil fertility development for resource-poor farmers in Zimbabwe: the research and development strategy beyond 2001. DRSS, Harare, 48 pp

Hilderbrand GL (1996) The status of technologies used to achieve high groundnut yields in Zimbabwe. In: Gowda GLL, Nigam SN, Johansen C, Renard C (eds) Achieving high groundnut yields: proceedings of an international workshop), 25–29 Aug 1995, Laixi City, Shandong, China. International Crops Research Institute for the Semi Arid Tropics. Patancheru, India. 300 pp (ISBN 92-9066-350-2)

Hilhorst T, Muchena F (2000) Nutrients on the move: soil fertility dynamics in African farming systems. International Institute for Environment and Development, London, UK, 146 pp

Keating BA, Carberry PS, Hammer GL, Probert ME, Robertson MJ, Holzworth D et al (2003) An overview of APSIM, a model designed for farming systems simulation. Eur J Agron 18:267–288. doi:10.1016/S1161-0301(02)00108-9

Mafongoya PL, Bationo A, Kihara J, Waswa BS (2006) Appropriate technologies to replenish soil fertility in southern Africa. Nutr Cycl Agroecosyst 76:137–151. doi:10.1007/s10705-006-9049-3

Mapfumo P, Giller KE (2001) Soil fertility management strategies and practices by smallholder farmers in semi arid areas of Zimbabwe. International Crops Research Institute for the Semi Arid Tropics (ICRISAT) and Food and Agricultural Organization (FAO), Bulawayo, Zimbabwe and Rome, Italy, 60 pp

Mazvimavi K, Twomlow S (2007) Conservation Farming for Agricultural Relief and Development in Zimbabwe. In: Goddard, T, Zoebisch MA, Gan YT, Ellis W, Watson A, Sombatpanit S (eds) 2008. No-Till Farming Systems, Special Publication No. 3, World Association of Soil and Water Conservation, Bangkok. ISBN: 978-974-8391-60-1. 169–178

Morris M, Kelly VA, Kipicki RJ, Byerlee D (2007) Fertiliser use in African agriculture: lessons learned and good practice guidelines. The World Bank, Washington DC, 144 pp

Moyo M (2001) Representative soil profiles of ICRISAT research sites. Chemistry and Soil Research Institute, Soils Report No. A666. AREX, Harare, Zimbabwe, 97 pp

Mushayi P, Waddington SR, Chiduza C (1999) Low efficiency of nitrogen use by maize on smallholder farms in sub-humid Zimbabwe. In: Maize production technology for future: challenges and opportunities. Proceedings of the sixth Eastern and Southern African maize conference, 21–25 September 1998. CIMMYT and EARO, Addis Ababa, pp 278–281

Ncube B, Dimes JP, Twomlow SJ, Mupangwa W, Giller KE (2007) Participatory on-farm trials to test response of maize to small doses of manure and nitrogen in smallholder farming systems in semi-arid Zimbabwe. Nutr Cycl Agroecosyst 77:53–67. doi:10.1007/s10705-006-9045-7

Ncube B, Twomlow SJ, Dimes JP, van Wijk MT, Giller KE (2008) Farm characteristics and soil fertility management strategies in smallholder farming systems under semi-arid environments in Zimbabwe. Soil Use Manage (in press)

Nhamo N, Mupangwa W, Siziba S, Gatsi T, Chikazunga D (2003) The role of cowpea (Vigna unguiculata) and other grain legumes in the management of soil fertility in the smallholder farming sector of Zimbabwe. In: Waddington SR (ed) Grain legumes and green manures for soil fertility in southern Africa: taking stock of progress. Proceedings of a conference held 8–11 October 2002 at the Leopard Rock Hotel, Vumba, Zimbabwe. Soil Fert Net and CIMMYT-Zimbabwe, Harare, Zimbabwe, 246 pp

Pender J, Place F, Ehui S (eds) (2006) Strategies for sustainable land management in the East African highlands. International Food Policy Research Institute, Washington, DC. doi:10.2499/0896297578

Robertson MJ, Sakala W, Benson T, Shamudzarira Z (2005) Simulating response of maize to previous velvet bean (*Mucuna pruriens*) crop and nitrogen fertiliser in Malawi. Field Crops Res 91:91–105

Rohrbach DD (1999) Linking crop simulation modeling and farmers participatory research to improve soil productivity in drought-prone environments. In: Risk management for maize farmers in drought-prone areas of Southern Africa. Proceedings of a Workshop, 1–3 October 1997. Kadoma Ranch, Zimbabwe, CIMMYT, Mexico. pp 1–4

Rohrbach D, Mazvimavi K (2005) Assessment of the 2004/05 Seed and Fertiliser Relief and Recovery Programs of ECHO, DFID and GTZ. ICRISAT and FAO, Bulawayo, Zimbabwe. http://www.prpzim.info/download-docs/index.php. Accessed April 2007

Rohrbach D, Charters R, Nyagweta J (2004) Guidelines for agricultural relief programs in Zimbabwe. ICRISAT, Bulawayo, Zimbabwe. http://www.prpzim.info/download-docs/index.php. Accessed April 2007

Rohrbach D, Mashingaidze AB, Mudhara M (2005) Distribution of relief seed and fertiliser in Zimbabwe: lessons from the 2003/04 season. ICRISAT and FAO, Bulawayo, Zimbabwe

Rusike J, Dimes JP, Twomlow SJ (2003) Risk-return tradeoffs of smallholder investments in improved soil fertility management technologies in the semi-arid areas of Zimbabwe. Paper presented at the 25th conference of the international association of agricultural economists. Durban, South Africa. 16–22 Aug 2003

Rusike J, Twomlow SJ, Freeman HA, Heinrich GM (2006) Does farmer participatory research matter for improved soil fertility technology development and dissemination in Southern Africa. Int J Agric Sustain 4(3):176–192

Ruthenberg H (1980) Farming systems in the tropics, 3rd edn. Clarendon Press, Oxford, 424 pp

Shamudzarira Z, Robertson MJ (2002) Simulating the response of maize to nitrogen fertiliser in semi-arid Zimbabwe. Exp Agric 38:79–96

Tittonell P, Vanlauwe B, Leffelaar PA, Shepherd KD, Giller KE (2005) Exploring diversity in soil fertility management of smallholder farmers in western Kenya. II. Within farm variability in resource allocation, nutrient flows and soil fertility status. Agric, Ecosys and Environ 110: 166–184

Twomlow SJ (2004) Increasing the role of legumes in smallholder farming systems—the future challenge. In: Serraj R (ed) Symbiotic nitrogen fixation: prospects for application in tropical agroecosystems. Science Publishers, NH, USA, pp 29–46

Twomlow S (2006) New partnerships boost impact from agricultural relief programmes. Landwards (IAgrE Journal) 61(4):2–5

Twomlow SJ, Ncube B, (eds) (2001) Improving soil management options for women farmers in Malawi and Zimbabwe. Proceedings of a collaborators workshop on the DFID-supported project 'Will Women Farmers Invest in Improving their Soil Fertility Management? Experimentation in a risky environment. 13-15 September 2000, ICRISAT Bulawayo, Zimbabwe. PO Box 776, Bulawayo, Zimbabwe: International Crops Research Institute for the Semi-Arid Tropics. 150 pp

Twomlow S, Rohrbach D, Rusike J, Mupangwa W, Dimes J, Ncube B (2007a) Spreading the word on fertiliser in Zimbabwe. In: Mapiki A, Nhira C (eds) Land and water management for sustainable agriculture. Proceedings of the EU/SADC land and water management applied research and training programmes inaugural scientific symposium, Malawi Institute Management. Lilongwe, Malawi, 14–16 February 2006. paper 6.21

Twomlow S, Rohrbach D, Hove L, Mupangwa W, Mashingaidze N, Moyo M, Chiroro C (2007b) Conservation farming by basins breathes new life into smallholder farmers in Zimbabwe. In: Mapiki A, Nhira C (eds) Land and water management for sustainable agriculture. Proceedings of the EU/SADC land and water management applied research and training programmes inaugural scientific symposium. Lilongwe, Malawi. 14–16 February 2006, Paper 7.2

UN Millennium Project (2005) Investing in development: a practical plan to achieve the millennium development goals. Earthscan, New York

Vincent V, Thomas RG (1960) An agricultural survey of southern Rhodesia, part I:agro-ecological survey. Government Printers, Salisbury

Vincent V, Thomas RG (1961) An agricultural survey of southern Rhodesia (now Zimbabwe), part 1: agro-ecological survey. Government Printers, Salisbury (now Harare)

Whitbread A, Braun A, Alumira J, Rusike J (2004) Using the agricultural simulation model APSIM with smallholder farmers in Zimbabwe to improve farming practices. In: Whitbread A, Pengelly BC (eds) Tropical legumes for sustainable farming systems in Southern Africa and Australia, ACIAR Proceedings, no. 115. Australian Centre for International Agricultural Research, Canberra, pp 171–80

Woolcock R, Mutiro K (2007) Cost benefit analyses of the protracted relief programs agricultural interventions. Report No. 33 June 2007. Technical Learning and Coordination Unit, Harare, Zimbabwe, 63 pp. http://www.prpzim.info/download-docs/index.php. Accessed April 2007

Zingore S, Murwira HK, Delve RJ, Giller KE (2007) Influence of nutrient management strategies on variability of soil fertility, crop yields and nutrient balances on smallholder farms in Zimbabwe. Agric Ecosyst Environ 119(1–2): 112–126

The Dryland Eco-Farm: A Potential Solution to the Main Constraints of Rain-Fed Agriculture in the Semi-Arid Tropics of Africa

D. Fatondji, D. Pasternak, A. Nikiema, D. Senbeto, L. Woltering, J. Ndjeunga, and S. Abdoussalam

Abstract This chapter presents the results of studies on a production system called Dryland Eco-Farm (DEF) that addresses a range of constraints to agricultural productivity in dryland Africa. It combines the use of live hedges and alleys of *Acacia colei*, "demi-lunes" in which are planted domesticated *Ziziphus mauritiania*. Annual crops like pearl millet (*Pennisetum glaucum* (L.) R.Br.), cowpeas and roselle (*Hibiscus sabdariffa*) are planted in rotation. This trial tests the effect of the system on (1) soil erosion control, soil fertility and water use efficiency, (2) crop yield and biomass production, and (3) improving income generation and diversification. Average pearl millet yields in the DEF were twice the control (880 vs. 430 kg ha^{-1}) when no mineral fertilizer was applied. With the application of NPK, millet yields were almost similar under both conditions (950 vs. 780 kg ha^{-1}). Cowpea yields were on average seven times higher than the control without NPK (1,400 vs. 200 kg ha^{-1} total biomass) and three times with NPK (1,850 vs. 650 kg ha^{-1} total biomass). Roselle yield increased four times on average without NPK (205 vs. 60 kg ha^{-1} calices yield) and two times with NPK (234 vs. 114 kg ha^{-1}). Therefore, the system has the potential to produce yield response similar to that of the recommended rate of 100 kg of the 15-15-15 fertilizer per ha. The return to land is estimated at US $224 for the DEF compared to US $77 for the traditional millet–cowpea system. This system has the potential to improve productivity and rural livelihood in the drylands of Africa while sustaining the natural resources base.

D. Fatondji (✉)
International Crops Research Institute for the Semi-Arid Tropics (ICRISAT) Sahelian Center, Niamey, Niger
e-mail: d.fatondji@cgiar.org; d.fatondji@gmail.com

Keywords Demi-lune · Domesticated *Ziziphus* · Dryland eco-farm · Millet–cowpea system · Pomme du sahel · Roselle · Sahelian region

Introduction

The dryland of Africa is the home to the poorest of the world, where about 90% of the population live in rural areas and sustain their livelihood from subsistence agriculture (Bationo et al., 2003). In this zone of Africa, low and erratic rainfall, its poor distribution within the growing period, prolonged dry spells, which usually occur during the season, and the lack of adequate water supply due to soil physical degradation (soil crusting) and nutrient shortage often adversely affect crop growth and yields (Zougmoré, 2003). Indeed, only 2% of the arable lands are irrigated, which means that rain-fed agriculture is the main source of food for the increasing populations in that zone (Parr et al., 1990).

Soil erosion (wind and water), low soil fertility (mining agriculture), low income (vicious circle of low crop yields), low water use efficiency, insufficient supply of animal feed (low biomass production and poor pasture) and poor distribution of the labour force (farmers are busy only 4 months of the year) are some of the constraints to agricultural production in Africa and particularly in the dryland zone. Research work conducted by the International Crops Research Institute for the Semi-Arid Tropics (ICRISAT), the International Fertilizer Development Centre (IFDC), the International Institute for Tropical Agriculture (IITA) and national research institutions has led to the development of technologies that

A. Bationo et al. (eds.), *Innovations as Key to the Green Revolution in Africa*,
DOI 10.1007/978-90-481-2543-2_114, © Springer Science+Business Media B.V. 2011

have proven to be efficient in addressing these constraints individually. But further work in the dryland has shown that to address these constraints a multi-component and multi-disciplinary research is required to increase the productivity of the arable land (Mando, 1997). The Dryland Eco-Farm (DEF) is an innovation that was recently developed at the ICRISAT Sahelian Center (ISC) in Niger in collaboration with the National Agricultural Research and Extension Services (NARES) partners. The DEF combines the use of live hedges of *Acacia colei*, earth bunds that turn into micro-catchments or "demi-lunes", high-value trees such as the domesticated *Ziziphus mauritiania* planted inside the "demi-lunes", and annual crops, each planted in half or a third of the field in rotation each year.

The purpose of the experiment was to test the effect of the DEF system on (1) soil erosion control, soil fertility and water use efficiency, (2) plant biomass production, (3) income generation and diversification and (4) labour productivity compared to the traditional millet–cowpea system.

Materials and Methods

The Experimental Site

This experiment was conducted at the ICRISAT research station at Sadore located at 13° 15′ N, 2° 17′ E, approximately 40 km southwest of the capital city Niamey. The long-term average annual rainfall at this site is 550 mm. The mean monthly temperature varies between 25 and 41°C (Sivakumar, 1993). The soils are classified as Psammentic Paleustalf (West et al., 1984). The experiment started in 2002 on a soil that was left fallow for 10 years, but the evaluation of the yield of the annual crops started in the rainy season in 2004 when the perennials were pruned for the first time at the end of the dry season.

Experimental Layout

The experiment tested the following treatments: land management – DEF vs. no-DEF (also called traditional system or control); cropping system – rotation vs. continuous cropping; mineral fertilizer management – +NPK vs. –NPK; organic amendment management – +mulch vs. –mulch. One hectare DEF was surrounded by a live hedge made out of Australian acacias planted at 2-m intervals. Two rows of *Acacia colei* were planted down the slope from the hedge at a spacing of 10 × 5 m. These two rows were followed by a row of the domesticated *Ziziphus mauritiana* called Pomme du Sahel (PdS) planted at 10 × 5 m spacing. Pomme du Sahel was planted inside 3 × 3 m micro-catchments (demi-lunes) joined by earth bands that diverted the runoff water into the demi-lunes. This configuration of two *Acacia* rows followed by a PdS row was repeated three times in the 1 hectare (100 × 100 m) field. Each of these bands was planted with annual crops. The annual crops were pearl millet – variety Composite Inter-Varietal de Tarna (CIVT); cowpea (*Vigna ungiculata*) (L.) Walp – variety Ecoh-Mali; roselle (*Hibiscus sabdariffa*) – variety Wankoy.

In the no-DEF, three bands were also considered but without tree components and rotation. In both DEF and no-DEF, treatments were laid in a randomized complete block design (RCBD) in four replications. DEF and no-DEF treatments were laid in two separate but adjacent fields. The fertilizer NPK was applied at the rate of 100 kg ha^{-1}, the recommended rate of the Sahel. The plot size was 24 × 5.75 m.

The time table of the cultural practices for both annual and perennial crops are given in Table 1. The trees are pruned every year almost at the same time. The annual crops were planted with the first significant rain. They were kept free of weeds with a traditional hoe called hilaire and harvested at maturity.

Data Collection and Analysis

Characterization of Soil Chemical Properties

Initial soil sampling was done in the field prior to installing the experiment using a stratified random sampling method. For chemical analyses (C, N, P, pH, total N), three cores were sampled, 0–15, 15–30 and 30–60 cm. The present chapter will present the first core, where most of the change in soil chemical properties is expected (Geiger et al.,

Table 1 Activities in the DEF year around

			Time period											
			January	February	March	April	May	June	July	August	September	October	November	December
Annual crops		Planting												
		Care annuals												
	Harvest and processing	Pearl millet												
		Cowpea												
		Hibiscus sabdariffa (roselle)												
Perennials	Acacia sp.	Planting												
		Harvest of seed												
		Prunning												
	Ziziphus	Planting												
		Fertilization (cattle manure)												
		Phytosanitary protection												
		Harvest												
Soil fertility management		Mulch application												
Water harvesting technology		Confect/repare of half-moons												

1992). Soil organic carbon content was determined by the Walkley-Black (1934) method; available P was determined using the Bray P1 method (Bray and Kurtz, 1945); total nitrogen was determined using the Kjeldahl method (Bremner and Mulvaney, 1982); and soil pH was determined in a 1:2.5 suspension with water. At the end of 2004, a second set of soil samples was collected to study the effect of the treatments on the soil characteristics. These samples were analysed for N, P and pH. To study the changes in the levels of these elements over the first 3 years of experimentation, samples were also collected in 2006, but are not reported in the present chapter because these are not yet analysed.

Yield Data

To study the performance of the various annual crops, yield data were collected at harvesting in a plot of 138 m^2. The straw and millet heads were sun-dried and weighed to determine dry weight. The heads of millet were threshed to determine the grain weight per plot which was extrapolated to kg ha^{-1}. The pods of cowpea were collected three times during the season, sun-dried, weighed on plot basis and threshed. The haulm was harvested at the end and weighed. The final yield was determined in the same way as for millet. The calices of *Hibiscus* are used to produce drinks and the grains are used by women to make local ingredients called "sumballa" for sauce. In addition to these products, the fallen leaves and the stalk serve as organic amendments to replenish soil fertility. Calices yield was determined in the following way: the whole fruit was collected to a working area. The calices were separated from the rest of the fruit and both of them were sun-dried. Calices dry weight was recorded on a plot basis and extrapolated to yields per ha. Thereafter, the fruits were threshed to get the grains, whose weight was obtained on a plot basis and then extrapolated to ha. This chapter will focus on grain yield of millet, total biomass of cowpea and calices yield of *Hibiscus* over 3 years. The other data will be reported in another paper.

The data were analysed for Analysis of Variance (ANOVA) using the Genstat v9 statistical package. In the process of statistical analysis, the DEF and no-DEF treatments were considered as nested blocks and there was combined analysis for all the 3 years.

Analysis of Contribution by Trees

On the perennials, firewood and total mulch production data were collected on the *Acacia* trees after pruning. Pomme du Sahel (improved *Ziziphus*) fruits were harvested in December–January. Only mean values of the data of the perennials are presented in the present chapter. Investment, labour and production costs were recorded and served as a basis of economic analysis of the system. Data on soil erosion as well as water use efficiency will be a subject of another paper.

Economic Assessment

An enterprise budget combined with an investment analysis was used to analyse the economic parameters of the DEF. Potential yields and biomass production of the trees had to be estimated since they have never been planted under existing climatic conditions. The lifetime of the PdS is the basis for the economic calculation period, assumed 50 years. Revenues were generated from the annuals (millet and cowpea grains and stover or haulm) and the trees (mainly PdS (improved *Ziziphus*) fruits and *Acacia* firewood). Expenditures included annual costs and fixed costs. Annual costs include seed and seedling costs, and organic and inorganic fertilizers, pesticides and labour costs. Labour costs included all activities from land preparation to harvesting. Fixed costs are mainly depreciation on assets such as PdS and *Acacia colei* trees and some agricultural equipment. The economical analysis of the DEF was compared to a traditional millet–cowpea production system under average conditions in Niger based on expert opinions. Crop prices were those collected by the Systèmes d'Information sur les Marchés Agricoles (SIMA) of Niger. Input prices were gathered from the Central d'Approvisionnement (CA) of Niamey. Wage was proxied by the opportunity cost of labour in rural areas. Data on fruit trees and wood were estimated and weighted using comparable products in Niamey's market. It was assumed that 100% of production was sold.

Results and Discussion

Rainfall Distribution

The cumulative rainfall was similar for the 3 years with a total of 562 mm in 2004, 527 mm in 2005 and 545 in 2006. These values are comparable with the long-term average of 550 mm for Sadore. However, rain distribution varied widely between years. In 2004, rain started in April and continued until the end of June with short dry spells of about 10 days. Another short dry spell of 1 week occurred at the end of August. In 2005, the rain started at the end of May with a long dry spell of 3 weeks from 9 to 29 June. Another dry spell of 10 days occurred in the third week of July. In 2006, the rain started in mid-June with a cumulative rain of 18 mm from 8 June to 6 July which delayed planting during this year. Since millet and cowpeas are short-day species, any delay in planting (as was the case in 2006) negatively affected yields. The effect of water shortage on millet yield was recorded by Kanitkar (1944). Philips and Norman (1967) reported that millet grain yield is adversely affected if rain ceases during the reproductive stage even though the total amount of water received during the cropping season may be adequate (Jensen et al., 1990).

Soil Characteristics

As reported earlier, the experiment was conducted in a field that was under fallow for 10 years. Prior to planting, the P-Bray N° 1 in the top 15 cm soil layer was 20 mg/kg of dry soil, total N was 236 mg/kg and organic carbon was 2.9% (Table 2). This fertility level was better than the level typical to the region (P-Bray N° 1 = 2.1 mg/kg, Org-C 1.7%) reported in Sinaj et al. (2001). From the analysis of the samples collected in October 2004 after 3 years of cropping, Bray N° 1 P as well as total N level decreased in both the DEF and the traditional system. However, the decrease was more pronounced in the control (Table 3). Thus the DEF system maintains a higher level of soil fertility than the traditional system.

Effect of Dryland Eco-Farm on Yield of Annual Crops

Effect on Millet Grain Yield

Average millet grain yield over 3 years in the DEF was about 1,000 kg ha^{-1} (Fig. 1a), which is three times the average grain yield in Niger (Bationo et al., 2003). Grain yields in the DEF were 2–3 times higher than in the traditional system independent of the year and the application of mulch or NPK (Table 4). The application of *Acacia* mulch did not affect grain yield production. This could be due to slow decomposition of the applied mulch. In a separate study (unpublished) at Sadore research station, it was observed that after 1 year, only 50% of the *Acacia* mulch was decomposed even in the presence of termites. Application of NPK in the DEF increased grain yield slightly but this was not statistically significant. In most of the years, grain yield in the DEF without NPK was similar or higher than that in the traditional system with NPK application, which leads to a preliminary conclusion that DEF without fertilizers can produce as much yield as that obtained with the recommended rate of NPK application for the region.

Effect on Cowpea Total Biomass Yield

Cowpea fodder is an important product both as a livestock feed and for income generation. Over the 3 years of experimentation, average cowpea total biomass yield of 2,000 kg ha^{-1} was produced in the DEF which is far above the yield in farmers' field

Table 2 Chemical status of the experimental soil before layout in 2002

Depth (cm)	pH (H_2O)	Bray P1 (mg-P/kg)	Total N (mg-N/kg)	Organic C (%)	Sand (%)	Clay (%)
15	5.2	20.2	236	0.29	93.9	6.1
30	5.0	9.9	126	0.15	93.2	6.9
60	4.9	2.6	87	0.10	92.5	7.6

Table 3 Chemical status of the experimental soil after three seasons in 2004

Depth (cm)	pH	Bray P1	Total N
The dryland eco-farm			
15	5.1	18.7	175
30	4.8	12.5	93
60	4.7	4.7	67
Traditional system (no-DEF)			
15	5.3	10.2	122
30	4.9	6.3	81
60	4.7	3.9	69

(Fig. 1b). Biomass yield in all years in the DEF was 2–3 times higher than in the control, regardless of the year (Table 4).

On average, mulch application did not affect total biomass production. NPK application increased total biomass production, particularly in the control. In the absence of NPK in the DEF as well as in the traditional system, mulch application depressed biomass production whereas with the application of NPK, a 25% yield increase was obtained which was statistically significant.

Effect on *Hibiscus* Calices Yield

Average calices yield of 250 kg ha^{-1} was produced in the DEF, which is far beyond 110 kg ha^{-1} reported by Robert S. McCaleb in *Market Survey: Hibiscus sabdariffa*; http://www.herbs.org/africa/hibiscus.html.

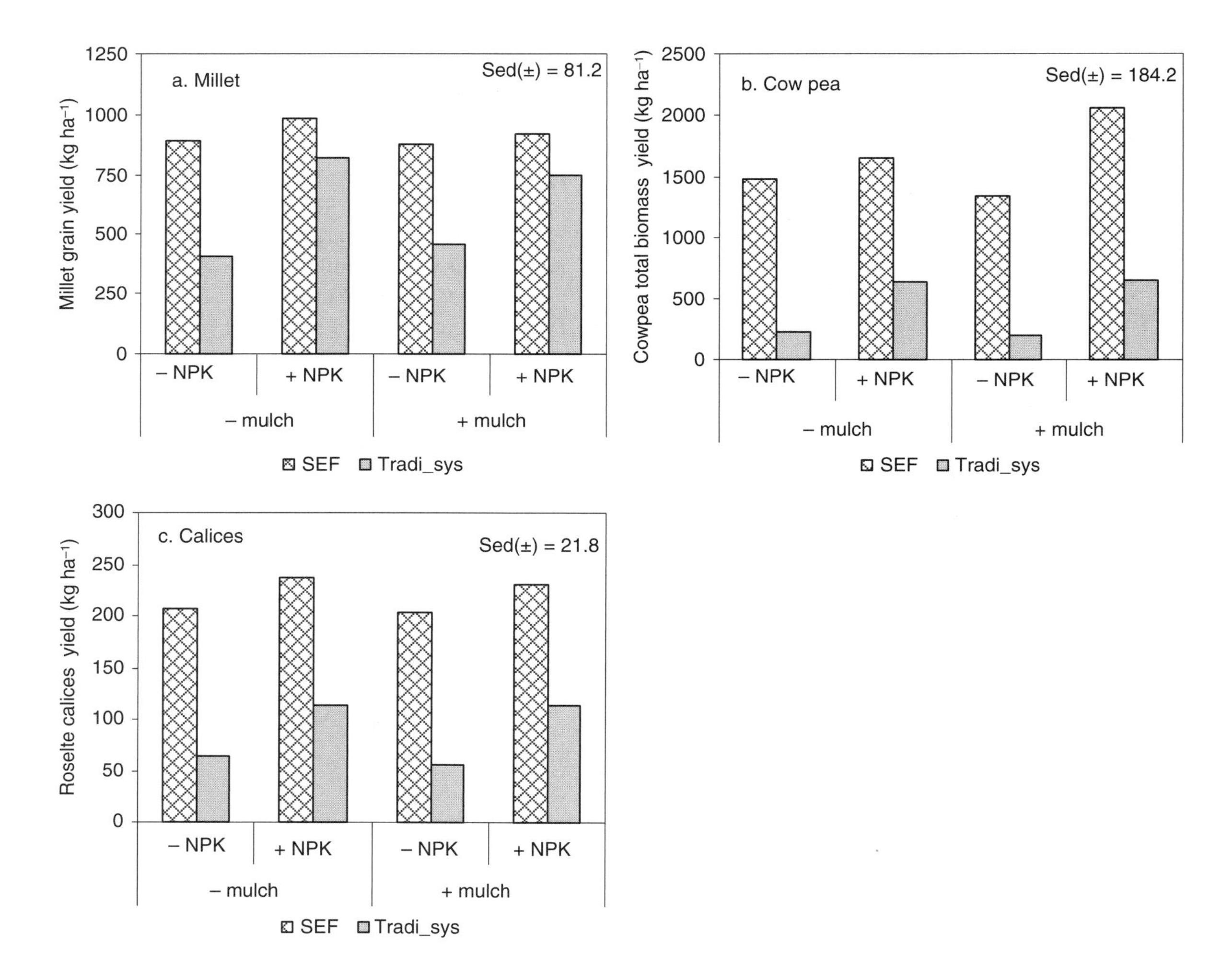

Fig. 1 Effect of the dryland eco-farm system on millet grain, cowpea total biomass and roselle calices yield; mean over 3 years 2004–2006; Sadore ICRISAT Research Station. Sed is standard error of difference between means. Tradi_sys = traditional system

Table 4 Ratio of DEF to traditional system in terms of annual crops for 3 years of cropping

	–Mulch		+Mulch	
Year	–NPK	+NPK	–NPK	+NPK
Millet grain yield				
2004	2.4	1.5	1.9	1.4
2005	2.5	1.1	1.7	1.0
2006	4.1	1.3	3.3	2.6
Cowpea total biomass yield				
2004	6.5	2.0	6.4	2.2
2005	10.0	4.3	10.4	6.5
2006	4.1	1.5	4.4	2.2
Hibiscus calices yield				
2004	2.9	2.4	3.2	2.5
2005	6.4	1.9	2.6	1.5
2006	16.0	14.3	13.6	5.3

This yield level was significantly higher than the control (50–100 kg ha^{-1}) (Fig. 1c). Mulch and NPK application did not significantly affect calices and grain production in the DEF but application of NPK significantly increased yields as compared to the control. Calices yield was 2–16 times higher in the DEF compared to the traditional system, regardless of mulch or NPK application (Table 4).

Effect of Dryland Eco-Farm on Yield of Trees (*Acacia* spp.)

The purpose of planting the *Acacia* trees was to produce firewood for energy and mulch to improve soil fertility, protect the soil against erosion and produce seed which can be used as source of protein for poultry. In the period 2004–2006, firewood production from *Acacia* trees was 2 t ha^{-1} and mulch production was about 2.5 t ha^{-1} (Table 5).

Table 5 *Acacia* firewood and *Ziziphus* fruit production in the DEF

Acacia sp.		
Year	2004	2006
Firewood	1,787	2,137
Mulch	2,575	2,693
Improved *Ziziphus*		
Year	2005	2006
Fruit	241	417

Improved *Ziziphus mauritiana* (PdS)

Pomme du Sahel is the second perennial component of the system providing both food and income. In 2006, the trees were 4 years old and produced 417 kg of fruit ha^{-1} (Table 5). Fruit yields are likely to double over the coming years since the trees had not reached full maturity.

Economical Analysis

The results of the investment analysis of millet and cowpea grown in the DEF are presented in Table 6. The setup costs of US $248 are mainly comprised of the trees, tools and labour. Fruit of the full-grown Pomme du Sahel brings the most income (average 228 US $/year) followed by the pearl millet grains. Annual costs are around 240 US $/year including labour costs at 2 US $/day (Table 6). The results indicate that with a loan with less than 41% as the interest rate, it is advisable to invest in the DEF with millet and cowpea. Credit schemes can help build confidence in the money market by making investing in the DEF less risky for the farmers.

The investment in the traditional system consists mainly of some tools and is so low (US $14) that it is not possible to compare the Internal Rate of Return (IRR) and Net Present Value (NPV) with the investment needed for 1 hectare of DEF. The benefit–cost ratios and return to labour and land can be compared. From Table 7 it can be seen that the benefit–cost ratio is not much higher for the DEF; however, return to labour and capital is much higher than in the traditional system.

Table 6 Set-up costs and Net Present Value (NPV) with different discount rates for millet and cowpea from 1 hectare of DEF

		DEF millet + cowpea
Set-up costs	(US $/ha)	$248
IRR (50 years)		41%
NPV at discount rate (50 years)	10%	$1,673
	20%	$462
	30%	$133
	40%	$5
	50%	–$54

Table 7 Comparing growing millet and cowpea traditionally and in the DEF

	Traditional	DEF
Benefit/cost ratio	1.71	1.93
Return to labour per year	$1.63	$3.35
Return to capital per year	$94	$275

Traditional = no-DEF

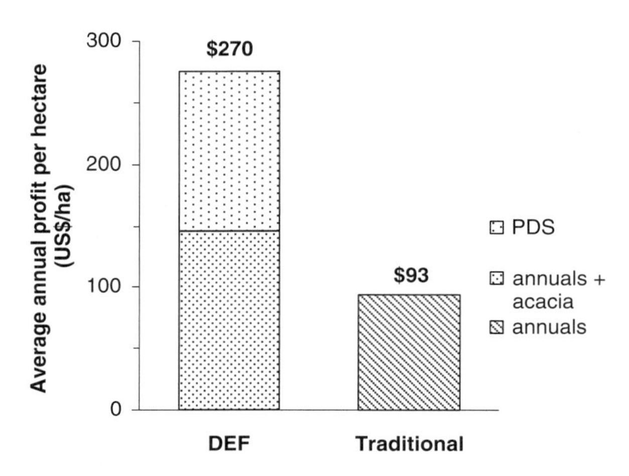

Fig. 2 Average annual profit per ha, contributions from PDS and annuals and acacias separated in the SEF. Traditional = no-DEF

In Fig. 2, the contributions to the average annual profit of the PdS are separated from the acacias and annuals. It can be seen that the incorporation of the PdS in the production system significantly increases the total profit. Besides that, higher yields of the annuals in the DEF translate directly into higher profit from cultivating millet and cowpea.

Conclusions

The Dryland Eco-Farm is an integrated trees–crops–livestock system under development. So far the performance of this system is outstanding in comparison with the traditional rain-fed production system. The incorporation of the Australian acacias results in improved soil fertility and in a corresponding increase in yield. The incorporation of a drought-tolerant fruit tree species significantly increases the profitability of the system. A system that is based on trees needs to be tested over a period of at least 10 years before concrete conclusions can be drawn. It should also be tested in farmer's fields to identify constraints for adoption. Therefore there is a need for further study and development of the system before it can be recommended for mass dissemination. All criteria used to evaluate the economic viability of the DEF in the Sudano-Sahelian zone suggest that it is worthwhile investment as an alternative to the traditional millet cowpea production system. This can be mainly attributed to the consistent fruit and firewood production of the PdS, the most dominant income-generating element of the DEF. From the analysis it was clear that accesses to credit for investment and markets that can absorb the product are very important provisions for viable investments. Institutional financing with fair interest rates and long repayment periods can play a major role in facilitating private farm investments in more intensive rain-fed production systems in the region. Social benefits from decreased erosion, water harvesting, forage production or less wood logging were not taken into account in the economic analysis.

Acknowledgements The research work on the DEF was supported by the Finnish Ministry of Foreign Affairs through its contribution to IPALAC (the International Program for Arid Zone Crops) and by USAID-West Africa. We wish to thank our technical staff Mr. Moustapha Amadou and Mr. Saidou Abdoussalam for their dedicated and professional work.

References

Bationo A, Mokwunye U, Vlek PLG, Koala S, Shapiro BI (2003) Soil fertility management for sustainable land use in the West African Sudano-Sahelian zone. In: M.P. Gichuru, A. Bationo, M.A. Bekunda, H.C. Goma, P.L. Mafongoya, D.N. Mugendi, H.K. Murwira, S.M. Nandwa, P. Nyathi and M.J. Swift (eds.) Soil fertility management in Africa: a regional perspective. Academy Science Publisher & Tropical Soil Biology and Fertility, Nairobi, Kenya, pp 253–292

Bray RH, Kurtz LT (1945) Determination of total, organic, and available forms of phosphorus in soils. Soil Sci 59:39–45

Bremner JM, Mulvaney CS (1982) Nitrogen–total. In: Page AL (ed) Methods of soil analysis. Part 2, 2nd edn. Agron, Monogr. 9. ASA and SSSA, Madison, WI, pp 595–624

Geiger SC, Manu A, Bationo A (1992) Changes in a sandy Sahelian soil following crop residue and fertilizer additions. Soil Sci Soc Am J 56:172–177

Jensen ME, Burman RD, Allen RG (eds) (1990) Evapotranspiration and irrigation water requirement. ASCE manual and reports on engineering practices No. 70. ASCE, New York, NY, 332pp

Kanitkar MV (1944) Dry farming in India. ICAR Sci. Monograph no. 15. New Delhi

Mando A (1997) Effect of termites and mulch on the physical rehabilitation of structurally crusted soils in the Sahel. Land Degrad Devel 8:269–278

McCaleb RS Roselle production manual (*Hibiscus sabdariffa*). In: Market survey: *Hibiscus sabdariffa*; agribusness in sustainable natural African plant productions. http://www.herbs.org/africa/hibiscus.html

Parr JF, Stewart BA, Hornick SB, Singh RP (1990) Improving the sustainability of dryland farming systems: a global perspective. In: Singh RP, Parr JF, Stewart BA (eds) Advances in soil science, vol 13. Dryland agriculture strategies for sustainability. Springer-Verlag, New York, NY, pp 1–8

Phillips LJ, Norman MJT (1967) A comparison of two varieties of bulrush millet (*Pennisetum typhoides* S. & H.) at Katherine. NT Melbourne, CSIRO Aust Div Land Res Tech Memo No 67/18

Sinaj S, Buerkert A, El-Hadjj G, Bationo A, Traore H, Frossard E (2001) Effect of fertility management strategies on phosphorus bioavailability in four West African soils. Plant Soil 233:71–83

Sivakumar MVK (1993) Agroclimatology of West Africa: Niger. Information bulletin no. 5. ICRISAT, Patancheru, AP, p 108

Walkley A, Black IA (1934) An examination of Degtjareff method for determining soil organic matter and a proposed modification of the chromic acid titration method. Soil Sci 37:29–37

West LT, Wilding LP, Landeck JK, Calhoun FG (1984) Soil survey of ICRISAT. Sahelian Center, Niger

Zougmoré RB (2003) Integrated water and nutrient management for sorghum production in semi-arid Burkina Faso. Wageningen University and Research Centre, Wageningen

Effect of Zai Soil and Water Conservation Technique on Water Balance and the Fate of Nitrate from Organic Amendments Applied: A Case of Degraded Crusted Soils in Niger

D. Fatondji, C. Martius, P.L.G. Vlek, C.L. Bielders, and A. Bationo

Abstract Experiments were conducted on degraded crusted soils to study water status and nitrogen release in the soil during the dry seasons of 1999 at ICRISAT research station and on-farm during the rainy seasons of 1999 and 2000 in Niger. Zai is a technology applied on degraded crusted soil, which creates conditions for runoff water harvesting in small pits. The harvested water accumulates in the soil and constitutes a reservoir for plants. The organic amendment applied in the Zai pits releases nutrients for the plants. Soil water status was monitored through weekly measurement with neutron probe; access tubes were installed for the purpose. Nutrient leaching was measured as soil samples were collected three times throughout the cropping season. A rapid progress of the wetting front during the cropping period was observed. It was below 125 cm in the Zai-treated plots 26 days after the rain started versus 60 cm in the non-treated plots. Applying cattle manure leads to shallower water profile due to increased biomass production. Total nitrate content increased throughout the profile compared to the initial status, suggesting possible loss below the plant rooting system due to drainage, which was less pronounced when cattle manure was applied. This study shows that the system improves soil water status allowing plants to escape from dry spells. However, at the same time it can lead to loss of nutrients, particularly nitrogen.

Keywords Drainage · Dry spells · Organic amendment · Water harvesting · Wetting front · Zai

D. Fatondji (✉)
International Crops Research Institute for the Semi-Arid Tropics (ICRISAT) Sahelian Center, Niamey, Niger
e-mail: d.fatondji@cgiar.org; d.fatondji@gmail.com

Introduction

Land and soil degradation is one of the major problems facing agricultural production nowadays. Sundquist (2004) reported that desertification along the Sahara desert proceeds at an estimated area of 1,000 km^2/year, which is in line with the findings of the Global Land Assessment of Degradation (GLASOD) (Oldeman et al., 1990), which reported that in Africa, 65% of the cropland is degraded to some extent. In the Sahelian zone, soil fertility restoration through the vegetative fallow system is becoming ineffective due to population pressure, which leads to shorter fallow periods or simply to land abandonment (Amissah-Arthur et al., 2000). Experience revealed that due to the mounting population pressure and the limited availability of fertile land, farmers in the desert margin are forced to rely on marginal or degraded lands for agricultural production.

Zai is one of the several techniques available for the rehabilitation of marginal lands. The Zai is prepared during the dry season as farmers dig small pits in the soil to collect water, wind-driven soil particles and plant debris around the plant. About two handfuls (equivalent to 300 g) of organic amendments such as millet straw, cattle manure or their composted form are added to the pits right after digging (Roose et al., 1993; Ouédraogo and Kaboré, 1996). Nutrient released from amendment added is used by crops sown in the pit. The soil excavated from the pit is put down the slope of the pit to act as water catchment area. Likewise, runoff water is collected in the pit to help the plant escape dry spells that are frequent in the Sahel.

In Burkina Faso, it was found that on the zipele (lateritic soil) it is mostly the hardpan that hinders

A. Bationo et al. (eds.), *Innovations as Key to the Green Revolution in Africa*,
DOI 10.1007/978-90-481-2543-2_115, © Springer Science+Business Media B.V. 2011

water infiltration and also limits crop production. In the Sahel of Niger, in addition to the crust, soil fertility is also a limiting factor (Hassan, 1996; Rockstom et al., 1999; Fatondji et al., 2006). Under both conditions, breaking the crust would increase water infiltration and deep percolation favoured by termite holes. The Zai technique thus combines water harvesting with nutrient management practices (Roose et al., 1992). The main investment required by the technology is manpower for digging the Zai holes, but the work is done during the dry period of the year when the farmers can invest more time to Zai making. According to Fatondji et al. (2006), Zai alleviates the effect of dry spells during plant growth and improves rain use efficiency by a factor of 2 compared to traditional flat planting, effects that are not only due to the water harvesting but also due to the amendments, and can be increased when using high-quality amendments. The use of Zai enables runoff water to be collected in small water pockets. The water accumulates and infiltrates in the soil profile and constitutes a reservoir for the crop. The crop planted in the Zai uses the nutrients released from the organic manure applied, but the nutrients can also be leached into deeper soil layers. However, no attempts have been made to study the pattern of water movement in the profile in Zai-treated plots and also to estimate the potential nutrient losses that can occur under these conditions. Therefore, experiments were carried out in the Sahelian zone of Niger, on-station under controlled water supply in 1999 and on-farm at Damari during the rainy seasons of 1999 and 2000 to address this problem. The objective of the on-station experiment was to determine the optimum application rate of organic amendments for pearl millet (*Pennisetum glaucum*) production as a function of the type of amendment. In the on-farm experiment, we studied resource use efficiency of millet under rainfed conditions in the Zai as compared to planting on flat soil. In the present chapter, the emphasis will be on water status of the soil throughout the cropping period and the effect on possible nutrient losses through the study of nitrate content in the soil profile at different sampling dates.

Materials and Methods

Site Description

The on-station experiment was conducted under controlled water supply at the ICRISAT research station at Sadoré (13° 15′ N, 2° 17′ E) in Niger from March to May 1999. Long-term average annual rainfall at this site is 550 mm, which falls between June and September. Monthly temperature varies between 25 and 41°C. The soils are classified as Psammentic Paleustalf (West et al., 1984). It is acidic with relatively high Al saturation and very high sand content (Table 1). The experiment was conducted on a field that had been subjected to severe wind and water erosion for a period of 4 years and that had developed extensive erosion crusts (Casenave and Valentin, 1989).

The on-farm trial was conducted during the rainy seasons of 1999 and 2000 at Damari (13°12′N and 2°14′E). Long-term average annual rainfall and monthly temperature amplitudes at Damari are similar to those at the ICRISAT research station. The soil at Damari is classified as Kanhaplic Haplustult (Soil Survey Staff, 1998). It is acidic, with 84% sand content

Table 1 Selected initial soil properties of the experimental fields at Sadoré, Damari and Kakassi (0–20 cm soil depth)

Soil characteristics	Sadoré	Damari	Kakassi
pH (H_2O)	4.5	4.2	6.4
pH (KCl)	3.9	3.9	5.4
Exchangeable base (cmol/kg)	0.4	1.7	7.9
Exchangeable acidity (cmol/kg)	0.7	1.1	0.04
ECEC[a] (cmol/kg)	1.0	2.8	7.9
Al saturation (%)	47	29	0
Base saturation (%)	37	61	99
P-Bray 1 (mg/kg)	2.3	2	0.8
C org (%)	0.1	0.2	0.2
Total N (mg/kg)	120	116	169
Bulk density (kg/m)	1.5	1.6	1.8
Sand (%)	92	84	69
Silt (%)	3	3	6
Clay (%)	5	13	25

Adapted from Fatondji et al. (2006)
[a]Effective cation exchange capacity

and relatively low effective cation exchange capacity (ECEC). The vegetation was an open bush with scattered trees. The selected field had been left fallow for 3 years prior to the experiment. In addition to small patches of loose sand deposits, which were cropped by the farmer, the field contained large patches of bare crusted soil, which were selected for installing the experimental plots.

Experimental Layout

On-station (at Sadoré)

Effects of amendment type (millet straw and cattle manure) and rate of application (1, 3, and 5 t/ha) on dry matter production of millet (*P. glaucum* L. R. Br) were evaluated in Zai pits under controlled irrigation. The field was sprinkler irrigated uniformly throughout the growing period at a weekly rate of 20 mm, with a total of 220 mm of water applied to harvest. The experimental design was a randomized complete block design (RCBD) with four replications. The control treatment was a non-amended pit. A local millet variety "Sadoré local" (120 days growing cycle) was sown on 17 March and the stover harvested on 25 May before grain production not only to avoid interference of rain with the treatments but also due to the photosensitivity of the crop. Zai holes were dug in all the plots.

On-farm

Effect of planting technique (planting on flat versus planting in Zai pits) and amendment type (millet straw and cattle manure) on millet yield was studied for over 2 years. In both years, the experimental design was an RCBD with four replications. The control plots received no organic amendment. The millet variety "Sadoré local" was sown on 29 June in 1999 and 26 June in 2000 and harvested at maturity. On-station as well as on-farm planting density was 10,000 pockets/ha and the crop thinned to three plants per pocket approximately 3 weeks after planting.

In both years, rain started towards the end of June (Fig. 1a and b). Cumulative rainfall was 499 mm in 1999 and 425 mm in 2000, which was below the long-term average of 550 mm. The same field was used in both years; therefore the pits dug in 1999 were used for 2000 but renewed.

Data Collection

Soil moisture profiles were measured weekly at 15-cm intervals down to 240 cm depth using a Didcot neutron probe (Didcot Instrument Company Limited, Wallingford, UK) starting from the day of planting. The first measurement was done before the first irrigation or rainfall. For that purpose, two 48-mm inner diameter aluminium access tubes were installed in each plot, one tube between the pockets and the other on the pocket close to the plant.

The depth of the shallowest tube was restricted to 45 cm due to the presence of a lateritic layer, while the deepest reached 300 cm. The probe had been calibrated in situ for the soils of the experimental sites applying the gravimetric method. Data of the tubes installed between the pockets are reported in this chapter.

From the neutron probe data, the volumetric soil water content was calculated. It is expressed here as

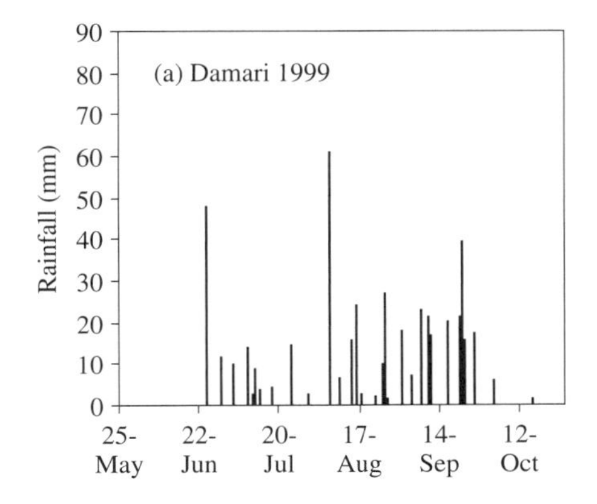

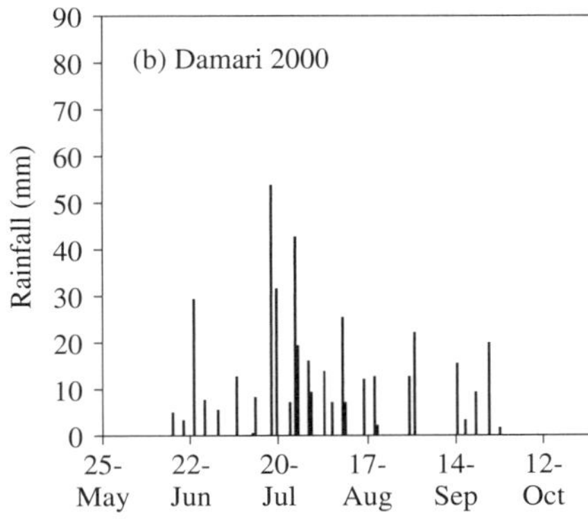

Fig. 1 Rainfall pattern at the experimental site in two seasons of 1999 and 2000

dimensionless ratio (cm^3/cm^3). In the same experiment, amendment decomposition was studied with litterbags and related samples collected three times during the cropping period. Along with these samples, soil samples were collected to estimate nitrate content. For each sampling, the maximum depth was determined in accordance with the progress of the wetting front. The level of the profile wetting front was determined as a function of soil water content in a given layer at a given time of measurement compared to its level before the rain or irrigation started. The samples were collected in three replications out of four. Prior to installation of the experiment in 1999, on-station soil samples were collected to evaluate the initial status of the soil with regard to nitrate content. Sample collection was done with an aluminium tube of 7.5 cm diameter. They were collected in/on the hole or the pocket after the plants had been removed. From 0 to 20 cm, samples were taken at 10-cm interval. Below 20 cm, samples were taken at 20 cm increment. The samples were kept in sealed bags and stored in a freezer until use.

Nitrate content was determined semi-quantitatively using nitrate test strips (Reflectoquant) and an RQflex reflectometer as described by Merck KgaA (64271 Darmstadt, Germany). KCl solution (50 ml) was added to 70 g of soil sample and the mixture shaken for 10 min. A sub-sample of the mixture was put in a test tube and left to elutriate, and then the nitrate content was read. The test strips have two reaction zones, which turn red-violet on contact with solutions containing nitrate, and the colour intensity depends on the nitrate concentration. To convert the reading into nitrate N, the values obtained must be multiplied by 0.226. At harvest, crop yield data were collected. Total dry matter for both on-farm and on-station experiments as well as seed dry weight and harvest index (on-farm only) was estimated.

Results and Discussion

Rainfall Pattern in Both Rainy Seasons at the Experimental Site

As mentioned earlier, in both years, rainfalls started on-farm in June. In 1999, the first important rainfall event occurred on 28th June (Fig. 1). The cumulative rainfall to planting was 60 mm. Thirty-four events were recorded out of which nine were above 20 mm, giving 57% of the total rainfall. In 2000, the rain started earlier with smaller rainfall events. The first important rain with which crops were planted occurred on 25 June. The cumulative rainfall to planting was 38 mm. A total of 30 events were recorded out of which seven were above 20 mm, giving 48% of the total rainfall. This year was also characterized by two dry spells of which the first of 9 days occurred during grain filling at the end of August and the second of 14 days at the beginning of September. This shows better rainfall distribution in 1999 than in 2000, but also higher cumulative rainfall in 1999.

On-station Under Controlled Water Supply

Effect of Amendment Type and Rate on Soil Profile Wetting Front

Figure 2 presents the wetting front on the respective dates of soil sampling, which was done purposely to study the relation between the progress of the wetting front and potential nutrient losses.

In all treatments, the wetting front was below or close to 200 cm 36 DAS (days after sowing). Figure 2 indicates a rapid progress favoured by the breakage of the soil crust but also the sandy loose structure of the soil on the station. According to Fatondji et al. (2006), the sand content of the experimental soil was 92% with 5% clay. Under such conditions, Fatondji (2002) observed 70% drainage of 220 mm irrigation applied during 72 days. Under similar conditions, Rockstom et al. (1999) also observed infiltration rates ranging from 15 to 182% of individual rainfalls depending upon the measurement position on a toposequence of 1–3% slope. The wetting front was still close to 200 cm at harvest in all treatments except for manure-treated plots, particularly the largest rate of application (5 t/ha) where the front was around 60 cm deep at harvest. This could be due to increased water consumption under this treatment because of higher biomass production. According to yield data collected in this experiment, 5 t/ha of manure produced the highest yield of 4,500 kg/ha total dry matter (Fatondji et al., 2006).

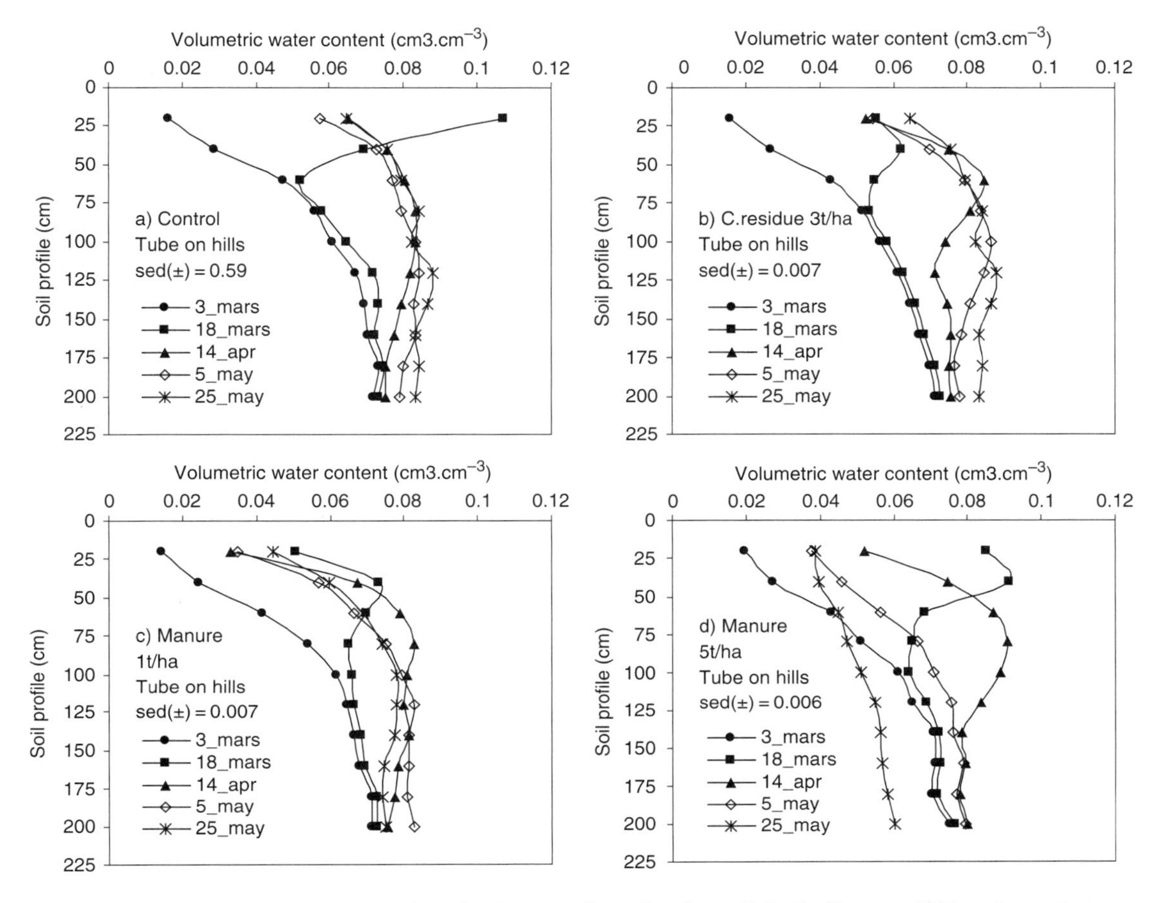

Fig. 2 Effect of amendment type and rate of application on soil wetting front; Sadoré off-season 1999, sed, standard error of difference between means; c.residue, crop residue

On-farm Experiment

Effect of Planting Technique on Soil Profile Wetting Front

In all cases in 1999, the wetting front was already below 200 cm on the day of planting in the Zai-treated plot (Fig. 3a, c and e), which confirms the trend observed on-station under the same conditions, while the wetting front was shallower on the same date in the non-Zai-treated plots (Fig. 3b, d and f).

The results indicate that despite the sandy structure of the experimental soil, breaking the surface crust and digging the pits was highly favourable for water infiltration compared to the flat treatment. Volumetric soil water content (VWC) was still very high at the deeper layer in the Zai than in the flat even towards the end of the season. In 1999, in the Zai control, for instance, at 200 cm depth, VWC was 0.084 cm^3/cm^3 compared to the initial level of 0.060 cm^3/cm^3, while in the flat control, it was 0.059 cm^3/cm^3 compared to the initial level of 0.055 cm^3/cm^3. The same trend was observed in 2000 but it was less pronounced (Fig. 4).

Effect of Amendment Type on Soil Profile Wetting Front

In both years, soil water profile was shallower in the manure-treated plots than the other treatments. Towards the end of the cropping season, in the Zai as well as on flat with cattle manure, soil water content decreased significantly compared to plots treated with millet straw, indicating high water consumption of the

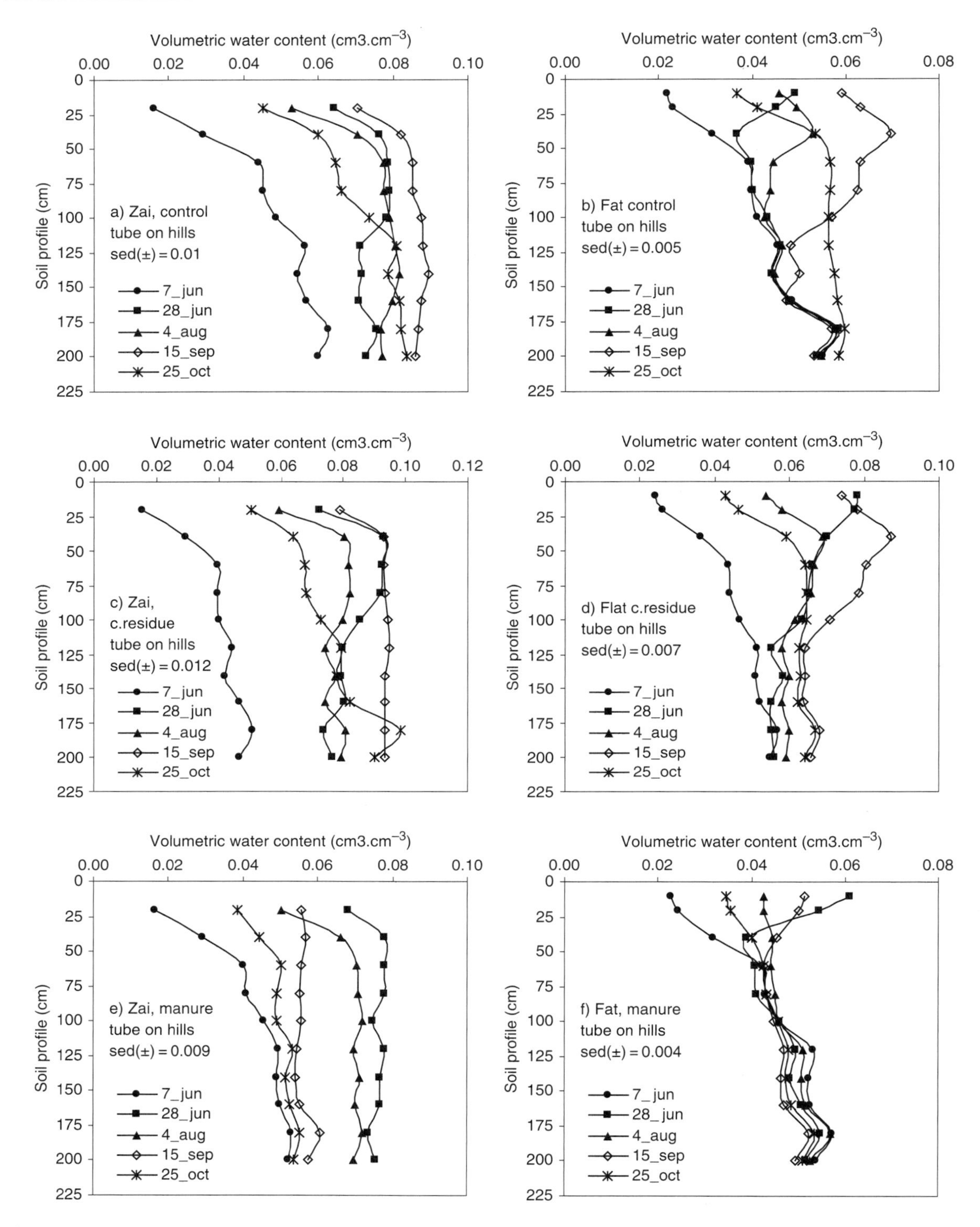

Fig. 3 Effect of planting technique and amendment type on soil water profile; Damari 1999. sed, standard error of difference between means; c.residue, crop residue

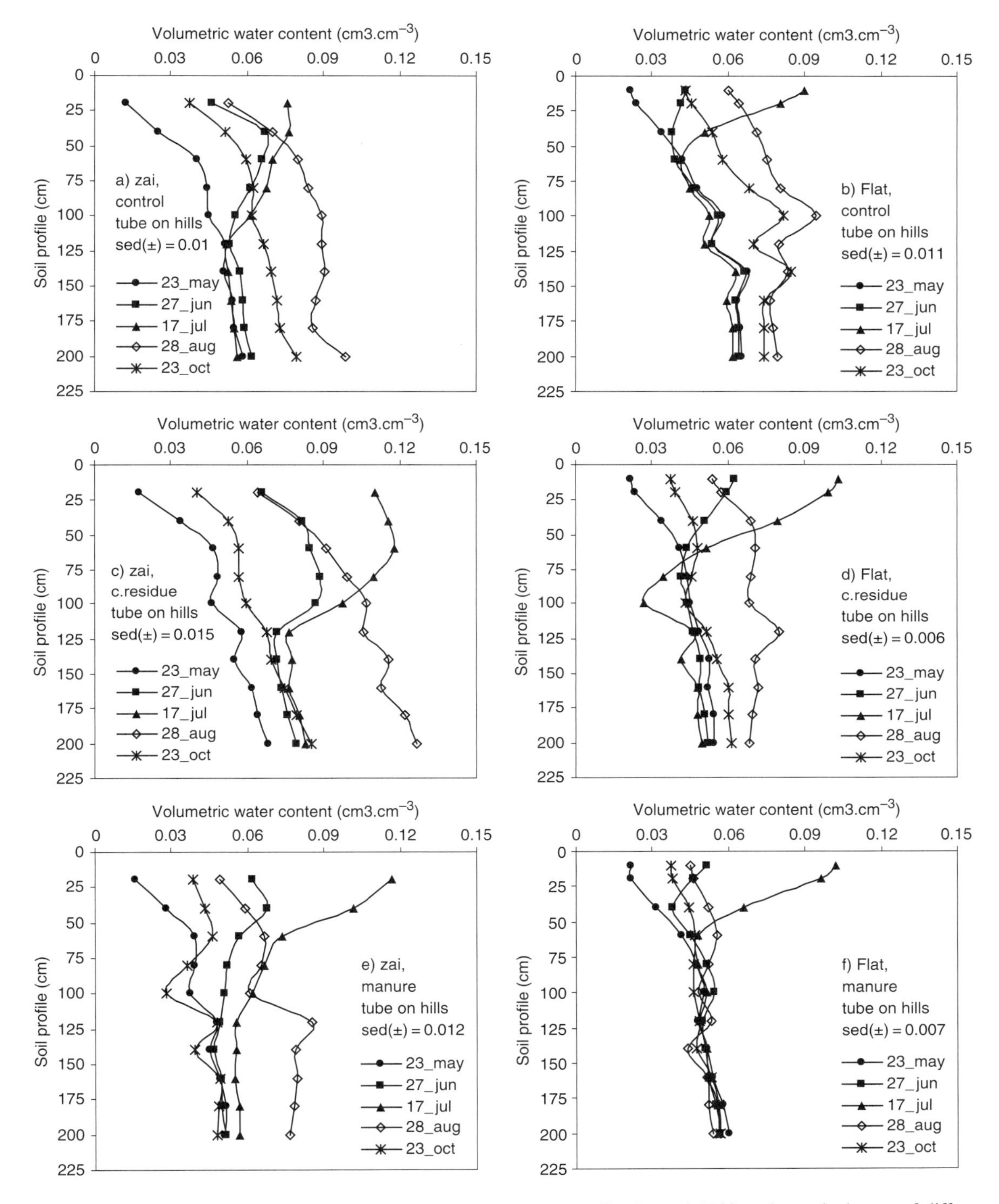

Fig. 4 Effect of planting technique and amendment type on soil water profile; Damari 2000. sed, standard error of difference between means; c.residue, crop residue

crop due to increased biomass production. Fatondji et al. (2006) reported rain water use efficiency of 8 kg/mm in manure-treated plot versus 2 kg/mm for millet straw on average calculated on total dry matter basis under similar conditions, which was accompanied by high yield. Particularly in non-Zai-treated plots amended with cattle manure, the wetting front remained at 60 cm during the whole growing period, which is due to not only the presence of crust that hampers infiltration but also increased crop uptake as reported by Payne et al. (1996) and Zaongo et al. (1997), who found that manure application increases soil water retention and favours root development and water uptake. All these resulted in increased crop yield.

Effect of Amendment Type and Rate of Application on Nitrate Content in the Profile

Compared to the initial nitrate content in the profile, nitrate level increased in all treatments during the growing period but more in plots with crop residue (Fig. 5). In these plots, the level increased at deeper layer, which presumes possible loss beyond the rooting depth. Nevertheless, in all treatments, nitrate content was close to or less than the initial level, particularly in the manure-treated plots, where it is observed that below 60 cm in the profile, nitrate content was lower than the initial level after 67 days of plant growth, indicating not only increased crop uptake following increased biomass production but also possible percolation in deeper layer that was compensated for by nitrate movement for upper layers. The observed trend

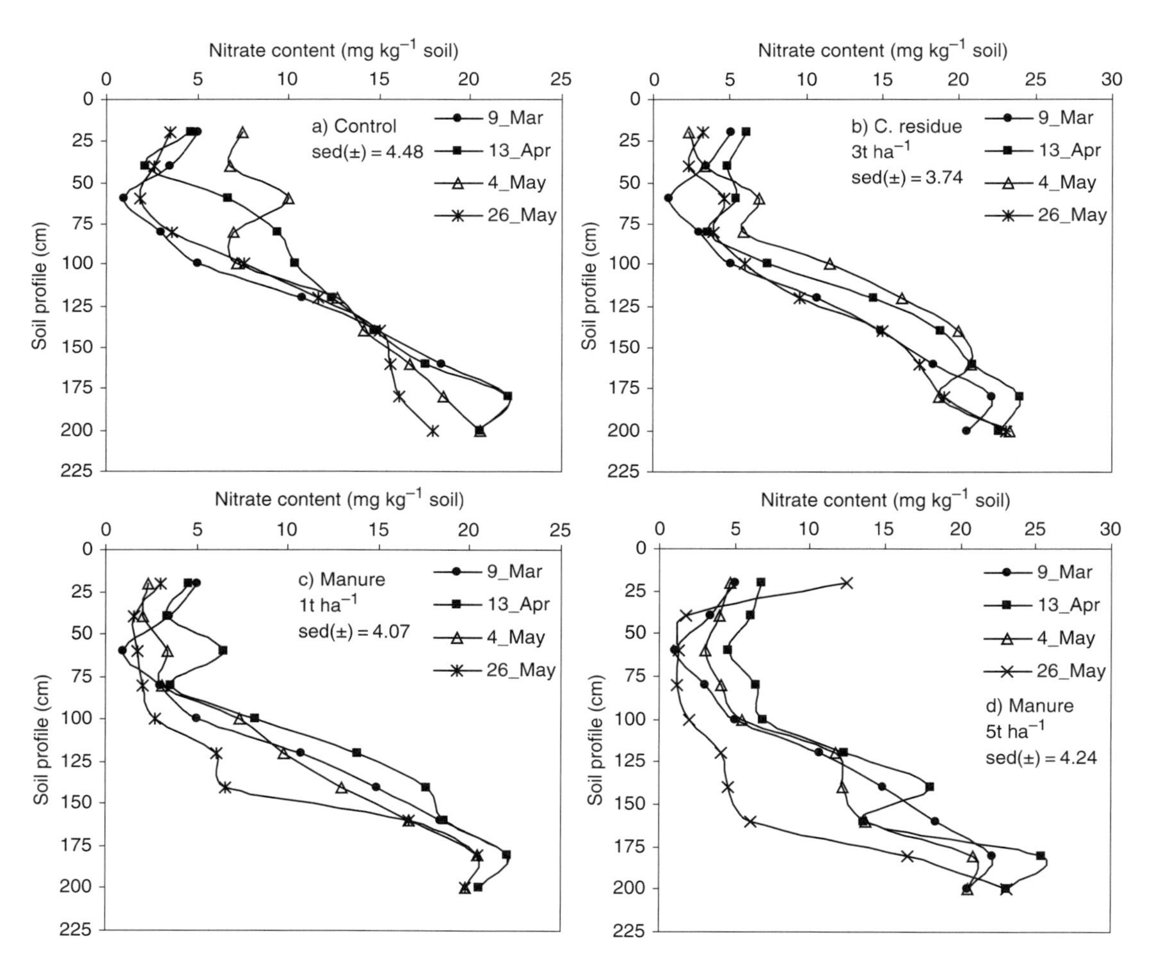

Fig. 5 Nitrate content in the profile as affected by amendment type and rate of application; Sadoré off-season 1999. sed, standard error of difference between means; c.residue, crop residue

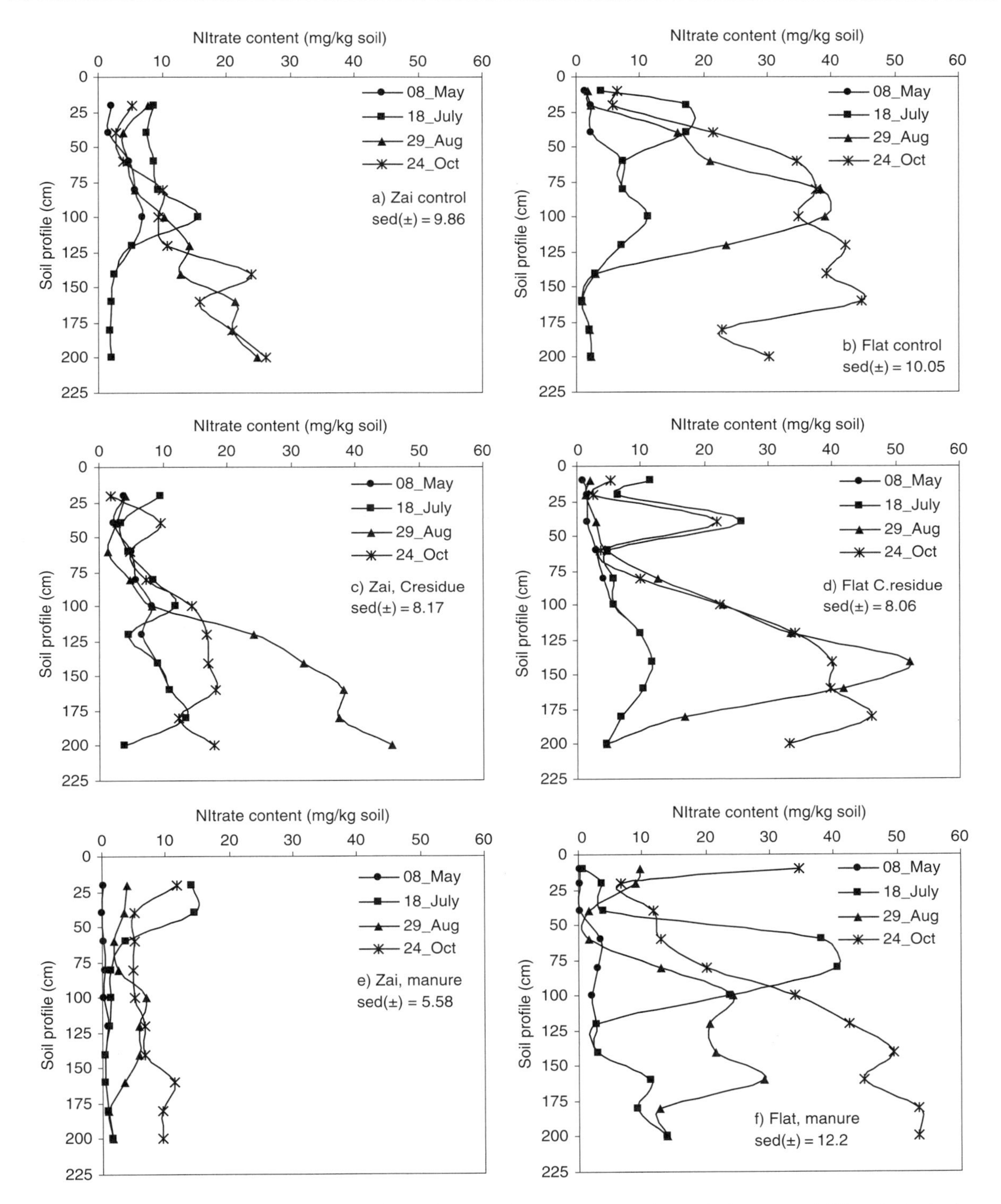

Fig. 6 Profile nitrate content as affected by planting technique and amendment type; Damari rainy season 2000. sed, standard error of difference between means; c.residue, crop residue

confirms the findings of Addiscot et al. (1991), who reported 5 kg/ha nitrogen uptake in a fast-growing crop.

Effect of Planting Technique on Nitrate Content in the Profile

Nitrate content in the soil of the Zai-treated plots was lower than the level observed in flat planted plots. Possible reason for this could be deeper percolation due to water harvested in the pits as opposed to the flat planting conditions. As reported earlier, water infiltration and percolation beyond the rooting zone was very fast in the Zai, whereas we observed that on flat, the wetting front was shallower throughout the growing period. Under these conditions, nitrate losses through drainage have also occurred (Fig. 6b, d and f) but in a limited proportion, which could explain the high concentration of nitrate observed in the measured profile, particularly in deeper layers.

The overall low level of nitrate content in the Zai pit could also be due to better plant development and the resulting increased plant nutrient uptake, particularly in the plots amended with cattle manure. Fatondji et al. (2006) reported 3–4 times grain yield increase and N and P uptake improvement in the range of 43–64 and 50–87%, respectively, due to Zai application in similar conditions.

Effect of Amendment Type on Nitrate Content in the Profile

In Zai-treated plots, nitrate content in the upper layer was slightly higher or similar to the initial level in the control as well as in plots with crop residue. Nitrate content was higher in manure-amended plots in the same layer. In deeper layer, nitrate content increased slightly, particularly under crop residue, compared to the initial level. This was more pronounced 3 months after amendment application, while in the plots with cattle manure it was relatively low throughout the cropping season. This could be due to increased crop uptake.

In non-Zai-treated plots (flat), nitrate content remained high in lower layers at the end of the cropping period compared to the initial level under all amendment management practices.

The trend observed with regard to soil water profile and nitrate content indicates that following crust breaking with the Zai pits, water infiltration increases. Therefore, water percolates to deeper layer in the Zai-treated plots. This water could recharge the water table. To improve soil fertility and plant growth and yield, organic amendment was applied in the Zai, but part of the nutrients released by this amendment may be drained to deeper layer and possibly not available to the crop. Nevertheless, not all but small proportion of this nutrient may be lost as reported by Addiscott (1996), who observed that 6–8% of nitrogen applied at a rate of 190 kg/ha was lost by leaching in winter wheat experiment. Even though this experiment was not conducted under similar condition with the actual study, it does give an indication of the potential loss. In plots amended with cattle manure, the crops developed higher biomass; therefore, water and nitrate losses were limited but not prevented. According to Fatondji et al. (2006), 4,500 kg/ha of dry matter was produced on-station and 5,000 kg/ha was produced on-farm with cattle manure in the Zai in 1999 and 3,000 kg/ha in 2000, which has resulted in higher nutrient uptake, limiting possible loss of nutrient by drainage.

Conclusion

From the above results and discussion, the following conclusions are made. Zai technology offers numerous advantages with regard to rehabilitation of degraded land and also facilitates sustainable cropping on marginal lands. Studies have shown that the most important benefits lie in the water-harvesting feature and also the concentration of nutrients in the rooting zone of crops (application of manure in Zai pit), which may favour crop root development and hence yield increase. Results further indicated that surface crust breakage improves water harvesting in the planting pits which later percolates to deeper layer together with part of the nutrient applied, particularly the mobile forms of nitrogen like nitrate, which when in excess can become a pollutant to the underground water. It is also concluded that plant uptake due to increased vegetative mass can limit this percolation. The results of the study enable to make a recommendation that amendments applied should be of good quality to draw the best from the technology and reduce nutrient losses due to improved plant water uptake resulting from increased vegetative growth. Under experimental conditions, water harvesting through the use of Zai technology is necessary to provide and store water in the soil for crop use, as it was shown that in plots amended with manure and not treated with Zai, the wetting front was limited to 60 cm. This would avoid shortage of water for the plants due to dry spells that are common during the growing season in the Sahel.

References

Addiscot TM (1996) Fertilizer and nitrate leaching. In: Hester RE, Harruson RM (eds) Agricultural chemicals and the environment. The Royal Society of Chemistry, Cambridge, pp 1–26

Addiscot TM, Whitmore AP, Powlson DS (1991) Farming, fertilizers and the nitrate problem. CABI Publ., Wallingford, UK

Amissah-Arthur A, Mougenot B, Loireau M (2000) Assessing farmland dynamics and land degradation on Sahelian landscapes using remotely sensed and socioeconomic data. Int J Geogr Inf Sci 14(6):583–599

Casenave A, Valentin C (1989) Les e ́tats de surface de la zone sahelienne; influence sur l'infiltration. In: les processus et les facteurs de reorganisation superficielle (ed) ORSTOM. institut Franc ̧ais de recherche scientifique pour le developpement en coope ́ration, collection didactiques, Paris, pp 65–190

Fatondji D (2002) Organic amendment decomposition, nutrient release and nutrient uptake by millet (*Pennisetum glaucum*) in a traditional land rehabilitation technique (zaï) in the Sahel. PhD thesis, Ecological and development series No 1. Center for Development Research, University of Bonn, Cuvillier Verlag, Gottingen, Germany

Fatondji D, Martius C, Bielders C, Vlek PLG, Bationo A, Gérard B (2006) Effect of planting technique and amendment type on pearl millet yield, nutrient uptake, and water use on degraded land in Niger. Nutr Cycling Agroecosyst 76:203–217

Hassan A (1996) Sustaining the soil. Indigenous soil and water conservation in Africa. In: Reij C, Scoones I, Toulmin C (eds) Improved traditional planting pits in the Tahoua department, Niger. An example of rapid adoption by farmers. Earthscan Publisher, pp 56–61

Oldeman LR, Hakkeling RTA, Sombroek WG (1990) World map of the status of human-induced soil degradation: an explanatory note, revised ed. International Soil Reference and Information Centre, Wageningen, United Nation Environment Program, Nairobi

Ouedraogo M, Kaboré V (1996) The "Zaï": a traditional technique for the rehabilitation of degraded land in the Yatenga, Burkina Faso. In: Reij C, Scoones I, Toulmin C (eds) Sustaining the soil. Indigenous soil and water conservation in Africa. Earthscan Publisher, pp 80–92

Payne WA, Brück H, Sattelmacher B, Shetty SR, Renard C (1996) Root growth and soil water extraction of three pearl millet varieties during different phonological stages. In: Dynamics of roots and nitrogen in cropping systems of the semi-arid tropics: proceedings of an international workshop. ICRISAT Asia Center, India, 21–25 Nov 1994

Rockstrom J, Barron J, Brouwer J, Galle S, De Rouw A (1999) On-farm spatial and temporal variability of soil and water in pearl millet cultivation. Soil Sci Soc Am J 63:1308–1319

Roose E, Dugué P, Rodriguez L (1992) La G.C.E.S. Une nouvelle strategie de lutte anti-erosive appliquee a l'amenagement de terroirs en zone soudano-sahelienne du Burkina Faso. Revue Bois et Forêts des Tropiques 233:49–62

Roose E, Kaboré V, Guenat C (1993) Le"zaï": fonctionnement, limites et amélioration d'une pratique traditionnelle africaine de réhabilitation de la végétation et de la productivité des terres dégradées en région soudano-sahelienne (Burkina Faso). cahier de l'ORSTOM, serie pedologie XXVIII(2): 159–173

Soil Survey Staff (1998) Keys to soil taxonomy, 8th edn. USDA/NRCS, Texas A&M University, USA

Sundquist B (2004) Land area data and aquatic area data; a compilation, 1st edn. Mar 2004. http://home.Alltel.net/bsundquist1/la0.html (checked 5.12.2005)

West LT, Wilding LP, Landeck JK, Galhoun FG (1984) Soil survey of the ICRISAT Sahelian Center, Niger. Soil and Crop Science Department/Tropsoils, Texas A&M System, College Station, Texas in cooperation with the International Crop Research Institute for the Semi-Arid Tropics

Zaongo CGL, Wendt CW, Lascano RJ, Juo ASR (1997) Interaction of water mulch and nitrogen on sorghum in Niger. Plant Soil 197:119–126

Counting Eggs? Smallholder Experiments and Tryouts as Success Indicators of Adoption of Soil Fertility Technologies

M. Misiko and P. Tittonell

Abstract The aim of this chapter is to analyse how successfully smallholders test–apply new soil fertility concepts and to understand the diverse adaptive strategies they rely on. Through in-depth interviews and participant observation in western Kenya, we analysed *success* stories in the form of use-as-you-learn applications, i.e. "experiments", following a participatory research initiative. The nature and prevalence of smallholder use-as-you-learn "experiments", referred to here as *tryouts,* is a useful indicator of future application of research concepts and technologies, whether they can be gainful and sustainable. Smallholder experimentation can work best when integrated into research agendas or in cases when researchers dedicate themselves to participatory research full-time and long enough. This chapter concludes that co-research initiatives are crucial for successful soil fertility research and also shows that the hunt for signs of success of research among smallholders is a crucial beginning point for any scaling-out initiative.

Keyword Co-innovation · Convenience · Soil fertility management

Introduction

Eighty percent of Kenyan people live in the rural areas and engage in subsistence farming (and pastoral activities) as the most available means of survival: fighting poverty and struggling to achieve food security (Republic of Kenya, 2005). To achieve food security or to reduce poverty among smallholders, the dependency on agriculture requires the application of functional soil fertility management technologies (Misiko, 2007a).

Smallholder application of soil fertility technologies is usually constrained by factors such as labour and poverty, which minimise success stories. Application of new technologies by smallholders can be of transitory nature, especially during or soon after the life of a research project. Smallholder tryouts are mainly inspired by economic and environmental reasons, and tend to epitomise social conditions such as labour capital constraints (Bentley, 2006). This chapter demonstrates that much of the application of new technologies is usually on tryout basis. Smallholder tryouts are usually the real practice, as they do not have chances to set up separate or elaborate experiments. When a technology *seems* practical to smallholders, they try it out under their household's social and farm-level ecological conditions (Misiko, 2007a). The challenge is how to translate these tryouts into long-term application (i.e. successful adoption) and to distil clear lessons from them for wider applicability for many more farmers.

Smallholder Tryouts as Experiments

Farmers constantly experiment. Nevertheless, we often do not pay enough attention to them (Bentley and Baker, 2005). Farmer experiments are organised in remarkably different ways from those of formal research (Table 1).

M. Misiko (✉)
Tropical Soil Biology and Fertility Institute of the International Centre for Tropical Agriculture (TSBF-CIAT), Nairobi, Kenya
e-mail: m.misiko@cgiar.org

A. Bationo et al. (eds.), *Innovations as Key to the Green Revolution in Africa*,
DOI 10.1007/978-90-481-2543-2_116,

Table 1 Differences between the research style of smallholder farmers and scientists

Characteristics	Scientist	Farmer
Shape	Square or rectangular	Irregular
Size	The same for each treatment	Different for each treatment
Repetitions	A must	Not used
Numbers (quantification)	Important	Visual analysis, with few numbers
Planning	Absolutely essential	Sometimes used
Serendipity	Less often	More often
Who is it for?	Others	For that farmer (specific)
Replicability	Always important	Not always
Capital cost	More	Less
Control plots	Included for comparisons	Not included – conditions "known"
Research	Main purpose	Relevance (apply as you learn)
Conclusions	From specific data measurements	Generality of suitability – many variables under scrutiny, memory comparisons, etc.

Adapted from Bentley and Barker (2005)

By learning from smallholder experiments, however, less scientific researchers can engage farmers in initiatives that aid better understanding of practical scientific concepts. Such an initiative was done by the Tropical Soil Biology and Fertility Institute of the International Centre for Tropical Agriculture (TSBF-CIAT) through the project "Strengthening 'Folk Ecology': Community-Based Interactive Farmer Learning Processes and Their Application to Soil Fertility" (henceforth the "'folk ecology' initiative" or FEI). One of the principles of the FEI is that an appropriate technology is a locally adapted end-product of inputs and practices that smallholders themselves derive from many sources, a perception supported by earlier findings (Warren et al., 1993; UNDP, 1991). This project sought to strengthen farmer practices through integrating research and folk knowledge. Co-research on soil fertility management concepts and technologies was a key component of the FEI.

Several years of the FEI resulted in changes at the farm level. The question however is: Are such changes that happen in the course of a project dependable? The assumption in this study was that the critical analysis of farmers' tryouts can help gaining understanding on the drivers of sustainable technology adoption, both technically and socially, and deriving lessons that can be of use in designing a farmer-to-farmer promotion of "green revolution" (co-innovations).

The objective of this chapter is to review *success* among smallholder farmers through presenting and analysing selected cases of farmer tryouts or "successes" that followed co-managed experimentation under the FEI.

Materials and Methods

In 2003, farmers in two sites and researchers at TSBF entered into the FEI as a collaborative learning programme that was implemented until 2007. This partnership involved farmers providing labour, land, knowledge and security for experimental plots, whereas scientists provided technical knowledge, seed and fertiliser. Throughout this exercise, experimental plots in each site were centrally located on farmer-identified farms to facilitate interactive learning.

These experiments included the following:

(i) Cereal–legume rotations – involving maize (*Zea mays* L.) as the cereal and *Mucuna pruriens*, soya bean (*Glycine max* [L.] Merr.) and groundnut (*Arachis hypogaea*, as farmer practice);
(ii) The organic resource quality concept illustrated through biomass transfer – using *Tithonia diversifolia, Calliandra calothyrsus*, maize stover and farmer-managed farmyard manure (as farmer practice);
(iii) Mineral fertiliser responses (most limiting nutrients N and P were chosen);
(iv) Mineral fertilisers and organic manure (i.e. farmyard manure – FYM) combinations;
(v) Screening (for selection) of promiscuous multipurpose soya bean varieties; and
(vi) Soil nutrient test strips – adjacent to all demonstrations.

Host farms were selected by farmers to ensure that they were representative in terms of soil type and history of cultivation (host farms had been cultivated continually from as far back as between 1951 and 1980). Participating farmers originated from several villages of Chakol Division (Teso district) and Matayos Division (Busia district), western Kenya (Table 2).

The farming systems at both sites are predominantly smallholder, i.e. rural cultivators practising intensive, permanent, diversified farming on relatively small land in areas of dense population and low incomes (cf. Netting, 1993). They have serious soil fertility problems (Ojiem et al., 2004; Ayuke et al., 2004), amid a long history of project work (Misiko, 2001), typified with low adoption of new practices (TSBF, 2001; Republic of Kenya, 2005).

Common crops include maize (*Z. mays* L.), common bean (*Phaseolus vulgaris* L.), cassava (*Manihot esculenta* Crantz), sorghum (*Sorghum bicolor* (L.) Moench), sweet potato (*Ipomoea batatas* (L.) Poir.), cowpea (*Vigna unguiculata* (L.) Walp.), finger millet (*Eleusine coracana* (L.) Gaertn. ssp. *africana*), green gram (*Vigna radiata* (L.) R. Wilczek), banana (*Musa* spp. L.), avocado (*Persea americana*), many species of vegetables, mango (*Mangifera indica* L.), cotton (*Gossypium hirsutum* L.) and tobacco (*Nicotiana tabacum* L.) (Busia and Teso districts), (Tittonell et al., 2005; Republic of Kenya, 1997). Common livestock include poultry, cattle, indigenous goats and sheep (Republic of Kenya, 1997).

Collective experiments consisted of single replicates of each treatment combination. Farmers managed and evaluated plots, and took notes about the legumes and effect of mineral fertilisers, during a 4-year span of an interactive learning process. Farmers and researchers jointly harvested each experiment. Harvested plants were weighed at the plot, and grain and stover subsamples were taken, weighed, sun dried at Maseno Research Station and weighed again to determine dry weights. During the on-farm weighing, yields were evaluated by farmers for quality. Legume residue was incorporated into the farm. Some or all of the maize stover was usually taken away by farmers to be fed to livestock.

Participatory monitoring and evaluation (PM&E) visits to trial sites and discussions were regularly held by both farmers and researchers. Visits to plots were also regularly undertaken by farmers alone to carry out independent participatory monitoring and evaluations (PM&E), i.e. to record observations free from researchers' influence and even-ranked treatments. At the end of the experiment process, yields of cereals and legumes were presented to participants for final rankings, discussions and lesson construction.

Participant observation was used to observe and detail illustrative incidents. We interacted intensively and regularly with about 165 farmers on the collective trials. Forty farmers were purposively identified. These smallholders were selected based on their

(i) experiments or tryouts resulting from the FEI;
(ii) participation in the activities of the Tropical Soil Biology and Fertility Institute of the International Centre for Tropical Agriculture (TSBF-CIAT);
(iii) gender (10 out of 20 in each site were women); and
(iv) related farming activities that changed because of their association with FEI.

Participant observation was critical in understanding tryouts from an *emic* perspective because to understand each practice we needed

(i) time (long period 5 years to make adequate follow up);
(ii) to understand physical and social environments within which farmers lived;

Table 2 Main biophysical characteristics of the study sites

Site	Altitude (m.a.s.l.) (s.d.)	Rainfall (mm yr^{-1})[a]	Coordinates (experimental sites)	Topography	Dominant soil types
Chakol	1155	1270–1600	Lat. 00°57′99.8″N Long. 034°19′00.7″E	Hilly	Dystric and Humic Cambisols
Matayos	1214	1020–1270	Lat. 00°34′34.3″N Long. 034°16′04.3″E	Hilly	Ferralic Cambisols, lithic or petroferric phase and Lithosols (rock outcrops)

[a]Bimodal distribution (i.e. first and second rainy seasons); study data (s.d.) of experimental farms
Source: FURP (1987); Republic of Kenya (1997)

(iii) to gain awareness of farmer understanding and experiences on the collective trials; and
(iv) through understanding local terminology, to appreciate social openings and barriers, within the study sites. This enabled the reliance on first-hand information with high face validity of data.

In-depth interviews among the 40 farmers were done to understand their conception of the trials, and to share key experiences. They were sampled purposively following detailed knowledge from participant observation (cf. Frankfort-Nachmias and Nachmias, 2005) and with the reliance of resource farmers in each site (Misiko, 2007b).

This study also relied on other secondary data published as TSBF-CIAT's reports for these two sites.

Results and Discussion

Tryouts Underline Micro-environments

During sampling procedures, participant observation and informant interviewing, many tryouts were identified and documented. These tryouts were so varied, but they generally followed patterns in Table 3.

Case Study 1: Soya Bean–Maize "Intercrop Rotations"

Promiscuous soya bean varieties have been promoted in Matayos as soil fertility technology, to be mainly grown in rotation with non-legumes, especially maize. Between 2004 and 2006, the nature of growing soya bean among smallholders was studied. Farmers grew soya bean in an irregular fashion, i.e. varying spacing, with low or no use of P fertiliser, mixing different varieties, intercropping, etc. The shift from rotation as done on co-managed experiments was inspired by small acreage, labour constraints, insufficient capital to buy fertiliser or by certain cultural meanings of farming, e.g. "to be like the rest", one has to plant maize.

Farmer Janet accepted soya bean farming when she observed lowered incidence of *Striga hermonthica* on the specific section where she had been rotating maize and soya bean in 2003 and 2004. Initially, she grew soya bean in pure stand for home consumption and cash. In the first rainy season of 2005, she planted an intercrop (see Fig. 1) and "rotated" the lines of soya bean and maize in the second season. The nature of the intercrop was inspired in trying to find a replacement of the common bean (*P. vulgaris* L.). Unlike the popular common bean, soya bean did not yield well when closely spaced with maize, for which she kept (and still continues) trying out different plant spacings.

Case Study 2: Implementing the Organic Resource Quality Concept and Using "Hot Compost"

The implementation of organic resource quality (ORQ) concept relied on the use of simple indicators of the intrinsic quality of organic materials as soil amendments. The implementation of ORQ was found to vary from farm to farm and done in styles that did not "conform" to the original research idea (in which,

Table 3 Observed farmers' "tryouts" in Chakol and Matayos

Technology	2002	2003	2004	2005	2006
Resource quality *(maize–bean test crop)*	0 C	2 C 20+ i			
Resource quality *(vegetable test crops)*		6 i	2 C 28 i	5 i	5 i
Resource quality *(addition to compost)*		8+ i	5 i	8 i	7 i
Improved farmyard manure	2 i	1 C 10 i	17 i	30 i	33+i
Legume–cereal rotations		2 C	2 C 20 i	2 C 10 i	2 C 10+ i
Legume (soya)–cereal intercropping			23+ i	70+ i	90+ i
Inorganic fertiliser comparisons	1 C 15+ i	2 C 20+ i	19+ i	17+ i	15+ i
Striga control *(legumes, IR maize)*				2 C 10 i	2 C 300+ i

Key: Observations were documented only among households in the villages where FEI experiments were located; they are not exhaustive (+) but indicative. C, Collective experiment (# of farmer research groups); I, individual experiment (# of households). The following selected tryouts on soya bean "rotations", biomass transfer technologies and soya bean "screening" show the different constraints smallholder face at local level

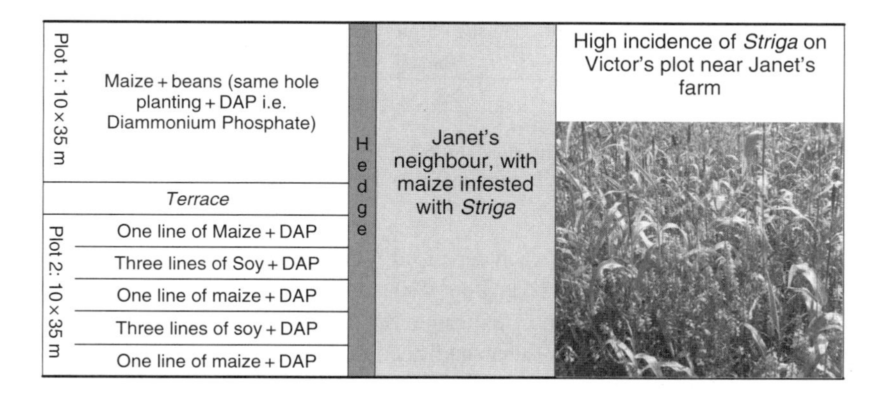

Fig. 1 An amended "tryout" by farmer Janet of Matayos in 2005 (Misiko, 2007a)

for instance, organic resources were harvested fresh and immediately applied). The researched application of the ORQ concept was appreciated but interpreted as impractical if done in the form in which it was presented, for instance, it involved selection, harvesting and transportation of bulky materials. Participating farmers therefore applied both research and local knowledge to make "quality composts" rather than adopting biomass transfer as was done on co-managed experiments (Fig. 2).

The *convenience* of applying ORQ concept was perceived from an *emic* perspective. *Convenience* refers to how variably and loosely a concept or a technology can be gainfully applied and was shaped by such factors as labour, land and capital constraints. In this case, *Tithonia* was picked "little by little" (in the language of farmers) and incorporated into traditional "composts" to "boost" quantity and quality and to speed up decay ("cooking") due to increased temperature – a *hot* compost "cooks" faster (Misiko, 2007b). Participating farmers made composts that comprised resources that easily crushed to powder when dry (c.f. Giller, 2000); such materials took shorter periods to "cook well" (Giller et al., 2007). These were *emic* labour saving tryouts of the ORQ and do not imply adoption.

Case Study 3: Soya Bean Continuous "Screening"

When new soya bean varieties were screened and evaluated in Chakol, western Kenya, through participatory approaches in 2003, farmers acquired and planted their preferred varieties on their individual farms. There were many agreements between scientists and farmer assessments, with respect to yield ranking and value of biomass (Misiko, 2007a). However, because soya bean was "new" in Chakol, the preferred varieties were not necessarily adopted. The preferred varieties were "screened" further under the learn-as-you-use tryouts.

The screening experiment conducted at Dismus and Wilmina Omalalu's farm in 2006 is an example from Chakol that demonstrates how soya bean screening was extended to other farmers' plots. All Omalalu's plots were located on same soil type (Assing'e – sandy soils), planted on the same day and weeded twice.

Likely action
i) Plant crops where it grew e.g. *Tithonia, Vernonia*
ii) Add to cattle manure (*Tithonia*, to increase compost quantity)
iii) use as cattle fodder (*Calliandra*)

Expanded practices
iv) Incorporate fresh at planting (selected crops)
v) Improve compost quality
vi) "heat up" composts (*Tithonia*)

High quality organic resource ⇦ ⇨ **Low quality organic resource**
Easier to tear ⇦ ⇨ Harder to tear
Darker green ⇦ ⇨ Light green
Bitter taste ⇦ ⇨ Menthol taste

Strengthened local knowledge
Fast decomposition ⇦ ⇨ Slow decomposition
Soil fertility indicator ⇦ ⇨ low fertility indicator
Lesser succulence ⇦ ⇨ Higher succulence

Usual action
i) fodder (maize stover)
ii) Deposit materials on 'traditional' compost (e.g. litter)
iii) make conservation strips, leave resource in the farm (e.g. prunings, stover)

Derived practices
iv) Make cattle "beddings", then (add to) compost
v) Make mulches, e.g. stover

Fig. 2 Common cases of expanded ORQ knowledge and application: the researcher and smallholder perspective (adapted from Misiko, 2007a)

Table 4 Yield of different soya bean varieties from Dismus' screening plots

Variety	Mksoy	Namsoy	SB3	SB20	SB17	SB19	SB15
Yield (kg)	3	2.3	0.5	1	1.2	1.8	1.5

Source: Misiko (2007b)

NB: the FEI screening sites and previous farmer tryouts had shown that without P the yields of all these varieties were lower. Dismus and Wilmina therefore did not plant any replicates and did not have P plots

The plots were 4 m × 6 m, each fertilised with triple superphosphate at one glass (i.e. about 200 g) per plot. In this farmer-type "screening", experiment SB20 was the least prolific contrary to what had been observed in co-managed experimental findings (Table 4). This underscored the fact that it is complicated for farmers to make decisions over which variety to trust and sets hurdles for any scaling out based on few season's or site's lessons.

Many more farmers planted only two or three varieties (see Misiko, 2007a:51). The continuous "screening" under farmer contexts revealed new criteria for selection of varieties. In 2005, wind was an important problem. In 2006, rain pattern and amount were very critical. As challenges evolved, so did coping strategies of farmers. For instance, farmers varied their dates of planting common bean and soya bean, rotated plots and changed seed variety more often than before.

Tryouts vs. Adoption: The Lessons

1. Tryouts were usually performed on a temporary basis. The nature of tryouts, therefore, necessitates two-phased projects. The first phase should mainly be focused on interactive practical learning and analysing smallholder tryouts, while the second phase focuses on applying learnt lessons, through scientific research with smallholders.
2. While adoption hinges on acceptance and wide approval, tryouts of soil fertility concepts are done by individuals principally on learn-as-you-use basis. This means that smallholders learn new ideas that scientists may have missed or overlooked. Tryouts are therefore statements of "correct" or "poor" learning among practising farmers.
3. Because tryouts are done on learn-as-you-use basis, use of a technology for purposes besides or instead of for what it was promoted was devised. Understanding new and latent functions can be a very useful opening for scientists to grab the initiative and fine-tune a technology for wider approval and acceptance.
4. Tryouts are not standardised experiments or application of a technology across sites and seasons. Lessons among farmers are drawn depending on each season and by relying on memory, decisions are made over whether more fertiliser "scorches" (i.e. *huchoma* in Swahili), or if the problem is germplasm, *Striga*, witchcraft, etc. These necessitate a more proactive scientific role in improving smallholder lesson distillation process. This may be through simplified co-experimentation processes within farmer contexts, e.g. on P or on "hot" composts in modified experiments. This may for instance "prove" whether "hot" composting is meaningful application of ORQ concept, or help understand the significance of "localised" forms of intercrops.
5. While tryouts, e.g. on ORQ concept, are rarely done uniformly across the study sites, adoption of artefacts or singe-unit material technologies like hoe and oxen plough is clear to measure. The adoption of the latter can result in easy-to-measure results, such as reduction in labour and cost. Adoption of soil fertility concepts is not all that is required. Soil fertility concepts require knowledge intensive undertakings, and even when technologies are adopted, it does not necessarily mean that soil fertility will be enhanced. For instance, legumes (such as *Crotalaria ochroleuca*, soya bean and *Calliandra*) or farm yard manure are used in ways that may compromise their soil fertility worth. This may indeed mean that we need to concurrently scale in the benefits of our work just as we scale them out.

Experimentation as Farmer Practice

1. *No control – "we know control, simply look around..."*

Because tryouts are meant to test for results under farmer-specific contexts, they are forms of experiments. Nonetheless, they do not have any control. As Dismus (Chakol) puts it, "...we know control; simply look around if you wish to know the poor plant growth we are trying to reverse". Many farmers already know that low crop yield results from non-use of appropriate technologies. Before any new technology is tried out, it has to be first favourably considered, but will be trusted only after its application under specific farmer contexts. In Janet's language, "to know if this thing (soil fertility status) improves, and whether soya treats my soil's other diseases (i.e. Striga), then I'll decide".

2. *For how long? Farmer tryouts vs. complex scientific experiments*

Smallholder tryouts were observed to last only up to 2 years. Even so, they were never redone but rather revised. When farmers were taken on learning trips, such as to on-station trials, they were perplexed that an experiment had been running for 4 years, with same high level of complexity and detail. The participants picked only closer-to-familiar elements from this trial, such as seed variety. Such complex experimentation may be more beneficial to scientists than smallholders as an on-farm interactive learning tool. On the one hand, scientists can gain from farmer tips, but on the other hand, however long a complex trial exists in a village, lessons will not meaningfully build up on their own, and non-participating smallholder's interest will, for instance, likely be on how much rent the *outsider* pays.

3. *No replicates – hard to make reliable conclusions, but aren't tryouts valid?*

The three tryout cases illustrated in this chapter do not rely on any replication. The role of replication is fundamental in scientific research (Frankfort-Nachmias and Nachmias, 2005). Smallholder experiments are rarely replicated by their neighbours, or simply given their nature – specific contexts – we cannot relocate a farmer to repeat her experiment in her kitchen garden. We cannot re-create exact farmer contexts as last season, etc. Nonetheless, since tryouts are smallholder practice, their results are simply valid depending on each farmer's circumstances. What we learn is that farmers do not specifically do soil fertility, rather they do life itself. They do not rehearse for better yield, they search it now against many independent variables by doing tryouts based on what they can manage. This is exactly why they may not adopt a seemingly worthy concept if their trust in it is inadequate, given their complex social contexts.

Conclusions

Given the nature of tryouts, it may be convenient to ignore them rather than incorporate any lessons into regular science. Because tryouts are the real practice, even though tedious to analyse, there is no avoiding the very subjects our research is meant to lead. Our hunt for signs pointing to success or illustrating circumstances that throttle smallholder steam out of soil fertility management is key to enable us to resolve the issue at hand.

Tryouts can easily indicate future success, because they are founded on the idea that a farming system's properties (i.e. social, economic and ecological) cannot necessarily be accurately understood independently of each other. The question we may all ask is, What are the guarantees for successful scaling out of soil fertility concepts or technologies that are only positively perceived or tried out while the project runs, i.e. are positive perceptions and/or tryouts enough promise for acceptance, wide approval and sustainability? We will need to be cautious while we count "un-hatched eggs".

Acknowledgements Special gratitude goes to all farmers of western Kenya, especially those who made this study a success, and to the invaluable financial support of the International Development Research Centre (IDRC) and TSBF-CIAT.

References

Ayuke FO, Rao MR, Swift MJ, Opondo-Mbai ML (2004) Effect of organic and inorganic nutrient sources on soil mineral nitrogen and maize yields in western Kenya. In: Bationo A (ed) Managing nutrient cycles to sustain soil fertility in sub-Saharan Africa. Academic Science, Nairobi, pp 66–76

Bentley J (2006) Folk experiments. Agric Hum Values 23(4):451–462

Bentley JW, Baker PS (2005) Understanding and getting the most from farmers' local knowledge. In: Gonsalves J,

Becker T, Braun A, Campilan D, De Chavez H, Fajber E, Kapiriri M, Rivaca-Caminade J, Vernooy R (eds) Participatory research and development for sustainable agriculture and natural resource management: a sourcebook. Volume 1: understanding participatory research and development. International Potato Centre-Users' Perspectives with Agricultural Research and Development, Laguna. Philippines and International Development Research Centre, Ottawa, Canada, pp 58–64

Fertiliser Use Recommendation Project (1987) Final report, Annex III, vol 5. Busia District. Ministry of Agriculture, Nairobi (Annex)

Frankfort-Nachmias C, Nachmias D (2005) Research methods in the social sciences, 5th edn. Arnold, London

Giller KE (2000) Translating science into action for agricultural development in the tropics: an example from decomposition studies. Appl Soil Ecol 14:1–3

Giller KE, Misiko M, Tittonell P (2007) Managing organic resources for soil amendment. LEISA LEUSDEN 22(4): 16–17

Misiko M (2001) The potential of community institutions in dissemination and adoption of agricultural technologies in Emuhaya, Kenya. MA thesis, University of Nairobi, Nairobi (Chapter 4)

Misiko M (2007a) Fertile ground? Soil fertility management and the African smallholder. PhD thesis, Wageningen University, Wageningen, The Netherlands

Misiko M (2007b) Participatory analyses of the "Strengthening 'Folk Ecology' Project" activities: a report on the January–March 2007 fieldwork. TSBF-CIAT, Nairobi, Kenya, 14 Mar 2007

Netting RMcC (1993) Smallholders, householders: farm families and the ecology of intensive, sustainable agriculture. Stanford University Press, Stanford, CA

Ojiem JO, Palm CA, Okwuosa EA, Mudeheri MA (2004) Effect of combining organic and inorganic phosphorus sources on maize grain yield in humic-nitisol in western Kenya. In: Bationo A (ed) Managing nutrient cycles to sustain soil fertility in sub-Saharan Africa. Academic Science, Nairobi, pp 347–357

Republic of Kenya (1997) Busia district development plan 1997–2001. Government Printers, Nairobi

Republic of Kenya (2005) Ministry of Agriculture Strategic Plan 2005–2009. Ministry of Agriculture, Nairobi

Tittonell P, Vanlauwe B, Leffelaar PA, Rowe EC, Giller KE (2005) Exploring diversity in soil fertility management of smallholder farms in western Kenya: I. Heterogeneity at region and farm scale. Agric Ecosyst Environ 110:149–165

TSBF (2001) Integrating soils, systems and society, strategic directions for TSBF. Tropical Soil Biology and Fertility Programme, Nairobi, Kenya

UNDP (1991) Agricultural extension. United Nation Development Program, New York, NY

Warren DM, Brokensha D, Slikkerveer LJ (1993) Indigenous knowledge systems: the cultural dimension of development. Kegan Paul International, London

Improving Smallholder Farmers' Access to Information for Enhanced Decision Making in Natural Resource Management: Experiences from Southwestern Uganda

K.F.G. Masuki, J.G. Mowo, R. Sheila, R. Kamugisha, C. Opondo, and J. Tanui

Abstract For more than 9 years the African Highlands Initiative (AHI) has been working in Kabale District, Uganda, on integrated natural resource management (INRM). During implementation of activities, it was learnt that farmers and their institutions lacked access to appropriate information necessary for applying methods and technologies to enhance productivity and ensure sustainable NRM. Access to information was hampered by limited communication services. Despite active involvement by farmers' institutions in mobilizing and sharing information, it was felt that additional information sources could improve farmers' livelihoods. The AHI-ACACIA project was designed to address this shortcoming. A road map to institutionalize information exchange between farmers and information providers was designed. The project involved key stakeholders including the NAADS Secretariat, IDRC (the donor) and the Community Wireless Resource Centre. A collective re-strategizing involved connectivity issues and renewal of research to align learning with some of the programmatic changes, to ensure that activities were closely aligned with the overall goal of improving farmers' livelihoods through better access to information. The findings show that priority farmers' information needs include market information (100%), post-harvest handling (98%), improved seed management (93%), natural resources management (bylaws, trenches, grasses; 71%), soil sample analysis using local indicators (63%), crop and animal pests and disease control (62%) and fertilizer management and application (55%). A total of 252 different information products of leaflets, pamphlets, booklets and poster were developed in both English and the local language (Rukiga) and disseminated to 6 village information centres (VICEs) and 2 telecentres. About 20% of the farmers preferred the brochure because it contained more detailed information and 30% preferred the booklet on post-harvest handling because it also contained more detailed information. Generally it was observed that the reading culture of farmers was very low because very few used the library at VICEs. Use of mobile phones to access information was found to differ from one parish to another. It was revealed that in general male farmers (59.3%) and female farmers (40.7%) use the phones; however, more female farmers (66.7%) request information on NRM and agriculture compared to male farmers (33.3%). Farmers were more excited about the phone than either the radio or paper products. Radio announcements did not seem to be of much help to the farmers; however, farmers preferred programmes rather than announcements. This calls for more needs assessment for information dissemination and flow for the betterment of the communities.

Keywords Information sharing · NRM · VICEs · Telecentres

Introduction

The African Highlands Initiative (AHI) has been working in Kabale District, Uganda, on integrated natural resource management (INRM) for the past 9 years. During AHI's phase II and III activities, it was found that farmers' institutions were lacking access

K.F.G. Masuki (✉)
African Highland Initiative, Kampala, Uganda
e-mail: k.masuki@cgiar.org

A. Bationo et al. (eds.), *Innovations as Key to the Green Revolution in Africa*,
DOI 10.1007/978-90-481-2543-2_117,

to appropriate information necessary to apply methods and technologies for increasing productivity and improving the management of the natural resource base. Farmers' needs were found to be continuously changing, with such changes requiring access to certain information that was not available due to limiting communication services. In response to these needs, information has then been captured from farmers' preferred sources, packaged according to farmers' preferred format and mode of delivery, and delivered to farmers for use and evaluation. AHI-ACACIA project was designed to address this shortcoming and to introduce an effective information chain that would connect farmers to appropriate information providers through telecentres. These telecentres would act as information hubs or centres where information needs were articulated by farmer institutions, information was gathered in response to these needs and products developed based on the information collected. AHI in partnership with NAADS and local-level farmer institutions has been developing a system for farmers to articulate their information needs. Needs assessment focuses on agricultural production, natural resource management and market information.

To ensure sustainability one of the project's strategy involved targeting the National Agricultural Advisory Service (NAADS) as the most promising partner for institutionalizing demand-driven information provision given their national mandate in demand-driven agricultural extension in Uganda and the complementarity of technology 'hardware' and information. The NAADS which replaced the old government extension system has a communication strategy that draws on a variety of techniques and media. In some cases, NAADS was already working with telecentres at the time of research; thus, access to accurate and timely market information had been identified as a priority area for NAADS. The NAADS focuses on the 'demand side' of agricultural extension services by organizing farmer groups that can collectively demand and buy information related to agriculture and markets. NAADS helps the groups to pay in the beginning, on the assumption that once they are putting new farm practices and market knowledge into practice, they will be successful enough to buy the information themselves, while the collective activity reduces the per-farmer cost. Thus, NAADS is helping to create effective demand for the kind of services that telecentres provide.

AHI-ACACIA sub-county telecentres liaise with parish-level village information centres (VICEs) to coordinate the flow of information needs from the farmers, to the source of information, and back to the farmer. A telecentre by definition is an integrated information and communication facility that houses a combination of information and communication technologies (ICTs). On the other hand, ICTs include those instruments, modes and means through which information and/or data are transmitted or communicated from one person to another or from place to place. The current facilities at Rubaya and Kabale fall short of constituting telecentres. The rural telecentre in Rubaya had been working with NAADS since 2001, since they shared many goals. The telecentre staff had been contracted by NAADS to help organize local farmer groups. This relationship appeared to be of great mutual benefit. It might increase the telecentre's value to the community, if the NAADS farmer groups learned to seek and apply the different kinds of information the telecentre might provide. AHI and NAADS co-developed an Information Needs Protocol, combining both NAADS criteria and AHI inputs and with consultations with farmer institutions at the district and sub-county level. NAADS has their own procedure for assessing information needs on agro-enterprises, setting the basis for determining more specific information needs focusing on agriculture production, NRM and markets and cross-cutting issues. Although farmer institutions were active in the district in mobilizing farmers and sharing information informally during meetings, it was felt that additional information sources from Kabale town and from active NGOs working in the district could greatly improve farmers' livelihoods.

The research dimension of the project was also expanded to encompass active learning at diverse levels such as action research on the overall design of the 'system' of demand-driven information provision in Kabale (information and communication technologies, local institutional structures and processes for articulating demand, district institutional arrangements to respond to demand and M&E systems); action research to validate instruments for farmers to articulate their information needs equitably, for information management (demand and supply) at telecentres and for monitoring of final information products and media; and empirical research on outcomes and impacts, including use of mobile phones, access to diverse information products and media and livelihood impacts.

The overall goal of the AHI-ACACIA II project is to improve the livelihoods of farmers through enhancing information access and use through improved flow of information between farmers, service providers, community members and NGOs on agriculture, market information and natural resource management.

Literature Review

Past research has firmly established the importance of knowledge and information sharing in R4D settings. Knowledge and innovation are now widely regarded as key drivers of economic growth. But what enables knowledge flow and innovation? While the answer is obviously complex, it is clear that ICTs are deeply implicated (Verlaeten, 2002). Access to appropriate information and knowledge is paramount to the success of NRM planning, implementation and evaluation processes. It is known to be among the single largest determinants of agricultural productivity (AHI-ACACIA II, 2006). Agriculture and the harvesting of natural resources provide livelihoods for over 70% of the African population. Farmers must have access to information about new technologies before they can consider adopting them. It is in working with and improving these information and communication systems that ICTs can be used to enhance the delivery of these services (ACACIA, 2006).

In East Africa agricultural production for local consumption and export plays a critical role in national economies, making up 40% of the sub-region's GDP. Nearly 80% of the people depend on agriculture for their livelihoods, most as small-scale farmers. Fifty percent of the population in Kenya, 35% in Uganda and 36% in Tanzania live below the poverty line. Agricultural information is a key component in improving small-scale agricultural production and linking increased production to remunerative markets, thus leading to improved rural livelihoods, food security and national economies. However, these farmers look up to research and extension agents as sources of new technologies; yet, the traditional approach to providing agricultural information through extension services has had several shortcomings mainly because the extension service is overstretched and under-resourced. New approaches of promoting access to agricultural information are being explored. Etta and Parvyn-Wamahiu (2003) observe that telecentres have the potential to transform the lives and livelihoods of many in the developing world and especially those in remote locations in developing countries. Many authors share the view that ICTs can be used to deliver agricultural information that could stimulate increased production by linking farmers to remunerative markets (Asingwire, 2003; Harris, 2002; Mayanja, 2002). The Comprehensive Africa Agricultural Development Programme (CAADP) of the New Partnership for Africa's Development (NEPAD) identifies agricultural research, technology dissemination and adoption through improved delivery systems to farmers and agribusinesses as a primary action necessary for long-term agricultural productivity and competitiveness (CAADP-NEPAD, 2002). Access to information is being increasingly recognized as central to helping poor communities have a stronger voice in matters affecting their well-being (World Bank, 2002). The rapid spread of information and communication technologies enables anyone with access to publish information and access vast amounts of information and millions of other users and creators of ideas. The abundance of information creates a pressure to innovate (Verlaeten, 2002). An important component of any agricultural system is the flow of information and the strength of the information systems that are managed by governments, farmers and NGOs. While a few large-scale, commercial farmers on the continent have used some of the decision support tools that ICTs are providing, relatively little attention has been paid to the potential benefits of the broader use of ICTs in this (mainly informal) sector, one of the few in which women often predominate (ACACIA, 2006). The NSGRP (2005) asserts that problems of access of ICTs include low literacy rates, low incomes and limited number of service providers. The recent rise in the use of ICTs including mobile telephony is still dominantly urban oriented and needs to be directed to rural areas as well.

The Acacia Initiative is an international effort to empower sub-Saharan African communities with the ability to apply information and communication technologies to their own social and economic development. The relevance to Africa of emerging, new market models that are enabled through ICTs needs to be understood in order to strategically integrate ICTs into national planning. African governments are expected to make large leaps of faith in shaping their nation's ICT futures with scarce knowledge resources to navigate this constantly changing sector (ACACIA, 2006). Appropriate models need to be developed that arise from a bottom-up assessment of the demands, the needs and the realities of African ICT users. However, for the African governments to embed this in a sustainable way there is a need to look into favourable conditions to support the move. Among the conditions are favourable policies. There is insufficient dialogue around how changing and shifting policies can assist these governments to meet their mandate to serve their people, to deliver services and to improve the public services through leveraging the benefits of ICTs. It is interesting to note that good policy and regulation is a necessary but not sufficient condition for improving access in Africa. Countries like Uganda – held up as a model of good policy and regulation – still have remarkably low penetration of ICT infrastructure and services. Income levels and ICT usage continue to show strong correlation. However, there is insufficient information about the behaviour and use the poor in Africa make of ICTs with regard to their livelihoods and social networking. In extremely poor countries such as Uganda (where 40% of the population live below the poverty line), further decreases in infrastructure and access costs will be required by the network if the effects of expanded access are ever expected to be felt.

Uganda is a landlocked country that occupies 270,000 km^2. It has a population of nearly 20 million, 88% of whom live in rural areas. With a per capita income of $255 Uganda is one of the poorest countries of the world. Fifty-five percent of her people live below the poverty line. Illiteracy levels are 38 and 55% for 15-year-old boys and girls, respectively. School enrolment rate is 35% although this is expected to increase rapidly through the Universal Primary Education policy which promotes free primary education for four children per family. Only 42% of the rural population have direct access to health care. The country has a telephone capacity of 70,000 lines and an annual rate of increase of 5%. It still has one of the lowest teledensities in sub-Saharan Africa. This situation will, however, change as a result of policy reforms introduced in 1996 liberalizing the telecommunication and broadcast sectors. The government has delinked itself from direct participation in the development of ICTs and opened the door for private sector participation and competition under the management of the Uganda Communications Commission, an independent regulatory body. There is a rapidly growing ICT market in the country with several Internet Service Providers, privately owned radio and television stations, several ICT training institutions and many donor-funded ICT initiatives. Prior to the growth of inexpensive telecommunications and the Internet, information was costly to acquire and disseminate. These costs act as a natural restraint on the spread of ideas and knowledge (ACACIA, 2006).

By contrast, the mobile phone companies have seen considerable growth in Africa. Mobile service providers have been able to reap rapid returns on their investments, working with newer, cheaper technologies that are easier to roll out. These services are eagerly absorbed by people whose pent-up demand for telephone services is so great that they are prepared to pay up to 4% of household expenditure (Ureta, 2005) to make voice calls. While the number of Internet users in Africa has grown 400% in the last 5 years to 22 million (Internet World Stats, 2005) and Africa is experiencing faster Internet growth than North America, most Africans have never used a computer or sent an e-mail. Problems of access of ICTs include low literacy rates, low incomes and limited number of service providers. The recent rise in the use of ICTs including mobile telephony is still dominantly urban oriented and needs to be directed to rural areas as well.

Mobile telephony has spread rapidly across the continent. Mobile phone lines in Africa outnumber fixed lines by more than two to one. In sub-Saharan Africa the ratio is closer to three to one. In 2004, Africa became the fastest growing mobile telephony market in the world. Recent results from the Acacia-funded *African e-Index* (LINK Centre, 2005) study have shown that Africans are willing to pay a higher proportion of their income than in developed countries for access to telephony; this indicates, among other things, a significant, unfulfilled demand for telephony. A recent study of telephone use in rural areas in Botswana, Uganda, and Ghana (McKemey et al. quoted in Batchelor and Scott, 2005) shows that 80% of residents had made at least one phone call in the last 3 months. There is emerging evidence of the economic impact of the spread of mobile telephony (Vodafone, 2005). Very little is known about the behaviour and consumer demands of African ICT users; individual consumers or public sector institutions; the poor; or those trying to deliver services.

Methodology

The project was implemented using participatory action research methodology whereby meetings were conducted between several stakeholders involved in the project, especially communities, service providers, NAADS, NGOs and AHI staff. Interviews with community leaders were also used to collect information, and various training sessions were used to equip stakeholders with the current issues related to information sharing.

A systematic action research and process documentation was employed to monitor the use of various ICTs to facilitate communication and information flow between the various levels of farmer institutions (farmer group, parish level, sub-county and district), between farmer institutions and the telecentres, and between information sources and the telecentres or information hubs. The progress of the integration process of an information delivery system with district NAADS actors and stakeholders was assessed through identified performance targets and indicators which cover the project's aspects such as technical service delivery, financial, managerial and equipment (maintenance). Information needs assessment protocol was used to identify farmers' information needs. The initial needs assessment protocol was developed in May 2006 and tested with farmer groups in Rubaya sub-county. This protocol determined the first priority information needs of farmers and was used as a basis for developing information sharing products at the Kabale telecentre. The protocol was then improved to enable collecting information needs on agro-enterprise, agriculture, NRM and markets. Monitoring and evaluation process was used to evaluate final products at VICEs level; this included keeping an eye on usage of mobile phones through Village Phone Usage Tracking Forms, and monitoring the spread of farmer products was done through VICE monitoring books.

Findings

Information Delivery System

The information delivery in the project area has undergone several changes over a period of time resulting in a number of lessons which have evolved from developing a demand-driven information delivery system. It was learnt that there exist several district-level NGOs and research organizations which work to provide information for farmers. Thus ACACIA needs to complement to existing projects and activities in the communities rather than working as an independent entity because this will ensure the sustainability and continuity of the project's activities.

Existing Structure at Kabale

The existing structure at Kabale (Fig. 1) illustrates the farmer institutions and the management structure relative to the telecentres. VICE (village information centres) are the parish-level telecentres where farmer

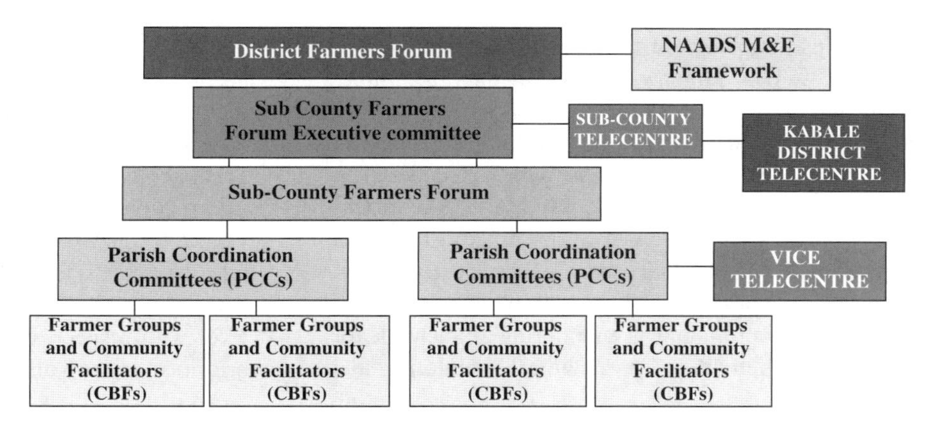

Fig. 1 Farmer institutions and management structure through NAADS

groups from the village level meet and discuss their information needs.

The following is a breakdown of the various levels of farmer institutions through NAADS, the roles and processes involved and links to the three levels of telecentres.

Parish Coordination Committees (*PCCs*) consist of two representatives from each farmer group (male and female) from each village. They meet monthly to discuss any issues or concerns.

Community Facilitators work with farmer groups in their parish. They are hired by NAADS to facilitate meetings and liaise with farmers to ensure that the NAADS service providers are fulfilling their contracts and farmers are benefiting from the services they provide. They meet monthly to assess progress and identify any challenges which might have emerged and how they might be addressed.

The *Sub-county Farmers Forum* (*SFF*) consists of two representatives from every parish in the sub-county. In Rubaya, there are six parishes, where two representatives from each parish (one male and one female) form the sub-county farmer's forum. They meet twice a year (mid-financial year and at the end of the financial year) to review progress.

The *Sub-county Farmers Forum Executive Committee* consists of 10 farmers, where 3–4 members are elected from the Sub-county Farmers Forum and each parish is represented by their chairperson. The Executive Committee meets quarterly at the sub-county level.

The *District Farmers Forum* consists of 17 farmer representatives from the 12 sub-counties in Kabale district and additional 5–6 representatives from the NAADS technical committee (based on enterprises). They meet every 3 months, to discuss issues raised from the sub-county level and coordinate district-level farmer institutional development and activities.

The *NAADS Monitoring and Evaluation* framework in the district has recently transformed into two management bodies: administrative and technical. The monitoring of local political activities is undertaken by existing district and sub-county NAADS representatives, including the NAADS coordinator, the CAO representative and other specific service providers. The key responsibilities include implementing M&E policy and coordinating a monitoring framework for all NAADS activities. The technical monitoring and evaluation of service provision is coordinated at the district level, through a specified committee who are technical specialists. The AHI-ACACIA II project has integrated its M&E activities into these two NAADS structures through the incoming service provider and existing NAADS coordinators for the administrative aspects and the Quality Assurance Committee (QAC) for the technical aspects of information delivery.

Structure to Enhance Information Access to Farmers

One of the project strategies is to make use of the available organizational structure to foster access to information through appropriate ICTs. Figure 2 represents the various levels of farmer institutions and their communication flows with the telecentres. The figure shows a structure and schematic representation of information flow and communication mechanisms.

The following is a breakdown and inventory of the communication mechanisms and ICTs available presently at each level.

(i) Village/farmer group to parish level
Farmer groups at the village level meet with the CBFs to discuss any issues or concerns they are currently facing in their farming practices. The CBFs use tools such as a 'needs tree' to generate discussions on current information needs. These needs are then validated through the Information Needs Protocol, whereby needs are categorized into three main subject matters – agriculture, natural resources management and markets. The groups select enterprise which is then identified parallel to the three subject matters. The CBFs then deliver this information to parish level where the PCCs compile the information and prioritize needs from the farmer groups. The ICTs available include radio and mobile phones.

(ii) Parish level to sub-county telecentre (Rubaya)
Feedback is received from the CBFs on farmers' information needs during monthly meetings and via pre-formatted NAADS forms. The PCCs compile the results from all the farmer groups and determine the information priorities at the parish level with respect to enterprise selection. These priorities at the parish level are then submitted

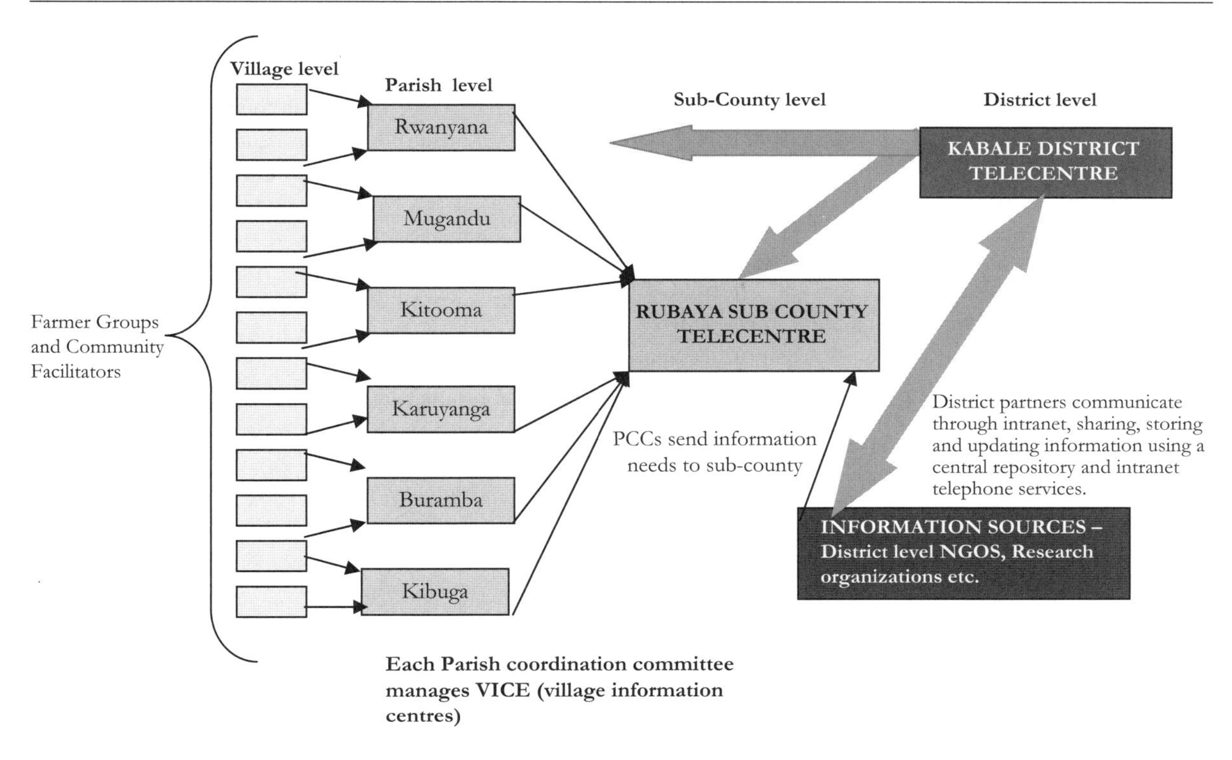

Fig. 2 AHI-ACACIA communication schematic

to the sub-county farmer forum. There are two modes of communication at the parish level: mobile phones and radio. The CBFs liaise with the PCCs on articulated information needs.

(iii) Sub-county to district
The sub-county telecentre collates all the information received from the six parishes and responds by (i) distributing existing information available at the telecentre and/or (ii) sending information needs to the district-level telecentre for identification and packaging of the information. Currently, the Rubaya telecentre requires further technical support to improve the communication channels between the parish group (which ideally should be via mobile phones) and the Kabale telecentre [ideally via Internet or mobile phone using text messages (for market information)].

(iv) District-level partners and information sources (Fig. 3)

At the district level, NGOs and development organizations are engaged in facilitating information delivery on natural resources management, agriculture and market information for farmers. Most of the ICTs in use include wireless connectivity between partners, mobile phones, Internet and intranet. These organizations can improve the level of information sharing at the district level to allow an efficient collection and dissemination system to sub-county and parish levels. One of the partners, IT+46, of the CWRC is in the process of planning a VOIP intranet system whereby partners in the district will be connected via wireless intranet that enables free phone calls and central intranet space to upload, share and exchange information.

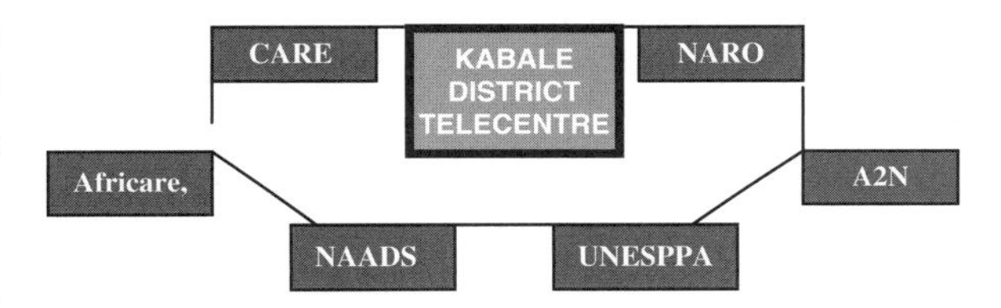

Fig. 3 Detailed AHI-ACACIA communication schematic with district-level partners

Information Needs Assessment and Priority

Table 1 summarizes the results of farmers' information needs assessment and their priority in all six parishes based on the initial Information Needs Protocol developed in May 2006. These priorities were later used to identify sources of information for identified subject matters. The assessment showed that market information was a higher priority and need (100%) for the farmers in Kabale district. The result suggests that priority information needs are in order of importance as market information (100%), post-harvest handling (98%), improved seed management (93%), natural resources management (bylaws, trenches, grasses) (71%), soil sample analysis using local indicators (63%), crop and animal pests and disease control (62%) and fertilizer management and application (55%).

Further analysis on market information revealed that farmers need to know information sources, marketing and communications. However, they need to know information on preparation for marketing, promotion/advertisement and market source and prices. In the case of natural resources management the farmers showed that they need information related to tree planting (matching species to their correct niche, nursery management, trees for soil erosion control, trees to protect springs or other), soil and water conservation (terracing, trenches, composting, enhancing spring recharge, vegetative contour), bylaws (how to create your own bylaws and how to improve enforcement of bylaws), water (water use, water quality control, water harvesting), participatory planning and watershed management (reducing NRM conflicts, improving management of farm boundaries and drainage, controlling grazing).

Communication Methods/Tools

The communication methods/tools in this regard refer to phones, paper products and use of radio. This section presents the extent to which these three were received and utilized by the AHI communities in Kibuga, Karujanga, Rwanyana and Mugandu parishes.

Phone: Phones (mobile and table sets) were distributed to different parishes around November 2006. Community-based facilitators (CBFs) were trained on how to use the phones and thereafter expected to (a) make the rest of the farmers in the parish aware of the existence and use of the phone and (b) keep, maintain and make available the phone to the farmers. AHI and the service provider (private sector) were expected to provide technical backup.

Paper products: These were assorted written materials in the form of leaflets, manuals, textbooks kept in a bookshelf: some written in English and others in the local language. The paper products were mainly on animal production, crop management and natural resource management. The library was located at the CBFs place.

Radio: This was in the form of announcements about prices or markets of different foodstuffs that included potatoes and beans.

Information Sharing Products

A number of information sharing products were produced in response to farmers' information needs. The products were made in both English and the local language (Rukiga). Table 2 presents a number of preliminary products that were prepared for farmers. In general a total of 252 different information products were developed and disseminated to 6 village information centres (VICEs) and 2 telecentres.

Farmers' Evaluation of Information Sharing Products

The preliminary results of the evaluation of information sharing products for one parish in Rubaya are presented in Table 3. During this evaluation, it was found that most farmers evaluated the products based on content and the relevance to the type of farming they do. The parish is particularly interested in Irish potato production and therefore interested in products about clean seed management of Irish potato. Two types of products on clean seed were developed: a one page flyer and a brochure. Farmers preferred the brochure (20%) because it contained more detailed information. Farmers also preferred the booklet (30%) on post-harvest handling, because it also contained more detailed information.

Table 1 Summary of farmer priority information needs in six sub-counties of Kabale District ($N = 55$ parish groups)

Topic	Parish							
	Mugandu	Karujanga	Buramba	Rwanyena	Kitooma	Kibuga	Total	%
Market information	6	10	9	10	10	10	55	100
Post-harvest handling	6	10	9	10	10	9	54	98
Improved seed management	6	10	9	10	10	6	51	93
Natural resources management (bylaws, trenches, grasses)	3	3	8	10	6	9	39	71
Crop and animal pests and disease control management	6	1	6	6	8	7	34	62
Soil sample analysis using local indicators	6	1	2	8	9	7	33	60
Fertilizer management and application (source, quality, where?, when?)	3	2	3	8	9	5	30	55
Local farmyard manure	2	4	5	6	1	8	26	47
Credit institutions/schemes	1	1	2	6	6	5	21	38
Beekeeping	0	4	2	1	2	4	13	24
Organic manure	1	2	1	1	4	3	12	22
Work plan development	0	1	0	4	0	6	11	20
Soil sample using local indicators	2	0	3	1	0	4	10	18

Table 2 Summary of products developed in one parish of Rubaya sub-county

Type of product	Quantity	Source	Title
Leaflet	2 English and Rukiga	NGO	Post-harvest handling
Poster	1 English and Rukiga	NGO, farmers	Local farmyard manure
Flyers	1 English and Rukiga	Farmers	Clean seed
Pamphlets	2 English and Rukiga	Farmers	Organic fertilizer

Table 3 Farmer evaluation of knowledge products

Product	% of farmers who selected	Major enterprise	Reasons for selection
1. *Leaflet* (seed management)	20	Irish potatoes	• Lack of seeds during planting season • Expensive to buy seeds from the market • Has detailed information
2. *Booklet* (post-harvest handling and storage of crops)	30	Beans Irish potatoes	• Lack of bean seeds for planting, they only get for consumption • Want to produce enough potato for consumption, seed and for sale • Contains detailed information
3. *Brochure* (seed management)	20	Irish potatoes	• Contains relevant information • Irish potatoes help to raise household income
4. *Leaflet* (how to make compost and farmyard manure using locally available materials)	20	Beans Irish potatoes	• It is more understandable to farmers • Artificial fertilizers are expensive • Poor yields of beans due to soil infertility
5. *Poster* (seed management)	10	Irish potatoes	• Increase potato yields for consumption and for sale

Evaluation on the Use of Communication Methods

Creation of awareness in the communities was handled by the CBF who kept the phones and library. Each CBF devised a way to carry out the awareness creation. In the case of Rwanyana parish, for example, the CBF called the parish and sub-county chiefs to a function where priority enterprise selection process was being launched. Village representatives were reportedly present at the function. The function provided an opportunity for the CBF to convey to people present to inform others about the existence of communication strategies, especially the phone. Other CBF mentioned that they called meetings to inform the communities about the VICES and others moved around informing groups. Let us see what happened in the different parishes.

The Phone

Kibuga parish mobile phone set was kept by the CBF who is a liaison between service providers working within the parish and the communities. The analysis showed that on average, about seven people were using the phone every week. However, frequency of use intensifies during planting and harvesting time. It was also noted that the phone was very useful in linking with the market, reaching veterinary services and keeping in touch or communicating with relatives. The link to market was mainly related to acquisition of inputs (seed and pesticides) from stockists, accessing price and market information so as to coincide harvesting (and bulking), especially of potatoes and beans with better prices. On the other hand the phone was very helpful when there was an outbreak of foot and mouth disease whereby about five farmers raised concern to the CBF who called the district veterinary officer (DVO). The DVO responded and vaccinated all animals in the parish. To ensure the sustainability of the strategy, the parish marketing committee/coordinating committee set up a modest rate of UShs 500 as the cost per unit used by any person who used the phone. Money generated from the phone is used for different purposes: buying airtime, saving to repair in case of any problem and also to buy another phone.

In Karujanga parish the cost per call on the phone was more expensive (UShs 500) than pay phones around the trading centre (UShs 100) which prompted most farmers to go for a cheaper phone. This has resulted in low usage of the phone. The analysis showed that on average 1 person was using the phone per week which was far low compared to 15 people per day using the general pay phones. The project phone gets seasonality of necessity, for example, when seeking for seeds, produce markets (as compared to those in office or business) and cost. Also one factor that might have influenced the low usage of the phone is the number of farmers who have their own personal mobile phones (estimated at about one phone per five people) in the parish.

Rwanyana parish had a village table set phone that was used by an average of five farmers per week. In this parish, the story was different. The CBF was the greatest asset, trustworthy, sober and always available. The project phones had more users (at least five users a week) because the commercial pay phones nearby had very poor network and were more expensive (charging UShs 1000 per unit as compared to the rate of the 'project' phone of Ushs 400). This phone was well promoted whereby the committee put up a sign post to inform people about the phone and its cost; however, the competitors kept removing the sign posts. About 80–90% of the people in the village are aware of the availability of the VICE. This parish has recorded the airtime loaded and calls made. It was found that they had saved about UShs 42,000 from phone calls. The savings are intended to buy another phone and therefore enhance a micro-financing scheme whereby members in need will be able to acquire 'soft' loans. This is in line with the study by Harris (2002) who has reported that there are some remarkable examples emerging that show that financial sustainability is possible, if built into the structure of a telecentre from the outset. On the other hand, a study by Mukhebi et al. (2004) observed that the financial sustainability of the centre is achieved when the centre provides basic services, namely Internet/e-mail, phone, fax and access to agricultural production and market information (electronic and hard copies). It charges placement fees for commodity offers to sell and bids to buy and market rated charges for communication services. Also sustainability can be achieved if the range of services offered by the centre was expanded beyond the basic services to include fixed and mobile phones, fax, photocopying, document preparation including identification, business and wedding cards, Internet and e-mail, library services including print and video and farm input supply delivery, especially seed and fertilizer.

It was found that the phone in Mugandu parish is used by about 20 people on average because of its cheap rates (UShs 300 per unit); however, the problem facing the phone is unstable network. About 60% of the people in the parish know about the phone. This means if the network was good, many people would benefit from the use of this phone.

The Radio Message

An analysis of the radio broadcasting accessing pattern showed that farmers in Kibuga parish accessed two radio stations, which were Voice of Kigezi (VOK) and Uganda Broadcasting Corporation (UBC). Farmers listened to different programmes; however, as a way of boosting marketing information, an announcement of commodity prices in Kabale and Katuna was made on VOK and broadcasted at 7:30 pm. Prices of the commodities such as potatoes, meat, beans, matooke, chicken, cabbage, carrots and wheat were included in the programme. In Karujanga parish two radio stations, VOK and Radio West, were accessed. The most favourite radio station was VOK because of clarity and the popular programme '*enkubito*' that talked about HIV/AIDS and family planning. The timing of the programme was said to be within the schedule of farmers. AHI made price/market announcements at 7:30 pm. Rwanyana received most radio stations including Rwanda stations.

The analysis showed that the information broadcasted in the radios was too general which could not be specifically applied by the farmers. Much as it was about prices one would prefer targeted information like price, availability and characteristic of market (place, quality and quantity of demanded commodity) and how to access such a market. During the discussion with farmers they voiced their concern that they would prefer a full programme that provided information package, especially on soil and water conservation, pests and diseases for newly introduced crops like apples. Farmers preferred if such programmes were presented by a renowned farmer on an active and interactive basis to allow live calling in for clarifications, questions and sharing of experiences.

The Paper Products

Generally the reading culture of farmers was observed to be very poor. Very few farmers used the library. In Kibuga parish, the library was well stocked; probably due to the reading culture of the CBFs, just as they brought information for personal interest, they were able to bring in information other that which was supplied by the AHI. Some were translated into local language, e.g. potato production by Steven Tindimubona (UNASPPA). In Karujanga parish, the library/VICE had very little material and these were never used by anyone (not even the custodian!). It was found that Rwanyana farmers made good use of the library, more than the rest of the parishes. However, the library is not well stocked to meet the interests of some readers, e.g. young farmers. In some cases, school pupils also accessed and used the library, though mostly they sought novels (which were not part of the collection). To make the library attractive, the CBFs put in their own book collections for others to use. The main snag of using library in Rwanyana is low publicity and accessing more books/materials.

The analysis showed that more than 100 people in Mugandu have visited the library/VICE but limited its use due to lack of diversity of material in the library. Farmers from this parish could use the main library in Rubaya sub-county headquarters which is 300 m away. It was found that farmers preferred information on crops of economic importance and/or with new technologies. Farmers were less willing to look for information about sorghum because it is considered as an indigenous crop with limited use (local consumption in the form of Bushera and Muramba), whereas potato has a market. Exposure to new/improved technologies increases the likelihood of seeking written information. This implies that creation of demand for paper products requires exposure of valued technologies. Analysis of farmer needs and prioritization followed by training and demonstrations would enhance the use of paper products. Materials with illustrations enhanced readability and understanding more than text material.

In general it was found that the information centres are better utilized by the contact farmers who usually demonstrate the technologies on their farms. So increasing the number of trained contact farmers who could demonstrate the technologies would be a better way of increasing information utilization as compared to use of paper products alone.

Gender Factor on Use of Parish-Level (VICE) Mobile Phones

One of the objectives of AHI-ACACIA project is to develop effective processes that are both gender sensitive and equitably managed for farmers to articulate their information needs. During the study the project staff at Kabale implemented a tracking system for each phone to assess who had been using the phones and for what purpose between September and October 2006 for three parish groups.

The results of the tracking are presented in Figs. 4, 5 and 6. It was found that the general phone usage was mostly in Rwanyana parish (48.1%) followed by Kitooma (33.3%) and Mugandu (18.5%) (Fig. 4). Higher usage at Rwanyana is attributed to cheap rates per calling as compared to other pay phones in the parish. The low usage in Mugandu is attributed to network problem which renders the phone unusable most of the time. On the other hand the overall gender analysis of usage of the phone showed that 59.3% of farmers who use the phone are males while 40.7% are the female farmers. Similar findings were reported by the ACACIA-funded research that in West Africa women have 36% fewer ICT-related opportunities and benefits than men (Mottin-Sylla and Marie-Hélène, 2005 as quoted by ACACIA, 2006).

The analysis of gender in the usage of phones across the parishes found that more male farmers made use of the phones in both Rwanyana (76.9%) and Mugandu (60.0%). On the other hand more female farmers made use of the telephone in Kitooma parish (66.7%) than in other parishes (Fig. 5).

The general analysis of the purpose of using the phone showed that women used the phone to request information on NRM and agriculture more than men across the three parishes, while more men used it for personal reasons and to search for market information (Fig. 6).

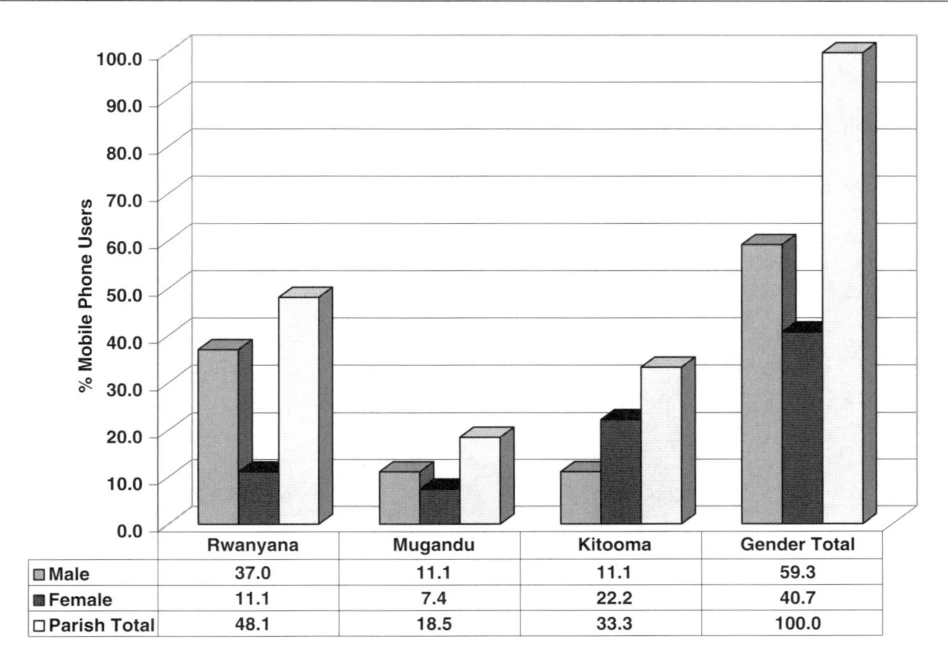

Fig. 4 General use of the mobile phone across the three parishes and gender

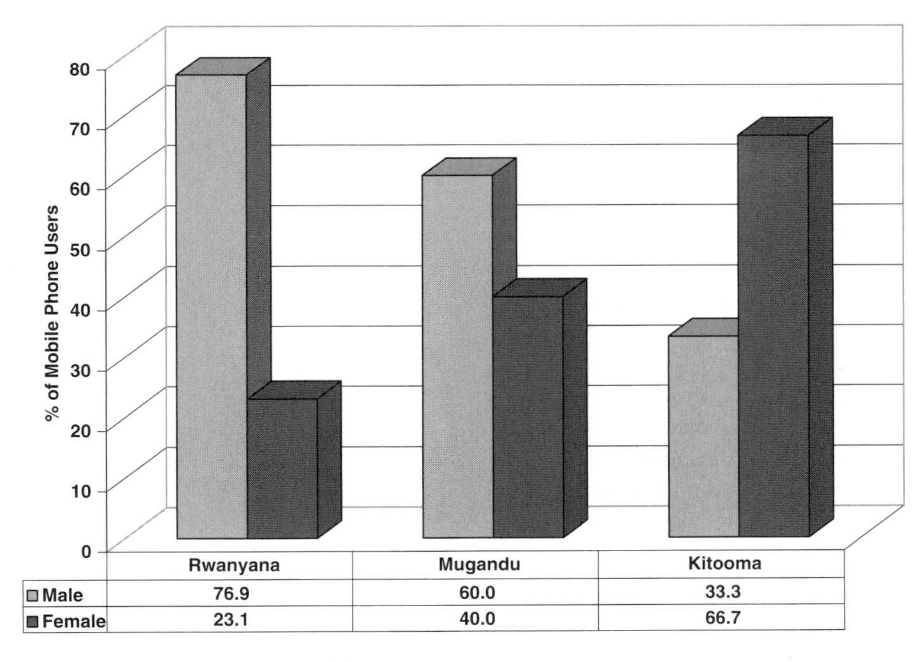

Fig. 5 Mobile phone usage by gender for each parish

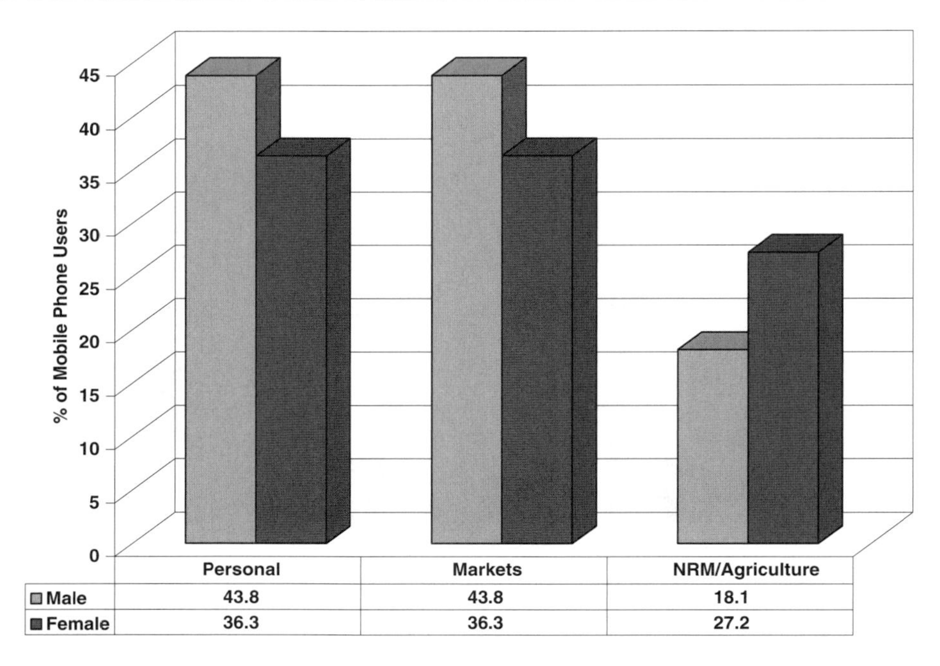

Fig. 6 Purpose of mobile phone usage by gender

Preliminary Conclusions and Recommendations

From the discussions and observations made, it can be concluded that farmers were more excited about the phone than about either the radio or paper products. However, in most situations, the phones appeared more to be assets of the committee than a resource to help the community. Materials in the libraries were rarely used, which raises concern about publicity and interests of the communities. Radio announcements did not seem to be of much help to the farmers. Preference was for programmes rather than announcements; this calls for more needs assessment for information dissemination and flow for the betterment of the communities.

There is a need for information delivery systems to be owned and managed by communities as it will create a sense of ownership in the communities as against the way it is now. There is a need for involvement of local government (at sub-county and parish levels) in developing the delivery system and as part of the decision-making process. This will ensure some sort of sustainability of the planned activities after the project tenure. To appropriately implement the use of ICTs at community levels, proper training and monitoring for information and communication should be enhanced. Farmers' information needs change according to time of the year, type of projects implemented in the district and level of communication between farmer groups. For that matter capacity building (managerial and technical) is required at all levels of telecentres in order to institutionalize the system.

References

ACACIA (2006) ACACIA prospectus 2006–2011. http://www.idrc.ca/acacia/ev-113431-201-1-DO_TOPIC.html

AHI-ACACIA II (2006) Improved access to information for development. African highlands initiative (AHI), Kampala, Uganda. Annual report Dec 2006

Asingwire N (2003) End of term evaluation report: electronic delivery of agricultural information to rural communities in Uganda. IDRC project No. 100206

Batchelor S, Scott N (2005) Good practice paper on ICTs for economic growth and poverty reduction. DAC J 6(3):1–69

CAADP-NEPAD (2002) NEPAD's comprehensive Africa agriculture program. http://www.fao.org/documents/show_cdr.asp?url_file=//docrep/005/y6831e/y6831e-06.htm

Etta FE and Parvyn-Wamahiu S (ed) (2003) Information and communication technologies for development in Africa: Vol 2. The experience with community telecentres. International Development Research Centre (IDRC) and the Council for the Development of Social Science Research in Africa (CODESRIA): Ottawa, Canada. Available at http://www.idrc.ca/openebooks/006-3/

Harris R (2002) A framework for poverty alleviation with ICTs. http://www.communities.org.ru/ci-text/harris.doc
Internet World Stats (2005) Internet usage statistics for Africa. http://www.internetworldstats.com/stats1.htm
LINK Centre (2005) Towards an African e-index – household and individual ICT access and usage across 10 African countries [Homepage of LINK Centre, Wits University], [Online]. http://www.researchictafrica.net/
Mayanja M (2002) The African community telecentres: in search of sustainability. http://www.google.co.ke/search?hl=en&q=Mayanja+and+ICT&btnG=Google+Search
Mottin-Sylla, Marie-Hélène (2005) Une Inquiétante Réalité: Fracture Numérique de genre en Afrique Francophone. Dakar, Senegal
Mukhebi A, Asaba JF, Day R (2004) A business model of a financially sustainable agricultural knowledge centre. CAB International, Nairobi
NSGRP (2005) National Strategy for Growth and Reduction of Poverty. Dar es Salaam
Ureta S (2005) Variation on expenditure on communications in developing countries – a synthesis of the evidence from Albania, Mexico, Nepal and South Africa (2000–2003). London School of Economics, Media and Communications Department, London
Verlaeten MP (2002) Policy frameworks for the knowledge based economy: ICTs, innovation and human resources – an OECD global forum. OECD, Paris, Available: http://www.oecd.org/dataoecd/49/11/1961062.pdf
Vodafone (2005) Africa: the impact of mobile phones, The Vodafone policy paper series, number 2. London
World Bank (2002) Empowerment and poverty reduction—a sourcebook. World Bank, Washington, DC

Market Access: Components, Interactions, and Implications in Smallholder Agriculture in the Former Homeland Area of South Africa

A. Obi, P. Pote, and J. Chianu

Abstract While insufficient market access is recognized as a key institutional constraint to smallholder development in Africa, the generalities that characterize much of recent research on the subject mean that the mechanisms by which market access exerts influence are not well understood. Drawing on household-level data from the former "independent homeland" of South Africa, this chapter employs the logistic model to isolate key components of market access, including access to market/price information, productive inputs, and infrastructure. Differences in the extent to which these factors constrain smallholder crop and livestock farmers buttress the expectation of greater policy impact from research that takes a wider view of market access. The chapter fits the foregoing finding against the backdrop of South Africa's troubled past that continues to negatively impact on its agricultural economy. How this history has influenced intra- and inter-sectoral relationships and coordination is discussed. The chapter further presents results that shed light on how policy and smallholder support measures can be better targeted to address the problems of limited market access in the communal/rural areas in order to increase the use of agricultural inputs such as mineral fertilizers, enhance agricultural productivity and equity, as well as improve overall rural livelihoods. Results will be extrapolated to other rural areas of sub-Saharan Africa which, in many respects, are similar to the former "independent homelands" of South Africa.

A. Obi (✉)
Department of Agricultural Economics and Extension, University of Fort Hare, Alice, Eastern Cape, South Africa
e-mail: aobi@ufh.ac.za

Keywords Agricultural productivity · Institutional and technical constraints · Market access · Rural livelihoods · Smallholders

Introduction

With democratic rule in South Africa, policies were introduced to redress the extreme wealth inequalities engendered by apartheid rule. There was expectation that enhanced access to productive resources such as land and technical support would quickly translate into increased agricultural productivity for these farmers, leading to increased incomes and reduction of poverty (Makhura and Mokoena, 2003). Recent studies suggest that there has rather been a growing pauperization of the citizens, especially the black population, manifesting mostly in deteriorating numbers for unemployment rates and poverty levels (Klasen, 1997; UNDP, 2007). For instance, while the broadly defined unemployment rates stood at 31.2% in 1993, they had deteriorated to 37.6% by 1997. The provincial data are equally disturbing with the Department of Labour data suggesting that unemployment rates in the Eastern Cape in 2003, for instance, were in the order of 30–70% (May et al, 1998; Department of Labour, 2003).

As many studies show, poverty is closely related to unemployment, among other factors (Klasen, 1997). Expectedly, the Eastern Cape province which has the highest unemployment rate also has the highest poverty rates, estimated at 71% in 1998 (May et al, 1998), and obviously worse today by all accounts (Department of Labour, 2003; Development Bank of Southern Africa, 2005). According to the available

A. Bationo et al. (eds.), *Innovations as Key to the Green Revolution in Africa*,
DOI 10.1007/978-90-481-2543-2_118, © Springer Science+Business Media B.V. 2011

data, this high poverty rate is accompanied by the highest level of income inequality in the world (Klasen, 1997; Lam, 1999). According to the UNDP (2007), the Gini coefficient estimated for South Africa for 2006 stood at about 0.59 which is among the most unequal in the world. Such a result is consistent with the fact that among the medium human development countries to which South Africa is placed by the UNDP, it is one of the few whose Human Development Indices actually deteriorated since the early 1990s, having fallen from 0.735 in 1990 to 0.653 in 2004 (UNDP, 2006).

Analyses based on comparable consumption aggregates from the Income and Expenditure Surveys (IES) suggest that over the 5- to 6-year period between 1994 and 2000, consumption growth has slowed to less than 1% per capita per annum, overall poverty headcount has remained unchanged, and the poverty gap has escalated (DLA/DoA, 2005). At the presentation of the 2004 Budget, the South African Finance Minister, Mr. Trevor Manuel, bemoaned the emergence of a "... *second economy* characterized by poverty, inadequate shelter, uncertain incomes and the despair of joblessness..." in which many South Africans are currently "trapped" (Government of South Africa, 2004).

The former "independent homeland" areas, namely Transkei, Ciskei, Venda, and Bophuthatswana, which were granted "independence" by the apartheid regime (Berry, 1996; Raeside, 2004), have exhibited these problems much more than any other part of the country. According to Van Zyl and Binswanger (1996) and Government of South Africa (2003), these former "homelands" were characterized by inadequate market access, infrastructure, and support services. A study conducted in 1999 showed that 60,000 commercial farmers controlled 86% of the land and an estimated 1.25 million blacks operated at subsistence levels on about 14% of the available land (Eicher and Rukuni, 1996).

It is crucial for both national and regional development that the food marketing system is functioning optimally (Eicher, 1999; Van Tilburg, 2003). As the South African Department of Agriculture has observed, access to markets is imperative as a mechanism for integrating new and emerging farmers into the country's agricultural economy (DoA, 2000). This view is shared by several experts, including Jayne et al. (1997), World Bank (2002), and Magingxa (2006). The Government of South Africa has accordingly initiated far-reaching trade liberalization and market deregulation processes targeting the agricultural sector (Makhura and Mokoena, 2003).

But to date, these measures have produced little or no improvement in the circumstances of the rural small-scale producers of South Africa whose conditions have either stagnated or actually become worse. Many researchers go as far as concluding that, for all practical purposes, the measures introduced to liberalize the domestic food market and integrate the country into the international system may actually have hurt rather than helped the small-scale farmers within the former homelands of South Africa (Makhura and Mokoena, 2003; Orthmann and Machethe, 2003).

It is against the foregoing background that this study aimed to examine the extent to which the smallholder's predicament can be linked to constraints at the technical and institutional levels which mediate access to profitable marketing opportunities, among other difficulties, as well as determine by what mechanisms such insufficient market access hinders smallholder development.

Materials and Methods

Against the foregoing background, a sample of 80 farming households was drawn from 3 towns in the former Ciskei "homeland" of South Africa, now included in the Nkonkobe Municipality of the Amatole District of the Eastern Cape Province. The three towns, Fort Beaufort, Seymour, and Balfour, were drawn randomly from the six main towns of the municipality.

Nel and Davies (1999) explain that racial conflicts in the early days of white settlement and ensuing apartheid regime resulted in laws that limited the access of the black population to the means of production, typically agricultural land. This situation has not changed much today. Where the black population has moved out of the traditional villages into the towns, they are more commonly found in the new settlements on the fringes where low-cost, poorly designed housing are being constructed by the new government under the Reconstruction and Development Programme (RDP).

General information on the institutional setup was obtained by open-ended interviews of community leaders and focus groups to complement information obtained through literature study. Subsequently, a single-visit household survey was undertaken using structured questionnaires which covered a wide range of issues, including demographic information, costs and returns, marketing arrangements, and access within a broad definition. The key variables utilized for the logistic regression are market access, equipment ownership, access to information, distance to market outlet, asset value, availability of infrastructure, and total income. The demographic data were not included in the logistic regression since market access actually has more to do with such factors as the physical conditions of the infrastructure, access to production and marketing equipment, and the way the marketing functions are regulated (Killick et al., 2000; IFAD, 2003).

The binary choice model was employed to estimate the probability that a smallholder farmer in the enumerated localities would have unsold produce in the current farming season on the basis of the reported market performance for 2006. Since only two options are available, namely "produce sold" or "produce not sold," a binary model is set up which defines $Y=1$ for situations where the farmer sold all produce and $Y=0$ for situations where some or all produce was not sold. Assuming that x is a vector of explanatory variables and ρ is the probability that $Y=1$, two probabilistic relationships can be considered as follows:

$$\rho(Y=1) = \frac{e^{\beta'\chi}}{1+e^{\beta'\chi}} \tag{1}$$

$$\rho(Y=0) = 1 - \frac{e^{\beta'\chi}}{1+e^{\beta'\chi}} = \frac{1}{1+e^{\beta'\chi}} \tag{2}$$

Since Eq. (2) is the lower response level, that is, the probability that some or all farm produce would not be sold, this will be the probability to be modeled by the logistic procedure by convention. Both equations present the outcome of the logit transformation of the odds ratios which can alternatively be represented as

$$\begin{aligned} \text{logit}\,[\theta\,(x)] &= \log\left[\frac{\theta\,(x)}{1-\theta\,(x)}\right] \\ &= \alpha + \beta_1\chi_1 + \beta_2\chi_2 + \cdots + \beta_i\chi_i \end{aligned} \tag{3}$$

thus allowing its estimation as a linear model, and for which the following definitions apply: θ is the logit transformation of the odds ratio; α the intercept term of the model; β the regression coefficient or slope of the individual predictor (or explanatory) variables modeled; and χ_i the explanatory or predictor variables.

In line with Hosmer and Lemeshow (1989) and Agresti (1990), the right-hand term in Eq. (3) is the natural logarithm of the modeled variables. A goodness-of-fit test, following Hosmer–Lemeshow, was conducted by examining the Pearson chi-square outcomes calculated from the table of observed and expected frequencies as follows:

$$X^2_{\text{HL}} = \sum_{i=1}^{g} \frac{(O_i - N_i\pi_i)^{\ 2}}{N_i\pi_i\,(1-\pi_i)} \tag{4}$$

where N_i is the total frequency of the items in the ith group, O_i the total frequency of obtaining particular event outcomes in the ith group, and π_i the average estimate of the probability that a particular event outcome in the ith group would be realized.

The foregoing operations were feasible within standard Minitab and SPSS packages. In relation to Eq. (3), the operation generated the odd ratios using the maximum likelihood procedure. The goodness-of-fit test examined the displayed results for the Pearson, Deviance, and Hosmer–Lemeshow methods, all of which gave high enough ρ-values to dispel doubts about the model fitting the data. It was therefore not necessary to consider alternative estimation procedures.

Results and Discussion

The summary statistics of the variables comprising demographic and production/marketing data is presented in Tables 1 and 2. In terms of the demographic characteristics of the sample, the summary statistics shows that the majority of the farmers were male, married, and aged about 57 years on average, with the youngest farmers being about 27 years while at least one farmer was as old as 91 years. Household size ranged from 2 to 21 but averaged about 6.7 persons. The sample suggests that the majority of the farmers had some education, mostly up to 7 years of primary school education although some did not have any education at all. A few of the sample farmers had post-secondary education. There was also evidence that some farmers supplemented their income by undertaking non-farm activities. It was also clear

Table 1 Summary statistics of demographic and socio-economic variables

Variable	Minimum	Maximum	Mean	Std deviation
Age	27	91	57.5	14.63
Household size	2	21	6.73	3.04
Education (years)	0	17	8.39	3.65
Area cultivated (ha)	0	25	2.59	4.95
Years of farming experience	1	55	21.5	10.38
Current enterprise experience	1	50	14.8	8.1
Total asset value (Rands)	0	240, 300	23,126	45,622
Crop income (Rands)	0	157, 575	8199.9	2,4790.3
Livestock income (Rands)	0	135, 000	9416.6	19,989.12
Total income (Rands)	0	157, 575	17,616.5	29,287.2
Market distance (km)	0	300	18.9	36.04

Table 2 Number of households as a proportion of total number of households

Modeled variable	Frequency	%
Fertilizer use	19	24
Farm planning	67	84
Assets	30	38
Infrastructure availability	28	35
Information availability	35	44
Credit need	58	73
Credit access	12	15
Equipment ownership	38	48
Unsold produce	9	11

from the data that the majority of the surveyed farm households had been in the farming business for some time, with length of experience of up to half a century. The data also picked up a few new entrants into the farming business.

Regarding the distribution of assets and income across the sample, the study reveals a pattern that closely mirrors the situation with respect to the overall population. For one thing, the data demonstrate pronounced inequities in terms of gross earnings from both livestock and crop production and the ownership of tangible/valuable assets in the study area.

Table 1 presents the picture for both income and assets among other variables. As Table 1 shows, the gross value of farm produce ranged from nothing at all to as much as R158,000 (equivalent to US $23,000) in 1 year, and the market value of assets ranged from those with negligible valuable assets to those with as much as R240,000 (equivalent to US $35,000) at current prices. These data were not analyzed for purposes of estimating household incomes but merely as a means for placing the households into rough socio-economic categories. However, they do show that about half of the survey households lacked the possibility to earn much more than R20 per day (or about US $3) on the basis of their reported gross value of farm income. The very high standard deviations of both asset value and gross farm income variables further confirm the huge disparities in socio-economic status even within the smallholder class, suggesting that it is by no means a homogenous category. The results similarly reveal other areas of inequalities in the smallholder sector and the serious constraints that this segment of the population still faces (Table 2).

The results of the analysis are presented in Tables 3, 4, and 5 with respect to the maximum likelihood estimates, the goodness-of-fit tests, and the associations between the response variable and the set of predictors modeled. At the initial fitting of the model, it was found that both equipment and infrastructure were not statistically significant even though the goodness-of-fit tests suggested that the model could be a good description of the data ($p > 0.05$). Since significance was required for efficient prediction, both variables were dropped by backward elimination to obtain a final structure for the model as presented in Table 3. Although the log-likelihood ratio in this case was somewhat higher than in the earlier run (Table 4), the fact that it was declining with successive iterations provided sufficient grounds to accept the model as a good description of the data.

These results could mean that the most important factors determining the chances that a smallholder in the survey area would face marketing constraints were access to information, asset ownership, and value of agricultural production. The picture is not ambiguous with respect to information since availability of information about such important market variables as

Table 3 Results of logistic regression analysis

	Maximum likelihood estimates						
						95% CI of odds ratios	
Parameter	Coeff.	SE Coeff.	*Z*-value	*p*-value	Odds ratio	Lower	Upper
Intercept	–7.640	2.173	–3.52	0.000			
Information	4.174	1.558	2.68	0.007	65.00	3.06	1378.80
Total assets	4.209	1.350	3.12	0.002	67.32	4.77	949.36
Total income	2.768	1.226	2.26	0.024	15.93	1.44	176.05

Table 4 Results of log-likelihood iterations

	Log-likelihood values	
Steps	First model run	Final model run
0	–28.137	–28.137
1	–18.118	–19.099
2	–13.109	–14.902
3	–11.809	–14.121
4	–11.466	–14.008
5	–11.433	–14.004
6	–11.433	–14.004
7	–11.433	–14.004
8	–11.433	

Table 5 Goodness-of-fit tests

Method	Chi-square	Degrees of freedom	*p*-value
Pearson	3.936	4	0.415
Deviance	4.530	4	0.339
Hosmer–Lemeshow	2.580	3	0.461
Brown			
General alternative	0.092	2	0.955
Symmetric alternative	0.058	1	.810

prices, supply, and demand is crucial to marketing performance. But the situation with respect to asset ownership and farm income may reflect the complex environment in which the smallholder operates and the interrelationships among the farm business and the rest of the farm household.

The results regarding infrastructure can also reflect the special circumstances of the smallholder farm family as well as the nature of the interactions among the explanatory variables. This constraint would normally only kick in when there is an output to sell and would have no significance to a smallholder who does not have anything to sell. If the farm is already so severely constrained in terms of insufficient information flow and lack of assets that production is limited, availability or otherwise of infrastructure would have little chance of influencing extent of marketed surplus. The same would be true if there is marketable surplus available but information is lacking about opportunities to sell them profitably. The results in Table 3 demonstrate that the odds ratios for information and assets are so high (at 65.0 and 67.3, respectively) that they most probably choked off other effects. For instance, gross farm income, although highly significant at ρ=0.024, was about four times less likely to constitute a binding constraint to marketing of farm produce than the other two explanatory variables, namely access to information and asset ownership. It would therefore have made no difference at all whether or not the farmer was served by a well-paved road network and had affordable power and water at his/her disposal.

Table 5 presents the results of the goodness-of-fit tests which show that, on the basis of alternative criteria, the model fitted the data very well. For instance, as Table 5 shows, the ρ-values were consistently higher than the chosen probability level for the logistic regression modeling (ρ=0.05).

Similarly, the test of the degree of association between the response variable and the predicted probabilities gave encouraging results regarding the predictive power of the model. In the light of the foregoing results, the logistic regression model can be summarized in the form of the following equation:

$$\log(Y) = -7.64 + 4.174\,X_1 + 4.209\,X_2 + 2.768\,X_3 \qquad (5)$$

where the modeled variables are assigned algebraic notations such that the unsold produce is Y, information is X_1, total assets is X_2, and total income is X_3.

Equation (5) will have the straightforward interpretation that reflects the strong positive influences of information, asset ownership, and level of farm production. It is also clear that there are chances that marketing will still occur even when these factors are frozen, but in that case a negative relationship is predicted.

Conclusions

This study aimed to test some generally held notions about the nature and pace of post-apartheid development and quantify the results in a way that would be relevant for policy. More specifically, it was intended to test the extent to which the farmers' market access is constrained by the technical and institutional environment in which they operate. A total of 80 farming households were surveyed in three towns of the Nkonkobe Municipality of the Eastern Cape Province of South Africa.

The descriptive analyses of a wide range of demographic, asset ownership, resource use, production, and marketing data confirm that serious livelihood challenges still remain. For one thing, income disparities are still substantial, even within a group that ordinarily would be considered homogenous. Thus, what one observes within the smallholder community is a reflection of the gross inequalities that characterize the wider South African society. Despite considerable farming experience and a commitment to farming demonstrated by their long-term planning, the majority of smallholders are unable to access the credit they need and therefore cannot acquire and use modern inputs such as mineral fertilizers and mechanical technologies. Access to publicly provided infrastructure such as good road network, water, and electricity is also limited and many farmers do not have access to the right amount of information and technical support that would make a difference in their farming business.

The results show that access to information, asset ownership, and level of output of farm produce, as proxied by the gross value of agricultural production, had the most chance of influencing the extent to which the smallholder could sell marketable surplus during the year. When modeled separately, ownership of farm equipment, access to physical infrastructure such as good roads, water, and electricity, and distance to the market did not seem to have had much influence. However, when these are combined with other assets of the household, regardless of whether they are used for the home or farm, the results showed very high statistical significance. These results probably reflect the complex interactions that take place within the smallholder environment.

Of course, this does not mean that, in and of themselves, infrastructure and equipment are not crucial to both the production and the marketing aspects of smallholder economic life. Rather, the result probably only highlights the relative strength of access to information and asset ownership vis-à-vis infrastructure and equipment taken separately. The result equally provides insights about the complex intra-household processes with respect to decision making about resource use. It is also crucial to appreciate the fact that a factor such as infrastructure and distance to market can only be critical when a farmer knows about profitable marketing opportunities beyond the farm gate and has the means to transfer produce to take advantage of such opportunities. Even with the best infrastructures, a farmer may still be unable to sell if there is insufficient information about profitable opportunities. Similarly, the physical distance between the farm and market is irrelevant to situations where the farmer does not have anything to sell. Taking a more holistic view of the smallholder's socio-economic and cultural context and defining market access in the widest possible sense will no doubt improve the policy relevance of research into the constraints to smallholder development.

These results have other important practical implications for policy, especially with respect to conclusions regarding access to information and issues related to the strengthening of smallholder production capacity as a precondition for enhanced livelihoods through market participation. Much of the problems that smallholders confront have to do with knowledge. These issues can be easily addressed by putting at the disposal of the farmer easily accessible information. The gap created by the dismantling of technical services to smallholders as part of the agricultural restructuring program following multi-party elections is now beginning to hurt and the time has come to reinstate these services and align the agricultural extension services to the needs of smallholders. This is true not only for the former "independent homelands" of South Africa, but for the rest of rural Africa where conditions are identical.

Acknowledgments Some aspects of the research for this chapter were financed out of a grant by the South Africa–Netherlands

Research Programme on Alternatives in Development (SANPAD) and this support is gratefully acknowledged.

References

Agresti A (1990) Categorical data analysis. Wiley, New York, NY

Berry B (1996) Divisions of South Africa 1910–1994. www.allstates-flags.com

Department of Labour (2003) Speech by the minister of labour at gala dinner, Cicira College, Umtata 16 August 2003. SA Department of Labour, Pretoria

Department of Land Affairs/Department of Agriculture (2005) Land and agrarian reform in South Africa: an overview in preparation for the land summit, Pretoria, Ministry of Agriculture and Land Affairs, South Africa, 27–31 Jul 2005

Development Bank of Southern Africa (2005) Development report 2005 – overcoming underdevelopment in South Africa's second economy. Midrand (Johannesburg), Development Bank of Southern Africa (DBSA), Human Sciences Research Council (HSRC), and United Nations Development Programme (UNDP)

Eicher CK (1999) Institutions and the African farmer, Issues in agriculture No. 14. Consultative Group on International Agricultural Research (CGIAR), Washington, DC

Eicher CK, Rukuni M (1996) Reflections on agrarian reform and capacity building in South Africa, staff paper No. 96–3, Department of Agricultural Economics, Michigan State University

Government of South Africa (2003) Programme operating manual for the municipal infrastructure grant (MIG) – draft 4. Department of Provincial and Local Government, Pretoria

Government of South Africa (2004) Budget speech by minister of finance Mr. Trevor Manuel. Department of Finance, Pretoria

Hosmer D, Lemeshow S (1989) Applied logistic regression. Wiley, New York, NY

IFAD (2003) Promoting market access for the rural poor in order to achieve the millennium development goals, Discussion Paper. International Fund for Agricultural Development, Rome

Jayne TS et al (1997) Improving the impact of market reform on agricultural productivity in Africa: how institutional design makes a difference. Michigan State University International Development Working Paper No. 66, East Lansing, Michigan State University

Killick T, Kydd J, Poulton C (2000) Agricultural liberalization, commercialization and the market access problem in the rural poor and the wider economy: the problem of market access. International Fund for Agricultural Development, Rome

Klasen S (1997) Poverty, inequality and deprivation in South Africa: an analysis of the 1993 SALDRU survey. Soc Ind Res 41:1–3

Lam D (1999) Generating extreme inequality: schooling, earnings, and intergenerational transmission of human capital in South Africa, Research paper. University of Michigan, Ann Arbor

Magingxa LL (2006) Smallholder irrigators and the role of markets: a new institutional approach. Unpublished PhD thesis, University of the Free State, Bloemfontein

Makhura M, Mokoena M (2003) Market access for small-scale farmers in South Africa. In: Nieuwoudt L, Groenewald J (eds) The challenges of change: agriculture, land and the South African economy. University of Natal Press, Natal, pp 137–148

May J et al (1998) Poverty and inequality in South Africa. Report prepared for the office of the executive deputy president and the inter-ministerial committee for poverty and inequality, Pretoria, South Africa

National Department of Agriculture (2000) Agricultural marketing – a discussion document. Pretoria, South Africa (mimeo)

Nel EL, Davies J (1999) Farming against the odds: an examination of the challenges facing farming and rural development in the Eastern Cape Province of South Africa. Appl Geogr 19:253–274

Orthmann G, Machethe C (2003) Problems and opportunities in South African agriculture. In: Nieuwoudt L, Groenewald J (eds) The challenges of change: agriculture, land and the South African economy. University of Natal Press, Natal, pp 47–62

Raeside R (2004) Flags of the world. www.allstates-flag.com/fotw/flags/za

UNDP (2006) Human development report 2006 – beyond scarcity: power, poverty, and global water crisis, New York, United Nations Development Programme

UNDP (2007) Country programme outline for South Africa 2007–2010. New York, Executive board of the united nations development programme and the united nations population fund (UNDP/UNFPA)

Van Tilburg A (2003) Framework to assess 'worldwide' the performance of food marketing systems. Paper presented at the brown bag seminar of the Department of Agricultural Economics, Michigan State University, 11 Nov 2003

Van Zyl J, Binswanger HP (1996) Market-assisted rural land reform: how will it work? In: van Zyl J, Kirsten J, Binswanger HP (eds) Agricultural land reform in South Africa. Oxford University Press, Cape Town, pp 413–422

World Bank (2002) Empowerment and poverty reduction – a sourcebook. International Bank for Reconstruction and Development/The World Bank, Washington, DC

Improving African Agricultural Market and Rural Livelihood Through *Warrantage*: Case Study of Jigawa State, Nigeria

M.A. Adamu and J. Chianu

Abstract Low agricultural commodity prices are the key causes of poverty in many sub-Saharan African countries. Efforts to improve rural livelihoods must improve agricultural produce marketing. This study was carried out to ascertain how *Warrantage* (a micro-credit scheme) can be used to improve agricultural marketing and livelihoods in *Madana* community, *Jigawa* State, Nigeria. The design was an action research approach, based on supervised enterprise project framework of the Department of Agricultural Extension and Economics of the University of Cape Coast, Ghana. Data were collected using questionnaires, interviews and group and personal discussions. Analysis was carried out using qualitative and quantitative methods. Prior to intervention, farmers in the area sold their produce at rock-bottom prices immediately after harvest to meet urgent cash needs, resulting in low returns on investment and limited use of improved inputs. A pilot phase led to the observation that farmers could overcome the above problem if offered the opportunity to hold on to the produce for a few months after harvest to take advantage of high prices during the lean season. Through the *Warrantage* system, farmers have been able to timely access subsidised farm inputs, increase production, store their produce and sell during the lean period when prices are high. Farmers are increasingly able to meet their cash and other needs. The impact of the project has generated widespread interest among other farmers even outside the *Madana* community who are adopting the *Warrantage* as a model for sustainable self-help and a robust means of improving their livelihoods.

Keywords *Warrantage* · Improved rural livelihoods · Improved farm commodity prices · Nigeria

Introduction

Agriculture plays an important role in the economy of Nigeria. The sector remains the largest contributor to the country's gross domestic product (GDP), accounting for over 38% of the non-oil foreign exchange earnings and employing about 70% of the active labour force (Federal Ministry of Agriculture, 2001). Low producer price, marketing restrictions and incidences of drought are some of the major problems that beset the agricultural sector of Nigeria. Besides, the sector has suffered much neglect by the government since the discovery of petrol in commercial quantity in 1958. The neglect has led to various problems including low farm produce marketing price (especially during crop harvest period) and inadequate or a complete non-availability of credit facilities and other essential farm inputs for agricultural production and marketing. This state of affairs has led to widespread poverty among the rural population, about 70% of whom are farmers.

If poverty is considered as low levels of annual cash income per household, then reducing it is about raising average annual cash income levels of households. If a particular level of annual cash income per person (or cash income per capita) is used as a poverty line below which one is considered to be poor, then the number or proportion of people who cross that line or stay above the line (or who are promoted out of

M.A. Adamu (✉)
Green Sahel Agro Venture, Gumel, Jigawa State, Nigeria
e-mail: muhddansista@yahoo.com

A. Bationo et al. (eds.), *Innovations as Key to the Green Revolution in Africa*,
DOI 10.1007/978-90-481-2543-2_119,

poverty) could be used to measure poverty reduction attainment. Providers of financial services who aim to empower and enable people to cross over and stay above the poverty line have focused on credit, especially credit for small enterprises such as agricultural production and agricultural commodity marketing. The high incidence of poverty among the rural population has fuelled ethnic and religious conflicts in parts of Nigeria in recent years. Indeed, any attempt to reduce the underlying causes of these conflicts must necessarily address the issues of poverty among the rural population (Adamu, 1985). However, general attempts in this direction have not yielded the right impact. Solutions at first have been piecemeal and ad hoc, lacking clear focus.

For the rural poor, improved access to credit facilities is one of the critical things needed for better livelihoods and other related impacts. No matter what is done, if poverty is not properly tackled, no development programme can be successful in Nigeria. In most developing countries (including Nigeria), credit has been an important ingredient for development. Finance ranks top among the inputs required to embrace modern agriculture in Nigeria. This is because funds are needed to pay for other farm inputs (e.g. seeds, mineral fertilisers, farm tools, equipment).

However, the resource-poor subsistence and peasant farmers are often unable to expand their production of different crops due to lack of access to credit facilities. Those who are able to produce more often encounter problems with the marketing of the extra productions. The most pronounced of these is low market prices at the time of harvest and immediately post-harvest. If it is possible for farmers to wait a little, they stand to reap better prices since most often the prices of such commodities soar 2 or 3 months from the time of harvest. Unfortunately, most farmers cannot store their farm produce to take advantage of the possible high prices during the lean season. They need to attend to some urgent household financial needs. Since most farmers have the habit of selling their farm produce during the harvest period or immediately post-harvest when prices are at rock-bottom levels, they often encounter income poverty. Consequently, many of them are often not able to purchase and use appropriate farm inputs during the following cropping season. Several measures taken by both the state and the national governments to resolve this problem have not yielded the right impact. It has therefore become very necessary to introduce a more innovative approach that is more community initiated, centred and owned in order to deal with the problems.

The *Warrantage* System

The *Warrantage* system was developed by the International Crops Research Institute for the Semi-Arid Tropics (ICRISAT) in collaboration with the Food and Agriculture Organization of the United Nations (FAO), the International Fertilizer Development Centre (IFDC), some commercial banks, several NGOs and donors (ICRISAT, 2001). The *Warrantage* credit approach bears all the hallmark of a community-centred approach that is working 'miracle' in several communities in the Sudano-Sahelian region of West Africa. The *Warrantage* is a community-based micro-credit scheme that allows the community or depositors (rural farmers or farmers' groups) and the lending agency (e.g. financial institutions, traders, NGOs) to hold full responsibility for the safekeeping of a merchandise (or farm produce) delivered by depositors (rural farmers or farmers' groups) into special warehouses. The merchandise is then used as collateral for the financial and commercial loans given to the depositors by the bank, traders, government or non-governmental financial sector/organisations to the depositors. Farmers stock their produce at harvest with local entrepreneurs or farmer organisations, with financial support from non-governmental organisations and/or commercial banks. The farmers then receive credit. The farmers and the local entrepreneurs sell the produce at an agreed time in the future (anywhere between 4 months after harvest and the next planting season in the case of Nigeria). Through this arrangement, farmers are able to earn 40–45% additional profit. This margin of profit is possible because at harvest time in October, prices are generally low but start to go up by March. *Warrantage* allows the farmers the opportunity to get good prices for their farm produce while at the same time enhancing their (farmers') access to adequate nutrition and increased income in order to meet other urgent social needs. It is estimated that improving the marketing system through *Warrantage* credit facility will increase the production of rain-fed crops in the study area, which will also increase feed for livestock and manure for soil fertility enhancement (ICRISAT, 2001).

The 'Warrantage' is an excellent credit title, which does allow not only for the purchase of raw materials but also the financing and stocking of farm inputs such as mineral fertilisers and the seeds of improved varieties of various crops. The *Warrantage* credit facility allows the farmers opportunities of being at an advantage to sell their farm produce at times of the year when prices are better. This refers to the lean as opposed to the post-harvest period when prices are often at rock-bottom levels.

Objective of the Study

This study aims at ascertaining how the *Warrantage* (a micro-credit scheme) system could be leveraged to improve agricultural commodity marketing and livelihoods of farm households in *Madana* community, *Jigawa* State, Nigeria. The target was resource-poor farmers with the aim of helping them to invest in their own small businesses.

Materials and Methods

Study Area

The study was conducted in *Madana* area of *Maigatari* Local Government Area of Jigawa State within the framework of the Supervise Enterprise Project (SEP) of the Department of Agricultural Extension and Economics, University of Cape Coast, Ghana. SEP is an action research, which involves trying out ideas in practice as a means of improving knowledge. *Madana* area is well known for the production of millet, sorghum, sesame, groundnut and cowpea.

Madana community was selected based on their need for this type of project and also for their record support and compliance with agricultural extension programmes. *Madana* is a rural area with human population of about 3800. It is located within the *Maigatari* Local Government Area of Jigawa State, Nigeria. Jigawa State is one of the 36 states of the Federal Republic of Nigeria, located at the extreme central northern part of the country sharing border with Niger Republic. Jigawa State has soil favourable for the cultivation of both rain-fed and dry season crops. The state covers a total land area of 2,315,400 ha (equivalent to ~23,154 km^2). The vegetation of Jigawa State is Sudan savanna. The annual rainfall varies between 635 and 1100 mm. The heart of Jigawa state economy is agriculture. This made the government of the state to put into practice some efforts to encourage private investment in agriculture and agro-industries by providing incentives, extension services and financial credit but without much success.

Madana has about 99% of its total population engaged in agriculture. The climate is characterised by 7 months dry season (from November to May) during which harmattan is experienced, 4 months rainy season (June to September) and 1 month transition between the two seasons (October). The annual rainfall in *Madana* ranges between 700 and 950 mm.

Study Methods and Instruments

A socio-economic analysis of the area was conducted prior to commencement of project implementation. This was done through focus group discussion, group meetings, observations and survey using interview and questionnaire. This provided the baseline on which to base assessment of the various achievements (e.g. increase in farm income, improvements in rural livelihoods) of the project. Data on the economic conditions of the project participants were among the baseline data collected using questionnaire. Monitoring and evaluation methods were used to underscore the lessons learnt and the experiences gathered during implementation. These were critical in informing about possible modifications in the design of the scheme as it expands to other areas.

Farmer Group Formation and Training

Four farmer groups of 15 individual farmers each were initially formed in order to facilitate group management. A central working committee (called an Apex committee) was formed with two members from each of the farmers' group. This committee served as the officials of the associations. The groups were registered both at the state and at the local government levels. Farmers who often encounter surplus farm produce voluntarily formed these groups to seek for ways of improving their own livelihoods in a sustainable fashion. The groups were used to train the other farmers on the *Warrantage* scheme as well as to undertake all the other activities of the scheme to ensure its smooth implementation. The researchers' role in the groups was to provide all the technical support as well as linking them with financial institutions (to access credit) and agro-input dealers (to access modern farm inputs).

Training was organised by Sasakawa Global 2000 (SG2000) Nigeria Project with the researchers as the main resource persons and facilitators. Co-facilitators from Ahmad Bello University and the Nigeria Agricultural and Rural Development Bank (NARDB) were invited to train the farmers on farm management and the roles of financial institutions in agricultural commodity marketing, respectively. Staff of Jigawa State Agricultural and Rural Development Authority (JARDA) was also invited to teach on post-harvest management issues and the importance of crop storage.

Sampling

The study used purposive sampling and selected 60 farmers that formed 4 groups (15 farmers each). Forming the members into four groups was only for easy handling of the participants. Farmers with land area ≥3 ha were selected for the study based on the assumption that this was the minimum farm size under which it is possible to generate farm produce surpluses that could be fed into the *Warrantage* system. Below a farm size of <3 ha, the assumption was that the level of food production would only be for feeding members of the household. It will not likely be possible for such farm households to have surpluses that could be fed into the *Warrantage* system.

Data Sources, Collection and Analysis

Both primary and secondary data were used for this study. Using questionnaires, interviews and participant observations as well as group discussions, primary data were collected from the 60 farmers sampled. Secondary data were collected from ICRISAT, SG2000, JARDA and various publications. Quantitative and qualitative data analyses were carried out.

Results and Discussion

Monthly Price Trend for Millet in the Study Area

Five-year historical data covering 1998–2002 clearly show the difference between the price of millet at harvest time and the price 4 months after harvest time (see Fig. 1). This clearly shows the need for the *Warrantage* system that would enable farmers to take advantage of this price difference and improve both their income security and their overall livelihoods.

Performance of Survey Farm Households with Respect to Millet Production

Most (about 65%) of the survey farm households used to produce 20–40 bags (of 100 kg) of millet annually. The remaining farm households were almost equally distributed with respect to the number that produces 41–50 bags (12%), 51–60 bags (10%) and ≥61 bags (13%) per annum (Table 1).

Sources of Funding for Farm Operations

Family sources (sale of family labour and own savings) accounted for about 87% of the sources of funding for

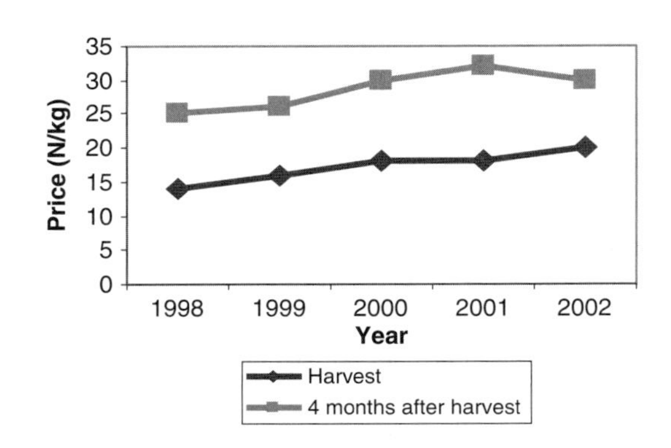

Fig. 1 Trend in temporal variation in the prices of millet in Nigeria: 1998–2002 (*Source*: Fieldwork, 2003)

Table 1 Distribution of survey farm households based on annual average production of millet

No. of bags usually obtained	No. of farm households	% of total survey farm households
20–30	10	17
31–40	29	48
41–50	7	12
51–60	6	10
61+	8	13
Total	60	100

Source: Field survey, 2003

Table 2 Distribution of sources of funding for farm operations in the study area

Source	No. of farm households	% of total survey farm households
Sale of family labour	41	69
Own savings	11	18
Money lender	8	13
Bank loans	0	–
Community association	0	–
Total	60	100

Source: Field survey, 2003

farm operations in the study area. It is clear that formal financial institutions (banks) and collective action were not available options for the farmers to source farm credit from (see Table 2). The lack of collective action was particularly worrisome, especially given its role in social capital formation and in responding to the constraints created by market and other institutional failures.

Table 3 Trend in average monthly price per bag (of 100 kg) of millet: 2003

Month	Market value (Naira)	Price difference (Naira)	% increase
January	1650	–	–
February	1800	150	9.2
March	2400	750	45.5
April	2600	950	57.6
May	2800	1150	58.9
June	3000	1350	81.8
July	3000	1350	0.0
August	2600	950	57.6
September	2000	350	21.2
October	1600	−50	−0.30
November	1600	−50	−0.30
December	1650	0	0

Source: Field survey, 2003

Awareness of the Warrantage System

The result of our investigation about the awareness of the survey farm households of the *Warrantage* system reveals that not a single one of them was aware of the *Warrantage* system prior to the present intervention. This further throws some light as to why while only 13% of the farmers sourced funding for farm operations from money lenders, none of them sourced farm credit from community associations or even banks, key organs for a successful operation of the *Warrantage* system (see Table 2).

Trends in the Price of Millet in the Study Area

This helps in exposing and demonstrating fluctuations in the price of millet and the real essence of the *Warrantage* system in the farming systems of the *Madana* community (see Table 3). The level of price difference is an important guide in determining the time to hold the produce up for better prices. Incidentally, the trend is more or less the same each year and can really be tapped into to improve the livelihood of the farm households participating in the *Warrantage* system.

According to Table 3, millet was sold at the lowest price of ₦1600 per bag in the immediate months (especially October and November) after harvest time. The trend in price increase continues till the price hits the highest of ₦3000 per bag in the months of June and July (the leanest months in this case). This implies that if farmers can hold on to their farm produce till around June and July, they will likely hit the highest price and make a gain of over 81% of what they would have secured selling during the harvest and/or post-harvest months (October and November in this case) (see Table 3).

Cost–Benefit Analysis of Millet Storage in Madana Community of Nigeria

The overall cost of the *Warrantage* system in *Madana* community was analysed. All items of cost were included, beginning from warehouse renovation4 through to and including the storage costs (see

Table 4 Cost–benefit analysis of the *Warrantage* system in *Madana* community, Nigeria

Month A	No. of bags B	Price/bag (Naira) C	Total storage cost (Naira)[a] D	Total revenue (Naira) E [B×C]	Financial benefit of waiting F	% increase in value of stored produce (from base) G	Benefit–cost ratio H [F/D]
January	305	1650	33,350	503,250	–	–	–
February	305	1800	33,350	549,000	45,750	9.1	1.4
March	305	2400	33,350	732,000	228,750	45.5	6.9
April	305	2600	33,350	793,000	289,750	57.6	8.7

[a]Cost of storage was assumed to be the same irrespective of the number of months the millet was stored which may not necessarily be so

Source: Computed from field data, 2003

Table 4). Table 4 shows the benefits of the *Warrantage* system in terms of what farmers stand to gain by storing their millet produce for some months after harvest instead of disposing it of immediately after harvest. Depending on how long the farmer is able to wait, the associated benefit–cost ratio for waiting ranges from about 1.4 (for 1 month waiting) to about 8.7 (for 3 months waiting). The increase in the value of the stored millet ranges from about 9.1% (for storing for 1 month or selling in February instead of January) to about 57.6% (for storing for 3 months or selling in April instead of January).

Conclusions and Recommendations

This study has demonstrated that the *Warrantage* system is one of the most effective modern tools that could be used to address several of the farmers' practical constraints including income poverty (due to low or rock-bottom prices for their produce when sold immediately after harvest) and limited access to and command of modern farm inputs (mineral fertilisers, improved varieties of crops, agro-chemicals, etc.).

Although the survey farmers were aware of the possible higher prices if one sells during the lean period, most of them could not wait due to urgent financial needs. This explains the almost immediate acceptance by farmers in Madana community of the *Warrantage* model which opens up opportunity for them to have access to farm inputs following the higher income opportunities that the model engenders. Forming farmers' groups helped to sensitise them on the benefits of collective action including improved ability to bargain for better prices for their farm produce. Through the group collective action, it was also easier to link the farmers to financial institutions that helped to facilitate the *Warrantage* system. The study also agrees with Adegeye and Dittoh (1981) who noted that adequate storage is the most practical marketing tool a farmer has. These authors observed that if storage is adequate, a farmer can hold on to his crop until such a time that prices become favourable. According to them, storage facilities must be made available for safekeeping of surplus production (needed during the lean period) after harvest.

In view of the outcome of this study, it is recommended that the Warrantage credit model or adapted forms of it be institutionalised and scaled up throughout the various farming systems of northern Nigeria and similar environments in sub-Saharan Africa. All stakeholders (NGOs, CBOs, faith-based organisations, financial institutions, donors, etc.) keen on seeing improvements in the welfare of farm families and other rural dwellers need to rally round and be involved in the scaling up of this noble intervention. This is an important way of not only reducing poverty but also bringing about sustainable agricultural production systems, devoid of soil mining.

Acknowledgements Our appreciation goes to Almighty God for his grace and mercy towards us that enabled successful completion of this work. A special thank goes to the Sasakawa Fund for Agricultural Education (SAFE) for financing the project. The work reflects the thoughtful guidance of Dr. John Adugyamfi, then head of ICRISAT Nigeria, who was the main initiator of the *Warrantage* project in *Madana* community. We are impressed by his sensitivity to the well-being of the resource-poor farmers in Nigeria and other African countries. Much of whatever integrity and quality this work has is a direct result of the contribution of Mr. William Ghartey of the University of Cape Coast, Ghana, who travelled to Nigeria during the project work and also reviewed the initial drafts of the manuscript. Lastly, we will like to thank our co-researchers, farmers and the people of *Madana* for their invaluable cooperation and support that made the outcome possible.

References

Adamu M (1985) Tackling poverty our own way. A paper presented at the Muslim–Christian dialogue at College of Education Gumel, Jigawa State

Adegeye J, Ditto SAA (1981) Essentials of agricultural economics. Impact Publishers Nigeria Limited, Ibadan, Nigeria, pp 166

Federal Ministry of Agriculture (2001) Annual Report. Abuja, Nigeria

ICRISAT (2001) ICRISAT Report, ICRISAT web site, www.icrisat.org

The Desert Margins Programme Approaches in Upscaling Best-Bet Technologies in Arid and Semi-arid Lands in Kenya

A.O. Esilaba, M. Okoti, D.M. Nyariki, G.A. Keya, J.M. Miriti, J.N. Kigomo, G. Olukoye, L. Wekesa, W. Ego, G.M. Muturi, and H.K. Cheruiyot

Abstract Kenya's land surface is primarily arid and semi-arid lands (ASALs) which account for 84% of the total land area. The Desert Margins Programme (DMP) in Kenya has made some contribution to understanding which technology options have potential in reducing land degradation in marginal areas and conserving biodiversity through demonstrations, testing of the most promising natural resource management options, developing sustainable alternative livelihoods and policy guidelines, and replicating successful models. In extension of sustainable natural resource management, two types of strategies were used: (i) strategies for the promotion of readily available technologies and (ii) approaches for participatory learning and action research. Thus DMP-Kenya initiated upscaling of four 'best-bet' technologies. Under the rangeland/livestock management options, scaling-up activities include improvement of rangeland productivity, rangeland resource management through community-based range resources monitoring/assessment, and fodder conservation for home-based herds. Restoration of degraded lands included rehabilitation of rangelands using the red paint approach in conservation of *Acacia tortilis*, control of *Prosopis*, planting of *Acacia senegal* trees in micro-catchments, and rehabilitation of degraded areas through community enclosures. Improved land, nutrient, and water management involved upscaling water harvesting and integrated nutrient management (INM) technologies. Activities under tree-crop/livestock interactions included upscaling of *Melia volkensii* and fruit trees (mangoes) and enhancing biodiversity conservation through support of beekeeping as a viable alternative livelihood. Participatory learning and action research (PLAR) was used for technology development and dissemination. Capacity building and training was a major component of upscaling of these best-bet technologies.

Keywords Approaches in upscaling technologies · Arid and semi-arid lands · Biodiversity · Kenya · Land degradation

Introduction

The problem of desertification and land degradation in Kenya presents a major threat to all facets of land productivity (Kilewe and Thomas, 1992). Land degradation, either natural or induced by humans, is a continuing process. It has become, however, an important issue through its adverse effects on national natural resources, food security, and livelihoods in sub-Saharan Africa (Nabhan et al., 1999). The potential scale of land degradation in Kenya is serious as 84% of the total land area is arid and semi-arid. Causes of land degradation are numerous and include decline of soil fertility, development of acidity, salinization, alkalization, deterioration of soil structure, accelerated wind and water erosion, loss of organic matter, and biodiversity (Nabhan et al., 1999). Kilewe and Thomas (1992) and Nandwa et al. (1999) reviewed the forms of soil degradation (both chemical and physical) occurring in Kenya. The absence of quantitative data necessary

A.O. Esilaba (✉)
Desert Margins Programme, Kenya Agricultural Research Institute (KARI), Nairobi, Kenya
e-mail: aoesilaba@kari.org; aesilaba@gmail.com

Olukoye (Deceased)

A. Bationo et al. (eds.), *Innovations as Key to the Green Revolution in Africa*,
DOI 10.1007/978-90-481-2543-2_120, © Springer Science+Business Media B.V. 2011

for predicting land degradation and detecting critical management alternatives has curtailed the development of conservation practices applicable to Kenyan conditions. This has also prevented objective evaluation and adaptation of models and experiences developed elsewhere and hindered the rehabilitation of degraded land.

The Desert Margins Programme (DMP) is a project in the Sub-Programme of Environmental Science and Research of the United Nations Environment Programme that covers nine countries in Africa (Burkina Faso, Botswana, Mali, Namibia, Niger, Senegal, Kenya, South Africa, and Zimbabwe). The DMP was developed in response to a recommendation made to the international research community at UNCED to consider specific contributions for implementation of the three international conventions on biodiversity, climate change, and desertification. The overall objective of the DMP is to arrest land degradation through demonstration and capacity-building activities developed through unravelling the complex causative factors of desertification and conservation and restoration of biodiversity through sustainable utilization in the desert margins in Africa. The Desert Margins Programme (DMP) in Kenya was implemented from 2003 to 2007 by the Kenya Agricultural Research Institute in collaboration with NARS [Kenya Forestry Research Institute (KEFRI), National Environment Management Authority (NEMA), the National Museums of Kenya (NMK), University of Nairobi (UoN), and extension agents], NGOs (ITDG and ELCI), and international institutions (ILRI, ICRISAT, TSBF, and ICRAF) resident in Kenya. DMP has local objectives of mitigating land degradation in the desert margins in Kenya. While addressing its primary objective of alleviating land degradation, DMP is expected to make a significant reduction in the negative impacts associated with other factors that cause degradation processes. During the past 4 years (2003–2006), the DMP project in Kenya has made some contribution to understanding which technology options have potential in reducing land degradation in marginal areas and conserving biodiversity through demonstrations, testing of the most promising natural resource management options, developing sustainable alternative livelihoods and policy guidelines, and replicating successful models. Thus DMP-Kenya has initiated upscaling of four 'best-bet' technologies that include various rangeland/livestock management options; restoration of degraded lands; improved land, nutrient, and water management; and tree-crop/livestock interactions.

Materials and Methods

Biophysical and Socioeconomic Characteristics of the Study Sites

The Kenyan component of DMP was implemented in three benchmark sites in Turkana/Turkwel, Marsabit, and the Southern Rangelands.

Turkana/Turkwel Site

The Turkana/Turkwel site is a river basin ecosystem in northwest Kenya. The ecosystem consists of the Turkwel riverine forest that supports hydropower production at the gorge, pastoralism, irrigated agriculture, production of Doum palms, fuel wood including charcoal production, and human settlements. Currently, *Prosopis juliflora* has become an obnoxious invasive weed within the Turkwel ecosystem and is also choking shores of Lake Turkana. Mean annual rainfall is approximately 500 mm in the upstream section and less than 200 mm downstream. Temperatures range between 24 and 38°C with a mean of 30°C. The forest is largely dominated by *Acacia tortilis* with *Faidherbia albida*, *Acacia elatior*, *Cordia sinensis*, and *Hyphaene compressa* as sub-dominants on the riverbanks and *Balanites pedicellaris*, *Boscia coriacea*, and *Acacia nubica* on the dry edge of the riverine zone. The soils are alluvial deposits of deep sandy or silty loams classified as calcaric Fluvisols (Van Bremen and Kinyanjui, 1992). The major source of livelihood is livestock sector, which contributes to 60% of the district population. However, crop production is mainly practised along major rivers and contributes to about 16% of the local economy. Despite the increasing dependency on agriculture most Turkana still keep herds of livestock (cattle, camels, goats, sheep, and donkeys) and are either nomadic pastoralists or agro-pastoralists. Poverty levels within this region are generally high with an overall poverty of 74% and food poverty of 81% (GOK, 2002).

Marsabit Site

Marsabit site is in northwest Kenya and comprises Kargi, Ngurunit, Korr, and Kalacha sub-sites. Pastoral livestock production with a bias on camels, sheep, and goats is the main livelihood, with settlement playing a significant role in degradation. Marsabit represents a gradient of ecological differentiation: from the oasis ecosystem in an otherwise desert zone at Kalacha through a very arid zone at Korr and Kargi to the montane semi-arid ecosystem at Ngurunit. The site has a diversity of plant species that are endemic.

Kargi is a Rendille settlement about 70 km northwest of Marsabit town at an altitude of approximately 600 m.a.s.l. with mean rainfall of 250 mm, mean monthly temperature of 27–29°C, and a mean annual potential evapotranspiration of about 2,500 mm. The vegetation is dominantly *Acacia* shrub land with *Duosperma*/*Indigofera* understory.

Ngurunit is an Ariaal/Rendille settlement situated at an altitude of approximately 700–800 m.a.s.l. The mean annual rainfall is about 600 mm and diurnal temperatures range between 22 and 27°C. The potential annual evapotranspiration is about 2,000 mm.

Korr is a Rendille settlement at an altitude of 500 m.a.s.l. The mean rainfall is 180 mm.

Kalacha is inhabited by the Gabbra pastoral community on the northern edge of the Chalbi desert. The settlement lies at an altitude of 386 m.a.s.l. The mean annual rainfall is less than 200 mm. The area surrounding the settlement is typical of most arid areas of Marsabit District except that there is an oasis with plenty of water and a big grove of Doum palms. The settled pastoralists do some irrigated farming using water from this oasis. The area is sparsely wooded with *A. tortilis* being the dominant species, sparsely interspersed by the salt-tolerant bush *Salsola dendroides*.

Southern Rangeland Site

This site is at the intersection of Kajiado, Makueni, and Taita-Taveta Districts with land degradation impacts emanating from farming, pastoralism, and wildlife conservation. The climate is semi-arid to arid with the mean annual rainfall ranging between about 600 and 750 mm as opposed to potential evaporation varying between 2,100 and 2,150 mm (Achieng and Muchena, 1979). In the national park, however, rainfall varies between 300 and about 900 mm and increases towards the south.

The altitude lies between 150 and 1,000 m.a.s.l. The natural vegetation is dominated by wooded bushland consisting of *Enteropogon macrostachyus* and *Chloris roxburghiana* (30% of grasses) with *Acacia brevispica*, *Combretum exallatum*, *Commiphora* spp. and grasses such as *Premma holstii*, *Ocinum basilicum*, and *Grewia* spp. plus other wooded species.

All soils in the district are developed on sandstones rich in ferro-magnesium minerals. They are well drained, deep to very deep, red to dusky red, and sandy clay to clay (Ekirapa and Muya, 1991). These are represented around Kiboko and west of Kibwezi by *chromic* Luvisols and interfluves of *orthic* and *xanthic* Ferralsols (Ministry of Agriculture, 1987). Soils are variable in depth depending on the parent material and slope and are generally low in organic matter and deficient in nitrogen, phosphorus, and potassium (Van Wijngaarden and van Engelen, 1985).

The population of Makueni District consists of some 43,377 households with a density of 10–150 people/km^2 and farm sizes averaging 3.5 ha per household (De Jager et al., 2006). Of the total farmland an average of 48.5% is under cultivation and up to 71% of the households live below poverty line.

Farm sizes are on average 3.3 ha per household of which about 1.8 ha is utilized to grow crops, with the rest of the land under pasture. Livestock are also kept but their quality is low due to poor breeds and poor feed supply and feed rations. Apart from rainfed agriculture there are extensive areas where irrigation is practised.

Methodologies/Approaches

Extension of the sustainable natural resource management (NRM) component of DMP was to foster improved and integrated soil, water, nutrient, vegetation, and livestock management technologies to achieve greater productivity of crops, trees, and animals to enhance food security and ecosystem resilience. In Kenya, extension of sustainable natural resource management was started in the second year of DMP. Two types of strategies were used: (i) strategies for the promotion of readily available technologies and

(ii) approaches for participatory learning and action research.

The research component started by developing inventories and characterizing available technologies ready for promotion, defining potential user groups/systems–farm typologies, and matching technologies with people. A bottom-up, social, and experiential learning approach was needed to foster technological change. Such a learning approach aimed at making the best use of locally available resources, knowledge, and decision making in combination with research-based understanding and analysis of the underlying processes, leading to technology adaptation, innovation, and change based on social interaction.

The methodologies/approaches used to implement DMP activities included reconnaissance surveys, diagnostic surveys, soil surveys, stakeholder workshops, on-farm trials/demonstrations, establishment of demonstration plots on forage germplasm, indigenous fruit trees, tree nurseries, medicinal plants, commercial fruit trees, multipurpose trees, woodlots, horticultural crops, drought-resistant crops, and other potential biodiversity activities and upscaling of 'best-bet' technologies. Best-bet technologies were upscaled through establishment of small-scale demonstration plots, on-farm trials, training and stakeholder workshops, information dissemination and exchange, and raising awareness to foster adoption of sustainable land use practices. The programme participated in a number of initiatives to generate a policy and regulatory environment to encourage sustainable use of natural resources. Community-based management of natural resources and monitoring and evaluation of the natural resource base with the pastoral communities was started in the DMP sites.

This was implemented in close collaboration with the relevant stakeholders involved in extension work. The existing and new farmer networks formed the basis for the set-up of stakeholder platforms for all relevant partners involved in research and development activities in the sites. They included extension services; international, national, and local non-governmental organizations (NGOs); and private sector who directly work with farmer groups and community-based organizations (CBOs) as primary beneficiaries. Detailed aspects of methods for each project in the programme are summarized under major activities for each project in the various DMP sites.

Results and Discussion

Rangeland/Livestock Management Options

Improvement of Rangeland Productivity

Range condition and trend assessments over the years have often pointed at worsening productivity of natural pastures in both the arid and semi-arid areas of East Africa (Coughenour et al., 1990; McPeak, 2003; Coughener, 2004). Wandera et al. (1996) highlighted the need for pasture improvement for greater meat production from indigenous livestock in pastoral areas of Kajiado and Baringo Districts in Kenya.

Range and livestock technologies can be used intensively to improve the productivity of the marginal areas in the rangelands of Kenya. Low levels of production in these areas are attributed not only to scarce and erratic rainfall but also to low levels of technical skills by farmers. Proper exploitation of the range resource potential through improved management can easily change the living standards of these farmers. The potential rangeland management practices that can be used to increase production in the semi-arid rangelands include range, plant, and livestock-related technologies.

Soils and soil moisture form the basis of rangeland productivity and therefore their conservation and management are of paramount importance. These can be achieved in grazing lands through pitting, contour terracing, and revegetation of watersheds and pastures. Forage species adapted to grazing and drought stress have been identified in ASALs of Kenya. Subsequently, field experiments were carried out to test and demonstrate the usefulness of the selected genotypes in the reclamation of degraded rangelands and production of good quality forages in the semi-arid zones.

During the period under review, DMP worked in five farmer clusters in the Southern Rangelands. In each cluster, five farmers were selected as contact demonstration farmers representing each village composed of about 50–60 active farmers. The demonstrations were carried out at the village level where all the farmers participated through participatory adaptive research and development approaches. In each village one farm was identified where demonstrations were carried out. Three grass species (*Cenchrus ciliaris, Eragrostis superba, and E. macrostachyus*) were established on a 0.25 ha of land. Other farmers were each allowed to carry grass seeds of one species of their choice for establishment in their farms. At the end of the first season a total of 21 farmers from 2 villages had well-established pastures. Their pastures ranged from 0.1 to 0.3 ha and the demand for grass seeds had increased, with farmers who had harvested grass selling all their stocks. After seed harvesting about 40% of the farmers harvested and baled the grass as hay while the rest allowed animals to graze during the dry season. Those who baled later sold them at a profit.

Results from participatory evaluation of trials on pasture reseeding and related trials indicated that farmers were encouraged by the increased forage for their livestock. Farmers realized increased income from increased milk yield, hay, and sale of grass seeds. The hay and grass seed sale resulted in diversification of income sources to the community. Other sources of increased incomes were due to sale of animals as a result of improved livestock conditions with improved availability of pastures in restricted areas. Farmers have therefore expanded the areas under pastures and forages. A community-based seed multiplication approach has been adopted to meet the high demands for seed. This group approach to on-farm trials and demonstrations has resulted in more rapid scaling up of technologies in the DMP sites within the Southern Rangelands through promotion and support of farmer groups and community-based organizations (CBO) that were registered during the EU/ARSP and ARIDSAK projects.

With the adoption of improved range management technologies, it is expected that primary and secondary rangeland productivity and the living standards of the communities will be improved as well as reduction in rangeland degradation.

Improvement in Rangeland Resources Management Through Community-Based Range Resources Monitoring and Assessment in ASALs

Over the last two decades community-based natural resource management (CBNRM) has become an important strategy to conserve and sustainably use biodiversity in Africa. Community resource management institutions and organizations are now receiving greater attention as a viable alternative to regulation by the state or privatization as a means of rectifying inefficiencies caused by attenuated property right systems and externalities (Fabrius et al., 2005; Gemedo-Dalle et al., 2006). There is growing recognition that biodiversity and indigenous knowledge are interrelated phenomena (Balard and Platteau, 1996; Warren 1993). Once the diversity of floral and faunal resources disappears, the knowledge associated with these resources also disappears. It was with this background that the community, in collaboration with KARI-Marsabit research scientists, initiated a community-based range resources monitoring and assessment activity. This is expected to harmonize the traditional methods of range resources monitoring and assessment with the conventional methods.

The main objective was to actively involve the community in biodiversity conservation through jointly documenting changes in natural resources over time by using traditional and conventional methods of range resources monitoring and assessment. The community was involved in monitoring the state of resources within their area using traditional indicators in collaboration with the research scientists who used conventional methods.

Conventionally, the technical staff described the various resource attributes within established permanent transects and also clipped the herbaceous vegetation falling at specified intervals within transects. Data were collected along three 500 m long transects and at intervals of 100 m; grasses, herbs, forbs, shrubs, and trees were sampled using the nested quadrat method. Descriptions were also made of the vegetation species, and the soil conditions and the results were compared with the community description. Two transects fall within the community livestock home range and the third transect outside the livestock home range.

For the community, data collected gave a general/subjective description of the area from their traditional indigenous knowledge perspective. Their method of recording and analysis depended more on memory and discussion than on written records. They describe the environment and their observations which are documented in terms of vegetation suitability, cover, soil conditions, and general observations on use or misuse of resources.

This approach was received with enthusiasm by the community members and promises to yield great impact in enhancing environmental conservation efforts at the community level. To sensitize the wider community on the importance of biodiversity conservation, the results from the studies are continuously made available to the community through workshops and outreach to institutions like schools. Since the communities took the necessary initiatives, the results from these studies will feed back into the community policy structures for conservation of resources. Communities in other areas have requested that the same structures for conservation be introduced in their areas. Hence, collaboratively with the GEF-funded Indigenous Vegetation Project, six more transects have been established for monitoring natural resources dynamics.

The project goal of conserving and restoring biodiversity in the desert margins is being met since the project's aim was to create awareness on resource use dynamics and influence conservation policies at the grassroots level. The greatest impact will be improvement in biodiversity within the areas that are already under pressure due to resource over-utilization. Ecosystem products and services such as vegetation resources, reduction in soil erosion, and improved water infiltration will improve. The total area targeted is over 30,000 ha and this will benefit over 1,200 households.

Fodder Conservation for Home-Based Herds

Because of many factors, there are many changes that have taken place within the drylands of Kenya which have led to changes in the vegetation structure. Several communities in the northern part of Kenya point to a good past in terms of vegetation resources endowment (Milimo et al., 2002). However, they indicate that there has been a downward trend in vegetation attributes over the years. This has led to scarcity of forage to livestock, especially the home-based herds, since the other livestock species migrate in search of pasture when conditions allow. To minimize forage scarcity for the home-based herds, fodder bank technology was introduced to the community. The technology involves fodder establishment at the community level/individual farmer's plots. Interested farmers who demand for grass seeds for fodder establishment are given technical advice on fodder establishment and hay bulking. This is a community-driven initiative that receives given technical support from research and extension agents.

The challenges of the recent drought, that made the government provide hay as part of relief efforts, have made the community take the initiatives of fodder conservation more seriously. The only challenge will be provision of grass seeds, which imply timely harvesting of grass seeds or sourcing of fodder germplasm.

The direct benefit of the initiative was provision of fodder for the home-based herds and securing food security at the household level. By implication, the provision of fodder reduces the pressure on vegetation resources, hence allowing regeneration and conservation of biodiversity.

Restoration of Degraded Lands

Rehabilitation of Degraded Rangelands

Rangelands have a relatively low production potential, are fragile, and are easily degraded through over-utilization or use of inappropriate technologies (Herlocker, 1999). The most common types of degradation in the rangelands are degradation of the soil, vegetation, and animal species.

The demand on the rangelands is high due to increasing livestock and human populations. The pressure is from within the rangelands and immigration from neighbouring densely populated high-potential areas.

There is need therefore to develop and/or source, adapt, and upscale appropriate rangeland rehabilitation technologies and to support efforts by research and other development agencies in providing technologies that will sustainably improve rangeland productivity.

The technologies being upscaled were mainly in soil and water management and improvement of primary productivity. The soil and water management technologies are as follows:

(i) Runoff harvesting for improving soil moisture
(ii) Construction of water and soil micro-catchments
(iii) Construction of modified terraces
(iv) Restoration of gully eroded rangelands

Technologies for improvement of primary productivity are as follows:

(i) Plant species enrichment that involves introduction or increase of the germplasm to improve ground cover and production
(ii) Bush management technologies that involve reduction of the woody vegetation to allow for increased herbaceous production

The technologies were implemented using on-farm trials and demonstrations carried out at the village level where farmers were involved through participatory adaptive research and development approaches.

Red Paint Approach in Conservation of *A. tortilis*

This involves the painting of the stems of *A. tortilis* trees with red paint. Red paint within the Rendille and Samburu pastoral communities is revered and anything painted red is disused or used judiciously. By marking young tree seedlings with red paint, one marks them out for disuse by the community and with this they can be allowed to establish into mature trees.

The process of identifying the desirable areas/plants for conservation involved discussions with the community. Areas with high *A. tortilis* regeneration, especially in old abandoned bomas (Kraals), were identified and mapped out. Paint was then provided to the community members to paint tree stems of the identified trees. However, not all trees were painted, to allow access and use of some of the resources.

This technology or approach was tried by various projects including IPAL, GTZ, and the EU/ARSP II with very good success (Lusigi, 1981; Goldsmith, 2000; Ndung'u et al., 1999). It has worked with the community's approval since the costs involved are minimal.

Increased biodiversity will lead to improved ecosystem services to the community. The target species, *A. tortilis*, which is well adapted to this area, is of high value in terms of provision of livestock feed (pods are a main dry season feed source), wood fuel, building poles, and also fencing materials, among others. Hence the impact will be felt at the household level, with a total population of over 10,000 persons benefiting from the technology.

Rehabilitation of Deserts Through Planting of *Acacia senegal* Trees in Micro-catchments

The Acacia Operation Project (AOP) is a project supporting food security and rural development of gums and resins in sub-Saharan African countries (Burkina Faso, Chad, Kenya, Niger, Senegal, and Sudan). This project aims to rehabilitate degraded land by planting *A. senegal* using novel water harvesting technologies and improving livelihoods through promotion of gum and resin production.

Micro-catchments for water harvesting were made using a specially designed tractor and plough (Delphino plough) designed to make micro-catchments. *A. senegal* seeds are planted in the micro-catchments to utilize harvested runoff. Grasses were also planted on the micro-catchments for soil conservation and as a source of fodder.

The approach used created synergy between nature and modern technology in improving management of natural resources. The local communities were involved in the identification of potential sites for development of plantations for gum and resin production, before micro-catchments were ploughed. There was also capacity building on markets and income diversification as well as training in dryland food production and utilization to realize immediate benefits from the plantations.

The potential of production of gums and resins as alternative livelihoods and diversification of the production systems formed the basis for the success of these technologies. About 342.7 ha was planted by the AOP project with *A. senegal* seeds/seedlings and drought-tolerant crops at Serolipi, Laisamis, and Ngurunit in Marsabit and Samburu Districts. The results indicated that *A. senegal* can be successfully established in degraded sites. However, more trials need to be conducted before concrete recommendations on integration of crops with *A. senegal*. Challenges include a cultural bias towards pastoral livestock keeping as opposed to agro-silvopastoralism and communal land ownership. The technology will contribute to sustainable development, food security, and combating desertification through the promotion and integration of gums and resins in rural economies as an alternative livelihood.

Rehabilitation of Degraded Areas Through Community Enclosures

Community enclosures are ideal for improving the overall ecological conditions of degraded areas so that they can provide better socioeconomic benefits and environmental services to the local people. Previous studies in ASALs show that enclosing land with fencing had many benefits that included increased livestock feed, fuel wood, more water, higher land value, and increased livestock production. The method is cheap, quick, and cost friendly in this region. Due to low and unreliable rainfall, afforestation programmes and grass reseeding have resulted in low survival rates of germplasm. Reid and Ellis (1995) reported that *A. tortilis* was successfully recruited in South Turkana through abandoned livestock enclosures. Goats which form the majority of livestock population in the district normally eat the *A. tortilis* pods. However, the seeds are not digested but are excreted in the droppings which will later germinate, but if the young seedlings are not enclosed they will be browsed. Emiru et al. (undated) and Olukoye et al. (2003) demonstrated successful cases in Tigray in Ethiopia and Noth Horr in Kenya, respectively.

An evaluation of one successful case study of established community enclosure at Kalatum in Turkana was undertaken to assess the effectiveness of tested technologies, community mobilization approaches, and opportunities/challenges of upscaling this technology. The success of enclosures was tested by assessing the adoption of fencing technology by community members and comparison of vegetation status between the fenced and unfenced plots. Adoption of fencing technology entailed assessment of fence expansion, while impact of other range rehabilitation was undertaken by comparing vegetation variables between fenced and unfenced areas, using a total of 25 sample plots. Sampling was done during the wet and dry seasons because of variation of seasonal impacts on vegetation status. Results of this study showed an increase of the fenced area from the initial plot of 5 ha to an extensive area of about 25 ha. The community adopted the use of locally available plant materials and deterrent rules to keep off livestock from the areas earmarked for range regeneration. Ecological assessment of the impact of fencing and water harvesting on range resilience revealed a higher density of important fodder species such as *A. tortilis* of 124 trees and 14 trees per hectare in the fenced and unfenced areas, respectively. The average percentage cover of the dominant grass *Chrysopogon plumulosus* was 36 and 4% in the fenced and unfenced areas, respectively. In total, 15 plant species were recorded in the enclosures compared to only 7 in open rangelands.

The enclosure has been a model used by the DMP project for exposing more community members to this technology. This has been accomplished by community sensitization and formulation of community action plans which led to initiation of three more enclosures with a total area of 10 ha in different localities within the district. Increased vegetation cover through enhanced natural regeneration while conserving biodiversity will lead to improved standard of living of the target communities.

Improved Land, Nutrient, and Water Management

The current high population pressure on land necessitates exploitation of arid and semi-arid lands (ASALs) for crop production. However, agricultural production in ASALs is limited by low and erratic rainfall, high transpiration rates, and generally fragile ecosystems which are not suitable for rainfed agriculture (Jaetzold et al., 2006).

Rainwater harvesting using tied ridges and open ridges is one of the cheap methods of mitigating dry spells in areas where farmers have inadequate resources to invest in irrigation. On-farm trials in Makueni District and other studies in ASALs in Kenya and sub-Saharan Africa have indicated that tied ridges increase maize yields by more than 50% above the conventional flat tillage practised by farmers (Njihia, 1977; Kipserem, 1996; Ngoroi et al., 1994; Jensen et al., 2003; Itabari et al., 2004; Miriti et al., 2007; Itabari et al., 2007). There were significant increases in maize yield when tied ridges are combined with integrated nutrient management (Jensen et al., 2003; Miriti et al., 2007; Itabari et al., 2007). However, tied ridging has not been widely adopted by small-scale resource-poor famers in the semi-arid lands. There have been contradictory reports on the effects of tied ridging due to the variation in soil and climatic characteristics among sites and between years. In addition, the net effect of tied ridging includes both positive and negative effects (Hudson, 1987; Lal, 1995). The research hypothesis was that combining water harvesting techniques with improved soil fertility will result in higher efficiency of resources and increase in crop yields in the ASALs.

Based on the above evidence from the ASALs in Kenya, DMP initiated upscaling water harvesting and INM in the benchmark sites in the Southern Rangelands and Marsabit District in northern Kenya. Researcher-farmer managed mother–baby on-farm trials and demonstrations were conducted to study the effect of different water harvesting technologies and integrated nutrient management practices on crop yields. The trial treatments included three water harvesting techniques [farmers' practice (flat tillage), contour furrows, and tied ridging] and five integrated nutrient management practices [control (without fertilizers), farmyard manure at 5 t/ha, farmyard manure at 5 t/ha + 20 kg N/ha + 20 kg P/ha, farmyard manure at 10 t/ha, and farmyard manure at 10 t/ha + 20 kg N/ha + 20 kg P/ha]. The treatment arrangement was split-plot and each treatment was replicated four times in a randomized complete block design (RCBD). Trials were established in five clusters in the Southern Rangelands. These clusters were at Kimana in Kajiado District, Kambu, Kamboo, Kathyaka, and Kalii in Makueni District. During the long rains in 2006 there were 5 mother trials and 15 baby trials, whereas during the short rains these were 4 and 18, respectively. Cowpeas K80 and green grams N26 were used as the test crops during the long rains and short rains, respectively. In Marsabit District four trial sites were selected in four different sub-locations [Songa, Galqasa/Goro Rukesa, Dirib Gombo, and KARI Marsabit (Majengo sub-location)]. Partners in the establishment of the trials were DMP research scientists from collaborating institutions that included the TSBF Institute of CIAT, KEFRI, KARI, extension staff from the line ministries, NGOs, and community-based organizations from the DMP sites.

Preliminary results indicated that there were differences between the tillage treatments at Katumani but not at Kiboko and Marsabit, and integration of manure and inorganic fertilizers within the treatments increased maize yields (Table 1). Application of manure 10 t/ha plus 20 kg N plus 20 kg P_2O_5/ha had the highest yields under tied ridging. The highest yield under contour furrows was at 5 t/ha with 20 kg N+20 kg P_2O_5/ha manure whereas under the farmers practice it was at 10 t/ha manure. The results show that tied ridging and contour furrows coupled with integrated nutrient management have potential as a viable option for improved crop production in arid and semi-arid lands. However, there is need for long-term trials as well as economic analysis prior to making any concrete recommendations.

Table 1 Preliminary results on the effect of water harvesting and INM on maize grain yields on farmers' fields in the Southern Rangelands (Katumani and Kiboko)

Treatments	Tied ridging	Contour furrows	Farmers' practice
Control	983	1,036	903
Manure 5 t/ha	1,194	1,123	963
Manure 5 t/ha + 20 kg N+20 kg P_2O_5/ha	1,139	1,180	1,018
Manure 10 t/ha	1,211	1,170	1,065
Manure 10 t/ha + 20 kg N+20 kg P_2O_5/ha	1,375	1,054	1,046
Mean	1,180	1,133	999
LSD (P< 0.05)	489	509	566

Tree-Crop/Livestock Interactions

Tree-crop/livestock integration offers a promising opportunity for intensifying agricultural production and increasing ecological integrity so as to have a positive impact on livelihoods and NRM in mixed farming systems (Karugu, 2004). For smallholders who have limited access to external inputs, research has documented that introduction of rotations of various crops, forage legumes, trees and use of manure help maintain soil biodiversity, minimize soil erosion, and conserve water. Activities under tree-crop/livestock interactions include upscaling of *Melia volkensii* and mangoes and enhancing biodiversity conservation through support of beekeeping as a viable alternative livelihood.

Upscaling of M. volkensii in the Southern Rangelands

M. volkensii Gürke has received significant research attention because of its socioeconomic importance in the drylands. *Melia* is a multipurpose tree species endemic in arid and semi-arid lands of eastern Africa. It is used as construction timber, fuel wood, fodder (fruit and leaves), medicine (bark), bee forage, mulch, and green leaf manure (Kamondo et al., 2006).

The Southern Rangelands located in the semi-arid areas of southeast Kenya form part of the drylands where *Melia* is endemic. The tree is heavily exploited in the region and very little remains in nature, calling for concerted efforts to conserve the natural stands as well as encourage farmers to plant *Melia* on their farms. In this region, DMP has responded to these needs and is currently upscaling *Melia* through the following:

(a) Capacity building on seedling production for farmers: The technology developed by KEFRI to raise *Melia* seedlings is an elaborate process that involves seed extraction from the nut, pre-treatments, and management of the seedlings. Capacity building is necessary for farmers' adoption and adaptation of the technology. Therefore, through support of DMP, KEFRI staff conducts on-station and on-farm training on requisite processes for successful seedling production through seeds. The training focuses on timing of seed maturity, depulping of *Melia* fruits, seed extraction, seed pre-treatment, sowing, pricking out, and tending of pricked out seedlings. Training is necessary because the species is highly sensitive to weather and environmental conditions.

During 2005–2006, members of four group nurseries were trained and they raised over 20,000 *Melia* seedlings that were either planted on their farms or sold off (Table 2).

In addition, over 200 farmers were trained in their individual capacity in raising *Melia* seedlings.

(b) Silvicultural management: Over 50 farmers supported to establish *Melia* woodlots in the Southern Rangelands site were trained on tree management to improve production of various *Melia* products. For timber production, pruning and spacing were the major training aspects. Tree-crop interactions/competition was also addressed through appropriate farm planning for optimum land productivity.

(c) On-station seedling production: On-station seedling production was initiated during phase II to meet the rising demand for *Melia* seedlings. This will be scaled down as farmers are trained to produce seedlings on farm. Over the past 3 years about 20,000 *Melia* seedlings were raised for out-planting (Table 3).

Therefore DMP is upscaling M*elia* through capacity building on seedling production and silvicultural management. By providing alternative source of forestry products like timber, the pressure on highland forests is expected to reduce in the long term, and this contributes to biodiversity conservation since highland forests are more endowed with biological diversity. The gains from increased *Melia* hectarage and conservation of ASALs will eventually lead to increased carbon sequestration. In addition to the above benefits, the

Table 2 Farmer groups trained on *Melia* seedlings production

Name of group	Number of members trained	Number of seedlings raised
Utheu wa Aka	21	4,500
Kituku's group	5	6,000
Nzonkolo group	15	1,500
Masongaleni group	5	5,000
DWA Sisal plantation	1	5,000

Table 3 Out-planted *Melia* seedlings

Division	No. of farmers	No. supplied	Farmers' own supplies	No. surviving	Average height (m)
Kibwezi	50	9,054	1,164	8,570	2.3
Kajiado	1	200	–	192	3.5
Marsabit	–	4,500	–	2,500	–
Makindu	2	2,000	–	1,600	3.0
Mtito Andei	3	3,720	–	1,970	2.2
Total	56	19,474	1,164	14,832	–

sale of its raw and processed products will contribute to household incomes as an alternative livelihood for local communities.

Upscaling Mango (Mangifera indica) Production and Marketing in the Arid and Semi-arid Lands in Kenya

Mango is the most important fruit in the tropics. The fruit marketability, especially for the export market, is high where the fruit fetches high prices. Mango which has been naturalized in Kenya grows well from sea level up to an altitude of 1,800 m and requires mean annual temperatures between 20 and 30°C. Once established the plant can tolerate a wide range of soils but prefers deep fertile well-drained soils with pH ranging from 5.5 to 7.5.

The Kenya Forestry Research Institute (KEFRI) and KARI through the ARIDSAK project were involved in the promotion of mango orchard establishment as a sustainable alternative livelihood source for the communities in the drylands. The key challenge to mango production is timely production of quality fruits in quantities that justify selling in lucrative markets located far away from the drylands. In addressing production and marketing challenges, different techniques and methodologies have been identified. Suitable types and varieties of mangoes for each region, grafting techniques, integrated soil water and fertility management practices, and pest and disease management options have been identified. In addition, there have been proposals by stakeholders on marketing and value added strategies through collective action, networking, packaging, and transportation of the fruits.

Promotion of mango production has been constrained by lack of quality planting materials. This is further exacerbated by lack of know-how by farmers on appropriate production and marketing practices. During 2003–2006, DMP responded to these needs in the Southern Rangeland site by the following:

(a) Raising and provision of quality germplasm and capacity building of farmers.

During 2005–2006, 57 farmers were provided with 5,308 mango seedlings and trained on planting and management, acquisition/production, and distribution of germplasm materials; formation of mango farmers' associations; and networking with potential marketing agencies (Table 4). After identification of farmers for mango orchard establishment, suitability of the site is assessed and field training of the targeted farmers undertaken on their farms. The farmers are given briefing on different mango types, grafting techniques, pitting, spacing, planting, and management of the orchard. They are also introduced to book-keeping and enterprise development as a complementary activity.

Table 4 Mango seedlings supplied and out-planted on farmers' fields

District	Division	Farmers recruited	Seedlings planted
Makueni	Kibwezi	43	3,898
	Mtito Andei	6	550
	Makindu	1	60
	Kathonzweni	5	500
Kajiado	Mashuru	2	300
Total		57	5,308

To produce quality germplasm materials, private nurseries were supported through training and provision of nursery materials. On-station seedlings production was done on a very low scale to complement the private nurseries.

(b) Improvement of mango fruit quality and quantity. Studies done by KEFRI show that better mango establishment is achieved when grafting is done in the field rather than in the nursery. During 2005–2006, DMP supported project staff to carry out grafting of mangoes on 30 farms. Farmers were trained on how to carry out grafting. To improve fruit quantity, production, and timeliness, DMP commenced a field study on nutrient management, watering, and pesticide application under a strict annual schedule. The study that commenced in 2005 has demonstrated that physiological manipulation of the mango tree through nutrient management and watering could alter its flowering pattern that impacts on the marketability of the fruits. Under treatment, mangoes flowered and yielded fruits 2 months earlier than those under normal conditions. The study is being documented and guidelines developed for dissemination to mango producers.

(c) Networking and marketing of products in the mango products value chain. Linkages were established by networking farmers with exporters and other marketers in potential outlets with better prices for products. The targeted ones were producers and traders of quality germplasm and buyers of the fruits. The mango seedlings' producers were linked up to Horticultural Development Authority (HCDA) for certification and cataloguing. The key buyers of the fruits from the site were the exporters and supermarkets mostly located in major urban centres and institutions and retail traders located in all marketing centres. All these were linked up by encouraging dialogue between them and farmers.

Empirical evidence has shown that mango production is highly profitable. The prices in the international market are highly competitive. Thus through capacity building, provision of quality germplasm, value adding, and improved marketing strategies, incomes of dryland communities will improve. This will help divert the farmers' focus from destructive and environmentally unfriendly practices such as charcoal production. Increased incomes will also help reduce over-reliance on livestock keeping, thus lessening pressure on the environment due to overstocking in ASALs.

Enhancing Biodiversity Conservation Through Support of Beekeeping as a Viable Alternative Livelihood

Beekeeping has several potential impacts and is particularly appropriate in the resource-poor ASALs because it does not compete directly for resources with other agricultural activities; it requires little space (50 hives can be accommodated in a tenth of a hectare) and land can be of poor quality. Thus, beekeeping is a viable and sustainable income generating activity in the drylands and has enormous environmental, social, and economic benefits for the ASALs in Kenya. Kenya's ASALs have a potential to produce between 80,000 and 100,000 t of pure honey per year, earning 20 billion shillings from the domestic and foreign markets (Nyariki et al., 2005).

To broaden their livestock base and diversify livelihood sources and income generation, settled pastoralists in ASAL ranges practise traditional beekeeping as an alternative livelihood source, which is highly compatible with biodiversity conservation. About 20% of pastoral households residing near the Ndotto Mountain ranges in Marsabit District rely on bees as a source of livelihood (Lengarite and Okoti, 2004). Traditional beekeeping is compatible with natural resource conservation since beekeepers strictly utilize fallen dead wood from 12 different woody species for making hives. *Commiphora* and *Euphorbia* spp. are the best preferred plants which have an average life span of over 10 years. Traditions discourage beekeepers from cutting down live trees for log hives and harvesting of wood products on trees bearing hives. In the upper slopes, families and clans control the use of resources in the apiary site and through routine surveillance control fire outbreaks and harvesting of woodland resources in the apiary sites. The increased interest in beekeeping has meant that many areas have been put under natural resource conservation, which contributes to biodiversity conservation.

The purpose of the project was to establish the potential of beekeeping in the Ndotto Mountain ranges and to describe development within the sector with a view to optimally exploit beekeeping as a sustainable livelihood strategy and for biodiversity conservation in this fragile ecosystem. Priority areas for training and appropriate beekeeping technologies were identified in collaboration with target communities. To improve the traditional system, several interventions were initiated that included capacity building of traditional beekeepers blending modern beekeeping technologies with indigenous skills to sustainably manage bees and hive products, social control measures to curb hive pillaging, and integrating women in beekeeping production. To improve the quality of hive products, simple and low-cost technologies on harvesting and post-harvest handling of hive products were initiated. To refine the results and discuss on survey findings, feedback workshops were conducted. Traditional beekeepers and representatives of women groups and local traders were taken for an external study tour to visit and learn from other beekeepers and create market linkages. The introduction of modern beekeeping technologies was carried out through participatory trainings and demonstrations.

The impact of beekeeping interventions in the study area showed that log hive population had increased by 12% and the volume of crude honey traded in the local market had increased by 20%. The increase in the amount of crude honey delivered to the market was attributed to more pastoralists venturing into beekeeping. The improved capacity of local markets to buy honey also stimulated many beekeepers in the Ndotto ranges to deliver honey to the local market, including

Marsabit town. The quality (absence of impurities) and shelf life of processed honey by the women groups have improved, due to better processing techniques and women now process strictly ripe honey for the market. Since hive products were less perishable, more women groups are investing on hive products like making of wax. This diversification of products has meant increased incomes to the various households.

Conclusions

The main conclusions from the implementation of best-bet technologies in ASALs in Kenya are that the Desert Margins Programme (DMP) in Kenya has made some contribution to understanding which technology options have potential in reducing land degradation in marginal areas and conserving biodiversity that will benefit ASAL communities. Several hectares of degraded lands and rangelands have been reclaimed through various land degradation control technologies. However, land ownership is critical to farmers' investment in land degradation control and better rangeland management strategies. Policy advocacy will be necessary to promote appropriate land ownership by rural communities hoping that this will lead to greater adoption of some of the best-bet technologies that address land degradation control. Good policies will promote sustainable livelihoods, while bad policies will discourage local initiatives and result in resource degradation, poverty, and conflict (Munyasi et al., 2007: IIRR, 2002).

Combating land degradation and conserving biodiversity in the ASALs require diversification of production and cropping systems. Hence, technologies that provide alternative livelihood options to the local communities are essential to the success of land degradation control measures. Technologies that have obvious and immediate benefits to communities are more readily accepted and adopted. This is illustrated by the increased income generation from technologies such as beekeeping and honey production and fodder conservation for sedentary livestock herds in Kenya. Thus community-based natural resources management is key to sustainable resource use and derivation of benefits by the local community.

Training of stakeholders is an important component of upscaling; therefore, many stakeholders (farmers, women, and technicians) have been trained during the implementation of the best-bet technologies. Training provided included grafting techniques, use of drip irrigation, and community-based monitoring and assessment of rangelands. However, the high cost of scaling up of some of the technologies, especially those that concern rehabilitation of degraded lands, is a major constraint to large-scale dissemination of these technologies. Another constraint to rehabilitation of degraded lands by tree planting and pasture production is protection of the land being reclaimed from livestock grazing, especially if it is communally owned. Moreover, sustainability of some of the technologies, especially tree planting and reseeding of rangelands, depends heavily on rainfall, which is often erratic.

Generally many of the technologies implemented have great potential to arrest land degradation and biodiversity loss in ASALs. However, larger scale upscaling of these technologies will necessitate influencing a change in policy.

Acknowledgements We acknowledge the financial support from the Desert Margins Programme for Africa that is funded by GEF, implemented by UNEP, and executed by ICRISAT in collaboration with NARS in Africa. We thank the various collaborative partners in Kenya [KARI, KEFRI, University of Nairobi, NMK, NEMA, ITDG (Practical Action–East Africa), Government of Kenya extension services and other extension agents, etc.] for their technical support and the Director, Kenya Agricultural Research Institute (KARI), for permission to publish this chapter.

References

Achieng NJM, Muchena FN (1979) Land utilisation types of the Makueni area. Miscellaneous Soil Paper no. M20. Kenya Soil Survey, Kabete Nairobi

Ballard JM, Plateau JP (1996) Halting degradation of natural resources: Is there a role for rural communities? Clarendon Press, Oxford

Coughener MB (2004) The Ellis paradigm – humans, herbivores and rangeland systems. Afr J Range Forage Sci 21(3): 191–200

Coughenour MB, Coppock DL, Ellis JL (1990) Herbaceous forage variability in an arid pastoral region of Kenya: importance of topographic and rainfall gradients. J Arid Environ 19:147–159

De Jager A, van Keulen H, Maina F, Gachimbi LN, Itabari JK, Thuranira EG, Karuku AM (2006) Attaining sustainable farm management systems in semi-arid areas in Kenya. Few technical options, many policy challenges. Int J Agric Sustainabil 3(3):189–205

Ekirapa AE, Muya EM (1991) Detailed soil survey of part of the University of Nairobi Dryland Field Station, Kibwezi. Machakos District. Detailed Soil Survey Report No. D75. Kenya Soil Survey

Emiru B, Demel T, Pia B (Undated) Contribution of enclosures to rehabilitate degraded drylands of Northern Ethiopia

Fabrius C, Koch E, Magome H, Turner S (eds) (2005) Rights, resources and rural development: community-based natural resources management in Southern Africa. Earthscan, London

Gemedo-Dalle, Mass B L, Isselstein J (2006) Rangeland condition and trend in the semi-arid Borana lowlands, southern Oromia, Ethiopia. Afr J Range Forage Sci 23(1):49–58

Goldsmith P (2000) Natural resource management in Marsabit District. Case studies of pastoral settlements. Analysis and recommendations. Project No. ACP. KE.061 – Kenya Agricultural Research Institute (KARI) Report. Kenya Agricultural Research Institute, Nairobi, Kenya

Government of Kenya (GOK). (2002) National Development Plan (2002–2008). Ministry of Planning and Development, Government Printer, Nairobi

Herlocker D (ed) (1999) Rangeland resources in Eastern Africa: their ecology and development. GTZ, German Technical Cooperation, Nairobi

Hudson NW (1987) Soil and water conservation in semi-arid areas. FAO Soils Bulletin No. 57. Food and Agriculture Organization of the United Nations (FAO), Rome

IIRR (2002) Managing dryland resources-An extension manual for eastern and southern Africa. International Institute of Rural Re-construction (IIRR), Nairobi

Itabari JK, Kwena K, Esilaba AO, Kathuku AN, Mohammad L, Mangale N, Kathuli P (2007) Land and water management research and development in arid and semi-arid lands of Kenya. A paper presented at an International Symposium on Innovations as Key to the Green Revolution in Africa: Exploring the Scientific Facts. African Network for Soil Biology and Fertility (AfNet) of TSBF-CIAT in collaboration with the Soil Fertility Consortium for Southern Africa (SOFECSA), Arusha, Tanzania, 17–21st Sept 2007

Itabari JK, Nguluu SN, Gichangi EM, Karuku AM, Njiru EN, Wambua JM, Maina JN, Gachimbi LN (2004) Managing land and water resources for sustainable crop production in dry areas: a case study of small-scale farms in semi-arid areas of Eastern, Central and Rift Valley Provinces of Kenya. In: Crissman L (ed) Agricultural Research and development for sustainable resource management and food security in Kenya. Proceedings of the agriculture/livestock research support programme, Phase II. End of programme conference. Kenya Agricultural Research Institute (KARI): Nairobi, Kenya. pp 31–42, 11–12 Nov 2003

Jaetzold R, Schmidt H, Hornet ZB, Shisanya CA (2006) Farm management handbook of Kenya, vol II. Natural Conditions and Farm Information, 2nd edn, vol II/ Part C Eastern Province. Ministry of Agriculture/German Agency for Technical Cooperation (GTZ), Nairobi

Jensen JR, Bernhard RH, Hansen S, McDonagh J, MØberg JP, Nielsen NE, Nordbo E (2003) Productivity in maize based cropping systems under various soil-water-nutrient management strategies in semi-arid alfisol environment in East Africa. Agric Water Manage 59:217–237

Kamondo BM, Kimondo JM, Mulatya JM, Muturi GM (eds) (2006) Recent Mukau (*Melia volkensii* Gurke) research and development. Proceedings of the First National Workshop, KEFRI Kitui Regional Research Centre. Kenya Forestry Research Institute, Nairobi, Kenya

Karugu WN (2004) The potential of agroforestry for subsistence farming in arid and semi-arid areas of Kenya. In: Temu AB et al (eds) Rebuilding Africa's capacity for agricultural development: the role of tertiary education. World Agroforestry Centre: Nairobi, Kenya, pp 170–175

Kilewe AM, Thomas DB (1992). Land degradation in Kenya. A framework for policy and planning. Food production and Rural Development Division, Commonwealth Secretariat, Marlborough House, Pall Mall, London SWIY 5HX

Kipserem LK (1996) Water harvesting on entric Fluvisols of Njemps Flats of Baringo district. A paper presented at the Dryland applied Research and Extension Project (DAREP) workshop at Embu from 4–6th June 1996

Lal R (1995) Tillage systems in the tropics. Management options and sustainability implications. FAO Soils Bulletin No. 71. Food and Agriculture Organization of the United Nations (FAO), Rome

Lengarite MI, Okoti M (2004) Survey report on the potential of honey production and marketing in Ngurunit and East of Ndoto Mountain ranges. Desert Margins Programme (DMP) Report. Kenya Agricultural Research Institute, Nairobi, Kenya

Lusigi WJ (1981) Combating desertification and rehabilitating degraded production systems in Northern Kenya. Integrated Project for Arid Lands (IPAL) Tech. Rep. A–4

McPeak JG (2003) Analyzing and addressing localized degradation in the commons. Land Econ 78(4):515–536

Milimo B, Olukoye GA, Moindi D (2002) The human dimension of desertification with special reference to people's participation in dune stabilisation measures in North Horr, Northern Kenya. Deutsche gesellschaft fur Technische Zusammenarbeit (GTZ) GmbH, TOB Publication No: TOB F-1/8e ISBN: 3-9801067-14-1, Eschborn 2002

Ministry of Agriculture (MoA) (1987) Description of the first priority sites in the various districts. Fertilizer use Recommendation Project (FURP) 26, Machakos District., NARL-Kabete

Miriti JM, Esilaba AO, Kihumba J, Bationo A (2007) Tied-ridging and integrated nutrient management options for sustainable crop production in semi-arid eastern Kenya. In: Bationo A, Waswa B, Kihara J, Kimetu J (eds) Advances in integrated soil fertility management in sub-Saharan Africa: challenges and opportunities. Springer, Netherlands, pp 434–441

Munyasi J, Esilaba AO, Wekesa L, Ego W (2007) Policy issues affecting integrated resource management and utilization in ASALS. A paper presented at an International Symposium on Innovations as Key to the Green Revolution in Africa: Exploring the Scientific Facts. Organized by AfNet of TSBF-CIAT in collaboration SOFECSA, Arusha, Tanzania, 17–21 Sept 2007

Nabhan H, Mashali AM, Mermut AR (1999) Integrated soil management for sustainable agriculture and food security in southern and east Africa. Proceedings of the Expert Consultation, Harare, Zimbabwe, 8–12 Dec 1997. Agricultural Technical and Extension Services (Agritex) and

Food and Agriculture Organization of the United Nations (FAO), Rome
Nandwa SM, Gicheru PT, Qureshi JN, Kibunja C, Makokha S (1999) Kenya country report. In: Nabhan H, Mashali AM, Mermut AR (eds) Integrated soil management for sustainable agriculture and food security in southern and east Africa. Proceedings of the Expert Consultation, Harare, Zimbabwe, 8–12 Dec 1997. Agricultural Technical and Extension Services (Agritex) and Food and Agriculture Organization of the United Nations (FAO), Rome
Ndung'u JN, Nyamori BO, Kuria SG, Njanja JC (1999) Participatory rural appraisal on livestock management practices of the Rendille pastoral community at Korr in Marsabit District. Kenya Agricultural Research Institute, (KARI) Report. Kenya Agricultural Research Institute, Nairobi, Kenya
Ngoroi EH, Gitari J, Njiru EN, Njakuthi N, Mwangi I, Muriithi C (1994) Effect of moisture conservation methods and fertilizer/FYM combinations on grain yield of maize in low rainfall areas. Annual Report, KARI Regional Research Centre, Embu
Njihia CM (1977) The effect of tied ridges, stover mulch and farmyard manure on rainfall abstraction in a medium potential area, Katumani, Kenya. Paper presented at the International Conference on role of soil physical properties, December 1977. International Institute of Tropical Agriculture (IITA), Ibadan, Nigeria
Nyariki D, Ikutwa C, Musimba N (2005)Capacity building through training of beekeepers in the southern rangelands of Kenya. Desert Margins Programme (DMP) Report. Kenya Agricultural Research Institute, Nairobi
Olukoye GA, Wamicha WN, Kinyamario JI (2003) Assessment of the performance of exotic and indigenous tree and shrub species for rehabilitating saline soils of Northern Kenya. Afr J Ecol 41(2):164–170
Reid RE, Ellis JE (1995) Impacts of pastoralism on woodlands in South Turkana, Kenya: Livestock Mediated Tree Recruitment. Ecological Appl 5:978–992
Van Bremen H, Kinyanjui HCK (1992) Soils of the Lodwar area: An inventory of soils, an evaluation of present land use and recommendation for future land use Reconnaissance Soil survey Report No R17. Kenya Soil Survey, Nairobi
Van Wijngaarden W, van Engelen VWP (1985) Soils and vegetation of the Tsavo area (Mtitu andei-Voi). Reconnaissance soil survey report no. R7. Kenya Soil Survey, Nairobi, Kenya
Wandera FP, Sitienei J, Mnene WN, Ahuya CO (1996). PRA report for Kajiado and Baringo Districts
Warren DM (1993) The role of indigenous knowledge in adaptation and development. Background paper for keynote address 'Pithecanthropus Centennial 1893–1993: human evolution in its ecological context'. University of Leiden, Leiden, The Netherlands

Soil Organic Inputs and Water Conservation Practices Are the Keys of the Sustainable Farming Systems in the Sub-Sahelian Zone of Burkina Faso

E. Hien, D. Masse, W.T. Kabore, P. Dugue, and M. Lepage

Abstract Rapid population growth and climatic change threaten the sustainability of natural resources in the sub-Sahelian region of West Africa. Environmental changes and degradations can be mitigated by the adaptation of improved farming practices. In Ziga, located in Yatenga region, a research program was implemented between 1980 and 1987. The aim of this research was to describe and to analyze manure practice management and to present their determinants, for deducing their effects on farming system sustainability. In 2005, a survey in the same village was carried out to assess the evolution of farming practices. According to the inquiry made, two practices, called "zaï" and "djengo," were largely used in cereal crop production. The characteristics of "zaï" and "djengo" practices were assessed and their effects on grain crop yields measured. The "zaï" characteristics depend on the farm manure availability. Zaï practice is a complex soil restoration system using manure localization and runoff capture in micro-watersheds on degraded soils to improve their productivity. In addition, another practice called the "djengo" that has not been described in previous works was noticed in Ziga. Like the "zaï," "djengo" is a technique of soil and water conservation characterized by localized supply of organic matter. In the case of "djengo," the micro-basin is dug after the first rain. The "djengo" is less expensive in time. These two practices revealed a strategy of farming system intensification by localization of organic and mineral fertilization, as well as a better management of rainwater. Tree regeneration occurs where "zaï" or "djengo" practice is used. This study highlights the necessity of better controlling soil, water, and organic matter to improve agrosystem viability as a key for the success of the Green Revolution in sub-Saharan Africa.

Keywords Djengo · Farming practices · Soil · Sustainability · Yatenga · Zaï

Introduction

Maintaining soil fertility is vital for sustainable productivity, particularly in resource-poor countries (Yang, 2006). Sub-Saharan African (SSA) countries have registered a continuous demographic growth in the last decades. This growth is still about 3% per year, which results in a twofold increase of the population every 25 years. In this regard, what is the capacity of natural resources such as soils to support this population growth and how can agriculture contribute to food security? Past studies have shown an increase of farmland and animal production at the expense of the natural vegetation and long-term fallows, which was considered to be the fulcrum of soil fertility management in West African savannahs. This resulted in a reduction of fallow duration over the past few decades from more than 20 years to less than 6 years (Bilgo et al., 2006). Consequently, soil has degraded and soil fertility has declined, leading to an increase of the vulnerability of the farming systems.

E. Hien (✉)
Université de Ouagadougou, UFR/SVT, 03 BP 7021, Ouagadougou 03, Burkina Faso
e-mail: edmond.hien@ird.fr

A. Bationo et al. (eds.), *Innovations as Key to the Green Revolution in Africa*,
DOI 10.1007/978-90-481-2543-2_121,

In the Yatenga region, degraded soils are named "zipellé" (Marchal, 1983), meaning "without vegetation" and characterized by a superficial crust. Because of natural (climate) and anthropogenic phenomena, the degradation of soil in Yatenga involved the formation of large stripped glacis whose importance is worrying (Dugué, 1986). During years of drought, many peasant families are compelled to leave their villages to settle in regions of higher rainfall elsewhere in Burkina Faso (McMillan et al., 1990) or in coastal countries, particularly in Ivory Coast, while others are attracted to urban centers (Reij et al., 2005). These dynamics illustrate the relations between population and the environment such as those conceptualized by the neo-Malthusians thesis, contrary to those developed by Boserup. The first indicates that the increase in population leads to land degradation and loss of soil productivity, which are solved by migration.

However, according to Boserup (1970), another option for solving soil productivity degradation is to adapt farming practices to natural resources (soil, vegetation) and new pedoclimatic and socio-economic conditions. These changes of practices, based on soil and water conservation and intensification of the farming systems, make it possible to maintain or to restore agricultural production to support the rural population.

Many research institutes carried out studies in Ziga village in Yatenga. This village also benefited from NGOs' actions from 1980 to 1985 (Dugué, 1989). It was interesting to evaluate the importance of changes in husbandries a few years later, results of which will contribute to promoting debate on population growth and dynamics of the environment. The first conclusion is that soil and water conservation (SWC) practices such as "zaï" observed on degraded soils and "djengo" on sandy soils were considerably developed by farmers. Second, these practices were combined with water conservation and organo-mineral fertilization. The hypothesis was that these practices have improved soil productivity, but also led to a change in field monitoring, working force management and organic matter flux in farming systems at various territory scales.

This chapter presents some observations on the Ziga farming systems particularly on the "zaï" and "djengo" practices and their environmental effects on natural resources.

Materials and Methods

Study Area

This study was conducted in Ziga (13°25′N, 2°19′W), a village located in the Yatenga province in the north of Burkina Faso (Fig. 1). This village is densely populated (70–100 dwellers/ha) by Mossi, Fulani, and Dogon ethnic groups. The region is characterized by an erratic rainfall distribution with an annual average rainfall between 400 and 800 mm. Since 1921, annual rainfall averages have decreased continuously up to 1990 and a small improvement has been registered. Soils are mainly classified as ferric leptosols (FAO, 1998). They are generally not very deep and poor in organic matter.

Selection of Study Farms

At the initial state of the study, a survey was conducted with a sample of 44 farmers, selected randomly in the three neighboring villages of Bossomboré-Yakin (BY), Biingwéogo (BW), and Légoum (LG). Then a distinction between farms was done, according to their input use. Selection criteria were mainly (i) plot area and manpower; (ii) herd size and manure availability; and (iii) farm equipment. Then 18 farms were selected for in-depth interviews. In this sample, we distinguished nine well-tended (input) farms and nine unkempt farms. Farms in-between were excluded.

Surveys

In implementing surveys, several visits (two, three, or four) were required. The main topics treated are summarized in the following three points:

- Farm assets, e.g., crop plots, equipments, livestock;
- Manuring and other SWC practices;
- Farm yield assessment in 2005 according to each manure practice.

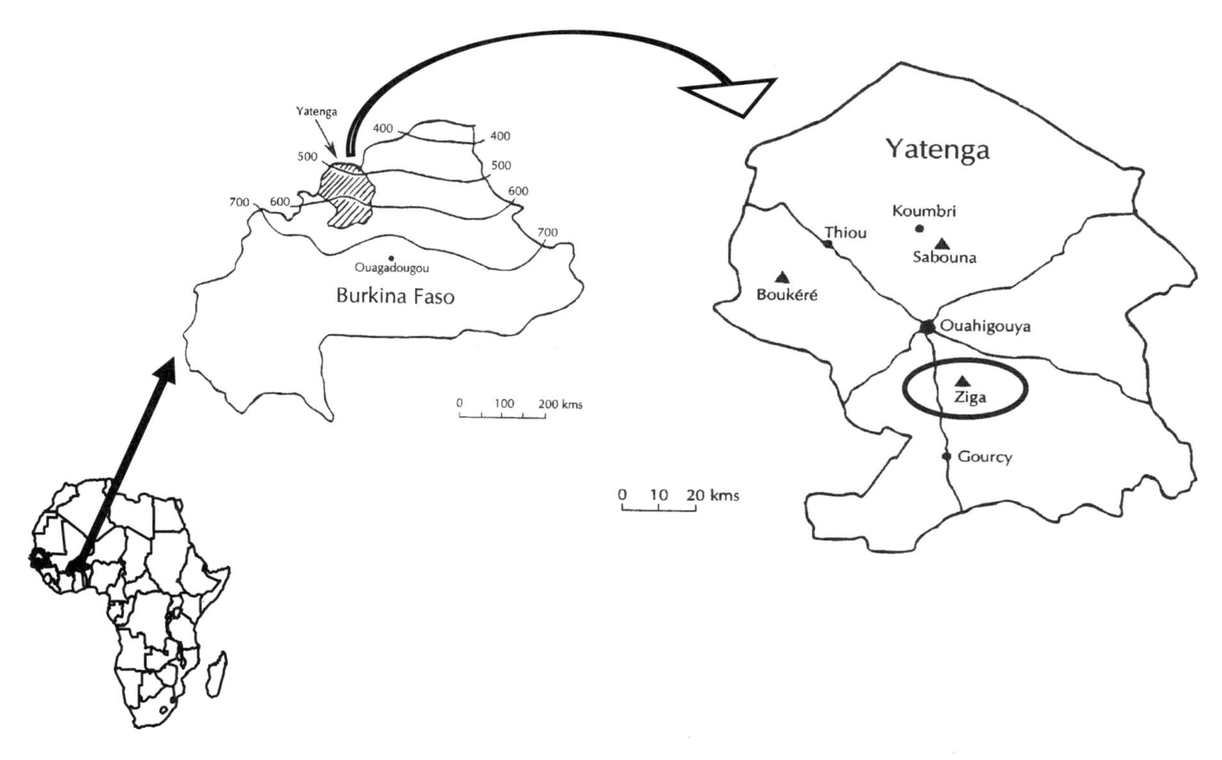

Fig. 1 Study area location

Assessment of Manuring Practices

To assess the manuring practices, we used squares of 9 m^2 (3 m × 3m) placed on various plots representative of different crop management sequences with manuring and SWC practices, such as zaï, djengo, till, or no-till practices with direct sowing. The following parameters on zaï and djengo practices were considered:

- Zaï or djengo hole shape (width, depth, diameters);
- Manure quantity (dry matter/ha);
- *Zaï* or *djengo* hole density and the time of work to settle 1 ha of these practices.

The effects of the different cultural practices on crop production were assessed by yields and its component measurement.

Results and Discussion

Farm Characteristics

Results show that large farms had an average of 6.3 ha against 3.7 ha for the small ones. Field dimensions vary according to the districts: it decreased gradually from LG to BW and finally to BY where cultivated areas were tiny. That indicates a stronger land pressure in BY. The number of workers per farm varied from 9 to 14 for large farms against 3 to 6 for small ones. On average, the large farms had two or three times more workers than did the small ones. Cultivated area per worker was higher on small farms than on large ones, irrespective of the agricultural district. Considered from the large farms to the small ones, these values were as follows: 0.5 and 0.93 ha (BY), 0.63 and 1.15 ha (BW), 0.57 and 0.72 ha (LG). Large farms with their workers and equipment did not try to extend their cultivated areas. Workers on these farms worked on reduced areas. This observation would be a sign of agricultural intensification.

Large farms were those which had more animals as was observed in the BY district where each farm had more than 15 cattle and 15 sheep and goats (Fig. 2). Small farms have no cattle but only some sheep and goats (less than five). We assume that one cattle was the equivalent of one tropical bovine unit (TBU) and a small ruminant 0.1 TBU. We also considered that 1 TBU produced 1 t of dry matter of dung per year. The potential manure per farm was very variable, according to cattle availability (Fig. 2). Some farms had a potential production of almost 22 t per year, while others had less than 1 t per year. If we consider a homogenous distribution of these manure on all the cropped area, the farms which could supply their fields with 5 t per ha of organic manure, which is the recommended dose by agricultural extension service, were few.

By comparing cattle dung production on the farm and quantities of organic manure brought this year on the field, we noticed that all the farms use more manure than they can obtain from their own cattle manure. Farms have other sources for organic matter. In fact, most of the farmers collected organic matters produced on their farms and stored them in manure pits. The organic resources used were the manure of stalling animals, dung collected on grazing areas, the household wastes, the crops residues, and even some grass from the fallow area. Some farmers brought water in the manure hole and regularly mixed the organic matter to produce a better compost. Composting is still rare.

The "Zaï" Practice

In the Moore language, "zaï" was derived from the word "zaïegré," which meant "wake up early to prepare the seedbed." The baseline principle of zaï is that the holes were dug before the onset of the rainy season and manure was put into the hole at the end of the dry season. The hole design on the field was done to catch the maximum of the runoff rain. Fifty-three percent of farmers practiced the zaï according to these rules. However, large variability occurred relative to the date of the hole digging and of sowing, to supply it or not with manure (Fig. 3). The availability of work force and of manure was another factor that influenced the diversity of zaï practice.

Another factor of variability of the zaï practices was the frequency of the zaï technique in the cropping system:

- 37.5% of farmers practiced the zaï every year on the same plots. The holes of the previous year were not necessarily the same as the next year. The holes were randomly positioned.

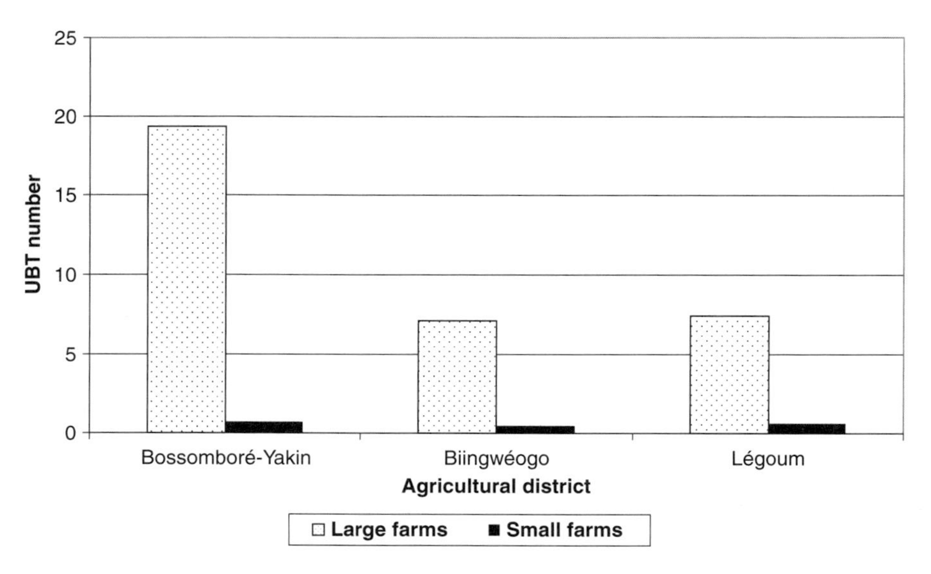

Fig. 2 Manure potential and livestock number

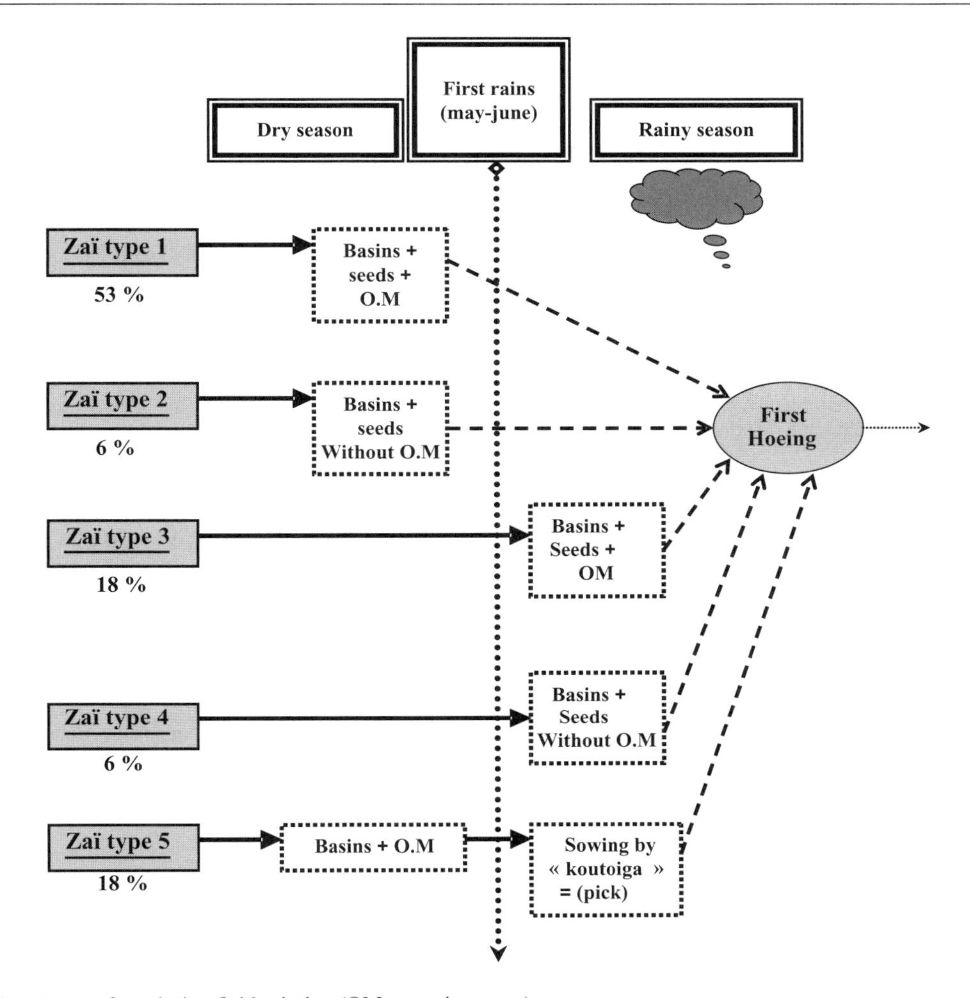

Fig. 3 Various types of zaï during field priming (OM, organic matter)

- 37.5% of surveyed peasants made zaï every year in a permanent way by using continuously the same basins for several years. In this case, each year, year *n*-1 basins were re-opened and manure brought again.
- 12.5% of farmers made zaï every 2 years conserving the same holes on the field. After the zaï practice on the year *n*, the farmers used the direct sowing in year *n*+1.
- 6.25% of peasants practiced the zaï in 2 out of three years; conserving the same holes over the 2 years of zaï. After 2 years of successive zaï practice, they proceeded on the third year of a direct sowing for 1 year and reverted to zaï practice for 2 years and so on.
- The last type of practice (6.25% of surveyed farmers) concerned farmers who practiced the zaï every year and used the crop residues produced in the year *n*–1 as mulch on the space between the holes that was not tilled with the zaï practice. Then, the objective of the farmers was to promote the soil restoration of the field. In this case, the localization of the holes remained fixed from one year to another.

Table 1 describes some parameters of plots cultivated with sorghum according to zaï practice. Hole density was 23,210 ha^{-1} on large farms against 33,889 ha^{-1} for small ones. Sixty-eight days and fifty-two days were necessary for a worker to dig 1 ha of zaï on large and small farms, respectively. In addition, zaï basin holes were significantly different from large to small farming systems. Indeed, the small diameter measured 32.9 and 23.5 cm for large and small farms,

Table 1 Characteristics of zaï and djengo practices

Practice	Farms	Hole density (number/ha)	Time for laying out 1 ha (days/ha)	Diameter 1 (cm)	Diameter 2 (cm)	Basin depth (cm)	Manure quantity per hole (g)	Manure quantity (t/ha)
Zaï	LF	23,210	68	32.9[a]	36.4[a]	10.6[a]	542[a]	12.57
	CV (%)			7.0	6.1	9.8		
	SF	33,889	52	23.5[b]	26.6[b]	10.3[b]	230[b]	7.77
	CV (%)			5.9	6.3	9.5		
Djengo	LF	41,481	15	26.9[a]	26.7[a]	7.0[a]	211.1[a]	8.76
	CV (%)			11.1	9.0	15.5	12.8	
	SF	38,889	24	22.5[a]	27.1[a]	8.3[b]	173.4[a]	6.74
	CV (%)			8.9	5.5	9.0	6.6	

LF, large farms; SF, small farms; CV, coefficient of variance; numbers followed by the same letter in a column are not statistically different ($P < 0.05$) for the same practice

respectively. Large diameter was 36.4 and 26.6 cm for large and small farms, respectively. The depth of the hole was 10.6 cm for large farms and 10.3 cm for the small ones. Large farms tend to accommodate big size holes, contrary to small ones. As much like hole dimensions, manure quantity in zaï basins was higher for big farmers who had high production of manure. On average, 542 g per hole was applied on large farms against only 230 g per hole on small-scale farms, which corresponds to 12.6 and 7.8 t ha^{-1}, respectively, in both cases.

Djengo, Another Practice in the Use of Manure

The other SWC practice, called djengo, similar to zaï was observed in Ziga. This technique was not noticed in the precedent studies in Yatenga region. The djengo is a hand tool, a hoe with long handle, used to carry out this practice. It was practiced on sandy soils and exclusively after the beginning of the rainy season. There were also several alternatives to the "djengo" practice. The majority of farmers (60%) did not till and waited for the growth of the first weeds to dig the holes, while other farmers plowed the soil before digging large holes. The quantity of manure applied also varied between farmers.

As the hole was not maintained in sandy soils during the rainy season, the localization of the holes in the djengo practice changed every year. The number of holes per hectare was relatively constant between farmers' fields (38,889 and 41,481 basins ha^{-1} for large and small farms, respectively) and the time of the field preparation was 15 and 24 days ha^{-1} for large and small farms, respectively (Table 1). In the same way, the hole size in djengo was not significantly different between the two types of farms: (i) small diameters: 26.7 and 22.5 cm; (ii) large diameters: 26.9 and 27.1 cm; and finally (iii) depths: 7.7 and 8.3 cm. The quantities of applied manure in djengo reached 8.8 t ha^{-1} in large farms and 6.7 t ha^{-1} in small ones.

The differences between zaï and djengo are given in Table 2. Zaï and djengo differ mostly by the used tools. The tool used for digging zaï holes is called "boam-boara," which has a short curved handle with a blade, while the tool used for digging djengo holes is called djengo, which has a longer handle. This difference in handle size induces a difference in the worker position: to dig zaï holes, the worker bends down, while to dig djengo holes, the worker is in a stand-up position. Zaï holes are larger than those of djengo, whatever be the nature of the soil, which implies that more manure is applied in zaï than in djengo (Table 1). Zaï is generally carried out during the dry season on massive and encrusted soils, while djengo is exclusively made on sandy soils after the beginning of the rainy season.

Assessment of the Sorghum and Millet Yields

Sorghum and millet yield is correlated to the number of grains per panicle or per ear (Table 3). There are no significant differences ($P = 0.05\%$ with SNK test) between grain yields produced with zaï, djengo, and

Table 2 Summary of key differences between zaï and djengo practices

Elements of difference Tools	Zaï practice Boamboara	Djengo practice Djengo
Basin density	27,481[a]	39,753[b]
Time for laying out 1 ha	62[a]	21[b]
Basin diameter 1 (cm)	31.3[a]	25.8[b]
Basin diameter 2 (cm)	35.1[a]	26.8[b]
Basin depth	10.9[a]	7.4[b]
Manure quantity (g/basin)	417.3	185.7
Manure quantity (t ha^{-1})	10.7	7.4
Soil type	Mostly loamy soils	Exclusively on sandy soils
Period	Dry and rainy seasons	Exclusively in rainy season

Numbers followed by the same letter on a row are not statistically different ($P < 0.05$)

Table 3 Pearson correlation test (parametric test) between yield and its components

	Sorghum ($n = 52$)		Millet ($n = 15$)	
Yield component	Correlation coefficient	*P*-value bilateral	Correlation coefficient	*P*-value bilateral
Bunch number ha^{-1}	–0.164	0.246	–0.801	0.0001
Stem number bunch^{-1}	0.293	0.036	0.497	0.056
Ear number stem^{-1}	0.630	<0.0001	0.103	0.715
Grain number ear^{-1}	0.610	<0.0001	0.753	0.001
Grain mass (g)	0.277	0.048	0.38	0.162

simple sowing (Table 4). However, the straw yields were significantly higher under direct sowing, both with zaï and with djengo practices. Because millet is produced on sandy soils, we compared only the millet yield between the djengo and the direct sowing practices. There is a slight increase of straw and grain yield under djengo compared with the direct sowing (Table 5).

Zaï is usually practiced on marginal soils where it gives more grains and straw than those obtained

Table 4 Sorghum yield (grain and straw) and its components (mean ± SE) for different agricultural practices at Ziga, Burkina Faso

Variables	Direct sowing ($n = 20$)	Djengo ($n = 9$)	Zai ($n = 23$)	F-test	*P*-value
Bunch number ha^{-1}	$31,107 \pm 341a$	$29,336 \pm 200b$	$28,884 \pm 328b$	13.4	< 0.001
Stem number bunch^{-1}	3.09 ± 0.15	3.51 ± 0.22	3.69 ± 0.23	2.52	0.091
Ear number stem^{-1}	0.57 ± 0.05	0.47 ± 0.03	0.49 ± 0.04	1.522	0.228
Grain number ear^{-1}	$716 \pm 69b$	$920 \pm 127ab$	$997 \pm 77a$	3.54	0.037
Grain mass (g)	0.023 ± 0	0.021 ± 0.001	0.021 ± 0	2.68	0.079
Grain yield (Mg ha^{-1})	0.93 ± 0.13	0.92 ± 0.16	1.1 ± 0.13	0.52	0.599
Straw yield (Mg ha^{-1})	$1.82 \pm 0.19b$	$2.32 \pm 0.49ab$	$2.96 \pm 0.35a$	3.69	0.032

ANOVA and Newman–Keuls mean comparison test (means with the same letter belong to the same group)

Table 5 Millet yield (grain and straw) and its components (mean ± SE) for different agricultural practices at Ziga, Burkina Faso

Variables	Djengo ($n = 6$)	Direct sowing ($n = 9$)	F-test	*P*-value
Bunch number ha^{-1}	$31,423 \pm 802$	$32,943 \pm 102$	5.39	0.037
Stem number bunch^{-1}	3.33 ± 0.13	2.78 ± 0.14	7.33	0.018
Ear number stem^{-1}	0.53 ± 0.05	0.62 ± 0.05	1.39	0.26
Grain number ear^{-1}	1350 ± 293	939 ± 79	2.62	0.13
Grain mass (g)	0.011 ± 0.001	0.012 ± 0.001	0.32	0.579
Grain yield (Mg ha^{-1})	0.8 ± 0.17	0.59 ± 0.06	2.04	0.177
Straw yield (Mg ha^{-1})	1.68 ± 0.5	1 ± 0.11	2.61	0.13

with direct sowing and djengo, which are often carried out on better soils. Some argue that the zai technique favors what is called extensification because it favors marginal soil cultivation. At the cultivated plot, however, the zaï practice is more an intensification of the farming system as previously unproductive soils give a crop production that could be higher than the yield obtained in "more easily cultivable soils." The stone line building and the hole digging, as the organic matter and mineral fertilizer application and water pouring, require additional labor and capital intensification.

To be efficient, the zaï as well as djengo need large quantities of manure. It is necessary to associate livestock production to agriculture production. Farmers should not only invest in rainwater management but also increase their manure production by animals. However, zaï production on marginal plots can reduce forage production. In Ziga, the animals stalling in each farm were developed during the last 20 years certainly to intensify animal production in response to the increasing urban demand for meat. More forage will be required in the future.

First works on soil and water conservation began in Ziga in the 1960s with the GERES program (1960–1964). This program had a limited success in comparison to the authorized investment and the negative effects of ridges made with soil. After the drought of 1984–1985, research and development (R&D) program was reoriented toward regional planning in the use of stone lines and grass bands. Some 18% of the village cropped area was protected with stone lines between 1987 and 1990 (Dugué et al., 1993) with logistic and technical support of R&D program. According to the CORAF/CRDI project (2002), 35% of the cropped area was conserved using stone lines. In 2005, 65% of the cropped area was managed with stone lines. All these figures testify to a steady progress and fast change of practices to fight against water runoff and erosion.

There was no zaï practices in the 1980s (Marchal 1983). The traditional practice was manual ridging intended to bury weeds and to increase soil surface roughness to reduce runoff. In 1989, Dugué observed a small extension of zaï in Ziga but no djengo on sandy soils. R&D was focusing on mechanization and the use of mineral fertilizers to increase the crop production. Twenty years later, our works show the importance of zaï in Ziga, confirmed by other studies such as the R3S project (2002), which indicated that 35% of the cultivated fields were tilled by zaï practice.

In Ziga, tree regeneration is visible due to the changes in farmers' practices and the various SWC interventions. Reij and Thiombiano (2003) have reported similar results in other Yatenga villages and shown an improvement of the agricultural outputs and a better management of soil fertility and tree regeneration with the development of soil and water conservation techniques. It appears that there was a significant change in agricultural practices in the sub-Sahelian village of Ziga that allow the maintenance and even an improvement of the crop production. SWC associated with an organic and mineral fertilization seems decisive in these changes. This evolution of the farming systems indicated that the village of Ziga would present a Boserupian type of dynamic. Boserup (1970) considers that rural population growth in developing countries is favorable for food security because population pressure triggers agricultural intensification of labor and capital and innovation.

The diversity of the zaï and djengo techniques pointed out in this study tends to confirm the capacity of farmers to improve their agricultural practices in response to climatic changes and population increase. As Mazzucato and Niemeyer's (2000) works show in Eastern Burkina Faso, the adaptation of the social organization also contributes to a great extent in the improvement of the agricultural production after a phase of degradation following the demographic and environmental changes. These social aspects of the evolution of natural resource management by populations have to be assessed in future studies in Ziga.

Conclusions

The results of this study show that the zaï and djengo practices based on the control of water and organic fertilizer supply were largely adopted in Ziga. The zaï and djengo practices allow a small increase in grain and straw yield per hectare, compared to traditional practices. It is possible to cultivate soils that were usually not used under traditional farming. Farm intensification of the farming systems using zaï and djengo practices is possible. In naturally poor tropical soils, the local supply of water and organic fertilizer to the soil plant system is the main determinant of crop production under erratic rainfalls. To be efficient, these practices need to produce manure which supposes a better association of cropping and animal production systems. Rural population growth is not necessarily a curse. It can be a favor to agricultural intensification and innovation. These results also show that innovations of agricultural practices occur and lead to an increase of agricultural yields and environmental regeneration that contribute to poverty reduction. Green Revolution in Africa is still possible by taking into account indigenous knowledge.

Acknowledgments This work was supported by the Desert Margins Program (UNEP-GEF-ICRISAT) and the French *Institut de Recherche pour le Developpement* (IRD). The authors thank the farmers from Ziga for their participation and contribution to the interviews and the experiments.

References

Bilgo A, Masse D, Sall S, Serpantié G, Chotte J-L, Hien V (2006) Chemical and microbial properties of semiarid tropical soils of short-term fallows in Burkina Faso, West Africa. Biol Fertil Soils. doi:10.1007/s00374-006-0107–4

Boserup E (1970) Évolution agraire et pression démographique. Flammarion, Paris

CRDI/CORAF (2002) Activités de recherche conduites dans le cadre du projet R3S: intégration agriculture-élevage et gestion des ressources naturelles (campagne 2001–2002). Rapp Tech 50p

Dugué P (1986) Programme de Recherche-Développement au Yatenga, programme d'agronomie. Rapp de synthèse 1984:29

Dugué P (1989) Possibilités et limites de l'intensification des systèmes de culture vivriers en zone soudano-sahélienne: le cas du Yatenga (Burkina Faso). Extrait de thèse de l'ENSAM, Collection « Documents Systèmes Agraires » N° 9 du CIRAD. 267p

Dugué P, Roose E, Rodriguez L (1993) L'aménagement de terroirs villageois et l'amélioration de la production agricole au Yatenga (Burkina Faso): Une expérience de recherche-développement. Cah Orstom sér Pédol XXVIII(2): 385–402

FAO (1998) World reference base for soil resources. World soil resources reports

Marchal JY (1983) Yatenga, nord Haute-Volta: la dynamique d'un espace rural soudano-sahélien. ORSTOM, Paris

Mazzucato V, Niemeijer D (2000) Rethinking soil and water conservation in a changing society: a case study in eastern Burkina Faso. Tropical resource management papers, 380p

McMilland D, Nana JB, Sawadogo K (1990) Settlement and development in river basin control zones: case study Burkina Faso. World Bank technical paper no. 200

Reij C, Tappan G, Belemvire A (2005) Changing land management practices and vegetation on the Central Plateau of Burkina Faso (1968–2002). J Arid Environ 63: 642–659

Reij C, Thiombiano T (2003) Développement rural et environnement au Burkina Faso: la réhabilitation de la capacité productive des terroirs sur la partie nord du Plateau Central entre 1980 et 2001. GTZ-PATECORE, USAID, rapport de synthèse, 82p

Yang HS (2006) Resource management, soil fertility and sustainable crop production: experiences of China. Agric Ecosyst Environ 116:27–33

Intercropping Grain Amaranth (*Amaranthus dubius*) with Soybean (*Glycine max*) for Sustainability and Improved Livelihoods in Western Kenya

M.N. Ng'ang'a, O. Ohiokpehai, R.M. Muasya, and E. Omami

Abstract This study was carried out in 2006 to determine the effect of different intercropping systems on grain yield, land use efficiency and economic returns in soybean (*Glycine max*) and grain amaranth (*Amaranthus* spp.) intercrops for two seasons in Teso district of western Kenya. The experiment was carried out on-farm at two sites (A and B) using randomized complete block design with three replicates. Results showed that intercropping using single and double rows significantly increased ($P < 0.05$) amaranth grain yield by 32–33% in site A and 47–46% in site B compared to monocropping. High average land use ratios of LER = 1.5 and 1.6 in single-row intercrops and LER = 1.8 and 1.9 in double-row intercrops were recorded in sites A and B, respectively, in both seasons. Intercropping using double-row intercrops also gave maximum benefit–cost ratio (BCR) values of 4.6 in site A and 4.2 in site B. Intercropping amaranth and soybean can therefore be an ideal farming system for increased food security, enhanced nutrition and income generation through efficient use of land and reduced cost of inputs towards improving livelihoods in this part of Kenya.

Keywords Benefit–cost ratio (BCR) · Food security · Land equivalent ratio (LER) · Monocropping

M.N. Ng'ang'a (✉)
Department of Seed, Crops and Horticultural Sciences, Moi University, Eldoret, Kenya
e-mail: marionnduta@yahoo.com

Introduction

Intercropping is one of the many farming systems that hold great potential in solving future food and economic problems in developing countries (Tsubo et al., 2001) like Kenya. Intercropping legumes and non-legumes is a common practice in western Kenya, with maize, beans and sorghum being the most popular crops grown. The Tropical Soil Biology and Fertility, Institute of CIAT (CIAT-TSBF) and its partners have been working in this region to develop, evaluate and disseminate integrated soil fertility management practices in which legumes are a prominent component (Vanlauwe et al., 2004). Among the legumes being promoted in western Kenya is the improved dual-purpose *promiscuous* soybean (*Glycine max*) variety, since farmers appreciate its ability to provide food, cash and improve soil fertility when grown as an intercrop in rotation with the major staple, maize (Vanlauwe et al., 2004).

In Kenyan rural areas, amaranth is known for its traditional vegetable, which grows in open field. However, in the 21st century, researchers are breeding improved varieties for both leaves and grains some which can fit in the vast Kenyan Arid and Semi Arid zones to curb food insecurity and malnutrition problems. The crop is drought resistant, has few pests and diseases and contains high nutritional levels like carotene, vitamin C, iron and calcium. Similarly, grain amaranth is rich in lysine, an essential amino acid that is lacking in diets based on cereals and tubers (African Executive Magazine, November 2005). However, grain amaranth is not widely consumed in Kenya due to lack of knowledge on its nutritional benefits and limited research on optimal field conditions and farming

A. Bationo et al. (eds.), *Innovations as Key to the Green Revolution in Africa*,
DOI 10.1007/978-90-481-2543-2_122, © Springer Science+Business Media B.V. 2011

systems ideal for its production. Besides, majority of research on intercrops and/or rotation have been based on legume/cereal systems with little documented on legume/pseudo-cereal (amaranth) production systems in this part of Kenya. Hence, there is need for research to establish the suitability of growing these two high-value crops together. The objective of this study therefore is to assess the grain yields, land use efficiency and economic returns of soybean and grain amaranth when the two crops are intercropped using single and double rows in two seasons in western Kenya.

Materials and Methods

Site Characteristics and Soil Measurement

The study was carried out on-farm in two sites A and B of Teso district in western Kenya. The specific location was Akiites, having a medium altitude between 1219 and 1295 m.a.s.l., average bimodal annual rainfall of 900–2000 mm and annual mean temperatures between 21 and 22°C (Jaetzold and Schmidt, 1982). Soils from this region are predominantly highly weathered Acrisols, Ferralsols and Nitisols that are low in nitrogen (N), phosphorus (P) and organic matter content (Sanchez et al., 1997). To obtain soil fertility baseline data before the experiment, soil samples were collected from nine randomly chosen spots from each site at a plough depth of 0–15 cm. The composite soil samples were air-dried and sieved prior to chemical and physical analyses at Moi University Soils Laboratory. Soil pH (water), total nitrogen (N), organic carbon (C), extractable phosphorous (P) and soil particle size were determined using the procedures given in Okalebo et al. (2002). The analysis revealed that soils from site A had moderate organic carbon content (1.85% C), high total nitrogen content (>0.25% N) but low available phosphorous content. The texture was sandy clay loam. On the contrary, soils in site B were low in organic carbon and total nitrogen but slightly high in available phosphorus (>10 mg P kg^{-1}) with a texture of sandy loams. The pH (water) levels of both soils varied between 5.53 and 6.20 reflecting their acid to near-neutral status (Table 1).

Experimental Design

The experimental design for each site was a randomized complete block with four treatments replicated three times. The treatments were amaranth and soybean as sole crops and amaranth/soybean intercrops using single and double rows grown in two seasons. Crop varieties planted were the dual-purpose *promiscuous* soybean variety (TGx1448-2E or SB 20) which was best preferred by farmers in the region (Chianu et al., 2006) and Simlaw select, *Amaranthus dubius* for grain amaranth. Certified soybean and amaranth seeds were obtained from CIAT-TSBF at Maseno and Simlaw Seeds (Kenya Seed Company), respectively. In double-row intercrops, planting was done at a distance of 50 cm between two soybean rows and 33.5 cm between two amaranth rows. A distance of 33.5 cm was left between two paired rows of soybean and amaranth. Inter-row interval of 33 cm between each soybean and amaranth row was used in single-row intercrops with a plant-to-plant (intra-row) spacing of 5 cm for soybean and 10 cm for amaranth being used in both intercropping systems. This gave equal plant densities of 266,667 plants ha^{-1} and 133,333 plants ha^{-1} for soybean and amaranth crops, respectively, in both intercropping systems. In sole crops, soybean was planted at an interval of 45 cm × 5 cm and amaranth

Table 1 Pre-plant soil analysis in sites A and B of Teso district in western Kenya

	pH (H_2O)	C (%)	N (%)	Olsen P (mg kg^{-1})	Clay (%)	Silt (%)	Sand (%)
Site A	5.53	1.85	0.41	6.1	22	13	65
Site B	6.20	1.39	0.23	10.2	11	11	78

at an interval of 30 cm × 10 cm, resulting in plant densities of 466,667 plants ha^{-1} and 333,333 plants ha^{-1} for soybean and amaranth crops, respectively. Planting was done manually with sowing dates being 17 March 2006 for first season crop (long rains) and 2 August 2006 for second season crop (short rains). The individual plot area was 3.0 m × 3.0 m (9 m^2). Prior to sowing, all plots were given a basal application of 30 kg N ha^{-1} as urea and 60 kg P ha^{-1} as triple superphosphate (TSP) by evenly broadcasting and manually incorporating into the top 20 cm of the soil (Frazen, 1999). Manual hoe weeding was done twice at 2 and 4 weeks after planting during each cropping season. Topdressing with urea at the rate of 20 kg N ha^{-1} was done immediately after second weeding, resulting in a total of 50 kg N ha^{-1} in season one. This falls within the reported range of nitrogen (N) fertilizer needed for amaranth production, i.e. 50–200 kg N ha^{-1} (Subhan, 1989), and forms a proper starter dose of N for soybean, i.e. 20–50 kg N ha^{-1} (Bohner, 2007). However, only P fertilizer was applied as TSP at a rate of 30 kg P ha^{-1} during the second growing season mainly to encourage biological N_2 fixation from soybean and observe the effect of intercropping amaranth with soybean without additional N fertilizer. Thinning out and gaping was done at 3–4 leaf stages to achieve optimum plant densities in both seasons.

Data Collection and Analysis

Four middle rows of every cropping treatment were identified for grain yield determination and 1 m of each row was hand harvested at physiological maturity. The harvested crop samples were sun dried, manually threshed and grain dry weight determined at 13% moisture level. Yield values were then scaled up to per hectare basis (kg ha^{-1}). Post-harvest soil fertility status was also determined at the end of each cropping season. Data were evaluated by analysis of variance (ANOVA) and means tested by least significant difference (LSD). The P values of <0.05 were considered as significant and statistical analyses performed using GenStat 4.24 analytical software. Grain yield mean values were then used to calculate the productivity of the intercropping system using land equivalent ratio (LER) as described by Willey (1979):

$$\text{Land equivalent ratio } (L_a + L_b) = Y_{ab}/Y_{aa} + Y_{ba}/Y_{bb}$$

where Y_{aa} is the pure stand yield of crop a, Y_{ab} is the intercrop yield of crop a, Y_{bb} is the pure stand yield of crop b, Y_{ba} is the intercrop yield of crop b and L_a and L_b are LERs of individual components of system.

Economic efficiency of intercropping over sole cropping was determined as described by CIMMYT (1988):

Adjusted mean yield = actual mean yield × 0.9
Gross income = adjusted mean yield × prevailing market prices of sole crops and intercrops
Net income = gross income less total variable costs of production
Benefit–cost ratio (BCR) (return per shilling invested) = gross income/total variable costs

The prevailing market prices at harvest were Kenyan shillings (Ksh) 65 for soybean and Ksh 40 for amaranth.

Results

Grain Yield (kg ha^{-1})

Intercropping using single and double rows significantly increased amaranth and soybean grain yields compared to their respective sole crops in both seasons (Fig. 1). Double-row intercrops recorded the highest average yields of 2526 kg ha^{-1} in site A and 2445 kg ha^{-1} in site B. However, these were 21 and 18% higher than yields obtained from single-row intercrops in sites A (1997 kg ha^{-1}) and B (1969 kg ha^{-1}), respectively, and 47 and 46% higher than sole crops in sites A and B, respectively (Fig. 1).

In both seasons, site A recorded significantly (P<0.05) higher amaranth yield than did site B, while site B recorded significantly higher soybean yield than did site A both in sole stands and intercrops (Fig. 1).

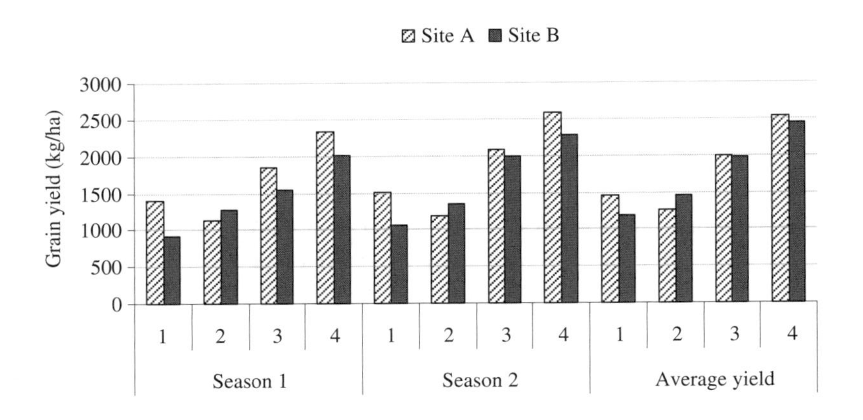

Fig. 1 Grain yield of amaranth and soybean (kg ha^{-1}) under sole cropping and two intercropping systems in western Kenya. 1, sole amaranth; 2, sole soybean; 3, single-row intercrops; 4, double-row intercrops

Similarly, higher amaranth and soybean yields were recorded in season two compared to season one in all cropping systems, although the difference was not statistically significant (Fig. 1).

Land Use Efficiency

Intercropping resulted in significant ($P < 0.05$) yield advantages compared to the respective sole crops in both seasons (Table 2). Double-row intercrops gave the highest total land equivalent ratios (LERs) of 1.9 in site B and 1.8 in site A. Similarly, single-row intercrops gave higher LER values than did sole crops with site A having higher LER value of 1.4 than site B (1.3) in season one but lower LER value (1.5) than site B (1.6) in season two (Table 2).

Economic Efficiency

Intercropping gave higher net returns and benefit–cost ratios (BCR) than did sole crops (Table 3). Double-row intercrops gave the highest net income of Ksh 88,487 with BCR values of Ksh 4.6 in site A and net income of Ksh 79,924 with BCR value of Ksh 4.2 in site B. Similarly, single-row intercrops also gave high net income of Ksh 63,755 and 60,734 with BCR values of Ksh 3.6 and Ksh 3.5 in sites A and B, respectively. However, sole soybean in site B had higher BCR value (3.6) than did single-row intercrop (3.5) in the same site (Table 3).

Post-harvest Soil Fertility Status

Intercropping amaranth with soybean using single and double rows led to improvement of soil fertility level in both seasons (Table 4). Plots under double-row intercrops and sole soybean had significantly ($P<0.05$) higher levels of residual soil N and organic carbon at the end of season two. However, soil N fertility and organic carbon content significantly declined in sole amaranth, while residual phosphorus (P_2O_5) declined in all cropping systems except in sole amaranth with maximum reduction occurring in sole soybean and single-row intercrops (Table 4).

Discussion

Soybean and amaranth grain yields were generally higher in intercrops compared to sole crops. Double-row intercrops gave the highest yields of these two crops in both seasons and sites (Fig. 1). These results agree with findings by Ayew (2003), who reported significant yield increase of grain amaranth when intercropped with soybean. Similarly, Mohta and De (1980) found a 31% increase in soybean yield in a maize/soybean intercrop system and a 26% increase in sorghum/soybean system in double alternate rows compared to single alternate rows. Potential for higher productivity of intercrops when inter-specific competition is less than intra-specific competition for a limiting resource has also been reported (Francis, 1989). In this study, intercropping soybean with amaranth has resulted in better utilization of existing resources that

Table 2 Mean grain yield (kg ha^{-1}) and land equivalent ratio (LER) values of soybean and grain amaranth intercrops in two seasons in western Kenya. Means with different letters within the same column are significantly different ($p < 0.05$)

	Season 1				Season 2			
Treatment[1]	Yield (kg ha^{-1})	Amaranth LER	Soybean LER	Total LER	Yield (kg ha^{-1})	Amaranth LER	Soybean LER	Total LER
Site A								
1	1395c	1.0	–	1.0	1506c	1.0	–	1.0
2	1128d	–	1.0	1.0	1187d	–	1.0	1.0
3	1842b	0.8	0.6	1.4	2081b	0.8	0.7	1.5
4	2329a	0.9	0.9	1.8	2593a	0.9	0.9	1.8
Site B								
1	915d	1.0	–	1.0	1053d	1.0	–	1.0
2	1271c	–	1.0	1.0	1352c	–	1.0	1.0
3	1545b	0.6	0.7	1.3	1997b	0.9	0.7	1.6
4	2019a	1.0	0.9	1.9	2286a	1.0	0.9	1.9

[1] 1, sole amaranth; 2, sole soybean; 3, single-rows intercrop; 4, double-row intercrops

Table 3 Economic analysis of soybean and amaranth intercrops using single and double rows for two seasons in western Kenya

Treatment[a]	Average yield (kg ha^{-1})	Adjusted yield[b] (kg ha^{-1})	Gross income[c] (Ksh)	Variable costs (Ksh)	Net income (Ksh)	Benefit–cost ratio[b] (Ksh)
Site A						
1	1451	1306	52240	22000	30240	2.4
2	1158	1042	67730	21400	46330	3.2
3	1962	1766	88355	24600	63755	3.6
4	2461	2215	113087	24600	88487	4.6
Site B						
1	984	885	35400	22000	13400	1.6
2	1312	1181	76765	21400	55365	3.6
3	1771	1594	85334	24600	60734	3.5
4	2153	1938	104524	24600	79924	4.2

[a] 1, sole amaranth; 2, sole soybean; 3, single-row intercrops; 4, double-row intercrops
[b] Adjusted yield was obtained by multiplying average yield by 0.9, while benefit–cost ratio (BCR) values were calculated by dividing gross income by variable costs (CIMMYT, 1998)
[c] Prices of Ksh 65 kg^{-1} and Ksh 40 kg^{-1} were used for soybean and amaranth grains, respectively, to calculate gross income

ultimately increased the overall crop yields especially when double rows were used.

However, intercropping using single rows increased overall yield but these were lower than yield obtained from double-row intercrops in both sites and seasons (Fig. 1). These results are in close conformity with Mohta and De (1980), who recorded a decrease in soybean yield when intercropped with maize using single rows. Legume yield was also found to be suppressed by cereals (Hauggaard-Nielsen and Jensen, 2001) when grown as intercrops. Similarly, Fisk (1993) and Henderson et al. (1993) reported excessive plant competition for growth resources like nutrients (especially N), water, light and space when amaranth was planted using narrow row spacing.

Overall grain yield of intercropped amaranth and soybean was higher in site A compared to site B (Fig. 1). Similarly, site A recorded higher amaranth yields, while site B recorded higher soybean yields. The results agree with those of Elbehri et al. (1993), who reported high response of amaranth to nitrogen fertilizer, and Myers (2003), who found a 43% increase in amaranth grain yield with increased nitrogen fertilization. Site A had higher soil fertility level (especially in N) than did site B which might have favoured amaranth growth, while site B had low N but higher P levels than did site A which might have favoured soybean growth.

Season two recorded higher grain yield than did season one in both intercropping systems. These

Table 4 Post-harvest soil fertility status after soybean and amaranth intercrops in western Kenya

	Site A			Site B		
Treatment[1]	C (%)	N (%)	Olsen P (mg kg^{-1})	C (%)	N (%)	Olsen P (mg kg^{-1})
Season 1						
1	1.81 (–2.16)[2]	0.39 (–4.88)	6.3(+3.28)	1.31 (–5.76)	0.20 (–13.04)	10.5 (+2.94)
2	1.87 (+1.08)	0.43 (+4.88)	5.2 (–14.7)	1.42 (+2.16)	0.26 (+13.04)	8.2 (–19.61)
3	1.86 (+0.54)	0.42 (+2.44)	5.6 (–8.20)	1.41 (+1.44)	0.24 (+4.35)	2.9 (–12.70)
4	1.88 (+1.62)	0.43 (+4.88)	5.7 (–6.56)	1.43 (+2.88)	0.25 (+8.70)	9.2 (–9.80)
Season 2						
1	1.82 (–1.62)	0.40 (–2.44)	6.5(+6.56)	1.36 (–2.16)	0.21 (–8.70)	10.9 (+6.86)
2	1.91 (+3.24)	0.45 (+9.76)	5.6 (–8.20)	1.43 (+2.88)	0.28 (+21.74)	8.8 (–13.73)
3	1.88 (+1.62)	0.43 (+4.87)	5.7 (–6.56)	1.45 (+4.32)	0.25 (+8.70)	9.1 (–10.80)
4	1.90 (+2.70)	0.44 (+7.32)	5.9 (–3.28)	1.47 (+5.76)	0.26 (+13.04)	9.6 (–5.9)

[1] 1, sole amaranth; 2, sole soybean; 3, single-row intercrops; 4, double-row intercrops

[2] Values in parentheses represent percentage (%) increase (+) or decrease (–) over original values, which were 1.85, 0.4 and 6.1 for C, N and P, respectively, in site A and 1.39, 0.23 and 10.2, respectively, in site B

results agree with those of Remison (1980), who reported the ability of companion crops in restoring soil fertility. Similarly, Clement et al. (1992a) reported an increase in N availability through application of N fertilizer, hence increasing productivity of the legume/non-legume intercrops. Yield increase in season two might have been due to residual effect of nutrients and biomass accumulation from season one. Moreover, there might have been reduced competition for N due to greater N fixation efficiency of the intercropped legume and transfer of fixed N to the associated non-legume component.

Irrespective of sites, the land equivalent ratios (LERs) indicated significant advantages in the intercropping systems (Table 2). The high LER values obtained agree with results of Martin et al. (1990), who obtained high LER in corn/soybean intercrops. In this study, intercropping amaranth with soybean may have encouraged more efficient utilization of growth resources, thus leading to yield advantages and increased stability compared to sole cropping (Okpara, 2002). A number of other studies have indicated that the LER of intercrops tends to be higher under low N conditions (Ofori and Stern, 1987). Low soil fertility in site B might have resulted in increased N_2 fixation from soybean especially in the second season from which amaranth might have benefited, hence resulting in higher land use efficiency compared to site A (Table 2).

Willey (1979) observed that practical significance of productivity in intercropping could be fully assessed only when related to the actual economic or monetary returns. The economic benefits of intercropping soybean and amaranth using double and single rows were generally greater than those of respective sole crops (Table 3). These findings are in close conformity with those of Ahmad et al. (1993), who recorded high economic returns in mash/soybean intercrops, and of Abou-Hussein et al. (2005), who found positive returns in green bean intercrops. Similarly, Polthanee and Kotchasatit (1999) observed clear monetary advantage of intercropping mung bean and cassava over monoculture. The high net returns in the intercrops may have been attributed to the increase in soybean and amaranth yields especially in the second season and the fairly high prices of soybean crop.

Overall soil fertility increased at the end of each cropping season (Table 4). The soils (especially in site B) were initially low in N content due to the cropping history of continuous cereal production with little or no N fertilizer application. Nitrogen fertility improved under amaranth/soybean intercrops with maximum increase being observed in double-row intercropping system. Improvement of nitrogen content due to inclusion of leguminous crops in cropping systems is well documented (Legard and Steel, 1992). However, N fertility decreased in sole amaranth, while phosphorous was low in single-row intercrops and least in sole soybean. The results agree with those of Koraddi et al. (1990), who found greater depletion of soil P_2O_5 after harvesting legumes. In addition, increase in soil N in site A was minimal compared to site B after the second growing season (Table 4). These results are consistent with findings from Clark and Myers (1994), who reported a decrease in soil nitrogen fixation by legumes with an increase in soil fertility as site A had generally higher soil fertility level than did site B even before the experiment.

Conclusions

An overall conclusion from this study is that grain amaranth and soybean can be effectively intercropped as a variation to the maize/bean intercrop system common in many parts of western Kenya. This is primarily due to the fact that higher yields and land use efficiency were obtained in intercrops compared to sole crops. In addition, the feasibility of implementing intercropping systems depends heavily on the profitability of the technology of which amaranth/soybean intercrops gave high economic returns. Post-harvest soil analysis also showed that amaranth/soybean intercrops are likely to increase soil fertility levels in the long term. Intercropping using double rows was found to be optimum since it produced highest average yields, land use efficiency, economic returns and significantly improved soil fertility levels, while single-row intercrops gave inconsistent but potentially promising results. More research should be done to investigate the effects of other farming systems like rotation and conservation tillage on yield and nutritional composition of these high-value crops that hold great potential in providing food security and improving livelihoods through enhanced nutrition and income generation.

Acknowledgements This chapter forms part of the M.Phil. horticulture thesis submitted to Moi University by the first author. Financial support from The Rockefeller Foundation through Tropical Soil Biology and Fertility, Institute of CIAT (CIAT-TSBF), is greatly acknowledged. The authors also appreciate active participation by farmers especially those whose farms were used for this experiment.

References

Abou-Hussein SD, Salman SR, Abdel-Mawgoud AMR et al (2005) Productivity, quality and profit of sole or intercropped green bean (*Phaseolus vulgaris* L.) crop. J Agron 4(2): 151–155

African Executive Magazine (2005) http://www.africanexecutive.com/modules/magazine/sections.php?magazine=41§ions=24

Ahmad R, Anwer MA, Nazir MS et al (1993) Agronomic studies on mash–soybean intercropping systems. Pak J Agric Res 14(4):304–308

Ayew SH (2003) Intercropping yield of *Amaranthus hybridus* and *Glycine max*. J West Afr Agron 232:22–29

Bohner H (2007) Soybean specialist. Soybean nodulation. Ministry of Agriculture, Food and Rural Affairs, Government of Ontario. http://www.gov.on.ca/MBS/english/index.html

Chianu J, Vanlauwe B, Mukalama J et al (2006) Farmer evaluation of improved soybean varieties being screened in five locations in Kenya: Implications for research and development. Afr J Agric Res 1(5):143–150. http://www.academicjournals.org/AJAR

CIMMYT (1988) From agronomic data to farmer recommendations. An economics training manual. Completely revised edition. Mexico DF, Crookston RK and Hill DS 1979. Grain yields and land equivalent ratios from intercropping corn and soybean in Minnesota. Agron J 71:41–44

Clark KM, Myers RL (1994) Yield response in intercropping of pearl millet, amaranth, cowpea, soybean, and guar. J Agron 86(6):1097–1102

Clement A, Chalifour FP, Bharati MP et al (1992a) Effects of nitrogen supply and spatial arrangement on grain yield of a maize/soybean intercrop in a humid sub-tropical climate. Can J Plant Sci 72:57–67

Elbehri A, Putnam DH, Schmitt M (1993) Nitrogen fertilizer and cultivar effects on yield and nitrogen use efficiency of grain amaranth. Agron J 85:120–128

Fisk JW (1993) Cover crop effects on establishment, growth, and N needs of amaranth. M.Sc. thesis, University of Missouri, Columbia

Francis C (1989) Biological efficiencies in multiple-cropping systems. Adv Agron 42:1–42

Franzen DW (1999) Soybean soil fertility. SF-1164. Extension soil specialist county commissions, North Dakota State University (NDSU) and US Department of Agriculture Cooperating. Duane Hauck, Director, Fargo, North Dakota

Hauggaard-Nielsen H, Jensen ES (2001) Evaluating pea and barley cultivars for complementarity in intercropping at different levels of soil N availability. Field Crops Res 72: 185–195

Henderson TL, Schneiter AA, Johnson BL (1993) Production of amaranth in the northern Great Plains. In Alternative crop research: a progress report. North Dakota State University, Fargo, pp 22–30

Jaetzold R, Scmidt H (1982) Farm management handbook of Kenya, vol II – Natural conditions and farm management information – Part A West Kenya (Nyanza and Western Provinces)

Koraddi VR, Rao GG, Guggari AK et al (1990) Overlapping versus simultaneous planting of intercrops in cotton, Karnataka. J Agric Sci 3:189–194

Legard SF, Steel KW (1992) Biological N fixation for sustainable agriculture. Kluwer, The Netherlands, pp 137–153

Martin RC, Voldeng HD, Smith DL (1990) Intercropping corn and soybean for silage in cool temperate region: yield, protein and economic effects. Field Crop Res 23: 295–310

Mohta NK, De R (1980) Intercropping maize and sorghum with soya bean. J Agric Sci (Cambridge) 95:117–122
Myers VT (2003) Intercropping of amaranth with conventional crops in Missouri. J Agron 333:233–239
Ofori F, Stern WR (1987) Cereal–legume intercropping systems. Adv Agron 41:41–90
Okalebo JR, Gathua KW, Woomer PL (2002) Laboratory methods of soil and plant analysis, 2nd edn. SACRED-Africa Press, Nairobi, 128 pp
Okpara CO (2002) Evaluation of cassava, soybean-intercropping system as influenced by cassava genotypes. Niger Agric J 28:3311–3318
Polthanee A, Kotchasatit A (1999) Growth, yield and nutrient content of cassava and mung bean grown under intercropping. Pak J Biol Sci 2(3):871–876
Remison SU (1980) Introduction between maize and cowpea at various frequencies. J Agric Sci 94:617–621
Sanchez PA, Shepherd KD, Soule MJ, Place FM, Buresh RJ, IzacA MN, Mokunywe AU, Kwesiga FR, Ndiritu CG, Woomer PL (1997) Soil fertility replenishment in East Africa: an investment in natural resource capital. In: Buresh RJ et al (eds) Replenishing soil fertility in Africa. SSSA special publication 51 SSSA, Madison, WI, pp 1–46
Subhan (1989) Effect of dosage and application time of nitrogen fertilizer on growth and yield of amaranth (*Amaranthus tricolor* L.) (Indonesian English summary). Penelitian Hortikultura 17:31–40
Tsubo M, Walker S, Mukhala E (2001) Comparisons of radiation use efficiency of mono–inter cropping systems with different row orientations. Field Crops Res 71(1):17–29
Vanlauwe B, Rotich E, Okalebo JR (2004) Integrated soil fertility management in practice in western Kenya. In: Bationo A, Sanginga N, Kimetu J, Kihara J, Ogola J (eds) The communitor. Tropical Soil Fertility Institute of CIAT (TSBF-CIAT), Nairobi, Kenya, pp 1–8
Willey RW (1979) Field Crops Res 32:1–10

Soil Conservation in Nigeria: Assessment of Past and Present Initiatives

B. Junge, O. Deji, R.C. Abaidoo, D. Chikoye, and K. Stahr

Abstract Soil degradation is one of the most critical environmental problems in sub-Saharan Africa (SSA). There is an urgent need to develop a system of effective soil resource management that can reverse the trend and sustain soil productivity and enhance food security. Earlier initiatives on soil conservation have resulted in a range of on-farm and off-farm technologies. But an evaluation of existing conservation strategies is still required to establish comprehensive practices for the West African savanna. In 2006, a soil scientist and a rural sociologist started an assessment study to identify past and present soil conservation initiatives and their effectiveness, including the sociological, technological, and economical aspects. An extensive literature review was conducted, including scientific and digital resources, to get information about past and present research and the performance of soil conservation in different agro-ecological zones of Nigeria. In 2007, villages with different types of conservation technologies were visited and farmers were interviewed to study the adoption of these technologies. The study provides information on the level of awareness of farmers about soil degradation, their attitude toward erosion and willingness to adopt adequate control measures, driving factors influencing the adoption of the technologies, adoption barriers, adaptation to the initiatives, and perceptions of impact. Based on the results, the most promising soil conservation technologies for the savanna are agronomic measures, such as mulching and cover cropping, as well as conservation tillage which can contribute to enhanced soil resource management in Nigeria.

Keywords Assessment of adoption · Nigeria · Soil conservation · Soil erosion control

Introduction

Erosion is one manifestation of the soil degrading processes that have been recognized for a long time as a serious problem in Nigeria (Stamp, 1938). Soil loss has accelerated as the traditional shifting cultivation system has been replaced by generally unstable systems with shorter duration fallows and the expansion of agriculture into marginal areas (Eswaran et al., 2001). Land use intensification was and still is necessary to augment food production to cover the high demand of the rapidly growing population. For example, the Nigerian population has increased from 115 million in 1991 to 140 million in 2006 (Federal Republic of Nigeria, 2007). But erosion has several environmental and economic impacts, especially in sub-Saharan Africa (SSA) where the resilience ability of the soil is limited (Lal, 1995). If accelerated erosion continues unabated, there will be low agricultural production, food insecurity, reduced income for the rural population, and rapid increase in levels of poverty. Hence, the avoidance of soil loss by improved management and conservation of the natural resource is important to maintain soil productivity and contribute to food security (Ehui and Pender, 2005).

B. Junge (✉)
University of Oldenburg, Germany
e-mail: birte.junge@uni-oldenburg.de

A. Bationo et al. (eds.), *Innovations as Key to the Green Revolution in Africa*,
DOI 10.1007/978-90-481-2543-2_123,

Research on soil conservation has already been done for many years in different parts of Nigeria. The existing initiatives have resulted in a range of on-farm and off-farm technologies. But an extensive review of initiatives has been done only for single erosion control measures up till now (Armon, 1980). These initiatives also raise questions about their efficiency, their adoption, and farmers' perception of impact as well as concerns about costs and benefits. The objective of this study was to make an assessment of former and present soil conservation initiatives in Nigeria.

Materials and Methods

Literature Review

An extensive literature review was conducted in 2006 and 2007. Scientific, governmental, and non-governmental institutions working on soil conservation were selected at the beginning, due to the size of the country and its numerous organizations. First of all, the resources of the International Institute of Tropical Agriculture (IITA), Ibadan, and the Internet were checked for appropriate literature. The most important National Research Institutes (NARS); the universities at Abeokuta, Ile-Ife, Nsukka, Maiduguri, and Zaria; and also the Lake Chad Research Institute, Maiduguri, were all visited. The Agricultural Development Program (ADP) in Maiduguri, Federal Environmental Protection Agency (FEPA) in Kaduna and Maiduguri, Agricultural Land Development Authority (NALDA) in Maiduguri, and the Rural Development Projects (RUDEP) in Kaduna and Oshogbo were contacted to get more information about the work of governmental organizations. Ministries such as the Federal Ministry of Environment and Water Resources, the Federal Ministry of Forestry, and the Federal Department of Agriculture in Kaduna, Minna, and Ibadan were also visited. In addition, some non-governmental organizations (NGOs) were contacted in Ibadan: Justice, Development, Peace Commission (JDPC) and the Nigerian Environmental Study/Action Team (NEST). A database of all references was generated. The literature was reviewed to identify the kind and location of installed soil conservation technologies and to select pilot villages for interviews with farmers.

Questionnaire

As sheet erosion is the most frequent kind of soil erosion in Nigeria (Igbozurike et al., 1989), study sites characterized by this type of process were selected for questionnaires. The pilot villages Esa-Oke (7°44′ N 4°50′ E) and Owode-Ede (7°42′ N 4°29′ E) are situated in the southwestern part of the country, an area with a gently undulating landscape and slopes up to 10%. The tropical climate is humid to sub-humid with a bimodal rainfall distribution (mean annual rainfall 1350 mm and mean annual temperature 26.8°C). The dominant soils in the study areas are Lixisols (Sonneveld, 1997). Farmers prepare the land by using hoes or hiring tractors and primarily cultivate food crops such as cassava and maize.

A total of 40 farmers in villages were interviewed in May 2006 to get information on the personal and socio-economic characteristics of farmers, land use pattern, environmental problems, and the characteristics of SCTs. Information such as awareness of the problem and willingness to participate in SCT projects was collected to explain levels of adoption. Various statistical data analyses, including analysis of variance (ANOVA) and regression and correlation analyses, were undertaken to examine possible relationships between different parameters.

Results and Discussion

Literature Review

The review generally has shown that most of the literature on soil conservation was found in scientific institutions such as IITA and NARS, and on the Internet. Records of work on soil erosion control in different governmental institutions were rare, even if staff members described the implementation of several initiatives in different areas. Another result of the study was that most research projects on soil conservation were performed on research farms and only a few on-farm with the participation of farmers.

Soil conservation has a long tradition in SSA and was already being performed in the pre-colonial era. Indigenous techniques focused on soil and water conservation by ridging, mulching, constructing earth bunds and terraces, multiple cropping, fallowing, and the planting of trees (Igbokwe, 1996). In colonial time, large-scale projects on soil loss control were started especially in areas of high agricultural potential, as the British Government was interested in the expansion of commercial farming enterprises. But many of the projects failed as the imported technologies had little relevance in the tropics. After independence, more emphasis was put on soil fertility issues. Decreasing funds at the end of the oil boom in the 1980s additionally restricted the performance of soil conservation schemes (Slaymaker and Blench, 2002). In 2007, the Federal Government of Nigeria planned to spend about half a million US dollars on soil erosion projects all over the country. This shows that the seriousness of this environmental problem still exists and is also recognized (FGN, 2007).

Recent research and implementation of soil conservation cover different kinds of strategies. This chapter includes selected references that focus on on-farm erosion control in Nigeria.

Agronomic measures use the effect of surface covers to reduce erosion by water and wind. The impact of mulching, the covering of the soil surface by residues from previous crops or other organic materials brought to the field, is demonstrated in many field experiments which were conducted on several research stations in Nigeria (Orimoyegun, 1988; Lal, 1993; Odunze, 2002). There are various investigations on the beneficial effects of the physical, chemical, and biological soil properties which influence the soil erodibility. Hulugalle et al. (1985) and Ogban et al. (2001) found out that the bulk density is decreased and infiltration capacity, hydraulic conductivity, and soil moisture are increased by mulching. The size and the stability of soil aggregates are increased because of higher activity of earthworms (Salako et al., 1999). The influence of mulching on the content of organic matter and nutrients was analyzed by Mbagwu (1991) and Olasantan (1999). They state that crop residues increase soil productivity, plant height, and crop yields. One limitation of mulching lies in the large quantity of material required (minimum 4 t ha^{-1} yr^{-1}) as 70–75% of the soil surface should be covered to achieve effectiveness. Significant changes in soil properties also require long periods of time, which makes this erosion control measure less attractive for farmers (Lal, 2000). Kirchhof and Odunze (2003) mention that providing enough crop residues is a problem, especially in the northern part of Nigeria where this material is often completely removed from the field for use as animal fodder, firewood, or construction material. Other possible disadvantages are the carryover of pests and diseases by residues from the previous crop and difficulties in controlling weeds. Extra costs for purchase and the transport of residues to the field, increased labor demand for distributing mulch on the farmland, and problems with planting through residues all make this strategy less attractive for farmers (Lal, 1995).

Prevention or reduction of soil loss can also be done by crop management. Multiple cropping involves different kinds of systems depending on the temporal and spatial arrangement of different crops on the same field (Morgan, 1995). It has been traditionally practiced and is still very common in Nigeria (Olukosi et al., 1991). Research preferentially focuses on increasing soil productivity and crop yields. For example, Carsky et al. (2001) and Singh and Ajeigbe (2002) investigated the intercropping of cereals with legumes, the common cropping system in northern Nigeria. Field trials with root- and tuber-based systems were performed by Van der Kruijs and Kang (1988). Agroforestry, a land use system in which woody perennials are integrated with crops and/or animals on the same land management unit, is another erosion control technology. Field trials with *Leucaena leucocephala* or *Gliricidia sepium* were made by Kang and Mulongoy (1987), Lal (1989), Okogun et al. (2000), and Vanlauwe et al. (2001). The results show that trees improve soil structure, maintain a high infiltration rate, and increase water holding capacity which reduces runoff. Investigations on alley cropping where multipurpose trees are planted as contour hedges between strips of cropland were part of the research program on many stations. Cover crops also play an important role in soil conservation. Research on *Pueraria phaseoloides* or *Mucuna pruriens* was done by Lal et al. (1979) and Tian et al. (1999). The investigations show that cover crops have positive effects on the soil, such as improving the structure, increasing nitrogen levels by the use of N-fixing legumes (Lal, 1989), and suppressing weeds such as speargrass (*Imperata cylindrica*) (Chikoye et al., 2002). Long fallow periods that were part of the traditional shifting cultivation system are no longer possible, due to the rapid increase in human and livestock populations. Taking soil out of use for some time is still an appropriate technology to encourage its recovery, so research on improved fallows is important. Appropriate investigations were made by Juo and Lal (1977) and Tarawali et al. (1999). The benefits of multiple cropping systems are various, as erosion is reduced and physical, chemical, and biological soil properties are improved. Additional benefits are a decreased risk of total crop failure and the suppression of weeds. As product diversification and higher crop yields help to ensure both subsistence and disposable income, polyculture is of huge economic value for the farmers (Kang, 1993). Carsky et al. (2001) recommend the cowpea–maize intercropping system as it can be a relatively productive low-input system for the savanna zone. But Diels et al. (2000) observed reduced yields of maize when grown in combination with leguminous trees and this might be caused by competition for light, water, and nutrients. Tian et al. (1999) got lower cassava yields on fields after a *Pueraria* fallow. Hence, special knowledge on the selection of species and good crop management is needed. Tarawali et al. (2000) state that some research on germplasm development, integrated management, and demonstration of multiple crop use to attract adopters is still required to improve cover cropping.

Soil management by conservation tillage includes different kinds of soil preparation methodologies. Minimum tillage describes a practice where soil preparation is reduced and 15–25% of residues remain on the soil surface. No-till or zero tillage is characterized by seeding a crop directly into the soil that has not been tilled since the previous crop was harvested (Morgan, 1995). Braide (1986) emphasized that these practices were appropriate for conserving Nigerian soils as their erodibility is reduced. Various field trials on the effect of different tillage methods on soil properties have been made in Nigeria by the research groups of Lal (1986), Ojeniyi (1986), Takken et al. (2001), and Chiroma et al. (2005). The results all show the benefits of reduced or zero tillage on erosion control through the long-term maintenance of the soil structure and an increase in water retention and hydraulic conductivity. Additionally, the organic matter content in the surface layer is increased and the hydrothermal regime for crop production and biological activity of the soil fauna is improved (Maurya and Lal, 1980). Lal (1986) also affirms that a continuous no-till system is applicable in tropical wetlands, the basis for increasing national rice production. But he also mentions disadvantages such as problems with diseases, pests, and weeds being carried over from the previous crops. Braide (1986) described a reduction in the yields of cereal crops and mentioned that reduced or no-tillage is not applicable to stem tubers and root crops which are planted on ridges.

Mechanical measures can also be installed on-farm to break the force of winds and to decrease the velocity of runoff to reduce soil erosion. Contouring means performing soil preparation and cultivation across the slope (Morgan, 1995). Contour ridging is common all over Nigeria, while ridge tying is primarily conducted in the northern part of the country to conserve both soil and water. Appropriate investigations were made by Malgwi (1992) and Kowal and Stockinger (1973).

Additional benefits of ridging are the accumulation of soil material and plant residues in the area of roots which increases their thickness and the fertility in the rooting space. Ridges also improve the aeration of roots during wet periods and this especially facilitates the growth of root and tuber crops (Kowal and Stockinger, 1973). But Lal (1995) states that this technology is not suitable for areas with long and steep slopes and high rainfall intensities, as contour ridges are easily destroyed by concentrated overflow. Permanent erosion control technologies are contour bunds made of earth or stones or terraces which consist of an excavated channel and a bank for cultivating crops. The first are installed on the slope of low gradients and the latter in hilly areas with steeper slopes. Research on banks was done by Couper (1995), who prepared an implementation guide for farmers. Terraces were built in Pankshin area, Jos Plateau (Slaymaker and Blench, 2002), and Maku near Udi-Nsukka (Igbokwe, 1996). There is little research on these issues nowadays as implementation and maintenance are often low due to the labor and material required and the fact that benefits are observed after a long period (Lal, 1995). Waterways, such as cut-off drainage, also aim to collect and guide runoff to suitable disposal points. They are primarily installed in areas with high rainfall rates and often covered with grass to prevent destruction (Morgan, 1995). Literature on investigations into drainage systems is rare in Nigeria. The implementation probably needs special knowledge on the water regime of the area and the construction of waterways (Lal, 1995).

Conducting studies focusing on better land use planning and proper soil management by using modern technologies such as GIS is another approach to improve soil conservation. For example, Igwe (1998) and Okoth (2003) made an assessment of soil degradation and risk modeling and suggested appropriate measures for land use.

Questionnaire

The results of the questionnaires in Esa-Oke and Owode-Ede are as follows.

Personal and Socio-economic Characters of the Respondents

The majority of the farmers are male (57.5%) and between the ages of 46 and 65 years. Thirty percent of the respondents had no formal education and 35% had post-secondary education (Table 1). The remainder had started primary or secondary schools. Households are often large: 22.5% had fewer than 6 persons, 65% had between 7 and 12 persons, and 12.5% had 13–18 members. The interviews also show that the majority (50%) of the farms in the area are about 16–26 acres (2.5 acres = 1 ha), whereas 35% are smaller than 5 acres and the remainder 6–15 acres or larger than 26 acres. Farmers acquire land most often by leasing (52.5%), followed by inheritance (35%), purchase (7%), and communal use (5%). Half of the interviewed farmers are additionally occupied in trading besides farming. Only a few have vocational jobs or are in the civil service. Hence, nearly all respondents have traveled within the country (55%), within Osun State (25%), or within the local government area (LGA, 17.5%). Families are the major source for credit (97.5%) and cooperative societies that lend money are rare. Nearly all farmers of

Table 1 Personal and socio-economic characteristics of the respondents

Subject	Freq.	%	Subject	Freq.	%
Sex			*Minor occupation*		
Male	23	57.5	None	18	45.0
Female	17	42.5	Civil service	1	2.5
			Vocational jobs	3	7.5
Age (yr)			Trading	18	45.0
<25	1	2.5			
26–45	8	20.0	*Extent of traveling*		
46–65	21	52.5	Never traveled anywhere	1	2.5
>65	10	25.0	Traveled within the LGA	7	17.5
			Traveled within the state	10	25.0
Level of education			Traveled within Nigeria	22	55.0
No formal education	12	30.0			
Primary school	9	22.5	*Major source of credit*		
Secondary school	5	12.5	Personal/family	39	97.5
Post-secondary education	14	35.0	Cooperative societies	1	2.5
Household size			*Membership in agricultural association*		
<6 persons	9	22.5	No	1	2.5
7–12 persons	26	65.0	Yes	39	97.5
13–18 persons	5	12.5			
			Average annual income (US $)		
Total farm size (1 ha = 2.5 acres)			<400	24	60.0
<5 acres	14	35.0	400–800	15	37.5
6–15 acres	4	10.0	800–1200	1	2.5
16–26 acres	20	50.0			
>26 acres	2	5.0	*Cost of input (US $)*		
			<40	16	40.0
Means of acquiring land			40–120	10	25.0
Communal	2	5.0	120–200	7	17.5
Purchase	3	7.5	200–275	2	5.0
Inheritance	14	35.0	275–350	2	5.0
Lease	21	52.5	>350	3	7.5

Freq., frequency; 1 ha = 2.5 acres

the villages of Esa-Oke and Owode-Ede are members of agricultural associations to receive more knowledge on farming issues. Sixty percent of the respondents have an average annual income of less than US $400, 37.5% earn US $400–800, and only 1 farmer out of 40 earned more than US $800. Accordingly, the majority of the respondents (40%) spend less than US $40 on inputs. The more the costs of input increase, the less the farmers spend.

Practiced SCTs and Comparison by Farmers

The majority of the respondents (82.5%) knew the meaning of mulching, cover cropping, and fallowing for controlling soil erosion, while very few (10.0%) were aware of intercropping and agroforestry as on-farm SCT. Very few (7.5%) of the respondents indicated that they knew nothing about any kind of erosion control.

Mulching, cover cropping, contour tillage, and cutoff drainage were the on-farm SCTs being practiced by farmers in the study area. The respondents compared these SCTs on the basis of labor intensiveness, compatibility, cheapness, complexity, availability of the input materials, and cost effectiveness. Mulching and cover cropping are regarded as not labor intensive, highly cost effective, compatible with the existing farming system, and easy and cheap to adopt. A disadvantage of mulching might be the required amount of grass, the main material used for reducing soil erosion and increasing fertility. Farmers recognize the value of cover crops as they are sources of food for man and animal and improve soil fertility through N fixation and decomposition. But they also see cover crops as competitors for soil nutrients and providers for insects, pests, and diseases which may be transmitted to the main crops and decrease the yield. Tillage along the contour line is also accepted as a compatible methodology which is easy and cheap to adopt and to practice, due to availability of the equipment, a common hoe. Farmers also installed cutoff drainage on the fields to get rid of surplus water. This SCT is highly labor intensive and costly as hired laborers have to maintain the channels regularly. Spades, the main tool for establishing and maintaining the drainage, are often not available. Another issue that reduces the adoption of this erosion control measure is the incompatibility with the culture. Digging the ground is associated with burying, which leads gradually to the death of people in the community without a cause until the dug ground is closed. Hence, farmers that practiced cutoff drainage were compelled to discontinue the method.

Source of Information

The farmers in the study areas regarded mulching, cover cropping, contour tillage, and agroforestry as indigenous farming practices inherited from their forefathers, even though the majority of the farmers did not practice these SCTs until the RUDEP made appropriate interventions to the communities. Cutoff drainage was an erosion control innovation through RUDEP.

Adoption of SCTs

Half of the respondents (51.4%) consequently rejected all SCTs, followed by 37.8% who adopted only one and 5.4% who accepted two or three technologies. However, the majority (72.2%) of the adopters were at the first stage of adoption (first installment), while 11.1% had completed installation and another 11.1% had discontinued implementation. Only 5.6% of the adopters were at the level of beginning to maintain adoption. The majority (66.7%) of the adopters had already continued to maintain the SCT for a period of 1–5 years, only a few for more years.

The mean number of SCTs adopted by the respondents was less than one. The average number of SCTs adopted was lower than the average number of SCTs that the respondents were aware of (2.0). This indicates a low adoption of SCTs among the respondents in the study area. The scenario that old people were in the majority among the respondents might have been a contributing factor to the low level of SCT adoption. The majority of the youth in Nigeria lack interest in rural livelihoods and in taking farming as a vocation, partly because of the poor condition that agriculture is assuming at the present time. This might also have contributed negatively to the low level of adoption in the study areas. In general, there was a positive average level of perception (4.2) toward soil erosion control among the respondents in the area of study, although there was a lower perception (2.7) toward soil erosion as a major factor influencing agricultural production.

Difference Between Adopters and Non-adopters of SCTs

About 56.5% of the adopters in the experimental group were female, while the majority (70.5%) of the non-adopters were male. Very few (16.7%) of the adopters had no formal education, whereas about 40.9% of the non-adopters had no formal education. The higher level of education among the adopters might have influenced their positive disposition toward SCTs, as literates are usually more aware and experienced about the significance of new technologies to livelihood than are the illiterates.

The majority of the adopters (94.4%) and non-adopters (95.5%) participated in training on SCTs conducted by RUDEP. The courses were free of charge and the extension approach was transferring the technologies by participation. All the adopters and the majority of the non-adopters (72.7%) indicated their willingness to continue participating in Soil Conservation Technology Development Project even if they have to contribute to the costs.

Most of the respondents were full-time farmers (adopters 72.2% and non-adopters 81.8%), commercial farmers (adopters 88.9% and non-adopters 68.2%), and engaged in individual farm practice (adopters 88.9% and non-adopters 95.5%). About half of the adopters (55.6%) and the non-adopters (59.1%) were cultivating only food crops, while the remaining 44.4% (adopters) and 40.9% (non-adopters) claimed to be cultivating both food and cash crops. None of the respondents cultivated only cash crops.

Conclusions

Both the literature review and the interviews show that mulching and crop management are likely to be the most useful erosion control strategies in Nigeria. The use of residues and polyculture is known to protect the soil from the impact of rain and runoff which are especially high in the tropics. Farmers adopt innovations on improved cropping systems more easily than mechanical erosion control measures as the first are more compatible with their system, easier, and cheaper to implement. Additional benefits of mulching and enhanced crop management are increased crop production that contributes to a better food supply and higher incomes of the farmers. Reduced tillage and tillage across the slope seem to be adequate in combination with crop management, especially in areas with gentle slopes and sheet erosion.

Acknowledgments The project has been financed by the German Federal Ministry for Economic Cooperation and Development (BMZ), Berlin, Germany, and the International Institute of Tropical Agriculture (IITA), Ibadan, Nigeria. We would like to acknowledge the support of Dr Odunze, B. Ukem, Dr Salako, Prof. Igwe, Dr Kwari, and Dr Oyedele from the universities in Zaria, Abeokuta, Nsukka, Maiduguri, and Ile-Ife for finding appropriate literature. We also thank A. Ogunyemi for assisting in reviewing publications, T. Baehrs for generating the reference database, M. Ibarahim for his work as enumerator, and Dr Thyen and R. Umelo for reviewing the manuscript.

References

Armon MN (1980) Effect of tillage systems on soil and water conservation and soil physical properties. MSc thesis, Department of Soil Science, University of Nigeria, Nsukka, 158pp

Braide FG (1986) Conservation tillage as a means of reducing erosion of erodible land. National workshop on ecological disasters in Nigeria: soil erosion, Owerri, Nigeria, 8–12 Sept 1986. Federal Ministry of Science and Technology, Lagos, Nigeria, pp 189–196

Carsky RJ, Singh BB, Oyewole B (2001) Contribution of early season cowpea to late season maize in the savanna Zone of West Africa. Biol Agric Hortic 18:303–315

Chikoye D, Manyong VM, Carsky RJ, Ekeleme F, Gbehounou G, Ahanchede A (2002) Response of speargrass (*Imperata cylindrica*) to cover crops integrated with hand weeding and chemical control in maize and cassava. Crop Prot 21: 145–156

Chiroma AM, Folorunso OA, Kundiri AM (2005) Effect of tillage and stubble management on root growth and water use of millet grown on a sandy loam soil. J Arid Agric 15:83–89

Couper DC (1995) No-till farming in the humid and subhumid tropics of Africa. IITA Research Guide 3. IITA, Ibadan, Nigeria, 26pp

Diels J, Lyasse O, Sanginga N, Vanlauwe B, Aihou K, Houngnandan P, Aman S, Tossah B, Iwuafor E, Omueti J, Manyong V, Akakpo C, Falaki AM, Deckers S, Merckx R (2000) Balanced nutrient management systems for maize-based systems in the moist savanna and humid forest Zone of West-Africa (BNMS). Annual report 2000, Jan 2000–Dec 2000 (Report No. 4). K.U. Leuven, Belgium, IITA, Ibadan, Nigeria, 95pp

Ehui S, Pender J (2005) Resource degradation, low agricultural productivity, and poverty in sub-Saharan Africa: pathways out of the spiral. Agric Econ 32(1):225–242

Eswaran H, Lal R, Reich PF (2001) Land degradation: an overview. In: Bridges EM, Hannam ID, Oldeman LR, Pening de Vries FWT, Scherr SJ, Sompatpanit S (eds) Responses to land degradation. In: Proceedings of the 2nd international conference on land degradation and desertification, Khon Kaen, Thailand. Oxford Press, New Delhi, India, 173pp

Federal Republic of Nigeria (2007) Official gazette: legal notice on publication of the details of the breakdown of the national and state provisional totals 2006 census. Government Notice Nr 21. Nr. 24. Vol. 94. Lagos, Nigeria

FGN (2007) 2007 Budget. http://www.budgetoffice.gov.ng/pub/2007budget.pdf

Hulugalle NR, Lal R, Okara-Nadi OA (1985) Effect of tillage system and mulch on soil properties and growth of yam (*Dioscorea rotundata*) and cocoyam (*Xanthosoma sagittifolium*) on an Ultisol. J Root Crops 1+2:9–22

Igbokwe EM (1996) A soil and water conservation system under threat. A visit to Maku, Nigeria. In: Reij C, Scoones I, Toulmin C (eds) Sustaining the soil: indigenous soil and water conservation in Africa. Earthscan Publication Ltd, London, UK, 260pp

Igbozurike UM, Okali DUU, Salau AT (1989) Profile on Nigeria: land degradation. NEST report submitted to Commonwealth Secretariat, London. NEST, Ibadan, Nigeria, 48pp

Igwe CA (1998) Land use and soil conservation strategies for potentially high erodible soils of central-eastern Nigeria. Land Degrad Dev 10:425–434

Juo ASR, Lal R (1977) The effect of fallow and continuous cultivation on the chemical and physical properties of an Alfisol in Western Nigeria. Plant Soil 47:567–584

Kang BT (1993) Alley cropping: past achievements and future directions. Agroforest Sys 23(2–3):141–155

Kang BT, Atta-Krah AN, Reynolds L (1999) Alley farming. Centre for Agricultural and Rural Cooperation (CTA), International Institute of Tropical Agriculture (IITA), Ibadan, Nigeria, 110pp

Kirchhof G, Odunze AC (2003) Soil management practices in the northern Guinea savanna of Nigeria. Paper presented at 16th triennial conference of International Soil Tillage Research Organization (ISTRO) soil management for sustainability, Brisbane, Australia, 14–18 Jul 2003, 6pp

Kowal J, Stockinger K (1973) Usefulness of ridge cultivation in Nigeria agriculture. J Soil Water Conserv 28(3):136–137

Lal R (1986) Effects of 6 years of continuous no-till or puddling systems on soil properties and rice (*Oryza sativa*) yield of a loamy soil. Soil Tillage Res 8:181–200

Lal R (1989) Agroforestry systems and soil surface management of a tropical Alfisol. Agroforestry systems 8. I Soil moisture and crop yield 7–29. II Water runoff, soil erosion, and nutrient loss 97–11. III Changes in soil chemical properties 113–132. IV Effects on soil physical and mechanical properties 197–215. V Water infiltrability, transmissivity and soil water sorptivity 217–238

Lal R (1993) Soil erosion and conservation in West Africa. In: Pimentel D (eds) World soil erosion and conservation. Cambridge University Press, Cambridge, pp 7–28

Lal R (1995) Sustainable management of soil resources in the humid tropics. United Nations University Press, New York, NY, 593pp

Lal R (2000) Mulching effects on soil physical quality of an Alfisol in western Nigeria. Land Degrad Dev 11:383–392

Lal R, Wilson GF, Okigbo BN (1979) Changes in properties of an Alfisol produced by various crop covers. Soil Sci 127(6):377–382

Malgwi GS (1992) Setting up contour guidelines on farmers field. In: Proceedings of monthly technology review meeting (MTRM) of Borno State Agricultural Development Programme (BOSADP), 6 Jun 2000, Maiduguri, Nigeria, 12pp

Maurya P, Lal R (1980) No-tillage system for crop production on an Ultisol. In: Lal R (ed) Soil tillage and crop production: research needs and priorities. IITA, Ibadan, pp 207–219

Mbagwu JSC (1991) Mulching an ultisol in southern Nigeria: effects on physical properties and maize and cowpea yields. J Sci Food Agric 57:517–526

Morgan RPC (1995) Soil erosion and soil conservation. Longman, Essex, 198pp

Odunze AC (2002) Mulching practice in a semi-arid zone of Nigeria for soil erosion control and grain yield of maize. J Sustain Agric 20(1):31–39

Ogban PI, Ekanem TP, Etim EA (2001) Effect of mulching methods on soil properties and growth and yield of maize in south-eastern Nigeria. Trop Agric (Trinidad) 78(2):82–89

Ojeniyi SO (1986) Effect of zero-tillage and disc ploughing on soil water, soil temperature and growth and yield of maize (*Zea mays* L.). Soil Till Res 7:173–182

Okogun JA, Sanginga N, Mulongoy K (2000) Nitrogen contribution of five leguminous trees and shrubs to alley cropped maize in Ibadan, Nigeria. Agroforest Syst 50(2): 123–136

Okoth PF (2003) A hierarchical method of soil erosion assessment and spatial risk modelling. PhD thesis, Department for Geoinformatic, University of Wageningen, The Netherlands, 211pp

Olasantan FO (1999) Effect of time of mulching on soil temperature and moisture regime and emergence, growth and yield of white yam in western Nigeria. Soil Till Res 50: 215–221

Olukosi JO, Elemo KA, Kumar V, Ogungbile AO (1991) Farming systems research and the development of improved crop mixtures technologies in the Nigerian savanna. Agric Syst Afr 1(1):17–24

Orimoyegun SA (1988) Influence of forest litter and crop residue on soil erosion. In: Proceedings of national workshop on ecological disasters in Nigeria: soil erosion, Owerri, Nigeria. Federal Ministry of Science and Technology, Lagos, Nigeria, 8–12 Sept 1986, pp 380–391

Salako FK, Babalola O, Hauser S, Kang BT (1999) Soil macroaggregate stability under different fallow management systems and cropping intensities in southwestern Nigeria. Geoderma 91:103–123

Singh BB, Ajeigbe HA (2002) Improving cowpea: cereals based cropping systems in the dry savannas of West Africa. In: Proceedings of the world cowpea conference III, Ibadan, Nigeria, 4–8 Sept 2000, pp 278–286

Slaymaker T, Blench R (2002) Volume I – country overview. In: Blench RM, Slaymaker T (eds) Rethinking natural resource degradation in sub-Saharan Africa: policies to support sustainable soil fertility management, soil and water conservation among resource-poor farmers in semi-arid areas. Cyber Systems, Tamale, Ghana

Sonneveld BGJ (1997) Dominant soils of Nigeria 1:1,300,000. Map. Stichting Onderzoek Wereldvoedselvoorziening van de Vrije Universiteit (SOWVU), Amsterdam

Stamp LD (1938) Land utilization and soil erosion in Nigeria. Geogr Rev 28:32–45

Takken I, Govers G, Jetten V, Nachtergaele J, Steegen A, Poesen J (2001) Effects of tillage on runoff ad erosion patterns. Soil Till Res 61:55–60

Tarawali G, Manyong VM, Carsky RJ, Vissoh V, Osei-Bonsu P, Galiba M (1999) Adoption of improved fallows in West Africa: lessons from *Mucuna* and *Stylosanthes* case studies. Agroforest Syst 47:93–122

Tarawali SA, Singh BB, Gupta SC, Tabo R, Harris F, Nokoe S, Fernandez-Rivera S, Bationo A, Manyong VM, Makinde K, Odion EC (2000) Cowpea as a key factor for a new approach to integrated crop-livestock systems research in the dry savannas of West Africa. In: Proceedings of world cowpea conference III, Ibadan, Nigeria, 4–8 Sept 2000, pp 233–251

Tian G, Hauser S, Koutika LS, Ishida F, Chianu J (1999) On-farm study of *Pueraria* cover-crop fallow system in the derived savanna of West Africa. In: Proceedings of conference on cover crops for natural resource management in West Africa, Cotonou, Benin. IITA, Ibadan, Nigeria, 26–29 Oct 1999, pp 185–193

Van der Kruijs ACBM, Kang BT (1988) Intercropping maize and cassava with leguminous shrubs on Ultisol. ITA Resource & Crop Management Program annual report 1986, Ibadan, Nigeria. IITA, Ibadan, Nigeria, 105pp

Vanlauwe B, Aihou K, Aman S, Tossah BK, Diels J, Sanginga N, Merckx R (2001) Leaf quality of selected hedgerow species at two canopy ages in the derived savanna zone of West Africa. Agroforest Syst 53(1): 21–30

Effect of Farmer Resource Endowment and Management Strategies on Spatial Variability of Soil Fertility in Contrasting Agro-ecological Zones in Zimbabwe

E.N. Masvaya, J. Nyamangara, R.W. Nyawasha, S. Zingore, R.J. Delve, and K.E. Giller

Abstract Variability of soil fertility within and across farms poses a major challenge for increasing crop productivity in smallholder systems of sub-Saharan Africa (SSA). This study assessed the effect of farmers' resource endowment and nutrient management strategies on variability in soil fertility and plant nutrient uptake between different fields in Gokwe South (average rainfall ~650 mm yr^{-1}; 16.3 persons km^{-2}) and Murewa (average rainfall ~850 mm yr^{-1}; 44.1 persons km^{-2}) districts, Zimbabwe. In Murewa, resource-endowed farmers applied manure (>3.5 t ha^{-1} yr^{-1}) on fields closest to their homesteads (homefields) and none to fields further away (outfields). In Gokwe the manure was not targeted to any particular field, and farmers quickly abandoned outfields and opened up new fields further away from the homestead once fertility had declined, but homefields were continually cultivated. Soil-available phosphorus (P) was more concentrated in homefields (8–13 mg kg^{-1}) of resource-endowed farmers than on outfields and all fields of poor resource farms (2–6 mg kg^{-1}) in Murewa. Soil fertility decreased with increasing distance from the homestead in Murewa, while the reverse trend occurred in Gokwe South, indicating the impact of different soil fertility management strategies on spatial soil fertility gradients. In both districts, maize nutrient uptake showed deficiency in nitrogen (N) and P, implying that these were the most limiting nutrients. It was concluded that besides farmers' access to resources, the direction of soil fertility gradients also depends on agro-ecological conditions, which influence resource management strategies.

Keywords Agro-ecology · Nutrient management strategies · Plant nutrient uptake · Resource endowment · Soil fertility variability

Introduction

Soil fertility status and nutrient management strategies are key factors that determine agricultural productivity with implications on food security and livelihoods of rural households (Mapfumo and Giller, 2001). Soil fertility variability on farms may be inherent resulting from differences in parent material and catenal position (Deckers, 2002). Farmers deliberately manage their fields differently, depending on inherent production potentials and other socio-economic factors such as availability of nutrient resources, household priorities and production strategies, farm size and labour resulting in the development of soil fertility gradients. As a result of the many possible combinations of the biophysical and socio-economic factors, the magnitude of the soil fertility gradients will also vary from farm to farm as well as across agro-ecological zones. Farmers typically apply most nutrient resources to fields closest to homesteads and manage these fields better, and this has led to the establishment of gradients of decreasing soil fertility from the homestead in some cropping systems (Vanlauwe and Sanginga, 2004; Tittonell et al., 2005; Giller et al., 2006). However, gradients of increasing soil fertility from homefields to outfields

J. Nyamangara (✉)
Department of Soil Science and Agricultural Engineering, University of Zimbabwe, Mount Pleasant, Harare, Zimbabwe; Chitedze Research Station,Tropical Soil Biology and Fertility Institute of the International Centre for Tropical Agriculture (TSBF-CIAT) Lilongwe, Malawi
e-mail: jnyamangara@yahoo.co.uk; j.nyamangara@cgiar.org

A. Bationo et al. (eds.), *Innovations as Key to the Green Revolution in Africa*,
DOI 10.1007/978-90-481-2543-2_124, © Springer Science+Business Media B.V. 2011

have also been reported in the central highlands of Ethiopia (Haileslassie et al., 2007).

Soil fertility variability between farms on similar soil types is mainly driven by differing access to nutrient resources between farmers of different resource endowment, e.g. through purchase and use of large amounts of fertilizers. Wealthier farmers also own more cattle and are able to import significant quantities of nutrients to their farms during the cropping season from grazing on open-access lands and during the dry season from grazing of crop residues on their fields and other farmers' fields (Swift et al., 1989). Therefore, nutrients accumulate on wealthier farms, often at the expense of the poorer farms (Zingore et al., 2007a).

Spatial variability in soil fertility associated with differential nutrient resource management at farm scale has largely been ignored when designing technological interventions in smallholder farming systems. For example, fertilizer recommendations used in Zimbabwe are blanket in nature and target application of major nutrients (N, P, K, S, Ca and Mg) and only differentiate agro-ecological regions (Nyamangara et al., 2000). In Malawi, the use of blanket fertilizer recommendations only based on N and P resulted in country-wide S deficiency and regional K, Zn and B deficiency (Wendt et al., 1994). Therefore fertilizer recommendations which do not take into account spatial variability in soil fertility, which also affects nutrient use efficiency, and farmer resource endowment will fail to allocate scarce fertilizer resources efficiently.

Although the occurrence of soil fertility gradients has been documented, this has been mainly in sub-humid conditions where production is relatively intensive and arable land for expansion limited. The aim of this study is to compare the effects of resource endowment and soil fertility management strategies by farmers on soil fertility variability and plant nutrient uptake in two smallholder areas, one located in sub-humid and the other in semi-arid agro-ecological conditions. It was hypothesised that gradients of decreasing soil fertility from the homestead occur irrespective of agro-ecology and farmer resource endowment.

Materials and Methods

Description of Study Sites

The study was conducted in two smallholder farming areas in Gokwe South (Njelele I and II and Nemangwe II and III wards) and Murewa (Cheuka ward) districts of Zimbabwe. Gokwe South district (18 and 19° latitude; 28 and 29° longitude) is located in agro-ecological region III and IV where rainfall is erratic and unreliable, while Murewa district (17 and 18° latitude; 31 and 32° longitude) is located in agro-ecological region II and rainfall is relatively well distributed and more reliable (Table 1).

Farmers in both study areas practise mixed crop–livestock farming with livestock providing draft power and manure for soil fertility improvement, while crop residues provide supplementary feed for livestock in winter when grazing is scarce and of poor quality. Grazing in both study areas is communal in open-access areas during the rainy season. Fields are individually owned and managed but are communally grazed in winter.

In Gokwe South, soils are mainly sands (Luvic Arenosols – FAO) of Kalahari origin, while in Murewa,

Table 1 Site characteristics of the study areas in Murewa and Gokwe South districts

		Gokwe South district (Njelele I and II and Nemangwe II and III wards	
	Murewa district (Cheuka ward)	Njelele I and II	Nemangwe II and III
Households sampled	23	19	15
Climate	Sub-humid tropical	Semi-arid tropical	Semi-arid tropical
Agro-ecological natural region	II	III	IV
Rainfall (unimodal) (mm annum^{-1})	750–1000	650–800	450–650
Soil type	Granitic sands	Kalahari sands	Kalahari sands
Population density[a,b]	44.1	16.3	16.3
Food security crop	Maize	Maize/sorghum	Sorghum/pearl millet
Cash crop	Maize	Maize/sweet potato	Cotton

[a]Statoids (2005)
[b]ZIMVAC (2005)

sandy soils (Haplic Lixisols – FAO) of granitic origin predominate and smaller areas (<1%) are covered by more fertile red clay soils (Chromic Luvisols – FAO) derived from dolerite intrusions (Nyamapfene, 1991; Zingore et al., 2007a). The sandy soils from both districts are inherently infertile, characterised by low soil organic carbon (SOC) and deficiencies of N, P and sulphur (S) (Grant, 1981; Ahmed et al., 1997). In advanced cases of nutrient mining and high rainfall such as in Murewa, nutrient imbalances and extreme acidity have been reported (Nyamangara et al., 2000).

Effect of Farmer Resource Endowment and Management on Soil Fertility and Plant Nutrient Uptake

Twenty-three and 34 farmers were selected in Murewa and Gokwe South districts to represent resource-constrained, intermediate and resource-endowed households according to researcher-developed criteria presented in Table 2 (Mtambanengwe and Mapfumo, 2005; Zingore et al., 2007a). The farmers in the different wealth categories were asked to indicate their most productive and least productive maize fields. The distance of each field from the homestead and cattle pen was measured and the field nearest to the homestead was designated 'homefield' and the one furthest 'outfield'. Cattle pens were located close to homesteads and therefore homefields were also nearest to the cattle pens. A structured questionnaire was used to collect information on soil fertility management practices employed by the selected farmers, including use of mineral fertilizer and the main crops grown.

At silking stage (~10–12 weeks after emergence), soil and maize cobleaf samples were taken to assess soil fertility status and plant nutrient uptake. Soil samples were air-dried, passed through a 2-mm sieve and analysed for clay content, pH (0.01 M $CaCl_2$), exchangeable bases, cation exchange capacity (CEC), SOC, total N and P, and available P (Anderson and Ingram, 1993; Okalebo et al., 2002). Cobleaf samples were oven dried (65°C), ground to pass through a 2-mm sieve, and analysed for total N, P, Ca, Mg, K, Fe, Mn, Zn and Cu (Okalebo et al., 2002). Plant nutrient contents were assessed using standard methods (Mengel and Kirkby, 2001).

Statistical Analysis

Questionnaires were analysed and socio-economic factors affecting nutrient management determined using counts and frequencies to separate farmers into different social groups using the Statistical Package

Table 2 Descriptive criteria for classification of farmers in Gokwe South ($N = 34$) and Murewa ($N = 23$) smallholder farming areas into different wealth classes generated from previous studies[a]

Farmer category	Description
Resource endowed	– ≥ 10 cattle and use own cattle for draught power – Field size usually > 3 ha
(Murewa, $n = 6$; Gokwe, $n = 6$)	– Own important farming equipment: scotch cart, plough, cultivator, harrow and wheelbarrow and all small implements – Housing type is usually brick under asbestos/iron roofing – Afford to hire labour
Intermediate	– Own between 4 and 10 cattle – Field sizes 1–3 ha
(Murewa, $n = 9$; Gokwe, $n = 12$)	– Own a plough and an ox-drawn cart – Housing type usually brick under iron roofing – Rarely afford to hire labour
Resource constrained	– Own 0–3 cattle – Fields < 1 ha
(Murewa, $n = 8$; Gokwe, $n = 12$)	– Own small implements such as hoes – Brick under thatch – Cannot afford to hire labour

[a]Mtambanengwe and Mapfumo (2005) and Zingore et al. (2007a)

for the Social Sciences (SPSS). Analysis of variance (ANOVA) tables were generated for soil and plant nutrient status across resources groups and between field types using GENSTAT 7.1 to test for significant differences at $P < 0.05$.

Results

Soil Fertility Resources and Management in the Field

The average cattle ownership in both study sites was low (5.4 and 5.6 cattle per household in Murewa and Gokwe, respectively) but was higher than the national average (2.8 cattle per household) for smallholder areas (Gambiza and Nyama, 2000). There was a significant ($p < 0.05$) relationship between quantity of manure applied and resource endowment, with wealthy farmers applying more manure (3.5–9 t ha^{-1}) to their fields in Murewa, compared with the intermediate (up to 1.5 t ha^{-1}) and resource-constrained (<1 t ha^{-1}) farmers. Resource-constrained farmers in Murewa applied significantly less NPKS fertilizer (<100 kg ha^{-1}) compared with resource-endowed (150–200 kg ha^{-1}) and intermediate (up to 175 kg ha^{-1}) farmers. Other nutrient resources used, but on a limited scale, were compost, leaf litter and anthill soil; these were targeted to homefields in Murewa, whilst in Gokwe South, very few farmers (6%) used them.

Farmers in Murewa owned small farms (1–3 ha) and continuously cultivated them, while farmers in Gokwe South owned larger farms (5–10 ha) and frequently fallowed them. Fields in Murewa had been under cultivation for longer periods (~30 years) compared with Gokwe (~15 years). Outfields were generally larger than homefields and the latter constituted about 17 and 32% of the farm area in Murewa and Gokwe South respectively (Table 3). There was a significant ($p < 0.05$) difference across resource endowment groups in the ability to hire labour with resource-endowed farmers being able to hire more casual labour compared with their less resource-endowed counterparts.

Soil Fertility Status

Soil pH was higher in homefields compared with outfields in Murewa but the difference was significant only for resource-endowed farmers where soil pH was extremely acidic in outfields (Table 4). Cation exchange capacity and all exchangeable bases were also higher in homefields compared with outfields and were largely similar for resource-endowed and intermediate farmers and much lower for the resource-constrained farmers. In Gokwe South, soil pH and exchangeable bases, except Mg, showed no specific trend. CEC decreased with decease in resource endowment in both homefields and outfields, and the trend was similar for Mg in homefields (Table 4).

There were differences in total soil N and SOC across resource endowment classes and field types in each wealth category but the differences were not

Table 3 Characteristics of homefields and outfields in Murewa ($n = 22$ farms) and Gokwe South ($n = 34$ farms) districts

	Field type			
	Murewa ($n = 23$)		Gokwe ($n = 34$)	
	Homefield	Outfield	Homefield	Outfield
Average field size (ha)	0.54 ± 0.061[a]	0.63 ± 0.106	1.13 ± 0.160	1.49 ± 0.244
Distance from homestead (m)	29 ± 12.7	159 ± 36.4	51 ± 14.3	763 ± 132.0
Range (m)	5–200	50–700	3–200	30–2500
Distance from cattle pen (m)	28 ± 12.8	134 ± 38.4	96.1 ± 16.5	431 ± 101.8
Period under cultivation (years)	30 ± 2.7	30 ± 2.7	16 ± 1.9	15 ± 1.9
Main crops grown (frequency %)				
Maize	39.3	46.4	26.5	38.3
Cotton	0	0	35.3	32.4
Groundnuts	31.4	21.4	17.6	8.9
Sorghum	0	0	5.9	2.9
Others	29.3	32.2	14.8	17.8

[a]Values given represent the standard error of the mean

Table 4 Selected soil properties across resource groups on sands of Murewa and Gokwe South districts

Resource group	Study site	% Clay		pH (0.01 $CaCl_2$)		CEC ($cmol_c\ kg^{-1}$)		Exch. Ca ($cmol_c\ kg^{-1}$)		Exch. Mg ($cmol_c\ kg^{-1}$)		Exch. Na ($cmol_c\ kg^{-1}$)		Exch. K ($cmol_c\ kg^{-1}$)	
		HF	OF	HF	OF	HF	OF	HF	OF	HF	OF	HF	OF	HF	OF
1	Murewa	3.5	3.3	6.1	4.4	6.83	5.20	3.10	1.70	0.50	0.30	0.02	0.00	4.00	2.90
2	Murewa	4.7	3.1	5.4	5.1	6.83	4.43	5.53	3.15	1.47	1.06	0.14	0.09	1.98	1.06
3	Murewa	3.0	3.3	5.2	4.8	4.17	2.95	2.95	1.97	0.65	0.17	0.08	0.07	1.10	0.65
SED		*0.63*		*0.30*		*0.95*		*0.98*		*0.31*		*0.08*		*0.98*	
1	Gokwe	4.3	4.0	5.7	5.7	15.2	14	16.75	4.2	5.75	1.10	0.02	0.08	1.67	0.51
2	Gokwe	3.3	4.0	5.3	5.0	6.4	10.9	2.70	10.7	1.95	1.85	0.11	0.12	0.22	0.51
3	Gokwe	5	4.0	4.8	5.8	3.4	4.8	4.00	1.95	0.40	0.45	0.05	0.09	0.51	0.56
SED		*0.86*		*0.22*		*2.601*		*3.05*		*1.04*		*0.02*		*0.27*	

significant (Fig. 1a, b). However, total SOC and N were higher in homefields compared with outfields in Murewa, and the opposite trend was observed in Gokwe South. Available P was particularly responsive to management and decreased sharply from >20 mg kg^{-1} in the resource-endowed group to less than 5 mg kg^{-1} in resource-constrained group in Murewa (Fig. 1c). In Gokwe South, available P was significantly higher in fields of the resource-endowed farmers than those of the intermediate and resource-constrained farmers.

Plant Nutrient Uptake

Although plant nutrient status varied across wealth classes in both study sites, differences were significant only for N and P in Murewa (Fig. 2), where cobleaf N and P concentration increased with decreasing resource endowment. However, in Gokwe, the reverse was observed (Fig. 2). The concentration of cobleaf N in resource-constrained farmers' fields in Murewa and resource-endowed farmers' fields in Gokwe South was low (2.0–2.5%), while the rest were in the deficient (<2.0%) category according to the criteria by Mengel and Kirkby (2001). Phosphorus availability was generally higher in Murewa, where cobleaf P concentration was adequate (0.20–0.50%) in the resource-endowed and resource-constrained groups (Fig. 2). In Gokwe South, cobleaf P concentration indicated P deficiency (<0.10%) in the resource-endowed group and low P (0.10–0.20%) in the intermediate and resource-constrained groups (Fig. 2). Maize cobleaf K, Mg, Ca and micronutrients (Fe, Zn, Cu and Mn) were in the adequate ranges (K 1.5–3.0%, Mg 0.2–1.0%, Ca 0.4–1.0%, Fe 1–400 mg kg^{-1}, Zn 20–100 mg kg^{-1}, Cu >5 mg kg^{-1} and Mn 20–200 mg kg^{-1}) at this growth stage in both study areas and generally tended to decrease with decrease in resource endowment (Table 5).

Discussion

Cattle ownership is a form of wealth in smallholder farming systems that can be converted into cash when the need arises (Rufino et al., 2006) justifying the use of cattle ownership to differentiate farmers. Livestock ownership explained the use of higher application rates of manure in farms of resource-endowed farmers in Murewa compared to resource-constrained farmers. However, the amount of manure applied to the preferred fields had declined compared with the 1980s where rates up to 80 t ha^{-1} were applied, mainly due to persistent droughts since the early 1990s that have reduced cattle ownership levels per household (Mugwira and Murwira, 1997). The majority of the resource-constrained farmers co-owned few cattle, e.g. with their sons, and the manure produced had to be spread in all fields, hence the low application rates. Farmers in Gokwe applied manure to their fields once every 2–3 years, similar to findings of Ahmed et al. (1997), who reported that smallholder farmers in semi-arid areas of Zimbabwe applied manure once every 3–5 years to their maize crop. Farmers cited the manure scarcity due to low cattle ownership (approx.

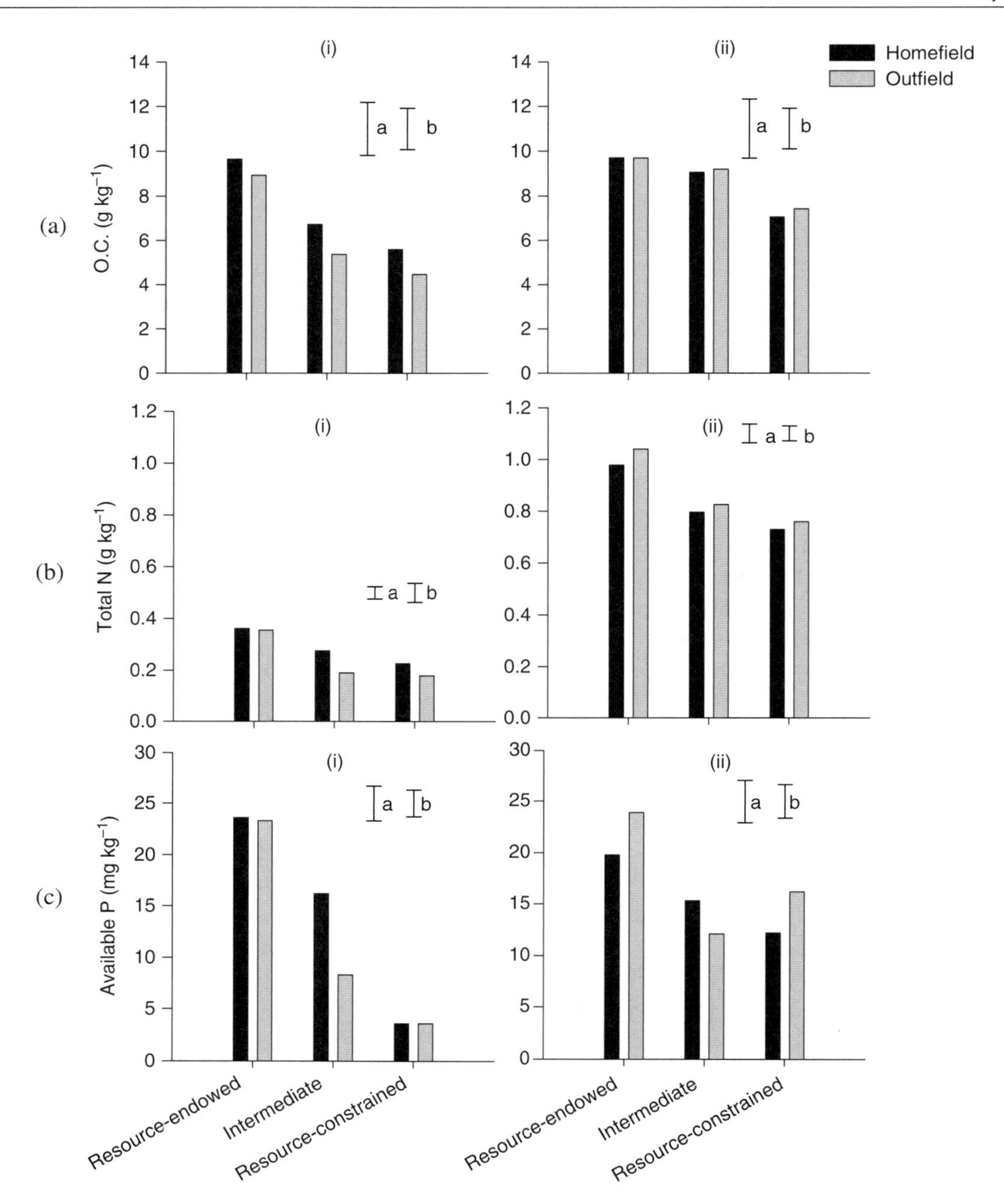

Fig. 1 Differences in (a) soil organic carbon, (b) total N and (c) soil-available P with resource endowment and field type in (**i**) Murewa and (**ii**) Gokwe districts. Bars represent SEDs for factors (a) resource endowment group and (b) field type

six cattle per household in Gokwe) as the main reason for rotating manure application in their field.

The observed decrease in SOC, total N and available soil P in soils belonging to poorer farmers in sub-humid conditions has been reported elsewhere (Mtambanengwe and Mapfumo, 2005; Zingore et al., 2007a) and attributed to differences in the nutrient resources available to the different classes of farmers. As the poorer farmers add little or no fertility amendments to their soils, fertility is likely to decline very rapidly within a few years of continuous cultivation on granitic and Kalahari sands (Zingore et al.,

Table 5 Maize cobleaf nutrient contents across farms of different resource endowments in Murewa and Gokwe South districts

		%K		%Ca		%Mg		Zn (mg kg^{-1})		Mn (mg kg^{-1})		Fe (mg kg^{-1})		Cu (mg kg^{-1})	
Resource group	Study site	HF	OF	HF	OF	HF	OF	HF	OF	HF	OF	HF	OF	HF	OF
1	Murewa	1.37	1.31	0.54	0.43	0.13	0.11	21	17	62	49	93	79	7	8
2	Murewa	1.21	1.08	0.55	0.45	0.12	0.12	15	13	69	68	80	79	7	6
3	Murewa	1.41	1.30	0.53	0.43	0.15	0.13	15	14	34	33	125	90	7	7
SED		*0.09*		*0.04*		*0.02*		*1.50*		*12.0*		*17.0*		*0.61*	
1	Gokwe	2.35	2.36	0.37	0.31	0.48	0.50	54	57	103	93	390	373	6	7
2	Gokwe	2.31	2.16	0.39	0.38	0.52	0.49	60	54	84	69	422	437	7	6
3	Gokwe	2.11	1.99	0.40	0.39	0.51	0.47	55	51	91	82	399	397	6	5
SED		*0.12*		*0.05*		*0.04*		*11.1*		*16.1*		*59.2*		*1.17*	

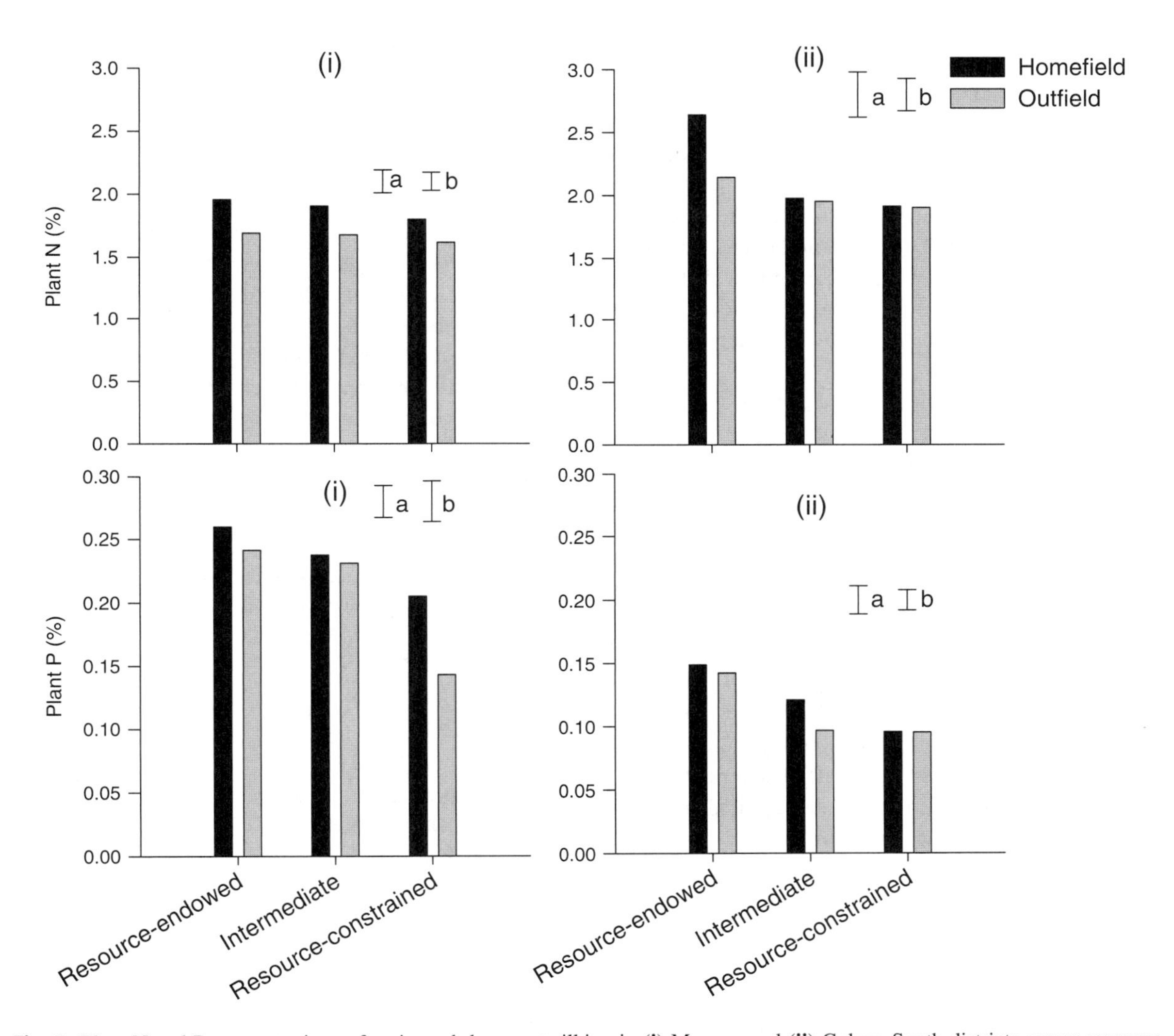

Fig. 2 Plant N and P concentrations of maize cob leaves at silking in (**i**) Murewa and (**ii**) Gokwe South districts across resource endowment groups. Bars show SEDs for differences in (a) resource endowment and (b) field type

2005). Resource-endowed farmers often have access to livestock manure and resources to purchase mineral fertilizer.

The higher soil fertility status in homefields in Murewa and outfields in Gokwe South implied that the farmers in the contrasting agro-ecological zones used different nutrient management strategies. In Murewa, where land holdings are small and land for expansion unavailable, farmers concentrated their nutrient resources in relatively smaller homefields, a practice that has also been reported elsewhere (Tittonell et al., 2005; Mtambanengwe and Mapfumo, 2006; Vanlauwe et al., 2006; Zingore et al., 2007b). In ward 5 of Shurugwi smallholder area, located in a semi-arid region but with a much higher population density than Gokwe, a study showed that farmers allocated organic fertilizer (animal manure, compost and leaf litter) to homefields, while farmers allocated mineral fertilizer (NPKS) to outfields, and in all cases, maize was grown (Nyagumbo et al., 2007). Since homefields are much smaller compared with outfields, it implies that significant increases in yield output, and hence food security, can only be realised if crop productivity in the outfields is also increased.

In Gokwe, where population density is relatively low (16.3 persons per km^2), land for expansion is available (miombo forest) and farmers quickly open up new fields further away from the homestead once fertility has declined (Mapedza et al., 2003). Soil fertility decline in the Kalahari sands is rapid due to poor physical protection of organic matter. Consequently, soil fertility in Gokwe was higher in the relatively younger or newly opened up outfields compared with the older homefields which are continually cultivated (e.g. N and P; Fig 1a–c). Table 3 shows that farmers did not target particular fields for cotton production, and therefore both homefields and outfields benefited from the relatively higher fertilizer rates applied to cotton compared to other crops. A similar trend of increasing soil fertility from homefields to outfields has also been reported in the East African highlands in Ethiopia (Haileslassie et al., 2007). Therefore, agro-ecology and farming system are important, in addition to farmer resource endowment and socio-economic conditions, in order to understand soil fertility spatial variability in the smallholder areas of sub-Saharan Africa.

Plant N and P concentrations were critical limits (Mengel and Kirby, 2001) at silking, implying that nutrient additions even by resource-endowed farmers were also low. Nitrogen is still the most important nutrient limiting crop production in Africa (Sanchez et al., 1997). Plant P was particularly low in Gokwe ($<0.15\%$), implying a higher potential response if the nutrient was applied.

Conclusions

Understanding variability in soil fertility in smallholder farming systems is required for improved targeting of soil fertility management interventions. Access to resources, farmers' management strategies and agro-ecology are important determinants of variability in soil fertility within and across farms. In intensive cropping systems in Murewa, fields closest to homesteads were more fertile than fields further away, following gradients of intensity of nutrient resource use. However, in extensive cultivation systems under semi-arid conditions, fields closets to homesteads were less fertile than fields further away, as fields further from homesteads are frequently fallowed, allowing them to maintain higher soil fertility status than fields closest to homesteads.

Acknowledgements AFRICARE-Zimbabwe and the Regional Universities Forum (RUFORUM) provided funding for this work. We are grateful to the farmers in Gokwe South and Murewa districts for their cooperation, the Department of Agricultural Technical and Extension (AGRITEX) in Gokwe South district for facilitating this study, and Francis Dzvene for field assistance in Murewa district.

References

Ahmed MM, Rohrbach DD, Gono LT, Mazhangara EP, Mugwira L, Masendeke DD, Alibaba S (1997) Soil fertility management in communal areas of Zimbabwe: current practices, constraints and opportunities for change: results of a diagnostic survey. Southern and eastern Africa region working paper no. 6, ICRISAT Southern and Eastern Africa Region

Anderson JM, Ingram JSI (eds) (1993) Tropical soil biology and fertility: a handbook of methods, 2nd edn. CAB International, Wallingford

Deckers J (2002) A systems approach to target balanced nutrient management in soilscapes of sub-Saharan Africa. In: Vanlauwe B, Diels J, Sanginga N, Merckx R (eds) Integrated plant nutrient management in sub-Saharan Africa. CAB International, Wallingford, CT, pp 47–61

Gambiza J, Nyama C (2000) Country pasture/forage resource profile, Zimbabwe. http://www.fao.org/ag/agp/agpc/doc/counprof/zimbabwe/zimbab.htm

Giller KE, Rowe EC, de Ridder N, van Keulen H (2006) Resource use dynamics and interactions in the tropics: Scaling up in space and time. Agric Syst 88:8–27

Grant PM (1981) The fertilization of sandy soils in peasant agriculture. Zim Agric J 78:169–175

Haileslassie A, Preiss JA, Veldkamp E, Lesschen JP (2007) Nutrient flows and balances at the field and farm scale: exploring effects of land-use strategies and access to resources. Agric Syst 94:459–470

Mapedza E, Wright J, Fawcett R (2003) An investigation of land cover change in Mafungabusi forest, Zimbabwe, using GIS and participatory mapping. Appl Geophs 23:1–21

Mapfumo P, Giller KE (2001) Soil fertility management strategies and practices by smallholder farmers in semi-arid areas of Zimbabwe. ICRISAT, FAO, Patancheru

Mengel K, Kirkby EA (2001) Principles of plant nutrition, 5th edn. Kluwer, The Netherlands

Mtambanengwe F, Mapfumo P (2005) Organic matter management as an underlying cause for soil fertility gradients on smallholder farms in Zimbabwe. Nutr Cycling Agroecosyst 73:227–243

Mtambanengwe F, Mapfumo P (2006) Effects of organic resource quality on soil profile N dynamics and maize yields on sandy soils in Zimbabwe. Plant Soil 281:173–191

Mugwira LM, Murwira HK (1997) Use of cattle manure to improve soil fertility in Zimbabwe: past, current research and future needs. Network research results – working paper no. 2, DR and SS, CSRI, Harare, Zimbabwe

Nyagumbo I, Nyamangara J, Rurinda J (2007) Scaling out integrated soil nutrient and water management technologies through farmer participatory research: experiences from semi-arid central Zimbabwe. Paper presented at the TSBF-AfNET symposium, Arusha, Tanzania, 17–21 Sept 2007

Nyamangara J, Mugwira LM, Mpofu SE (2000) Soil fertility status in communal areas of Zimbabwe in relation to sustainable crop production. J Sust Agric 16:15–29

Nyamapfene K (1991) Soil of Zimbabwe. Nehanda, Harare

Okalebo JR, Gathua KW, Woomer PL (2002) Laboratory methods of soil and plant analysis: a working manual, 2nd edn. TSBF-CIAT and SACRED Africa, Nairobi

Rufino M, Rowe EC, Delve RJ, Giller KE (2006) Nitrogen cycling efficiencies through resource-poor African crop–livestock systems. Agric Ecosyst Environ 112:261–282

Sanchez PA, Shepherd KD, Soule MJ, Place FM, Buresh RJ, Izac AMN (eds) (1997) Soil fertility replenishment in Africa: an investment in natural resource capital. Replenishing soil fertility in Africa. Soil Science Society of America Special Publication No. 51, Madison, WI, pp 1–46

Statoids (2005) Districts of Zimbabwe. Gwillim Law. http://www.statoids.com/yzw.html

Swift MJ, Frost PGH, Campbell BM, Hatton JC, Wilson KB (1989) Nitrogen cycling in farming systems derived from savannah. In: Clarholm M, Bergstrom L (eds) Ecology of arable land. Dev Plant Soil Sci 39:63–67

Tittonell P, Vanlauwe B, Leffelaar PA, Rowe EC, Giller KE (2005) Exploring diversity in soil fertility management of smallholder farms in western Kenya I. Heterogeneity at region and farm scale. Agric Ecosyst Environ 110: 149–165

Vanlauwe B, Sanginga N (2004) The multiple roles of organic resources in implementing integrated soil fertility management. In: Delve RJ, Probert ME (eds) Modelling nutrient management in tropical cropping systems. Australian Centre for International Agricultural Research (ACIAR) proceedings no. 11, pp 12–24, Pirion Printing: Canberra, Australia. http://aciar.gov.au/files/node/542/pr114.pdf (Accessed 21 April 2011)

Vanlauwe B, Tittonell P, Mukalama J (2006) Within-farm soil fertility gradients affect response of maize to fertiliser application in western Kenya. Nutr Cycling Agroecosyst 76: 171–182

Wendt JW, Jones RB, Itimu OA (1994) An integrated approach to soil fertility improvement in Malawi, including agroforestry. In: Craswell ET, Simpson J (eds) Proceedings ACIAR/SACCAR workshop on soil fertility and climatic constraints in dryland agriculture, Harare, 1993. ACIAR, Canberra, pp 74–79

ZIMVAC (2005) Zimbabwe rural food security and vulnerability assessments – June 2005. www.zimrelief.info/files/attachments/doclib/0003.pdf

Zingore S, Manyame C, Nyamugafata P, Giller KE (2005) Long-term changes in organic matter of woodland soils cleared for arable cropping in Zimbabwe. Eur J Soil Sci 56: 727–736

Zingore S, Murwira HK, Delve RJ, Giller KE (2007a) Influence of nutrient management strategies on variability of soil fertility, crop yields and nutrient balances on smallholder farms in Zimbabwe. Agric Ecosyst Environ 119: 112–126

Zingore S, Murwira HK, Delve RJ, Giller KE (2007b) Soil type, historical management and current resource allocation: three dimensions regulating variability of maize yields and nutrient use efficiencies on African smallholder farms. Field Crop Res 101:296–305

Empowering Farmers in Monitoring and Evaluation for Improved Livelihood: Case Study of Soil and Water Management in Central Kenya

F.M. Matiri and F.M. Kihanda

Abstract There has been a paradigm shift in terms of research planning, implementation, monitoring and evaluation, and upscaling in agricultural research and development. The previous procedure of little involvement of wider key stakeholder participation in this process ended up entrenching inherent biases that also led to entrenching rural poverty. Therefore, there has been a need to invert the paradigm from top-down to bottom-up approaches to enhance farmers' participation so that they become more proactive in agricultural research and development process. Also, the indicators that are used at the grassroots rarely incorporate interests and/or knowledge of the communities. Therefore, this study was designed in Central Kenya to understand the farmers' perception of monitoring and evaluation and build their capacities in development of monitoring and evaluation indicators that would empower them to assess success or failure of projects that they jointly implement with other stakeholders. This process was subsequently meant to influence research through feedback mechanisms and increase adoption and development of appropriate technologies that would create impact at the grassroots level for improved livelihood and poverty reduction. A farmers' group involved in soil and water management was selected as a case study. The results indicate that wider stakeholder participation empowers them to create relevant feedbacks that inform research to develop appropriate technologies and accelerate impacts to improve livelihoods. The results also show that farmers have knowledge and capacity to develop frameworks for monitoring and evaluating projects, making them more relevant to the communities' reality.

Keywords Empowerment · Evaluation · Monitoring · Participation and poverty

Introduction

There has been a strong recognition of a need for alternative approaches to conventional agricultural research processes. This is because of low levels of adoption of new technologies developed without farmers' participation, hence failing to address the issues of rural poverty appropriately due to inherent biases that tend to obscure the extent of rural poverty and deprivation (Pretty et al., 1995). On the basis of this there has been a dramatic shift from top-down to bottom-up approaches, from centralised and standardised to local diversity and from blueprint to learning processes.

Different experts have defined participatory monitoring and evaluation (PM&E) differently. Participatory monitoring and evaluation is described by Guijt (1998) as *assessment of change through processes that involve many people or groups, each of whom is affecting or affected by the impacts being assessed.*

The soil and water management (SWM) programme was among the first ones in KARI to adopt participatory methodologies in designing and implementing its research activities. Hence, it was selected in this case study. In KARI, there was limited information on farmers'/wider stakeholders' participation in development of PM&E indicators. There was also limited evidence

F.M. Matiri (✉)
Kenya Agricultural Research Institute (KARI), Embu, Kenya
e-mail: francis_matiri@yahoo.com

A. Bationo et al. (eds.), *Innovations as Key to the Green Revolution in Africa*,
DOI 10.1007/978-90-481-2543-2_125,

on whether there are changes in monitoring and evaluation with the ongoing shift towards participatory research. In addition, there is evidence in stakeholders' participation in project planning but not in the monitoring and evaluation stages. Most reports indicate that they are more involved in planning and implementation but less in monitoring and evaluation and how they are involved in these stages is not clear (Matiri, 1999). Also, evaluation has been the domain of consultants using indicators that are developed without main stakeholders' participation. Hence, the process has always been for passing "subjective judgement" about a project's success or failure but not an empowering and social learning process. Therefore, the setting of monitoring and evaluation (M&E) criteria, deciding on indicators and who should participate in M&E has been the domain of scientists and consultants. Thus, though the process is proclaimed to be "PM&E", it may be operating like the conventional M&E in many ways (Guijt, 1998). Hence, a strategy/framework that gives appropriate feedback to all the stakeholders in order to improve the research process may be lacking. This therefore creates a problem of measuring "what it is against what is expected". In addition, leaving out key actors in development of monitoring criteria (indicators) may make the results irrelevant in terms of assessing the performance of participatory research. Therefore, there is need to build the main stakeholders' capacity to participate in indicators' development and entire M&E process. This would consequently assist in developing a participatory monitoring and evaluation strategy that takes the diverse perspectives of the key stakeholders. It would also inculcate PM&E as a learning process and not as a tool for assessing outputs/outcomes against a set of predetermined indicators.

Objectives

The objectives of the study included the following:

1. To facilitate main stakeholders in identifying the main M&E variables at the grassroots
2. To build main stakeholders' capacity to develop more appropriate and relevant M&E indicators for SWM
3. To provide guidelines that can be used to develop SWM PM&E framework
4. To contribute to an efficient feedback mechanism for M&E to enhance technologies' adoption, upscaling and improve communities' livelihoods
5. To contribute to developing negotiated participatory monitoring and evaluation guidelines for programmes using experiences in SWM (action research)

Methodology

The study was carried out in Meru South, Kirege location, with a farmers' group. The group was called for the meeting through the extension staff, the objective of the study explained and meeting for developing M&E indicators set. It was also agreed that key local leaders need to be invited whether they are group members or not because they are important opinion shapers in the community and are key in upscaling of technologies and information.

The first and second meetings were attended by 17 members (9 men and 8 women) and 31 farmers (13 men and 18 women). The second meeting was to document the implementation of M&E process as well as participatory development of M&E indicators. The first meeting was held during the short rains (October 2003), while the second meeting for development of M&E indicators was held in the long rains (April 2004).

Through participatory facilitation and dialoguing, the team jointly came up with statement/s that assisted the group to discern the implementation activities in SWM that would assist in eliciting M&E issues. The question that was agreed upon was phrased as detailed later and presented to both women and men.

Results

The group identified the key activities/steps that need to be undertaken and was presented as in Table 1.

From the presentations, it appeared that men came up with a more detailed catalogue of activities that are undertaken in the implementation process of SWM project. The group attributed this to men having more interest on issues of soil conservation than women. Despite the men's presentation being more detailed or richer, differences in identification of implementation

Table 1 Activities undertaken in SWM grassroot implementation process at Kirege, Meru South

Men's group	Women's group
Problem identification and ranking	Identification of ways to improve soils
Options identification and ranking	Observe for change in soil erosion in different fields/plots over time
Selection of SWC interventions for evaluation	Repair soil erosion control structures when necessary
Selection of soil fertility improvement interventions	Check time taken for MPTs to take up
Selection of the plots for evaluations	Check on growth behaviour of MPTs
Observe for change in soil erosion in different fields/plots over time	Replacing of dead MPTs
Reinforce SWC structures when necessary	Check whether MPTs are affected by inadequate rain
Adjust the spacing of SWC structures where necessary	Observe for pest infestations of MPTs
Observe duration of time before establishment of MPTs	Pest control measures
Observe the performance of different MPTs	Check for when to prune
Filling of the gaps of MPTs	Check time required for pruning different MPTs
Observe for breakages on conventional SWC structures	Assess ability of MPTs to provide fuel wood
Repair of conventional SWC structures	Identify pruning intervals
Observe for moisture stress for different MPTs at different stages of establishment	Observe how fast MPTs regrow after cutting
Continuous observation for pest infestations of MPTs, e.g. psyllids in lucaena	Check for ability of MPTs for frequent harvesting
Application of pest control measures	Assess amount of material harvested from different MPTs
Seeking expert intervention if no solution to pest infestation	Observe how effective the material spread in the plot is able to suppress weeds
Observing for appropriate time of cutting/pruning of MPTs – for livestock or spreading in the plots	Time for gully clearing depending on conservation measure
Pruning of MPTs	Check for increase of soil fertility
Observe duration of time before pruning of MPTs	Assess crop production over time
Check for capacity of MPTs to regenerate	Compare milk production due to feeding with MPTs
Check for capacity of MPTs to withstand pruning	
Assess biomass production for different MPTs	
Establishment uniformity of MPTs	
Period of time for gully filling as per the intervention (MPTs and conventional SWC structures)	
Observe for soil fertility change in different plots	
Assess the amount of prunings required to bring change in soil fertility	
Yields of the test crop over time/SWC measure	
Labour requirement for establishing layouts of MPTs and conventional SWC technologies	
Assess milk yield change due to feeding on different MPTs	
Assess animal's live weight change due to feeding on different MPTs	

steps between men and women were found to be insignificant. The small differences that were noted include the following:

Men perceived implementation process to start from constraint identification, while women perceived it to start with identification of options to improve soil productivity. This is probably due to the fact that very few women participated in the constraint identification and ranking at the beginning of the project.

Women are more concerned with labour allocated to managing/pruning MPTs than are men, while men are more concerned with labour allocated to managing conventional SWC measures – this was highly attributed to gender allocation of labour, roles and responsibilities. This implied that the role of feeding livestock is mostly for women, while establishment and management of conventional SWC measures like terracing is mostly done by men.

The capacity of Multi Purpose Trees (MPTs) biomass to suppress weed was taken more seriously by women than men – this was another indication of gender division of labour where women are more involved in weeding than are men.

Women are more concerned with the capacity of MPTs to generate fuel wood – another indication of gender division of labour. Women are concerned with cooking than are men because the latter rarely cook.

Women did not consider change in animal's live weight to be an important evaluation factor – this was attributed to gender access and control of household resources. Livestock control and access are almost exclusively for men.

With the assistance of the facilitator, the group polished the separate lists presented by men and women to come up with a comprehensive one incorporating men's and women's views as shown in Table 2.

Table 2 Comprehensive list of activities in the SWM implementation project at grassroots level in Kirege, Meru South

Soil conservation	Soil fertility
Problem identification and ranking	Problem identification and ranking
Option identification and ranking	Option identification and ranking
Selection of the plots for evaluations	Selection of soil fertility improvement interventions
Selection of SWC interventions for evaluation	Establishment of soil fertility interventions (organic and inorganic)
Establishment of SWC interventions	Establishment of the test crop
Monitor change in soil erosion/plot	Monitor change in soil fertility/plot
Repair/reinforce soil erosion structures	Monitor appropriate pruning time of MPTs
Adjust the spacing of soil erosion structures	Pruning of MPTs
Monitor establishment time of MPTs	Evaluate biomass production of MPTs
Monitor performance of MPTs	Evaluate establishment uniformity of MPTs
Gapping of MPTs	Monitor test crop's pest infestation
Monitor conventional soil erosion structure breakages	Control pests in test crop
Repair conventional soil erosion structures	Monitor soil fertility change/plot
Monitor moisture stress of MPTs	Monitor biomass effectiveness in weed control
Monitor pest infestations of MPTs	Evaluate pruning quantities due to change in fertility
Application of pesticides	Evaluate crop yields/intervention
Feedback for intervention	
Monitor appropriate pruning time of MPTs	
Pruning of MPTs	
Evaluate regenerative capacity of MPTs	
Evaluate pruning frequency of MPTs	
Evaluate establishment uniformity of MPTs	
Assess ability of MPTs to provide fuel wood	
Gully filling duration/intervention	
Yields/intervention	
Layout labour requirement of MPTs	
Conventional SWC technologies' labour requirement	
Evaluate milk yields/MPT	
Live weight gain/MPT	

Differentiating Various Activities in the Implementation Process

Looking at the checklist of the activities in the implementation process, the participants' different activities in the implementation process can be grouped together depending on their similarities. In order to clearly perceive and conceptualise this, question or statement was posed to the farmers.

Through facilitation by the researcher, extension staff and one member from a local NGO, the group dialogued at length and agreed that there are differences because establishment of MPTs is planting of the MPTs, which is a physical/cultural activity, while checking whether there is pest infestation is monitoring so as to come up with a way of stopping the damage to the MPTs.

Hence, the implementation process activities were categorised as "physical"/cultural/agronomic activities, observatory activities (monitoring), action/remedial activities and assessment activities (evaluation for outputs and outcomes).

From the catalogue of implementation activities developed as in Table 1, the participants were facilitated to group them into four categories. This process led the team to come up with the classification of activities as shown in Table 3.

Monitoring and evaluation was further conceptualised by the groups by taking them through the graphics in Fig. 1.

Hence, the implementation of an SWM project can be perceived to be as in the process shown in the flow diagram (Fig. 2).

Participatory Development Process of M&E Indicators for SWM at Grassroots

The participants noted that the project had two interrelated components whose indicators for success or failure need to be developed/identified. These are

Table 3 SWM implementation process classification of activities

Cultural	Monitoring	Action/remedial	Evaluation
Establishment of SWC interventions Pruning of MPTs Establishment of soil fertility interventions (organic and inorganic) Establishment of the test crop Pruning of MPTs	Monitor change in soil erosion/plot Monitor establishment time of MPTs Monitor performance of MPTs Monitor conventional soil erosion structure breakages Monitor moisture stress of MPTs Monitor pest infestations of MPTs Monitor appropriate pruning time of MPTs Gully filling duration/intervention Monitor change in soil fertility/plot Monitor moisture stress of the test crop Monitor test crop's growth uniformity Monitor appropriate pruning time of MPTs Monitor test crop's pest infestation Monitor biomass effectiveness in weed control	Repair/reinforce soil erosion structures Adjust the spacing of soil erosion structures Gapping of MPTs Repair conventional soil erosion structures Application of pesticides Control pests in test crop	Evaluate regenerative capacity of MPTs Evaluate pruning frequency of MPTs Evaluate establishment uniformity of MPTs Assess ability of MPTs to provide fuel wood Evaluate yields/intervention Evaluate layout labour requirement of MPTs Evaluate conventional SWC technologies' labour requirement Evaluate milk yields/MPT Live weight gain/MPT Evaluate biomass production of MPTs Evaluate establishment uniformity of MPTs Evaluate pruning quantities due to change in fertility Assess amount of inorganic/organic fertilizer/plot Evaluate crop yields/intervention

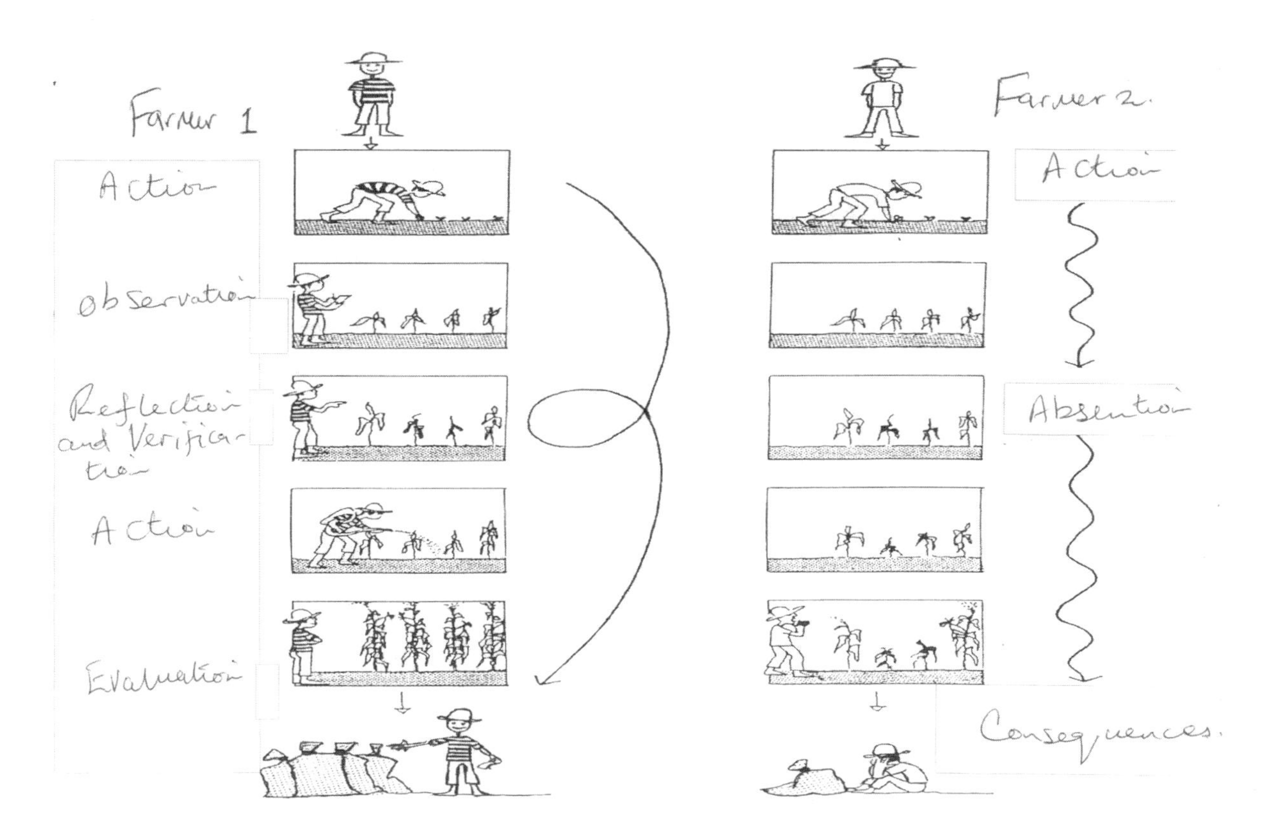

Fig. 1 Two farmers cultivating maize (German and Gohl, 1996)

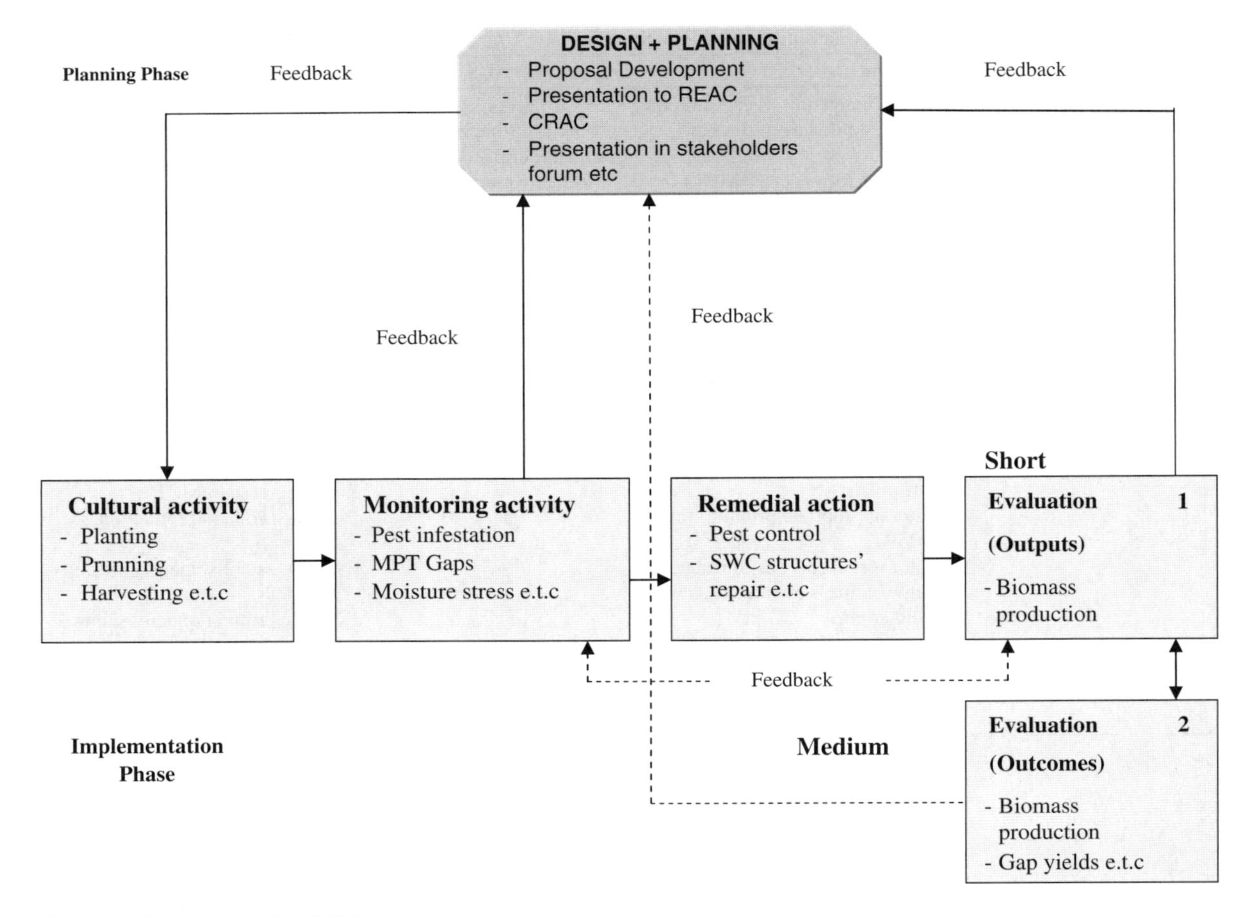

Fig. 2 Implementation of an SWM project process

soil and water conservation (SWC) and soil fertility improvement (SFI) indicators. Hence, participants were divided into two groups composed of equal representation of men and women. Each group was requested to develop SWC and SFI indicators by use of sort cards. Through facilitation, one group developed SWC, while the other developed SFI M&E variables and indicators. The results were as presented in Tables 4 and 5.

The participants noted that SWC and SFI implementation processes in SWM were different but with interrelationships that influence one another. Hence, there is no dichotomy between monitoring and evaluation for either SWC or SFI activities, implying that there are overlaps and both monitoring and evaluation may be carried out simultaneously. Therefore, the monitoring and evaluation processes for SWM and water management at the grassroot level can be presented as in the flow diagram (Fig. 3).

Result Sharing Among Key Stakeholders

The farmers were facilitated to hold a stakeholder forum to share the results of this process. The participants in the forum included researchers, extension staff, representatives of NGOs working in the region, local leaders and some CBO representatives. This forum was an important platform for influencing research and development process so that technologies development became more responsive to the needs of the farmers, hence increasing adoption and impact at the grassroots.

Conclusions

This study shows that farmers have the capacity to develop M&E indicators that are more relevant and

Table 4 Participatory catalogue of M&E indicators for SWM at the grassroots

Soil and water conservation indicators

Monitoring variable	Indicators	Evaluation variable	Indicators
Monitor change in soil erosion/plot	– No. of visible gullies – Soil colour – Soil build on lower hedges – Types of weed	Evaluate regenerative capacity of MPTs	– Time taken between prunings – Time taken for MPTs to be "unproductive"
Monitor establishment time of MPTs	– Time taken from planting to pruning	Evaluate pruning frequency of MPTs	– Time taken between prunings
Monitor performance of MPTs	– Time taken to be out of "drying risk	Evaluate establishment uniformity of MPTs	– Plant height over time
	– Amount of forage in the first harvest – Lifespan of MPTs		– No. of dead MPTs/variety
Monitor conventional soil erosion structure breakages	– Time span before first repair	Assess ability of MPTs to provide fuel wood	– Height of the MPT – Woodiness of the stem and branches – Branching capacity
Monitor moisture stress of MPTs	– Colour of leaves of MPTs	Evaluate yields/intervention	– Yields/test crop/intervention
Monitor pest infestations of MPTs	– Height/time – Vigour of MPTs – No. of plants infected – Type of pests infesting – Pest count/plant	Evaluate layout labour requirement of MPTs	– Man days/MPT/plot
Monitor appropriate pruning time of MPTs	– Height of MPTs – Forage production of MPTs	Evaluate conventional SWC technologies' labour requirement	– Man days/conventional SWC structure/plot
Monitor gully filling duration/intervention	– Period of gully filling/intervention	Evaluate milk yields/MPT	– Litres/animal/MPT fed
			– Animal's lactating period
		Live weight gain/MPT	– Time taken to maturity
		Evaluate biomass production of MPTs	– Livestock price/age
		Evaluate establishment uniformity of MPTs	– Amount of forage produced/MPT/season – No. of man loads/harvest/MPT – Area of plot spread with the pruning of MPTs – No. of seedlings used for gapping

appropriate in their environment and situation. It also shows that the process is empowering and enhances ownership of the M&E process. The results created awareness of communities in identification of M&E indicators and built their confidence in participating in PM&E process, discarding the past myth that M&E is a responsibility of the "consultants and outsiders with very little/no involvement of the local communities". Hence, this process could be used in development of more appropriate and community-friendly M&E indicators that the communities can use to monitor their own projects without necessarily waiting for outsiders to do it for them. It also creates synergies among participants making it more effective and efficient in monitoring and evaluating SWM projects at the grassroots. The results also show that stakeholders' participation in this process provides relevant feedbacks to inform research, hence enhance development of appropriate technologies, accelerate impacts and improve household and community livelihoods. In addition it also shows that there is room for upscaling of these results at the programme level and hence develop negotiated PM&E frameworks at the institute level.

Table 5 Catalogue of monitoring and evaluation indicators for SFI variables/issues at the grassroots

Monitoring variable	Indicators	Evaluation variable	Indicators
Monitor change in soil fertility/plot	– Soil colour – Type of weeds – Crop vigour	Assess amount of inorganic/organic fertilizer/plot	– Kilograms of inorganic fertilizer/plot – Man loads of green manure/plot
Monitor test crop's pest infestation due to low soil fertility	– No. of plants infested with low-fertility-related pests – Types of low-fertility-related pests identified	Evaluate green manure biomass production	– Man loads/plot – No. of biomass harvests/season or cycle
Monitor moisture stress of the test crop	– Crop size/height – Crop vigour – No. of plants stunted due to stress – Crop colour/yellowing	Evaluate pruning quantities due to change in fertility	– Amount of MPTs produced due to change in soil fertility – Amount of crop residues produced due to change in soil fertility – Amount of green manure produced due to change in soil fertility
Monitor test crop's growth uniformity	– No. of hills gapped – Crop height uniformity – Cob size uniformity – Harvest time uniformity	Evaluate crop yields/intervention	– Bags/test crop/soil fertility intervention – Cobs/plot/intervention
Monitor biomass effectiveness in weed control	– Types of weeds identified/plot – Weed intensity/type of biomass – Weeding frequency/type of biomass		

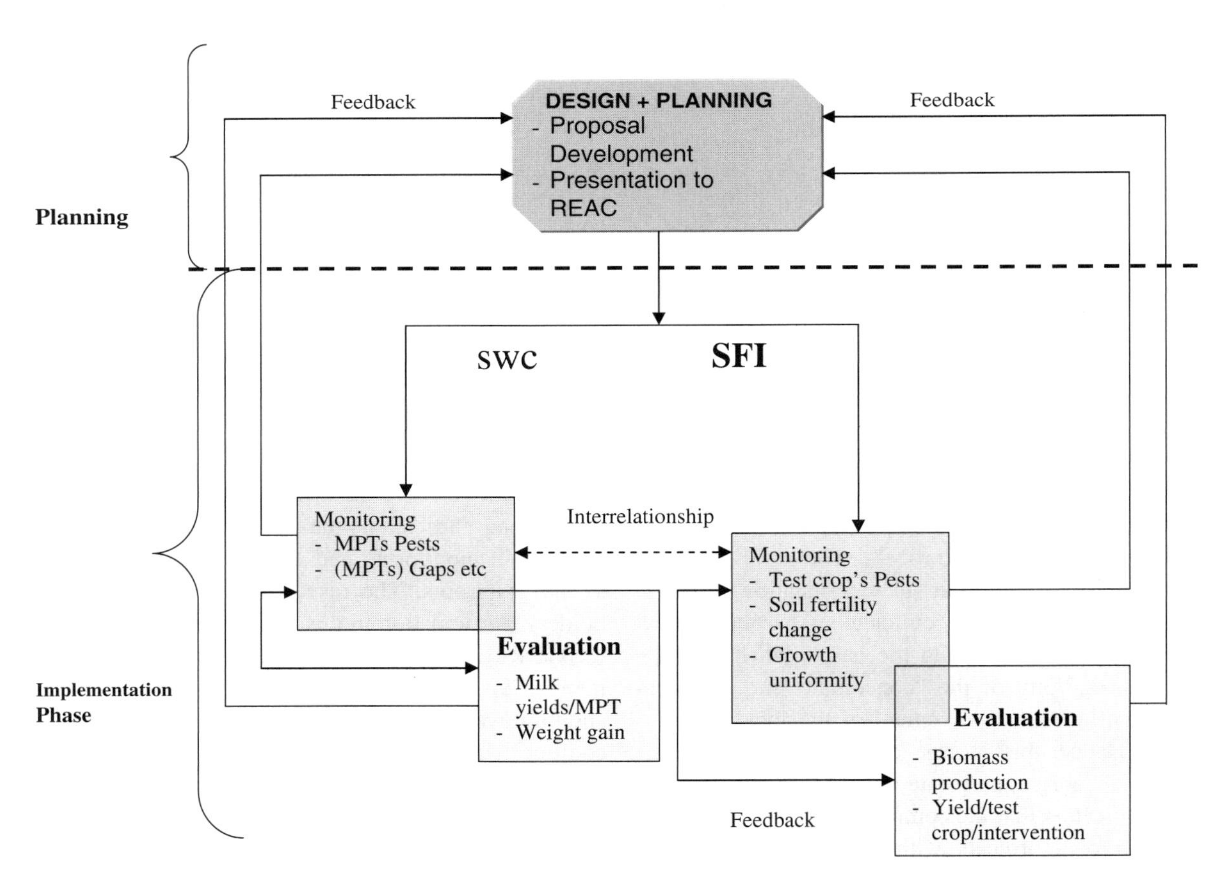

Fig. 3 Participatory development process of PM&E indicators for SWM

References

German D, Gohl E (eds) (1996) Participatory impact monitoring booklet 4: the concept of participatory impact monitoring. GATE/GTZ, Eschborn

Guijt I (1998) Participatory monitoring and impact assessment of sustainable agricultural initiatives: an introduction to the key elements. IIED, London

Matiri FM (1999) How participatory is farmer participatory research: critical review of KARI DFID NARP II. MA dissertation, School of Development Studies, UEA, UK

Pretty JN, Guijt I, Thompson J, Scoones I (1995) A trainer's guide for participatory learning and action. IIED participatory methodology series. IIED (HT2700 PRE), London, pp 54–71

Effectiveness of "PREP-PAC" Soil Fertility Replenishment Product on Performance of the Diversified Maize–Legume Intercrops in Western Kenya

E.J. Rutto, J.R. Okalebo, C.O. Othieno, M.J. Kipsat, and A. Bationo

Abstract Western Kenya region is densely populated, with population densities ranging from 500 to 1200 persons/km^2. This has resulted in reduced land sizes, continuous cropping, high rates of soil nutrient depletion, and food insecurity. PREP-PAC was tested on a small-scale farm in western Kenya for three continuous seasons. Seven legume varieties were intercropped with maize (*Zea mays*) using MBILI intercropping system. The treatments were in a 7×2 factorial arrangement in a randomized complete block design, with each treatment replicated four times. PREP-PAC application contributed to a significant increase ($p < 0.01$) in soil available P and soil pH, with a maximum increase in soil-available P of 8.20 in maize–yellow gram intercrop and pH increase of 0.72 in maize–dolichos bean obtained in 2004 LR. A significant positive change ($p < 0.01$) in maize grain yields across all intercrops was also recorded, with a maximum change in yield of 2616 kg/ha obtained from maize–bambara nut intercrop in 2004 LR. A positive change in increase in legume grain yield was obtained for most legumes, with beans giving a maximum significant increase ($p < 0.01$) of 658 kg/ha in 2004 SR. Economic analysis indicated a significant increase ($p < 0.01$) in improved farm income. Maize–yellow grams gave the highest net profit of 89,506.70 Ksh/ha/year and maize–cowpeas gave the lowest net profit of −19,403.20 Ksh/ha/year. This study therefore concluded that intercropping yellow grams, soybeans, beans, groundnuts, and bambara nuts with maize was profitable and is recommended for adoption toward nutrient replenishment and food security in western Kenya.

Keywords PREP-PAC · Soil pH and available P · Maize-legume MBILI intercropping system · Grain yields

Introduction

Sub-Saharan Africa (SSA), including Kenya, continues to experience constraints of food insecurity, which largely contribute to overall poverty and poor health in rural areas. The highlands of western Kenya occupy only 15% of the country's land area but contain 40% of the country's population (500–1200 persons/km^2) (ICRAF, 1999). This has resulted in reduced farm sizes of less than 0.2 ha/household (Swinkels et al., 1997), which arises from subdivisions of rather small pieces of land amongst household members. The outcome of this is the widespread food insecurity and poverty in western Kenya, where 90% of the population earns less than US $1/day (Obura et al., 2001). In an effort to produce more food, farmers in this region cultivate continuously about 500,000 ha of widely N- and P-depleted soils (Okalebo, 2000), thus exposing western Kenya to high rates of phosphorus losses of 3–13 kg/ha/year (Woomer et al., 1997). Therefore, the region has experienced several decades of low maize (less than 1.0 t/ha) and bean yields (200–500 kg/ha) (Sanchez et al., 1997).

Western Kenyan soils are acidic, classified mainly as Ferralsols, Acrisols, and Nitisols (FAO classification), because of high weathering and heavy leaching of basic cations into underlying horizons. Under this condition, P is fixed by Al and

E.J. Rutto (✉)
Department of Soil Science, Moi University, Eldoret, Kenya
e-mail: emyruto@yahoo.com

A. Bationo et al. (eds.), *Innovations as Key to the Green Revolution in Africa*,
DOI 10.1007/978-90-481-2543-2_126,

Fe compounds or sesquioxides. The presence of Al toxicity in these acid soils also limits root elongation (Kanyanjua et al., 2002).

Maize/common bean conventional intercropping is a common cropping system in western Kenya (Obura et al., 1999). This intercropping system has been found to favor build up of pests and diseases such as bean root rot (Otsyula et al., 1998) and to reduce light penetration to beans growing under the maize, thereby leading to reduced bean yields of 200 kg/ha and maize grain yields of below 500 kg/ha per season.

However, MBILI intercropping system (planting staggered two rows of maize alternated with two rows of a chosen legume) has been proven to increase both maize and legume grain yields (Tungani et al., 2002) The system also improves *P*- and *N*-use efficiencies and uptake through enhanced root density (Thuita, 2007).

In the quest to detect nutrient limitation in soils and solutions to correct these constraints in western Kenya, several nutrient replenishment strategies have been studied. These include the use of inorganic fertilizers (FURP, 1994; Allan et al., 1972); the use of manures, compost, and their combinations with inorganic fertilizers (Okalebo et al., 1999); and the use of agroforestry shrubs capable of fixing atmospheric N into the soil (Jama et al., 1997). However, adoption of these technologies is low because inorganic fertilizers are expensive (Smaling, 1993) and organics are generally inadequate and are frequently of low quality (Probert et al., 1992). The use of Minjingu phosphate rock (MPR) has been studied and tremendous results in terms of maize, legume, and agroforestry shrub yields have been reported (Waigwa et al., 2002; Ndungu et al., 2006).

Earlier findings of PR effectiveness led to the development of PREP-PAC package, at Moi University, Eldoret Kenya, in 1997. PREP-PAC consists of 2 kg Minjingu PR (source of *P*), 0.2 kg urea, 125 g food grain legume seed (to enhance biological N fixation), the legume seed inoculant (*Rhizobium*), and gum arabic adhesive and lime for pelleting (to make the soil reaction favorable for the growth of rhizobia), along with instructions for use written in Kiswahili and a few other local languages. The legume seeds in the PREP-PAC are inoculated with 100 g of rhizobium inoculants and 30 g of gum arabic adhesive, which is enough for 5 kg of legume seeds before planting immediately. PREP-PAC replenishes the fertility of soils in patches of 25 m^2 (Nekesa et al., 1999).

Considering the nutrient depletion problem, food insecurity, and poverty, this study was undertaken to quantify (1) the effect of affordable PREP-PAC package on the yields of maize and seven legume grains; (2) the effect of PREP-PAC on soil properties; and (3) the economic impact on maize and legume production.

Materials and Methods

Site Description

The study was conducted in the Nyabeda area, Siaya district in western Kenya. Siaya district lies between latitude 0°03′N and longitude 34°25′E. The altitude is between 1140 and 1141 m.a.s.l. The average annual rainfall ranges between 800 and 2000 mm, with temperature of about 27–30°C and annual minimum temperature of 15–17°C (Jaetzold and Schmidt, 1983). The predominant soil types according to FAO/UNESCO classification in the district are Orthic Ferralsols (Oxisols), Dystric Nitisols (Alfisols), and Acrisols (Ultisol) (Republic of Kenya, 1994), which are generally acidic and depleted of important plant nutrients.

Experimental Design and Management

A field experiment was established on 46 m × 33 m land with each experimental plot unit 4.5 m × 4.0 m. The experiment was in a 2 × 7 factorial arrangement in a randomized complete block design, whereby there were two levels of soil fertility treatments: with and without PREP-PAC. The treatments were then randomly assigned to the seven maize–legume intercrops replicated four times. The maize (*Zea mays*) variety WH 502 was intercropped with seven legume types, namely bambara nuts (*Voandzeia subterranea*), soybean (TGx 14482E), yellow grams (*Phaseolus aureus*), groundnuts (*Arachis hypogea*), beans (*Phaseolus vulgaris*), cowpeas (*Vigna unguiculata*), and dolichos bean (*Lablab purpurens*), using the MBILI spatial arrangement (maize rows are spaced as 50-cm pairs that are 100 cm apart and two rows of legumes are planted within the gap with 33-cm row spacing). Plant population for this intercropping system is 44,000 and 88,000 for maize and legumes, respectively, just as in conventional intercropping system. PREP-PAC was applied once per season in each marked plot. In 2003 SR, PREP-PAC (100 kg P/ha + 40 kg N/ha) was applied with no addition of N as a topdress. This led to yellowing of some plants in PREP-PAC plots, suggesting a possible N deficiency (J.R. Okalebo, *personal observation); hence* the N rate in the PREP-PAC package was increased to 60 kg N/ha in 2004 LR, applied in two equal splits of 30 kg N/ha at planting and as topdress at 6 weeks after sowing. The residual effect of the applied PREP-PAC in 2003 SR and 2004 LR together with N added at the rate of 60 kg N/ha to PREP-PAC plots was studied in 2004 SR.

The experiment was weeded two times per season and pests and diseases were controlled by spraying the legume crops with "Ridomil" and "Duduthrin."

At maturity, maize was harvested in 11.5 m^2 effective area and weighed. The maize cobs and stover were separated, weighed, and sample maize cobs were taken for sun drying. Later the maize cobs were shelled to obtain the grains and weighed to determine the yield.

The cowpeas, beans, yellow grams, and soybeans were uprooted from the effective area of 11.5 m^2 and total grain and trash weights were taken. The pods were then separated from trash, weighed, and a sample for trash and grain was taken for sun drying. Later, the sample dry weights of legumes, grain and trash, were taken to determine the total grain and trash weight.

The bambara nut and groundnut legumes were carefully dug out using a hoe within the effective plot area of 10.85 m^2 for each treatment. The pods were separated from total biomass by hand and pod fresh weight was taken. Sample pods of 200 g per plot were sun dried and shelled to determine the total grain yields per hectare.

Soil Sampling, Laboratory, and Statistical Analyses

Soil sampling was done before the application of treatments for site characterization and at the end of every season (at harvesting) to monitor the changes in soil chemical properties as affected by the treatments and their residual effects. Soil samples from all plots were taken at random, using a soil auger from 12 points per plot and mixed well to get a composite soil sample for each plot. Soil samples were analyzed for pH, organic carbon, total N, and available P according to standard procedures outlined in Okalebo et al. (2002). Plant tissues were also analyzed for total P using procedures in Okalebo et al. (2002).The results obtained for the study parameters were statistically analyzed using GLM procedures of SAS system version 8 software (SAS, 2003) to determine treatment effects on yields, soil pH, and available P after every harvest and Microsoft excel was used to separate means. The results obtained were analyzed per season.

Economic Analysis

Costs and benefits associated with intercropping system in land use were determined. Prices for legumes and maize were established through direct survey of the prevailing market prices within the western Kenyan region. The PREP-PAC price of Ksh 67 per pack (Obura et al., 1999) was adopted. Labor requirement for land preparation, planting, weeding, pest and disease control, and harvesting was obtained from field management records and valued using the prevailing wage rates. Prices of inputs such as maize, bean and groundnut seeds, Biofix, DAP, CAN, stalk borer dust, "Duduthrin" and "Ridomil" were obtained from nearby market centers, based on prevailing market prices. Opportunity cost of capital was estimated as 10% per season, which is the commonly used rate for studies involving resource-poor smallholder farmers (Jama et al., 1997). The production cost was the product of the prevailing prices of each crop in the particular season. The economic indicators that were used comprised of net present value (NPV), enterprise budgets, and marginal rate of return (MRR).

Results and Discussion

Effect of PREP-PAC Application Under MBILI Intercropping System on Soil Properties

Soil pH

The results indicated significant increases ($p < 0.01$) in soil pH due to PREP-PAC application throughout

Table 1 Mean values of soil pH (H_2O) and their positive changes from PREP-PAC addition, measured during maize–legume cropping at Nyabeda, Siaya district

	2003 SR soil pH (H_2O)			2004 LR soil pH (H_2O)			2004 SR soil pH (H_2O)		
Legumes	Control	PREP-PAC	pH change	Control	PREP-PAC	pH change	Control	PREP-PAC	pH change
Bambara nuts	5.17	5.76	0.59	5.22	5.70	0.48	4.97	5.49	0.52
Beans	5.57	5.58	0.01	5.43	5.59	0.16	5.26	5.57	0.31
Soybeans	5.38	5.57	0.19	5.24	5.58	0.34	5.14	5.54	0.40
Yellow grams	5.37	5.74	0.37	5.45	5.71	0.26	5.21	5.53	0.32
Cowpeas	5.52	5.54	0.02	5.55	5.73	0.18	5.18	5.71	0.53
Groundnuts	5.28	5.62	0.34	5.39	5.60	0.21	5.15	5.59	0.44
Lablab	5.33	5.56	0.23	5.17	5.89	0.72	5.20	5.68	0.48

Mean, 5.46; SE, 0.13; CV, 4.58

the cropping season. However, significant decrease ($p < 0.05$) in soil pH was observed within control plots in 2004 SR season compared to the earlier season. The increased soil pH due to PREP-PAC addition is explained by the increased MPR dissolution due to low pH of the soil in the study (Table 1). This led to availability of calcium (CaO) which dissociates from the apatite rock during the solubilization process, thereby replacing Al^{3+} and Fe^{2+} ions at the exchange sites; hence increased soil pH (Okalebo et al., 1999). Rock phosphate (RP) is sparingly soluble in water; therefore, its complete breakdown toward the release of its components takes long time (Sikora and Giordano, 1993). This explains the slow but steady increase in soil pH due to the addition of PREP-PAC under MBILI system.

Available P

The results indicated a significantly increased ($p < 0.01$) available P in PREP-PAC plots compared to the controls plots, especially in 2004 LR and 2003 SR. Positive interaction of PREP-PAC with maize and legume intercrops and seasons indicated rise in soil-available P throughout the cropping seasons. A positive change in soil-available P was observed throughout the cropping season, with maize–yellow gram intercrop giving the highest value and maize–groundnut intercrop giving the least value on the average (Table 2). This could be attributed to positive responses of the soil to applied P in the form of MPR, since the soils in the area were acidic. The decline in soil-available P in the third cropping season (2004 SR) might have been due to the uptake of P by intercrops from previous cropping seasons, microbial immobilization over time, and possible fixation of P in the acidic soil (Nguluu et al., 1996).

Nevertheless, the residual effect of PREP-PAC in 2004 SR was noted; the values of available P in the soil still remained higher in PREP-PAC plots compared to the control plots, the finding also earlier made by Mnkeni et al. (1991).

Table 2 Means of available P (mg P/kg soil) in 0–15 cm soil, obtained from PREP-PAC and control plots in an experiment in Nyabeda farm at Siaya district

	Soil-available P (mg P/kg soil)								
	2003 SR			2004 LR			2004 SR		
Maize–legume intercrops	Control	PREP-PAC	Change in soil P	Control	PREP-PAC	Change in soil P	Control	PREP-PAC	Change in soil P
Bambara nuts	2.41	3.35	0.94	1.80	6.83	5.03	2.03	7.29	5.26
Beans	2.15	3.39	1.24	2.07	8.14	6.07	2.07	7.46	5.39
Soybeans	2.33	3.38	1.05	2.43	9.30	6.87	1.76	7.21	5.45
Yellow grams	1.83	4.10	2.27	1.87	10.07	8.20	2.34	8.41	6.07
Cowpeas	2.35	3.22	0.87	2.34	8.06	5.72	2.61	9.86	7.25
Groundnuts	2.15	2.57	0.42	1.95	4.85	2.90	2.18	6.28	4.10
Dolichos bean	2.67	3.18	0.51	2.41	7.44	5.03	2.99	5.86	2.87

Mean, 4.21; SE, 0.65; CV, 30.93

Effects of PREP-PAC Application to Maize–Legume Intercrops on Crop Yields

Maize Grain Yield

PREP-PAC, season, MBILI intercrops, and their interaction all gave significant increases ($p < 0.01$) in maize grain yield. The maximum grain yield was observed in 2004 LR, followed by 2004 SR and the least in 2003 SR, with mean increases in maize grain yields of 2002, 1881, and 735 kg/ha, respectively, above the control (Table 3). The maize–bambara nut intercrop gave the highest overall maize grain yields of 2616 kg/ha and maize–dolichos bean intercrop the least maize grain yields of 1152 kg/ha (Table 3). The lowest response in maize grain yield values observed in maize–dolichos bean intercrop was thought to have been caused by the lush vegetative growth of the legume, which choked the maize, thereby reducing its potential. The higher yields observed could mainly be attributed to the addition of both N and P in the PREP-PAC as RP and urea (Obura et al., 2001). The liming potential of MPR in PREP-PAC probably reduced P fixation (through soil pH rises), thus the recorded maize grain increase. The maize grain yields are above the farmers' grain yield levels of 0.5 t/ha, an indication that PREP-PAC under the MBILI system is effective.

Legume Grain Yields

The results indicated PREP-PAC significantly increased ($p < 0.01$) legume yields. Legume yields, ranging from 52 to 1171 kg/ha (Table 4), varied with the cropping season, maize–legume intercrop, legume species, and PREP-PAC application. The resultant yields were attributed to addition of N and P nutrients to the already depleted soils of Nyabeda through PREP-PAC application. Liming potential of MPR

Table 3 Increase in maize grain yields (kg/ha) due to PREP-PAC application to maize–legume intercrops in three seasons at Nyabeda, Siaya district, and western Kenya

	Maize grain yield (kg/ha)								
	2003 SR			2004 LR			2004 SR		
Maize–legume intercrops	Control	PREP-PAC	Change in yields	Control	PREP-PAC	Change in yields	Control	PREP-PAC	Change in yields
Bambara nuts	385	1474	1089	667	3283	2616	411	2237	1826
Beans	439	1174	735	830	3346	2516	578	2620	2042
Soybeans	553	1143	590	878	3033	2155	600	2485	1885
Yellow grams	513	1141	628	1054	2724	1670	300	2804	2504
Cowpeas	291	1287	996	1561	3357	1796	620	2061	1441
Groundnuts	307	854	547	1174	3285	2111	604	2193	1589
Dolichos bean	507	1070	563	1887	3039	1152	–	–	–

Mean, 1469.19; SE, 245.34; CV, 46.63

Table 4 Increase in legume grain yields (kg/ha) due to PREP-PAC application to maize–legume intercrops in three seasons at Nyabeda, Siaya district, and western Kenya

	Legume grain yield (kg/ha)								
	2003 SR			2004 LR			2004 SR		
Legume	Control	PREP-PAC	Change in yields	Control	PREP-PAC	Change in yields	Control	PREP-PAC	Change in yields
Bambara nuts	433	439	6	302	348	46	892	1171	279
Beans	409	685	276	290	743	453	208	866	658
Soybeans	114	247	133	69	369	300	52	457	405
Yellow grams	134	234	100	135	323	188	99	538	439
Cowpeas	158	336	178	–	–	–	–	–	–
Groundnuts	427	511	84	461	514	53	602	794	192

Mean, 417.79; SE, 103.48; CV, 49.54

in PREP-PAC contributed to the increased soil pH, leading to release of previously fixed P due to acidic soil, hence available to the growing crops. The MBILI intercropping system promotes high root density, light penetration, and reduces cases of pests and diseases. The legume grain yields from PREP-PAC plots were above the average farmers' value of < 200 kg/ha, suggesting the effectiveness of PREP-PAC toward soil fertility alleviation and food security.

Economic Analysis

Economic analysis results were carried out in order to calculate the gross margins of the maize–legume intercrops. This was done to quantify the use of PREP-PAC as a soil input to boost soil fertility and crop yields against the control treatment. The results (Table 5) indicated that the use of PREP-PAC on maize–legume intercrop was profitable. Maize–bambara nut intercrop gave the highest gross margin of 118,527 Ksh/ha/year, while maize–dolichos bean intercrop gave the lowest gross margin of –17,673.50 Ksh/ha/year (Table 5). However, the low gross margin of maize–dolichos bean intercrop is largely due to lack of harvest of dolichos bean, which was due to its indeterminate and hence continuous flowering and subsequent prolonged seed production. The logistics in this experiment did not provide for labor to continuously harvest dolichos beans.

The net change in profit analysis (Table 6) indicates that, with an exception of dolichos bean and cowpea intercrops, the use of PREP-PAC for yellow grams, soybeans, beans, groundnuts, and bambara nuts intercrops was profitable. The recorded negative net change in profit of the dolichos bean and cowpea intercrops was because of the missing grain data in 2004 LR and 2004 SR, which was attributed to poorly distributed rainfall (for cowpeas) and unable to continuously harvest dolichos bean due to its indeterminate growth behavior.

These economic results concurred with the results obtained by Obura et al. (2001), where PREP-PAC proved to be profitable in a maize–soybean conventional (soybean rows between maize rows) intercrop. Also, Mwaura (2003) in her study determined that, PREP-PAC gave a high return to land than the absolute control and other nutrient replenishment technologies.

Based on these findings, it is therefore recommended that maize (var. WH 502) successfully be intercropped with common beans, yellow grams, groundnuts, bambara nuts, and soybeans in western Kenya. However, this research did not obtain enough data on the yields of dolichos bean and cowpeas. Further studies are encouraged in order to find these facts.

Table 5 Results of gross margin analyses of maize–legume intercrop production under PREP-PAC experiment in Nyabeda, Siaya district in three cropping seasons

	Returns to factors of production			
Intercrop type	Gross margin (Ksh/ha/year)	Labor productivity (Ksh/Ksh) spent on labor	Capital productivity (Ksh/Ksh)	Value cost ratio
Maize–bambara nuts	118,527.75	4.30	1.62	2.45
Maize–groundnuts	118,398.00	4.20	1.61	2.44
Maize–yellow grams	97,754.40	3.48	1.32	2.18
Maize–beans	82,528.55	2.94	1.14	2.02
Maize–soybean	75,955.25	2.70	1.04	1.93
Maize–cowpeas	24,672.25	0.90	0.34	1.31
Maize–lablab	−17,673.50	−0.67	−0.26	0.83

Table 6 Partial budget for use of PREP-PAC on maize–legume intercrops in an experiment in Nyabeda Siaya District, in three cropping seasons

Intercrop type	Net change in profit (kshs/ha/year)
Maize–yellow grams	89,506.70
Maize–soybeans	76,594.95
Maize–beans	75,540.10
Maize–groundnuts	39,014.55
Maize–bambara nuts	25,236.90
Maize–lablab	–16,906.40
Maize–cowpeas	–19,403.20

Acknowledgments I would like to acknowledge TSBF-CIAT and Dr Martin Wood from the University of Reading for financing my research project, my supervisors Prof. J.R. Okalebo and Prof. C.O. Othieno for the support and constant guidance, and last but not the least, my family and colleagues for their constant encouragements.

References

Allan AY, Were A, Laycock D (1972) Trials of types of fertilizer and time of nitrogen application. In: Kenya Department of Agriculture, NARS Kitale. Ann. Rpt. Part II:90–94

FURP (1994) Fertilizer use recommendations project, vols 1–22. Kenya Agricultural Research Institute, Nairobi, Kenya

ICRAF (1999) International Center for Research in Agro Forestry, report to the Rockefeller Foundation on Preparation and Agronomic Evaluation of Ugandan phosphates (September 1998 to August 1999)

Jaetzold R, Schimdt H (1983) Farm management handbook of Kenya, vol II. Natural conditions and farm management information. Part A. Western Kenya (Nyanza and western provinces). Kenya Ministry of Agriculture, Nairobi, Kenya p 397

Jama B, Swinkles RA, Buresh RJ (1997) Agronomic and economic evaluation of organic and inorganic sources of phosphorus in Western Kenya. Agron J 89:597–604

Kanyanjua SM, Ireri L, Wambua S, Nandwa SM (2002) Acidic soils in Kenya: constraints and remedial options. Technical note number 11, KARI Headquarters, Nairobi, Kenya

Mnkeni PNS, Semoka JMR, Buganga JBS (1991) Effectiveness of Minjingu phosphate rock as a source of phosphorus for maize in some soils of Morogoro Tanzania. Zimbabwe J Agric Res 29(T):27–37

Mwaura HW (2003) Evaluation of soil fertility management options in a maize–bean production system in western Kenya. MPhil thesis, Moi University, Eldoret, Kenya

Ndung'u KW, Okalebo JR, Othieno CO, Kifuko MN, Kipkoech AK, Kimenye LN (2006) Residual effectiveness of Minjingu phosphate rock and fallow biomass on crop yields and financial returns in western Kenya. Exp Agric 42:323–336

Nekesa P, Maritim HK, Okalebo JR, Woomer PL (1999) Economic analysis of maize bean production using a soil fertility replenishment product (PREP-PAC) in Western Kenya. Afr Crop Sci J 7:585–590

Nguluu SS, Probert ME, Myers RJ, Waring SA (1996) Effect of tissue phosphorus concentration on the mineralization of nitrogen from stylo and cowpea residues. Plant Soil 191: 139–146

Obura PA, Okalebo JR, Othieno CO, Maritim HK (2001) An integrated soil fertility amelioration product intended for smallhold farmers in Western Kenya. Presented at the 5th conference of African Crop Science Society, Abuja, Nigeria, Oct 2001

Obura PA, Okalebo JR, Woomer PL (1999) The effect of PREP-PAC components on maize soybeans growth, yield and nutrient uptake in the acid soil of Western Kenya, PREP annual report

Okalebo JR (2000) Effects of short fallow species, *Tephrosia* and *Crotalaria* as soil fertility indicators. Proceedings of the 18th conference SSSEA, Mombasa, Kenya, p 47

Okalebo JR, Gathua KW, Woomer PLJ (2002) Laboratory methods of soil and plant analysis, 2nd edn. A working manual, KARI, SSSEA, TSBF, SACRED Africa, Moi University, Nairobi, p 128 Marvel EPZ (K)

Okalebo JR, Palm CA, Gischuru M, Owuor JO, Othieno CO, Munyampundu A, Muasya RM, Woomer PL (1999) Use of wheat straw soybean trash and nitrogen fertilizer for maize production in the Kenyan highlands. Afr Crop Sci J 7: 423–432

Otsyula RM,, Nderitu JH, Buruchara RA (1998) Interaction between bean stem maggot, bean root rot and soil fertility. In: Farrel G, Kibata GN (eds) Second biennial crop protection proceedings. National agricultural research project II, Kenya Agricultural Research Institute, pp 70–77

Probert ME, Okalebo JR, Simpson JR, Jones RK (1992) The role of 'boma' manure for improving soil fertility. In: Probert ME (ed) Sustainable dryland cropping symposium. Proceedings no. 41. ACIAR, Australia, pp 63–70

Republic of Kenya (1994) Siaya district development plan (1994–1996) office of the vice-president and minister of planning and national development. Government Printer, Nairobi

Sanchez PA, Shepherd KD, Soule MJ, Place FM, Buresh RJ, Izac AMN, Mokuwonye AU, Kwesigaa FR, Ndiritu CG, Woomer PL (1997) Soil fertility replenishment in Africa. An investment in national resource capital. In: Buresh RY, Sanchez PA, Calhoun F (eds) Replenishing soil fertility in Africa. SSSA special publication. 51. Soil Science Society of America, Madison, WI, pp 1–46

SAS Institute (2003) SAS/STAT users guide: release 9.1 edn. SAS Institute, Cary, NC

Sikora FJ, Giordano PM (1993) Future direction for agricultural phosphorus research. Presented at FAO/IAEA consultants meeting on "Evaluation of agronomic effectiveness of phosphate fertilizers through the use of nuclear and related techniques", Vienna, Austria, 10–12 May

Smaling EMA (1993) Soil nutrient depletion in sub-Saharan Africa. In: Ruelar VH, Prins WH (eds) The role of plant nutrients and sustainable food production in sub-Saharan Africa. Plonsen and Looijen, Wageningen, pp 53–67

Swinkels RA, Franzel S, Shepherd KD, Ohlsson E, Ndufa JK (1997) The economics of short-term rotation improved fallows: evidence from areas of high population density in Western Kenya. Agric Syst 55:99–121

Thuita M (2007) A study to better understand the 'MBILI' intercropping system in terms of grain yield and economic returns compared to conventional intercrops in Western Kenya. MSc thesis, Moi University, Eldoret, Kenya

Tungani JO, Eusebius JM, Woomer PL (2002) A Handbook of innovative maize–legume intercropping. SACRED Africa, Bungoma, 20 pp

Waigwa M, Othieno CO, Okalebo JR (2002) Phosphorus availability as affected by application of phosphate rock combined with organic materials in acid soils of western Kenya. Exp Agric 39:395–407

Woomer PL, Okalebo JR, Sanchez PA (1997) Phosphorus replenishment in western Kenya: from field experiments to an operational strategy. Afr Crop Sci Conf Proc 3(1): 559–570

Risk Preference and Optimal Crop Combinations for Smallholder Farmers in Umbumbulu District, South Africa: An Application of Stochastic Linear Programming

M. Kisaka-Lwayo

Abstract Using data collected from 200 rural farm households in Umbumbulu district of KwaZulu-Natal in South Africa, the stochastic linear programming model is used to model the farm family crop production enterprise incorporating risk with a view of developing the optimal enterprise combination that would enable households maximize their utility. The model incorporates farmers' risk preferences, revenue fluctuations and resource restrictions. The results show that (1) changes in risk preference do affect optimal crop combinations and (2) the typical cropping pattern is rational under the present level of farmer's risk preference estimated in the study site; however, slight differences exist between the three groups of farmers studied. Effective extension programmes that will educate the farmers on efficient allocation of resources are pivots upon which the various smallholders' development programmes initiated by the government and/or other stakeholders should be built.

Keywords Optimal enterprise combination · Risk preference · Stochastic linear programming

Introduction

The ability of smallholder farmers in South Africa to adapt crop and resource management strategies to production uncertainty is the basis for their survival in a high-risk agricultural environment. Returns to crop production are precariously low for these resource-poor farmers and show tremendous variability over the years. Hence efficient allocation of resources through optimal enterprise combination by rural smallholder farmers though important has been evasive. Food crop production remains a major component of the smallholder farm family's production activities in SA and in Umbumbulu district of KwaZulu-Natal, and is the major use of arable land (Gadzikwa et al., 2006; Mushayanyama and Darroch, 2006). The crops *amadumbes* (*traditional yams*), sweet and English potatoes and green beans are a major source of farm income for these low-income farmers (Mushayanyama and Darroch, 2006) with farming characterized by low level of operation and low literacy levels.

These households aim to maximize their welfare through the choices of farm production and marketing methods and allocation of farm resources to farm and non-farm activities. Previous studies by Lwayo et al. (2007) and Gadzikwa et al. (2006) have shown certified organic farming to bring in the highest proportion of farm income with studies by Mushayanyama and Darroch (2006) showing these set of farmers to have the highest gross margin across all produce. Notwithstanding these conditions, farmers have not rushed into certified organic farming, with only 48 farmers as fully certified and 103 as partially certified in Umbumbulu district who are members of the Ezemvelo Farmers' Organisation (EFO). These may be attributed to the risk and constraints associated with the adoption of new and/or improved technology such as the difficulty of obtaining quality seeds of high-yielding variety and credit, income implications and farmers' perceptions. To the author's knowledge, no study in South Africa has explicitly taken into account

M. Kisaka-Lwayo (✉)
Agricultural Economics Discipline, School of Agricultural Sciences and Agribusiness, University of KwaZulu-Natal, Pietermaritzburg, South Africa
e-mail: maggiekisaka@yahoo.com

A. Bationo et al. (eds.), *Innovations as Key to the Green Revolution in Africa*,
DOI 10.1007/978-90-481-2543-2_127, © Springer Science+Business Media B.V. 2011

the farmer's risk preference which may be one of the most important factors for the adoption of new technology and which may also inform the degree of diversification the farmer will be willing to engage in.

The study will seek to compile linear programming models incorporating risk for smallholder organic farming in Umbumbulu with the overall aim of (1) determining the optimal crop combinations of smallholder farmers that maximize household welfare, (2) determining the sensitivity of the crop combination with changes in risk preference and (3) analysing constraints faced by these smallholder organic farmers.

Theoretical Framework

Farmers make production and marketing decisions. These include the portfolio of crops, land allocation, input use for all crops as well as obtaining access to markets, allocating and selling output to relevant choices of marketing channel. These decisions are interrelated and should be simultaneously determined. The theoretical model for the smallholder farmer resource allocation has some basic assumptions concerning the objective function of the farmers: (i) farmers are assumed to have specific utility functions; (ii) the objectives or goals of the farmers are many and may or may not be in conflict; (iii) the objectives are in turn assumed to be functions of the model decision variables; and (iv) the objective of production is to achieve satisfactory levels of specified objectives subject to the limitations imposed by the system and the environment.

The theory of production economics is concerned with the optimization of the objectives or goals and optimization implies efficiency (Baumol, 1977). Decision makers are presumed to be concerned with the maximization of measure of achievement such as profit or utility. Resource allocation according to Heady and Candler (1969) refers to the technical concept of efficiency, which brings about great product to the society from the given resources. The principle of equal marginal returns is the neoclassical economic criterion for efficiency in resource use and allocation in multi-product firms such as smallholder economy. It states that for a product firm to be said to have allocated its resources optimally (highest output per unit input), among its feasible production enterprises, it must do it in such a way that the marginal value product (MVP) of every variable input is equal in all enterprises in which it is employed and also equal to the price of the output. A stochastic linear programming (SLP) model is used to establish the optimal enterprise mix of smallholder farmers in Umbumbulu so that farmers can maximize their welfare subject to a set of constraints.

The model will account for fixed costs associated with machinery, information and family labour. Also to be included among the fixed costs are certification costs for organic farmers. For the EFO, group certification has enabled them reduce the average fixed costs associated with certification. It may also be important to note that for these farmers, this cost has been subsidized in one way or another by various stakeholders in both monetary and non-monetary terms. The study will seek to identify the fixed costs associated with certification and establish sensitivity of the results in the stochastic model to this cost. The EFO has acquired a tractor for which each farmer pays a "rental" fee, hence machinery fixed costs are not incurred but the farmer is faced with rental costs. An attempt will be made to account for the annual renting costs. The model will account for labour costs, the prices of which are predetermined and relatively constant in Rands per day, with a working day equivalent to six working hours. Authors Berry and Cline (1979) and Bhalla (1979) argue that family labour is cheaper than hired labour incorporated in the model in hours per day. Remittances may be used by the farm household towards farm production activities and will be accounted for in the model.

Empirical studies show an inverse relationship between farm size and efficiency, hence a tendency to disregard the fact that the adoption and use of any technology involved fixed transaction and information costs (Lyne, 1996). Fixed information costs are included in this study by modelling them in terms of opportunity costs of time or hours per week spent on getting information on farming. In contrast to costs being scale free, returns are perhaps best described as being proportional to scale (Welch, 1978). The amount of information acquired is positively related to farm size, and owing to the larger area that they command, the benefits of marginal increment in knowledge are higher for larger farms than for smaller farms (Feder and Slade, 1984). Size is therefore expected to have a positive effect on farmers' demand for specialized information, but a negative effect on farmers' demand for general information (Jones et al., 1989). The farmer's choice of marketing channel is also included in the model taking into consideration both direct and indirect marketing options (Mushayanyama and Darroch, 2006) that are available to farmers in Umbumbulu district. Direct markets include the direct sale to consumers including farmer markets. Indirect markets refer to the sale to retailers and wholesalers. Transaction cost figures from Gadzikwa et al. (2006) for the different farmer groups are incorporated to establish their link to certification and marketing channel option. The Arrow–Pratt risk aversion (APARA) coefficient will be calculated and included in the model as the tuning factor to establish how the crop combination changes with change in the farmers' risk attitude.

Analytical Framework

The linear programming model based on simple assumptions of additivity, linearity, divisibility, finiteness and single-valued expectation (Olayide and Heady, 1982) has remained the most popular mathematical programming technique (Anderson et al., 1977) and has received a wide application in farm planning problems (Jeffrey et al., 1992). The solution to the linear programming formulation gives the optimal operating conditions for the modelled farms and guarantees the best result under the conditions specified in the constraints, hence its use as the basis for the model used in this study. Stochastic linear programming presents a suitable method for economic evaluation of the farming system under risk since it is able to simultaneously incorporate the major factors determining the relative advantage of the crops. These factors cover the farmers' risk attitudes, net return fluctuations and constraints of both land and labour. Fluctuations of net returns are determined by many elements such as yield and product prices. The labour coefficients as well as yield reflect technological factors. Both farmers' risk attitudes and skill in operation are related to human factors. The criterion for stochastic linear programming adopted in this study is that of maximizing expected utility empirically written as

$$E[u(r'x)], \quad \text{subject to} \quad Ax \leq b, x \geq 0$$

where E is the expectation, $u(r'x) = 1\text{–}\exp(-ar'x)$ is the utility function employed by Freud et al. (1996), parameter $a(a \geq 0)$ is the risk aversion constant which may be considered as a measure of the aversion to risk, r is the vector of stochastic net returns and costs, A is the matrix of resource requirements, b is the vector of resource availability and x is the solution vector. The stochastic linear programming is formulated to maximize utility from a combination of five enterprises (*amadumbe*, potato (I), potato (E), beans and maize) subject to constraints arising from land, labour, tractor use and fixed and operating costs among each of the three farmer groups.

The expected utility model is often employed in empirical studies for modelling and simulating the economic behaviour of the farmers under risk. However, it is not easy to estimate directly the level of farmer's risk aversion constant since farmers do not recognize their utility functions. In this chapter the level of farmers' risk aversion is estimated using the Arrow–Pratt absolute risk aversion coefficient using the *EXRISK* software. The linear interactive discrete optimizer (LINDO) programme was used to run the model.

Study Area and Survey

The selected study area is in the rural Umbumbulu magisterial district of KwaZulu-Natal in South Africa. KwaZulu-Natal is the province in SA with the largest concentration of people who are relatively poor, and social indicators point to below-average levels of social development (Development Bank Southern Africa, DBSA, 1998). A survey was conducted during October–December 2004. Farm-level information was recorded for the previous growing season of 2003/2004. Socio-economic, demographic, institutional data and household characteristics and activities were recorded in a questionnaire through interview sessions with the principal decision maker in the participating EFO households and among the non-organic farmers. Questions included information resources, socio-economic and farm characteristic data such as products grown, existing crop combinations, institutional data such as marketing options, constraints and challenges, organic certification, farm management and demographics example household characteristics.

Transaction cost information from Gadzikwa et al. (2006) is used in the analysis. Farmers' risk attitudes were quantified using direct interviews based on hypothetical questions regarding risk alternatives, and the APARA coefficient was calculated for each farmer. Information collected will assist in the investigation of the heterogeneity of the preferences and risk attitudes by individual characteristics such as gender, age, education as well as farm characteristics such as farm structure and land size.

The survey was stratified into three groups: fully certified organic farmers, partially certified organic farmers and non-organic farmers. In all, 200 farmers were surveyed: 151 organic farmers who were members of EFO were census surveyed and interviewed, while another sample of 49 non-organic farmers who were not EFO members was randomly selected within the same region from a sample frame constructed from each of the five neighbouring wards. The EFO farmers consisted of 48 fully certified members and 103 partially certified farmers.

Results and Discussion

Table 1 presents the summary statistics of the variables for the three groups of farmers: fully certified organic, partially certified organic and non-organic farmers. These are the variables that are used in the model. Quantitative variables are expressed as mean averages, whereas the qualitative variables are expressed as dummy variables.

The proportion of arable land planted was smaller for certified organic farmers compared to the partially certified and non-organic farmers. This can be attributed to the intensive nature of certified organic farming, hence farmers opted to plant manageable areas based on the availability of labour and other input requirements. Similar results were found by Gadzikwa et al. (2006). In his study, he found that certified organic farmers had smaller farms, more family labour and earned a greater share of total farm income from farming, having the highest net benefit score. The

Table 1 Summary statistics and technical coefficients matrix

Variable	Fully certified (mean, $n = 48$)	Partially certified (mean, $n = 103$)	Non-organic (mean, $n = 49$)
Farm size (ha)	0.52	0.62	0.81
Enterprises (ha)	0.19	0.20	0.19
Amadumbe	0.08	0.09	0.10
Potato (English, E)	0.06	0.07	0.01
Potato (Irish, I)	0.03	0.05	–
Beans	–	–	0.07
Maize	–	–	3.46
Sugarcane	–	–	0.01
Bananas	–	–	0.01
Chillies	–	–	0.12
Peanut			
Farm labour (h/day)	38.73	22.94	14.49
Hired labour	337.60	65.49	43.47
Family labour	8.71	8.22	6.85
Family size			
Household income (R/yr)	2428.17	2214.95	2018.18
Net off-farm income	957.77	1007.75	332.48
Net farm income	3385.94	4580.67	2374.87
Total net income	0.34	0.22	0.14
Proportion of income from farming			
Machinery costs (R)	251.83	87.14	106.79
APARA coefficient	–9.28	–15.45	–24.37
Transaction costs	1.77	1.18	1.30
Age	52.4	49.73	51.82
Education	4.81	4.10	3.59
Gender	0.83	0.77	0.79

representative farm household in the study area across the three farmer groups grew *amadumbe*, potato (E), potato (I) and beans; however, non-organic farmers also grew a large proportion of sugarcane. Non-organic farmers grew both organic and non-organic crops on different plots of land.

Due to the intensive nature of certified organic farming, the hours of labour spent on farming was highest for the certified organic farmers. The average family size across all the groups, however, was large: over seven for all the three groups. Large family sizes are an indication of availability of labour and provide the opportunity for the farm to develop the technical know-how required for certified organic farming. The potential to meet peak labour demand also highlights the importance of the availability of family labour. The proportion of income from farming was highest for the certified organic farmers compared to the other groups. The average sample farmer in the study area was classified as risk averse. This is because the overall APARA coefficient was negative for the study farmers. However, certified organic farmers were less risk averse than were their counterparts. The average age of the respondents is generally high, averaging 50 years across the three groups. This result is consistent with previous studies in KwaZulu-Natal by Matungul (2001), who found the average age of household heads to be 60 years. Female farmers were predominant in the study area, with the male members working as migrant workers in other towns. About 88% of the certified organic farmers were female, supporting findings by Marcus et al. (1995), who found that most de facto heads were female. The education levels in the study area are low across all groups, with average years of schooling as 4 years. These results support findings in the study area by Gadzikwa et al. (2006), who found a similar result of low literacy levels among all farmer groups.

Characteristic of Crop Farming in Umbumbulu District

Mixed farming is a typical cropping pattern in the study area. A majority of the farmers plant English and Irish potatoes, *amadumbes* and maize. During the previous growing season of 2003/2004, 10 crops were cultivated (*amadumbe*s, English potatoes, Irish potatoes, green beans, dry beans, maize, sugarcane, bananas, chillies and peanuts). Table 2 shows popular combinations of crops in the study area. The three major combinations based on allocated hectares (ha) are *amadumbe*–potatoes (E)–potatoes (I) (5.76 ha), *amadumbe*–potatoes (E)–potatoes (I)–beans (4.24 ha) and *amadumbe*–potatoes (E)–beans (3.00 ha) for the fully certified farmers; *amadumbe*–potatoes (E)–potatoes (I) (16.2 ha), *amadumbe*–potatoes (E)–beans (7.77 ha) and *amadumbe*–potatoes (I) (4.33 ha) for the partially certified farmers; and potatoes (I)–dry beans–maize (6.00 ha), *amadumbe*–potatoes (I)–dry beans–maize (4.50 ha) and *amadumbe*–potatoes (E)–potatoes (I)–dry beans (2.09) for the non-organic farmers. Consequently the *amadumbe*-based combinations account for over 80% of the total plots. Thus *amadumbe*s play a significant role as a basal crop for various cropping systems. There is an already existing formal market for *amadumbes* and baby potatoes with the current arrangement that the farmers have with the park house agent at the upmarket grocery store Woolworths in Durban. This enables these smallholder farmers to tap into the high-end market and get premium price for their produce. On the other hand, the non-organic farmers who are more risk averse than the other farmer groups grew the safer option of long shelf life beans and maize.

Table 3 gives the optimal solutions for various levels of risk aversion constant of the utility model. The risk aversion coefficient is set at 0.00 (solution I), assuming that the decision maker is neutral. The optimal combination for the organic farmers comprises of *amadumbe*, English and Irish potatoes and beans with no option for maize, while the combination for the non-organic farmers includes maize, sugarcane and banana. Progressively, higher levels of risk are analysed. As the risk increases, the combination among the fully and partially certified farmers remains the same; however, the hectarage increases for *amadumbe* and Irish potatoes but decreases for the English potato. The model also brings out maize as an option, with the hectarage under maize and beans increasing with risk. This change in crop combination is brought about by higher variance of the net return for the English potato compared to that of the other crops.

The mean of total net return for the certified organic farmers decreases with higher levels of risk. This can

Table 2 Popular crop combinations and their hectarage allocation in Umbumbulu district, KwaZulu-Natal, SA, 2004

	Fully certified		Partially certified		Non-organic	
Crop combinations	Number of plots	Hectares	Number of plots	Hectares	Number of plots	Hectares
Amadumbe/potatoes (E)	2	0.24	8	3.42	–	–
Amadumbe/potatoes (E)/potatoes (I)	13	**5.76**	19	**16.2**	–	–
Amadumbe/potatoes (E)/ potatoes (I)/beans	4	**4.24**	4	1.84	–	–
Amadumbe (monocrop)	4	0.09	11	1.53	–	–
Amadumbe/potatoes (I)	6	0.67	14	**4.33**	–	–
Amadumbe/potatoes (E)/beans	3	**3.00**	8	**7.77**	–	–
Potato (I) (monocrop)	–	–	1	0.06	–	–
Potato (E)/potato (I)	–	–	3	0.11	–	–
Potato (E)/beans	–	–	1	0.04	–	–
Total	*32*	*14*	*69*	*35.3*		
Combinations for non-organic farmers						
Amadumbe/potatoes (I)/dry beans/maize	–	–	–	–	7	**4.50**
Amadumbe/potatoes (I)/maize	–	–	–	–	5	0.80
Amadumbe/dry beans/maize	–	–	–	–	5	1.70
Amadumbe/potatoes (E)/dry beans/maize	–	–	–	–	8	1.70
Amadumbe/potatoes (E)/potatoes (I)/maize/banana	–	–	–	–	5	0.70
Amadumbe/potatoes (E)/maize	–	–	–	–	4	1.52
Potatoes (I)/dry beans/maize	–	–	–	–	5	**6.00**
Amadumbe/potatoes (E)/potatoes (I)/dry beans	–	–	–	–	7	**2.09**
Total					46	19.01

Bold values represents the largest acreage under cultivation.

be attributed to the fact that a risk averter would rather trade higher levels of net return with lower levels of variance. The decrease in aspiration level with higher risk indicates the probability that real net returns for the certified organic farmers exceed the aspiration level.

Table 3 also demonstrates that with higher levels of risk, family labour among all farmer groups increases, while hired labour gets negligible because this constraint is loose and with migrant labour rampant in the study area, availability of hired labour is not a reliable option. Family labour accounts for over 80% of the labour used by the three farmer groups.

Table 4 represents the constraints faced by the different groups of farmers based on a Likert scale of 1–3 with 1 being no problem and 3 being a severe problem. While non-organic farmers assigned more weight to crop damage due to uncertain climate, livestock damage and lack of cash and credit to finance inputs, fully certified farmers highlighted constraints that had a direct bearing on their production of certified inputs. These were uncertain climate and tractor unavailability. The latter can be attributed to the fact that there is one tractor that has been allocated to the members of EFO at an annual rental fee, hence posing challenges during the land preparation phase when most of the farmers will be preparing their field and hence farmers have to queue for the tractor. Other constraints included delays in payment by the packhouse agent through which they have a contract to supply the retail store Woolworths.

Conclusion and Recommendations

This study aimed at (1) determining the optimal crop combinations of smallholder farmers that maximize household welfare, (2) determining the sensitivity of the crop combination with changes in risk preference and (3) analysing the constraints faced by these smallholder organic farmers. A total of 200 smallholder farmers from Umbumbulu district in KwaZulu-Natal were sampled for the study. These consisted of 48 fully certified farmers and 103 partially

Table 3 Summary of optimal solutions for selected levels of risk aversion constant

		I	II	III	IV	V	Transaction costs
Risk aversion constant	–	0	0.001	0.010	0.020	0.030	
Reliability constant	–	0.600	0.866	0.913	0.958	0.977	
Aspiration level	R 1,000	3385.94	2652.88	2535.81	2408.46	2341.17	
Mean of total net return	R 1,000	3385.94	3373.41	3308.64	3281.84	3213.91	
SD of total net return	R 1,000	1022.89	196.585	1000.21	870.12	795.44	
Optimal planting areas							
Fully certified farmers							1.77
(a) *Amadumbe*	Hectare	0.03139	0.01807	0.01872	0.02448	0.02870	
(b) Potatoes (E)	Hectare	0.02301	0.00301	0.00251	0.00202	0.00193	
(c) Potatoes (I)	Hectare	2.14290	1.79189	1.82001	1.85230	2.44541	
(d) Beans	Hectare	0.06017	0.06017	0.06017	0.07121	0.11830	
(e) Maize		0	0.41686	0.41686	0.51221	0.68526	
Optimal labour							
Family labour inputs	Hours/days	337.60	344.35	361.57	362.62	353.91	
Hired labour inputs	Hours/days	38.73	13.32	4.58	1.58	0	
Total	hr/days	376.33	357.67	376.15	364.20	353.91	
Partially certified farmers							1.18
(a) *Amadumbe*	Hectare	0.1929	0.1929	0.25225	0.32793	0.42630	
(b) Potatoes (E)	Hectare	0.7500	0.7500	0.62541	0.52097	0.43241	
(c) Potatoes (I)	Hectare	0.5730	0.5730	0.58199	0.59363	0.60550	
(d) Beans	Hectare	0.0399	0.0399	0.04720	0.08496	0.88130	
(e) Maize	Hectare	0	0	0	0.0730	0.10384	
Optimal labour							
Family labour inputs	Hours/days	65.49	66.80	70.14	70.34	70.54	
Hired labour inputs	Hours/days	22.94	7.89	7.71	0	0	
Total	Hours/days	88.43	74.69	77.85	70.34	70.54	
Non-organic farmers							1.3
(a) *Amadumbe*	Hectare	0.1950	0.1950	0.1950	0.1950	0.1950	
(b) Potatoes (E)	Hectare	0.1018	0.1018	0.1018	0.1018	0.1018	
(c) Potatoes (I)	Hectare	0.1120	0.1120	0.1120	0.1120	0.1120	
(d) Beans	Hectare	0.1021	0.1021	0.1021	0.1021	0.1021	
(e) Maize	Hectare	0.0704	0.0704	0.0704	0.0704	0.0704	
(f) Banana	Hectare	0.0118	0.0118	0.0118	0.0118	0.0118	
(g) Sugarcane	Hectare	3.4625	3.4625	3.4625	3.4625	3.4625	
Optimal labour							
Family labour inputs	Hours/days	43.47	44.33	46.55	46.69	46.82	
Hired labour inputs	Hours/days	14.49	4.98	1.71	0	0	
Total	Hours/days	57.96	49.31	48.26	46.69	46.82	

Source: Gadzikwa et al. (2006)
Fully certified farmers incur the highest transaction costs. Interestingly the partially certified farmers incur the lowest transaction costs due to the fact that they have the two options of marketing through the formal channels and also the informal route, hence distributing their risks across the two channels

certified farmers who constituted the 151 members of the EFO and were purposefully selected for the study and 49 non-organic farmers selected randomly from a list comprising of farmers drawn from five of the neighbouring wards. The stochastic linear programming model was used to analyse the data using the LINDO programme. Results obtained showed mixed cropping as the optimal crop combinations among the fully certified, partially certified and non-organic farmers, albeit with different hectarages. Family labour costs increased with higher risk levels, while hired labour decreased. Where a change in the risk tuning factor gave different hectarage combinations for the organic farmers, the enterprise combination remained the same with higher risk for the non-organic farmers. Uncertain weather was highlighted as a constraint

Table 4 Constraints faced by farmers

Constraints	Fully certified	Partially certified	Non-organic
Uncertain climate	**2.98**	**2.83**	**2.80**
Tractor not available when needed	**2.92**	**2.75**	2.46
Delays in payment by packhouse	**2.88**	2.30	–
Inputs not available at affordable prices	**2.83**	2.52	2.51
Lack of cash and credits to finance inputs	2.77	2.60	**2.78**
Lack of affordable transport for produce	2.74	2.43	2.06
More work than the household can handle	2.63	2.31	**2.53**
Cannot find manure	2.63	1.94	–
Lack of proper storage facilities	2.56	2.47	–
Lack of telephones to negotiate sales	2.54	**2.68**	2.22
Livestock damage	2.51	**2.82**	**2.80**

Bold values represents the four main constraints experienced by farmers

by all farmer groups. This perhaps explains the reason for the crop diversification among the farmer groups. A diversified crop combination remains the best combination option for these smallholder farmers as it may cushion farmers against risk associated with weather uncertainty and other socio-economic conditions like market uncertainty, family subsistence requirements, marketing and storage needs. Adequate information needs to be made available to the farmers in order for them to make informed decisions and also effectively allocate resources among their competing needs. Effective extension programmes and smallholders' development programmes initiated by the government and/or other stakeholders will enable the farmers improve production.

References

Anderson JR, Dillon JL, Hardaker JB (1997) Agricultural decision analysis. Iowa State University Press, Ames, IA

Baumol WJ (1977) Economic theory and operation analysis, 4th edn. Prentice Hall International, London

Berry RA, Cline WR (1979) Agrarian structure and productivity in developing countries. The John Hopkins University Press, Baltimore, MD

Bhalla SS (1979) Farm size, productivity and technological change in Indian agriculture. In: Berry RA, Cline WR (eds) Agrarian structure and productivity in developing countries. The John Hopkins University Press, Baltimore, MD

DBSA (1998) KwaZulu-Natal development profile, development information business unit. Development paper 133

Feder G, Slade R(1984) The acquisition of information and the adoption of new technology. Am J Agric Econ 66:312–320

Freud EH, Phillipe P, Jacques R (1996) Innovations in West African smallholder cocoa: some conventional and non-conventional measures of success. In: Nur EM, Mahran HA (eds) Food security and innovations: successes and lessons learned. International symposium, Hohenheim, Germany, pp 131–146

Gadzikwa L, Lyne MC, Hendriks SL(2006) Collective action in small holder organic farming: a case study of the Ezemvelo farmers organisation in KwaZulu-Natal. S Afr J Econ 74:2

Heady EO, Candler W (1969) Linear programming methods. Iowa State University Press, Ames, IA

Jeffrey SR, Gibson RR, Famiwow WD(1992) Nearly optimal linear programming as a guide to agricultural planning. Agric Econ 8:1–19

Jones E, Batte MT, Schnitkey GD(1989) The impact of economic and socio-economic factors on the demand for information: a case study of Ohio commercial farmers. Agribusiness 5:557–571

Lwayo MK, Darroch MAG, Ferrer SRD (2007) A discriminant analysis of factors associated with the adoption of certified organic farming by smallholder farmers in KwaZulu-Natal, South Africa. Paper to be presented at the second annual conference of the African Association of Agricultural Economists, to be held in Accra Ghana, La Palm Beach Hotel, Accra, 20–22 Aug 2007

Lyne MC (1996) Transforming developing agriculture. Establishing a basis for growth. Agrekon 35(40):272–392

Marcus T, Mac Donald C, Maharaj P, Manicon D, Phewa R (1995) An analysis of land availability, social demographics, stakeholder and institutional arrangements, School of rural community development, University of natal, Pietermaritzburg, South Africa

Matungul MP (2001) Transaction costs and crop marketing in the communal areas of Impendle and Swayimana, KwaZulu-Natal. Dev S Afr 18(3):347–363

Mushayanyama T, Darroch MAG (2006) Smallholder farmers' perception of factors that constrain the competitiveness of a formal organic crop supply chain and marketing margins in KwaZulu-Natal. Master thesis, Agricultural economics discipline, University of KwaZu lu-Natal, 119 pp

Olayide SO, Heady EO (1982) Introduction to agricultural economics. Ibadan University Press, Ibadan, Nigeria

Welch F (1978) The role of investment in human capital in agriculture. In: Schultz TW (ed) Distortions of agricultural incentives. Indiana University Press, Bloomington, IN, 259 pp

Scaling Out Integrated Soil Nutrient and Water Management Technologies Through Farmer Participatory Research: Experiences from Semi-arid Central Zimbabwe

I. Nyagumbo, J. Nyamangara, and J. Rurinda

Abstract Key attributes driving the adoption of nutrient and water management technologies, farmer experimentation and relationships between innovativeness and resource endowment were explored in semi-arid Shurugwi district of Zimbabwe. The approach used participatory techniques from 2005 and field testing of new options by innovators in 2006/2007 season. A formal survey of 46 respondents involving 30 innovators and 16 non-innovators was conducted in June 2007. By 2007, various technologies of water management and soil fertility improvement were being tested by farmers. Innovator farmers tested significantly ($p < 0.05$) more technology options than did their counterparts. Resource-endowed farmers also innovated more compared to resource-constrained farmers particularly with regard to water management options ($p < 0.05$). Organic fertility resource allocations were found to be skewed in favour of homestead fields, while options with inorganic resources were used significantly ($p < 0.05$) more often in far-fields, a tendency attributed to transport and labour bottlenecks in ferrying organic resources over large distances. On the other hand, water management innovations, e.g. infiltration pits and potholing, were being applied significantly ($p < 0.001$) more to far-fields compared to homestead fields. The disparity between homestead and far-fields was considered a major bottleneck to integrated use of water and fertility management options, which results in loss of potentially realizable yields. Activities in the last two seasons generally had more impacts on innovations related to water than to nutrient management technologies. More emphasis on integration of fertility and water management options in further scaling out activities and stronger linkages with both input and output markets is recommended.

Keywords Fertility management · Innovators · Resource endowment · Scaling out · Water management

Introduction

Crop productivity in semi-arid smallholder farming areas of Zimbabwe is limited by two main factors, nutrients and water. Recurrent droughts and erratic rainfall patterns, particularly in the semi-arid zones, often result in reduced yields or complete crop failure (Barrow, 1987; Bratton, 1987). Efforts to address some of these issues in Ward 5 of Shurugwi district date as far back as the 1990s with development efforts introduced through the Smallholder Dry Areas Resource Management Programme (SDARMP), a development-oriented programme which operated in the area up to 1998.

Between 1998 and 2002, TSBF-CIAT initiated soil fertility management research programmes that sought to improve livestock manure utilization through improved handling methods (Nzuma and Murwira, 2000), legume intercrops and green manuring technologies. Further initiatives between 2002 and 2005 came through collaborative work between the Department of Soil Science and Agricultural

I. Nyagumbo (✉)
Department of Soil Science and Agricultural Engineering, University of Zimbabwe, Mount Pleasant, Harare, Zimbabwe
e-mail: inyagumbo@agric.uz.ac.zw

A. Bationo et al. (eds.), *Innovations as Key to the Green Revolution in Africa*,
DOI 10.1007/978-90-481-2543-2_128, © Springer Science+Business Media B.V. 2011

Engineering, the University of Zimbabwe and TSBF-CIAT. These initiatives attempted to address the constraints through the use of integrated soil nutrient and water management technologies. This mainly involved the use of pit-stored livestock manure (Nzuma and Murwira, 2000) banded in crop rows combined with various water-conserving tillage options such as post-emergence, tied ridging, and rip and potholing systems (Nyagumbo, 2002). These initiatives were later expanded in 2005 through funding from the Soil Fertility Consortium of Southern Africa (SOFECSA). This initiative sought to scale out identified soil nutrient and water management options through participatory technology development work involving field demonstrations and self-experimentation by farmers. This chapter presents preliminary findings from these experiences since 2005 with respect to scaling out of the various nutrient and water management options and seeks to draw lessons from these experiences. The objectives are to establish the extent to which farmers are experimenting with various soil nutrient and water management innovations since 2005, identify key factors driving the testing and adoption of innovations, and determine the relationship between technology innovativeness and farmer resource endowment.

Materials and Methods

The approach involved use of various participatory technology development tools such as participatory extension approaches (PEAs) as illustrated in Fig. 1 (Hagmann, 1998) and farmer field school concepts. To start the process, the PEA was used for problem analysis and identification of the most feasible options for both nutrient and water management, recognizing constraints faced by farmers in accessing fertility amelioration resources. Following a stakeholder workshop, key fertility management issues (access and affordability of inorganic fertilizers, risk, etc.) were analysed and options for further testing were identified and implemented on five farmer-selected demonstration sites across the ward with one in each village.

A mid-season field tour was later carried out in March 2006 during which over 200 farmers participated in touring the five demonstration sites as well as other initiatives in the ward. In September 2006 a workshop was carried out again to review the previous season and to replan activities for the following season. Farmers identified new sites on which they wanted new demonstrations to be set up. Farmers were also asked to volunteer and try out any of the promising options

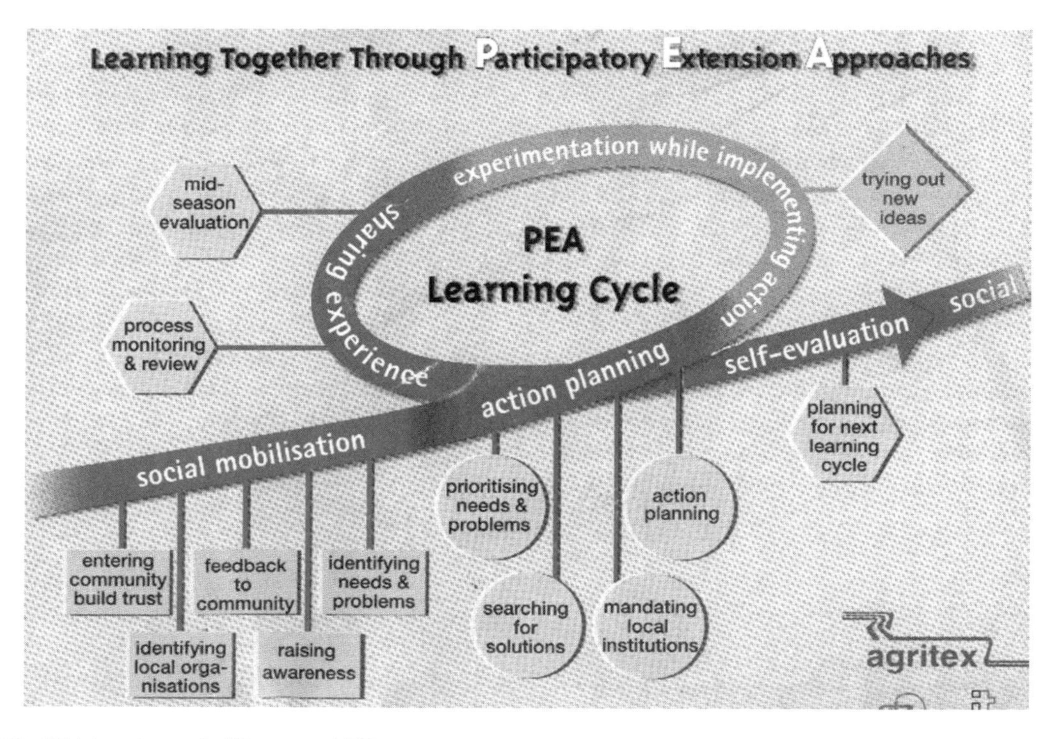

Fig. 1 The PEA learning cycle (Hagmann, 1998)

of their choice but would not receive any inputs for that. Another mid-season field tour was carried out in 2007 to assess performance of the demonstrations and innovations.

In June 2007 a technology adoption survey in which 30 innovator farmers who volunteered to test various innovations were interviewed together with 16 neighbouring non-innovators using a formal questionnaire. Key issues addressed by the questionnaire involved household sizes, full-time labour, land and livestock ownership, fertility and water management options in use and associated constraints. Respondent households were classified into resource categories (RGs) based on cattle ownership, with farmers owning more than 4 cattle being categorized as RG1, between 1 and 4 cattle as RG2 and no cattle as RG3. Innovativeness of each respondent was based on a dimensionless index computed from a reciprocal of the sum of fertility or water management options tried by each respondent where the value of 1 indicated a *yes* to that option and 2 denoted a *no*. This reciprocal was multiplied by 100 to give an innovativeness index whose value increased as the number of tried technology options of each respondent increased. Results from the survey were processed and analysed electronically and Pearson's χ^2 test was used to compare allocation of fertility resources or water management technologies between homesteads and far-fields.

Results

Performance of Nutrient and Water Management Technologies

Findings from year 1 (2005/2006) clearly demonstrated the benefits of water management techniques in the form of post-emergence tied ridging superimposed on various fertility management options (Fig. 2). Thus, all fertility ameliorants induced greater maize yield than did the control. Moreover, manure and fertilizer combinations gave the highest yields on all the five demonstrations as also reported in a sister paper (Nyamangara and Nyagumbo, 2010 unpublished). Thus over 30 farmers volunteered to try out various options on their own by the beginning of year 2 (2006/2007).

Use of Nutrient and Water Management Technologies

Findings from the survey showed that intercropping, compost, rotations, livestock manure and anthill materials were being used by over 70% of the respondents, while the use of inorganics was rather low with only

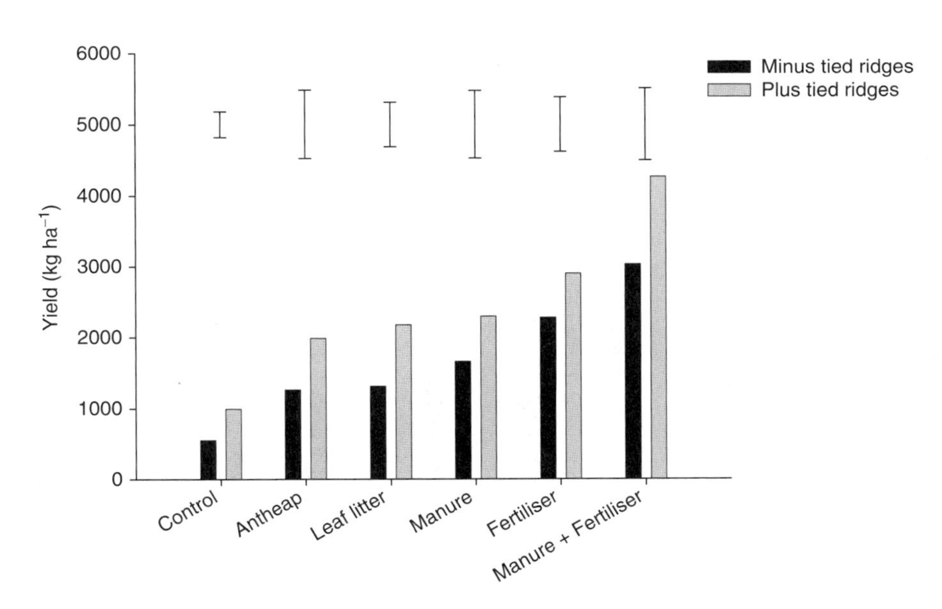

Fig. 2 Effect of different nutrient resources and tied ridging on mean maize grain yields from five demonstration sites in 2005/2006, Ward 5, Shurugwi district, Zimbabwe. *Source:* Nyamangara and Nyagumbo (unpublished)

50% using basal fertilizers (Fig. 3). With respect to water management, the oldest technique of contour ridging was used by 100% of the respondents, while infiltration pits were used by as much as 60% of the respondents (Fig. 4). Relatively new techniques such as tied ridging, fanya juus, deepened contours and potholing were being used by less than 30% of the respondents (Fig. 4). Zero-till and conservation farming basin techniques had not been introduced to the farmers and hence recorded nil in Fig. 4. On the other hand, the rip and pothole combination which also recorded nil was being tested by farmers either as ripping only or as potholing only. By June 2007, a total of 43 farmers were on the extension worker's register of users of water management techniques (Table 7).

Innovators also used significantly ($p < 0.05$) more fertility and water management options than did non-innovators based on a t-test for paired observations on each of the 11 possible fertility options and 10 water management options (Table 1). However, there were no significant differences in land ownership, household sizes and full-time labour between innovators and non-innovators (Table 2). On average, arable land size was 4.42 ha per household. In addition, the average household size was 5.89 with only 2.67 full-time labour persons per household (Table 2). This finding agrees with results from other surveys in Zimbabwe (CSO, 1992; Nyagumbo, 1992; MPSLSW, 1996).

Effects of Resource Endowment on Innovativeness

Based on livestock ownership (cattle, donkeys and goats) per household, there were no significant linear relationships between livestock ownership and innovativeness on fertility management technologies. Significant linear relationships ($p=0.0097$) were, however, found between livestock ownership and innovativeness to water management. In both cases, increased livestock ownership generally positively correlated with innovativeness, i.e. resource-endowed farmers tended to innovate more (Fig. 5a, b).

In terms of resource allocation, the survey showed contrasting allocations of fertility management resources to either homestead fields or far-fields. Organic resources such as livestock manure, pit-stored manure and leaf litter materials were being applied significantly ($p < 0.05$) more to homestead fields than to far-fields, while inorganic fertilizers were being applied significantly ($p < 0.05$) more in far-fields (Fig. 6). There were no significant differences between homesteads and far-fields in the allocation of anthill materials (Fig. 6). This tendency was attributed to transport and labour associated with moving organic materials and high risk of failure associated with inorganic resources. However, overall 23.35% of the

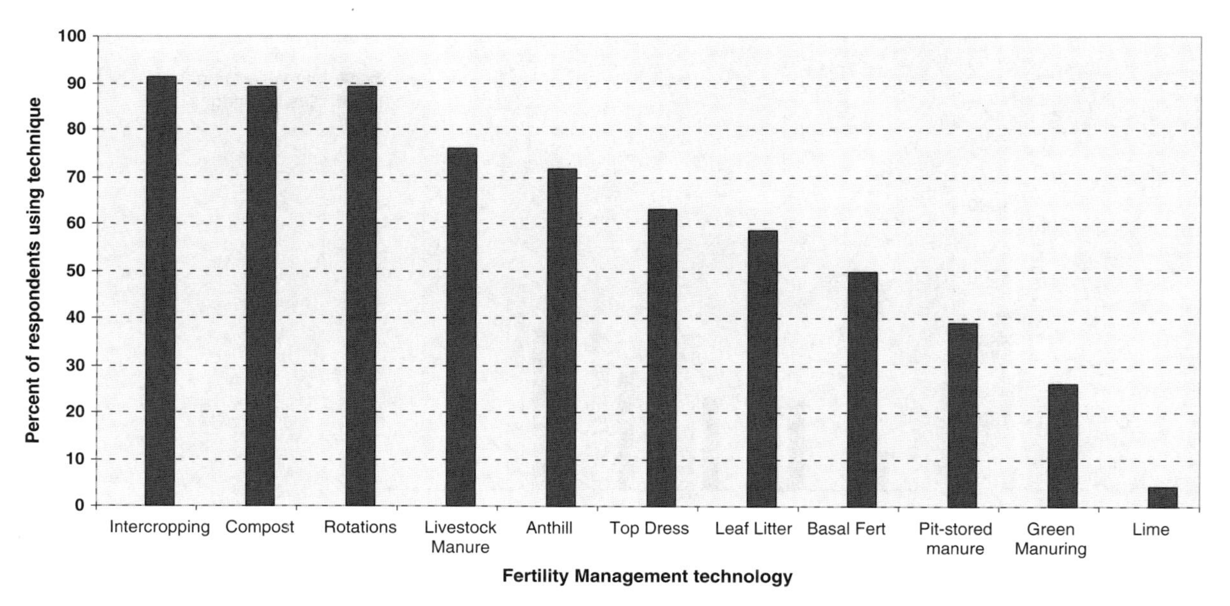

Fig. 3 Use of various soil fertility amelioration options by respondents in Ward 5, Shurugwi district, Zimbabwe, in June 2007 ($N=46$)

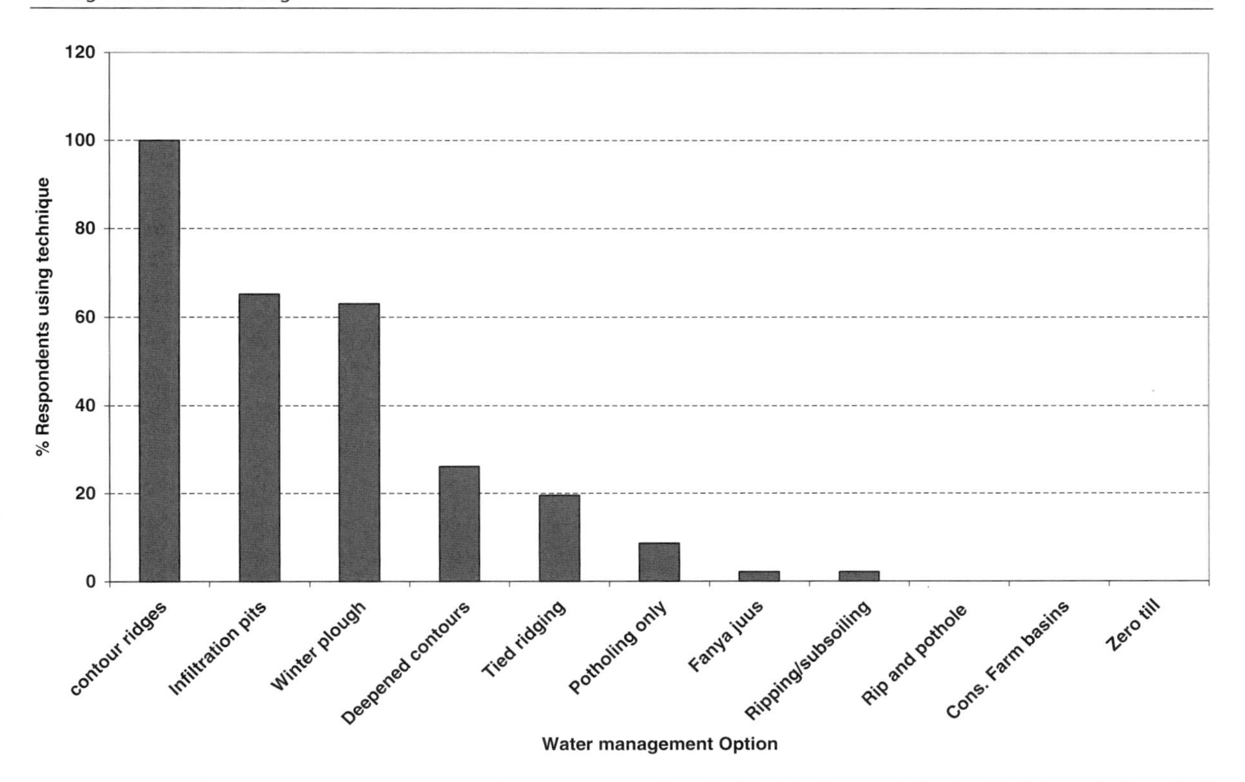

Fig. 4 Use of various water management options by survey respondents in Ward 5, Shurugwi district, Zimbabwe, in June 2007 (N=46)

Table 1 Comparison of fertility and water management options used by innovator and non-innovator respondents in Ward 5, Shurugwi district, Zimbabwe, in June 2007

	Option type	
Respondent category	Fertility management	Water management
Percentage use by innovator farmers	67.6[a**]	36.3[c*]
Percentage use by non-innovator farmers	45.5[b**]	14.4[d*]
Overall % use by all respondents	59.9	28.7
Statistics[1]	*N=11 paired observations, d.f.=10, p=0.003*	*N=10 paired observations, d.f.=9, p=0.04*

[1]Statistics: t-test for paired observations with innovator and non-innovator groups and each pair based on 11 fertility management options (Fig. 3) and 10 water management options (Fig. 4). Means in the same column followed by different letters are significantly different at * $p < 0.05$ and ** $p < 0.01$

Table 2 Land ownership, household size and full-time labour per household in Ward 5, Shurugwi district, Zimbabwe, in June 2007

Category	Land size per household (ha)	Household size (persons)	Full-time labour per household
Innovators (N=30)	4.42	5.50	2.60
Non-innovators (N=16)	4.44	6.63	2.81
Overall means	4.42	5.89	2.67
Statistics	*n.s.*	*n.s.*	*n.s.*

n.s., no significant differences between innovators and non-innovators based on an independent sample t-test for comparison of means

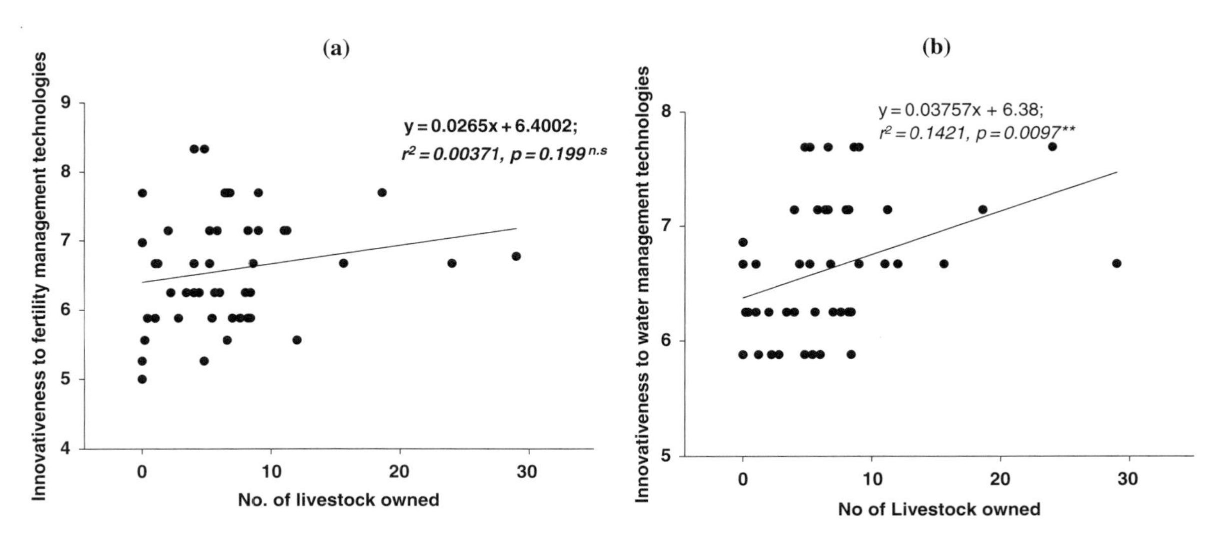

Fig. 5 Effects of resource endowment (based on livestock ownership) on innovativeness to (**a**) *fertility management* and (**b**) *water management* technologies

Fig. 6 Allocation of different soil fertility resources between homesteads and far-fields in Ward 5, Shurugwi district, Zimbabwe, in June 2007. For each fertility resource, $N =$ --; **, ***, n.s. denotes the total number of respondents using that resource and that differences between homestead and far-fields are significant at $p < 0.01$, $p < 0.001$ or not significant at $p > 0.05$, respectively, based on a χ^2 test. Thus bars in each fertility resource followed by the same alphabetical letters are not significantly different at $p < 0.05$

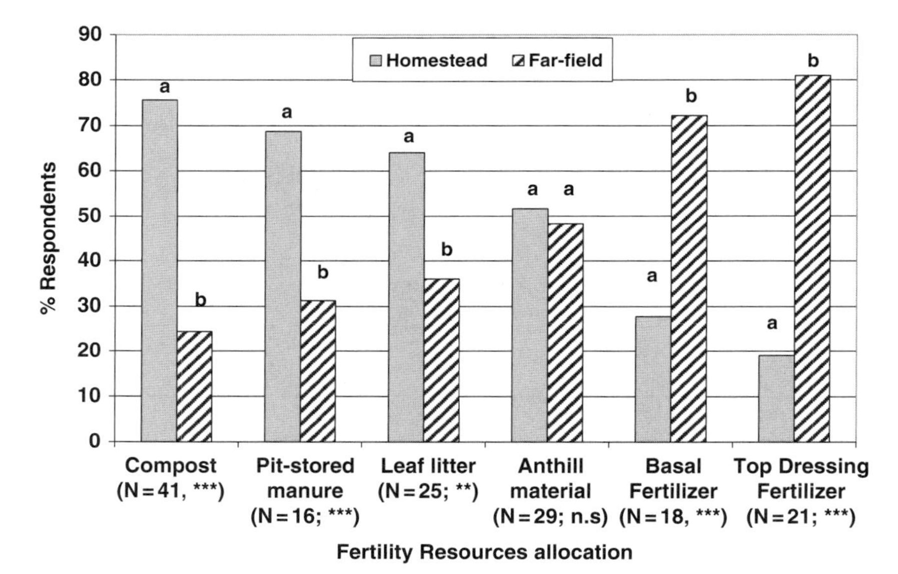

farmers allocated fertility management resources to homestead fields compared to 24.63% in far-fields. On the other hand, water management options were allocated more to far-fields (22%) than to homestead fields (10%). Hence, this resulted in lack of integrated/harmonized use of water and fertility management options in fields. Far-fields had more water management options, while soil fertility amelioration (considering both organic and inorganic) options were spread across resulting in loss of potentially realizable yields (Fig. 7).

Factors Impacting on Uptake of Nutrient Management Innovations by Smallholder Farmers

A wide range of constraints were identified as limiting farmers' capacity to scale out fertility management technologies (Table 3). Constraints to locally sourced organic materials related mainly to their scarcity, lack of livestock, labour and subsequent weed control problems when these are used. On the other hand, the main

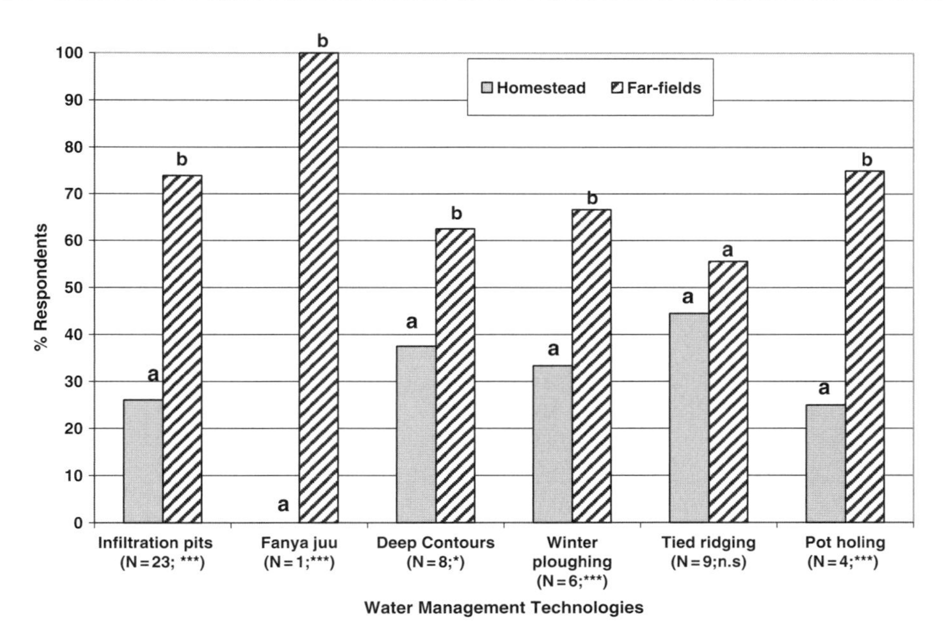

Fig. 7 Distribution of different water management technologies between homesteads and far-fields in Ward 5, Shurugwi district, Zimbabwe, in June 2007. For each technology, N = --; *, ***, n.s. denotes the total number of respondents using that technology and that differences between homestead and far-fields are significant at $p < 0.05$, $p < 0.01$, $p < 0.001$ or not significant at $p > 0.05$, respectively, based on a χ^2 test. Thus bars for each technology followed by the same alphabetical letters are not significantly different at $p < 0.05$

Table 3 Constraints to uptake of nutrient management innovations by smallholder farmers in Shurugwi, Zimbabwe

Constraints	Livestock manure	Compost	Pit-stored manure	Anthill materials	Basal fertilizer	Top dressing fertilizer, e.g. AN	Lime	Green manuring
Lack of livestock	71.7	30.4	26.1	19.6	2.2	2.2	0.0	0.0
Insufficient materials	54.3	47.8	15.2	26.1	0.0	0.0	0.0	21.7
Inadequate labour	41.3	47.8	21.7	37.0	0.0	0.0	0.0	6.5
Weed problem	39.1	34.8	2.2	6.5	4.3	4.3	4.3	0.0
High costs	10.9	8.7	2.2	8.7	60.9	69.6	10.9	0.0
Lack of free inputs	8.7	2.2	2.2	4.3	13.0	13.0	6.5	0.0
Poor yields	8.7	15.2	2.2	6.5	0.0	0.0	0.0	0.0
Availability of fertilizers	6.5	2.2	2.2	6.5	60.9	69.6	10.9	0.0
Discouragement	6.5	8.7	2.2	4.3	0.0	0.0	0.0	0.0
Inadequate training	6.5	6.5	2.2	2.2	0.0	0.0	2.2	0.0
Fear of victimization	4.3	6.5	0.0	4.3	0.0	0.0	0.0	0.0
Lack of confidence	0.0	6.5	0.0	2.2	0.0	0.0	2.2	0.0

bottleneck to the use of inorganic fertilizers related mainly to high cost and unavailability indicated by 60.9 and 69.6% for basal and top dressing fertilizers, respectively (Table 3). In addition, 21.7% of the respondents highlighted insufficient materials such as lack of seed as the major constraint to adopt green manuring technologies.

With regard to water management technologies, major constraints highlighted related to high labour demands associated with mechanical conservation structures, high cost of tillage equipment and its unavailability (Table 4). For example, labour was cited as the major constraint to contour ridges reinforced with infiltration pits and fanya juus by 47.8 and 15.2% of the respondents, respectively, while unavailability of equipment featured as the main constraint to tied ridging, ripping and making potholes. As shown in Table 2, there are

Table 4 Constraints to uptake of water management innovations by smallholder farmers

Constraints	Contour ridges with infiltration pits	Fanya juus	Deepened contours	Winter ploughing	Post-emergence tied ridging	Ripping/ subsoiling	Rip and potholing	Potholing only
High labour demands	47.8	15.2	19.6	30.4	6.5	6.5	2.2	0.0
Lack of equipment	15.2	2.2	2.2	28.3	6.5	13.0	10.9	2.2
High costs of equipment	13.0	0.0	2.2	26.1	2.2	6.5	2.2	0.0
Weed problem	4.3	0.0	2.2	2.2	0.0	0.0	0.0	0.0
High draft requirements	4.3	0.0	2.2	13.0	0.0	0.0	0.0	0.0
Discouragement	4.3	0.0	2.2	6.5	0.0	0.0	0.0	0.0
Inadequate training	4.3	6.5	8.7	2.2	4.3	2.2	2.2	2.2
Lack of yield incentive	2.2	0.0	2.2	8.7	0.0	0.0	2.2	0.0
Lack of free inputs	0.0	0.0	0.0	0.0	0.0	0.0	0.0	0.0
Poor yields	0.0	0.0	0.0	2.2	0.0	0.0	0.0	0.0
Fear of victimization	0.0	0.0	0.0	0.0	0.0	0.0	0.0	0.0
Lack of confidence	0.0	6.5	4.3	4.3	4.3	2.2	2.2	2.2

on average only 2.67 full-time labour persons per household.

Impacts of SOFECSA on the Scaling Out of Innovations

The survey results also showed that the period when various technologies were first used by the farmers synchronized with the emphasis placed on those technologies during the last 10 years by various extension or research institutions (Tables 5 and 6). Despite a wide range of soil fertility management techniques, farmers in recent years (1999–2002) have increasingly paid attention to the use of intercropping, pit-stored manuring techniques and green manuring which were promoted through TSBF-CIAT–UZ initiatives. Between 2005 and 2007, an increase in the use of leaf litter and anthill materials was noted, a period during which SOFECSA became more active in the area. There was also an apparent increase in the use of water management options since 2003, e.g. infiltration pits, tied ridges and potholing techniques which were promoted through TSBF–UZ and SOFECSA initiatives. Techniques such as tied ridging, potholing and ripping essentially started being used by the farmers from 2005 when the current initiatives took centre stage. Physical ground checks by the local extension worker in 2007 estimated the combined total number of farmers using pit-stored manure, leaf litter and anthill materials at 145, while the total registered number of innovators using infiltration pits, post-emergence tied ridges, deep contours and potholing was 43 (Table 7). The total number of farmers testing various innovations in Ward 5 by June 2007 was thus estimated at 188, although no data were available from the other years for a trend comparison.

Discussion

The study shows that a wide range of fertility management techniques have been tried by farmers as a means of alleviating nutrient deficiencies in soils. It became apparent that farmers were now testing and using a wide range of soil fertility and water management innovations. Although all the farmers were using standard contour ridges in their fields, their relevance in semi-arid conditions has been questioned (Hagmann and Murwira, 1996) as they are pegged to dispose of excess water rather than retain it (Elwell, 1981). Organic fertility resource (livestock manure, compost and leaf litter) allocations were found to be skewed in favour of homesteads, while inorganic resources (basal and top dressing fertilizers) were used more often in far-fields, a tendency attributed to transport and labour bottlenecks in ferrying organics over large distances. This confirms findings in other studies, which have shown steep fertility gradients between far and homestead fields, the latter showing higher fertility status due to more organic resource additions annually (Mutambanengwe, 2006; Zingore, 2006). Lack of significant differences in allocation of anthill materials

Table 5 Institutional contributions to uptake of various soil fertility management technologies in Ward 5, Shurugwi district, Zimbabwe, between 1998 and 2007

Fertility management option	Period during which majority of respondents started using technique	Magnitude of majority (% of respondents)	Institutions active in ward during period	Remarks
1. Livestock manure	Before 1998	60.9	Agricultural Research and Extension Services (AREX) SDARMP	A traditional soil fertility ameliorant promoted since the 1960s (Grant, 1967).
2. Compost	Before 1998	56.5	AREX	A traditional soil fertility ameliorant promoted since the 1960s but revived in recent years due to increased costs of inorganics (Mapfumo and Giller, 2001)
3. Pit-stored manure	1998–2002	34.8	TSBF-CIAT, Africare	A technology developed and promoted mainly through TSBF-CIAT initiatives (Nzuma and Murwira, 2000) and later by UZ
4. Leaf litter	2006–2007	17.4	SOFECSA	An indigenous soil fertility management technique revived due to high cost and poor access to mineral fertilizers since 2000
5. Anthill soil	Before 1998	32.6	AREX	A widely practiced indigenous technology for soil fertility amelioration
6. Basal fertilizer	Before 1998	39.1	AREX	An old extension recommendation
7. Top dressing	Before 1998	54.3	AREX	An old recommendation
8. Intercropping	1998–2002	69.6	TSBF-CIAT, Africare	Increased drive on legume intercrops by TSBF-CIAT and Africare
9. Green manuring	1998–2002	26.1	TSBF-CIAT, Africare	Increased drive on legume green manures by TSBF-CIAT and Africare

Notes: AREX, Agricultural Research and Extension Services; SDARMP, Smallholder Dry Areas Resource Management Programme; TSBF-CIAT, Tropical Soil Biology and Fertility Institute of International Center for Tropical Agriculture; SOFECSA, Soil Fertility Consortium of Southern Africa; UZ, University of Zimbabwe, Department of Soil Science and Agricultural Engineering

between homesteads and far-fields was attributed to their availability across both types of cropping fields. Thus, their use depended more on availability and spatial distribution than on preferential usage, although their bulkiness could still not ease their transportation.

In contrast, the results also suggested that water management options were more often tried in far-fields than in homestead fields. This could be due to the fact that some of them, e.g. infiltration pits, involve considerable structural changes in landforms, which may alter the outlook of the homestead if applied there or could be due to farmers' perceptions of the associated costs and benefits. Unfortunately it could not be ascertained within the scope of the study why this was so but a clear outcome of this finding is that there is disparity between fields allocated to water management options and those allocated to fertility management options. This effectively means farmers applying these technologies in different locations fail to fully capitalize and benefit from the integrated effects of both fertility and water management technologies (Bationo and Mokwunye, 1991).

The study also found that in terms of innovations, well-endowed farmers tended to innovate more than their resource-constrained counterparts particularly with regard to water management technologies. Most of the water management technologies tend to demand more labour, draft power and equipment which the resource-constrained farmers could not easily access, hence these differences. As a result, none of the respondents had tried the rip and potholing technique due to lack of equipment. This suggests that as expected, adoption of various innovations depended on farmer resource endowments and hence the need to carefully target technologies based on resource demands. This would help to define options for the

Table 6 Institutional contributions to uptake of various water management technologies in Ward 5, Shurugwi district, Zimbabwe, between 1998 and 2007

Water management option	Period during which majority of respondents started using technique	Magnitude of majority (% of respondents)	Institutions active in ward during period	Remarks
1. Contour ridges	Before 1998	95.7	Arex, SDARMP	An old extension recommendation constrained mainly by labour and draft power problems
2. Winter ploughing	Before 1998	56.5	AREX	An old extension recommendation constrained by draft power shortages
3. Infiltration pits	2003–2005	39.1	UZ-SSAE, TSBF-CIAT,	A recent drive towards water harvesting by NGO Africare and UZ research initiatives
4. Fanya juus	1998–2002	2.2	TSBF-CIAT, Africare	A recent drive towards water harvesting by NGO Africare and UZ research initiatives
5. Deepened contours	1998–2002	10.9	Africare TSBF-CIAT UZ-SSAE	A recent drive towards water harvesting by NGO Africare and UZ research initiatives
6. Post-emergence tied ridging	2006–2007	15.2	SOFECSA	Promoted widely since 2005
7. Ripping/subsoiling	2003–2005	2.2	UZ-SSAE SOFECSA	Introduced as a water conservation method through breaking plough pans
8. Potholing	2006–2007	6.5	SOFECSA	Promoted by SOFECSA to enhance water infiltration between crop rows since 2005

Notes: AREX, Agricultural Research and Extension Services; SDARMP, Smallholder Dry Areas Resource Management Programme; TSBF-CIAT, Tropical Soil Biology and Fertility Institute of International Center for Tropical Agriculture; SOFECSA, Soil Fertility Consortium of Southern Africa; UZ-SSAE, University of Zimbabwe, Department of Soil Science and Agricultural Engineering

Table 7 Registered number of farmers using various water management technologies in Ward 5, Shurugwi district, Zimbabwe, by June 2007

Water management technology	Number of farmers using technique
Infiltration pits	15
Post-emergence tied ridging	16
Deepened contours	3
Potholing	9
Total	43

different farmer resource categories but could also mean well-resourced farmers have a higher capacity to risk and experiment with new technologies.

In terms of dissemination impacts in the ward, SOFECSA activities mainly contributed towards increased uptake of water management technologies such as infiltration pits, potholing and tied ridging. The increased use of such techniques in 2006/2007 may have been driven by the long dry spells which induced severe water stress to most crops in the ward, hence farmers tried to address a felt need. SOFECSA activities since 2005 also revived the use of anthill and leaf litter options for fertility management. This revived use of the latter materials was attributed to two factors including the realization from field 2005/2006 demonstrations that leaf litter could yield better than the unfertilized controls and the increased scarcity and high cost of inorganic fertilizers. Such tendencies to use locally available fertility ameliorants have also been observed in other parts of the country (Chuma et al., 2000, 2001; Mapfumo and Giller, 2001).

Overall the study showed that farmers will try out new technologies that fit within their resource endowment as long as the potential of such technologies is demonstrated to them. Prevailing socio-economic circumstances also determined the use of the technologies by farmers at any particular time. For example, while farmers had started using inorganic fertilizers well before the 1990s, their current use had declined considerably due to both their high cost and unavailability. Thus, in terms of scaling out, the study showed that the prevailing socio-economic environment was brought about through the use of participatory dissemination approaches in the last 3 years which empowered farmers and increased farmers' willingness to try out new techniques through improved sharing of ideas, competitiveness and increased self-confidence. Although such dissemination tools have initially high set-up costs, they have been found to improve efficiency in technology development and dissemination among farmers in the long run (Rusike and Twomlow, 2006). However, major bottlenecks to increased dissemination related to poor access to input and output markets, e.g. seed, fertilizer and equipment suppliers. Further work in the ward should thus focus on fostering stronger linkages between the farmers and these input/output markets so as to facilitate increased productivity and economic benefits.

Conclusions

The approach used generally saw dramatic increases in the use of various nutrient and water management options. Generally innovators have tried out more soil nutrient and water management options than did their counterparts. Resource-endowed farmers were generally more innovative than their resource-constrained counterparts in terms of testing new technologies. Innovation among resource-endowed farmers related mainly to their ability to risk and diversify investments in new technologies which the poor-resourced farmers fail to afford. The results also showed that there is a disparity in fertility and water management resource allocation between homesteads and far-fields with organic resources being allocated more to homestead fields and inorganic resources more to far-fields, resulting in the commonly observed steep fertility gradients between homesteads and far-fields. Furthermore, water management techniques were mostly being used in far-fields, thereby resulting in farmers failing to realize full benefits of integration. It is clear from the study that farmers face a wide range of constraints which hinder the uptake of fertility and water management technologies. Notable among these constraints are labour, high cost of equipment and generally poor accessibility. This suggests that for increased productivity to take place, there is need for mechanizing some of the manual operations that the farmers have to do while at the same time improving issues of access. Finally, it is apparent that SOFECSA impacts have been most apparent with respect to uptake of water management options but has also benefited from previous efforts by other players.

Recommendations

Based on the findings of this study, it is apparent that for integration of water and fertility management technologies to take place, there is need for future scaling out activities to emphasize on superimposing fertility to water management options in order for farmers to benefit fully from these options. Current practices by the farmers suggest lack of this integration. Options for mechanizing some of the manual operations carried out by the farmers also need to be considered bearing in mind the prevailing labour constraints and resource endowments. There is need to continue supporting the current technology innovation momentum in Ward 5 probably through stronger linkages with equipment, seed and fertilizer input service providers who will help to ease existing constraints in terms of accessibility.

Acknowledgements The authors of the chapter would like to thank the Soil Fertility Consortium of Southern Africa supported by the Rockefeller Foundation for funding this research in the last 2 years.

References

Barrow C (ed) (1987) Factors affecting tropical agricultural development. In: Water resources and agricultural development in the tropics. Longman Scientific & Technical, Longman Group UK Limited, Essex, pp 3–55

Bationo A, Mokwunye AU (1991) Alleviating soil, water and nutrients management constraints to increased crop production in West Africa. Fertil Res 29:95–115

Bratton M (1987) Drought, food and the social organization of small farmers in Zimbabwe. In: Glantz M (ed) Drought and hunger in Africa denying famine a future. Cambridge University Press, New York, NY, pp 31–35

Chuma E, Mombeshora BG, Murwira HK, Chikuvire J (2000) The dynamics of soil fertility management in communal areas of Zimbabwe. In: Hilhorst T, Muchena F (eds) Nutrients on the move: soil fertility dynamics in African farming systems. International Institute for Environment and Development, London, pp 45–64

Chuma E, Mvumi B, Nyagumbo I (2001) A review of sorghum and pearl millet-based production systems in the semi-arid regions of Zimbabwe. SADC/ICRISAT Sorghum and Millet Improvement Program (SMIP), International Crops Research Institute for the Semi-Arid Tropics (ICRISAT) (Limited distribution), Bulawayo, Zimbabwe, 72 pp

CSO (1992) National population census 1992. Central Statistical Office, Harare, Zimbabwe, 83 pp

Elwell HA (1981) Contour layout design, 1981. Conex publication, Government Printers, Harare, 61 pp

Grant PM (1967) The fertility of sandveld soil under continuous cultivation. Part II. The effect of manure and nitrogen fertilizer on the base status of the soil. Rhod Zambia Mal J Agric Res 5(12):117–131

Hagmann J (1998) Learning together through participatory extension: a guide to an approach developed in Zimbabwe. Department of Agricultural Technical and Extension Services (AGRITEX; Integrated Rural Development Programme (GTZ/IRDEP) and Intermediate Technology Development Group Zimbabwe (ITDG), Harare, 59 pp

Hagmann J, Murwira K (1996) Indigenous soil and water conservation in Southern Zimbabwe: a study of techniques, historical changes and recent developments under participatory research and extension (part 1). In: Reij C, Scoones I, Toulmin C (eds) Sustaining the soil: indigenous soil and water conservation in Africa. Earthscan, London, pp 97–106

Mapfumo P, Giller KE (2001) Soil fertility management strategies and practices by smallholder farmers in semi-arid areas of Zimbabwe. Limited distribution, 2001. International Crops Research Institute for Semi-Arid Tropics and Food and Agriculture Organization of the United Nations, Bulawayo Zimbabwe, 60 pp

MPSLSW (1996) Poverty assessment study survey, 1995. Ministry of Public Service, Labour and Social Welfare, Government of the Republic of Zimbabwe, Harare, Zimbabwe, 80 pp

Mutambanengwe F (2006) Soil organic matter dynamics and crop productivity as affected by organic resource quality and management practices on smallholder farms. DPhil thesis, Department of Soil Science and Agricultural Engineering, University of Zimbabwe, Harare, 245 pp

Nyagumbo I (1992) The influence of socio-economic factors on potential adoption of no-till tied ridging in four communal areas of Zimbabwe. In: Kronen M (ed) Proceedings of the 3rd annual scientific conference, SADC Land and Water Management Research Programme, Harare, Zimbabwe, 5–7 Oct 1992, pp 319–329

Nyagumbo I (2002) Effects of three tillage systems on seasonal water budgets and drainage of two Zimbabwean soils under maize. DPhil thesis, Department of soil science and agricultural engineering, University of Zimbabwe, Harare, 270 pp

Nyamangara J, Nyagumbo I (2010) Interactive effects of selected nutrient resources and tied-ridging on plant growth performance in a semi-arid smallholder farming environment in central Zimbabwe. Nutr Cycling Agroecosyst 88:103–109

Nzuma JK, Murwira HK (2000) Improving the management of manure in Zimbabwe. Manage Afr Soils 15:1–20, Apr 2000

Rusike J, Twomlow SJ Freeman HA, Heinrich GM (2006) Does farmer participatory research matter for improved soil fertility technology development and dissemination in Southern Africa? Int J Agric Sustain 4(2):1–17

Zingore S (2006) Exploring diversity within smallholder farming systems of Zimbabwe: nutrient use efficiencies and resource management strategies for crop production. PhD thesis, Wageningen University, Wageningen, The Netherlands, 215 pp

Reducing the Risk of Crop Failure for Smallholder Farmers in Africa Through the Adoption of Conservation Agriculture

C. Thierfelder and P.C. Wall

Abstract Current degradation of the natural resource base calls for innovative approaches to sustainable agriculture in Africa. Conservation agriculture (CA) is a sustainable cropping system based on minimal soil disturbance, soil cover with crop residues and crop rotations. CA leads to soil organic matter accumulation and improves water harvesting, and therefore more stable yields and a reduction of the risk of crop failure. After several years, soil quality improvement results in greater crop productivity. However, smallholder, resource-poor farmers in Africa generally manage mixed crop/livestock systems and depend on crop residues for animal feed in the dry season. Strategies therefore need to be developed to convert the farm from conventional to conservation agriculture. Step-wise incorporation of CA into the farming system and concentration of plant nutrient resources will allow increased productivity of both food and crop residues. Once productivity is increased part of the crop residues can be used as animal feed while still leaving sufficient residues for soil cover and soil quality regeneration. Greater production stability and reduced labour requirements of CA make it possible for farmers to use part of the farm for higher value crops, thus generating additional income. Reduced labour requirements of CA allow farmers to involve in alternative activities. CA systems, however, are knowledge intensive and although the principles have very wide application, the actual techniques and technologies to apply these principles are site and farmer-circumstance specific, necessitating the development of multi-stakeholder "innovation networks" focused on adapting CA systems to local conditions.

Keywords Conservation agriculture · Crop residues · Diversification · Risk · Soil quality

Introduction

Current degradation of the natural resource base in many parts of Africa is a major thread for food security and sustainable agriculture intensification (Nana-Sinkam, 1995). The global annual loss of about 6 million ha of arable land due to various forms of soil degradation has raised concerns about the effects of conventional farming practices on the natural resource base on small-scale farms. In Africa, an estimated 500 million ha of land was affected by soil degradation between 1950 and 1990 (ISRIC-UNEP, 1991). Traditional farming practices in sub-Saharan Africa are characterized by land preparation using hand hoe or mouldboard plough. Crop residues are generally removed, used as animal feed or burned, although occasionally incorporated. Use of inorganic fertilizers is low, with an average annual fertilizer use of less than 10 kg N ha^{-1} year^{-1} across Africa (Wichelns, 2006). Organic fertilizers are generally scarce, although more widely utilized. Available water for crop production is often limited. Land holdings are generally small and often located on soils with inherently low fertility. Farmers face serious labour constraints due to low-input, labour-intensive cropping systems, a problem exacerbated by the effects of HIV/AIDS on labour availability. As a result of this, farmers in many parts

C. Thierfelder (✉)
CIMMYT Zimbabwe, P.O. Box MP 163, Mount Pleasant, Harare, Zimbabwe
e-mail: c.thierfelder@cgiar.org

A. Bationo et al. (eds.), *Innovations as Key to the Green Revolution in Africa*,
DOI 10.1007/978-90-481-2543-2_129, © Springer Science+Business Media B.V. 2011

of Africa are disadvantaged and face serious crop production risks.

This chapter aims to present a new and innovative approach to sustainable agriculture in Africa – conservation agriculture (CA) – in order to reduce the crop production risks for smallholder farmers. The chapter emphasizes the biophysical and economic benefits and challenges of CA (especially in the smallholder farmers' context), the strategies for farmers to implement this technology and new methods of promoting complex technologies such as CA.

Conservation Agriculture – A Sustainable System

The sustainability of an agricultural system can be measured in many ways, and many different indicators of sustainability have been proposed. Although economic sustainability is the longer term goal, it is impossible to achieve it if there is degradation of the natural resource base. One relatively simple measure of sustainability is the ability of a system to maintain or increase soil organic matter content and soil fertility/soil quality. Unless the decomposition and loss of organic matter are halted, the soil fertility continues to decline and the system is not sustainable (Wolf and Snyder, 2003). Many people concentrate on how the chemical aspect of soil fertility could be improved by mineral fertilization. Nevertheless, a higher chemical soil fertility does not necessarily mean that the overall cropping and soil system has improved. Physical and biological soil fertility, the often neglected parts of the system, are equally essential. Their destruction or deterioration through tillage leads to the breakdown of soil aggregates, compacted layers, reduced water infiltration and soil water-holding capacity, reduced root growth, and increased water run-off and soil erosion.

To counteract soil fertility decline in Africa two possible solutions are available. One is applying large amounts of manure but this is often in short supply and can therefore only be used on a limited area of land. The other solution is conservation agriculture (CA).

CA is a cropping system based on three principles, namely (a) minimum soil disturbance which basically means no soil inversion by tillage, (b) soil surface cover with crop residues and/or living plants and (c) crop rotations, an important component of the system to maintain sufficient crop residues and if problems such as pests and diseases occur (Calegari et al., 2005; Jones et al., 2006). The principles of CA appear to have extremely wide adaptation, and CA systems are currently employed by farmers on a wide range of soil types, in a wide range of environments and on numerous crops and cropping systems (Wall, 2007). CA is used on soils with over 80% clay in Brazil to around 90% sand in Australia. CA farmers can be found from the equator to at least 50° latitude (north and south) and from sea level to over 3500 m.a.s.l. Many crops can be grown under CA such as maize (*Zea mays* L.), wheat (*Triticum aestivum* L.), soybeans (*Glycine max* L.), cotton (*Gossypium hirsutum*, L.), sunflower (*Heliathus annus* L.), rice (*Oryza sativa* L.), tobacco (*Nicotiana tabaccum* L.) and even root crops, although the harvest of these crops causes considerable soil movement. However, the techniques to apply the principles are very varied and depend to a large degree on environment and farmer circumstances.

The short and longer term benefits of CA summarized by Wall (2007) include the longer term benefits on physical, chemical and biological soil quality (Unger and Fulton, 1990; Derpsch, 1999). However, reduced soil disturbance and residue retention lead to improved water infiltration, control of run-off and erosion, higher soil moisture content and reduced evaporation almost immediately (Derpsch et al., 1986; Dardanelli et al., 1994; Thierfelder et al., 2005). This is clearly illustrated by Fig. 1a, b, showing the effects of two CA-maize management systems (direct seeding and planting basins, with both surface residue retention) on water infiltration (in mm h^{-1} with a simulated rain of 100 mm in 1 h) compared to a conventionally ploughed maize treatment. These data are from two contrasting soil types: sandy soils (*Arenosols*) at Henderson Research Station (HRS) in Zimbabwe and clay-rich soils (*Lixisols*) at Monze Farmer Training Centre (MFTC) in Zambia. Infiltration measurements using a mini-rainfall simulator (Amézquita et al., 1999) were carried out for 2 years at MFTC and 3 years at HRS after trial establishment. Final water infiltration rate of the mean of the two CA treatments was 34% higher than the conventionally ploughed treatment at HRS. The significantly highest final infiltration rate (LSD 16.2 at $p \leq 0.05$ probability) was found in the direct seeded treatment (74.8 mm h^{-1}) and lowest in the conventionally ploughed (51.5 mm h^{-1}). At the MFTC, the mean infiltration rate of the two CA

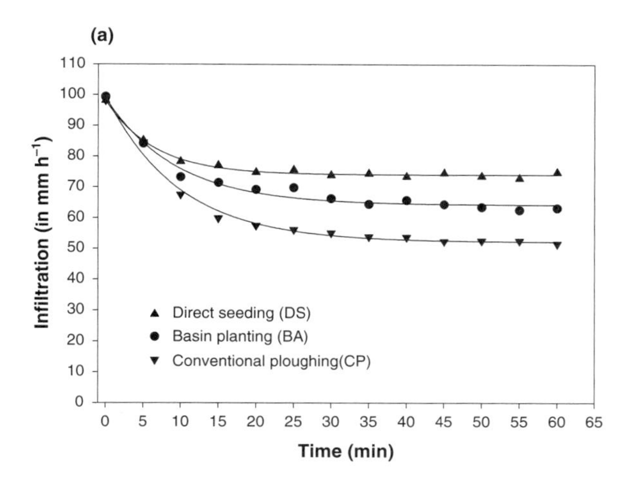

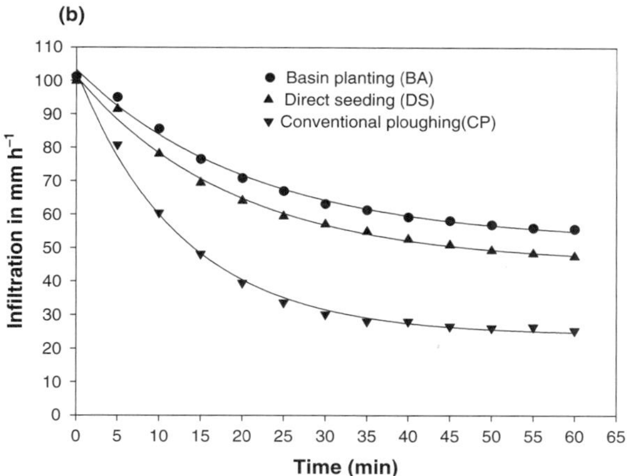

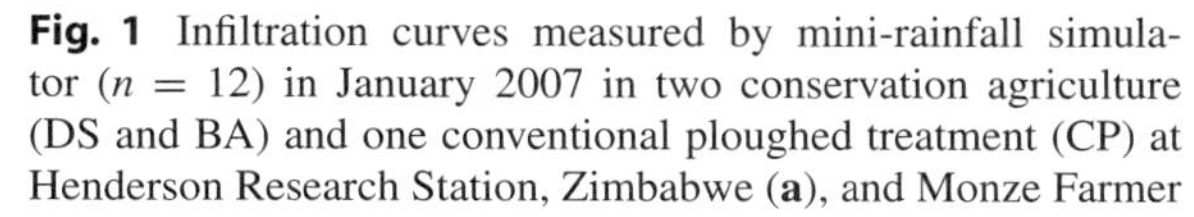

Fig. 1 Infiltration curves measured by mini-rainfall simulator ($n = 12$) in January 2007 in two conservation agriculture (DS and BA) and one conventional ploughed treatment (CP) at Henderson Research Station, Zimbabwe (**a**), and Monze Farmer Training Center, Zambia (**b**). Experimental design: completely randomized block with four replications, mean rain intensity 100 mm h^{-1}, duration of simulation 60 min (LSD at $p \leq 0.05$ probability)

treatments was 103% higher than the conventionally ploughed control. Significantly highest final infiltration rate (55.6 mm h^{-1}) was measured in the basin treatment followed by the direct seeded treatment (47.4 mm h^{-1}, LSD 19.4 at $p \leq 0.05$ probability) compared to only 25.3 mm h^{-1} in the conventionally ploughed treatment.

Residue retention on the surface protects the soil against climatic impacts and serves as an additional source of nutrients for both soil fauna and flora. Increases in the availability of many essential plant nutrients such as phosphorous, potassium, calcium, magnesium and sulphur have been observed on CA fields (Sidiras and Pavan, 1985). Leaving the residues on the surface leads to a higher biological activity of the soil (Power and Legg, 1978). In the longer term, the slowly decomposing residue material which is translocated by soil organisms leads to higher organic matter and available nutrient contents in the whole soil system. Key to maintaining SOM is the lack of inversion ploughing: not only does tillage lead to soil aeration and thus rapid oxidation of SOM but also it reduces fauna and flora populations by removing their source of food from the top layer to deeper and for microorganisms less favourable layers.

At the MFTC in Zambia, we investigated the biological activity of six CA treatments with residue retention (one basin treatment with maize and five direct seeded treatments with (a) sole maize, (b) cotton after maize, (c) maize after cotton, (d) sunn hemp (*Crotalaria juncea* L.) after maize and (e) maize after sunn hemp) and compared them with a conventionally ploughed treatment without residue retention. Results (Fig. 2) show that earthworm densities were, on average, 450% higher in the CA treatments than in the conventionally tilled treatment (45 earthworms/m^2 on average in the CA treatments and 10 earthworms/m^2 in the conventionally tilled treatment) and were highest (64 earthworms/m^2) in the maize treatment seeded after the sunn hemp green manure cover crop (Nhamo, 2007, unpublished).

Reduced soil disturbance and higher soil organic matter contents enable the soil matrix to build up a stable aggregate and pore system (Kladviko et al., 1986; Six et al., 2002; Simpson et al., 2004). Finally, rotations and green manure cover crops generate additional organic matter, improve the soil structure and reduce pests and diseases and may mobilize additional plant nutrients (Anderson et al., 2001). Longer term soil organic matter accumulation, the improved water harvesting through a better pore system and reduced surface run-off lead to more stable yields and reduce the risk of crop failure. Although the water conserving effect in normal years may have little conspicuous effect on short-term yields it is particularly beneficial in dry years, thus reducing production risk and yield oscillations (Erenstein, 2002). This is shown in Fig. 3 illustrating results of the Tarata Experimental Station,

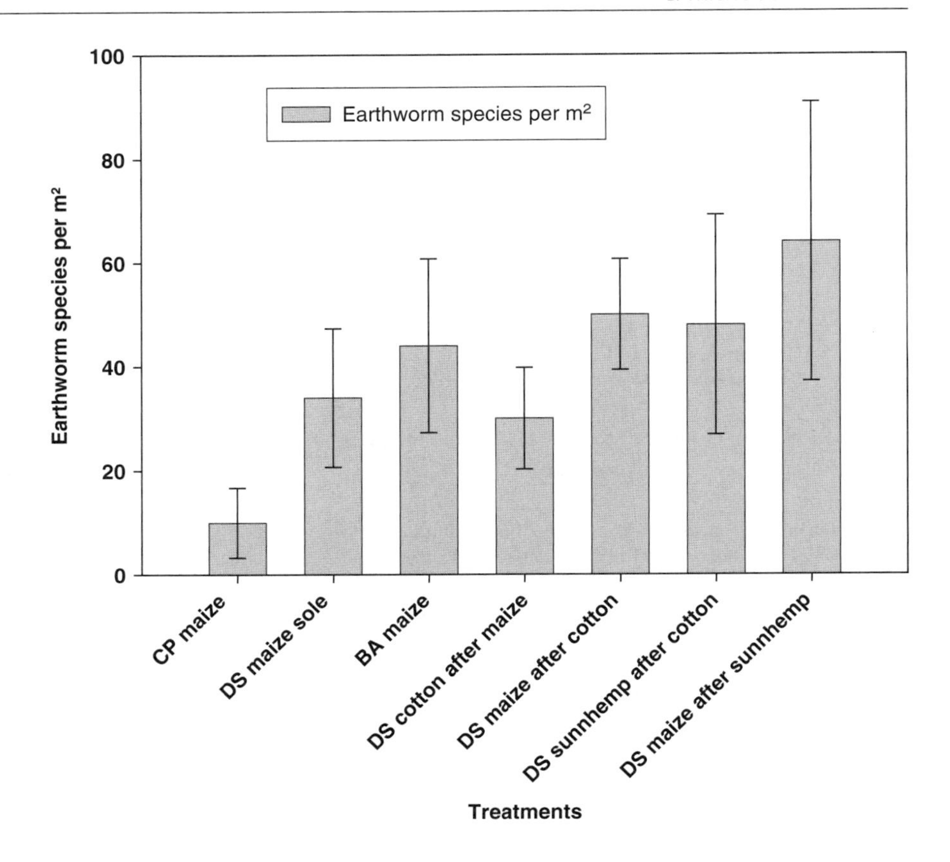

Fig. 2 Earthworm counts (counts per m^2 in the first 0–30 cm) in six conservation agriculture (DS maize sole; basin maize sole; DS cotton after maize; DS maize after cotton; DS sunn hemp after cotton; DS maize after sunnhemp) and one conventional ploughed (CP) treatment ($n = 10$) at Monze Framer Training Center, Zambia, January 2007 (Nhamo, 2007 unpublished)

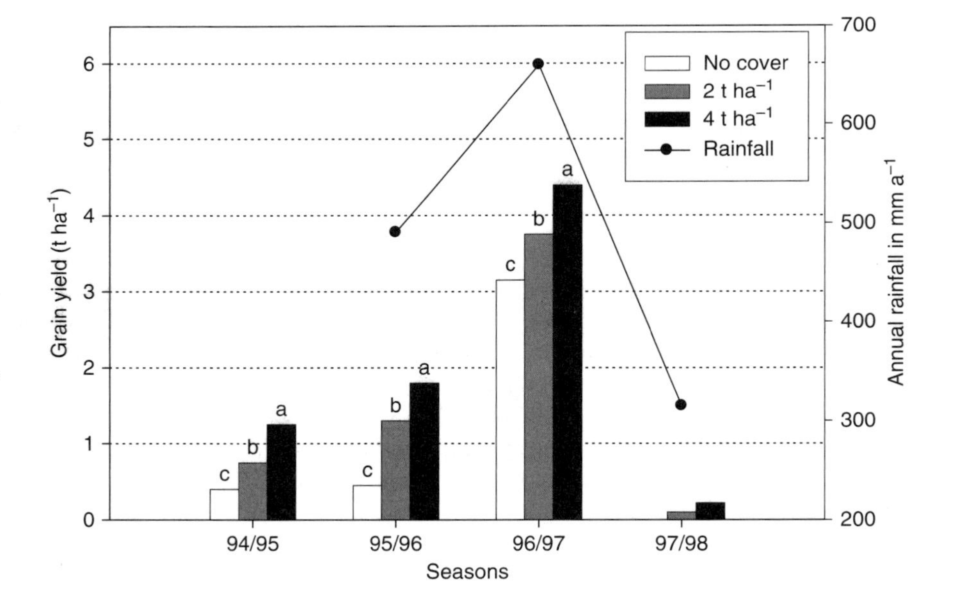

Fig. 3 Effects of residue levels on wheat grain yield in four consecutive seasons at the Tarata Experiment Station, Cochabamba, Bolivia. Rainfall data for 1994/1995 are not included here. Bars that do not share the same letter are significantly different (LSD at $p \leq 0.05$ probability) (adapted from Wall, 1999)

Cochabamba, Bolivia, summarized by Wall (1999). The aim of this study was to evaluate the effect of crop residue ground cover on yield. Grain yields of direct seeded wheat in four consecutive seasons and three levels of residue retention are compared. Rainfalls in the 1994/1995 (data not presented here) and 1995/1996 seasons were close to the long-term means. The 1996/1997 season was very wet while the 1997/1998 season was extremely dry. Grain yields in all years are directly related to the amount of residue retained from the previous crop. Even in the very wet season (1996/1997), there was a positive effect of mulching while in the very dry season, there was no yield obtained on plots without ground cover and 220 kg ha^{-1} on plots with 4 t ha^{-1} mulch. Over all seasons, yield was economically increased with ground cover, thus reducing the risk of crop failure (Wall, 1999).

Short- and long-term socio-economic benefits are observed due to the drastic reduction in farm inputs (fuel, machinery wear and drudgery) and optimized allocation of farm labour (Sorrenson et al., 1998). This is particularly true for large-scale commercial farms where the use of tractors and bigger farm implements are common. For smallholder farmers, if weeding is carried out manually, the labour saving through direct seeding technologies might be less significant as more time has to be spent on manual weed control, particularly in the first season. Nevertheless, farmers in rural areas of Zimbabwe opted for CA after the first season although they did not see any yield benefit compared to conventional treatments. They say, "It saves us labour on land preparation that we can use for other purposes."

Challenges and Conflicts for the Widespread Adoption of CA in Africa

Introducing CA to smallholder farmers involves challenges and, if they are not properly addressed, often leads to the rejection of the technology in the first season. CA does require better management skills compared to traditional farming practices as it requires a special set of cultural practices that may be different from traditional plough-based systems (Lal et al., 2000).

One of the major reasons that farmers till the soil is to control weeds. If soil tillage is omitted, weed control becomes very important. Different weed control strategies such as manual and/or chemical weed control, post-harvest weeding and suppression of weeds through cover crops are effective ways to overcome the problem of weeds and if carefully controlled in the first seasons, the weed pressure diminishes over time (Anderson, 1977; Wall, 2007).

Better fertility (nitrogen) management is required until the system and specifically the soil organisms have adjusted to the new agricultural method (Barreto, 1989). In conventionally tilled systems, farmers mineralize nitrogen by tilling the soil to stimulate SOM breakdown, which then results in a flush of N when soil fauna and flora die as the availability of the food source declines. However, there is a high risk of leaching of the mineralized N as plants cannot fully use it in the early stages of crop establishment and heavy rain may leach it below the rooting zone. In CA systems the soil is not turned and decomposition of organic matter is much slower (but also more regular) and the lack of large amounts of readily available N early in the season can lead to N stress, especially in the first years of conversion from tillage-based agriculture to CA. Furthermore, leaving residues on the soil surface can lead to other side effects. Most cereal stover consists of wide C:N ratios (>20–30) and low nitrogen concentrations (<1%), resulting in temporary immobilization of N(Mueller et al., 2001). This phenomenon has been observed mostly in the first seasons of transition before enough soil organic matter and available N has accumulated and more N becomes available (Barreto, 1989). Both restrictions can be addressed by using a different fertilizer management, i.e. slightly higher N doses in the early stages of transition. Models suggest that once SOM levels and biological activity increase as a result of the lack of tillage and permanent soil cover, a new equilibrium of SOM breakdown is achieved, and this addition of extra N (greater than in conventionally tilled systems) is not necessary. Although there is a paucity of research results to illustrate this, results from farmers' fields in Brazil show that after 20 years of CA, fertilizer applications to maize have been reduced by 30%, and at the same time maize yields have increased by 50% (Derpsch, 2005).

The management of crop residues is crucial, especially in the drier parts of Africa: CA does not work if the residues are not retained (Wall, 1999). Most of the benefits in CA are directly or indirectly linked to crop residues (Erenstein, 2002) but in most cases the primary use of residues for smallholder farmers remains as animal fodder to offset dry season shortages (Sain and Barreto, 1996). In Africa, smallholder farmers manage mixed crop–livestock systems where animals contribute to food security, reduce risk, provide draft power and add to capital. The trade-off between residues for retention on the field and for feed is therefore very important and can lead to serious conflicts between crop and livestock producers (Mueller et al., 2001).

In many areas of southern Africa, farmers plant large areas of land with maize to assure food security: sufficient area is planted to be able to harvest family requirements even in a dry year. However, in a normal or good year, management becomes difficult as labour as well as fertilizers are insufficient, and so productivity is reduced by nutrient stress or weeds. Consequently, the average maize and residue biomass yields are low. As a solution, step-wise incorporation of CA into the farming system will allow increased productivity of both food and crop residues. The main emphasis in this strategy is to concentrate both inputs and management on smaller farm areas and to increase the productivity on these areas first. In the process of increasing productivity, the intervention areas can be slowly increased so that the productivity of the whole farm will increase. Once output is increased part of the crop residues can be used as animal feed while sufficient residues remain for soil cover and soil quality regeneration. If farmers are assured of food security thus reducing the production risks, they can start producing higher value crops, which gives them additional income-generating options. Furthermore, if farmers restrict themselves to certain limited cropping areas, they will free up labour time for alternative activities such as off-farm labour or value addition to their products (i.e. make marmalade out of fruit or chutney out of vegetables, etc.).

Knowledge – A Key Component to Adoption of CA Technologies

Traditional management practices have been handed down through generations and are well understood by farmers. Introducing CA to farmers involves a complete change in their mind-set in approach to agriculture. The farmers have to acquire knowledge about weed control, seeding techniques and times, equipment, fertilization, residue management, crop rotations and intercropping, harvest technologies and much more. CA does not involve just a simple change in one production practice, but rather is a complex technology involving simultaneous change in many practices. At the same time it is a knowledge-intensive cropping system, where success is more dependent on what the farmer does, than on the levels of inputs she/he applies. Smallholder farmers generally do not have access to formal information systems such as books, periodicals, libraries or the internet: they depend to a large degree on both governmental and non-governmental extension systems for new agricultural information. However, often the extension agents themselves are poorly linked to information systems, suffer from lack of new information and promote technologies that are often outdated. Many of these factors have direct effects on the widespread adoption of CA. Agricultural universities and national research institutions do not have CA on their agenda because the change in mind-set among their professors and directors is a slow process. The mind-set that ploughing is necessary for producing good crops is well established in those institutions, and views about the "symbol of the plough" as the major tool for "proper" agriculture are difficult to overcome. It will take time before innovators in those institutions start to believe in CA which will enable a change in institutional thinking.

New Ways of Promoting Complex Agriculture Systems

In the Americas and Australia, the adoption of CA technologies was mainly driven by farmers and their organizations. The first innovative farmers started practicing CA technologies in the mid-1970s and slowly attracted other players including other farmers, researchers, extension agents, machinery manufacturers and input suppliers. Initially, implementation was slow but since the early 1990s, adoption has increased exponentially in these areas. To date over 95 million ha are under no-tillage worldwide, one of the principle components of CA (Derpsch, 2005). Most of the area under CA is on large-scale commercial farms in the Americas and Australia: only a few examples are available where CA adoption has advanced on small-scale farms. In Africa, some farmers in Ghana and Zambia have started practicing CA on smallholder farms and some few examples are reported from Zimbabwe, Malawi and Tanzania.

However, there are some conditions under which CA does not work – or at least we do not yet know how to manage CA under these conditions. Because of the increased water infiltration and reduced evaporation under CA, the system does not work well in soils with impeded drainage: the system increases the severity and frequency of waterlogging. Under extremely moisture-limited conditions it becomes impossible to produce sufficient biomass for effective ground cover, and therefore CA does not, as yet, work under these conditions. However, if ground cover can be achieved with other means (i.e. in China the use of plastic film was successfully tested), the principles of CA should work equally well under these conditions as under other dryland conditions.

The fact that the principles of CA have an extremely wide application does not allow for blanket recommendations or set instructions on how to put it into practice. Often only the general principles of CA (minimum soil disturbance, residue retention and rotations) are the same on different sites while the actual formulae are very site and farmer-circumstance specific. The need to fine-tune CA systems to farmers' circumstances makes the technology and its promotion different from traditional component technologies. Traditionally in Africa, component technologies (e.g. a new maize variety or a new crop) are investigated and promoted in a linear research and extension pathway, where researchers conduct basic and strategic research mostly on-station, do applied research on-farm to test the technological component under representative conditions and pass the results to the extension services. Extension agents then pass the information on to farmers hoping for adoption of the new technology (Ekboir, 2002; Wall et al., 2002). However, for a complex technology such as CA, involving numerous components, the linear model of knowledge development and transfer does not work. Multi-agent innovation networks have been proposed for complex change (Rycroft and Kash, 1994), and in Latin America and South Asia these have proved to be an efficient way for the development and adoption complex agricultural change – the uptake of CA and no-tillage agriculture.

Unlike the linear research and extension pathway, innovation networks make use of complex interactions between stakeholders, because the different players differ in their activities and abilities (Ekboir, 2002). In other words, researchers, extension agents and farmers work together in a participatory way to adjust the CA system to local circumstances (Fig. 4). Other players such as machinery manufacturers and input suppliers join the network to adjust machinery and input supply to the needs of the innovation network. The network is open for other players such as market facilitators, credit providers and politicians in order to interact, provide input and improve the network; the participation of all agents that can enhance the network is encouraged. New learning routines such as participatory research methods, multidisciplinary approaches, acceptance of information generated without experimental design, creation of a common language and open dissemination of information have proved to facilitate such networks leading to farmer-to-farmer exchange with strong support from researchers, exchange agents, machinery and input suppliers (Ekboir, 2002).

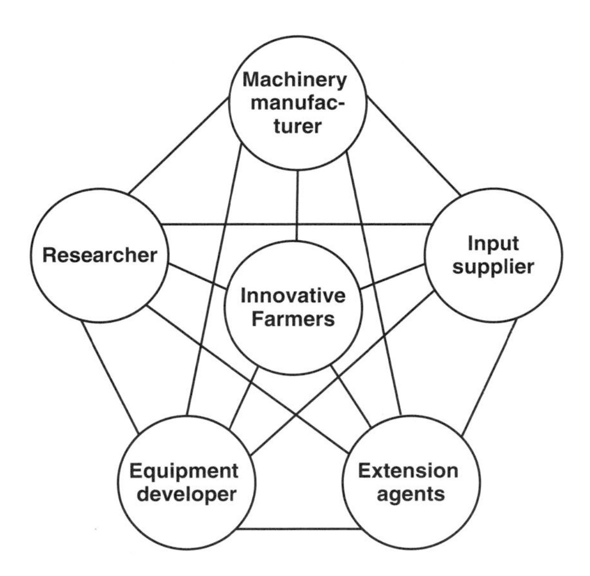

Fig. 4 Representation of a simplified innovation network of knowledge and information sharing based on Wall et al. (2002)

Conclusion

Conservation agriculture has proved its ability to alleviate soil and land degradation in many parts of the world in different agro-ecosystems. Proper use of the three principles of CA (minimum soil disturbance, crop residue retention and crop rotations) offers many benefits to the farming systems and to farmers. Through CA many aspects of physical, chemical and biological soil quality improve. In most situations water harvesting of the system increases and organic matter accumulates, thus increasing the crop productivity and reducing the risk of crop failure. Promotion of CA to communities involves biophysical and economic challenges (e.g. weed control, residue and nutrient management). Furthermore, changing the mind-set of people towards CA is difficult as for centuries the plough has been the symbol of agriculture. CA is a knowledge-intensive and complex technology and strategies for implementing such technologies have to be developed. Promotion of CA needs to follow a different modus operandi than in normal research, necessitating the development of multi-stakeholder "innovation networks" focused on participatory methods and adapting CA systems to local conditions.

Acknowledgements This work was carried out under the project "Facilitating the widespread adoption of conservation agriculture in smallholder maize-based systems in Eastern and Southern Africa" by the International Maize and Wheat Improvement Center (CIMMYT) in collaboration with multiple partners, including the University of Hohenheim, Germany. Financial support of the Ministry of Economic Cooperation (BMZ) of the German Government is gratefully acknowledged, as is the technical support of Tobias Charakupa and Sign Phiri from Zimbabwe and Mangwala Sitali and Patient Munakalanga from Zambia.

References

Amézquita E, Cobo QL, Torres EA (1999) Diseño, construcción y evaluación de un minisimulador de lluvia para estudios de susceptibilidad a erosion en areas de laderas. Revista Suelos Equatoriales 29(1):66–70

Anderson W (1977) Weed science: principles. West Publishing Company, St. Paul

Anderson S, Gündel S, Pound B et al (2001) Cover crops in smallholder agriculture. ITDC Publishing, London

Barreto HJ (1989) Cambios en probiedades quimicas, patrones de fertilizacion, y encalcimiento en suelos bajo labranza cero. In: Barreto HJ, Raab R, Tasistro A, Violic AD (eds) Labranza de Conservación en Maiz. CIMMYT, Mexico, DF, pp 43–70

Calegari A, Ashburner J, Fowler R (2005) Conservation agriculture in Africa. FAO, Rome

Dardanelli JL, Bachmeier OA, Salas HP et al (1994) Evaporación en un suelo Haplustol éntico bajo dos sistemas de labranza. Ciencia de Suelo (Argentina) 12:17–21

Derpsch R (1999) Expansión mundial de la SD y avances tecnológicos. In: Proceedings of the 7th national congress of AAPRESID, Mar del Plata, Argentina, 18–20 Aug 1999

Derpsch R (2005) The extent of conservation agriculture adoption worldwide: implications and impact. In: Proceedings on CD. III world congress on conservation agriculture: linking production, livelihoods and conservation, Nairobi, Kenya, 3–7 Oct 2005

Derpsch R, Sidiras N, Roth CF (1986) Results of studies made from 1977 to 1984 to control erosion by cover crops and no-tillage techniques in Paraná, Brazil. Soil Tillage Res 8: 253–263

Ekboir J (2002) CIMMYT 2000–2001 world wheat overview and outlook: developing no-till packages for small-scale farmers. CIMMYT, Mexico, DF

Erenstein O (2002) Crop residue retention in tropical and semi-tropical countries: an evaluation of residue availability and other technological implications. Soil Tillage Res 67: 115–133

ISRIC-UNEP (1991) World map of the status of human-induced soil degradation, An explanatory note. ISRIC, UNEP, Wageningen

Jones C, Basch G, Baylis A et al (2006) Conservation agriculture in Europe: an approach to sustainable crop production by protecting soil and water? SOWAP, Jealott's Hill, Bracknell

Kladviko EJ, Griffith DR, Mannering JV (1986) Conservation tillage effects on soil properties and yield of corn and soya beans in Indiana. Soil Tillage Res 8:277–287

Lal R, Eckert DJ, Fausey NR et al (2000) Conservation tillage in sustainable agriculture. In: Edwards CA, Lal R, Madden P, Miller RH, House G (eds) Sustainable agricultural systems. St Lucie Press, Ankeny, pp 203–225

Mueller JP, Pezo D, Benites J et al (2001) Conflicts between conservation agriculture and livestock over utilization of crop residues. In: Garcia-Torres L, Benites J, Martínez-Vilela A (eds) Conservation agriculture: a worldwide challenge. ECAF/FAO, Córdoba, pp 211–225

Nana-Sinkam SC (1995) Land and environmental degradation and desertification in Africa: Issues and options for sustainable development with transformation. ECA/FAO, Rome

Power JF, Legg JO (1978) Effect of crop residues on the soil chemical environment and nutrient availability. In: Lewis WM, Taylor HM, Welch LF, Unger P (eds) Crop residue management systems. ASA Special Publication No. 31. ASA, CSSA, SSSA, Madison, WI, pp 85–100

Rycroft RW, Kash D (1994) Complex technology and community: implications for policy and social sciences. Res Policy 23(6):613–626

Sain GE, Barreto HJ (1996) The adoption of soil conservation technology in El Salvador. J Soil Water Conserv 51:313–321

Sidiras N, Pavan MA (1985) Influencia do sistemas de manejo do sol no seu nivel de fertilidade. Revista Brasiliera de Ciencia Solo 9:244–254

Simpson RT, Frey SD, Six J et al (2004) Preferential accumulation of microbial carbon in aggregate structures of no-tillage soils. Soil Sci Soc Am J 68:1249–1255

Six J, Feller C, Denef K et al (2002) Soil organic matter, biota and aggregation in temperate and tropical soils – effects of no-tillage. Agronomie 22:755–775

Sorrenson WJ, Duarte C, Lopez Portillo J (1998) Economics of no-tillage compared to traditional cultivation on small farms in Paraguay, Asunción. MAG/GTZ soil conservation project

Thierfelder C, Amezquita E, Stahr K (2005) Effects of intensifying organic manuring and tillage practices on penetration resistance and infiltration rate. Soil Tillage Res 82: 211–226

Unger PW, Fulton LJ (1990) Conventional- and no-tillage effects on upper root zone soil conditions. Soil Tillage Res 16: 337–344

Wall PC (1999) Experiences with crop residue cover and direct seeding in the Bolivian highlands. Mt Res Dev 19:313–317

Wall PC (2007) Tailoring conservation agriculture to the needs of small farmers in developing countries: an analysis of issues. J Crop Improv 19:137–155

Wall PC, Ekboir J, Hobbs PR (2002) Institutional aspects of conservation agriculture. Paper presented at the international workshop on conservation agriculture for sustainable wheat production in rotation with cotton in limited water resource areas, Taschkent, Usbekistan, 13–18 Oct 2002

Wichelns D (2006) Improving water and fertilizer use in Africa: challenges, opportunities and policy recommendations. Background paper prepared for the African Fertilizer Summit, Abuja, Nigeria, 9–13 Jun 2006

Wolf B, Snyder GH (2003) Sustainable soils: the place of organic matter in sustaining soils and their productivity. The Haworth Press, New York, NY; London, Oxford

Dissemination of Integrated Soil Fertility Management Technologies Using Participatory Approaches in the Central Highlands of Kenya

D.N. Mugendi, J. Mugwe, M. Mucheru-Muna, R. Karega, J. Muriuki, B. Vanlauwe, and R. Merckx

Abstract Declining soil fertility and productivity is a critical problem facing smallholder farmers in the central highlands of Kenya. A study to improve soil fertility and farm productivity within the smallholder farming systems in the area was carried out from 2003 to 2006. The specific objectives were to identify farming system constraints, evaluate and disseminate potential integrated soil fertility management (ISFM) interventions using participatory approaches, assess achievements and impacts, and document learning experiences emanating from the methodologies used. The participatory approaches used were Participatory Rural Appraisal (PRA), mother–baby approach (with emphasis on demonstration), farmer groups, stakeholders planning meetings, village training workshops, cross-site visits and participatory monitoring and evaluation. The core problems identified were low crop and fodder yields that were caused by erratic rainfall, soil erosion, low soil fertility and small land sizes. There was high participation of farmers in all the partnership activities, and this possibly contributed to the high uptake of the technologies for testing by farmers whereby after only 2 years a total of 970 households were testing the new technologies. Maize yields at the farm level increased by more than 150% following use of the new ISFM interventions and about half of the farmers within the groups planted close to 500 trees propagated in the group nurseries. We recommend that pathways to reach more farmers should concentrate on demonstrations, farmer training grounds, field days and farmer groups and that a policy framework should be put in place to impart appropriate skills in ISFM to the extension workers.

Keywords Demonstration trials · Mother–baby trials · Field days · Farmer groups

Introduction

Declining soil fertility in sub-Saharan Africa, a phenomenon that has ultimately led to soil degradation and reduced per capita food production, has become a serious issue of global concern (Bationo et al., 2004). Studies indicate that high population growth rates in sub-Saharan Africa, resulting in abandonment of fallow periods and subsequent intensification of agriculture without proper land management and addition of nutrients, is the primary cause of nutrient depletion in this region (Smaling et al., 1997). As population grows, soil fertility continues to be depleted by crop-harvest removals, leaching and soil erosion, as farmers are unable to sufficiently compensate these losses by returning nutrients to the soil via crop residues, manure and mineral fertilizers.

The humid highlands of central Kenya are characterized by high population pressure with growth rate of about 2.9%. To feed this ever growing population, the available land is cropped continuously without adequate external inputs of both organic and inorganic nutrient supplements. This has led to soil nutrient depletion and land degradation evidenced by declining agricultural productivity, soil analytical indicators and soil nutrient balance studies. Use of inorganic fertilizers to replenish the soil nutrients is one of the

D.N. Mugendi (✉)
Department of Environmental Sciences, School of Environmental Studies, Kenyatta University, Nairobi, Kenya
e-mail: dmugendi@yahoo.com

A. Bationo et al. (eds.), *Innovations as Key to the Green Revolution in Africa*,
DOI 10.1007/978-90-481-2543-2_130,

major ways of counterbalancing this nutrient depletion but farmers in central Kenya use insufficient amounts of inorganic fertilizers (Woomer and Muchena, 1996). High costs of fertilizer, lack of know-how about their usage and lack of incentives to use fertilizer have individually or jointly constrained fertilizer optimal use. As a result, soil fertility continues to decrease and has been identified as one of the root causes of declining per capita food production in the area.

Over the years, integrated soil fertility management (ISFM) technologies have been developed by scientists to address the problem of soil fertility decline but the adoption in the farmers' fields remains minimal. Poor research extension linkages coupled with poor dissemination methodologies are some of the causes of low adoption (MoA, 2002). The successful dissemination and adoption of new knowledge-intensive practices such as ISFM technologies requires much more than the transfer of knowledge and germplasm; it involves building partnerships with farmers and a wide range of stakeholders (participatory), ensuring appropriateness of the practice, assisting local communities to mobilize resources and ensuring participation of farmers' groups and encouraging adoption (Franzel et al., 2002). Participatory research methods have been advocated as a means to improve relevance of the developed technologies and adoption by smallholder farmers (Cramb, 2000). According to its supporters, the benefits of this approach are twofold: locally adapted improved technologies and improved experimental capacities of farmers. Practical field experiences reveal that impressive results can be achieved when farmers and outsiders "join hands" (Haverkort, 1991).

Involvement of farmers in selection of soil fertility technologies has proved more problematic, with few successful models (Kanyama-Phiri et al., 1998). Highly variable performance of technologies is one challenge, and local adaptation is one of the ways of optimizing performance in the heterogeneous environment that exists in smallholder farming systems. However, gaps still remain on how this can be achieved as there are relatively few studies with farmer-participatory approaches. The purpose of this study was therefore to use the participatory approach (Chambers et al., 1987; Okali et al., 1994) to diagnose farming system constraints, involve farmers in ISFM technologies development and dissemination and document achievements as well as learning experiences emanating from the methodologies used. Farmers' involvement is essential because they (farmers) understand their complex biophysical, social-cultural and economic environment better than anyone else and are therefore better in adapting new technologies which are crucial in accelerating adoption.

Research in ISFM Technologies

Integrated soil fertility management (ISFM) is now regarded as a strategy that helps low resource endowed farmers mitigate problems of poverty and food insecurity by improving the quantity of food, income and resilience of soil productive capacity. According to Kimani et al. (2003), the ISFM approach advocates for careful management of soil fertility aspects that optimizes production potential through incorporation of a wide range of adoptable soil management practices. It also entails the development of soil nutrient management technologies using inputs that meet the farmers' production goals and circumstances as a way of enhancing adoption of improved soil management options (Palm et al., 1997). According to these authors, there has been a major change in the research and development paradigms over the years, which has led to adoption of the integrated soil nutrient management paradigm.

Integrated nutrient management (INM) is the backbone of ISFM approach, which entails integrated use of organic and inorganic sources of plant nutrients, as well as the entirety of possible combinations of nutrient-adding practices. INM emphasizes on improving the agronomic efficiency of the "external inputs" that are being used (Kimani et al., 2003). Many studies including long-term ones have shown that INM results in more yield and nutrient use efficiency than that expected from mere additive effects of sole applications (Bekunda et al., 1997).

One of the organic resources which has generated a lot of interest among the scientific community is *Tithonia diversifolia* (tithonia). Tithonia is a herbaceous roadside weed which is also commonly used as a boundary hedge and has been shown to improve soil fertility and increase crop yields substantially in western and central Kenya (Jama et al., 1997; Mugendi et al., 2001). The green leaf biomass of tithonia is high in nutrients, in the order of 3.5–4.0% N, 0.35% P, 4.1% K, 0.59% Ca and 0.27% Mg on a dry matter basis (Rutunga, 2000), produces large quantities of biomass and tolerates regular pruning (Buresh and Niang, 1997), which is a favourable characteristic for its use in soil fertility improvement. Application of tithonia biomass has been shown to double maize yields without application of mineral fertilizers in a wide variety of farmer-managed trials (Sanchez and Jama, 2002). In many cases, maize yields were higher with application of tithonia biomass than with commercial mineral fertilizer at equivalent rates of N, P and K.

Fast growing leguminous trees/shrubs have also been identified to have potential to alleviate farmers' problems of soil infertility and soil erosion (Mugwe and Mugendi, 1999). These trees have root nodules that biologically fix nitrogen from the atmosphere, making them available to plants via soil incorporated prunings (Odee, 1996). In addition, when planted on contour hedgerows, they reduce soil erosion through the hedge effect and enhance formation of natural terraces (Mugwe et al., 2004). *Calliandra calothyrsus* (calliandra) and *Leucaena trichandra* (leucaena) are some of the key leguminous trees that have shown positive results in western Kenya and central highlands of Kenya (Jama et al., 1997; Mugendi et al., 1999). For example, results from Embu (central Kenya) showed that the treatment plots that received biomass of calliandra and leucaena (with or without fertilizer) collected outside the plots gave the highest maize yields (Mugendi et al., 1999).

Researchers have also shown that herbaceous legumes, also known as legume green manure crops, are an effective means of sustaining soil fertility (Abayomi et al., 2001). They have been found to contribute significant quantities of N (30–110 kg N ha^{-1}) to the associated or succeeding crops (Constantinides and Fownes, 1994). Notable examples include *Mucuna pruriens* (mucuna or Velvet bean), *Crotalaria ochroleuca* (Tanzanian sunnhemp) and *Lablab purpureus* (Dolichos or Lablab bean) that have successfully been intercropped with maize in many parts of the world with resultant increase in grain yields of the subsequent maize compared to continuously grown maize (Giller, 2001).

Cattle manure is another very important resource that farmers have been using for many years to effectively alleviate the problem of soil fertility decline (Probert et al., 1995). Manures have the advantage of supplying the essential plant nutrients either directly or indirectly (Palm et al., 1997), add to soil organic matter (humus) and improve soil physical and chemical characteristics (Kimani et al., 2004). Manure has also been demonstrated to restock the particulate organic matter fraction better than fresh residues (Kapkiyai et al., 1998) and to exhibit a long-term benefit with its ability to release nutrients slowly, thereby guarding against loss through leaching.

There was a realization however that even after many years of research a large yield gap remained between the yields in the farmers' field and the on-station experiments. Therefore, in 2000, a demonstration trial aiming at offering farmers ISFM technologies for replenishing soil fertility was established in Kirege School, Meru South District. Farmers from Kirege started taking up the technologies and by 2003, about 203 farmers were trying the ISFM technologies in their farms and they indicated that their yields were increasing. With this information it was realized that there was need to expose the technologies to more farmers in the region; therefore, a project aimed at increasing the dissemination and adoption of the technologies to farmers was started in 2003.

Materials and Methods

Description of the Study Area

The study was conducted in Chuka division of Meru South District of Kenya. Meru South District is situated between latitudes 00°03′47″N and 0°27′28″S and longitudes 37°18′24″E and 28°19′12″E. The district covers an area of 1032.9 km^2 and Chuka division covers an area of 169.6 km^2. According to agro-ecological conditions (based on temperature and moisture supply), the area is in the Upper Midland Zone (UM2–UM4) (Jaetzold et al., 2006) on the eastern slopes of Mt. Kenya with an annual mean temperature of 20°C and a total bimodal rainfall of 1200–1400 mm. The rainfall comes in two seasons: the long rains (LR) lasting from March through June and short rains (SR) from October through December.

The area is dominated by farming systems with a complex integration of crops and livestock and smallholder farms that are intensively managed. Land sizes are small ranging from 0.1 to 1.5 ha with an average of 1 ha (GOK, 2001). The main cash crops are coffee and tea while the main staple food crop is maize, which is cultivated from season to season mostly intercropped with beans. Other food crops include potatoes, bananas and vegetables that are mainly grown for subsistence consumption. Livestock production is a major enterprise in the region, especially of dairy cattle mostly consisting of improved breeds. Other livestock in the area include sheep, goats and poultry.

The study was carried out in four sites. These were Muthambi Division (Murugi), Chuka Division (Mucwa and Kirege) and Magumoni Division (Mukuuni). These sites were chosen using GPS coordinates to try to cover the district as much as possible. Kirege (00°20′07.1″S; 37°36′50.8″E) is located in upper midland 3 with an altitude of approximately 1473 m above sea level. Mucwa (00°18′48.2″S; 37°38′38.8″E) is located in upper midland 3 with an altitude of approximately 1373 m above sea level. Mukuuni (00°23′30.3″S; 37°39′33.7″E) is located in upper midland 4 with an altitude of approximately 1287 m above sea level. Murugi (00°23′30.3″S; 37°39′33.7″E) is located in upper midlands 3 and 4 with an altitude of approximately 1287 m above sea level.

The soils in Kirege, Mukuuni and Mucwa are Rhodic Nitisols while in Murugi they are Humic Nitisols. These are deep, well-weathered soils with moderate to high inherent fertility but over time soil fertility has declined due to continuous cropping and minimal use of fertilizers. They have generally low concentrations of organic carbon (<2.0%), nitrogen (<0.2%) and available phosphorus (<10 ppm) and are moderately acidic (pH ranges from 4.8 to 5.4). All these conditions are unfavourable for crop production and result in low crop yields.

Dissemination Approaches Used

Participatory Rural Appraisal (PRA)

Participatory Rural Appraisal (PRA) is a growing family of approaches and methods to enable local people to share, enhance and analyse their knowledge of life and conditions and to plan, act, monitor and evaluate their circumstances (Chambers, 1997). The goal of PRA is to implement a project that is socially acceptable, ecologically sustainable and economically viable. PRA helps the community to mobilize its human and natural resources, define prevailing problems, prioritize opportunities, evaluate local institutional capacities and develop a systematic site plan of action for the community to adopt and implement in their farms. This project used the PRA to evaluate the current state of affairs regarding soil management in the target area in a participatory manner. The PRAs were conducted in all the four sites during the month of February 2004 and involved two phases. Phase one involved problem diagnosis while phase two involved the research team providing feedback of the diagnosis phase to the farmers and identification of potential solutions to the constraints.

Mother–Baby Approach

The mother–baby trial model is an upstream participatory research methodology designed to improve the flow of information between farmers and researchers about technology performance and appropriateness under farmer conditions (Snapp, 1999). The methodology was initially developed and implemented to test legume-based soil fertility management technologies in Malawi and was later expanded to Zimbabwe. The trial design consists of two types of trials: mother and baby. The mother trial is researcher-designed and conforms to scientific requirements for publishable data and analysis. A baby trial consists of a single replicate of one or more technologies from the mother trial. Each farmer manages their baby trial on their land. According to Johnson et al. (2003) the mother–baby trial methodology has three goals. The first is to generate data on which to assess the technology performance under realistic farmer conditions. The second is to complement the agronomic trial data with farmers' assessment of the adoption potential technologies. The third is to encourage farmers to actively participate in the trials and is expected to stimulate farmer experimentation with and adoption of new technologies and practices. In this approach, all the farmers within the vicinity of the "mother" sites were given equal opportunities to participate in the study (Franzel et al., 2002). The methods used to disseminate and scale up ISFM technologies included (a) field days and (b) individual farmer visits to the demo sites. The objectives of the field days and the visits were to conduct a field visit for farmers in the demonstration site and to get feedback from farmers concerning the performance of the various treatments.

Training Workshops

The project highly borrowed on the methodology described by Bunch (1982) that advocates a combination of 80% practical training and 20% theory. Trainings were conducted for both farmers and extension staff. The objectives of the trainings were to train farmers and extension staff on the principles of the ISFM strategies; to practically demonstrate on the use of technologies in the mother trial; to train farmers on the best usage and application methods of the different resources into the soil; to advice farmers on research benefits, record keeping and group dynamics; and to train farmers who in turn can train others. Each training workshop was organized into six workstations, (i) Organic materials (*tithonia*, *calliandra* and *leucaena*): sole or in combination, (ii) Manure management, (iii) Mineral fertilizers, (iv) Herbaceous and grain legumes (v) Community mobilization and technology transfer and (vi) Soya bean utilization. In the manure and fertilizer training, appropriate agronomic procedures (timing, rates and application procedures) were emphasized. For example, some of the farmers who had prior experience of using some of the technologies acted as trainers.

Farmers' Groups

With the realization that it would not be possible to reach every individual farmer in the project working area, the farmers were encouraged to form groups in each village. Farmer groups are formed on the principle that when farmers come together, they develop effective working relationships and have synergy, defined as the ability to combine the perspectives, resources and skills of a group of people. Farmer groups or community-based organizations have been used successfully in western Kenya to promote improved fallows technology for soil fertility improvement (Place et al., 2004). In central Kenya, Wambugu et al. (2001) reported success in scaling up of calliandra for fodder using farmer groups.

Stakeholders' Workshops

A strong partnership with stakeholders that includes institutional support has been advocated as an important factor for a successful scaling-up strategy especially with soil fertility-related technologies (Snapp et al., 2002; Sanchez, 1999). The stakeholders included researchers from Kenyatta University, Kenya Forestry Research Institute (KEFRI), Katholieke Universiteit Leuven, Kenya Agricultural Research Institute (KARI), and Tropical Soil Biology and Fertility Institute of CIAT (TSBF-CIAT), extension staff from the Ministry of Agriculture, field technicians and farmers.

Participatory Monitoring and Evaluation

There is often a danger with all new approaches to rural development that this "promising idea" will fall out of favour before its practitioners have had sufficient time to evaluate results and improve on methods (Belshaw, 1997). Now it is increasingly recognized that there is a need for greater attention to the monitoring and evaluation of participatory research projects (McAllister and Vernooy, 1999). With this background a participatory monitoring and evaluation framework was developed where all the different stakeholders developed their own indicators for monitoring and evaluating the project (Mugendi and Mucheru-Muna, 2006). The indicators for technology evaluation that farmers developed include crop foliage colour, height, vigour, inflorescence size and general crop condition.

Results and Discussions

Participatory Rural Appraisal

A total of 1,428 farmers (831 males and 597 females) attended the problem diagnosis meeting while 2,118 farmers (1186 males and 932 females) attended the feedback meetings to rank the problems and suggest solutions. Table 1 shows the problem ranking and their solutions as done by the farmers.

The results indicate that the immediate or core problem is food shortage in all the sites in Meru South District. The farmers associated the food shortage to low yields from their farms. The causal factors for the low yields of both crop and fodder were also outlined as lack of finances to enable farmers allocate adequate resources to food production. Some also argued that sometimes seeds purchased from the local shops are forged and hence did not produce as expected. Low yields consequently led to low finances, and farmers were therefore unable to purchase farm inputs. There was also a problem of inability to connect with extension workers, and this also linked to lack of knowledge on what inputs to apply on the farms for efficient productivity. The farmers also outlined lack of adequate water and rainfall, presence of poor soils, soil erosion and small farms as underlying factors impeding solid production figures. On top of the limited natural resource base, aggravating factors include poor tillage methods, lack of proper knowledge on how to manage soils and lack of finances. Consequently, it emerged that all these problems have a cyclical nature and are all inter-related as shown in Fig. 1.

Since it was not practical to address all the problems identified by the farmers, the project agreed together with the farmers to focus on soil fertility and food security-related problems. During the second phase (feedback meetings), the farmers also identified locations where the demonstration sites (mother trials) would be situated. The farmers agreed that the trials should be (i) on public land, (ii) secure, (iii) accessible to all participants and (iv) at least 2–3 acres in size.

Mother–Baby Approach

Four demonstration trials (mothers) were established in Meru South in each of the study sites (Kirege, Mucwa, Mukuuni and Murugi). They were all situated in public schools except in one site (Mukuuni) where not enough public land was available. Here the farmers selected one farmer who agreed to give land for the demonstration. The mother trials addressed different ISFM themes; in Kirege and Mucwa, the theme was on biomass transfer using tithonia, calliandra, leucaena, manure and mucuna while Mukuuni focused on maize/grain legume intercrops (groundnuts, beans and cowpea) and Murugi on maize and grain legume (soya beans and green grams) rotations.

Farmers were invited to attend joint field days in the mother trials during the grain filling stage every season. The field days were attended by a wide range of stakeholders who included farmers, teachers of the local schools, students from the local secondary and primary schools, local administration, extension personnel from the Ministry of Agriculture and researchers and lecturers from Kenyatta University. The field days involved field visits to the demonstration plots and discussions where farmers interacted with the researchers and extension personnel. Due to the different themes in each of the sites, cross-site visits were also organized during the field days. This enabled farmers to move and share experiences with each other. During the field days, farmers gained knowledge on various technologies in learning by seeing what was being done in the demonstration (mother) plots and then started testing technologies of their choice on their farms (babies).

Four field days were held in Mucwa, Mukuuni, Murugi and Kirege with a total of 1,644 farmers attending. The data shows that more female farmers attended the field days during the 2nd (55%) and 4th (52%) field days compared to the 1st and 3rd field day where there were 35% and 50% females, respectively. Murugi recorded a consistent increase in the female farmers' attendance as time progressed (Table 2). The large number of farmers who participated in the field days can be an indication that they perceive soil fertility as a genuine problem. Versteeg and Koudokpon (1993) also reported an enhanced farmer participation in their study when farmers perceived soil fertility to be a serious problem.

Although the attendance by female farmers (48%) was less than that of males, their participation was relatively high compared to other activities that report poor participation by female farmers due to demanding family chores. Their participation is crucial at enhancing transfer of the ISFM technologies because women usually carry out the bulk of agricultural activities (Adiel, 2004).

Before the start of the project, farmers were using varied vegetative materials such as maize stover, grevillea, banana leaves as bedding material for livestock

Table 1 Problems diagnosed, ranked and suggested solutions by farmers in all the sites during the PRA in Meru South District, Kenya

Problem diagnosed by farmers	Rank by farmers	Solution suggested by the farmers
Poor soil fertility and soil erosion	1	• Practice soil conservation (bench terraces, contour hedges, crop rotation, grass strips and planting trees) • Government to help in the provision of demonstration and training on soil fertility improvement technologies • Practice early planting and mixed cropping • Provision of water for irrigation • Be assisted in soil testing
Animal and crop pests and diseases	2	• Increase extension staff to teach farmers • Use indigenous knowledge to control pests and diseases
Lack of knowledge and extension services	3	• Increase extension staff • Farmers be consulted so as to participate in decision making with extension staff
Inadequate finances to purchase fertilizer	4	• Use of organic materials plus mineral fertilizers • Be provided with credit facilities
Erratic rainfall	5	• Be provided with demonstration and training on soil fertility improvement • Practice soil and water conservation (bench terraces, contour hedges, crop rotation grass strips and planting trees) • Be provided with water for irrigation and practice mixed cropping
Poor markets	6	• Farmers to form marketing groups in order to eliminate middlemen in the marketing of farm produce • Government to ensure that farmers get quality seeds verified through KARI and other organizations • Government to ensure that farmers get markets and good market linkages through formation of marketing groups
Inadequate fodder	7	• Plant more fodder (Napier grass and improved fodder)
Lack of fertilizers and chemicals. Fake seeds	8	• Provide technical know-how through seminars, farmer field days and educational tours, etc. • Government to lower prices of inputs such as fertilizers and seeds • Government to ensure that farmers get quality seeds verified through KARI and other organizations
Small farms	9	• Improve production within small farms • Hire land for cultivation • Offer labour for sale to enable farmers buy food
Food insecurity	10	• Improve production within small farms by improving soil fertility • Use indigenous knowledge to control pests and diseases • Offer labour for sale to enable farmers buy food

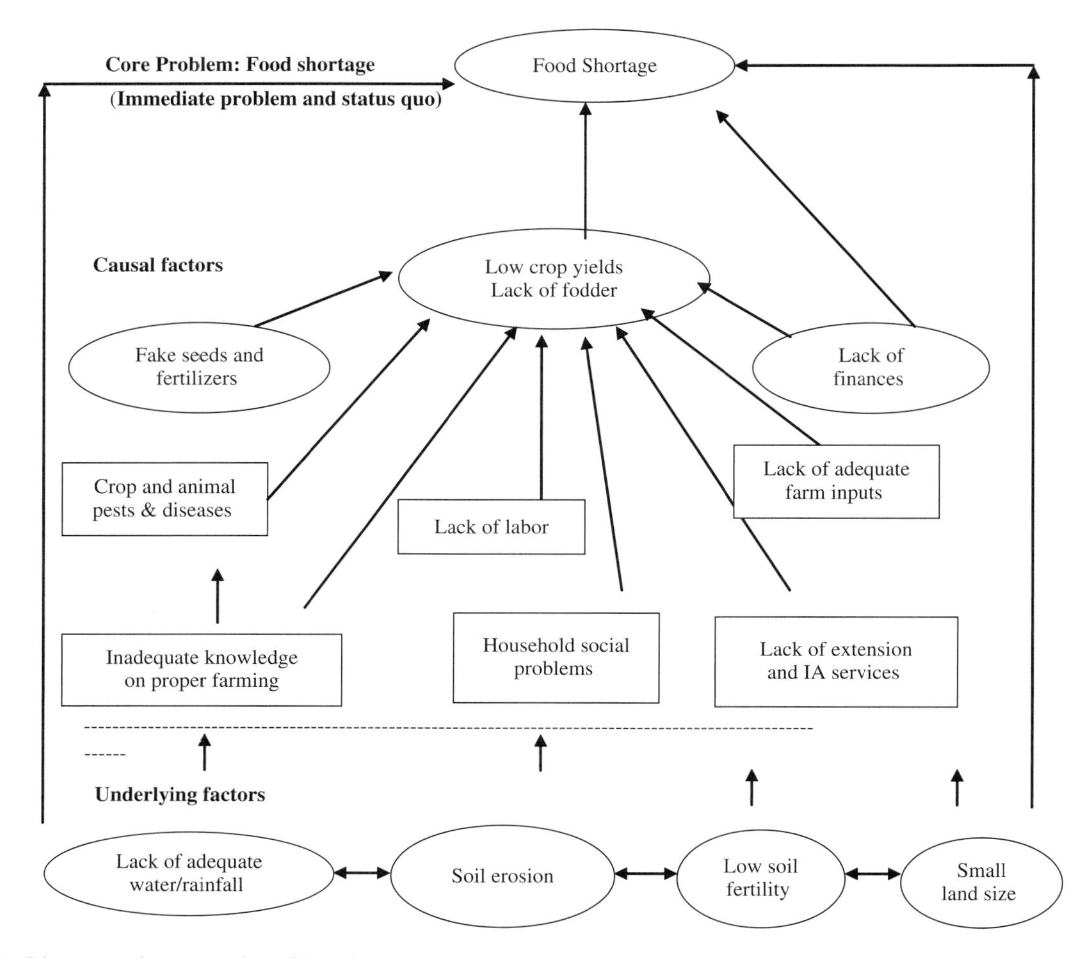

Fig. 1 Diagrammatic presentation of the primary problem status in Meru South District, Kenya, as outlined by farmers during PRA in 2004

and to increase quantity of manure. Manure made from such material was however of poor quality and farmers lacked adequate knowledge on its management and application rates. The use of mineral fertilizer was also common before the project but farmers lacked knowledge on types, dosage and management. During the field days and village training workshops, farmers were trained on proper manure management and appropriate use of mineral fertilizers. Calliandra, tithonia and leucaena were new organic materials introduced to the farmers. A farmers survey conducted during 2005 LR season, involving 563 farmers who

Table 2 Field day attendance by gender in the different sites during 2004 and 2005 in Meru South, Kenya

	First field day		Second field day		Third field day		Fourth field day	
Site	Male	Female	Male	Female	Male	Female	Male	Female
Kirege	64	38	–	–	–	–	–	–
Mucwa	45	37	20	31	46	40	–	–
Mukuuni	145	55	101	114	75	63	80	56
Murugi	81	53	91	109	46	62	77	115
Total	335	183	212	264	167	165	157	171

Table 3 Soil fertility replenishment technologies tested by farmers during 2005 LR in Meru South, Kenya

	Kirege	Mukuuni	Mucwa	Murugi	Total
Technology	$N = 73$	$N = 195$	$N = 62$	$N = 233$	$N = 563$
Tithonia	3	22	11	32	68 (12.1%)
Tithonia + fertilizer	0	10	0	14	24 (4.3%)
Tithonia + manure[a]	2	3	8	5	18(3.2%)
Tithonia + manure + fertilizer[a]	0	1	1	2	4 (0.7%)
Manure	15	63	22	37	137 (24.3%)
Calliandra	1	0	1	2	4 (0.7%)
Fertilizer	27	135	26	78	266 (47%)
Manure + fertilizer	38	69	31	108	246 (43%)
Mucuna + manure[a]	2	0	0	0	2 (0.4%)
Leucaena	0	0	0	1	1 (0.2%)
Leucaena + fertilizer	0	1	0	0	1 (0.2%)
Leucaena + calliandra + manure[a]	0	0	1	0	1 (0.2%)
Calliandra + manure + tithonia[a]	0	0	1	0	1 (0.2%)

[a]Farmers innovated technologies
Percentages add to more than 100% because farmers tried more than one technology

had registered as having tried the new technologies, showed that farmers were testing a wide range of inputs (Table 3). The majority of the farmers tried mineral fertilizer (47%), and a slightly smaller percentage tried manure plus mineral fertilizer (43%).

Most of the farmers tried more than one technology while others tried various combinations of the technologies, modifying them based on their own talent for innovation (Table 3). The modifications of the technologies mainly involved combining different organic materials. At the demonstration site single-type organic materials were solely applied or combined with mineral fertilizer. Farmers indicated that they combined different organic resources because the amount of biomass of an individual kind was mostly inadequate.

Farmers' modification of technologies has been reported by other authors. For example, Franzel and Scherr (2002) noted that small-scale farmers are rarely able to manage any single enterprise in the "optimal" manner prescribed by researchers and therefore adapt the technologies to their circumstances. According to these authors, farmers make compromises in the management of individual enterprises in order to reduce risk, alleviate constraints and increase the productivity of the entire household livelihood system. Modifications of agricultural technologies have similarly been reported by other authors (Adesina et al., 1999; Pisanelli et al., 2000). Adesina et al. (1999) argued that farmers make modifications to fit their managerial and production systems, and these modifications often lead to a final technological package for farmers which is adopted if it is technically feasible, profitable and acceptable to farmers.

Of the total number of farmers who were trying the technologies, 88% had attended field days, while 62% had attended village training workshops. This finding is against our hypothesis where we stated that the village training workshops will be more effective in reaching the farmers than the field days. An explanation for this could be that during the field days farmers were able to see, learn and evaluate technology performance while in the village training workshops, they only learnt how to apply the inputs.

The yields from the farms (babies) varied among the treatments during 2005 SR (Table 4). Maize grain yield from the plots with inputs was significantly higher ($P < 0.05$) than the control treatment during both seasons. However, during 2005 SR, the lowest yields were obtained in the control treatment followed by calliandra and fertilizer treatments. During 2006 LR, the highest yields were obtained from tithonia + manure + fertilizer and tithonia + fertilizer treatments, an indication that sole organic materials, their combinations or a combination of organic materials + fertilizer yielded more than the fertilizer alone treatment. These findings agree with other authors who reported that integration of organic and inorganic nutrient gave higher yield than inorganic fertilizer in the region (Mugendi et al., 1999;

Table 4 Maize grain yield from the different treatments during long rain 2005 and short rain 2006 in Murugi and Mukuuni, Meru South District, Kenya

	Grain yield, mg ha^{-1}			
	Short rain 2005		Long rain 2006	
Treatment	Mukuuni	Murugi	Mukuuni	Murugi
Control	1.2^{b}	0.9^{bc}	0.7^{c}	1.0^{cd}
Fertilizer	2.4^{a}	1.9^{abc}	1.5^{bc}	1.3^{cd}
Calliandra + fertilizer	2.6^{a}	2.0^{ab}	1.8^{ab}	2.1^{bc}
Manure	2.6^{a}	2.0^{ab}	2.2^{ab}	1.4^{cd}
Manure + fertilizer	2.7^{a}	2.1^{ab}	1.5^{bc}	2.3^{b}
Calliandra	2.7^{a}	1.8^{bc}	1.8^{ab}	1.7^{bcd}
Tithonia	3.0^{a}	2.3^{ab}	2.1^{ab}	1.9^{bc}
Tithonia + fertilizer	3.1^{a}	2.4^{ab}	2.3^{ab}	2.9^{a}
Tithonia + manure + fertilizer	3.2^{a}	2.5^{ab}	2.5^{a}	3.6^{a}
Tithonia + manure	3.4^{a}	2.9^{a}	2.6^{a}	2.5^{b}
LSD	1.0	1.0	0.9	0.8

Note: Figures in each column followed by the same letter are not significantly different at $P < 0.05$
LSD, average least significant difference in the means at $P < 0.05$

Mucheru-Muna et al., 2007). It has been reported that a combination of organic and inorganic nutrient sources ensures synchrony between nutrient release and plant uptake, increasing nutrient use efficiency (Vanlauwe et al., 2002).

During the fourth year (2006) farmers shared the experiences in testing of the technologies on their farms (Table 5). The farmers used yields, soil improvement seen through crop responses, soil colour change and labour as their criteria to rank the soil fertility technologies. Tithonia was ranked the best technology followed closely by manure + fertilizer and calliandra and leucaena. The least ranked was fertilizer and as indicated, the major reason was its high costs.

Table 5 Farmers' experiences and constraints in adopting soil fertility technologies in Meru South District, Kenya

Technology	Experiences	Constraints	Score	Rank
Manure	• Requires fertilizer • Must be fully decomposed • Gives high yields • Encourages soil micro organisms	• Labour intensive • Cannot be used directly from the cowshed	5.5	4
Fertilizer	• High yields • Required all seasons • May acidify soils	• Costly • Not readily available	5	5
Manure + fertilizer	• High maize vigour • High yields	• Fertilizer is expensive • No livestock, no manure • Labour intensive • Limited know-how on manure management	7	2
Tithonia	• Crops are not affected by nematodes • Improved soil fertility • High yields	• Biomass not readily available • Labour intensive	8.5	1
Leucaena/ calliandra	• Good fodder for livestock • Increase milk production • Increase yields	• Land is limited • Inadequate biomass	7	2

Training Workshops

Village training workshops were held at every site in September 2004 and July 2005 with a total of 1,971 farmers participating (Table 6).

The first village training workshop was well attended, and most of the farmers were males (58%). During the second village training, the number of female farmers attending increased to 55%. The village workshops provided a forum for learning and sharing experiences among farmers and between farmers and scientists. The village training strategy was extremely important in that it was more attractive to women who are not able to attend training venues far from their villages due to their gender roles and relations in the household.

Farmer Groups

A total of 40 farmer groups were formed and trained on tree nursery establishment and management (Mugwe et al., 2007), group dynamics and book keeping. The group membership consisted of 60% females and 40% males. A survey carried out in June 2005 indicated that in Kirege, 43% of the female-headed households who were trying the technologies were group members while they were 42,73 and 78% in Mukuuni, Murugi and Mucwa, respectively. Male-headed households who were group members were 70% in Murugi, 56% in Mukuuni, 55% in Mucwa and 27% in Kirege. This shows that both male- and female-headed households found group membership an important avenue to try the new technologies. In Murugi and Mucwa, most of the farmers who were trying the technologies were group members indicating the importance of groups as far as the uptake of the technologies was concerned.

Table 6 Farmers village training workshops attendance by gender in Meru South District, Kenya

	First training		Second training	
Site	Male	Female	Male	Female
Kirege	337	235	62	50
Mucwa	100	33	68	78
Mukuuni	220	185	107	151
Murugi	96	103	72	94
Total	753	556	309	373

One of the major achievements of these groups was the ability to propagate seedlings in their nurseries and plant them on-farm. During the three seasons that they produced seedlings (2004 SR, 2005 LR and 2005 SR) farmer groups from Murugi produced the largest quantity with 89,631 seedlings followed by Mukuuni farmer groups with 45,196 seedlings. Kirege and Mucwa farmer groups produced the least number with 32,343 and 24,130 seedlings, respectively. Difference in the quantities produced was mainly attributed to the number of groups in each site and membership per group. For example, Murugi and Mukuuni had 10 groups each but had more members per group than Mukuuni, and this could explain why Murugi produced more seedlings than Mukuuni. The use of group nurseries as a means of availing propagation materials to farmers was effective, and considerable success was registered as noted in the number of seedlings farmers received within the short time period. At the end of the three seasons, about a third of all members in the group nurseries received close to 500 seedlings, an amount enough to supplement animal protein for one dairy cow.

Before 2004, only a few farmers had calliandra and leucaena on their farms. An on-farm survey carried at this time showed that Kirege had 8 farmers out of 46 (17%), Mucwa had 7 out of 26 (26%), Murugi had 6 out of 63 (10%), while there was none at Mukuuni who had calliandra or leucaena. In addition, the mean number of trees per farmer was few with Murugi having a mean of 3 trees, Kirege and Mucwa having 11 and 44 trees, respectively. The relatively higher number of trees at Kirege and Mucwa was attributed to past contact of these sites with project activities and other institutions such as the Ministry of Agriculture that promoted planting of calliandra and leucaena. By 2006, the mean number of trees surviving and being utilized had increased to more than 300 trees per farmer, an indication of a positive impact of the group nurseries.

Though the major objective of forming the group nurseries was to enable propagation of calliandra and leucaena seedlings and consequently increase on-farm planting, farmers had an opportunity to interact and know each other better and gained confidence of remaining united. Apart from raising calliandra and leucaena seedlings, and other tree and vegetable species, the groups started other extra activities. For instance, the groups started trying the different technologies in their group farms and also started activities revolving around generating income for the group. Merry-go-round was the most common with more than 50% of the groups having merry-go-round activities as a way of raising funds for the group and also as a group welfare activity. This was reflected by some of the groups transforming themselves to a more formal grouping through registration in the Ministry of Culture and Social Services. These groups are therefore likely to continue even after the end of the project.

Stakeholders' Workshops

The partnerships between the different stakeholders were formalized through the signing of memorandum of understanding (MOU) between the research organizations, farmers and extension personnel. Stakeholders' workshops were held once every year to evaluate and review the progress of the project. For the farmers, each group was represented by two farmers selected by the group, and these representatives were different each year to enhance full participation by the members of the groups. A total of 280 representatives attended the stakeholders' workshops during the 4 years. The stakeholders highlighted the progress they had made during each year and the challenges they had experienced. They also planned together on activities for the coming year and shared responsibilities.

During the second annual stakeholders' workshop (January 2005), farmers suggested another approach of "farmers' training ground" (Fig. 2) to complement the "mother–baby" approach. Farmers indicated that they would be more confident to take technologies to their farms if they first practiced together in small groups (arrow 3, Fig. 2). They indicated that they would be able to confirm whether the good performance they observed at the demonstration site would be transferable to their farms. However, they also indicated that some farmers who were willing to test straight from demonstration were free to do so (arrow 1, Fig. 2). In the "farmers' training ground" approach, the farmers form groups, and one of the farmers avails a piece of land where the farmers practice some of the technologies demonstrated at the mother trial site. The farmers agreed on few technologies to test at the training grounds.

The farmers noted that the approach was more practical and realistic as it gives the farmers a hands-on opportunity to do what is done on the demonstration site. The farmers planned together on which day to come and prepare the land, plant, weed, water and so on. Farmers selected a few technologies from the demonstration trial which they tried within their groups and then eventually started trying the technologies in

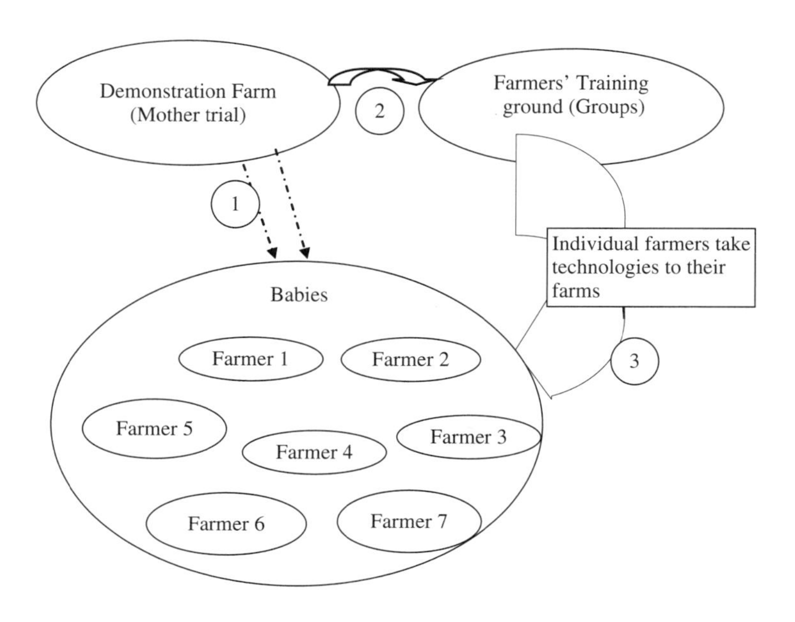

Fig. 2 The adopted new mode of participatory technology transfer by farmers in Meru South District, Kenya

their individual farms. Development of this approach by farmers was one of the major outcomes from the annual evaluation and planning meetings. This is an indication of farmers' interest, full participation and a sense of owning the project activities. According to Franzel et al. (2002) farmers' workshops can be an important means for farmers to discuss issues related to new practices, exchange opinions and lessons, and come to consensus or clarify differences.

Participatory Monitoring and Evaluation

Farmers identified four major outcomes which they expected to achieve by the end of the project and then identified indicators that they would use to measure the extent of progress. The first outcome was higher and better quality yields, and the indicators to measure this were (i) presence of technologies on farms, (ii) increased food supply at household level and (iii) increased food for sale. The second outcome was they gained knowledge on better and improved farming methods, and the indicators to measure this were (i) presence of technologies on farms, (ii) increased knowledge about the technologies, (iii) confidence in teaching other farmers and (iv) confidence in inviting other experts to train them and in asking questions. The third outcome was improved soil conservation, and the indicators to measure this were (i) reduced soil erosion and (ii) formation of bench terraces. The fourth outcome was improved animals, and the indicators to measure this were (i) presence of more healthy animals and (ii) more milk production. The farmers continuously used these indicators to evaluate the progress of the technologies in their farms and the project.

During the third year (2005), farmers in Mukuuni and Murugi sites scored and ranked technologies they were testing in their farms in terms of overall benefits accrued. Results showed that in both sites, the tithonia + manure treatment was ranked highest receiving the highest possible score of 10 in both male and female groups (Tables 7 and 8). Calliandra alone or combined with mineral fertilizer was ranked low in both sites.

Table 7 Farmers scoring and ranking of the technologies on basis of benefits in Mukuuni, Meru South District, Kenya

	Group mean scores				
Technology	Male	Female	*t*-test, *P*	Total mean	Rank
Calliandra	2.0	3.3	0.016	2.7	9
Fertilizer	3.8	2.0	0.038	2.8	8
Calliandra + fertilizer	4.0	4.3	0.768	4.2	7
Tithonia	5.3	5.0	0.768	5.2	6
Manure	6.0	5.3	0.609	5.7	5
Tithonia + fertilizer	8.0	7.3	0.561	7.7	4
Manure + fertilizer	7.3	8.3	0.251	7.8	3
Tithonia + manure	7.67	8.3	0.519	8.0	2
Tithonia + manure + fertilizer	10.0	10.0	Na	10.0	1

Table 8 Farmers scoring and ranking of the technologies on basis of benefits in Murugi, Meru South, Kenya

	Group mean scores				
Technology	Male	Female	*t*-test, *P*	Total mean	Rank
Calliandra	2.7	2.3	0.678	2.5	9
Manure	3.3	4.0	0.374	3.7	8
Calliandra + fertilizer	3.0	4.3	0.116	3.7	7
Fertilizer	6.7	3.3	0.024	5.0	6
Manure + fertilizer	6.7	7.0	0.768	6.8	5
Tithonia + manure	5.3	8.7	0.002	7.0	4
Tithonia	8.3	6.3	0.013	7.3	3
Tithonia + fertilizer	8.3	8.0	0.815	8.2	2
Tithonia + manure + fertilizer	9.7	10.0	0.374	9.8	1

There were significant differences in the scoring of some technologies by males and females.

The results showed that scores allocated to calliandra by the female farmers were significantly higher than those allocated by males (Table 6). This implies that female farmers had a higher preference for calliandra than the male farmers in Mukuuni. This relates to the fact that women are responsible for feeding the animals in the evening, after milking, and calliandra serves as a fodder supplement that is cheap. Since most of the women do not control a lot of cash within the household, this is a better option for them. On the other hand, scores allocated to mineral fertilizer by male farmers were significantly higher than those allocated by female farmers in both Mukuuni and Murugi (Tables 7 and 8), implying that male farmers had a higher preference for mineral fertilizer than the female farmers. This could be because the male farmers can easily afford to purchase fertilizers since they control most of the cash in the households.

Preference for tithonia + manure by female farmers was significantly higher than by the male farmers, while male farmers significantly preferred sole tithonia. This could be attributed to the fact that the female farmers are the ones responsible for collecting and carrying the biomass and would therefore prefer a technology that is not very labour intensive (manure). The male farmers on the other hand being heads of households can be able to solicit for labour from their wives and children.

Stakeholders' workshops are also an important means to find out farmers' views on the technologies, their potential impacts and general project performance in terms of achieving its objectives. Therefore, during the last stakeholders meeting held in March 2007, farmers used the indicators they had developed during the PM&E activities and evaluated the extent to which the stipulated outcomes had been achieved (Table 9). Generally, increased knowledge about the technologies, confidence in teaching others and more healthy animals scored highest (80%). This could be attributed to intensive training that was conducted throughout the project period in terms of field days and village training workshops. Using the knowledge gained, farmers were able to use the technologies on their farms (68%) and consequently increase food supply at the household level to about 68%. Increase in milk production (78%) as a result of improved health of the animals (80%) could be explained by farmers feeding fodder trees to the animals. At the start of the project, farmers had on average less than 10 fodder trees per farmer but after 3 years, the number of trees per farmer had increased to about 300 trees.

Table 9 Expected outcomes, indicators and farmer assigned percentages of achievements in Meru South

Expected outcome	Indicators	(%)
High yields	Presence of technologies on farms	68.1^{ab}
	Increased food supply at household level	68.7^{ab}
	Increased food for sale	33.7^{c}
Knowledge on better and improved farming methods	Presence of technologies on farms	68.1^{ab}
	Increased knowledge about the technologies	80.0^{a}
	Confidence in teaching others	80.0^{a}
	Confidence in inviting other experts and asking questions	74.5^{a}
Improved soil conservation	Reduced soil erosion	63.8^{ab}
	Formation of bench terraces	61.3^{b}
Improved animals	More healthy animals	80.0^{a}
	More milk production	77.7^{a}
	LSD	*16.3*
	P	*<0.001*

Percentage means with the same superscript are not significantly different

Table 10 Farmers' evaluation of the different approaches used in the project in Meru South District, Kenya

Approach	Mean (%)	Rank
Training grounds	77.5[a]	1
Demo farms	76.2[ab]	2
Nurseries	71.2[abc]	3
Field days	68.7[abc]	4
Groups	58.7[abc]	5
Training workshops	55.0[bc]	6
Cross-site visits	50.0[c]	7
LSD	*22.2*	

Percentage means with the same superscript are not significantly different

The farmers also evaluated the participatory approaches that the project had been using in disseminating the technologies using a score of 10 for most effective and 1 for least effective. Training grounds were ranked the best overall followed by demonstration farms with 77.5 and 76.2%, respectively, while the least scored were cross-site visits (50%) and village training workshops (55%) (Table 10). The training grounds were significantly more effective in the dissemination of the technologies than cross-site visits and village training workshops (Table 10).

Possible explanation for this could be the fact that cross-site visits only involved a few farmers due to transport costs. The reason why the training grounds were significantly more effective than the village training workshops could be due to the fact that the farmers could see the crop growing and they were also involved in the actual management of the technologies, while in the village training workshops they were observers and not doers. In addition the farmers were the inventors of the training ground concept indicating that they actually owned the idea. We had earlier anticipated that the village training workshops would have a higher contribution to dissemination compared to the field days, but this was not the case as far as the farmers were concerned. This agrees with Franzel et al. (2002) who noted that whereas in many cases the information provided by farmers in such workshops is what the researchers might have anticipated, it is possible that new information may be obtained.

Lessons Learnt

A number of lessons were learnt during the project, and this agrees with Franzel et al. (2002) who noted that participatory research offers researchers, extensionists, policy makers and farmers themselves an opportunity to learn important lessons about achieving effective dissemination of agricultural technologies, as well as feedback on further research priorities. Some of the lessons learnt include the following:

(i) The use of "mother–baby" approach to scale up the technologies may not be very effective for all technologies. The use of "farmers' training ground" approach where the farmers form groups and one farmer avails a piece of land where they practice some of the technologies from the "mother trial" is a better option. This gave the farmers a hands-on experience to do what is done on the mother trial. The farmers plan together on which day to come to carry out various activities on their plot like prepare the land, plant, weed, water.

(ii) Working in partnership is very constructive and enhances sustainability as well as impact. For instance, the project activities have awakened some retired agricultural officers who are now being invited by the groups to come and discuss agricultural matters with the farmer groups. In this way there is a renewal of synergy and recognition and appreciation of the knowledge of other professional community members. In other areas, the farmers have made an effort to forge links with the Department of Forestry and are being taught various other methods of handling seedlings. This is an interesting experience given that only 3 years ago the very farmers were accusing the Ministry staff for not being available to assist them.

(iii) The framework of participatory interaction has enabled farmers to view the "outsiders" as partners in development and with whom they do engage in development discourse. This indicates that they are in a position to interact very well with other partners who use the participatory approaches used in this project.

(iv) The use of participatory monitoring and evaluation has been a major eye opener for all stakeholders on how the farmers would like to direct the course of a project for their optimal benefit.

(v) Keeping a vibrant and an effective working partnership is very challenging and more expensive than had been anticipated. Constant meetings and regular field team visits must be put in place.

Conclusions and Recommendations

The farmers' participation was quite high in most of the joint activities, and this possibly contributed to the high rate of farmers' acceptance to test the technologies. The immediate problems identified by the farmers were low crop yields and lack of fodder which were caused by inadequate finances to purchase inputs, inadequate knowledge on farming and inadequate extension services among others. The underlying constraints to the causal factors were erratic rainfall, soil erosion, low soil fertility and small farm sizes. The current study, however, addressed only two of them (low soil fertility and soil erosion) and realized positive impacts. The problem of erratic rainfall is critical and there is need to address it through exploration of strategies to conserve moisture and optimize its usage. As for the problem of small farms, options of optimizing efficiency in agricultural production need to be sought.

Adoption of the introduced ISFM technologies has potential to increase crop yields by over 100%, and the high yields observed from innovations of mixing organic materials developed by farmers is a reflection that farmers are modifying the technologies to fit their needs and circumstances. This is a lesson that researchers should facilitate the process of farmer learning, experimentation and modification of technologies as what is likely to remain on the farmers' fields is what farmers have modified to suit them as this is what they can cope with. In addition, the gender differences observed in the participation of project activities and preferences of technologies is also a lesson that gender issues are critical and need to be put into consideration in dissemination of ISFM technologies.

This study has provided pathways for future scaling up of ISFM activities. It is recommended that future efforts aimed at reaching farmers should concentrate on demonstrations, farmer training grounds, field days and farmers' groups. Farmer groups are particularly effective in activities that involve bulking of planting materials as revealed by the success in propagating seedlings. Since ISFM technologies are management and information intensive and require skills, there is need to build capacity for field extension workers. A policy framework should therefore be put in place to impart appropriate skills to the extension workers.

It was evident that income generation activities among the groups were instrumental in keeping the groups cohesive and individual members interested in the group activities. Researchers and the other stakeholders should therefore channel their efforts towards empowering farmers to engage in income-generating strategies as these are likely to improve resilience of the groups.

Acknowledgements Financial support was provided by the Belgium Government through the Flemish Inter-university Council (VLIR), the Rockefeller Foundation (RUFORUM) and FARM AFRICA Technology Transfer Maendeleo Fund. We also appreciate the contributions of collaborators from the Kenya Agricultural Research Institute, Kenya Forestry Research Institute, Kenyatta University, Katholieke Universiteit Leuven, Tropical Soil Biology and Fertility Programme of CIAT and the Ministry of Agriculture of Kenya. This work could not have been possible without the cooperation of farmers who participated and are highly appreciated. The local administration (chiefs and sub-chiefs) played a crucial role in mobilizing the farmers.

References

Abayomi A, Fadayomi O, Babatola JO, Tian G (2001) Evaluation of selected legume cover crops for biomass production, dry season survival and soil fertility in a moist savanna location in Nigeria. Afr Crop Sci J 9:615–627

Adesina AA, Coulibaly D, Manyong VM, Sanginga PC, Mbila D, Chianu J, Kamleu G (1999) Policy shifts and adoption of alley farming in West and central Africa. IITA and Meg Communication, Ibadan, Nigeria

Adiel RK (2004). Assessment of adoption potential of soil fertility improvement technologies in Chuka Division, Meru south district Kenya. MSc thesis, Department of Environmental Foundations, Kenyatta University, Nairobi

Bationo A, Kimetu J, Ikerra S, Kimani S, Mugendi DN, Odendo M, Silver M, Swift MJ, Sanginga N (2004) The African network for soil biology and fertility: new challenges and opportunities. In: Bationo A (ed) Managing nutrient cycles to sustain soil fertility in sub-Saharan Africa. Academy Science Publishers and TSBF of CIAT, Nairobi, pp 1–24

Bekunda MA, Bationo A, Saali H (1997) Soil fertility management in Africa. A review of selected research trials. In: Buresh RJ, Sanchez PA, Calhoun F (eds) Soil fertility replenishment in Africa: an investment in natural resource capital, SSSA special publication no. 51, Madison, Wisconsin, SSSA, pp 63–79

Belshaw D (1997) 'And fourteenthly': is operational rural development theory too difficult and the paradigm shift too easy? Paper presented at Annual Conference of the Development Studies Association, University of East Anglia, Norwich, UK

Bunch R (1982) Two ears of corn: a guide to people-centered agricultural development. Oklahoma City, World Neighbors

Buresh RJ, Niang AI (1997) *Tithonia diversifolia* as a green manure: awareness, expectations and realities. Agroforest Forum 8:29–31

Chambers R (1997) Whose reality counts? Putting the first last. Intermediate Technology Publications, London. Institute of Development Studies, Brighton, UK

Chambers R, Pacey RA, Thrupp L (1987) Farmer first: farmer innovation and agricultural research. Intermediate Technology Publications, London

Constantinides M, Fowness JH (1994) Nitrogen mineralisation from leaves and litter of tropical plants: relationships to nitrogen, lignin and soluble polyphenol concentrations. Soil Biol Biochem 26:49–55

Cramb RA (2000). Processes affecting the successful adoption of new technologies by smallholders. In: Stur WW, Horne PM, Hacker JB, Kerridge PC (eds) Working with farmers: the key to the adoption of forage technologies. ACIAR Proceedings No. 95. Australian Centre for International Agricultural Research, Canberra, pp 11–22

Franzel S, Scherr SJ (2002) Trees on the farm: assessing the adoption potential of agroforestry practices in Africa. CAB International, Wallingford, CT

Franzel S, Scherr SJ, Coe E, Cooper PJM, Place F (2002) Methods of assessing agroforestry adoption potential. In: Franzel S, Scherr SJ (eds) Trees on farm: assessing the adoption potential of agroforestry practices in Africa. Wallingford, UK, CAB International, pp 11–36

Giller K (2001) Nitrogen fixation in tropical cropping systems, 2nd edn. CAB International, Wallington

GOK (2001) Poverty reduction strategy paper. Central Bureau of Statistics, Nairobi

Haverkort B (1991) Farmers' experiments and participatory technology development. In: Harverkort B, Van der Kamp J, Waters-Bayer A (eds) Joining farmers experiments: experiences in participatory technology development. Intermediate Technology Publications, London, pp 3–16

Jaetzold R, Schmidt H, Hornet ZB, Shisanya CA (2006) Farm management handbook of Kenya. Natural conditions and farm information, vol 11/C. Eastern Province, 2nd edn. Ministry of Agriculture/GTZ, Nairobi

Jama BA, Buresh RJ, Ndufa JK, Shepherd KD (1997) Agronomic and economic evaluation of organic and inorganic sources of phosphorus in western Kenya. Agron J 89:597–604

Johnson NL, Lilja N, Ashby JA (2003) Measuring the impact of user participation in agricultural and natural resource management research. Agric Syst 78:287–306

Kanyama-Phiri G, Snapp SS, Minae S (1998) Partnerships with Malawi farmers to develop organic matter technologies. Outlook Agric 27:167–175

Kapkiyai JJ, Karanja NK, Woomer P, Qureish JN (1998) Soil organic carbon fractions in a long term experiment and the potential for their use as a diagnostic assay in highland farming systems of central Kenya. Afr Crop Sci J 6:9–28

Kimani SK, Macharia JM, Gachengo C, Palm CA, Delve RJ (2004) Maize production in the central highlands of Kenya using cattle manures combined with modest amounts of mineral fertilizer. Uganda J Agr Sci 9:480–490

Kimani SK, Nandwa SM, Mugendi DN, Obanyi SN, Ojiem J, Murwirwa HK, Bationo A (2003) Principles of integrated soil fertility management. In: Gichuru MP, Bationo A, Bekunda MA, Goma HC, Mafongoya PL, Mugendi DN, Murwirwa HM, Nandwa SM, Nyathi P, Swift MJ (eds) Soil fertility management in Africa. A regional perspective. Academy Science Publishers, African Academy of Sciences, Nairobi, pp 51–72

McAllister S, Vernooy R (1999) Action and reflection: a guide for monitoring and evaluating participatory research. International Development Research Centre, Ottawa

Ministry of Agriculture (2002) Annual report for Meru south district. District Agricultural Office, Meru south, Nairobi

Mucheru-Muna MW, Mugendi DN, Micheni A, Mugwe JN, Kung'u JB, Otor S (2007) Effects of organic and mineral fertilizer inputs on maize yield and soil chemical properties in a maize cropping system in Meru South District, Kenya. Agroforest Syst 69:189–197

Mugendi DN, Mucheru-Muna MW (2006) FARM Africa technical MATF end of year 2 report. Kenyatta University, Kenya

Mugendi DN, Mucheru MW, Micheni A, Mugwe JN, Kung'u JB, Otor R (2001) Soil fertility strategies for improved food production in the central highlands of Kenya. Afr Crop Sci Conf Proc Lagos Nigeria 5:897–904

Mugendi DN, Nair PKR, Mugwe JN, O'Neill MK, Woomer P (1999) Alley cropping of maize with calliandra and leucaena in the subhumid highlands of Kenya. Part 1: soil fertility changes and maize yield. Agroforest Syst 46:39–50

Mugwe JN, Mugendi DN (1999) Use of leguminous shrubs for soil fertility improvement in sub-humid highlands of Kenya. In: Tenywa JS, Zake JYK, Ebanyat P, Semalulu O, Nkalubo ST (eds) Proceedings of the 17th soil science society of East Africa conference, 6th–10th Sept 1999, Kampala, Uganda, pp 194–198

Mugwe JN, Mugendi DN, Mucher-Muna M, Kungu JB, Wakori S (2007) Promotion of leguminous trees for soil fertility management through group tree nurseries in the central highlands of Kenya. Unpublished. VLIR Annual report. Kenyatta University, Kenya

Mugwe JN, Okoba BO, Mugendi DN, Tuwei PK (2004) Soil conservation and fertility improvement through use of leguminous shrubs in central highland of Kenya. In: Bationo A (ed) Managing nutrient cycles to sustain soil fertility in sub-Saharan Africa. Academy Science Publishers, Nairobi, pp 277–297

Odee DW (1996) Nitrogen-fixing trees in agroforestry systems: myths and realities. East Afr Agr Forest J 62(1):139–149

Okali C, Sumberg J, Farrington J (1994) Farmer participatory research: rhetoric and reality. Intermediary Technology Publication, London

Palm CA, Myers RJK, Nandwa SM (1997) Organic-inorganic nutrient interactions in soil fertility replenishment. In: Buresh RJ, Sanchez PA, Calhoun F (eds) Replenishing soil fertility in Africa. Soil Science Society of America Special (SSSA) Publication no. 51. Soil Science Society of America, Madison WI, pp 193–217

Pisanelli A, Franzel S, Dewolf J, Romnelse R, Poole J (2000) Adoption of improved tree fallows in Western Kenya. Farmer practices, knowledge and perception. International Centre for Research in Agroforestry (ICRAF), Nairobi

Place F, Kariuki G, Wangila J, Kristjanson P, Makau A, Ndubi J (2004) Assessing the factors underlying differences in farmer groups: methodological issues and empirical findings from the highlands of central Kenya. Agric Syst 82:257–272

Probert ME, Okalebo JR, Jones RK (1995) The use of manure on smallholder farms in semi-arid eastern Kenya. Exp Agr 31:371–381

Rutunga V (2000) Potential of *Tephrosia vogelii* Hook F and *Tithonia diversifolia* (Hemsley) A Gray short duration fallows for improving the productivity of maize in Western Kenya. PhD thesis, Faculty of Agriculture, University of Nairobi, Nairobi

Sanchez PA (1999) Improved fallows come of age in the tropics. Agroforest Syst 47:3–12

Sanchez PA, Jama BA (2002) Soil fertility replenishment takes off in East and southern Africa. In: Vanlauwe B, Diels J, Sanginga N, Merckx R (eds) From Integrated plant nutrients management in sub Saharan Africa: concept to practice. CABI, Wallingford, pp 23–46

Smaling EMA, Nandwa SM, Janssen BH (1997) Soil fertility is at stake. In: Buresh RJ, Sanchez PA, Calhoun F (eds) Replenishing soil fertility in Africa, Special publication no. 51. American Society of Agronomy and Soil Science Society of America, Madison, WI, pp 47–61

Snapp SS (1999) Mother and baby trials. A novel trial design being tried out in Malawi. Target Newsletter of the Southern Africa Soil Fertility Network 17:8

Snapp S, Kanyama-Phiri G, Kamanga B, Gilberts R, Wellard K (2002) Farmer and researcher partnerships in Malawi: developing soil fertility technologies for the near term and far-term. Exp Agric 38:411–431

Vanlauwe B, Palm CA, Murwira HK, Merckx R (2002) Organic resource management in sub-Saharan Africa: validation of a residue quality driven decision support system. INRA, EDP Sciences, www.edpsciences.org, pp 839–846

Versteeg MN, Koudokpon V (1993) Participative farmer testing of four low external input technologies, to address soil fertility in Mono Province (Benin). Agric Syst 42: 265–276

Wambugu C, Franzel S, Tuwei P, Karanja G (2001) Scaling up the use of fodder shrubs in central Kenya. Development and agroforestry. Dev Practice 11(4):487–494

Woomer PL, Muchena FN (1996) Recognizing and overcoming soil constraints to crop production in tropical Africa. Afr J Crop Sci 4:503–518

Success Stories: A Case of Adoption of Improved Varieties of Maize and Cassava in Kilosa and Muheza Districts, Eastern Tanzania

C.Z. Mkangwa, P.K. Kyakaisho, and C. Milaho

Abstract The project of Eastern Zone Client Oriented Research and Extension (EZCORE) was designed to improve the capability of researchers to better respond to the needs of the farmers and empower the districts to obtain and implement relevant research results for the benefit of the rural populations. In the process, farmers in Mkobwe village of Kilosa district reported that they were getting very low maize yields due to problem which was later identified as maize gray leaf spot. Similarly, farmers of Tongwe village in Muheza district had a disease problem in cassava which was later identified as cassava brown streak disease. Researchers were commissioned to solve these two problems which were severely limiting crop yields. In Mkobwe village, a maize variety called TMV-2 was tested and found suitable, while in Tongwe village, a cassava variety *Kiroba* was the best. The maize Var. TMV-2 and cassava Var. *Kiroba* were then provided to the villagers of Mkobwe and Tongwe, respectively. The monitoring and evaluation of project activities for the last 3 years indicated that all the farmers in Mkobwe village and more than 50% of the farmers in the nearby villages are growing TMV-2. With regard to *Kiroba* variety, the variety is planted by all the farmers in Tongwe village, and now the variety has spread to over eight villages in Muheza district. Apart from Muheza district, it is also planted in many villages in the nearby districts. The source of cassava cuttings in all the villages planted with variety Kiroba is Tongwe village.

C.Z. Mkangwa (✉)
Ilonga Agricultural Research Institute, Kilosa, Tanzania
e-mail: mkangwa@yahoo.co.uk

Keywords Improved varieties · Technology adoption · Kilosa

Introduction

The Eastern Zone Client Oriented Research and Extension (EZCORE) Project was established in 1999 and aimed at improving food security and livelihood of the rural poor and reducing poverty through policies and programmes which empower the beneficiaries. The programme strives to attain improved food security and sustainable management of natural resources, therefore contributing directly to national goals of poverty reduction and hence improving livelihoods. In Tanzania, client-oriented research (COR), sometimes also referred to as client-oriented research and management (CORMA), has been adopted as a strategy for delivery of research services in future. Until now, COR has been implemented in the northern and lake zones. Basically, COR strives to forge closer links between the demand side (farmers, governments, NGOs, traders and businesses) and the supply side (research stations, NGO and training institution). But in the case of EZCORE, the supply side included extension services, which include government and service providers from the private sector.

EZCORE made deliberate efforts to forge links with extension services by involving district councils on behalf of the beneficiaries to manage the research resource as well as manage the entire research process, from problem identification to contracting out and using the outcomes of research findings.

In order to achieve the set objectives, EZCORE had to support many activities which eventually changed

A. Bationo et al. (eds.), *Innovations as Key to the Green Revolution in Africa*,
DOI 10.1007/978-90-481-2543-2_131, © Springer Science+Business Media B.V. 2011

the mindset of the farmers. Farmers were organized into groups and they were trained on group dynamics and on skills to interact with local government authorities on various matters related to their development initiatives and were eventually able to identify their problems, formulate researchable issues and forward to the district level. On the other hand, village extension officers (VEOs) were trained on participatory extension methods so that they can be able to assist the farmers in identifying the problems using participatory approaches.

Through such skills, farmers in Mkobwe village, Kilosa district were able to identify a problem mentioned as drying off of maize leaves. This problem was causing yield decline of up to 90% (Mbiza et al., 2002). Similarly, in Tongwe village, Muheza district, farmers identified cassava root rot problem, which caused yield decline of up to 74% (Muhana and Mtunda, 2002). In Mkobwe village, farmers reported that maize yields used to be about 1250 kg ha^{-1} but gradually declined to less than 500 kg ha^{-1} in January 2002. After infection, the yields of both maize and cassava could not support the family for food even for 3 months after harvest. Most families were therefore food insecure and poverty levels kept on increasing because they could not get cash by selling maize and from cassava produce.

These two disease problems were forwarded to the respective district office, and then to researchers. The solutions to these two diseases were not available and therefore researchers were commissioned to solve these two problems. Different maize and cassava varieties were evaluated in the farmers' fields by researchers in collaboration with farmers and identified tolerant varieties of TMV-2 for GLS and *Kiroba* for CBSD, and thereafter the varieties were adopted by farmers. This chapter explains the extent of adoption of maize variety TMV-2 and cassava variety *Kiroba* after being found to resist grey leaf spot and cassava brown streak disease, respectively.

Materials and Methods

Mkobwe village is in Kilosa district, found in Uluguru mountain ranges of eastern arc mountains, and located at an elevation of about 1700 m above sea level (m.a.s.l.), while Tongwe is found in Muheza district coastal area at an elevation of 500 m.a.s.l.

Commissioned researchers from Ilonga Agricultural Research Institute and Kibaha Sugarcane Research Institute collected samples from affected plant parts of both maize and cassava plants for identification of the diseases. The plant parts collected were affected leaves from maize plants and leaves and tuber from cassava plants. After identification, seven maize varieties were compared for resistance/tolerance against GLS in Mkobwe village. The varieties used were Tuxpeno, TMV-1, TMV-2, Staha, Kito-ST, Kilima-ST and Ngida. The plot size used was 10 m × 10 m and assessed for severity of GLS using scores of 1–5. The lowest score represented less affected, and highest score represented most affected plants. This was then followed by establishing four demonstration plots, each located in one sub-location using three varieties. These were TMV-2, Kilima-ST and Ngida. These varieties were chosen because of being more tolerant to GLS than were the other varieties. The basic seed obtained from seed farms was used in all cases. The demonstration plots were laid in farmers' fields, and farmers were involved in the GLS assessment and in the field days. Two field days were conducted, the first at vegetative phase and the second at physiological maturity. The evaluation was administered by researchers of Ilonga Agricultural Research Institute in close collaboration with farmers and extension staff of Kilosa district. It was important to conduct the second field day at physiological maturity because the GLS-infected plants normally have dry leaves as if the crop has matured, but if examined, the cobs are still green and if husks are removed, grain filling is poor. Normal maize physiological maturity starts with the maize cob by drying of the husks.

In the case of cassava, eight varieties were compared in Tongwe village. These include Kiroba, Namikonga, Kitumbua, Naliendele, Kigoma-red, Kikombe, Kibaha, Mahiza and Kibanda-meno. In the first year of 2003, the varieties were planted on four farmers' fields on a plot size of 15 m × 15 m. In the second year of 2004, the varieties were planted on 14 farmers' fields. The plot size used was the same as that of the first time. In each of these two plantings, farmer field- and cassava days were conducted and attracted more farmers, were also used to make some assessments. Apart from assessing the resistance of these varieties to CBSD, farmers also evaluated other qualities which are important in cassava industry. These qualities include yield level, taste and time to maturity. Farmers gave scores starting from 1 to 5 for each quality. Apart from evaluation, cassava day was used to popularize the variety by distributing the cuttings to the farmers who showed interest. The evaluation was administered by researchers of Root and Tuber Programme based at Kibaha in close collaboration with extension staff of Muheza district.

Results and Discussion

Table 1 gives the scores of severity of maize varieties to GLS as scored by farmers. Maize varieties which were most susceptible to GLS were TMV-1, Kito-ST and Tuxpeno with the highest score of 4.0, followed by Kilima-ST and local variety Ngida with comparable scores of 3 and 3.5, respectively. The most resistant variety to GLS was TMV-2. Similar observations were obtained in the southern highlands of Tanzania, which include Mbeya, Iringa and Ruvuma regions. In these regions, Mbiza et al. (2002) reported that TMV-2 was also found to be resistant to GLS.

Table 1 GLS severity score of maize varieties in Mkobwe village in year 2002

Maize variety	Score
Tuxpeno	4.0
TMV-1	4.0
TMV-2	1.0
Staha	4.0
Kito-ST	4.0
Kilima-ST	3.5
Ngida	3.0

Score scale 1–5, where 1, clean leaves; 5, severely affected

Initial symptoms of GLS disease can be observed as small lesions surrounded by a yellow halo confined between the veins, which are easily visible when the infected leaf is held against light. In severely infected maize plant, leaves are completely blighted and cob filling is very poor or there is no formation of seeds. The GLS pathogen (*Cercospora zeae-maydis*) causes extensive leaf blighting. Reduction in green leaf area of the upper leaves may cause significant yield reduction of 75–90% of the photosynthates required by the ear during grain filling (Ward et al., 1997).

The scores given by farmers with regard to tolerance of different cassava varieties to CBSD are given in Table 2. With the exception of *Kiroba*, the other varieties which were evaluated were susceptible to CBSD at varying degrees. The local varieties (Mahiza and Kibanda-meno) were more susceptible than those which were introduced by researchers. Apart from being less susceptible to CBSD, *Kiroba* variety had better qualities than did other varieties. For instance, the yield per plant was 12–18 kg, which is equivalent to 76–115 t ha^{-1} of fresh cassava, and matures only within 9 months. These qualities made most farmers prefer planting variety *Kiroba* in Tongwe and later in neighbouring villages.

In the year 2003, the variety which showed resistance to GLS started to be demanded by many farmers

Table 2 Farmers' assessment of cassava resistance to CBSD

	Scores			
Varieties	CBSD susceptibility	Yield	Taste	Time to maturity
Kiroba	1	1	2	1
Namikonga	2	2	1	1
Kitumbua	3	3	1	3
Kibaha	2	3	3	4
Naliendele	2	3	2	3
Kigoma-red	3	3	2	4
Kikombe	2	3	1	1
Mahiza	5	3	3	3
Kibanda-meno	3	3	2	4

Scoring: CBSD susceptibility: 1, not susceptible; 5, very susceptible. Yield: 1, 12–18 kg $plant^{-1}$; 2, 10–14 kg $plant^{-1}$; 3, 6.5–12 kg $plant^{-1}$. Taste: 1, very sweet; 5, very biter. Time to maturity: 1, 8–9 months; 2, 9–12 months; 3, 12–18 months; 4, 18–24 months

Table 3 Adoption of maize variety TMV-2 by December 2006

Village	Farmers planting TMV-2 (%)	Average grain yield (t ha^{-1})
Mkobwe	70	3.2
Masenge	55	3.0
Lufikiri	50	3.2
Nongwe	60	3.8
Mandege	62	2.9
Rubeho	65	3.0
Kisitwi	40	2.6

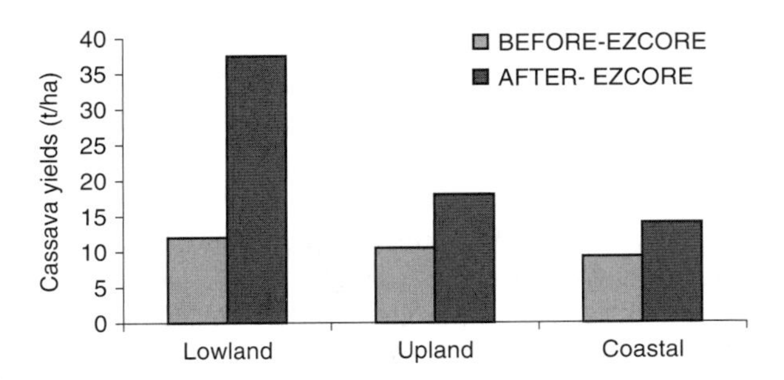

Fig. 1 Changes in cassava production (t ha^{-1}) in different zones of Muheza district

of Mkobwe village, and in the following year, the farmers from the nearby villages also showed interest of planting TMV-2.

Monitoring and evaluation mission of year 2006/2007 indicated that maize variety TMV-2 has been adopted by other villagers in the villages found along Ukaguru mountain ranges. These villages include Masenge, Kisitwi, Rubeho, Mandege, Nongwe and Lufikiri. Most farmers in these villages have adopted the variety (Table 3). The adoption rate of this maize variety ranged from 40 to 70%. This was possible because the seed was available from the farmers who were commissioned to produce quality-declared seed and sell the seed at affordable price equivalent to US $0.3. The seed is still demanded by more farmers, as the seed produced in the village has not satisfied the demand of all the farmers. It is worth noting that the problem of GLS has disappeared in these villages. Apart from tolerating the disease, maize yield obtained from this variety was in the range of 2.6–3.8 t ha^{-1}. This is equivalent to 420–660% increase if compared to the yield of 0.5 t ha^{-1} obtained by GLS-infected maize varieties.

With regard to *Kiroba*, the variety has spread very fast. Within 2 years, *Kiroba* cuttings were sold and adopted by farmers in eight different villages within Muheza district (Table 4). This figure is according to the village records, but it may be higher since some farmers may have given the cuttings to their relatives without reporting to the village authority. The selling of the cuttings was also to six districts within the same period.

According to the monitoring and evaluation report of year 2007/2008, cassava production in Muheza district has increased by 40–60% due to the introduction and adoption of high-yielding *Kiroba* variety (Figure 1). The increase in cassava production varied from 68% in the lowland, 41% in the upland, to 34% in the coastal zone. This high cassava production has improved food security and the surplus was sold and improved farmers' income. Some farmers revealed that increased income from selling of cassava cuttings in Tongwe village enabled them to pay school fees, purchase radio, mobile phones and iron sheets.

At district level, the adoption of *Kiroba* variety featured high compared to other technologies extended within the same period (Table 5). This could be due to the fact that farmers are best rewarded by planting this variety compared to other technologies.

Table 4 Purchases of Kiroba variety from Tongwe village

Year	Villages within Muheza district	Districts
2005/2006	Mtimbwani, Mavovo, Maramba, Amani, Bombani	Tanga, Handeni, Kilindi
2006/2007	Kwemhosi, Nkumba, Ubembe	Same, Pangani, Korogwe

Table 5 Overall adoption of technologies in Muheza district

Types of technologies	Adoption (%)
Maize production technologies	18
Orange improvement technologies	9
Cassava improvement technologies	36
Chicken improvement technologies	36
Milk improvement technologies	18
Cashew nut improvement technologies	27
Soil and water conservation/soil fertility improvement	5

Conclusions

Empowering small-scale farmers with knowledge can improve their participation in technology testing and adoption of improved technologies. This approach normally leads to the change of the mindset of the farmers and can be used to break with business as usual syndrome, and contribute to green revolution in Africa.

Acknowledgements The authors would like to thank the farmers for their commitment in identifying the problems and who were involved in evaluation and eventual adoption of the varieties. We also thank AfNet for providing avenue to present this success story from Tanzania.

References

Muhana M, Mtunda K (2002) Report on the study of cassava root rot problem in Muheza district, Tanga region-Tanzania, Muheza District Council. Unpublished report, p 42

Mbiza ABC, Mbwaga AM, Chilagane A (2002) An outbreak of grey leaf spot disease in Mkobwe village Nongwe Division, Kilosa district, Unpublished report, p 30

Ward JMJ, Laing MD, Riijikenberg FHJ (1997) Frequency and timing of fungicide applications for control of grey leaf spot in maize. Plant Dis 81:41–48

The Role of Forest Resources in the Strategies of Rural Communities Coping with the HIV/AIDS Epidemic in Sub-Saharan Africa

J.B. Kung'u

Abstract The HIV pandemic has had dramatic effects on rural livelihoods in Africa. In rural communities only a few people have access to treatment due to high prices of medicine, poor health infrastructure, and long distances to the nearest health centers. These have led to a greater dependence on the natural resources by the rural communities to alleviate the problems. In the rural areas, the natural resource products are easily accessible to most people and their use has increased over the years. There has been higher demand for wood, to prepare food for increasingly frequent funerals and for making coffins. Similarly, HIV has complicated existing livelihood crises resulting from droughts, high prices of fertilizer, and poor marketing services. The impact of HIV and AIDS on household labor has also intensified the dependence on tree products like fruits, roots, tubers, and vegetables. This chapter examines the role of forest resources in the responses to HIV and AIDS, particularly in terms of herbal medicines, energy, and food. The chapter shows that HIV and AIDS epidemic has tremendously increased the dependence on wood resources and that the pandemic has environmental and natural resource management implications. Some policy and program interventions that might help lessen the impact of the pandemic on natural resources and the role forest resources can play in response to HIV and AIDS have been highlighted.

J.B. Kung'u (✉)
Department of Environmental Sciences, School of Environmental Studies, Kenyatta University, Nairobi, Kenya
e-mail: kungu_james@yahoo.com

Keywords Forest resources · HIV/AIDS · Labor · Health · Food · Livelihoods

Introduction

It is reported that in the year 2004 alone, the global HIV/AIDS epidemic killed more than 3.1 million people and that an estimated 4.9 million acquired the human immunodeficiency virus (HIV), bringing the number of people living with the virus around the world to 40 million (UNAIDS, 2005). Sub-Saharan Africa (SSA) remains by far the worst affected. For a long time, HIV/AIDS has been seen as purely a health issue, and most financial resources have passed through Ministries of Health toward reducing infection rates, buying antiretroviral (ARV) drugs, educating and awareness raising, and, more recently, increasing home-based care (UNAIDS, 2001; FAO/WHO, 2002). There is need to realize that HIV and AIDS have implications that reach far beyond health which include great impacts on agricultural and food production systems and natural resource management (FAO, 1995; 2001; FAO/UNAIDS, 2003; FAO/WHO, 2002).

In sub-Saharan Africa, most infected people live in the rural areas and HIV and AIDS have become mostly a rural problem (UNDP, 2002). Sub-Saharan Africa has largely a rural-based economy and it is unlikely that the epidemic can be controlled without the effective support of agricultural and natural resource management (du Guerny, 1999). This sector is in a strong position to assist in both the prevention and the mitigation of HIV and AIDS (Gari and Villareal, 2002). Equally, there are limitations to the extent to

A. Bationo et al. (eds.), *Innovations as Key to the Green Revolution in Africa*,
DOI 10.1007/978-90-481-2543-2_132,

which the health or the agricultural sector can operate, and for this reason a multi-sectoral approach is crucial (FAO, 2005). Impacts experienced by people living with AIDS (PLWA) include health constraints, labor shortages and a weakened labor force, social isolation, monetary shortages, redistribution of tasks, and paying more attention to the sufferer at the expense of other necessary activities. PLWA provide less labor, have less capital, and are more in need of risk management strategies. As they struggle to pay increased medical bills, they lose their earning capacity and at the same time their financial wealth decreases which may lead to their assets such as livestock, tools, or seed reserves being sold. Cash crop production is often abandoned due to its excessive financial and labor requirements. All of these aspects contribute to a decline in production in rural communities and to farm degradation in terms of a decrease in the use and conservation of agrobiodiversity, a decrease in food quality and quantity, and an abandonment and disinvestment in land. Since PLWA are less able to grow crops, they increasingly shift to gathering trees and scrubs for their daily subsistence needs for food, medicine, and other products. This chapter examines the role of tree and scrub resources in the response to HIV and AIDS, particularly in terms of herbal medicines, energy, and food. Some policy and program interventions that might help lessen the impact of the pandemic on natural resources and the role farm trees and shrubs can play in the multi-sectoral response to HIV and AIDS are highlighted.

Impact of HIV/AIDS on Population Structure

Demographic projections of the impact of HIV/AIDS on population structures reveal dramatic changes in the size, age, and sex compositions (Fig. 1). Not only does HIV/AIDS reduce the total population but also the age and sex structure, resulting in a population dominated by the elderly and the youth. In some countries (Fig. 1), AIDS is erasing decades of progress made in improving mortality conditions and is reducing life expectancies. The average life expectancy in sub-Saharan Africa is now 47 years when it would have been 62 years without AIDS. In Botswana, for example, life expectancy at birth has dropped to a level not seen in that country since 1950 (UNAIDS, 2002).

Impact of HIV/AIDS on Household Labor

HIV/AIDS poses a direct threat to household food security as it affects the most productive household members. When a person is sick, the household has to manage not only without their labor inputs but with the loss of labor from those who have to care for the sick. AIDS is characterized by recurrent periods of sickness, and so recurrent loss of labor, which eventually erodes agricultural production and food security. Much of rural agricultural production is highly labor dependent and often labor demands are concentrated in specific periods of the year. For instance, sickness or funeral attendance may mean that the planting season is missed and with it a full crop. Gross agricultural production is also affected by labor shortages. The FAO has estimated that in the 25 hardest hit countries in Africa, AIDS has killed around 7 million agricultural workers since 1985 and it could kill 16 million more before 2020 (FAO, 2001). It is reported that the most affected African countries could lose up to 26% of their agricultural labor force within a few decades leading to severe impacts on the national economy (FAO, 2001). In Bondo district, Kenya, the number of female-headed households (FHH) and the number of orphan-headed households increased dramatically due to HIV/AIDS between 1992 and 2002 (Fig. 2). This is unlike the married-headed households (MHH) which decreased dramatically.

As Table 1 shows, the projected loss of total population due to HIV/AIDS in some countries by 2020 is expected to be as high as 30%, while the loss in agricultural labor force is expected to be as high as 26%. This will have a very big impact on agricultural production since households will be mainly headed by orphans and old people (Fig. 2).

The death of women and men in their economic prime is also accompanied by the loss of their skills and knowledge base. Many are dying before they have time to share these essential livelihood skills with their children, thereby reducing the range of livelihood options for the next generation. Further disruptions appear when the increased need for medicine and

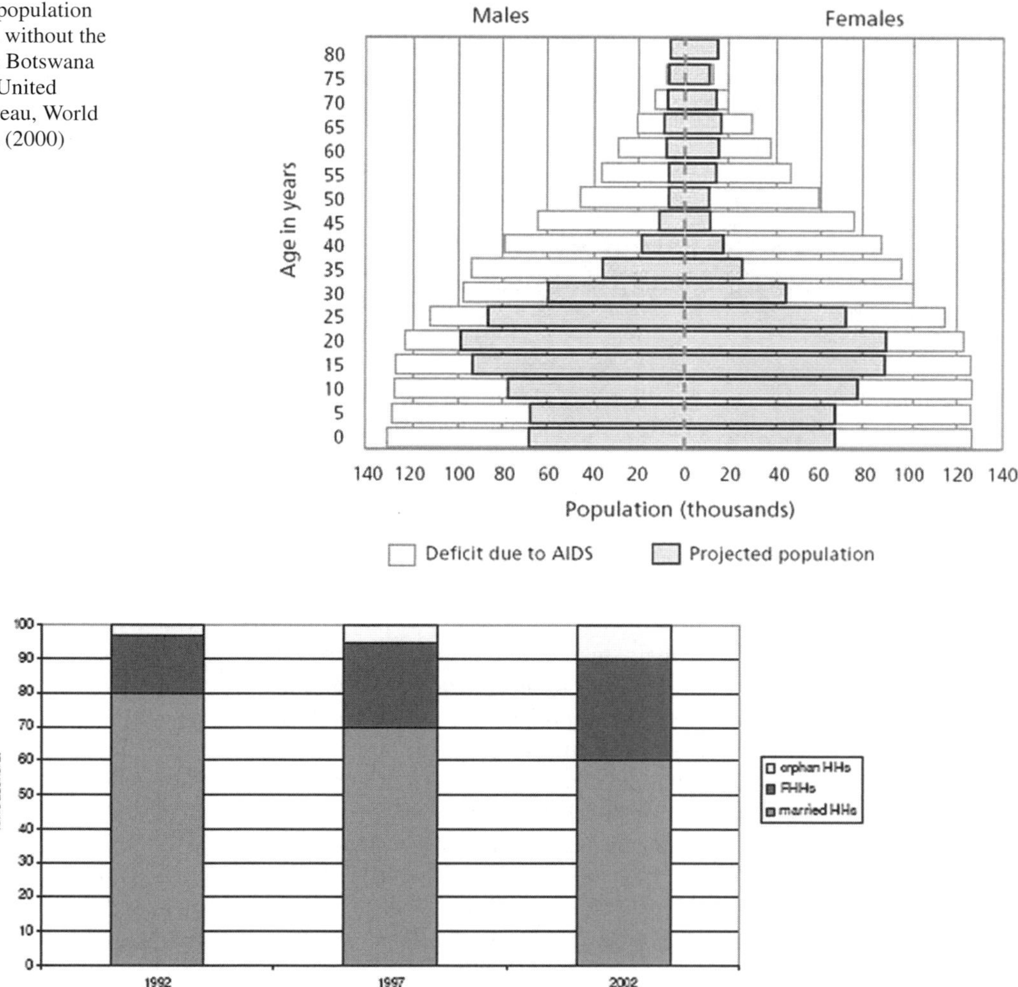

Fig. 1 Projected population structure with and without the AIDS epidemic in Botswana by 2020. *Source:* United States Census Bureau, World Population Profile (2000)

Fig. 2 Changing composition of households in Bondo district, Kenya, between 1992 and 2002. *Source*: Bishop-Sambrook, 2003

Table 1 Projected loss in total population and agricultural labor force due to AIDS, 1985–2020, in some SSA countries

Country	Total population (%)	Agricultural labor force (%)
Namibia	–17	–26
Botswana	–30	–23
Zimbabwe	–23	–23
Mozambique	–16	–20
South Africa	–27	–20
Kenya	–16	–17
Malawi	–17	–14
Uganda	–8	–14
Tanzania	–7	–13

Source: FAO (2001)

treatment forces productive family members to leave their farms in search of paid employment elsewhere. Children are often taken out of school because their labor is needed on the farms and because there is no longer money to pay for education. The stress of HIV/AIDS on the social capital within communities also erodes the traditional transmission of knowledge between households and communities and contributes to the demise of local seed exchange systems.

Farming households affected by HIV/AIDS have labor shortages and are using various coping strategies (Engh et al., 2000; Egal and Vastar, 1999). The household health care represents 25–50% of the net annual income of most small farms in developing countries. High prices of drugs and recent market orientation of health-care systems limit access to medical treatment (Nnko et al., 2000; Farah, 2001).

Role of Tree and Scrub Resources in the Prevention, Care, and Treatment of HIV/AIDS

It is estimated that about 1.6 billion people in the world, which make up more than 25% of the world's population, rely on forest resources for their livelihoods. Of these, almost 1.2 billion live in extreme poverty (World Bank, 2001a). People residing in and near forests typically obtain a considerable and variable amount of nutritious foods from forests. Forest ecosystems contribute to the diets and subsistence of forest dwellers, and in increasingly market-oriented economies, they provide a significant portion of the food and medicines consumed by urban populations. Indigenous people's foods form part of rich knowledge systems. Traditional food systems typically draw on local biodiversity and are based on local production and management of land and specific environments (Johns, 2006).

Individuals infected with HIV/AIDS are recommended to eat more food as their bodies require more nutrients (FAO, 2001). Low-income and HIV/AIDS-affected households often rely upon tree and forest products to complement their diets (i.e., wild food plants, bushmeat, nuts, leaves, and roots). Forest foods are often good sources of micronutrients (vitamins and minerals which are essential for good nutrition and health) and are essential to HIV/AIDS-affected households. As Table 2 shows, some fruits in the forest contain high amounts of vitamins and nutrients.

Wild foods continue to provide the major portion of the animal fats, proteins, and minerals in the diets of millions of people. In 62 developing countries, it is reported that people obtain more than 20% of their protein from wild meat and fish (Bennett and Robinson, 2000). In the Congo basin alone, more than 1 million tons of wild meat is consumed yearly which is equivalent to 4 million cattle (Wilkie, 2001), while in the Amazon basin, 67,000–164,000 tons of wild meat per year is consumed (Bennett et al., 2002).

Wild supplies of food comprise much more of the diet of subsistence populations than is often realized (Hoskins, 1990). HIV/AIDS-affected households tend to attach more importance to forest product collection than do non-affected households (Barany et al., 2005). Approximately 1,500 species of wild plants have been reported as being collected for consumption in Central and West Africa (Chege, 1994). In some parts of Africa, diets based on staple grains depend largely on tree products to provide essential vitamins.

There are many trees that produce oil seeds, edible leaves, and fruits that are rich in important vitamins and nutritional elements (Hoskins, 1990; Ogden, 1990). After the oil palm, the shea butter tree (*Butyrospermum paradoxum*) has been reported to be the second most important source of fat in African diets (FAO, 1995). Dietary supplementation with forest and tree products can play an important role in community nutrition, given growing evidence that malnutrition is a major underlying cause for the rapid expression of AIDS in Africa's HIV-infected individuals (Enwonwu and Warren, 2001). There is evidence that agricultural labor and cash shortages amongst HIV/AIDS-affected households have led to the reversion and increased consumption of wild foods, including fruits, nuts, leafy vegetables, fungi, and protein sources such as bushmeat and insects (UNAIDS, 1999; Kengni et al., 2004). Wild plants are a principal source of traditional medicines that may help to treat many of the symptoms of opportunistic infections that are associated with AIDS (Kung'u et al., 2006). Indigenous medicinal plants (including cultivated tree nuts and wild fruits) may also boost the immune system of HIV/AIDS patients.

Role of Forest Resources as Source of Food for Rural Communities

People residing in and near forests typically obtain a considerable, although variable, amount of nutritious foods from forests – with poor people generally more dependent on such food. Forest resources contribute to the diets and subsistence of rural communities and in increasingly market-oriented economies, they provide a significant portion of the food and medicines consumed by urban populations. Most societies recognize

Table 2 Some neotropical fruits in Brazil that are excellent sources of provitamin A

Fruit	Portion analyzed	α-Carotene (μg/g)	β-Carotene (μg/g)	β-Cryptoxanthin (μg/g)	Other carotenoids (μg/g)	Vitamin A activity in mixed foods (retinol activity equivalents/100 g)
Mauritia vinifera	Pulp	80.5	360		γ-Carotene, 37	3,050
Astrocaryum vulgare	Pulp		107	3.6		930
Eugenia uniflora	Pulp		9.5		γ-zeacarotene, 5.9	830
Acrocomia makayayba	Pulp		55			490
Bactris gasipaes	Boiled pulp	3.2	22		γ-Carotene, 18	270
Malpighia glabra			26	3.6		230
Mammea americana	Pulp		14		β-Apo-10′-carotenal, 5β-apo-8′-carotenol,11	195
Spondias lutea	Pulp and peel		1.4	17.0		93
Cariocar villosum	Pulp		1.2	4.4		30

Source: Adapted from Rodriguez-Amaya (1996)
Note: By comparison, mango (*Mangifera* spp.) and papaya (*Carica papaya*) provide 38–257 and 25–150 retinol activity equivalents per 100 g, respectively (USDA-ARS, 2004)

that food, medicine, and health are interrelated. Food is typically associated with cultural identity and social well-being. Indigenous people's foods form part of rich knowledge systems. Traditional food systems typically draw on local biodiversity and are based on local production and management of land and specific environments (Johns, 2006).

Ethnobiological literature documents the historical and current importance of an array of resources consumed by communities living in and around the world's forests. It also demonstrates the richness of the traditional knowledge of indigenous and local communities related to the gathering and hunting of plant and animal foods and the medicinal value of forest species. From a wide range of ecosystems, some 7,000 of the earth's plant species have been documented as gathered or grown for food (Wood et al., 2005), and thousands more have medicinal properties (Napralert, 2006).

Forest-based societies draw on traditional knowledge for most of their subsistence needs and the use of a diversity of resources can be contributed to health. Although many traditional subsistence systems depend on one or more staples such as cassava, sago, rice, or maize, such diets are kept diverse and balanced through small but complementary amounts of animal source foods including birds, fish, insects, and mollusks, as well as sauces, condiments, snacks, and beverages obtained from plants. From a nutritional perspective, forest environments offer ample sources of animal (vertebrate and invertebrate) protein and fat, complemented by plant-derived carbohydrates from fruits and tubers and diverse options for obtaining a balance of essential vitamins and minerals from leafy vegetables, fruits, nuts, and other plant parts. Although many forest types have scant wild sources of carbohydrate, this lack can be overcome through forest-based agricultural production of cereals (e.g., maize), roots and tubers (e.g., cassava and yams), or bananas. Similarly, traditional cultivation systems drawing on agrobiodiversity can make adequate food available in spite of potential intermittent and seasonal shortages of many forest foods. Thus forest food resources can provide a valuable safety net in case of shortage of food crops. Undoubtedly, then, forest biodiversity is the basis for nutritional sufficiency for some populations.

Some forest products, such as the fruits of *M. vinifera* and other Brazilian palms that are rich in provitamin A (β-carotene and other carotenoids), are recognized as exceptional nutrient sources (Rodriguez-Amaya, 1996, 1999). However, the nutrient composition of most wild species and minor crops has been poorly studied (Burlingame, 2000).

Links between food and health are increasingly understood in terms of the functional benefits provided by phytochemicals, including numerous carotenoids and phenolics, apart from their value as essential nutrients (Johns and Sthapit, 2004). Stimulants of immunity and antioxidant, glycemic, and lipidemic agents can moderate communicable and non-communicable diseases such as diabetes, cancer, and cardiovascular disease. Guava, for example, is rich in the antioxidant lycopene, which has recognized anti-cancer properties. Many nuts have a high content of specific oils such as omega-3 fatty acids (walnut) and monounsaturated fatty acids (almond, macadamia, pistachio, and hazelnut), which reduce the risk of cardiovascular and other diseases. Argan nuts (*Argania spinosa*) from the southwestern part of Morocco offer similar benefits, but many forest species with commercial potential have not been characterized for their specific fatty acid composition (Leakey, 1999). Leaves of many plant species are rich sources of xanthophylls which contribute to optimal eye function. Examples include leaves of *Gnetum* spp. and *Adansonia digitata* (baobab), which are widely eaten in sub-Saharan Africa, and *Cnidoscolus aconitifolius*, which is locally important as a vegetable in Central America (Serrano et al., 2005).

While these kinds of functional properties of foods are seldom recognized by local communities without the benefit of scientific analyses, people often attribute to particular foods value in treating or preventing diseases. Indeed the distinction between food and medicine which characterizes scientific perspectives stands in contrast to traditional concepts of health which recognize the therapeutic and sustaining values of food more holistically.

The widespread use of roots, barks, and other forest plant parts as medicine appears to offer public health benefits, but these are difficult to validate scientifically. Ethnobotanical studies in tropical forest areas typically document knowledge of hundreds of species within local communities and widespread use of plants in primary health care. Much of the recorded data on the use of medicinal plants are anecdotal and idiosyncratic, and their specific contribution to the health of individuals cannot be effectively evaluated without controlled investigations. Ethnopharmacological research, including clinical studies, demonstrates the efficacy of many traditional remedies while failing to substantiate the pharmacological value of many others. Long-term epidemiological studies would be needed to confirm the contribution of specific remedies, phytomedicines, or foods to the health of populations. Even these remain inadequate to measure the efficacy and contributions of traditional healing practices to physical and mental health.

Forest Resources as Source of Medicine for Communities with HIV/AIDS

Many forest plants and animals produce poisons, fungicides, antibiotics, and other biologically active compounds as defense mechanisms, and many of these have medicinal uses. Compounds that have common medicinal uses such as cola nuts, caffeine, chocolate, chili peppers, and cocaine are also found in forest areas. Many western pharmaceutical products derive from tropical forest species, e.g., quinine from *Cinchona* spp.; cancer-treating drugs from rosy periwinkle (*Catharanthus roseus*); treatments for enlarged prostate gland from *Prunus africana*; forskolin, which has a variety of medicinal uses, from the root of *Coleus forskohlii*; medicine for treating diabetes from *Dioscorea dumetorum* and *Harungana vismia*; and several medicines based on leaves of the succulents of the Mesembryanthemaceae family. Some of these products are now synthesized, but others are still collected from the wild. The economic value of traditional medicines is considerable. Achieng (1999), for instance, reported that the bark of *P. africana* alone was worth US $220 million annually to the pharmaceutical industry.

Traditional health-care systems are based on significant local knowledge of medicinal plants in all major tropical areas. These health-care systems are important, particularly where formal health-care services are absent. The market for traditional medicines is large and expanding, and much of it is in the hands of women, particularly that involving less commercially valuable medicinal plants. There is also growing scientific evidence of the efficacy of some of these widely used traditional remedies. At the same time, medicinal plants are threatened globally. Some of the threats include slow growth patterns of desirable species, loss of traditional mechanisms that contributed to sustainable use, and competing uses of the same species, in tandem with growing commercialization and global markets. Certification of medicinal plants and better forest management techniques are two possible partial solutions.

Pharmaceutical companies have sometimes been charged with reaping unacceptably large benefits from forest peoples' knowledge, given the widespread poverty in forested areas. Issues relating to intellectual property rights, implications for cultural integrity, and amounts and recipients of benefits are complex. The Convention on Biological Diversity (CBD) aims to protect benefit-sharing rights, but adequate mechanisms for doing so are not in place, especially in many developing countries. Attempts to establish collaboration between the pharmaceutical industry and local communities in bio-prospecting have had mixed results (Kate and Laird, 1999). The forest sector plays an important role in the prevention, care, and treatment of HIV and AIDS and the mitigation of their impact. Tree products plays an important part not only in the care and treatment of HIV-related illnesses but also in income generation and other livelihood activities that can help alleviate the impact of the disease on households. To mitigate the impact of HIV and AIDS, one component of interventions in the tree sector should be directed toward supporting the sustainability of those tree benefits on which households and communities affected by HIV and AIDS rely on. Such interventions should also aim to alleviate those interactions that aggravate the impacts of HIV and AIDS on households (e.g., household labor reductions and scarcity of forest products to meet subsistence needs, in particular fuel wood).

More intensive forest management to increase productivity and accessibility of forest resources is in itself a mitigation strategy (FAO, 2004). Within communities, there is a need to improve the management of natural woodlands for multiple purposes. However, it is also necessary to lighten excessive pressure on forest resources, through either an increase in the supply of wood and non-wood forest products (through forest planting, cultivation of medicinal plants, and transport of wood from greater distances) or a decrease in the demand (through the use of more efficient wood stoves, possibly the switch to other fuel types, and alternative income-generating activities that may not be woodland based).

Conclusion

HIV and AIDS have dramatically changed rural life in sub-Saharan Africa, where close to 70% of the population is rural. Most households remain poor, with limited resources to fight the pandemic. Rural areas are also absorbing a significant part of the burden of urban AIDS cases as those who fall ill in urban areas return to rural areas to seek family care. Forests provide essential foods, medicines, health care, and mental health benefits to people in sub-Saharan Africa. The amount of these benefits generally increases with proximity to the forest. Persistent shocks of HIV and AIDS have long-term structural impacts on availability of labor and expertise, accumulation and distribution of capital, people's sense of long-term security, and outlook for the future. The advent of HIV and AIDS in rural sub-Saharan Africa is already in a precarious state of declining small-scale agriculture and increasing utilization of natural resources, particularly forests and woodlands. Use of forest resources for food and medicinal purposes has increased in many parts of sub-Saharan Africa.

References

Achieng J (1999) African medicinal tree threatened with extinction. Third world network. Internet document. Available at www.twnside.org.sg/title/1903-cn.htm

Barany M, Holding-Anyonge C, Kayambazinthu D, Sitoe A (2005) Firewood, food and medicine: interactions between forests, vulnerability and rural responses to HIV/AIDS. In: Proceedings from the IFPRI conference: HIV/AIDS and food and nutritional security, Durban, South Africa, 14–16 Apr 2005

Bennet T, Robinson JG (2000) Hunting of wildlife in tropical forests: implications for biodiversity and forest people. World Bank, Washington, DC

Bennett EL, Robinson ES, Eves H (2002) The scale of hunting and wild meat trade in tropical forests today. Presented at the society for conservation biology symposium on mitigating unsustainable hunting and the bushmeat trade in tropical forest countries: using science to change practices, Canterbury, UK, 18 July 2002

Bishop-Sambrook C (2003) Labour constraints and the impact of HIV/AIDS on rural livelihoods in Bondo and Busia districts Western Kenya. Working document for IFAD and FAO

Burlingame B (2000) Wild nutrition. J Food Compos Anal 13:99–100

Chege N (1994) Africa's non-timber forest economy. World Watch 7(4):19–24

Du Guerny J (1999) AIDS and agriculture in Africa: can agricultural policy make a difference? Food Nutr Agric 25: 12–19

Egal F, Valstar A (1999) HIV/AIDS and nutrition: helping families and communities to cope. Food Nutr Agric 25: 20–26

Engh I-E, Stloukal L, du Guerny J (2000) HIV/AIDS in Namibia: the impact on the livestock sector. FAO, Rome. http://www.fao.org/sd/WPdirect/WPan0046.htm

Enwonwu C, Warren R (2001) Nutrition and HIV/AIDS in sub-Sahara Africa. In: Watson RR (ed) Nutrition and AIDS, 2nd edn. CRC Press, Boca Raton, FL, pp 175–192

FAO (1995) The international expert consultation on non-wood forest products. FAO, Rome, Italy

FAO (2001) AIDS epidemic as a disaster which requires a response of an emergency nature. New challenges for FAO's emergency response in sub-Saharan Africa. TCOR, SDWP, ESNP. FAO, Rome, Italy

FAO (2004) Understanding the interface between natural woodlands and HIV/AIDS affected communities in Southern Africa. FAO, Harare, 1–2 April

FAO (2005) Agrobiodiversity and local knowledge for the mitigation of HIV/AIDS. A manual by Ard Lengkeek, Julia Wright, Yohannes Gebre Michael, and Mundie Salm. FAO, Rome, Italy

FAO/UNAIDS (2003) Addressing the impact of HIV/AIDS on ministries of agriculture: focus on eastern and southern Africa. FAO/UNAIDS, Rome, Italy

FAO/WHO (2002) Living well with HIV/AIDS. A manual on nutritional care and support for people living with HIV/AIDS. FAO, Rome, Italy

Farah D (2001) Seeking a remedy for AIDS in Africa: continent's woes limit reach of cheaper drugs. Washington Post, June 12, p A17

Garí JA, Villarreal M (2002) Agricultural sector responses to HIV/AIDS: agrobiodiversity and indigenous knowledge. Population and Development Service (SDWP), FAO, Rome

Hoskins M (1990) The contribution of forestry to food security. Unasylva 41(1):3–13

Johns T (2006) Agrobiodiversity, diet and human health. In: Jarvis DI, Padoch C, Cooper D (eds) Managing biodiversity in agricultural ecosystems. Columbia University Press, New York, NY, pp 382–406

Johns T, Sthapit BR (2004) Biocultural diversity in the sustainability of developing country food systems. Food Nutr Bull 25:143–155

Kate KT, Laird SA (1999) The commercial use of biodiversity: access to genetic resources and benefit sharing. Earthscan, London

Kengni E, Mbofung CMF, Tchouanguep MF, Tchoundjeu Z (2004) The nutritional role of indigenous foods in mitigating the HIV/AIDS crisis in West and Central Africa. Int Forest Rev 6(2):149–160

Kung'u JB, Kivyatu B, Mbugua PK (2006) Challenges to utilisation and conservation of medicinal trees and shrubs in Eastern Africa. In: Owuor BO, Kamoga D, Kung'u JB, Njoroge GN (eds) Some medicinal trees and shrubs of Eastern Africa for sustainable utilisation and commercialisation. World Agroforestry Centre, Nairobi, pp 13–25, 84p

Leakey RRB (1999) Potential for novel food products from agroforestry trees: a review. Food Chem 66:1–14

Napralert (2006) Natural products alert database. Available at http://www.napralert.org/

Nnko S, Chiduo B, Wilson F, Msuva W, Mwaluko G (2000) Tanzania: AIDS care-learning from experience. Rev Afr Pol Econ 86:547–557

Ogden C (1990) Building nutritional considerations into forestry development efforts. Unasylva 41(1):20–28

Rodriguez-Amaya DB (1996) Assessment of the provitamin a contents of foods-the Brazilian experience. J Food Compos Anal 9:196–230

Serrano J, Goñi I, Saura-Calixto F (2005) Determination of b-carotene and lutein available from green leafy vegetables by an in vitro digestion and colonic fermentation method. J Agric Food Chem 53:2936–2940

UNAIDS (1999) A review of household and community responses to the HIV/AIDS epidemic in the rural areas of sub-sahara Africa. UNAIDS, Geneva

UNAIDS (2001) The global strategy framework on HIV/AIDS. Available at http://data.unaids.org/publications/IRC-pub02/jc637-globalframew_en.pdf

UNAIDS (2002) HIV/AIDS, human resources and sustainable development. World summit on sustainable development, Johannesburg, 2002, UNAIDS, Geneva

UNAIDS (2005) AIDS epidemic update December 2005. Available at www.unaids.org

UNDP (2002) Agriculture and HIV/AIDS. FAO/Ease international/UNOPS/UNDP South East Asia HIV and development programme, Bangkok, Thailand

United States Department of Agriculture, Agricultural Research Service (USDA-ARS) (2004) USDA national nutrient database for standard reference, Release 16–1. Available at www.nal.usda.gov/fnic/foodcomp

Wilkie DS (2001) Bushmeat hunting in the Congo Basin – a brief review. In: Bakarr MI, da Fonseca GAB, Mittermeier RA, Rylands AB, Painemilla KW (eds) Hunting and bushmeat utilization in the African rain forest. Conservation International, Washington, DC, pp 17–20

Wood S, Ehui S, Alder J, Benin S, Cassman KG, Cooper HD, Johns T, Gaskell J, Grainger R, Kadungure S, Otte J, Rola A, Watson R, Wijkstrom U, Devendra C (2005) Food. In: Ecosystems and human well-being, vol 1. Current state and trends, Millennium ecosystem assessment. Island Press, Washington, DC, pp 209–241

World Bank (2001a) Medicinal plants: local heritage with global importance. www.worldbank.org

World Bank (2001b) Medicinal plants: rescuing a global heritage. World Bank technical paper no. 355, Washington, DC

Farmer Managed Natural Regeneration in Niger: A Key to Environmental Stability, Agricultural Intensification, and Diversification

M. Larwanou and C. Reij

Abstract A study to assess the extent of farmer managed natural regeneration (FMNR) and its impacts on agriculture and people's well-being as well as to determine what motivated farmers to practice natural regeneration at a larger scale was conducted in the southeastern part of Niger (Zinder). Farmer managed natural regeneration is so called when farmers deliberately and actively protect and manage sprouts and germinating plants in their fields in order to recreate tree vegetation. Most of the tree species are of economic value. This practice differs from tree plantation (for village woodlots, windbreaks, etc.) or management of natural stands in the forest outside farmer's fields. FMNR in Zinder could reach 1 million ha, with high dominance of gao (*Faidherbia albida*) and baobab (*Adansonia digitata*). Natural forests have almost disappeared. The farmers interviewed said that ecological crisis which occurred during the 1970s and 1980s motivated them to protect and manage more systematically and massively young trees than they did in the past. Other policy aspects went in favor of this farmer's innovation. The high pressure on natural resources has also motivated farmers toward agricultural intensification. The systematic protection of young *F. albida* has contributed to the creation of agroforestry parklands which help to maintain or improve soil fertility. This helps agricultural intensification and diversification in many villages. Despite this favorable evolution, some villages still have young parklands with little effect on soil fertility. These village territories are in the transitional phase of intensification. The production systems are being complex with a better integration of crop–livestock and trees. Despite the fact that the macroeconomic and macropolitic conditions in Niger were less favorable between 1985 and 2000, farmers have spontaneously continued to intensify the production system and at the same time improve their environment.

Keywords *Adansonia digitata* · *Faidherbia albida* · Natural regeneration · Production systems · Niger

Introduction

During the 1970s and 1980s, many articles were published on energy crisis in Sahelian countries and elsewhere in arid and semi-arid zones (Eckholm, 1975; Winterbottom, 1980). The author's principal argument was the important gap between the energy needs of the people that were exclusively wood-based and the annual growth of the forest. At the same time, the Sahel followed a period of crisis because of the recurrent droughts. The agriculture extended more and more on the marginal lands where the vegetation was strictly destroyed. The assumption was that in the near future, areas near big towns in the Sahel will be seriously denuded due to increasing needs of fuel wood by the population with increasing growth rate.

The actual perception is that in the Sahel, the vegetation is still being degraded because of overexploitation of fuel wood (PANLCD, 2000). The vegetation degradation in the Sahel is in fact undeniable, but at the same time, field observations, studies, and

M. Larwanou (✉)
Faculté d'Agronomie, Université Abdou Moumouni de Niamey, Niamey, Niger
e-mail: m.larwanou@coraf.org

A. Bationo et al. (eds.), *Innovations as Key to the Green Revolution in Africa*,
DOI 10.1007/978-90-481-2543-2_133,

satellite images showed important regreening zones (Olsson et al., 2005; Hermann et al., 2005). In Niger, for instance, an important regreening was observed, especially in Zinder region where the protection and management of natural regeneration by farmers are exceptional.

Farmer managed natural regeneration is defined as a technique wherein farmers actively protect and manage new sprouts and seedlings in their own farms in order to increase vegetation of trees. In most cases, the species are of economic value. It is called farmer managed natural regeneration to distinguish this practice from reforestation by planting trees in woodlots, windbreak, and the management of natural forest blocks outside farms zones.

It is surprising that even though natural regeneration in farm lands, protected and managed by farmers themselves, has been so spectacular many people are not aware of this phenomenon.

A study aimed to evaluate the evolution of agriculture and environment in Niger estimated that the natural regeneration covered more than 2 million hectares, but current studies estimate that the extent of this phenomenon is over 5 million hectares. In Zinder region, the extent of FMNR alone was more than 1 million hectare (Larwanou et al., 2006). If FMNR has taken this extent, it has to be unique in the Sahel and probably in all of Africa.

A study was conducted to identify and understand what motivated farmers to protect natural regeneration in their fields, how they manage this new resource, and what are the impacts of this on the environment, agricultural intensification, and diversification.

Methodology

Before going on field, the team had developed a list of points to be discussed in the villages. This list served as a guide for semi-structured interviews, and the majority of the points were discussed with villagers. Twenty-two villages were concerned with this survey in three districts, and the discussion was spontaneous at the selected villages. The strategy used was that no village was to be informed of the arrival of the team. Also, on the way to the villages, farmers were interviewed in their farms. In many cases, the group involved in discussion was large and included many socio-professional individuals; in the majority of villages, the group was composed of almost 100 persons with 10–25 participating in the debate.

Data were collected with the list of discussion points and were analyzed and interpreted.

What Motivated Farmers to Invest in Tree-Assisted Natural Regeneration?

Ecological Crisis

The droughts of the 1970s and 1980s remained in the memories of Sahelian farmers in general and in those of farmers from Niger in particular. Just after the droughts, the productive potential was weakened by the death of animals and trees. This had also led to soil fertility decline for agricultural crops.

In economic terms, droughts led to a strong recapitalization and many families experienced extreme poverty. Having been conscious of this situation, and due to the fact that it was urgent to act for survival, many actors (authorities, technical services, and farmers) (CND, 1984) came together in order to find alternative solutions that would help reverse the tendency (DE/DFPP, 2002). Alternative strategies were adopted such as the protection of natural vegetation by either improved cutting of tree stems or through identification, caring, and protection of newly germinated tree seedlings. Farmers were conscious of these environmental rehabilitation options and the need to have better ecological conditions. The sensitization developed by extension agents has succeeded in this part of Niger.

These droughts have seriously disturbed the existing ecological equilibrium, and thus, there is a need for a new environmental and political orientation. This has been translated into action by reinforcing the politic of natural resources preservation by making laws and regulations, forestry code which was modified in 1974, and the creation of new protected forests, especially *Acacia senegal* forest stands. Tree plantation in urban centers, houses, and villages (schools, markets, and other public places) was carried out. This period marked also the era of *reforestation projects called first-generation projects in Niger and the first green belts projects around the big cities* (CNEDD,

2003). *These actions were translated by the promotion of trees in agricultural fields, creation of village woodlots,* Acacia senegal, *and* Borassus aethiopum *forest management and urban green belts, dune fixation, and soil and water conservation actions.*

Farmer managed natural regeneration practice, even though institutionalized in the new political perception, existed but was generalized toward the end of the 1980s, just after the drought that caused a serious ecological crisis with wind and sand dune movement; FMNR was underlined as a local dynamic since decades and became spontaneous and supported by extension agents (Guengant and et Banoin, 2002; Joet et al., 1998; Larwanou, 1998; PAFN, 2005).

Demographic Pressure and Production System Evolution

The south of Zinder zone (Matamèye, Magaria, and Mirriah) is one of the most densely populated areas in Niger (more than 100 inhabitants per km^2). The rapid population growth has led, at the beginning of the twentieth century, to land colonization and saturation. This crop extension strategy (mostly used by majority of farmers to increase crop production) consequently led to the disappearance of pastoral and forest lands. It is probable that for a certain period of time, extensification had permitted meeting food needs for the population. But the combined effects of droughts, demographic pressure, and soil fertility decline had constrained farmers to adopt agricultural intensification. Trees are a fundamental element in the production systems (Raynault et al., 1997).

Certain societies developed close relationships with trees; the severe protection measures institutionalized by the sultans in Zinder since 1860 to protect trees like *Faidherbia albida* were somehow underlined to illustrate the integration of trees in African agriculture. Giffard cited by Hambally (1999) reported that in Zinder and at that period anyone who cut without authorization a *F. albida tree would have his head cut off; anyone who without any reason mutilates a tree will have his hand cut off* (Bonkoungou, 1993). The same author adds that for Zinder region *the importance of this tree species is well established in rural economy so that in land sharing in certain ethnic groups, the number of F. albida trees is taken into account.*

> by its conception and assigned role, tree stand, in agricultural lands, appears as a revealing strategy that each society adopts in area where it is inserted. Then, it is not only the needs and techniques that parkland means, but the nature of the society and its history, and somehow the structure it brings (Pélissier, 1980; Raison, 1988; Raynaut, 1997; Seignobos, 1982).

Demographic structure played an important role in agricultural intensification, and it appears that in this region, the density of *F. albida* trees is very high in areas with high population density. This is also true in other parts of Niger, where populated zones have high tree density.

Development Partner Actions in the Region

Many development projects did intervene in this region and among them is the 3 M project which was financed by USAID. It was an integrated project which conducted productive systems improvement actions (artificial plantations, sensitization and training of farmers in environmental rehabilitation). Another project for reinforcing agricultural technical services has worked in areas of training and sensitization on rural development aspects; polyvalent extension agents were involved in imparting such training and sensitization to producers. During the implementation phase of this project, farmers were well trained on improved ways of land clearing and caring and protection of assisted natural regeneration. It could be true that these project actions have contributed to the adoption of assisted natural regeneration techniques in Niger.

Farmer's Perception of the Impacts of Assisted Natural Regeneration

This part presents the results of interviews conducted with farmers in the various villages and also of individuals and families met in their farms.

Impacts on Revenues

FMNR has in many cases important positive impacts on farmers' revenues including those of women and young people; in the village of Ara Sofawa, for example, it was stated that rural exodus of young people had reduced, because more than 100 people could gain cash from transportation and sale of wood. In a big garden in Mirriah, we met young people collecting baobab leaves. They bought the leaves of baobab trees for 4000 FCFA after which they would sell them in some big cities in Niger like Diffa and Niamey.

Baobab leaves were not only consumed almost everyday in the majority of households, but they *constitute an important source of revenue for thousands of people* in this region. A bag is sold *for 1000–3000 FCFA* according to age and quality of leaves and *a big baobab tree could produce 5–6 bags*.

The abundance of *F. albida* trees in this region has an important impact on livestock, because its pods constitute a good fodder source in the area without natural pasture lands. The price of a bag of pods is variable but 500 FCFA a bag is a normal price. Removal of barks of some medicinal tree species like *Sclerocarya birrea* is a common practice in this region. Traditional healers are engaged in this activity either for their own use or for sale to others in Zinder. A bag is sold between 250 and 400 FCFA.

Impacts on the Environment

The people interviewed were well aware of the impacts of FMNR on the environment. Trees deliberately kept on farm provide various environmental services like the following:

- *microclimate improvement*: near the trees, humidity is high and heat is less, according to the farmers;
- *soil fertility improvement*: agricultural production increases with tree density;
- *erosion control*: natural regeneration, besides its role in soil fertility management, constitutes according to farmers windbreak, the effects of which were to bury young crops at the beginning of the rainy season;
- *well-being of the people*: farmers affirm that the presence of trees in their territories makes them happy in terms of landscape beauty and well-being.

FMNR scale in Zinder region has another important impact on soil fertility management. In the 1970s and 1980s, many women used animal dung as energy sources for making food. In many villages, farmers said that animal dung as energy source has now disappeared, because there is enough firewood from trees kept in the farms.

Impacts on Agricultural Production

The emblematic tree species in the 3 M region of Zinder is *F. albida*. It is predominant but in other territories, *Piliostigma reticulatum and Hyphaene thebaïca* parklands are more dense. Some other species like *Balanites aegyptiaca* were also present. Farmers adopt selective assisted natural regeneration. Two types of species were preserved: those that improve soil fertility and food tree species.

F. albida is considered as an excellent fertilizing tree species. For instance, all farmers interviewed confirm that soil fertility in their farms has improved and this is related to the high density of that particular tree species. It has been estimated that because of *F. albida* natural regeneration, farmers have maintained soil fertility in their farms at a certain level. Even though land fallow is absent and the use of mineral fertilizers is low, cropping lands are not always poor. In villages where the *F. albida* parkland has reached a good height, farmers said that the soil fertility is improving from year to year, but in villages with young parkland of this species, farmers perception is that soil fertility is still decreasing.

Impacts on Livestock

There is evolution in the modes of livestock rearing in many village territories because of lack of natural pastoral lands. In Haoussa village, livestock rearing in the house with animals tied at home for almost 8 months in a year is dominant. During the remaining 4 months of the year, animals are divagating during the day or are herded by someone. In rainy season, a large part of the animal herd from the village is led by a herder (Fulani or Touaregs), who will go for transhumance out of agricultural zones. *In most cases, the number of animals is significantly increasing in most of the villages.* The villages where there is a decrease in number of animals are those where there is water shortage; this is the case in the west, southwest and northwestern parts of Mirriah.

Impacts on Food Security

Trees have an important place in farmers' strategy for food security. Fertilizing tree species like *F. albida, P. reticulatum,* and *P. africana* contribute to increased agricultural crop production (millet and sorghum), which are the principal food sources of the people.

Food tree species like baobab, date tree, mango, *Parkia biglobosa*, and *B. aegyptiaca* (Sahel date tree) help to diversify food and constitute an important food reserve for the community. In the villages where milk production is insufficient, farmers used the baobab pulp and *Tamarindus indica* leaves for a local dish preparation (cream from millet grain) which is the principal food for rural people in Niger. Some tree species like baobab, date tree, and mango are also important sources of revenue for rural people to face food deficit. The baobab leaves, date tree fruits, and mangoes provide cash that is reinvested in buying cereals.

Impacts on Women

In social work division in Sahelian rural societies, women are in most cases responsible for fetching firewood and collecting some forest products and non-timber products. Natural regeneration near the villages has reduced the time for firewood collection for women.

Impacts on the Right and Access to Trees

Tree Individual Ownership

To the question, who owns the trees that are in the farms? the majority of farmers interviewed responded: us. Few responded that it was government property. Production systems evolution with the integration of tree as a production factor has changed its status. Trees are no more a common property, and, since the 1980s, with changes in forestry policies that became less repressive, the government has gradually been disengaged in tree management.

Natural regeneration practice has amplified the tendency toward tree privatization. Old trees have disappeared and replaced by younger ones from FMNR. There is an investment that authorizes farmers to own the trees. *When farmers talked of F. albida, they usually said "my F. albida" as to confirm their right to these trees.* This situation revealed a change in tree access conditions. Access to certain tree species and their products is subjected to first asking the owner of the farm.

Monetary Transactions and Arrangements on Trees

Access to trees is according to the different modalities of the farmer's objectives (collection, branch cutting, whole cutting, sales, etc.) and to species. One could obtain access to a tree through farmer authorization without any charge, monetary transaction between buyer and seller, forestry services permit, and stealing.

Farmers and the government control access to trees. Here, traditional chiefs and communities have lost the right of control of trees. Many actors do exploit trees: farmers, wood salers, traditional healers, livestock herders (local and in transhumance), women, children, etc. The sale of trees is a recent dynamic due to high tree regeneration in the farms.

Thus, *P. biglobosa* tree, date tree, and baobab could be sold. For example, a baobab tree could be sold according to its height and age for 9000–20,000 FCFA. Baobab trees are sold to individuals who exploit and sell the leaves and the fruits. We did not go in-depth to inquire who sells and who buys, but there are women among those who bought. A price of 20,000 FCFA for a baobab tree shows that there is a relationship between the price and the monetary value produced by the tree because of its leaves.

At the same time, their products (leaves and fruits) could be sold before they are collected from the tree. We saw just near Mirriah someone collecting leaves from a baobab tree by paying 4000 FCFA which can then be sold in Zinder and other cities of Niger. When it comes to sale of a farm, food tree species are subjected to another price.

"The land has its own price and the tree its own" said a farmer. The handcrafters and traditional healers pay before accessing the trees; the handcrafters who make mortar and other materials most often bought the trees.

Access of Women to Trees and Its Products

Access conditions differ with tree species and type of products. For food tree species, like baobab, women could own by inheriting or buying. They have free access to dead wood in the farms and other products like *F. albida* pods.

Access to wood is sometimes after animal herders cut it for fodder. Women do gather the dead branches left over by herders. In a certain manner, there was a compromise between herders and farmers; the herders can cut the branches to feed their animals with leaves and pods and the farmers could get the wood.

Conflicts on Trees

Tree exploitation sometimes leads to conflict between farmers, exploiters, and foresters and between farmers and herders. It must be stated that officially, one must get permit from forest services before exploiting the trees in his farms, but very few people do that.

Uncontrolled Tree Cutting

In some territories, especially very close to urban centers, where there is high demand, uncontrolled cutting of wood is high. It is normally done by wood exploiters who profit from the absence of farmers during the dry season and cut their trees. Sometimes, it is during the night that they cut the trees.

Conflicts with Transhumant Herders

Transhumant herders are considered as important tree cutters. During the dry season, *F. albida* is almost the only tree species that keeps its young leaves. There are two arguments here. Farmers consider *F. albida* as their own property whereas the herders consider it as a shared resource but protected by foresters. In many cases, the herders get official permit from foresters to cut in a private domain. In this case, even though tree ownership by farmers is real, the latter understand that it is a shared resource.

A Typology of Situation and Tendency

The typology of the parkland in this region is based on their composition. Another typology could be based on the tendencies and development perspectives. For example, in a certain situation, the presence of *F. albida* parkland has not yet transformed, according to farmers, soil fertility management and thus led to an increase in agricultural production, but elsewhere farmer's perception is that *F. albida* parkland has already had positive impacts on soil fertility and crop yields. Even with high density in the 3 M, the situation is not homogeneous. Four different situations are present: FMNR with legumes in depressions; FMNR with livestock; FMNR without agricultural intensification; and FMNR with transition toward an intensive production system.

Perspectives

FMNR in this region is impressive and could cover 1 million hectare. In comparing the production systems in this region 20 years earlier, it seems justified to talk of a mutation. But what could be its future in a context characterized by high population density with an annual growth rate of 3%? What if there were some years of consecutive droughts, which will make poor people to cut more trees to survive? Is it that the process of agricultural intensification has attained its limits?

FMNR in a Context of High Demographic Growth

The general observation in the visited sites is that FMNR is important in densely populated areas. In order to respond to increasing needs in wood and non-timber forest products, and due to the fact that trees in this part of the country exist only in private lands, farmers engage in assisted natural regeneration in an intensive way. In the past, land clearing was systematic as there were some reserves of wooded communal lands. With the increase in population, there is need for more agricultural land. Farmers were and are obliged to leave trees in their farms in order to satisfy their needs.

FMNR represents a type of recapitalization of agricultural lands, and it is probable that this process will continue; nevertheless, there are dangers that threaten this process. Increased demography has motivated local population to protect natural regeneration on their farms, but at the same time a rapid demographic growth is threatening the process. Almost everywhere, the farmers interviewed stated that they produce more now and ate better than 20 years ago, but have more people to feed, which diluted the gains. There is a general feeling that the high demographic growth is a problem, and census data confirm this. The population in Zinder has doubled between 1977 and 2001.

FMNR Exploitation

The assisted natural regeneration exploitation in all the villages is controlled by local people. In all the visited sites, it was reported that "no one could cut a tree in somebody's farm without the owner's permission." In Maradi region, there are village committees for FMNR surveillance because there are some villages that do not practice FMNR. In the 3 M, FMNR is quasi-general and everyone protects trees in their farms, which reduces the necessity to create village organizations to protect trees. The observed dynamic in this zone is systematic in terms of FMNR protection. It is difficult to find someone in a village who is not protecting trees. In terms of exploitation for various uses, the following were identified.

Exploitation for Firewood

It is generally women, children, and sometimes the household chiefs who are responsible for this activity. With increasing human density, the highest in the country, the wood used for energy is very important. As there is no dead wood, the wood sold and used in Zinder is fresh and the highest percentage is from *F. albida*. Data are lacking to evaluate the quantity of wood from the farms and used everyday to satisfy the needs of many families. But it is reassuring that the time for wood collection has reduced, especially for women.

Exploitation for Service Wood

House making and stencils for cooking and for agricultural works were done with wood coming from FMNR. The quantity is important to meet the needs of the population.

Exploitation for Selling

During a group interview in the village, farmers in Gaounaoua in Magaria district said that their revenue source was based on sale of wood in various markets (Bandé, Magaria, and Zinder). This affirmation has been made in various villages; this means that local population manages the natural regeneration not only for landscape beauty or ecological services but also for the immediate needs of the household. The cash obtained is used to meet various economic and social needs (contribution during naming ceremonies, marriage, death, etc.). It is the very poor people in the villages that exploit trees for selling wood.

Exploitation for Fodder

Aerial fodder from trees is an expanded practice in this region; as there is barely any natural pasture lands and with increasing number of animals in the villages, the only way to feed the animals is through leaves, pods, and fruits from many tree species in the territories.

Exploitation for Traditional Pharmacopeias

Because of insufficient health centers in most of the villages, coupled with cultural behavior, forest products are usually being used to cure various sicknesses for humans and animals. Blessed with different medicinal tree species, this area is the principal source of traditional pharmaceutical products of many other parts of the country.

From these various uses, it can be inferred that FMNR exploitation in this zone is a day by day activity; coupled with high population density, one could imagine that with actual rhythm of exploitation, continuous efforts are needed by these populations in order to maintain a balance between exploiting potential and need satisfaction.

Conclusions

Various perceptions from decision makers, national and international experts, and urban population are that most often Sahelian environment in general and Niger in particular are subjected to a continuous degradation process. This study, conducted in Zinder region, with highest population density in Niger, shows the contrary and confirms the Boserup hypothesis, which sees an increase in population pressure on available resources as a factor that induces agricultural intensification (Boserup, 1970).

The protection and management of assisted natural regeneration by farmers on their crop fields in this part of Niger is a real and spectacular phenomenon (1 million hectare). The importance of natural regeneration is more than that of artificial plantations. Only small portions of natural forests remain in the 3 M, as almost all the trees are in farmers' fields, forming amazing parklands with the dominance of *F. albida*.

It was mostly the ecological and economic crisis of the 1970s and 1980s that motivated farmers to invest systematically in protection and management of trees in their farms and have new and beautiful agroforestry parklands and complex and productive production systems constructed. These farmers' efforts were facilitated by positive evolution in national forestry politics (after the Maradi declaration in 1984) and some projects.

Farmers protect and manage trees in their farms because in their own perceptions, they have an exclusive right on them (individual ownership). Before, they considered that trees belonged to foresters (government), but this perception has changed since the 1980s; it is difficult to identify the causes. In some cases, the national politics became more exciting (high population in development activities and an effort to inform them about ecological crisis and technical solutions); on the other hand, there is weakening and disengagement of the government because of economic and political crisis.

The *F. albida* parklands support agriculture (maintain soil fertility) and livestock (providing fodder); thus, they allow agricultural intensification and a recapitalization of agricultural lands. Trees have reduced population vulnerability during drought years (leaves and fruits, sale of wood, and fodder for animals).

References

Bonkoungou EG (1993) Fonctions socioculturelles et économiques d'acacia albida en Afrique de l'Ouest. In: Vandenbelt RJ, Renard C (éds) Faidherbia albida dans les zones tropicales semi-arides d'Afrique de l'Ouest: comptes rendus d'un atelier, 22–26 avril 1991. ICRISAT, Niamey, Niger. Andhra Pradesh, Inde: ICRISAT et ICRAF, Nairobi, pp 1–6

Boserup E (1970) Evolution agraire et pression démographique. Flammarion, France, 222p

CND (1984) L'engagement de Maradi sur la lutte contre la désertification, Maradi du 2& au 28 mai 1984, Conseil national de developpement. Niamey, Niger

CNEDD (2003) Evaluation des actions menées au Niger dans le domaine de l'environnement pendant les 20 dernières années. Cabinet du premier ministre; République du Niger, 138p

DE/DFPP (2002) Rapports annuels des années 1984 à 2002

DRE Rapports annuels 1984 à 2002 (Maradi, Tillabéry, Dosso, Tahoua, DRE/CUN, Agadez, Diffa et Zinder)

Eckholm EP (1975) The other energy crisis: firewood worldwatch paper 1. Worldwatch Institute, Washington, DC

Guengant J-P, et Banoin M (2002) Dynamique des populations, disponibilités en terres agricoles et adaptation des régimes fonciers. CICRED, Paris

Hambally Y (1999) Modes d'intervention et pratiques paysannes eu Niger. Thèse de doctorat, EHESS, Marseille

Herrmann SM, Anyamba A and Tucker CJ (2005) Recent trends in vegetation dynamics in the African Sahel and their relationship to climate. Global Environmental Change Part A, 15(4):394–404

Joet A, Jouve PH, et Banoin M (1998) Le défrichement amélioré au Sahel: une pratique agroforestière adoptée par les paysans. Bois et Forêts des Tropiques 225(1):31–43

Larwanou M (1998) Rapport d'activités. INRAN, 57p

Larwanou M, Abdoulaye M, Reij C (2006) Etude de la Régénération Naturelle Assistée dans la région de Zinder (Niger): premiére exploration d'un phénoméne spectaculaire. International Resources Group, Washington, DC, 67p

Olsson L, Eklundh L, Ardo J (2005) A recent greening of the Sahel—trends, patterns and potential causes. J Arid Environ 63:556–566

PAFN (2005) Plan villageois d'aménagement des parcs agroforestiers dans 4 terroirs du Département de Magaria. Ministère de l'hydraulique de l'environnement et de la lutte contre la désertification, Niamey, République du Niger, 29p

PANLCD (2000) Programme d'Action National de Lutte contre la Désertification et de Gestion de Ressources naturelles (PAN/LCD-GRN). République du Nigar, 80p. **Available at** http://www.unccd.int/actionprogrammes/africa/national/2000/niger-fre.pdf

Pélissier P (1980) L'arbre en Afrique tropicale. La fonction et le signe (préface) In: Cahiers de l'ORSTOM. Sér Sci Hum XVII(3–4):127–130

Raison J-P (1988) Les «parcs» en Afrique: état des connaissances et perspectives de recherches

Raynault C et al (1997) Sahels. Diversité et dynamiques des relations sociétés-nature. Khartala, Paris

Raynaut CL (1980) Recherches multidisciplinaires sur la region de maradi; Rapport de synthèse. DGRST/ACC: lutte contre l'aridité en milieu tropical; Université de Bordeaux II

Raynaut CL (1997) Rapport de synthèse des études menées en 1977, Université de Bordeaux

Seignobos C (1982) Végétations anthropiques dans la zone soudano-sahélienne: la problématique des parcs. In: Revue de Géographie du Cameroun. Tome 3(1):1–23

USAID, CILSS et IRG (2002) Investir dans la forêt de demain: vers un programme de révitalisation de la foresterie en Afrique de l'Ouest. International Resources Group, Washington, DC, 35p

Winterbottom RT (1980) Reforestation in the sahel: problems and strategies. An analysis of the problem of deforestation and a review of the results of forestry projects in Upper Volta. Paper prepared for presentation at the African studies association annual meeting, Philadelphia, PA, 15–18 Oct 1980, 32p

Achieving a Green Revolution in Southern Africa: Role of Soil and Water Conservation in Enhancing Agricultural Productivity and Livelihoods in Semi-arid Areas of Malawi

A. Kabuli and M.A.R. Phiri

Abstract Soil and water management are essential ingredients for increased crop productivity and subsequent attainment of green revolution in sub-Saharan Africa. This is particularly significant in countries whose populations are heavily dependent on rain-fed agriculture. However, farmers in semi-arid areas often suffer from inadequate extension systems that do not function well due to the marginal nature of arid areas. Consequently, farmers' adoption of soil and water conservation innovations continues to remain a challenge. Using a combination of both qualitative and quantitative assessments on a sample of 200 farming families in Central Malawi, our research study showed that smallholder farmers, who are a key to achieving a green revolution, face a variety of socioeconomic and institutional bottlenecks that are both inherent in their farming systems and those caused by the broader macroeconomic environment. High on the list was the inadequate access to fertilizer and seed inputs by many farmers as well as declining land holding sizes which have impacted negatively on technology adoption and productivity. Similarly, increasing poverty and the prevailing HIV/AIDS pandemic coupled with inappropriate markets have significantly reduced communal investments in most of the resource-demanding soil and water conservation innovations. The study recommended that a broad-based green revolution in Southern Africa is achievable only if smallholder farmers adopt on a wider scale the use of improved seed, fertilizer and traditional low-cost best bet soil conservation strategies to increase yields as well as to sustain the natural resource base.

A. Kabuli (✉)
Soil Fertility Consortium for Southern Africa, Bunda College of Agriculture, Lilongwe, Malawi
e-mail: amonmw@yahoo.com

Keywords Green revolution · Soil · Water · Conservation · Adoption

Introduction

Land degradation is one of the fundamental problems confronting sub-Saharan Africa in its efforts to increase agricultural production, reduce poverty and alleviate food insecurity. For Malawi, problems of soil degradation and soil erosion have been recognized for many years now, yet land pressure and its consequences on loss of soil fertility continue to increase with rapid population growth rates and continuous cultivations (World Bank, 1995). Many conservation programs designed to address soil fertility problems in the traditional agricultural sector have fallen far short of expectations. Malawian farmers often resist adoption of recommemnded soil and water conservation technologies and farm practices and frequently abandon them once research and promotional programs have ended. This study was aimed at contributing to the emerging knowledge of how farmers perceive soil and water technologies and their impact on adoption, agricultural productivty and livelihoods.

In semi-arid areas of Southern Africa under which Malawi falls, climatic vulnerability is characterized by low mean annual rainfall, compounded by high variability in its spatial–temporal distribution. In addition, the past few decades have shown a definite trend for increasing aridity within the Sahelian region of sub-Saharan Africa. Hence, the livelihood strategies in these areas are primarily geared towards coping with high degree of uncertainty, minimizing risk and meeting subsistence needs rather than maximizing

A. Bationo et al. (eds.), *Innovations as Key to the Green Revolution in Africa*,
DOI 10.1007/978-90-481-2543-2_134,

production and profits (Scoones, 1996). In terms of the sustainable livelihoods, the vulnerability faced by poor people includes that brought about by uncertainties in climate, politics, markets and potential conflict situations. This has implications for individual time preferences and investment decision making amongst smallholder farmers. However, the relationship between vulnerability and soil and water conservation remains unclear in much literature on sub-Saharan Africa. It is often assumed that soil and water conservation is undertaken as a risk-reducing investment in the context of vulnerability, but case studies from Tanzania and Uganda (Hatibu et al., 1995) suggest that investment is reduced as perceived vulnerability increases.

The main objective of the project was to understand the overall factors that shape farmer perceptions, choice and subsequent adoption of soil and water productivity enhancing technologies in dryland areas of Malawi. Specific objectives include (1)assessement and characterization of existing soil and water conservation techniques available to farmers in dryland areas, (2) assessement of farmers' knowledge and evaluation of the recemmended soil management practices as they impact on both livelihoods and their farming systems, (3) determination of the influence of socio-economic variables in farmers' decision-making process on investment in soil and water productivity and (4) determination of the influence of farm and technology characteristics in subsequent adoption decisions by farmers.

Methodology

The study involved a broad spectrum of stakeholders comprising government research bodies, the university staff involved in similar work and a number of NGOs carrying out conservation work. Two rural development projects were purposively selected in Kasungu Agricultural Development Division (KADD), namely Dowa East and Madisi RDP, simply because of their previous experience with soil conservation projects, one initiated by CIMMYT's Risk Management Project and the other by the Ministry of Agriculture and Irrigation in collaboration with Malawi Agroforestry Extension Project. Sampling of farmers to participate in the study began with stratifying them into adopters and non-adopters using sampling frames obtained from local agricultural experts. The final sample for the quantitative survey was 200 farmers randomly sampled from two RDPs in Dowa district.

In order to be able to solicit farmer perceptions, knowledge and attitudes on soil and water conservation and how they influenced their adoption decisions, a total sample of 50 adopters and 50 non-adopters were interviewed per site (RDP) giving a total sample of 200 farmers for the two sites. The involvement of the two sites was also helpful in making comparisons on farmer perceptions, choice and adoption decision considering the different geographical and agroecological zones. Random samples were drawn from each stratum of the farmers to be interviewed for the study. A total of 100 farmers in each category were interviewed in the survey. In addition, a sub-sample of the farmers (key informants) were selected for a series of meetings to discuss further on the research questions posed in the project. The study used two data collection tools: a strucutured questionnaire as well as key informant interviews using a checklist to supplement the secondary data from literature reviews. Interviews were conducted with farm household heads or spouses including focus group discussions that were stratified into men, women and youth to capture any differences by gender. Data collected were cleaned and entered into a computerized database developed for this purpose using Microsoft Access. The information was finally analysed using the statistical package for social scientists (SPSS). The anaytical techniques applied included the *Chi-square* test to detect any systematic association between the dependent variable of interest and specific household characteristics. Frequency, means, ratios and percentages were computed to provide some descriptive statistics.

Results and Discussion

Household Socio-economics and Demography

In the final sample, there were almost an equal number of men and women. From the 200 survey households, about 46.5% were female and 53.5% male. The mean household or family size of adopters was 5.95 persons per household with a range of 1–13 persons, and for non-adopters they had an average of 5.32 persons ranging from 2 to 12 persons per household. The study showed that there was no statistically significant difference in household size between the two groups. The mean age for the adopting households was 42.57 years. Age was one of the demographic characteristics the study hypothesized to influence the retention of technologies by farmers. However, the study results showed that there was no statistically significant difference in age between adopters and non-adopters.

At least 68.0% of the sampled household heads were able to read and write. Similarly, another 33.5% had completed primary school. Only 2.0% of the total respondents had participated in adult literacy education. There was no difference in the literacy status between adopters and non-adopters. The *Chi-square* analysis showed that there was no systematic association between the literacy status of the household head and the adoption of conservation structures. Overall 57.0% of the respondents belonged to one or more social institutions existing in their community and of these 41% were farmer club members. The result from the *Chi-square* analysis also showed a strong relationship between belonging to a social group and adoption of conservation structures.

Household Wealth and Income Sources

Livestock was an important component of the farming system in the study areas. A vast majority (71.0%) of the sampled households included in this study owned livestock of different kinds. Both the mean monetary value of livestock and the mean family income from all sources last year varied between adopters and non-adopters. Overall, the survey households had a mean livestock value of MK 8023.29 and a mean family income of MK 14,995.45 (Fig. 1). The high mean income observed in adopters probably explains why they were able to invest in these technologies. Additionally, it reflected a positive outcome from using soil and water management technologies through realization of high crop yields in the 2005/2006 season. Furthermore, there was a strong association between income from all sources last year and adoption of soil conservation structures as shown in Fig. 1.

In the survey areas, more adopters (9.3%) cultivated land more than 3 ha compared to 3.3% only for non-adopters. This finding indicated that farmers with smaller pieces of farm land tended not to use or adopt soil and water conservation on any of their plots. Alemu (1999) noted that if land ownership and associated user rights can be alienated from the holder at any point in time by forces outside his/her control and without the consent of the individual farmer, the farmers will have little incentive to invest in structures for improving land quality. *Chi-square* statistics showed a no systematic relationship between adoption of conservation measures and how the farmer obtained his/her land. However, land was mostly obtained through sharing amongst family members (57.0%), and a very small fraction (2.5%) of farmers rented in land. The average number of fields per household was 1.97, with a range of 1–8 fields per household. The study showed that more adopters owned more fields than

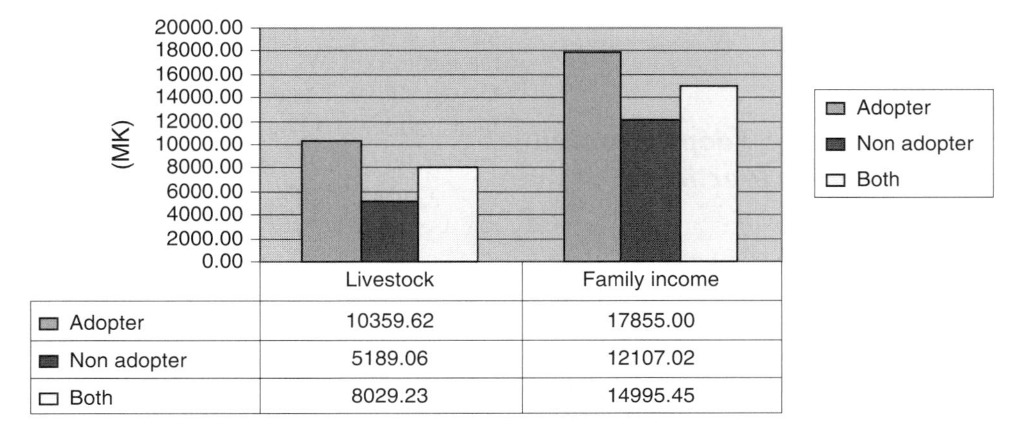

	Livestock	Family income
Adopter	10359.62	17855.00
Non adopter	5189.06	12107.02
Both	8029.23	14995.45

Fig. 1 Value of livestock and family income from all sources

non-adopting households with a mean of 2.02 fields per household compared to non-adopters who had an average of 1.92 fields.

About 66.0% of the households in the study areas had been farming independently for over 10 years, with adopters constituting for about 38.0% and non-adopters 28.0%. The higher number of years of farming for adopters indicated that they had acquired both enough experience and benefits of using the soil and water conservation technologies as opposed to their colleagues. A very small percentage (1.0%) of the non-adopting households reported to have been farming independently for less than 1 year.

Cropping Systems and Soil Management Strategies in Semi-arid Areas of Malawi

The soil and water conservation technologies employed on the farms varied between the fields with major and minor crops in the 2005/2006 cropping season. However, maize and tobacco fields shared a common soil conservation and management strategy with the majority of farmers using fertilizer to improve soil fertility. The study clearly showed that households who were more dependent on crop production for their livelihoods invested more in soil and water conservation. The most practised technologies in legume fields were observed to be traditional soil and water conservation structures such as box ridges, contour bunds and agroforestry technologies. The overall mean acreage for a soil management technology was 1.1483 acres, with a range of 0.10–5.00 acres. Sole cropping was very common in maize as well as in tobacco fields for the reason that farmers wanted to produce more yields from the unit area.

Extent of Soil Erosion and Adoption of Soil and Water Conservation Practices

Over half of the households (68.0%) reported to have experienced soil erosion/soil infertility problems in their fields, with non-adopters having a significantly ($p = 0.004$) higher rate than adopters. The most commonly experienced type of soil erosion at the time of the survey was sheet erosion, and the major cause was reported to be poor soil conservation. Gully erosion was mainly caused by up field runoff and fields located along steep slopes.

During the focus group discussions, farmers stated that the following factors were critical in determining which soil and water technologies should be adopted and practised on their farm: (1) number of people in the family, (2) climatic changes, particularly how the rainfall pattern was changing from one season to the other, (3) resources at one's disposal, (4) the type of soil: whether sand, clay or loam, (5) availability of inputs like seeds and labour at household level, (6) desire for immediate food security benefits versus the long-term benefits of soil conservation, (7) high and long-term discounting value – net return to investment outlay, (8) and effectiveness of the soil and water conservation technologies in restoring the degraded land. However, most of them also perceived the following benefits of adopting the current soil water conservation practices on their agricultural production and livelihoods. They noted that soil and water conservation technologies assisted them in soil/land conservation which helped both to beautify the landscape, e.g. Vetiver and agroforestry trees, and to increase crop yields. Despite the benefits or returns, farmers also encountered a number of constraints which reduced their ability to invest in large-scale soil and water conservation such as limited number of tools, pests for some agroforestry trees (e.g. Nsangu – *Faidherbia Albida*), poor growth habits of certain agroforestry trees, lack of sufficient training in agroforestry trees and inadequate seed/inputs of certain agroforestry species needed to establish effective conservation structures and climatic variability in areas where rainfall was not enough were some the most critical constraints to adoption. Furthermore, limited extension service support and lack of initial startup capital were also noted as additional constraints on the macrolevel. With respect to the survey on whether households encountered soil erosion effects in their fields, 80.88% of non-adopters said they had encountered a number of soil erosion effects in their fields compared to 45.59% of adopters. The high response rate from non-adopters could be attributed to the fact that many of them used traditional ridges which made their fields more susceptible to soil erosion compared to modern ridges. Similarly, of the 39.71% of households that reported very significant loss of maize (>5 bag of maize) in their field due to soil erosion, 24.26% were non-adopters (Table 1).

Table 1 Effects of soil erosion and perceived impact on maize yield

	Adopter	Non-adopter	Total
Effects of soil erosion encountered			
Loss of top soil	15.44	37.50	52.94
Loss of soil nutrients	3.68	12.50	16.18
Reduction in crop yield	20.59	22.06	42.65
Loss of soil water	5.15	5.88	11.03
Reduced vegetative cover	0.74	2.94	3.68
Total	45.59	80.88	**126.47**[a]
Perceived maize loss due to erosion			
Not significant	2.94	2.94	5.88
Significant	19.85	34.56	54.41
Very significant	15.44	24.26	39.71
Total	38.24	61.76	**100.00**

[a]Percentage above 100% due to multiple responses

The study showed that almost all of the surveyed households had heard of some soil and water conservation practices, but only a few of them were applying them in their fields. Of the households, 43.5% had been practising these soil and water control measures in their fields in order to arrest surface runoff, while a smaller number of farmers (4.0%) were practising the technologies to improve the landscape. The average size of land on which a household used soil and water control measures was 2.09 acres. The average distance of the plot with soil and water conservation structure from the homestead was found to be 25.32 min, with adopters covering more minutes than non-adopters and hence having plots further from homesteads than non-adopters (Fig. 2).

Farmers were asked to rate the conservation measures used on the basis of impact on crop yields particularly maize which is the staple food. Most of the respondents observed that they perceived increase in maize yields by more than the current yield as a major benefit. More than half of the respondents (55.9%) indicated that the introduced conservation practices performed better in containing surface runoff, while a very small fraction (12%) said that the introduced conservation measures also improved soil nutrient-holding capacity (Table 2).

The role of farmer perceptions of a technology attributes in enhancing or eroding adoption decisions is well acknowledged in literature. It was hypothesized that farmers' expectations of the effectiveness of conservation structures in retaining soil from erosion would have a positive effect on retention of soil conservation structures in the long run. Our findings revealed that there was a significant association

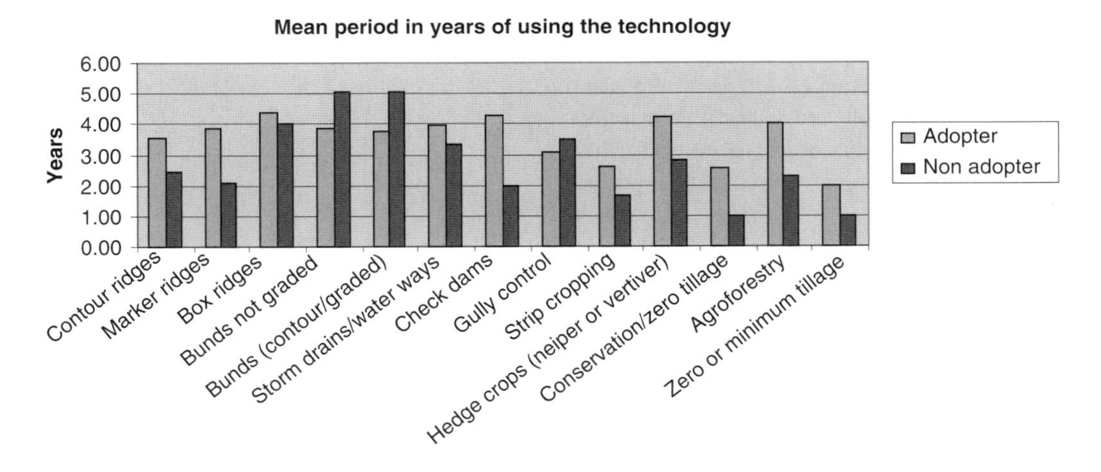

Fig. 2 Mean period of using soil and water conservation technologies

Table 2 Benefits from soil and water conservation technologies

	Adopter	Non-adopter	Total
Percentage of those that have tried any of the technologies	49.5%[1]	22%	71.5%
Estimated maize yield after using technologies			
By less than current yield	3.5%	9.0%	12.5%
By more than current yield	37.0%	9.0%	46.0%
By double the current yield	7.5%	0.5%	8.0%
By more than double current yield	1.5%	0.5%	2.0%
Other	0.0%	3.0%	3.0%
Average distance to soil and water conservation plots			
Mean (minutes)	26.02	23.72	25.32

[1]Percentage over 100 due to multiple response, *Chi-square*: 34.09; $p = 0.000$

between expected benefits from conservation structures and period of practising the technology (*Chi-square* = 18.91; $p < 0.000$). Furthermore, it was hypothesized that the further away the plots are from homestead the lesser the effort employed in maintaining the structures. However, it was found in this study that the adopters were the ones who had to travel a longer distance to the plot with soil conservation structure but still they were able to retain and adopt them. This was probably because of benefits they perceived were coming from the effectiveness of conservation structures in retaining soil from erosion which motivated them to continue using them despite the longer distance than non-adopters.

In terms of the challenges faced by farmers with the conservation activities, the most important one was that conservation practices competed for labour that could have been allocated for other activities, especially for production of food security crops. Others (15.4%) reported lack of inputs as a major challenge to adoption of these technologies, while 16.8% reported lack of technical knowledge as a challenge. Only 2.1% experienced land shortage problem. This agreed with the findings by Hailu and Runge-Metzger (1993), who stated that adoption of soil and water practices represented a decision by households to intensify their agricultural production through improving output per unit area through capital investment or an increase in labour inputs. They noted that it was essential to recognize that soil and water conservation measures impose an opportunity cost through their demands on labour, often at times of peak labour demand. As such it was important to consider looking at the cost and benefits and particularly on-farm costs versus off-farm costs of soil degradation.

Almost all the studied households have used some soil fertility enhancing practices in their fields except for a very small percentage (6.0%) of non-adopting households that have never ever used any of the soil fertility enhancing practices in their field. Legume–maize intercrops were the most commonly practised soil fertility enhancing practice amongst the surveyed households, and most of them have been practising them for a greater number of years than any other practice (average of 5.4 years).

Access to Credit and Inorganic Fertilizer Input Sources

It was found that only 9.0% of the respondents reported having received or taken loan in the 2005/2006 farming season. Whereas the majority 91.0% had not received

Table 3 Problems experienced during establishment of soil and water conservation technologies

	Adopter (%)	Non-adopter (%)	Total (%)
Land shortage	2.1	0.0	2.1
Labour constraints	16.8	4.2	21.0
Lack of inputs	9.8	5.6	15.4
Lack of technical knowledge	6.3	10.5	16.8
Lack of time	3.5	2.8	6.3
Limited financial resources	7.7	1.4	9.1

or taken any loan in the 2005/2006 farming season, it was found that 83.3% of adopters had obtained credit as against only 16.3% of non-adopters. The major source of credit in the study areas was Concern Worldwide, a non-governmental organization that was geared to uplifting the livelihoods of the rural communities through increased agricultural production by providing small-scale farmers with either agricultural or non-agricultural loans. Most of these loans were agricultural and were for the purchase of fertilizer and seed. According to Diagne et al. (2001), availability of non-farm income in the form of credit is an indicator of access to financial capital and has a positive influence on investment in soil and water conservation. Financial capital is manly used to pay for additional labour when investing in soil and water conservation. In some cases, it is needed to pay for cement for water diversion structures.

Over 75% of the surveyed households reported having applied chemical fertilizer in the 2005/2006 cropping season though most of them (73.0%) stated that the fertilizer was not enough to cover their fields. The average size of farm in acres, on which households applied the fertilizer, was 1.99 acres with more adopters registering a higher average acreage than non-adopters (Table 4)

Poor rural households in developing countries lack adequate access to credit. This in turn impinged negatively on technology adoption, agricultural productivity, nutrition, health and overall household welfare (Diagne and Zeller, 2001). This indicated that farmers who had access to credit had a higher probability of retaining conservation structures than those with no access. This may be explained by the fact that the requirement to pay back credits will motivate farmers to invest more on yield enhancing activities such as soil and water conservation and as a result greater effort will be put in maintenance and continued use of the structures already installed (Wogyehu and Drake, 2002).

Extension/Training in Soil Management Technologies

The study found that the most important source of agricultural information was through extension services as about 69.0% of the respondents reported working with agriculture extension officers on various technologies. The study also revealed that there was lack of regular interaction between farmers and agriculture extension staff who were the most reliable source of information to the farmers practising these new technologies. This in turn contributed negatively to the adoption decision of farmers. Again, the study found that farmers did not participate regularly in field days that were organized specifically to address soil and water management/soil fertility issues. This was because of lack of regular interaction between the farmers and extension staff. Less than half (45.0%) of the surveyed households reported that they had been participating in field days in their area, while the majority (55.0%) did not participate. Demonstration plot on new technologies was another important aspect in influencing farmers' decision to adopt a technology, as it offered practical experience of the technology. However, less than half of the survey population (23.5%) reported having a demonstration plot on soil and water conservation technologies mounted by extension workers in their area.

Institutional Support for Soil and Water Conservation in the Area

Through the focus group discussions with farmers, an institutional and stakeholder analysis of the community institutions as well as soil and water conservation stakeholders operating in the

Table 4 Fertilizer application by adoption on major crops (maize and tobacco)

		Adopter	Non-adopter	Total
Applied fertilizer last season		(82) 53.9%	(70) 46.1%	(152) 100.0%
Mean size of farm (acres) on which fertilizer was applied		2.08	1.88	1.99
Was the fertilizer enough?	Yes	13.2%	13.8%	27.0%
	No	40.8%	32.2%	73.0%

area was done. This was to see how much contribution they had made towards promoting soil and water conservation and the challenges being faced in the communities. It was found out that the following government and non-governmental organizations were already working in the area of environmental and natural resources management: The Malawi Environment Endowment Trusts, World Vision International; Concern Worldwide; Village Development Committees; Community Action Coordinating Committee; and CIMMYT Conservation Agriculture Project. Previously, Government, through the Initiative for the Development of Equity Efficiency in Agriculture funded by the Rockefeller Foundation, ICRISAT, and CIMMYT Risk Project had worked in the area on soil fertility management and soil and water conservation. A government project known as PROSCARP funded by the European Union had also been implemented in the area for more than 5 years and had constructed a number of soil and water conservation structures which the farmers interviewed were still using. There were no committees or organizations working specifically on afforestation in the area. The village headman had given the go-ahead to community members to form forestry committees for the village in order to facilitate afforestation activities which were crucial to the success of the soil and water conservation efforts. However, the current decentralization and community-based natural resources management committees mainly represented by the Village Development Committees were responsible for going around the surrounding villages to understand their problems related to environmental and natural resources management. These groups in turn reported the challenges to Community Action Coordinating Committees (CACCs) who in turn report to the District Assembly through the Area Development Committees. However, participants reported that the channel followed was not very effective as most communities normally did not get any feedback on the problems reported. Decentralization in Malawi has led to major coordination problems because district and local authorities continue to be supervised by their respective ministries. Village Development Committees also lack accountability and are corrupt. This has been reinforced by the primacy of political correctness over technical performance as a criterion for rewards. Corrupt and ineffective government institutions have led to the disappearance of those soil and water conservation practices that are dependent on the enforcement of the by-laws. The communities have also established some mechanisms of resolving conflicts related to land boundaries and water use. The chief manages and resolves all the conflict. No person is allowed to cut trees in the village except in his/her own land.

Conclusions and Recommendations

The empirical results show that certain socio-economic and institutional variables affect the farmers' decision on soil and water conservation technology adoption and subsequent impact on yields, food security and the general livelihoods of communities. It can be concluded that farmers belonging to a local organization had higher probability of maintaining conservation structures compared to those not belonging to any organization. Membership in a local organization reflects some certain level of social capital a farmer possesses in the community. Farmers with relatively larger land holdings had higher probability of maintaining conservation structures compared to the smaller ones. This necessitates intensification of agricultural production through provision of appropriate support services. To realize success in this regard, agricultural research, extension and provision of farm inputs have to be combined with soil conservation activities. Introduced soil conservation measures also need to be monitored regularly together with farmers so that problems can easily be singled out and appropriate improvements are forwarded. Hence, a sort of mechanism should be established to monitor farmers' plots where conservation structures have been installed. By doing so, we learn from experience and encourage farmers to take proper care to maintain the structures.

The results of the analysis of the factors influencing adoption of soil conservation practice indicated that the adoption behaviour of farmers is influenced by economic, institutional, physical as well as attitudinal factors. It was found from the analysis that farmers' perceived risk of soil erosion in the future, distance of a plot from the homestead, wealth status of the household and perception of benefits of conservation structure have a significant impact on farmers' retention behaviour. Therefore, it is reasonable to conclude that adequate consideration of these variables may greatly contribute to the increase of sustainable use and widespread adoption of introduced conservation structures and subsequent increase in maize yields, food security, income and livelihoods in dryland areas of Malawi and the whole of Southern Africa.

Acknowledgements The principle researcher would like to thank the International Foundation for Science (IFS) for the financial support that enabled him to undertake this fellowship in Malawi which has generated a lot of useful information which would help to influence policy and practice on soil and water conservation in Malawi and the whole of Africa. Similarly, many thanks go to the research assistants and all mentors for the support that was rendered during the fellowship period. To all the farmers who participated in the study, I wish to express my gratitude for the time and patience during the data collection period.

References

Alemu T (1999) Land tenure and soil conservation: evidence from Ethiopia. Göteborg University, Sweden

Diagne A, Zeller M (2001) Access to credit and its impact on welfare in Malawi. International Food Policy Research Institute (IFPRI), Washington, DC

Hailu Z, Runge-Metzger A (1993) Sustainability of land use systems. The potential of indigenous measures for the maintenance of soil productivity in Sub Saharan agriculture: a review of methodologies and research. Margraf, Weikersheim

Hatibu N, Mahoo HF, Kayombo B, Ussiri DAN (1995) Evaluation and promotion of rainwater harvesting in semi-arid areas of Tanzania research project. 2nd Interim Technical Report, SWMRP. Sokoine University of Agriculture, Morogoro

Scoones I (1996) Hazards and opportunities. Farming livelihoods in dryland Africa: lessons from Zimbabwe. Zed Books/International Institute for Environment and Development, London

Wogayehu B, Drake L (2002) Adoption of soil and water conservation measures by subsistence farmers in the Eastern Ethiopia. Presented at the 17th World Congress of Soil Science, Bangkok, Thailand

World Bank (1995) World development report. World Bank, Washington, DC

Managing Soil and Water Through Community Tree Establishment and Management: A Case of Agabu and Kandota Villages in Ntcheu District, Malawi

H.J. Kabuli and W. Makumba

Abstract Agriculture is the backbone of Malawi's economy, with over 90% of the country's population living in rural areas depending on agriculture for their livelihood. Consequently, deforestation is very pronounced because of the active agricultural industry which necessitates extensive clearing of woodlands. Trees are required in agriculture for both wood and non-wood products, protection of water catchments and preventing soil erosion thereby increasing agricultural productivity. In order to find out the technical and organizational aspects in community tree establishment and management, a research study was carried out in Agabu and Kandota villages in Ntcheu District. The study revealed that more than 55% of the people in both villages are involved in community forestry activities which include seed collection, seed processing, seed storage, nursery establishment, nursery management and out-planting of trees in communal village forest areas, gardens and around homes. More than 70% of the community members value indigenous trees over the exotic trees because of their multipurpose functions as a source of fuel, timber and poles, edible fruits and leaves, animal feed, medicine and as a source of manure which improves soil fertility and retains soil moisture. The study also identified that in both villages there are community forest clubs which are responsible for ensuring that members of the communities participate actively in the communal forest activities. The study recommends that government agencies, communities and other stakeholders must be involved in planning, implementation, monitoring and evaluation of projects and programmes to ensure sustainability in the management of natural resources.

Keywords Agriculture · Community · Trees · Soil · Water

Introduction

Malawi is among the poorest countries in the world, with a population estimated at 12 million and population growth of 3.2%. The country depends mostly on agriculture for its livelihood. Over-reliance on agriculture as a main engine of economic growth and inadequate development in the industrial and service sectors have promoted agricultural expansion in marginal land and extensive deforestation. The consequence of this is that an increasing proportion of the total cultivated land is exposed to degradation.

Although soil erosion is the major type of land degradation that poses the biggest threat to sustainable agricultural production, other forms of land degradation include wind erosion, rapid soil movements (e.g. the case of Phalombe disaster in 1991), increasing salinity (e.g. underirrigation), chemical, physical and biological degradation (e.g. through nutrient mining and leaching, loss of organic matter) that leads to reduced nutrient and water holding capacities and weakening of the soil structure. Results from spot trials of soil erosion under various cover and farming practices have shown that soil loss ranges from 0 to 50 t/ha/year (Amphlett, 1986). The national average soil erosion is estimated to be 20 t/ha/year. Taking into

H.J. Kabuli (✉)
Department of Agricultural Research, Chitedze Research Station, Ministry of Agriculture and Food Security, Lilongwe, Malawi
e-mail: hjinazali@yahoo.com

A. Bationo et al. (eds.), *Innovations as Key to the Green Revolution in Africa*,
DOI 10.1007/978-90-481-2543-2_135, © Springer Science+Business Media B.V. 2011

consideration different assumptions, the World Bank estimates that the above-soil loss leads to a mean yield loss of between 4 and 11% for Malawi as a whole.

Soil erosion is widespread and severe throughout Malawi, especially on land under cultivation. However, soil erosion also occurs on land that is not under cultivation such as stream banks, grazing land and forest land, and this poses a great threat to sustainable economic growth and food security in the country. As the soil fertility declines, there is reduction in crop yields, sedimentation and siltation of rivers and reservoirs, and fertile low-lying areas may become unproductive due to the deposition of infertile sand. Silt loads of surface water run-off lead to significant problems in downstream water quality, such as increased suspended solids and turbidity, water treatment costs and water flow problems. During the rainy season, virtually all rivers carry heavy loads of sediments. The turbid water is not good for human consumption. The majority of people in rural communities depend on untreated river water supply and chances of drinking unclean water are therefore very high (Government of Malawi, 1994).

Trees are important in soil and water conservation as they reduce erosion of the soil. Where there are more trees, leaf litter provides a soil surface mulch, which is important in the improvement of soil structure, infiltration of water into the soil and improvement in soil fertility. Trees also help to ensure supplies of fresh water, prevent flooding, protect crop from wind damage, stabilize soil and avoid excessive siltation of river beds and downstreams.

Community-based tree resource management, in its broader sense covering customary lands as well as forest reserves, is a concept with a major potential to realize the drive for participation of communities, non-governmental organizations and private sector alongside government in the management of woodland on customary land and forest reserves (Malawi Government, 2001). Community tree management promotes improved livelihood of communities and sustainability in the utilization of tree resources (Luhanga et al., 1993; Kayambazinthu, 2000), as the responsibility of planting and looking after the trees is taken by the community itself with the role of forest services being mainly a catalytic one. Community involvement in tree resource management is important in ensuring flexibility in the implementation of projects and programmes, active participation of communities and ensuring sustainability in the utilization of tree resources.

The success of most community reforestation programmes is dependent on collection of seed. A growing concern is to collect seed of high quality. The type of tree seed collected locally for establishment grows best in the nursery and also in that particular area as most tree species are adapted to a specific environment (Gowela and Masamba, 2002; Odera, 2004).

Methodology

In order to identify some of the participatory approaches that are aimed in empowering communities in establishment and management of tree resources, a research study was carried out in Agabu and Kandota villages in Ntcheu District, Central Region of Malawi.

Participatory Rural Appraisal (PRA) was used in data collection. The PRA tools which were used included resource mapping to gather information on the type of settlement and the key features in the community, transect walk to verify information on land use systems, household interviews to allow a representative number of people to give more detailed information and to gather information that was sensitive in nature and key informant interviews to gather specific type of information. Focus group discussions were also done to collect general issues from different gender categories of those involved in forestry activities and those who are not involved in forestry activities, respectively. Trend analysis studies were also carried out to show how resources were changing with time.

Results and Discussion

The study identified that most of the households were male-headed, and more than 55% of the people in both villages have land holding size of less than 3 acres. The major source of income in the area is agriculture. The crops that are commonly grown in the areas included maize, pigeon peas, groundnuts, millet, soybeans, bambara nuts, pumpkins, sweet potatoes, cowpeas, cotton, cassava, sorghum, barley tobacco, beans and pepper. Livestock that are commonly kept in the area include goats, sheep, chickens, cattle, pigeons, ducks, rabbits and pigs.

A total of 53.3 and 60.0% of the households in Kandota and Agabu villages, respectively, do not have enough food throughout the year, especially from October to March. There has been a decrease in crop production due to loss of soil fertility as a result of deforestation, decrease in land holding size located to individuals as a result of population growth because of births and immigration of people from other parts of the country in search of land for settlement, and unreliable rainfalls, which sometimes are heavy or inadequate.

The study also revealed that tree population in the area had been decreasing at an alarming rate since the 1970s as a result of opening up of new gardens for agriculture and settlement, increasing the demand for fuel wood, brick making, construction of tobacco barns and setting up of bush fires in search of wild animals, e.g. mice. Apart from reduced agricultural production in the area as a result of deforestation, the study also observed that there was an increase in siltation in some parts of the major tarmac roads and seasonal drying up of streams.

In order to combat the challenges of low agricultural production due to declining tree resources, more than 56% of the people in both villages are involved in tree management activities. More than 70% of the people in both villages value indigenous trees over the exotic trees because they have multipurpose functions, are well adapted to the soil and climate of the area and are not easily attacked by pests and diseases. But in most cases, exotic trees grow faster than indigenous trees and coppices very well at a faster rate.

Tree Seed Collection

Most of the people who are involved in tree management activities do collect tree seed from surrounding forests, around homes and from the gardens. Collection of tree seed is based on some conditions: the mother tree should be growing on fertile soil which is characterized by a lot of grass growing underneath; should have more leaves, not attacked by pests or diseases; should have no dead branches on top; should have good shape; should be tall and straight and should have more fruits which could be easily collected. There are various types of tree species that are commonly found in Agabu and Kandota villages, but few tree species are preferred for seed collection and these include *Trichitia emetica* (msikidzi), *Lonchocarup cappassa* (chimphakasa), *Senna spectabilis* and *Senna siamea* (mkesha), *Gmelina arborea* (malayina), *Carica papaya* (papaya), Eucalyptus species (bulugamu), *Mangifera indica* (mango), *Psidium guajava* (guava), *Pterocarpus angolensis* (mlombwa), *Acacia polyacantha* (mthethe), *Tamarindus indica* (bwemba) and *Alcacia albida* (msangu).

A. albida was spotted out to be a very important tree species because it is a source of fuel wood, it can be used in construction, and it has the capability to enrich the soil. Its distinctive feature in shedding of leaves during the rainy season enriches the top soil while permitting adequate light penetration to the crop below for normal photosynthesis. The tree litter and canopy also influence the micro-environment in terms of improved rainfall infiltration, soil structure improvement, reduced evapotranspiration and temperature extremes and increased relative humidity (Saka et al., 1994).

There are various ways by which communities determine seed maturity as outlined in Table 1. Seed of most tree species are collected from July to December, which coincides with some agricultural operations.

Most of the times, communities collect enough tree seed. If in excess, the seed are thrown away due to unavailability of reliable markets, given to friends, planted in the next season, or sometimes sold locally. In some cases, there is inadequate seed collection due to sickness and failure of trees to produce enough seed

Table 1 Determination of seed maturity

Determinant	Examples
Change in colour	*B. thoningii* – change of fruit from green to brown
Fruit falling from tree	*Gmelina arborea*
Hardness of fruit/pod	*B. thoningii*
Heaviness of seed and sinks in water	*L. carpassa*
Fruits can be consumed	*Mangifera indica*
Hardening of fruit hairs	*Pterocarpus angolensis*

due to inadequate rainfall or heavy rainfall as the fruits and seed tend to rot.

Seed Processing

Seed Drying

Drying of seed reduces the amount of water they contain for longer storage. The method of seed drying depends on the type of tree species, but for most trees, seed are sun-dried. Drying is done away from livestock, on top of roofs of granaries and houses, and on the rack (*chithandala*). Materials used in seed drying include sacks, baskets, mats, plastic plates, plastic papers and sometimes seed are dried directly on the ground.

Seed Extraction

Seed extraction is done to remove the seed from its pod or fruit, especially for those species that do not release their seed upon drying. In Kandota and Agabu villages, seed extraction is done mainly by use of hands to avoid damaging the seed thereby reducing its viability.

Seed Storage

Seed is stored in khaki papers, ragged clothes, empty tins, glass bottles, cartons, *zikwatu* (bags made from maize leaves), gourds and clay pots. Seed is stored inside the house by hanging on the roof of the house away from children. However, some people do not store seed, and seed is directly sown soon after collection to prevent the seed from damage caused by rats and fungi or bacteria. Preventive/cultural measures that are followed to protect seed from pests and diseases include ash dusting, storage in dry places to prevent the seed from rotting, taming cats as a biological control of rats and storing the seed for a short time to maintain viability.

Overcoming Seed Dormancy

There are various ways that local communities use to overcome seed dormancy and these include boiling the seed for 5 min before sowing, e.g. *S. siamea* and *B. thoningii*; soaking the seed in cold water before sowing, e.g. *G. arborea*; scarification of seed, e.g. *P. angolensis*; and soaking in hot water for 15 min, e.g. *Leucaena leucocephala*.

Nursery Establishment and Tree Planting

In Agabu and Kandota villages, the communities establish tree nurseries where tree seedlings are managed before out-planting the trees in the village forest areas, around homes and in the gardens. The conditions that are followed in the selection of the nursery site include closeness to water supply in order to reduce labour and time in watering the seedlings, gentleness of slope to reduce erosion and water logging and the vicinity of the community.

Organizational Aspects in Community Tree Management

The existence of forestry clubs in Agabu and Kandota villages ensures ownership and sustainability in communal tree management. Committee members in these forestry clubs ensure that there is equal sharing of roles and responsibilities, active participation and equal sharing of benefits among the communal members and that the community members abide to the laws defined by the community in tree resource management, e.g. no careless cutting down of trees and no setting of bush fires in search of mice.

Challenges of Community Tree Establishment and Management

Non-involvement of Some Community Members

Some members in Agabu and Kandota villages are not involved in communal tree management activities due to laziness, lack of knowledge on the importance of communal tree management, perception that communal tree management activities are more labour demanding and lack monetary profits. The villagers therefore give preference to temporary employment (*ganyu*) in order to source income and food.

Limitations in Seed Collection and Storage

The study revealed that in some cases seed is eaten by animals, e.g. rats; also, sometimes the seed is blown away by whirlwind during drying. It was also discovered that some seed get rotten in storage due to increased moisture content as a result of leaking houses during the rainy season as most houses are grass-thatched.

Conclusion

Communities have been empowered to play an active role in utilization and sustainable management of tree resources. This has been achieved though training of the communities in seed collection, nursery establishment and tree management.

Communal organizational structure through the existence of forestry committees ensures ownership and passing of knowledge in tree management from one generation to another.

If managed in a sustainable fashion, trees can provide long-term environmental and economic benefits for those willing to grow them. Trees absorb carbon dioxide, fertilize soil, prevent erosion, and help to absorb rainwater, while at the same time providing fuel, food, fodder, compost, building materials and even medicines from their wood, leaves and fruit.

Recommendations

Some community members who are not involved in community tree management activities, especially due to laziness, should be encouraged and trained to take part in different activities so that labour should not be a problem, especially during the rainy season when there are labour bottlenecks due to labour demands for tree management and field crop management operations.

Communities should be encouraged to plant multipurpose tree species in their gardens in order to increase or improve soil fertility levels, thereby increasing food productivity. There is also need to promote and encourage already existing communal manure clubs as households may not invest labour and time nurturing trees when there are more pressing needs for food security.

There is need for further research on how to establish tree seedlings apart from use of seed for faster tree establishment. Considering the long-term nature of the forest enterprise, degraded state of forest bases and people's short-term perspective for returns on investment, the need for a new research and development (R&D) order becomes an urgent task. Such an R&D programme should tackle areas of knowledge where research can add value to pave the way for greater innovation on an issue and problem-driven approach considering indigenous knowledge and skills for managing forest goods and services.

There are opportunities for diversifying income sources through development of forest-based enterprises and other innovative opportunities in developing new income sources from biodiversity, including consumptive and non-consumptive use of wildlife, guinea fowl rearing, beekeeping, fruit tree establishment, etc.

There is need to empower community members to collaborate with government and other stakeholders in managing tree resources, develop tree resource-based enterprises and carry out good management of tree resources on customary land based on clear mechanisms of access, ownership, responsibility and control.

Acknowledgements The author acknowledges the European Union Social Forestry Project for financially and materially supporting the study. The author would like to thank the staff of the Department of Forestry, Mr. Charles Gondwe, Mr. Namalima, Mr. Francis Nkungula, Mr. Kenneth Mandala, and all the staff who assisted in various ways for the study to be conducted.

References

Amphlett MB (1986) Soil erosion research project. Bvumbwe Malawi, Summary Report, Hydraulics Research, Wallingford, UK

Gowela JP, Masamba CR (2002) State of Forest and Tree Genetic Resources in Malawi. Prepared for the Second Regional Training Workshop on Forest Genetic Resources for Eastern and Southern African Countries 6–10 December 1999, Nairobi, Kenya; and updated for the SADC Regional Workshop on Forest and Tree Genetic Resources, 5–9 June 2000, Arusha, Tanzania. Forest Genetic Resources Working Papers, Working Paper FGR/27E. Forest Resources Development Service, Forest Resources Division. FAO, Rome (unpublished)

Kayambazinthu D (2000) Empowering communities to manage natural resources: where does the power lie? – The case of Malawi. In: Shackleton S, Campbell B (eds) Empowering communities to manage natural resources: case studies from Southern Africa. CIFRO, USAID, IUCN, WWF, Africa Resources Trust, CSIR, and IES: Lilongwe, Malawi

Luhanga JM, Coote HC, Lowore JD (1993) Community use and management of indigenous forests in Malawi: the case of Chemba village forest area. Forestry Research Institute of Malawi Report No. 93006, Zomba, Malawi

Malawi Government (1994) National environmental action plan. Lilongwe, Malawi

Malawi Government (2001) Malawi's national forestry programme: priorities for improving forestry and livelihoods. Department of Forestry, Lilongwe, Malawi

Odera J (2004) Lessons learnt on community forest management in Africa. In: Lessons learnt on sustainable forest management in Africa. Nairobi, Kenya: African Academy of Science (AAS) and FAO: Nairobi, Kenya

Saka AR, Bunderson WT, Itimu OA, Phombeya HKS, Mbekeyani Y (1994) The effects of *Alcacia albida* on soils and maize grain yield under smallholder farm conditions in Malawi. For Ecol Manage 64:217–230

Adoption and Up-Scaling of Water Harvesting Technologies Among Small-Scale Farmers in Northern Kenya

M.G. Shibia, G.S. Mumina, M. Ngutu, M. Okoti, and Helge Recke

Abstract In sub-Saharan Africa, especially in low-potential areas, agricultural production among small-scale farmers is constrained by declining soil fertility and unreliable rainfall. The "transfer-of-technology" model dominates most research and development approaches and it is important to understand drivers of technology innovation among small-scale farmers to guide future development of models that are effective in ensuring that farmers benefit more from research. This study was initiated in phases aimed to identify, test and develop appropriate water-saving technologies using Participatory Learning and Action Research (PLAR) approach. Combinations of four water-saving technologies were selected and tested with vegetable farmers; technologies included drip kit, compost basket, mobile sack and improved sunken bed. The improved sunken bed was more adopted compared to the other options. In the first crop-growing season, 100 farmers adopted technology from 9 trial farmers. The farmers adopted the technology due to modification to ensure that technology meets diverse socio-economic backgrounds of various households. Through adoption of improved sunken bed, farmers increased income and nutritional status at the household level with total net benefits of USD 1,500 per year for 10 m^2 land area. Among other factors, promotion of farmer innovation and partnership between researchers and extension agencies improved the technology adoption pathway. To meet food deficits due to increased population in marginal areas, it is important to consider farmer innovations as part of technology development.

Keywords Farmer innovation · Vegetable · Dry land farming

Introduction

In sub-Saharan Africa, especially dry land farming is increasingly constrained by problems of soil fertility and unreliable rainfall. Due to rapid increase in human population, most households are settling in marginal areas and advances in the development of drought-tolerant crops have brought more of marginal areas under crop cultivation. The farmers adopt farming practices suited to high-potential areas gained from interactions with the immigrants and with limited adoption of improved technology. Despite these challenges, small-scale farming remains a key livelihood option among the rural population.

The case of northern Kenya is no exception. Studies conducted show that farmers in the marginal areas of Marsabit Mountain experience crop failure every two to three seasons. The region is located within the Ecological Zone V–VI, characterized by a mean temperature of 30°C and an average rainfall of 250 mm that varies with altitude and on average 700 mm of rainfall is received on the selected mountain pockets. The soil is generally poor and not suitable for crop farming except in the selected arable pockets, which consist of 3% of the total land area. In comparison to the whole area, it is small but supports the livelihood of the rural community. For example, over 42% of the population around Marsabit Mountain is involved

M.G. Shibia (✉)
Kenya Agricultural Research Institute, National Arid Lands Research Centre, Marsabit, Kenya
e-mail: schibier@yahoo.com

A. Bationo et al. (eds.), *Innovations as Key to the Green Revolution in Africa*,
DOI 10.1007/978-90-481-2543-2_136, © Springer Science+Business Media B.V. 2011

in crop farming both for food and for income (Kioko et al., 2001). However, the study carried out by Ngutu and Kioko (2000) showed that the most limiting factor for crop farming around Marsabit Mountain was the amount of water available for crop production due to limited water and unreliable rainfall.

In this agro-ecosystem, diversity of farming practices exists and it is largely inhabited by agro-pastoralists. The population in Marsabit, including Marsabit town, consists of the poorest in Kenya with almost 80% living below the absolute poverty line of USD 1 per person per day (Barrett et al., 2003). Illiteracy among the female population in northern Kenya is as high as 94% and reaches over 70% in the male population (SID, 2004). They are diverse and homogeneous households almost do not exist as very minor changes in the socio-economic situation of a household may set it miles apart from a neighbour who shares a similar ecological environment. This means that any technology that is not tailor-made to fit the individual household needs will most probably not be adopted by that household. It is impossible for scientists to develop technologies that meet each and every household's requirement and so using participatory technology development approach, there is need to encourage the diverse farming households to experiment and develop technologies that suit their very specific needs.

In an effort to improve food security, various development agencies have implemented different interventions following the immediate needs by farmers to increase return from their farms. It is evident that most of the practices reduce land potential and farming practices are not recommended for the agro-ecosystem. Many farming systems are becoming unsustainable and in the long term, farm level productivity declines below the expected yields due to depletion of soil fertility and the farmers lack the initiative to replenish. The farms are abandoned by the opportunistic farmers as they respond to emerging economic activities that promise greater returns, including charcoal burning and firewood sale. The key to increasing and sustaining the productivity of this farming system is to understand the dynamics of this continuum and promote sustainable practices, which will not only conserve the fragile agro-ecosystem but also enhance food security among the settled agro-pastoralist communities.

Until recently, the "transfer-of-technology" model dominated most research and development approaches. The model essentially prescribes that new ideas to help farmers solve their problems are with the "experts" who must find innovative ways of passing these ideas on. The researchers have not been able to replicate the complex and marginal physical and socio-economic environment of the farmers and rarely is the innovation potential of the farmer sought or exploited (Haverkort et al., 1988) and only recently have steps been taken to acknowledge and harness indigenous knowledge in a link with science (Waters-Bayer and van Veldhuizen, 2005). Participatory Technology Development has been advocated recently for the types of farming systems found in semi-arid areas (Mubiru et al., 2004). If given the opportunity, it is possible that farmers will develop technological alternatives that take into account the complexity and diversity of prevailing circumstances at the household.

These findings draw on experience of projects implemented in the year 2002 to establish factors triggering farmers to intrinsically experiment with new ideas and how they decide to modify technologies to suit their specific needs. An understanding of the above will help to add to the body of knowledge on researcher–farmer interactions, understand drivers of adoptions and help in future development of models that are effective in ensuring that farmers benefit more from research.

Materials and Methods

The diagnostic survey was implemented using the Participatory Learning and Action Research (PLAR) approach as described by Defoer and Buddelmann (2000). The approach ensured that farmers are involved in all stages of the research process and includes the entire community in the selection of the test farmers. In the diagnostic phase, problem of water shortage and declining productivity were the main constraints ranked, and trial of high-value crop, specifically vegetable farming, was suggested to boost farm-level return. In the community workshop, trial farmers were selected based on three management levels (poor, medium and good) and the characteristics for each category were agreed upon in the same workshop according to the procedure outlined in Defoer and Buddelmann (2000).

Five study clusters were selected, including Karare, Gabra scheme, Gororukesa, Dirib Gombo and Kituruni. However, in the test clusters, agro-pastoralists are similar showing very small differences in practice and in each site four selected trial farmers had to represent similar management levels. The selected farmers were first taken on study tours to areas with a comparable climate, but with a much longer farming history, exposing the Marsabit farmers to what farmers in similar circumstances were practicing. The preferred improved technologies were agreed upon in community planning meetings and subsequently trials with those four different water-saving technologies were set up by the selected 16 farmers.

In participatory community meetings, the improvement of water utilization in small-scale vegetable production was selected as most likely to produce positive impact since this would improve available income (vegetables are still a scarce and pricey commodity around Marsabit Mountain), improve household food security and optimize the use of a scarce resource (water). The farmers developed the experimental designs, and the parameters to measure were agreed upon by the community. The trial farmers in the four clusters were monitored by a community research committee and the follow-up was done once a month by a team of scientists.

The four technologies selected by farmers included drip irrigation in the form of a simple drip kit as described by Arid Lands Information Network – Eastern Africa (2002), the compost basket, the sunken bed, the mobile sack and a control (Ngutu and Recke, 2006). For all the trials, kales were the test crop. A total of 16 farmers participated in the trials from 4 different clusters with each farmer trying at least 4 technologies. To determine if a certain technology had the ability to save on the amount of water required to sustain a crop, farmers decided that the best variables to measure included yields and the number of days it took before a plant started wilting. Equal amounts of water were given to all treatments and the resultant days to wilting and yields were recorded. Those farmers who gave recommended amounts of water for all technologies at the same interval recorded the yield and only medium- to large (approximately 60–100 cm long)-sized leaves harvested were recorded. The recommendation was approximately 20 l of water per m^2 per week.

Those farmers who participated in the study tours explained all new technologies demonstrated during the tour to interested farmers. The scientists explained the principles behind each of the technologies and detailed discussions on merits and disadvantages followed before a decision on the four technologies to be tested was reached. The trials were farmer-managed and the scientists did not seek to influence the farmers to allow them modify technologies to suit their immediate socio-economic environment. In different clusters, scientists held discussions with the community on the issues that related to technology adaptation. These included technology performance, crop performance, water utilization, crop management, labour requirements and other issues of concern to the farmer; modifications made on technology and the rationale behind it; technology adoption or reason for non-adoption and neighbour reactions through farmer-to-farmer interactions.

A cluster research committee was set up to coordinate the community research effort and to act as a linkage between scientists and the community. This was particularly useful as the committee guides any new farmers taking up the technology or modifications being made to the originally introduced technologies. Trial farmers were encouraged to monitor the progress of their trials and keep records of their observations. Scientists visited the trial farmers at least once a month and collected both qualitative and quantitative data on the performance of the various technologies. The approach encouraged farmers to do their own analysis based on suitability and sustainability of the various technologies. The data collected were subjected to statistical analysis using MS Excel.

Results and Discussion

Eighty percent of trial farmers ranked the compost basket as the better technology for small-scale vegetable gardening out of the four technologies tried. Farmers initially tried the technology as introduced but with time they began to make modifications (Table 1). The greater proportion of farmers adopting technology was women (53%). Initially, those farmers who had been in the exchange tour and selected farmers introduced technologies as they had been demonstrated in other parts of the country. It was

Table 1 Modifications made on the compost basket

Original design	Modifications	Rationale of modification
Circular hole 90 cm wide and 60 cm deep	Rectangular trench, depth was varied by different farmers, some dug holes only 35 cm deep	Utilizes more space efficiently Reduced labour requirements
Filling of hole with top soil and manure at a ratio of 4:1	No major modifications	–
Lining base of the hole with single new polythene (PE) sheet	Use of old pieces of PE overlaid on each other or reuse of home-used nylons	Long sheets of PE have to be acquired in shops. Initially. the PE used to line sugar sacks was given free by shopkeepers but with increased demand the PE was being sold, so farmers used smaller pieces of PE such as shopping bags and overlaid them
Installation of a round basket made of sticks at the centre of the hole with a diameter of 30 cm	Replacement of baskets with plastic containers	The basket made of twigs rot after some time and takes time to construct. Old plastic containers solved both problems
Filling of the basket with vegetative material (compost)	Replacement of vegetative material with stones and sand	Vegetative material may not be readily available on-farm (used for feeds, etc.). Stones allow water infiltration to lower levels of soil profile and prevent evapotranspiration from lower levels

expected that once the original technologies were tried the communities without modification would adopt the most promising technologies wholesome. However, while implementing the project, interesting trends in adaptation of technologies were observed. In the test clusters among the four technologies introduced, only the compost basket and sunken bed technologies allowed the farmer the opportunity to change. The modification allowed farmers to suit the technology to their socio-economic environment and it was more adopted across the cluster areas.

Individual farmers modified the technologies to suit their specific needs. Most of the farmers who wanted to go commercial opted for the more compact rectangular trenches, which they themselves had innovated while those interested in production for home consumption maintained the circular beds. Farmers near water sources mainly did away with the baskets, as water was not a problem while those far away maintained them and modified to utilize locally available material. It can be said that farmers were able to adapt the technologies to suit their specific needs. Scientists and development agencies would generally see this as undesirable because their recommendations are not implemented as planned. This means that they may not be able to show impact of their work, as they would like, as the effects of the modifications have not been tested in scientific ways. However, farmers who understood the principles of the technology appreciated positive benefits. The study did not seek to scientifically establish if the above farmer sentiments were true but all farmers with modifications said they had realized benefits. Farmers adopted compost basket technology at different levels of modifications (Fig. 1). In all the test clusters, the greater proportion of farmers adopted the technology with modification and not wholesome as introduced.

Triggers to Enhancing Farmer's Innovation

From working with the community during the implementation of the activity over the 3-year period, the following were observed relating to the triggers that encourage farmers to innovate:

- Exposure of farmers to new ideas in a dramatic and thought provoking manner
- Organizing farmers to follow up on the new ideas
- Allowing farmers to experiment by giving them the principles behind any new idea they have learnt and not overemphasizing the "package"

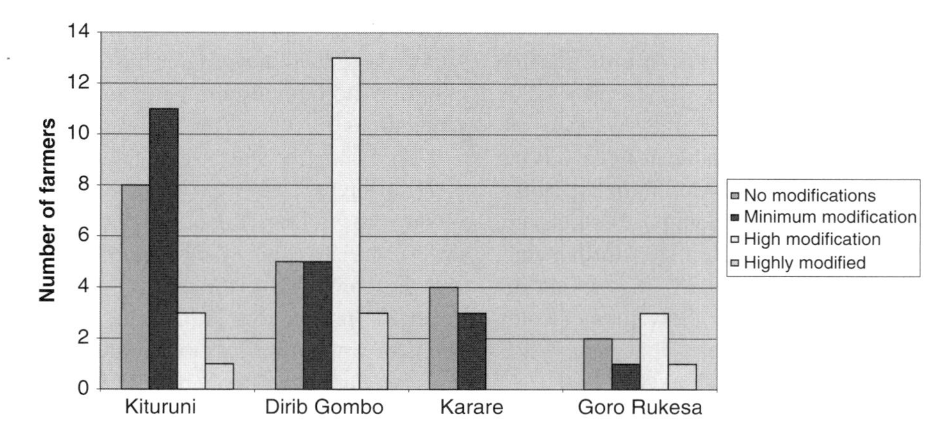

Fig. 1 Number of farmers making modifications to the compost basket

Exposure of Farmers to New Ideas

Exposure of farmers to new ideas can be done in many ways. However, methods that are dramatic and thought-provoking are important especially because farmers are averse to risk when it comes to deviating from what they have done for many years. One tool that was found to open up farmers was the exchange visit. Well-targeted and organized tours allow farmers the opportunity to see farmers in similar circumstances as their own trying out innovative ways of solving their problems. Once the farmers interact at the farmer level and honestly discuss the merits and challenges related to implementing a certain technology, the visiting farmer starts to feel that they too can do it. It is important to visit farmers who are truly implementing the technology in question and not the farmers who have the technology just to please the development agency they are working with (and these are many). The exchange tour therefore serves to among other things create a sense of ownership in the "discovery" of the technology.

Organizing Farmers

Farmers are so used to doing things in a certain way, largely considered as traditional and the old way by many researchers and development agencies. The experience of exposure to new proven ideas may not lead to immediate or instant translation to better practice on their farms even when they are convinced that the ideas can be implemented. Experience from the trials showed that before planning workshops were held in the cluster areas to consolidate the farmers' new idea, few farmers were implementing the technologies they had been most impressed with. However, if at most the packages are adopted there are no mechanisms of reaching and taking information regarding the packages beyond the farm or community where interventions were made. It is important after exposure to new ideas to follow up on farmers. This is achieved through development of participatory structures of communication within the community that will enhance exchanges of experiences related to the new ideas or proven practices. Such structures can simply be groups of test farmers meeting or more complex structure such as research committees to coordinate cluster-based research efforts.

Allowing Farmers to Experiment

Work done by the German development agency-supported Marsabit Development Project (MDP) around Marsabit Mountain showed that use of demonstration plots by development agencies made farmers feel that the technologies on display are expensive and alien to adopt. When demonstrations are done by development agents ("outsiders") the sense of ownership of the technology by the community is low (Kioko et al., 2001). The need to have communities have a sense of ownership for a technology is important if they are to fully adopt it.

The difficulty of making communities own a new way of doing things stems mainly from research and development agencies coming with a ready-made package or package options which they want to sell to the community as the most viable way to solving specific problems (Reijntjes and Moolhuijzen, 1995). For communities to own a technology, research and development agencies need to go a step further than just showing farmers "how it is done"; they should let farmers learn the principles and allow them to play around with these ideas to develop what they feel is most appropriate to their individual needs.

Adaptation of technologies by farmers will come naturally once the community realizes the rationale behind the technology and experimental benefits commensurate with the effort put in. To encourage this to happen, researchers must desist from insisting on farmers to follow strict recommendations and should appreciate different levels of adoption of technologies. Farmers may not implement technologies strictly as recommended but could modify them to suit their unique environmental or socio-economic situation. Encouraging such modification and going further to study the added benefit of the modifications made could be more important than sticking to rigid recommendations.

Concerning the water-saving ability of the technologies, different trials were compared, average number of days to wilting (Fig. 2) and average yields using different trials (Fig. 3). The mobile sack, sunken bed and the compost basket required minimum water for vegetable production. Farmers in areas prone to high levels of water scarcity observed that approximately 5 days after application of about 10 l of water per m^2 the kales under the above technologies did not show signs of wilting. In the control (farmer practice) and the drip kit, it only took between 1 and 2 days before wilting began.

In Fig. 3, the results show that the compost basket had the highest yield while the sunken bed, the drip kit and the control did not have any significant difference in yields.

Seventy-eight percent of the trial farmers reported a reduced pest and disease incidence in the compost

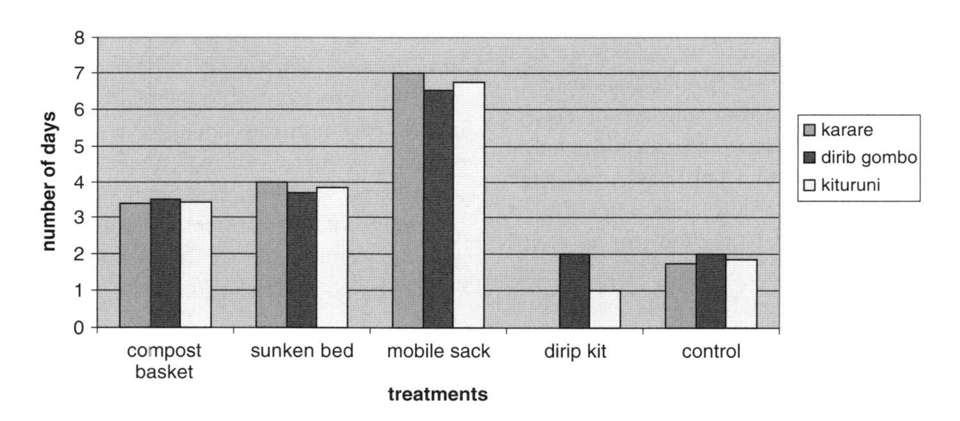

Fig. 2 Average number of days to wilting (August–September 2002)

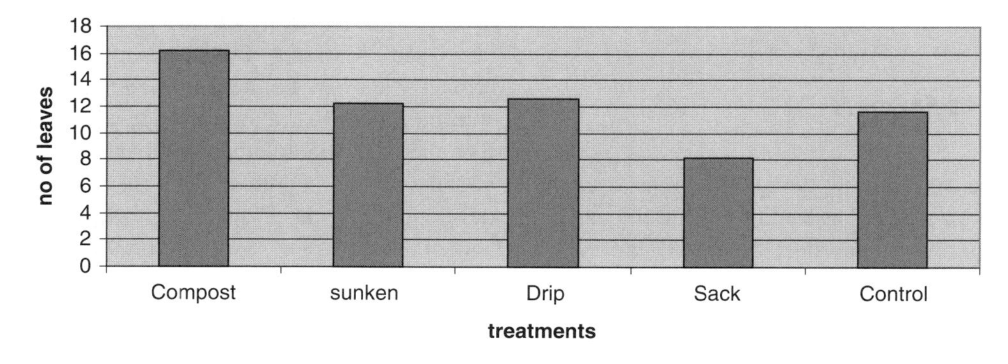

Fig. 3 Average yields for the 2002 long rain season (number of leaves/m^2/week)

basket and drip kit. The above situation may be attributed to the fact that water is not applied directly to the crop for the stated technologies. Weeding was considered to be easier for all introduced technologies.

Socio-economic and Environmental Dimensions

The adoption of the various technologies has been high, especially the compost basket; it was observed that over 40 non-test farmers from an initial 9 test farmers in one cluster area adopted it. This shows that the technology fits the community's socio-economic capability to implement. Small-scale vegetable production has been shown to put extra cash in the hands of women and so enable them to take more charge of their household needs. Farmers noted that adoption of the technology even at a very small scale (5 m^2) was economically rewarding, as they were able to sell vegetables and make at least Kshs 200 (USD 3 a day) during dry or drought season and on average they earn an equivalent to USD 1,500 per year. The technologies pay for the cost, especially for household increase in water needs (Fig. 4). The technologies while utilizing less water to produce more vegetables increased the overall demand for water in the households.

In the marginal areas to mitigate such effect, farmers fetch water using donkeys and bicycles to sustain the vegetables at the peak of the dry spell. The assumption had been that since the farmers would only be growing the vegetables at a very small scale, wastewater could be used to sustain vegetable production throughout the year. Since water demand increases, technologies that increase water availability at household level need to be investigated. Currently, trials on the household-based water pans are being tried to go hand in hand with the improved vegetable production technologies. All trial farmers said that vegetable production was more convenient for them than charcoal production. In this respect, it can be said that if the water issue can be solved, then this alternative source of income would be beneficial to the environment, as farmers would shift from environmentally destructive forms of livelihood sustenance to more friendly forms.

The weak research-extension–farmer linkages limited agricultural technology adoption. The national agricultural extension was inadequate to link farmer–researcher linkages, therefore allowing NGOs to fill the gap (Republic of Kenya, 2004). Among other factors, partnership between researchers and local extension organizations and introduction of community research committees helped to build synergy and enhance farmer innovations and knowledge and reach a large number of farmers. Presently, in total, 100 non-test farmers have adopted the technologies from 9 trial farmers. The figures represent minimum numbers, as they include only those areas where we have reliable estimates while more diffusion occurred through farmer-to-farmer transfers.

Further, the PLAR approach has proved an invaluable tool in ensuring that farmers not only take up the suitable technologies but also are encouraged to own the technologies to the extent that they continually experiment and modify them to suit their specific needs. As noted in the results, the rate of uptake of technologies was high and this is primarily attributed to farmers owning the research process. In community workshop, farmers would discuss the benefits of technologies and decide on the ones they believed were most appropriate, and the most interesting

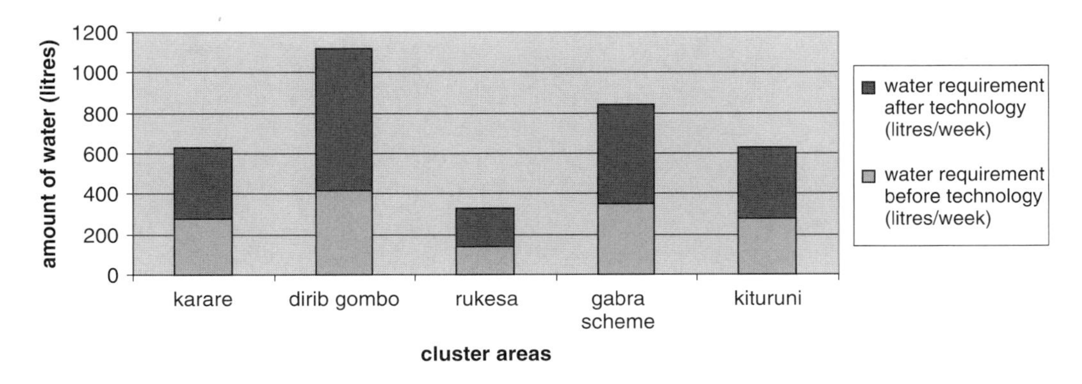

Fig. 4 Average household water requirement before and after technology introduction

development was the confidence the farmers gained as exemplified by their continuous modification of technologies to suit their needs.

Conclusion

Small-scale farmers and those living in close proximity live in diverse circumstances that make technologies developed by scientists inapplicable to each and every one of them. This means that individual farmers should be encouraged to experiment with new ideas so that they are able to find the level at which they can adopt a given technology. Farmers will generally want to continue doing what they have done for many years as they can ill afford to take chances and fail. New innovations that can help the small-scale farmer are best developed by the same farmer but with the guidance of scientists because this is the only way the farmers will truly own the technology. In addition, in the up-scaling of proven technologies, participatory approaches stimulate innovative spirit of farmers and enhance utilization of valuable technical knowledge to catalyse adoptions of the technology.

Acknowledgements The authors would like to acknowledge EU for the timely provision of financial support and farmers in Marsabit Mountain without which the project would not have been successful.

References

Arid Lands Information Network (Eastern Africa) (2002) Baobab: information on dryland agricultural practices. Nairobi, Kenya. Available from http://www.alin.or.ke

Barrett CB, Chabari F, DeeVon B, Little PD, Coppock DL (2003) Livestock pricing in the Northern Kenyan Rangelands. J Afr Econ 12(2):127–155

Defoer T, Budelman A (eds) (2000) Managing soil fertility. A resource guide for Participatory Learning and Action Research (PLAR). Royal Tropical Institute, Amsterdam, The Netherlands

Haverkort B, Hiemstra W, Reijntjes C, Essers S (1988) Strengthening farmers' capacity for technology development. ILEIA Newslett 4(3):3–7

Kioko TM, Ngutu M, Galgallo O, Kimani M (2001) Lessons learnt and experiences gained: the Marsabit Development Programme (MDP): sustainable crop production systems. Internal report, German Technical Cooperation (GTZ), Marsabit, Kenya

Mubiru DN, Ssali H, Kaizzi CK, Byalebeka J, Tushemereirwe WK, Nyende P, Kabuye F, Delve R, Esilaba A (2004) Participatory research approaches for enhancing innovations and partnership in soil productivity improvement. Uganda J Agric Sci 9:192–198

Ngutu M, Kioko TM (2000) Report on analyses of farming systems on the Mount Marsabit region of Marsabit districts. Kenya Agricultural Research Institute, Internal Report, Marsabit, Kenya.

Ngutu M, Recke H (2006) Exploring farmers' innovativeness: experiences with the adaptation of water saving technologies for small small-scale vegetable production around Marsabit Mountain in Northern Kenya. Exp Agric 42:459–474

Reijntjes C, Moolhuijzen M (1995) Searching for new methods. ILEIA Newslett 11(2):4

Republic of Kenya (2004) Strategy for Revitalizing Agriculture 2004–2014, Ministry of Agriculture & Ministry of Livestock and Fisheries Development. Government Printer: Nairobi, Kenya

SID (Society for International Development) (2004) Pulling a part-fact and figures on inequality in Nairobi, Kenya. Society for International Development, Eastern Africa Regional Office, Nairobi. Available from www.sidint.org

Waters-Bayer A, van Veldhuizen L (2005) Promoting local innovation: enhancing IK dynamics and links with scientific knowledge. IK Notes No. 76, January 2005, World Bank, Washington, DC. Available from http://www.worldbank.org/afr/ik/default.htm

Social and Economic Factors for the Adoption of Agroforestry Practices in Lake Victoria Catchment, Magu, Tanzania

A.J. Tenge, M.C. Kalumuna, and C.A. Shisanya

Abstract Environmental degradation is an issue of concern in the Lake Victoria catchment in Mwanza, Tanzania. Deforestation and improper soil management have lead to soil erosion, lake siltation, accumulation of C in the atmosphere and climate change. Agroforestry practices, soil conservation measures, manure application and retaining low-quality crop residues as soil organic inputs could reduce environmental degradation while increasing farm productivity and income. However, these options are not widely used by smallholder farmers in the Lake Victoria catchment area. Promotion and sustainable uses of these options require understanding of the prevailing socio-economic situation of the target farmers. This research investigated the social and economic factors that influence the adoption of agroforestry practices by smallholder farmers in Magu district, Mwanza region. Focused group discussions, household surveys and transect walks were used to collect data. A total of 120 households were interviewed and several fields visited during the transect walks. Data was analysed using cross-tabulation, cluster analysis and chi-square methods. Results indicate that involvement in off-farm activities such as fishing, livestock keeping, mini-business and insecure land tenure negatively influences the adoption of agroforestry practices. Contacts with extension agents, level of training, perception of the problem and farm sizes positively influence the adoption. Recommendations to facilitate adoption include integration of socio-economic factors into agroforestry plans, creation of more awareness among farmers and other stakeholders on environmental degradation problems and strengthening the extension services.

Keywords Adoption · Agroforestry · Environmental degradation · Socio-economic factors

Introduction

Environmental degradation is an issue of concern in the Lake Victoria catchment ecosystem in Mwanza, Tanzania. Deforestation and improper soil management practices have lead to soil erosion, lake siltation, accumulation of carbon in the atmosphere and climate change.

There are several land use options that could reduce environmental degradation while increasing farm productivity and farmers' income. Among such options include conversion of monoculture crop land to agroforestry in the form of orchards, live fences, scattered trees or woodlots.

Past research results have demonstrated that inclusion of agroforestry-based soil management practices can increase the grain yield to as much as 1.5 t ha^{-1} yr^{-1} (Ojiem et al., 2000). Agroforestry practice has an added advantage of reducing soil erosion and providing fruits, firewood, timber for construction, fodder for livestock and other woody products (Kaswamila and Tenge, 1998). Major environmental benefit of agroforestry is its ability to sequester atmospheric carbon, thus reducing greenhouse gases that contribute to the climate change.

Despite these benefits of agroforestry practices, farmers of the Lake Victoria catchment ecosystem do not widely use them. Factors that hinder the

A.J. Tenge (✉)
The University of Dodoma, Dodoma, Tanzania
e-mail: ajtenge@yahoo.co.uk

A. Bationo et al. (eds.), *Innovations as Key to the Green Revolution in Africa*,
DOI 10.1007/978-90-481-2543-2_137,

adoption of agroforestry in this area have not been thoroughly investigated in order to enhance adoption and sustainability.

The major objective of this research is to investigate the social and economic factors that influence the adoption of agroforestry practices in the Lake Victoria catchment ecosystem.

Materials and Methods

Description of the Research Sites

This research was conducted in Mayega and Chamgasa villages in the semi-arid area of Magu district, Mwanza Tanzania. Magu district is one of the seven districts in Mwanza region covering an area of 4800 km^2 with a total population of 371, 262 (URT, 2003). The district forms part of the Lake Victoria catchment to which several rivers originating in the northern part of Tanzania transverse before entering Lake Victoria. Magu district receives a bimodal rainfall pattern with an annual range of 700–1000 mm. The short rainy season is from mid-October to January and long rains from March to May. The two villages are part of the Kalemela ward and represent the downstream areas of the Lake Victoria catchment. The Kalemela ward is 60 km from Mwanza along the Mwanza–Musoma road near Lake Victoria on a plain with mainly sand soils locally known as *lusenyi*.

Data Collection

Data collection for this research involved discussions with key informants, farmers groups and administering pre-designed questionnaires to the representative households. Secondary data from scientific reports, maps and statistical abstracts were also used as additional sources of information. The type of data that was collected includes household characteristics, household perceptions on environmental degradation, farm household resources and tree planting practices. Other type of data was the external institution factors and their influences on the adoption of agroforestry practices.

Sampling Procedure

The sampling frame for household survey consisted of lists of heads of household obtained from the village leaders. The list was first stratified into male- and female-headed households.

From each stratified sample, a random sample was selected to make a total sample size of 120 respondents.

Data Analysis

Collected data was processed and analysed using the SPSS and Excel computer programs (Norusis, 1990). Data were analysed to answer the following research questions: (i) What are the types of households that reside in the research area? (ii) What are the major farming systems in the research area and how do they influence the adoption of agroforestry practices and (iii) What are the relationship between socio-economic characteristics and the adoption of the agroforestry practices.

Farm household types were distinguished on the basis of household characteristics such as age, sex, education, marital status and family composition and also on the basis of resource availability such as farm size, land tenure, possession of livestock, farm income, labour availability and involvement in off-farm activities.

Cluster analysis (Norusis, 1990) was used to group farmers and examine the conditions that can make them interested to undertake agroforestry practices. Farming system analysis involved the identification of major crops, economic activities, tree planting and land management practices. Households with any of the agroforestry practices (live fence, woodlot or scattered trees) were identified and grouped as adopters and those with none of these practices were identified and grouped as non-adopters.

Cross-tabulation was used to compare the household characteristics, resource endowment and external institutional factors between adopter and non-adopters. The statistical significance of the identified adoption factors was evaluated by the chi-square (χ^2) as the test statistics (Devore and Peak, 1993).

Results and Discussion

Household Characteristics

The type of households that reside and farm in Kalemela is indicated in Table 1. Based on the household characteristics and resource availability, farm household categories can be distinguished in several ways as shown in the following sections.

Male-headed households: Results indicate that 55% of households are married men. This group includes the most influential people and decision makers at the village and household levels. While it is important to consider this group, care needs to be taken during planning and implementation of the environmental conservation activities such that other groups particularly women are not marginalized.

Female-headed households: Survey results indicate that on average, 45% of the household heads are women. These women heading the households are either widow, divorced or separated. Women generally have limited access to information and other resources due to traditional social barriers. Women are also involved in many regular household activities than are men. These may have negative effects on the adoption of agroforestry practices. This group is likely to be motivated by agroforestry practices that provide fuel wood and relief their workload on search for firewoods.

Tribes: The results indicate that majority (93%) of the households that reside in Kalemela are Sukuma tribes who are native to the area and the largest tribe in Tanzania. Few (4.6%) are Jita and very few (2.4%) from other tribes. This implies that there has been a minimum integration with other tribes. Traditional believes may be strong and need to be considered. It may also imply richness in the indigenous knowledge on various issues related to agroforestry. This indigenous knowledge needs to be explored and utilized in promoting agroforestry practices.

Education-level groups: Four education-level groups can be distinguished in Kalemela: non-formal (<1 year in school), lower primary (1–4 years in school), upper primary (5–8 years in school) and secondary education (9–12 years in school). About 70% of the households have primary school education. Less than 1% of the households have secondary school education. Educated households are expected to understand environmental degradation problems, have more access to information related to environment and hence can easily adopt agroforestry practices to reduce degradation while enhancing productivity. The results also suggest that sustainability of agroforestry can be enhanced by awareness creation and training

Table 1 Household characteristics of respondents in the two villages of Kalemela, Magu

		Respondents by village (%)		
Characteristics	Variable	Mayega	Chamgasa	Mean
Sex	Female	40	50	45
	Male	60	50	55
Marital status	Married	74.3	79.4	76.9
	Divorced	0	2.8	1.4
	Separated	8.6	7.5	8.1
	Widow	14.3	9.3	11.8
	Single	2.9	0.9	1.9
Tribe/ethnic	Sukuma	94.3	91.6	93
	Jita	2.9	6.5	4.7
	Others	2.8	1.9	2.4
Education level	None (<1 year)	39.6	18.5	29.1
	Lower primary (1–4 years)	11.3	15.4	13.4
	Upper primary (5–8 years)	49.1	64.6	56.9
	Secondary	0.0	1.5	0.8
Age groups	Young (18–35 yrs)	6.7	21.7	14.2
	Middle (36–45 yrs)	22.2	39.1	30.7
	Old (46–60 yrs)	51.1	27.5	39.3
	Very old (> 60 yrs)	20	11.6	15.8

at primary school level where majority of future land users pass through.

Age groups: Four sub-groups were identified based on age: young (18–35), middle (36–45 years), old (46–60 years) and very old (> 60 years). About 55% of the households are in the age group of old to very old. This may imply labour shortage and threat to the sustainability of farming activities in the area. Also old farmers tend to be conservative sticking to their traditional ways of farming.

Resource Availability and Uses

Based on the major economic activities and resource availability the households in Kalemela are distinguished in three categories as shown in Table 2.

Crop growers (agriculture): Almost all (99%) households are involved in agriculture. Major crops are maize, rice, sweet potatoes, beans, groundnuts, cassava, pigeon pea and sorghum. Cotton is the main cash crop.

Few farmers grow vegetables such as tomatoes. Monocropping is the dominant cropping system. However, few farmers intercrop maize or cassava with beans and grow *Leucaeana* and pigeon pea on contours.

Livestock keepers: About 46% of the households in Kalemela are involved in livestock keeping. Important types of livestock are cattle, goat, sheep and chicken. These livestock graze freely on the communal land locally knows as *ngitili*. When no crops are growing, livestock are allowed to graze freely on farms. Livestock keepers can be stipulated to adopt agroforestry practices that provide fodder such as grass strips and multi-purpose trees.

Households with off-farm activities: Major off-farm activities are fishing, employment and mini-business. About 13% are involved in fishing, 3% are employees and 5% are involved in small businesses. Off-farm activities may have a negative effect on the adoption of agroforestry by reducing labour availability. On the other hand, off-farm activities can be a source of income to invest on farming and agroforestry activities.

Land tenure: Results in Table 2 indicate show that there are five major types of land ownership in the research area; inheritance, buying and allocation by the village government. Other land tenure systems are renting and borrowing. The results indicate that about half (51%) of the households own the land by buying. This was not expected because majority of the households are native to the areas and hence should own the land under the inheritance system. This indicates that the land from the inheritance system is becoming too small to make any meaningful division among family members. About one-quarter (26%) of the households own the land under the inheritance system, few (10%) have been allocated by the village government, renting (10%) and very few (4%) use a borrowed land. Security in land ownership determines household decisions on long-term investments. Households who do not own the land have little or no control over the types of technologies to adopt. Also they may be unwilling to implement agroforestry practices due to the lack of long-term responsibility on the land.

Table 2 Economic characteristics of households in Kalemela, Magu

		Percentage respondents by village		
Characteristics	Variable	Mayega	Chamgasa	Mean
Major economic activities	Agriculture	100	98.6	99.3
	Livestock	56.7	34.3	45.5
	Fishing	17.4	8.1	12.8
	Employment	2.2	2.7	2.5
	Business	2.2	8.1	5.2
Farm size (acres)		5.6	5.7	5
Land tenure	Bought	59	46	51
	Inherited	25	26	26
	Rented	8	12	10
	Borrowed	2	4	3
	From village government	6	13	10

Tree Planting Practices

Farmer's Reasons for Planting Trees

Major reasons for planting trees by farmers in Kalemela are indicated in Fig. 1. The reasons include provision of shade, fuel wood, wind break, demarcated farm bolder and control of soil erosion.

Other reasons are fertility improvement, construction and medicinal. These reasons determine the importance farmers attach to different tree species and hence the acceptance of the agroforestry practices. The results indicate that very few (10%) farmers have planted trees for the fertility improvement purpose. This is due to the inadequate knowledge on the tree species that can be combined with crops to improve the soil fertility.

Constraints to Tree Planting

Major constraints to tree planting include water shortage, pests and diseases, and livestock destructions (Fig. 2). There are also few farmers who are constrained by lack of sufficient knowledge and areas for tree nurseries. Capacity building through training on tree planting and management, and rainwater harvesting technologies can contribute towards solutions to these problems. There is also a need to review the by-laws that govern animal grazing on crop fields.

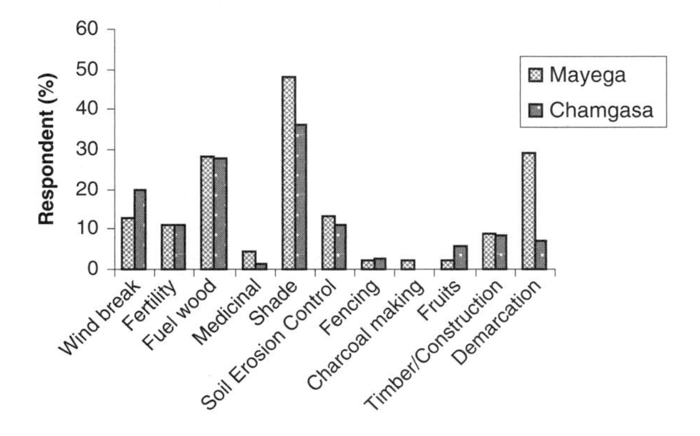

Fig. 1 Reasons for planting trees by farmers of Kalemela, Magu

Factors for Adoption of Agroforestry Practice

Household Characteristics

Gender of the household head: The results in Table 3 indicate that female-headed households have more adopters of agroforestry practices compared to male-headed households. The results indicate that the differences in adoption between male and female headed households is statistical significant at significant level of 0.1 ($\alpha = 0.1$). This could be due to their much involvement in farming activities and limited access to other economic activities. These results are similar to the findings by Tenge et al. (2004), who found that female-headed households in The West Usambara highlands, Tanzania, have adopted soil conservation measures more frequently than do male headed households. This observation was attributed to frequent visits to farms and limited access to other off-farm activities by women farmers.

Education level: The results in Table 4 show significant difference ($\alpha = 0.01$) in adoption levels between different education levels. The results (Table 4) show that almost all households with secondary school education (9–12 years in school) have adopted agroforestry practices. Although the sample size of the households with secondary school education was relatively small, these results imply that educated households are more likely to adopt agroforestry technologies because they have a better understanding of the environmental degradation problems and have more access to information related to conservation practices.

Age: Results indicate significant ($\alpha = 0.01$) difference in adoption between different age groups. The

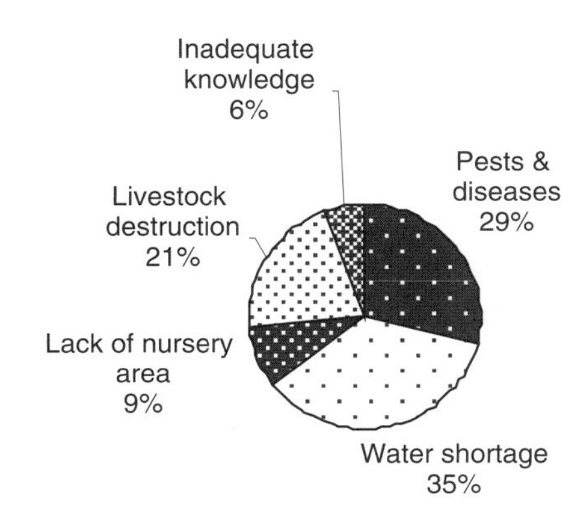

Fig. 2 Constraints to tree planting in Kalemela

Table 3 Household characteristics of adopters and non-adopters of agroforestry in Kalemela, Magu

Variable	Variable description	Adopters ($n = 44$)	Non-adopters ($n = 76$)	Chi-square	Significant level (α)
Sex	Sex of the head of household				*
	%Male	60.9	39.1	4.5	
	%Female	74.5	25.5		
Marital	Marital status				n.s.
	Married	66	34		
	Divorced	67	33		
	Separated	70	30		
Tribe	Ethnic group				n.s.
	Sukuma	68.2	31.8		
	Jita	66.7	33.3		
Birth	Place of birth				n.s.
	Within village	63.8	36.2		
	Outside village	71.4	28.6		

n.s., not significant; *significant at 0.1

Table 4 Education level and age characteristics of adopters and non-adopters of agroforestry in Kalemela, Magu

Variable	Variable description	Adopters ($n = 44$)	Non-adopters ($n = 76$)	Chi-square	Significant level
Education	Average school years			4.3	*
	<1	75	25		
	1–4	86.6	13.4		
	5–8	87.9	12.1		
	9–12	100	0		
Age	Average age of head of household (years)			6.2	*
	18–35	44.4	55.6		
	36–45	73	27.0		
	46–60	66.7	33.3		
	> 60	81.3	18.8		

n.s.: not significant; *significant at 0.1

young households showed a low adoption rate on agroforestry practices (Table 4). This was not expected as young farmers are supposed to have a longer planning horizon and hence be more eager to invest on environmental conservation activities. This low adoption rate among young farmers is due to their involvement in off-farm activities. Similar results were observed by Amsalua and De Graaff (2005) in the central Ethiopian highlands where adoption of soil conservation practices was more among older farmers than the young ones due to their differences in farming experiences and conservation practices.

Resource Availability and Uses

Off-farm activities: The results in Table 5 show significant difference ($\alpha = 0.01$) in the adoption rate between households that are involved in off-farm activities and those that are not involved. The results show that only 14–37% of households that are involved in off-farm activities have adopted agroforestry practices compared to 70% adopters from households that are involved in agriculture. These results provide sufficient evidence that involvement in off-farm activities negatively affects the adoption of the agroforestry practices. This is due to the competition in labour between agroforestry and off-farm activities. Lack of short-term benefits from agroforestry activities compared to off-farm activities also explains this observation. These results support the findings by Msuya et al. (2006), who observed that adoption of agroforestry practices by farmers in Shinyanga, Tanzania, was influenced by their production objectives and involvement in off-farm activities.

Land tenure: The results show that almost all the households with borrowed land have not attempted to practice agroforestry and only very few (11%) who

Table 5 Economic characteristics of adopters and non-adopters of agroforestry in Kalemela

Variable	Variable description	Adopters ($n = 44$)	Non-adopters ($n = 76$)	Chi-square	Significant level
FSize	Average farm size (acre)	5.8	5.1		n.s.
HHsize	Household members (no)	7	7		n.s.
Activity	Major economic activities			24.8	**
	Agriculture (%)	66.9	33.1		
	Livestock keeping (%)	37	63		
	Fishing (%)	28.6	71.4		
	Employment (%)	33.3	66.7		
	Mini-business (%)	14.3	85.7		
Land tenure	Type of land ownership			20.5	**
	Bought	67.7	32.3		
	Inherited	80.6	19.4		
	Rented	11.1	89.9		
	Borrowed	0	100		
	Allocated by village government	80	20		

n.s., not significant; **significant at 0.001

rent the land have adopted the agroforestry practices. This is because of the lack of security in the land ownership under these systems. These results provide sufficient evidence that security in land ownership significantly affects the adoption of agroforestry practices. The results are similar to findings by Msuya et al. (2006), who also observed an influence of the land tenure on the adoption of agroforestry practices in Shinyanga, Tanzania. However, the results are different from the findings by Posthums (2006), who found that insecure land tenure was positively influencing farmers in Peruvian Andes to adopt soil conservation measures because their rights to continue using the land were attached to the implementation of conservation measures.

External Factors

There is an influence of the external factors such as contacts with extension agents, training and availability of market. Results in Table 6 indicate that there is significant influence ($\alpha = 0.01$) of the external factors such as contacts with extension agents, training and availability of market. The results in Table 6 show that majority of households who have adopted agroforestry practices have contacts with extension agents (66.7%), have received training on tree planting (65%) and perceive market situation to be not a problem (77.3%). Membership in farmer groups was also found to be positively influencing the adoption of agroforestry practices.

These external contacts enable exchange of information and create great awareness among farmers on the environmental conservation issues. Farmer-to-farmer visits, farmer's field schools and linking the agroforestry and productivity to the market are likely to enhance adoption. Discussions with key informants revealed that there are few extension staff and not adequate facilitated. The results suggest the need for strengthening the extension services to enhance adoption. These results support the findings by Paudel and Thapa (2004) in Nepal, which showed that extension

Table 6 External factors for adoption of agroforestry practices in Kalemela, Magu

Variable	Variable description	Adopters ($n = 44$)	Non-adopters ($n = 76$)	Chi-square	Significant level
Extension	Contacts with extension agent	66.7	33.3	6.4	*
Group	Membership in farmer groups	55.8	44.2	3.2	*
Market	Market situation	77.3	22.7	5.2	*
Training	Received training	65	35	4.7	*

n.s., not significant; * significant at 0.1

visits and training were significant factors that influenced the adoption of soil conservation practices. Msuya et al. (2006) had similar findings in Shinyanga, Tanzania.

Conclusions

Results on the identification of social and economic factors for the adoption of agroforestry practices in Magu, Tanzania, lead to the following conclusions:

There is diverse economic activities by households that reside and farm in Magu. Major off-farm activities include livestock keeping, fishing and mini-businesses. These off-farm activities negatively affect the adoption of agroforestry practices.

Adoption of agroforestry practices in Magu is likely to increase with better security in land tenure, strong extensions services and higher level of education and awareness on the environmental degradation and conservation practices. Government commitment is required to strengthen extension services. There is also need for the policy and institution support to increase security in land tenure.

Physical factors such as frequent drought negatively influence the adoption of the agroforestry practices. Rainwater harvesting should be part of the activities in any agroforestry project in drought-prone areas like Magu.

Acknowledgements This research was conducted as part of the VicRes project titled "Transferring Soil Carbon Sequestration Best Management Practices to Smallholder Farmers in Lake Victoria Catchment Ecosystem". We wish to express our thanks to the Inter-university Council for East Africa for the financial support to carry out the work reported in this chapter under the VicRes grants programme. We acknowledge support from Kalemela farmers, village leaders, Magu district extension office and anonymous reviewers who read a first draft of this chapter for their constructive comments.

References

Amsalua A, De Graaff J (2005) Sustainable adoption of soil and water conservation practices in Barossa Watershed, central Ethiopia highlands. A paper presented at the EFARD (European forum for agricultural research development) conference in Zurich, Apr 2005

Devore J, Peak R (1993) Statistics. The exploration and data analysis. Duxbury Press, Pacific Grove, CA

Kaswamila AL, Tenge AJM (1998) The neglect of traditional agroforestry and its effects on soil erosion and crop yields. The case of West Usambara Mountains. Research report. REPOA, DSM, Tanzania

Msuya TS, Mndolwa MA, Bakengesa S, Maduka S, Balama C (2006) Factors affecting adoption and dissemination of agroforestry technologies in Tanzania: case of HASHI/ICRAF and Ruvu fuel wood pilot projects. In: Chamshama SAO, Nshubemuki L, Idd S, Swai RE, Mhando ML, Sabas E, Balama C, Mbwambo L, Mndolwa MA (eds) Proceedings of the second national agroforestry and environmental workshop: partnerships and linkages for greater impact in agroforestry and environmental awareness, Mkapa Hall, Mbeya, 14–17 March 2006, pp 199–907

Norusis MJ (1990) Advanced statistical guide. Statistical Package for Social Sciences (SPSS) International BV, Gorinchem

Ojiem JO, Mureithi JG, Okwosa EA (2000) Integrated management of legume green manure, farmyard manure and inorganic nitrogen for soil fertility improvement in western Kenya. In: Mureithi JG, Gachene CKK, Muyekho FN, Onyango M, Mose L, Magenya O (eds) Participatory technology development for soil management by smallholders in Kenya. Proceedings of the 2nd scientific conference of soil management and legume research network projects, Mombasa, Kenya, June 2000, pp 98–102

Paudal GS, Thapa GB (2004) Impacts of social institutions and ecological factors on land management practices in mountain watershed of Nepal. Appl Geogr 24:35–55

Posthumus H (2006) The adoption of terraces in the Peruvian Andes. PhD dissertation, Wageningen University, Wageningen

Tenge AJ, De Graaff J, Hella JP (2004) Social and economic factors for adoption of soil and water conservation in west Usambara Highlands, Tanzania. Land Degrad Dev 15: 99–114

URT (United Republic of Tanzania) (2003) Magu irrigation development plan, Mwanza Region. Government Printer, Dar es Salaam

Index

A. Bationo et al. (eds.), *Innovations as Key to the Green Revolution in Africa*,
DOI 10.1007/978-90-481-2543-2, © Springer Science+Business Media B.V. 2011

D

E